Applied Superconductivity 1995

Related titles

Superconductor Science and Technology (IOP Journal)

High Temperature Superconductivity, Proceedings of LT-19 Satellite Conference, Cambridge, 13–15 August 1990
J E Evetts

Critical Currents, Proceedings of 6th International Workshop, Cambridge, 8–11 July 1991
J E Evetts

Directory of Low Temperature Research in Europe
P McDonald

Superfluidity and Superconductivity, 3rd edition
D R Tilley and J Tilley

High Temperature Superconductivity, Proceedings of the 39th Scottish Universities Summer School in Physics, St Andrews, June 1991
D P Tunstall

Magnetism, Magnetic Materials and their Applications, Proceedings of the International Workshop, La Habana, Cuba, 21–29 May 1991
F Leccabue

Applied Superconductivity 1995

Proceedings of EUCAS 1995, the second European Conference on
Applied Superconductivity, held in Edinburgh, Scotland, 3–6 July
1995

Volume 1. Plenary talks and high current applications

Edited by D Dew-Hughes

Institute of Physics Conference Series Number 148
Institute of Physics Publishing, Bristol and Philadelphia

CODEN IPHSAC 148 1–1760 (1995)

British Library Cataloguing in Publication Data

A catalogue record for this book is available from the British Library.

ISBN 0 7503 0370 0 Vol. 1
     0 7503 0371 9 Vol. 2
     0 7503 0348 4 (2 Vol. set)

Library of Congress Cataloging-in-Publication Data are available

Published by Institute of Physics Publishing, wholly owned by The Institute of Physics, London
Institute of Physics Publishing, Techno House, Redcliffe Way, Bristol BS1 6NX, UK
US Editorial Office: Institute of Physics Publishing, The Public Ledger Building, Suite 1035, 150 South Independence Mall West, Philadelphia, PA 19106, USA

Printed in the UK by J W Arrowsmith Ltd, Bristol BS3 2NT

Contents

* Invited

Materials processing

* Invited

Critical currents

Bulk material and conductor properties

xvi

VOLUME 2. SMALL SCALE APPLICATIONS

Oral contributions

* Invited

Thin film deposition

* Invited

Thin film microwave properties

Microwave devices, circuits and applications

Junction fabrication, properties and arrays

Foreword

In early July the organizing committee for EUCAS '95 was delighted to welcome to Edinburgh over 650 delegates, together with 30 accompanying persons and the industrial exhibitors at 24 stands, who were attending the successor to the first European Conference on Applied Superconductivity held in Göttingen in October 1993. The attendance at Edinburgh was marginally greater than that at Göttingen, but there was a much wider representation of European countries and a gratifyingly large representation from other continents.

The 424 papers in these volumes reflect the themes and issues prominent in the 571 plenary, invited and contributed papers that were delivered during the three days of the meeting. As you read them you will be encouraged by the growth, since 1993, in the percentage of papers which refer to actual applications or their enabling science. This is important—the change in the funding climate of the last few years means that HTS must soon find real applications and marketable products. In fact these are clearly on their way, and some were already on sale at the EUCAS Industrial Exhibition.

Particular attention to this theme was highlighted in the Keynote Lecture given by Professor H Piel, the text of which is to be found on page 1, who as Vice-Chairman of CONECTUS (CONsortium of European Companies determined To Use Superconductivity) was very well qualified to address the topic. It also came through in the Technology Transfer Session in which six speakers, P Williams of Oxford Instruments Group plc, M Leghissa and B Gromoll of Siemens Research, S Hedges of GEC Marconi Research Centre, R Penco of Ansaldo and P Grant of the EPRI, gave refreshingly frank accounts of their organizations' perceptions of the economic and engineering challenges which must be faced in bringing HTS to the marketplace.

Eighteen months' preparation went into EUCAS '95 and many worked hard for it. I particularly acknowledge the work of the International Advisory Committee, the time and effort given by members of the UK Organizing Committee and the professionalism shown by the staff of Meeting Makers as they tackled both routine duties and wholly unexpected problems. Principal of the latter was the severe blow dealt by the sudden death of our first Treasurer, Dr Tony Appleton, on 15th December 1994. In fact the blow was to applied superconductivity as a whole, for Tony's contribution to our field was seminal. Our loss was spelt out by Dr Peter Komarek in the introductory remarks, to be found on page xlv, to his plenary talk, to which we are proud to give the title of *The Appleton Lecture*.

Within our three days we found time to hear about two Edinburgh graduates whose scientific contributions still influence everyone in superconductivity. They were Professor James Clerk Maxwell, whose boyhood home was visited by many delegates, and Professor Sir James Dewar, whose vacuum flask and other achievements are described by Dr Robert Soulen on page 23. However the conference scene which I will longest remember is that of the dinner, where 480 delegates found three helium-filled decorative balloons at every table. Enthusiastic experiments resulted in some 60 of these pressing against the high glass roof of the Royal Scottish Museum; advanced technology, though xxxviii

not of the superconducting kind, was applied in getting them down.

EUCAS '97 will be held in the Netherlands, organized by Professor Horst Rogalla and his colleagues. We all look forward to it, and to hearing of further great advances in applied superconductivity.

Gordon Donaldson, Conference Chairman

Acknowledgments

This meeting has benefited by support from the following organizations:

Ansaldo Energia
BICC Central Technical Unit
City of Edinburgh District Council
Cookson
Department of Trade & Industry
European Materials Research Society (EMRS)
Engineering & Physical Sciences Research Council (EPSRC)
GEC Alsthom NV
Lothian and Edinburgh Enterprise Ltd
Merck Industrial Chemicals
NKT Research Centre
Oxford Instruments
Quantum Design
The National Grid Company plc
The Royal Society
The Royal Society of Edinburgh

The meeting would also like to thank the following organizations for their assistance:

British Cryogenics Council
Institute of Electrical Engineering

Preface

These volumes contain the majority of those conference presentations for which a manuscript has been received. All of the published contributions have been rigorously refereed by members of the Programme Committee and their colleagues, to whom I am indebted for their efforts. In order to ensure rapid publication of the proceedings, deemed to be essential in an area of endeavour which is undergoing such rapid development and exciting change, it has not been possible to return the refereed manuscripts to the authors for revision. Typographical errors, spelling mistakes, non-colloquial or infelicitous use of English, provided the meaning is not obscured, and minor deviations from the prescribed format, have been allowed without alteration; their presence must not be interpreted as a reflection on the efficiency or integrity of the referees.

The proceedings have been divided into sections which are believed to be logical in the light of the material actually presented in the conference sessions. The papers have been assigned to these new sections in a manner which groups together similar topics, and may, therefore, not appear in the order in which they were presented at the conference.

David Dew-Hughes, Editor

Conference Organization

Conference President

Sir Martin Wood, FRS

EUCAS '95 Organizing Committee

Prof Gordon Donaldson	(University of Strathclyde)	Conference Chairman
Dr Colin Pegrum	(University of Strathclyde)	Conference Secretary
Dr David Caplin	(Imperial College)	Treasurer
Dr Archie Campbell	(IRC, Cambridge)	Programme
Dr David Dew-Hughes	(University of Oxford)	Publication
Dr Ian McDougall	(Oxford Instruments)	Exhibition
Prof Neil Alford	(South Bank University)	Exhibition
Prof Colin Gough	(University of Birmingham)	Student Workshop
Dr John Gallop	(National Physical Laboratory)	Programme
Dr Damian Hampshire	(University of Durham)	Programme
Dr Peter Hatton	(University of Edinburgh)	Local Arrangements
Dr John Macfarlane	(University of Strathclyde)	Programme

Dr Tony Appleton, who served as Treasurer during 1994, sadly died on 19 December 1994.

International Advisory Committee

A Barone (Italy)
W Blau (Ireland)
A Braginski (Germany)
Y Bruynseraede (Belgium)
T Claeson (Sweden)
A Dawson (USA)
J Ferreira da Silva (Portugal)
R Flükiger (Switzerland)
T Freltoft (Denmark)
H Freyhardt (Germany)
K Fossheim (Norway)
K Gundlach (France)
H Hayakawa (Japan)
H Koch (Germany)
R Navarro Linares (Spain)
P Komarek (Germany)

D Larbalestier (USA)
A Lauder (USA)
Y Liang (UK)
J Mannhart (Switzerland)
J Mooij (The Netherlands)
N Pedersen (Denmark)
W Phillips (UK)
H Peil (Germany)
H Freyhardt (Germany)
C Rizzuto (Italy)
H Rogalla (The Netherlands)
J Rowell (USA)
K Sato (Japan)
S Tanaka (Japan)
F Yndurain (Spain)

Eulogy

Dr Peter Komarek prefaced his plenary Appleton Lecture on 'Large Scale Applications of Superconductivity' with the following eulogy to Tony Appleton:

Ladies and Gentlemen,

It is a great honour for me to be invited as a speaker of this lecture, which is devoted to the late Dr Tony Appleton, one of the worldwide well known pioneers of large scale applications of superconductivity.

He believed in this technology so much that he felt motivated to work restlessly for its development and industrial implementation for many decades. In particular, the development of applications in electric power engineering was one of his primary goals. His early work on homopolar machines and fault current limiters has become well known. To remind you, I show a slide which I received from Tony in 1971—nearly 25 years ago—which shows the famous Fawley motor, a 7.5 MW superconducting homopolar machine, driving a power station water pump at 200 rpm. The machine was successfully built on Tony Appleton's ideas and initiatives.

In recent years Tony became a highly respected advisor and a driving force for the UK programme on applied superconductivity. He was charged by the DTI with many tasks in this area, among them the UK representation on the executive committee on an IEA implementing agreement on assessment of the potential of HTS in the electric power sector. I have the privilege of chairing this committee and enjoyed very much the substantial contributions by Tony Appleton in each of the meetings which took place in a sequence of about six months. The last time when we were sitting together in that frame was October 1994 in Karlsruhe.

Inst. Phys. Conf. Ser. No 148
Paper presented at Applied Superconductivity, Edinburgh, 3–6 July 1995
© *1995 IOP Publishing Ltd*

1

Future Prospects of Applied Superconductivity in Europe

Helmut Piel

Institute of Materials Science, University of Wuppertal, 42097 Wuppertal, Germany

Abstract. This contribution is a personal view of the current state and the future evolution of applied superconductivity with a specific focus on the European scene. It is approached by giving some examples from the history of applied superconductivity and by sketching the present market for products using superconducting components. Today's superconducting NMR, MRI and SQUID systems and the large scale projects of the near future, especially in the field of particle accelerators and energy technology, are exclusively based on metallic superconductors. The High Temperature Superconductors (HTS) still have a long way to go before they will have a noticeable impact on the superconductivity market, but quite a few practical uses of high Tc components are beginning to take shape. HTS bulk materials in the form of current leads are the first products which are used in commercial systems. The fabrication and application of HTS Josephson junctions is progressing and may find uses in the fields of biomedical research and non destructive evaluation. HTS wires and tapes are being tested in cryogen free or very high field magnets and in energy technology. One particular application area which is carrying much hope for the introduction of HTS materials to present technology will be described in more detail. It is the practical use of superconductors in high frequency fields. Epitaxial HTS films are expected to have their first wide spread use in mobile communication and rf signal sensing. In closing the early activities of a new European organization called CONECTUS are introduced. In this context an updated forecast of the size of the future superconductivity world market will be given.

1. Introduction

The last 150 years have certainly been the most exciting in the history of science. During this time our understanding of the laws of nature has progressed so much that we were able to create technologies which have changed our life. This century has seen a revolution in the communication, information and medical technologies. It may serve as an illustration that this year we celebrate the one hundreds anniversary of the discovery of the X-rays by C.F. Röntgen and of the invention of the radio by G.Marconi. There is a parallel between the discovery of the X-Rays and that of High Temperature Superconductivity. It lies in the excitement of the scientific world which both discoveries created. More than one thousand papers were published in the first year after Röntgen's discovery. Comparing the publication habits around the turn of the century with our times this number is quite comparable to the flood of papers following the discovery of the superconducting cuprates. The Meißner effect can now be demonstrated at every dinner party and superconductivity has gained the interest of the general public. Many more laboratories can participate in research on superconductivity as only liquid nitrogen, readily available almost everywhere, is necessary for cooling. If one asks the question if superconductivity will have a future similar to that of the X-Rays it is comforting to see that Nuclear Magnetic Resonance (NMR) spectroscopy plays an ever increasing role in creating new pharmaceutical products and that Magnetic Resonant Imaging (MRI) is challenging X-ray technology as a diagnostic tool. These two techniques have become the most important and visible successes of applied superconductivity. Both are used for structural analysis and are supplementing X-ray technology. One of the promising areas for the application of the High Tc Superconductors (HTS) is Marconi's world. The first wide spread use of HTS films is expected in communication and remote sensing by receiving, transmitting and analyzing radio frequency (rf) waves.

In the first part of this contribution I want to sketch past, presence and future of superconducting magnets and rf resonators with more emphasis on the latter. This is to show that all evolution, especially the one of a complex technology, needs its time. I will also describe shortly some important large scale European projects which are under construction or on the drawing board. In the second part I will introduce the Consortium of European Companies Determined To Use Superconductivity" (CONECTUS) by describing its early activities. Some of these are: a recent forecast of the future size of the world market for systems using superconductivity as an enabling technology, a listing of these products, an overview of the market today and an estimate of European funding.

2. Some Examples from the History of Applied Superconductivity or "Rome was not Built in one Day"

2.1 A condensed history of sc magnets.

Immediately after the discovery of superconductivity it was obvious that there would be important applications of this new phenomenon. In his Nobel lecture Heike Kammerlingh Onnes proposed a 10 Tesla magnet wound from lead wires. He then discovered the existence of the critical field and it took about half a century before such magnets could be built. The first commercially produced superconducting magnet made from NbZr in Europe was offered by Oxford Instruments in 1962 [1], half a century after the first proposal.

By far the most important success of applied superconductivity is NMR and MRI. In 1946 Bloch and Purcell performed their first experiments on nuclear magnetic resonance. About 25 years later the first NMR spectrometers using superconducting magnets were produced commercially. MRI developed out of NMR and in 1982 almost forty years after the basic discovery the American Food and Drug Administration approved the use of MRI systems for medical diagnostics.

In 1954 the first metallic high temperature superconductor Nb_3Sn was discovered [2] which may well serve as an historic example for the development of HTS wires. Its mechanical brittleness was a major stumbling block for its application. The first "wires" were Nb_3Sn coated niobium or steel tapes preceding in principle the recent success in thin HTS tapes at Los Alamos [3,4]. The first commercial Nb_3Sn magnet was produced by General Electric (GE) in 1965 [5] thirteen years after the material was discovered. In 1968 Nb_3Sn magnets were produced in Europe by Oxford Instruments (OI). In 1975 multilamentary Nb_3Sn wire became available from Imperial Metals Industries and the first filamantary Nb_3Sn magnet was delivered by OI in 1976. In 1994 the first commercial cryogen free magnets on Nb_3Sn basis were produced in Japan and by OI in Great Britain and in the same year GE introduced a cryogen free MRI magnet using Nb_3Sn tape as a conductor. This year, only 9 years after the discovery of Bednorz and Müller, a cryogen free HTS magnet reached a field of three Tesla at 21 K at Sumitomo [6].

Since the 1970's one has discussed the use of Nb_3Sn magnets in Fusion Reactors. Now these ideas become reality by the International Thermonuclear Experimental Reactor ("ITER") which will be discussed in P.Komarek's [7] contribution on the use of superconductivity in energy technology. ITER will certainly give a much wanted push to the continuation of the development of the Nb_3Sn wire technology.

The largest future project in applied superconductivity world-wide is the construction of the Large Hadron Collider ("LHC"). It was authorized in the beginning of this year at CERN. This projects requires the construction of approximately 1.300 dipole magnets of 13.5 m length with a field of 8.65T and about 600 quadrupole magnets of 3.2 m length and a field gradient of 220T/m [8]. This project will be the largest customer for NbTi wire in the near future.

From these examples one can see that the average delay between a fundamental discovery in superconductivity and its apparent commercial application is about 20 to 40 years. Nine years after the appearance of HTS we should therefore not be discouraged that there are not as yet applications which are visible commercially.

I do not want to discuss the cosequences of the wonderful quantum interference phenomena by Josephson, in the beginning of the sixties, because I am not very familiar with this promising field. Josephson junctions and their applications in active electronic devices and sensor technology will be reviewed by H. Rogalla in his contribution to this conference.

2.2 The Development of a New Technology - Superconducting Cavities.

In 1934 Heinz London [9] suggested that Joule losses should be observable in superconductors placed in a high frequency field. He expected these losses to be orders of magnitude lower than the losses due to the skin effect in normal conductors. With his considerations the application of superconductivity in radio frequency (rf) engineering was born. After the development of the Radar and microwave technology during World War II and after the invention of the first practical helium liquefier by Collins at MIT in 1949 rf superconductivity became a very active research field at Cambridge [10], Yale [11] and MIT [12]. After it was found that at 4,2K the surface resistance of superconducting niobium could be up to six orders of magnitude lower than the one of copper applications of superconducting cavities to particle accelerators were discussed. The first large scale system was the Stanford Superconducting Recyclotron which started operation in the beginning of the 1970's [13]. In 1977 the principle of the Free Electron Laser was demonstrated with the Stanford superconducting accelerator [14]. During this early large scale application of sc cavities anomalous field limitations like electron multipacting, electron field emission and thermal instabilities were observed not inherently related to the superheating field which is close to the thermodynamical critical field of a superconductor. In the work of the years from 1977 to 1985 at Stanford, CERN, Cornell and Wuppertal these obstacles were analyzed. Clean room work with superconducting cavities, a special shape to avoid electron multipacting, the improvement of the thermal conductivity of the niobium material and high temperature UHV treatment made the construction of the first large scale systems possible. In 1985 superconducting rf cavities were built commercially on a large scale (360 five cell cavities) for the superconducting linear accelerator of CEBAF in Newport News, Virginia. This accelerator is presently the largest superconducting rf installation and has reached its design energy of 4 GeV on May 9th of this year [15]. It will soon serve the nuclear physics community.

At CERN the worlds largest superconducting rf accelerating structure is presently being mounted into the LEP storage ring in order to increase its center of mass energy to aprox. 200 GeV [16]. 272 niobium accelerating cavities of 352 MHz with an accelerating field of 6 MV/m are now being installed. These structures will add 2 GeV of energy to the circulating electrons during each of their turns and thereby compensating for the energy lost by synchrotron radiation at a particle energy of 100 GeV. The upgraded LEP storage ring will mainly serve the purpose to study the weak boson production and may give first clues to the existence of the Higgs Boson. Without superconducting cavities this energy increase would have been inconceivable.

At present projects for very high intensity proton and H⁻ linear accelerators for energies in the GeV range and beam powers in excess of 1 MW are being discussed. An example for such a project is the European Pulsed Spallation Source (ESS) [17] were a sc linear accelerator for 1.3 GeV H⁻ ion of 5 MW is under consideration as one of the options [18].

4

At DESY in Hamburg the TESLA Test Facility (TTF) is built by an international collaboration in order to study the feasibility of a superconducting collider [19] were in the end 500 GeV electrons shall produce head on collisions with 500 GeV positrons. Such a machine will depend on approximately 40 km of superconducting accelerating structures. These 40.000 sc Cavities are the biggest future project of rf superconductivity as yet behind the horizon.

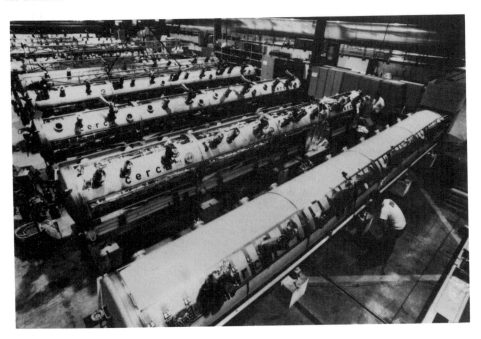

Fig 1. View over one part of the 200 superconducting accelerating sections for the energy upgrade of the LEP storage ring at CERN

3. A Step of Superconductivity into Marconi's world - Possible Applications in Mobile Communication

3.1 General Remarks

The only significant application of superconductivity at radio or microwave frequencies is today in the area of particle accelerators. Epitaxial films cannot be deposited so far on large (m²) curved surfaces and an economical procedure which would beat the metallic superconductors in accelerator applications is nowhere in sight. There are however other areas of high frequency applications in the communication and in the remote sensing technologies. The military world always wanted to see "what is on the other side of the hill" and at the same time wanted to hide whatever was possible to hide. I have no knowledge in this field although I know from colleagues that it is a very important application area for passive and active HTS electronics. Switchable filter banks, stable oscillators and chirp filters are the key components in this context. All are based on thin film planar technology. I only can comment on civilian applications of HTS materials which are based on planar microwave technology very suited to the two dimensionality of the HTS materials. There are two general areas of rf applications which may well belong to the first major application fields of HTS thin films. These are the base stations in mobile communication systems and the sensor coils in NMR and MRI systems. The latter will be described in the contribution of R.Withers to this conference. I will

5

describe an application to mobile communication, to give an example of our work in Wuppertal.

3.2 Mobile Communication

Mobile communication or cellular phone systems are a rapidly growing market. It is expected that by 2005 about 50 % of all communication terminals will become mobile. In addition to voice communication (8-30kb/s) other services like video conferences (500 kb/s), vehicle location and traffic information will become available. One envisaged system to handle such a flow of information is the Universal Mobile Telecommunication System (UMTS) with data transfer rates of up to 2Mb/s at a carrier frequency of 2 GHz. In a cellular phone system the participant communicates with a base station in the center of the cell in which he is located. The base station has a sensitive receiving system and high power transmitters to transfer the information to the receiving participant. It is confronted with a very complex signal scenario out of which it has to select the signal of the individual mobile unit in order to transfer it undisturbed to the receiver of the information. On its receiving side the base station has to suppress man made noise and narrow band frequency sources of possibly strong but unwanted signals (e.g. cochannel interference signals). On its transmission side many different carrier frequencies have to use a common antenna which requires a transmitter combiner filter bank. The individual frequency channels have to be narrow and non overlapping. They are composed of narrow band filters with steep skirts which should handle between 5 and 30 W of transmitted power. These filters are the most likely candidates for an early use of HTS elements in communication technology. They require a clever filter design to avoid high surface current densities on the HTS films and high quality epitaxial YBCO or TBCO films. Figure of merit is the surface resistance R_S and its power dependence.

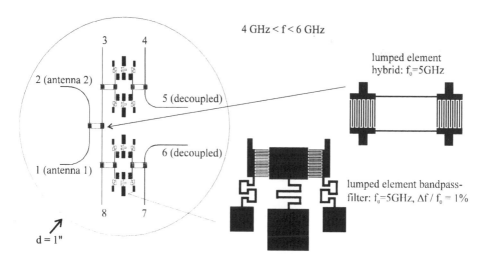

Fig.2: Example of an HTS microwave signal processing unit for two antennas of a receiving array including the beam forming network and 2 frequency multiplexers

The signals which are received by the base station may in future be analyzed by intelligent antenna systems which adjust their directional sensitivity according to the rf scenario and which suppress interference signals by an appropriate beam forming network for the required antenna array. Such beam forming networks need a very large number of low power microwave components like couplers, delay lines and filters. If made from normal conductors such a system would be very bulky and expensive. The existing high quality epitaxial HTS-films allow the design of a superconducting network composed of lumped element capacitors

6

and inductors which because of the strong reduction in Joule losses can be highly miniaturized. An example of such a microwave circuit is displayed in the lumped element beam forming network for two antennas in fig.2

It is characteristic for the work in NMR receiving antennas and in HTS electronics for mobile communication that much of the successful work, especially the one on high power filters, has already become proprietary. This is a development which does not ease our curiosity but it is a very good sign for the application of HTS in the future.

4. CONECTUS - ISIS and the Future.

Since August 1993 a group of 15 European Companies (Siemens AG, Daimler Benz AG, GEC Alsthom, Thomson-CSF/LCR, BICC plc, Oxford Instruments plc, Hoechst AG, ICI Superconductors, Ericsson, Merck plc, ACCEL GmbH, Ansaldo GIE , Noell GmbH, Preussag AG and Cryoelectra GmbH) came together for the foundation of the Consortium of European Companies Determined to Use Superconductivity (CONECTUS). The goals of CONECTUS are: to promote closer communication between European industries involved in the technology of superconductivity - to encourage closer liaison between European industry and publicly funded research laboratories which are interested in the commercial applications of superconductivity - to act as a focus for European industry in international discussions - to promote the flow of information on national and international projects involving superconductivity - to promote techno-economic studies in the field of superconductivity - to encourage the adoption of international standards within the field of superconductivity and to influence European and national governments for continued support for research, development and industrial activities involving superconductivity. The first chairman of CONECTUS is Sir Martin Wood.

One of the major tasks of CONECTUS is to represent European interests at the "International Superconductivity Industry Summit" ISIS. This institution is an international Organization founded by representatives of the American and Japanese Industry interested in the application of the technology of superconductivity.

Childhood and adolescence of a new technology needs public and corporate support. An attempt to look into the future of this technology, as difficult as it may be, is therefore very much wanted. It was therefore one of the first tasks of ISIS do perform such an attempt by carrying out a survey in its constituency in 1992. At ISIS 93 in Hakone these results were published and have since then been repeatedly referred to. At the time the European input to this endeavor was negligible and it appeared reasonable to us to attempt a forecast from a European perspective later. Ian McDougall from Oxford Instruments used the market knowledge of this experienced company to create a first European guess which later on was supplemented and modified by other experts from CONECTUS. It is shown in table 1. It is certainly daring to present such a forecast but EUCAS'95 is such a welcome environment for a discussion of this look into the future that it appears to be justified. Forecasts are generally to low and they often miss the really important application which mostly comes as a surprise. This forecast however gives an interesting outlook into the commercial world and the present technological expectations. It may also be something about which we can smile at a later time.

The superconductivity market of approximately 2.67 Bill. US $ is today (1995) dominated by the aprox.2 Billion in MRI systems. It is followed by about 0.5 Bill. $ in NMR systems and the rest distributes itself to laboratory magnets, accelerator components and SQUID sensor systems. The European share of the total market is about 45%.

In preparation of the next ISIS meeting and as one of its other activities CONECTUS has contacted representatives of more than 20 European countries in order to find out to what extend they were publicly supporting research in superconductivity. In summing up all the contribution we arrived at US $ 128 Mill. in 1994 compared to $ 276 Mill in the US and 226

Mill. in Japan. Taking into account the uncertainty in such an investigation the numbers indicate that Europe should do more to keep up its present technology position.

Table 1. Forecast of the market development by CONECTUS in June 1995 in billion US $

Market Sector	1995	2000	2010	2020
Electronics				
Computing			0,6	11
Passive		1	5	20
Active		1	1,5	7,5
Subtotal (Electronics)		2	7,1	38,5
Energy				
SMES		0,1	1,5	12
MHD			0,7	2
Transformers			1,5	13
FCL		0,1	3,5	17
Motors			0,1	1
Generators				
Cables			0,1	2
Subtotal (Energy)		0,2	7,4	47
Transport				
Motors				1
Levitation			0,1	4
Barings				0,1
Actuators			0,1	1
Subtotal (Transport)		0	0,2	6,1

Market Sector	1995	2000	2010	2020
Medicine & Science				
Science	0,1	0,2	1	2
MRI	2	3	5	10
Accelerators	0,05	0,2	0,5	1
NMR	0,5	0,6	1	1,5
Sensors	0,02	0,1	0,5	2
Subtotal M&S	2,67	4,1	8	16,5
Processing Industry				
NMR		0,1	1	5
Magnetic Separators		0,2	1,5	6
Actuators		0,1	0,5	2
Levitation			0,1	1
Reaction Control			0,1	1
Sensors			0,1	1
Subtotal		0,4	3,3	16

Conclusions

In concluding I want to summarize the impressions I gained during the preparation of this contribution. The economical future of applied superconductivity is resting on the metallic as well as on the oxide superconductors. The driving forces are the availability of cryogen free systems which are more and more incorporating HTS materials with Nb_3Sn as a possible forerunner. The future applications will depend strongly on inventions and developments in other fields and on the development of cheap and reliable refrigeration systems for the temperature range above 10 K. The future of applied superconductivity appears bright and justifies the substantial public funding which supports this field but it also will need this support for quite some time to come. The technology of superconductivity needs a vacuum shielded low temperature environment - this creates long development times. The remarkable achievements in the application of HTS superconductors already nine years after their discovery reflects the great public support during their infancy. The European financial engagement in this field however is considerably below the standards of the United States and Japan. This needs a readjustment if Europe wants to maintain its presently good position in the world market of products using superconductivity as enabling technology.

Acknowledgments

I have to thank Audrey and Martin Wood, M.Leenen, H.Heinrichs, W.Weingarten, P.Komarek, G.Bogner, Ian McDougall, K.I. Sato, P.Arendt, Richard Withers, Heinz Chaloupka and the colleauges in so many European countries for their support in collecting the information for this report. Without their help this contribution would not have been possible. I also want to thank the BMBF , Germany for supporting this work under grant nr. 13 N 6634

References

[1] M.Wood, Proceedings of the ASC 94
[2] B.T.Matthias, T.H.Geballe, S.Geller and E.Corenzwit, Phys. Rev. **95**,1435 (1954)
[3] Y.Ijima, K.Onabe, N.Futaki, T.Tanabe, N.Sadakata, O.Kohno, Y.Ikeno, J.Appl. Phys. **74** , 1905   (1993).
[4] S.R.Foltyn, P.N.Arendt, X.D.Wu, W.R.Blumenthal, J.D.Cotton, J.Y.Coulter, W.L.Hults, H.F.Safar, J.L.Smith, D.E.Peterson, Proceedings of the International Workshop on Superconductivity, 1995, Maui, Hawaii
[5] C.Rosner, private communication
[6] K. Sato,K.Ohkura, K.Hayashi, M.Ueyama, J.Fujikami and T.Kato, ubmiited to PHYSICA B, proceedings of the International Workshop on Advances in High Magnetic Fields, Feb. 1995, NRIM, Tsukuba, Japan
[7] P. Komarek, contribution to this conference
[8] D.Leroy, R.Perin, proceedings of the ASC 94
[9] H. London, Nature **133**, 497 (1934)
[10] A.B. Pippard, Proc. Roy. Soc. A 191, 371 (1947)
[11] W.M. Fairbank, Phys.Rev. 76. 1106 (1949)
[12] E.Maxwell, P.M.Marcus and J.C.Slater, Phys. Rev. 74, 1234 (1948)
[13] H.A.Schwettmann in "Near Zero", ed. by J.D.Fairbank,B.S.Deaver, Jr., C.W.F.Everitt, P.F.Michelson, (W.H.Freeman and Company, New York, 1988), p 376
[14] L.R.Elias, W.M.Fairbank, J.M..J. Madey H.A.Schwettmann and T.I.Smith, Phys. Rev. Lett. **36**, 71 (1976)
[15] H.Grunder, private communication
[16] G.Cavallari, E.Chiaveri, J.Tückmantel and W.Weingarten in Proceedings of the EPAC 94, ed. by V.Suller, Ch.Petit-Jean-Genaz, (World Scientific Publishing Co., 1994), p 2042
[17] H.Lengeler, ibid ref. [2.2.8], p 249
[18] H.Heinrichs, H.Piel and R.W. Röth, contribution to this conference
[19] M. Leenen, ibid ref.[2.2.8], p 2060

Inst. Phys. Conf. Ser. No 148
Paper presented at Applied Superconductivity, Edinburgh, 3–6 July 1995
© *1995 IOP Publishing Ltd*

THE APPLETON LECTURE

Large Scale Application of Superconductivity

P. Komarek

Research Centre Karlsruhe, Institute for Technical Physics, D-76021 Karlsruhe, Germany

Abstract. Large scale application is a prior topic since practical superconductors became available about 30 years ago. Many hopes did not materialize yet, others like magnets for small and huge research facilities and the commercial market for NMR-spectrometers and magnetic imaging systems became the dominating areas of present day applications.

With the discovery of the high T_c superconductors (HTSC) immediately all hopes for a much broader large scale application reappeared and indeed there exist a great deal of justifications for that mood. The key for breakthroughs is of course sufficient progress in conductor performance. After several years of hard work in this area it became rather clear now where at least near term applications can be established. At 77 K-cooling, power transmission cables, fault current limiters and even transformers can become attractive, at much lower cooling temperatures (about $4 \div 20$ K) all kind of magnet applications are looking promising. Thus, it is not surprising, that important development projects have been started worldwide in the last two years with good progress so far. Especially the attempt for a breakthrough in power engineering might be very important for the future of superconductivity, on the other hand this is also a very challenging field due to the severe boundary conditions for compatibility and reliability within the whole power grid system.

The new hopes have simultaneously consolidated further low T_c superconductor (LTSC) applications. Beside the above-mentioned ones this concerns e.g. small and medium size SMES. Thereby a bridge between LTSC- and HTSC-application is switched by HTSC-current leads reducing cryogenic power needs substantially.

The big fusion magnet activities can be considered as a special case. Here the most sophisticated technology in LTSC is required, challenging conductor and coil manufactures. Projects as the Large Helical Device, Wendelstein 7-X and especially ITER could on the other hand create a positive thinking in industry to invest also in long term projects with HTSC as mentioned above.

1. Introduction

For the purpose of reviewing the status of large scale application of superconductivity and comparing it with an earlier one, it is useful to treat separately different major areas of applications. As best moment for comparing the state of the art and its future prospects with the present ones, the year 1987 is chosen for obvious reasons, namely metallic superconductors were well in use, while HTSC were just discovered. The subdivision chosen to discuss the applications is in the order of the amount on activities, research, equipments, medical diagnostics, thermonuclear fusion, power engineering and finally, other

10

engineering techniques. To give a rather complete picture, it is of course necessary to consider the use of metallic superconductors as well as that of HTSC.

2. Superconductivity in research equipments

Looking back to the late 1960s, where bubble chamber magnets became one of the first large scale applications of superconductors, continuously since that time magnets for particle physics, laboratory magnets and very high field magnets for solid state physics and NMR spectroscopy respectively, as well as superconducting cavities have been an active driver for the technology at all. Thereby always challenging development goals could be met.

2.1. Particle detector magnets

At the time of 1987 large thin wall detector magnets for modern, complicated particle detector arrays have been built successfully. These were solenoids with typically 5 m bore, a length of 6 - 7 m and a flux density $\geq$ 1.5 T, produced by a one layer winding of a NbTi/Cu-conductor with an additional Al-stabilizer cross section. The largest ones were ALEPH and DELPHI for the LEP-accelerator at CERN, Geneva, still in operation now [1].

The design of new accelerators called also for new detector types and a bigger detector volume. The first new type was the toroidal coil system for CEBAF, with a kidney shaped contour of the toroidal coils, a volume similar to that of the earlier solenoids, but now an induction of 3.5 T at the winding. This magnet was brought into operation recently [2]. An extraordinary example of a new detector magnet system is that of the ATLAS detector, under development for LHC at CERN, sketched in Fig. 1 [3]. It consists first of a very big toroidal

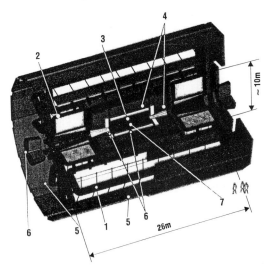

Fig. 1   Layout of the biggest particle detector so far, the ATLAS detector for LHC, under design and engineering development now [3]. (1 large torus coils, 2 end torus coils, 3 thin wall solenoid, 4 hadron calorimeters, 5 ion detectors, 6 particle calorimeters, 7 central detectors)

magnet system with nearly rectangular shaped coils of about 26 m length, an inner torus diameter of 9.4 m and an outer one of 19.5 m. In the centre is a thin wall solenoid of 5.3 m length and 2.44 m inner bore for a central field of 2 T. The ends are plugged by two "smaller" toroidal coil systems. The maximal field at the conductor of the torus coils is 4.2 T. These conductors consist - as nearly typical for the detector magnets - of a NbTi-conductor cable embedded in an Al-bar (7 x 70 mm2).

2.2. *Particle accelerator magnets*

The beam guidance in ring accelerators by dipols and the refocussing by quadrupols is one of the few superconducting magnet applications where a series production could already take place in industry. The successful completion and reliable routine operation of the Energy Doubler at FNAL, USA, and of HERA at DESY, Germany, since the end of the 1980s, demonstrated the availablility of the technology with NbTi-cable conductors, cooled at about 4.2 K [3]. It also provided a positive mood for the development of advanced magnets for new accelerators. Out of them the LHC (Large Hadron Collider) at CERN succeeded to get approval. It will be installed in the already existing LEP-tunnel with 27 km circumference. More than 1 800 dipoles (~ 9 m length) and 400 quadrupoles (~ 4 m length) will be needed. To match the developed magnet technology with NbTi-cable conductors with the required high field needs, cooling with He II at 1.8 K is foreseen. As already demonstrated by many model magnets and even first industrial full size magnets, this will savely provide a rated flux density of 8.6 T [4]. Another feature of the development was the twin-aperture, meaning that the field coils for the two counter-rotating proton-antiproton beam lines are placed in a common iron yoke and cryogenic vessel. This is saving a lot of space in the LEP tunnel. The completion of LHC is scheduled for the time between the year 2002 and 2008, depending on the annual funding.

First dipoles for even higher field are subject of basic technology developments and just recently a Nb3Sn-dipole (1 m long) has been tested successfully up to 11.5 T [5].

2.3. *RF cavities*

At CERN and CEBAF, the availability of good quality Nb-cavities coming from industrial series productions has been demonstrated. Acceleration voltages of 4 - 5 mV/m can be provided routinely.

The development of cavities with much higher voltage capability (~ 20 mV/m) has been started for a new generation of linear accelerators. Focus is so far the "Tesla Test Facility" (TTF) at DESY, Germany.

2.4. *High field and usual laboratory magnets*

Laboratory magnets, especially solenoids, are a commercial product since many years. The progress concerned in earlier years mainly an increase of the maximum flux density in the magnet bore. Most advanced Nb3Sn-conductors enables now the fabrication of magnets up to 21 T in a bore of about 5 cm if cooled at ≤ 2 K.

Recently new development lines have been started too, based on the advances in HTSC and cryocoolers. First line are the so called "cryogenfree" magnets. Starting point have been magnets with Nb_3Sn-conductors for a moderate field (a few Tesla), operated at about 10 K. The winding is cooled by heat conduction only, so that with low loss HTSC current leads an operation with a cryocooler can become feasible. Of course, the strong hope is to apply HTSC at 20 K or even higher in near future. Indeed recently a small solenoid with a 60 mm bore, wound with a Bi(2223) tape conductor achieved a flux density of 3 T at 20 K (4 T at 4.2 K) [6].

The second development line concerns the enhancement of the ultimate limit for superconducting high field magnets by HTSC. Highlights so far have been the recent tests of insert coils with Bi(2223) tape conductors in both a 21 T superconducting solenoid and a 23 T Hybrid-magnet. At 4.2 K up to 1.46 T additional flux density could be achieved, at 27 K still 0.9 T [6]. These experiments carried out by the NRIM/Tsukuba and Sumitomo El., are a very positive sign that Bi-conductors can complement the metallic superconductors to build mangets with a flux density level substantially higher than 21 T. This will have an important impact also on NMR-spectrometer magnets discussed below.

2.5. NMR spectrometer mangets

NMR spectrometers are a highly sophisticated market product. In the time since 1987 the maximum frequency could be increased from 600 MHz to 750 MHz for commercial systems. This requires solenoidal magnets with 17.6 T in the bore. Of specific importance are the field homogeneity and the timely constancy of the magnetic field. To achieve the required spatial homogeneity of $\Delta B/B_0 < 10^{-7}$ in about 1 cm^3 sphere, a special winding arrangement, a winding fabrication accurately fulfilling the computed specifications and additional outside shiming coils are required.

Even more difficult to meet and sometimes unclear in discussions is the request for timely constancy. For NMR-investigations in the solid state physics area already $\Delta B/B_0 < 10^{-5}$/h might be sufficient, while for high resolution spectroscopy in chemistry and biology $\Delta B/B_0 \leq 10^{-8}$/h is required. This can only be achieved by a persistent mode operation with all joints fully superconducting and a rated current significantly below the critical current.

As already mentioned above, the progress in high field magnet technology opens the chance for NMR spectrometers with even higher frequencies. 850 MHz systems corresponding to 20 T magnets with advanced Nb_3Sn-conductors might become a reality very soon, the only question thereby could be the timely resolution. 1000 MHz systems correspond to 23.5 T, which looks now achievable with Bi-tape conductors as innermost coils. Sufficient spatial homogeneity and expecially timely constancy (superconducting joints and resistivity of the Bi-conductor) remain as challenging development targets.

3. Magnetic resonance imaging

As known to everybody, this is still the major market for superconductivity. Developments in the recent years concerned mainly industrial improvements, like the integration of recondensing cryocoolers and recently the use of HTSC current leads.

A very interesting new approach is a cryogenfree split coil system and a cryostat providing in sito access to the patient. The coils are wound with a Nb_3Sn tape (cheap) conductor, providing 0.5 T when operated at 10 K, using a Gifford-McMahon refrigerator [7].

It is for sure that based on the continual improvements MRI will be a stable market for superconducting magnets also in the next future.

4. Superconductivity in thermonuclear fusion

Large scale fusion experiments with magnetic plasma confinement and of course later reactors need superconducting magnet systems. Two approaches complementing each other are taking place, namely the use of superconducting coils in selected large plasma experiments and the mission oriented development for the first generation of nuclear systems, at present focussed on ITER (International Thermonuclear Experimental Reactor).

4.1. Plasma experiments with superconducting magnets

The classic approach of ring coils with monolythic NbTi-conductors cooled at 1.8 K has successfully been applied for the toroidal field (TF) coils of TORE SUPRA (CEA-Cadarche) in the 1980s. This tokamak is in smooth routine operation now since 6 years [8]. Another tokamak of very similar size, T15, has been constructed with a Nb_3Sn-forced flow conductor at the Kurchatov Institute in Moscow, but for several reasons it has not been operated with full performance yet. At present two plasma experiments of the stellarator type are under construction with superconducting coils. Firstly, the "Large Helical Device" of the Japan National Institute for Fusion Science in Toki [9] and secondly the Wendelstein W7-X of IPP Garching [10]. Fig. 2 shows as example the coil system of W7-X and a winding pack cross section. The nonplanar coils (50 around the torus circumference) have an equivalent diameter of about 3 m and a flux density of 6 T at the conductor. This conductor is of cable in conduit type with NbTi/Cu strands. The conduit is made from an Al-alloy which experience a hardening process during the curing of the epoxy impregnation of the winding. The coil development for W7-X is well proceeding, the final EURATOM approval of the whole W7-X machine is expected for autumn of 1995.

4.2. The coil developments for ITER

The biggest activity at present concerns the development for the "International Thermonuclear Experimental Reactor" (ITER). This effort is a joint one of many laboratories and companies of all ITER partners (EURATOM, Japan, Russian Federation, USA). The TF coil system will consist of 20 D-shaped coils with 10.9 x 16.5 m^2 at the two main axes of the D. The maximum flux density at the conductor will be 12.5 T. The poloidal coil system is consisting of 7 outer ring coils and the central solenoid (CS). In the later one a field swing of 13 T within about 60 s will be required [11]. These specifications call for very sophisticated conductors which are shown in Fig. 3 [12]. Of course, due to the high field strength, Nb_3Sn has to be used as superconductor. The large size and the suitability for field changes call for a high rated current and a multistage cable.

14

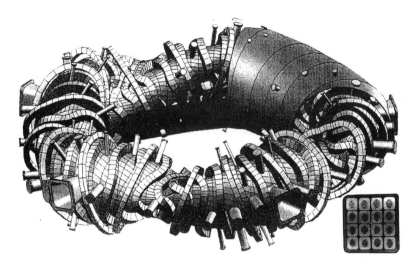

Fig. 2   The coil system for the stellarator W7-X of IPP-Garching [10]. On the lower
          right corner a cross section of the superconducting winding pack can be seen.

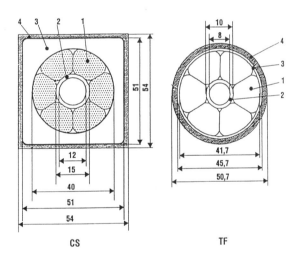

Fig. 3   The conductors for ITER (1 subcable, 2 central tube for supercritical He-flow,
          3 conduit, 4 turn insulation) [12].

This current is chosen as 60 kA (critical current about 120 kA). The TF-conductor
has a thin wall round conduit (because it will be embedded into grooves of steel
plates), the PF-conductors a squared thick one, made of Incoloy 908 or steel
(316 LN). The conductors are developed and 6 km of cable (26 tonnes of strands)
are under fabrication, to be used for a CS- and a TF-model coil, which will be
tested in 1998. This can be seen as a big boost for Nb_3Sn conductor
manufacturing. If ITER finally will be built, about 350 000 km on strands (1 600
tonnes) will be needed, extending present annual world production by about two
orders of magnitude.

5. Superconductivity in electric power engineering

Since the beginning of large scale application developments, electric power engineering equipments have been considered as a long time prime target. However, in this field superconductivity has to compete technically and economically with well established conventional equipments proven and continuously improved since many decades. Thus, it is not surprising that in spite of remarkable technical progress a real market implementation has not taken place yet. However, there exist strong indications that with the present progress on HTSC this might change within the next decades.

5.1. Generators

Due to a minimal useful size, mainly turbogenerators have been investigated so far. The progress in the recent years was not very fast. The biggest effort is still seen in Japan, where the so called "Super GM" project is executed. It will be featured by the test of 3 generators with about 70 MVA each, in 1996/97. They differ mainly in the type of NbTi conductor in the rotor field winding, suited for different fast transient field behaviour.

Unfortunately, the present tendency on the market is not bigger, but smaller generators ($\leq$ 300 MVA). In accordance to recent studies by Siemens [13] it will be not easy even for generators with LN_2-cooled HTSC rotor windings (if suited conductors would be available) to compete. Fig. 4 shows the reasons. If the costs of a water cooled turbogenerator for 1 000 MVA are set as one unit the costs of other systems in dependence of the rated power P_n follow as sketched. From the investment costs of the s.c. generator one can deduct the energy saving costs due to the improved efficiency. As can be seen, even LN_2 cooled systems might hardly have a breakeven point below 500 MVA.

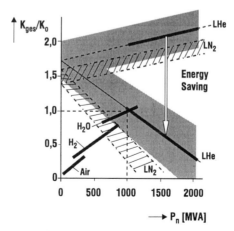

Fig. 4   The breakeven margin for superconducting turbogenerators in respect to conventional generators with H_2O, H_2 or air cooling respectively. The cost ratio K_{ges}/Ko, taking into account the energy saving of superconducting generators is plotted versus the rated power P_n [13].

Such considerations are certainly the reason that industry is not agressively active in this field.

5.2.   Motors

In the late 1970s and early 1980s technical feasibility of homopolar machines was demonstrated. However, the prospects were not convincing enough for further projects.

With HTSC, studies predict economic benefits for large synchronous motors (1 - 10 MW) in comparison to present day induction motors, due to the energy saving, if operated at 77 K. Based on that, a programme has been started by Reliance El. and ASC in USA, to develop such motors with Bi(2223) conductors in the rotor winding. So far, a 3.75 kW model of a rather simple design has been operated successfully and a 100 kW model is in progress [12]. However, for full scale application HTSC wires with a $j_c \sim 10^5$ A/cm^2 at 5 T and 77 K will be required, which looks still very challenging.

5.3.   Power transmission cables

Twofold economic chances for cables with HTSC, cooled at 77 K, have been identified. Firstly, a breakeven for systems ≥ 1 GVA transmission power and secondly cost benefits and no additional space requirements for the retrofit of existing cables, enhancing the transmission capacity from typically 100 MVA to 300 - 400 MVA. However, for both cases the benefits will only materialize if a critical current density of the HTSC material of about 100 kA/cm^2 can be taken into account (at the fortunately small flux density of ~ 0.2 T). Other constraints concern the a.c. losses, which should be kept below 0.1 W/m (conductor losses only). On the other hand, comparing with the present state of the art of Bi(2223) conductors, such values are challenging, but seem to be feasible with appropriate development effort.

Indeed, three development projects have been started, one in Germany, Japan and USA each, with similar aims in the first project phase until 1997/98 (test cable of several ten meters relevant for 66 - 115 kV), as reported elsewhere in this conference.

5.4.   Transformers

Modern power transformers have already very high efficiency, namely ≥ 100 MVA and already $\sim 99.4\%$ for systems of 1-10 MVA. However, even the remaining small losses, integrated over a continuous operation period of 30 - 40 years represent an energy amount where further saving would pay off. Therefore, it is worth to consider superconducting power transformers.

There exist two major difficulties for superconducting transformers. Firstly, the conductor must have very small a.c. losses. Such small losses (~ 20 μW/Am at 0.1 T/50 Hz) could be demonstrated in recent years by NbTi/CuNi conductors with very thin filaments [16] and resulted in the construction of model transformers up to 330 kVA with such conductors [17]. The second problem comes from the peak short circuit current conditions. Present day transformers are designed such, that they can withstand large short circuit

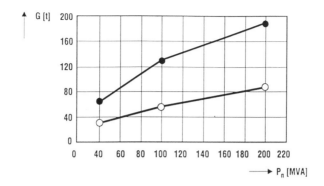

Fig. 5   Gain in weight G for s.c. HTSC transformers (o) in comparison to conventional
            transformers (•) for a rated power P_n, based on an ABB study [18].

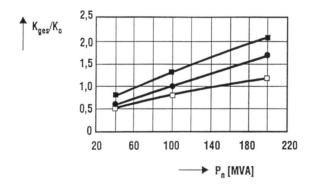

Fig. 6   Cost advantages K_{ges}/K_0 for s.c. HTSC transformers in comparison to
            conventional transformers (•) versus the rated power P_n. Cost advantages
            would not be given for present HTSC conductor cost (■), but significant for
            costs which are a factor of 5 lower (□). These are results of an ABB study [18].

currents, which can reach values up to 20 times the rated current. If a
superconducting transformer would be designed in the same way, the required
amount on superconducting wires would become too large, increasing the a.c.
losses and the costs such, that no benefit would remain. Therefore, a
superconducting transformer can only be designed for a few times the rated
current and must be protected by a superconducting fault current limiter (FCL), a
component discussed in the next subchapter. It is also considered, that the
primary winding (or a part of it) acts as integrated FCL.

The model transformers with NbTi superconductors confirmed the
expected principle performance and the predictable savings in weight and
volume. Thus, such transformers could already be beneficial for certain
application areas as railway engineers, where weight and volume saving are a
major issue. Indeed, a French-German project (GEC/Alsthom - Siemens) for a
railway transformer with NbTi/CuNi wire has been started.

Of course, the ultimate goal would be the application of HTSC conductors to gain volume/weight and energy savings. These benefits have been identified for the very interesting power range from 40 - 200 MVA by an ABB-study [18]. Fig. 5 and Fig. 6 show the key results concerning saving in weight and costs. As already indicated in Fig. 6, the cost of the HTSC material is significant to earn the benefits. A cost reduction by a factor of 4 - 5 in comparison to present prices will be required. In additon, the study results are based on technical data for the material, which are a challenge too, namely an engineering current density $j_c >$ 10 kA/cm^2 at 0.3 T and 77 K and a.c. losses of < 0.25 m W/A.m at 0.3 T. Based on the study results a first demonstration project has been started by ABB (conductor from ASC) in Switzerland, aiming for a 630 kVA distribution transformer (18.7 kV//420 V) to be operated in 1997 at the local utility (SIG) in Geneva [16] (further observing partner is the French EdF).

5.5. Superconducting fault current limites (FCL)

The stringent need for a FCL has already been mentioned above for superconducting transformers. Not stringent, but probably still required will be a FCL for superconducting power transmission cables. Beside these needs, the increasing power of transmission and distribution networks calls for additonal measures to limit short circuit currents. Utilities worldwide are becoming more and more interested in FCL and research was growing rapidly in the recent years. The work started already with the NbTi/CuNi wires to investigate fast resistive limiters. This work is proceeding successfully and the present state of the art will be reviewed by a separate talk in this conference. The major critical problem might be the needed cryogenic power. This of course, will be a much smaller problem with HTSC at 77 K. Thereby, the material capabilities so far opened up a quite different type of FCL too, with a primary winding over an iron core, shielded by a HTSC cylinder. This results in a combined inductive/resistive current limitation and the state of the art in this research is again reviewed in another talk in this conference.

Due to the premature state of the art on HTSC wires or thick films with resistive matrix, investigations on pure resistive FCL with HTSC are in a rather early stage.

5.6. Superconducting energy storage systems

While magnetic energy storage units (SMES) are investigated since many years, flywheel systems with HTSC bearings became an additonal option in the very recent years, especially for small storage capacity ($\leq$ 100 kWh).

The attractiveness of small and medium size storage systems, especially SMES with immediate response capability, was recently the motivation for several successful model experiments in the 1 MWs range. In parallel, in USA the idea of using solenoidal SMES systems in the 1 MWs/1 MW range as uninterrupted power supply (UPS) was explored and successfully deployed for the commercial market [19]. It consists of a solenoid with a lossless NbTi-wire, a helium pool boiling cryostat and a sophisticated power electronic converter system. A further improvement step are HTSC current leads now. The initiative was taken by the small company Superconductivity Inc., now also IGC became involved with a 3 MWs/0.75 MW system. It will be interesting to see how the market chances in comparison to conventional UPS systems will develop in the near future.

In Japan and Europe the aim for small SMES (and flyweel systems) is simply speaking "power quality". In all cases for fringe field minimization, only toroidal coil systems are designed and built (with NbTi-wires). In Japan the major activity is concentrated in a national programme for a 100 kWh/40 MW system with a broad spectrum of different design principles (for conductors etc.) under investigation. In Europe, at present mainly in Germany activities are running with experiments in the MWs-range and a detailed design activity of utilities and industry to confirm the estimated economic advantages for SMES as spinning reserve of the utility power grid [20].

Power quality is also the issue of the largest SMES under construction so far, a 0.5 MWh/30MW system for the Anchorage Municipal Light and Power, USA, built by Babcock & Wilcox. It consists of a 18 m long solenoid with 2.8 m diameter, with a NbTi-conductor winding [21].

The investigation of very large SMES systems (> 500 MWh) for peak load shaving is not seen as an urgent task now. Different studies showed that economic benefits in comparison to pumped hydro or pressurized air storage can hardly be seen, so that mainly limited availability constraints for these systems would call for a SMES. Thereby the potential use of HTSC with 77 K cooling would not change the situation substantially. This is different for small SMES, where HTSC application, already as "cryogenfree" magnets at 20 K would represent an additional economic benefit and is also not out of scope with the expected progress in conductor technology. Coils for about 5 - 6 T, wound with several kilometers of conductors with about 1 kA rated current and a conductor current density of ≥ 10 kA/cm^2 would be needed. This defines the conductor development goal for that application which could be very promising.

Conventional flywheel generator/motor systems are well developed and in use up to rather large units (GWS). The appeal of superconductivity is a reduction of friction losses by several orders of magnitude in using passive superconducting bearings. These bearings consist so far out of melt textured Y(123) bulk pieces face to face with a powerful permanent magnet. Due to the intrinsic high critical current density in melt textured Y(123), sufficient high levitation forces could alraeady be demonstrated. A few experimental systems up to about 100 Wh and a power in the few kW-range have been operated succesfully. Larger systems are planned and should demonstrate competitiveness in certain application areas (e.g. solar and wind power stations) [22].

6. Other engineering applications

In this area mainly levitated trains, magnetic separation and synchrotron radiation sources can be mentioned at present.

Concerning levitated trains, large activities with superconducting magnets can still be reported only from Japan. They will be subject of a separate talk in this conference. However, in any case it should be recognized that the principle political and economical decision for or against such train systems is not depending on the type of the magnet system and its possible coolant.

Synchrotron radiation sources need superconducting bending magnets if they should be very compact or they need wiggler magnets for specific synchrotron radiation beam lines. However, since the completion of the compact system of

Oxford Instr. not many further activities with superconducting systems have been reported yet.

Similar seems to be the situation in magnetic separation. Well functioning prototypes (NbTi conductor coils) for open gradient and for high gradient magnetic separators have been operated in the late 1980s. Not much has been reported on further systems and a real market in the last few years. No clear reasons are known for that stagnation. However, just recently cryogenfree magnets for high gradient magnetic separators in caolin industry started as a perhaps convincing product now.

7. Conclusions

Large scale applications are already at present and will also be in future a major market for superconductors. NbTi and Nb_3Sn wires are the foundation for the present market products MRI and NMR and for large research projects in elementary particle physics and nuclear fusion. HTSC would be able to immerse the different mentioned applications only step by step, depending on improved properties. As a start, HTSC current leads are already becoming an important component to improve equipments built with LTSC windings, e.g. to construct cryogenfree magnets operated with the simultaneously improved cryocoolers.

The big area of electric power engineering has not been conquered by superconductivity yet. However, small SMES units for power quality improvement, especially for UPS and for spinning reserve are now on a good way to be considered seriously by the utilities, or already being in market deployment, as UPS systems. In several other areas, as e.g. power transmission cables, HTSC cooled at 77 K are essential for a breakthrough. It is good to see that relevant development projects for cables, transformers, fault current limiters (FCL) and even motors, have been started. Especially the FCL became a strongly desired device by the utilities, increasing the development effort everywhere. Thereby both, LTSC and HTSC systems can have big market chances.

For all these applications the properties of HTSC wires must improve substantially, perhaps in several milestones to meet certain application needs. Such milestones could be first the needs for transmission cables and transformers, where the good properties only in flux densities of 0.1 - 0.3 T would be requested, while for winding applications a field level of typically > 5 T has to be taken into account. The critical current densities must approach 100 kA/cm^2, the rated current of final conductors has to be several kA and in a.c. applications the 50/60 Hz losses have to be typically $\leq$ 0.25 mW/Am. Beside these already challenging specifications, mechanical strength and much lower prices will be essential.

Meeting all these constraints for HTSC will take time and patience will be needed. For the meantime the capabilities with LTSC should be used as agressively as possible to strengthen the superconductivity market.

References

[1] Baze J M and et al. 1988 *IEEE Trans. on Magn.* **24** *1260-63*

[2] Ross J S H and et al. 1995 *IEEE Trans. on Magn.*, in print
     (Proc. MT 14 Tampere, Finland, June 1995)

[3] Wiik B H 1988 Bericht DESY HERA 88-05

[4] Bona M, Perin R, Acerbi E and Rossi L 1995 *IEEE Trans. on Magn.*, in print
     (Proc. MT 14 Tampere, Finland, June 1995)

[5] Den Ouden A, et al. 1955 *IEEE Trans. on Magn.*, in print
     (Proc. MT 14 Tampere, Finland, June 1995)

[6] Togano K 1995 *IEEE Trans. on Magn.*, in print
     (Proc. MT 14 Tampere, Finland, June 1995)

[7] Laskaris E T, et al. 1955 *IEEE Trans. on Appl. Superconductivity* (in print)

[8] Turck B 1995 *IEEE Trans. on Magn.*, in print
     (Proc. MT 14 Tampere, Finland, June 1995)

[9] Satow T, et al. 1993 *IEEE Trans. on Appl. Superconductivity*, **3** 365-68

[10] Sapper J 1995 VDI Berichte 1187 233-47

[11] Green B J and Huguet M 1995 *IEEE Trans. on Magn.*, in print
      (Proc. MT 14 Tampere, Finland, June 1995)

[12] Mitchell N 1994 *IEEE Trans. on Magn.* **30** 1602-7

[13] Intichar L 1995 VDI-Berichte Nr. 1187 165-75

[14] Joshi H C 1995 *IEEE Trans. on Appl. Superconductivity* (in print)

[15] Ashworth S P, Metra P and Slaughter R J 1994 ETEP,
      Vol. 4 293-300

[16] Fevrier A, et al. 1988 *IEEE Trans. on Magn.* **24** 1477-80

[17] Hörnfeld S, König F and Bonmann D 1994 ABB-Technik 1/94 13- 9

[18] Demarmels A 1995 VDI-Berichte Nr. 1187 195-203

[19] Daugherty M A 1993 *IEEE Trans. on Appl. Superconductivity* **3** 204-6

[20] Jüngst K P, Komarek P and Maurer W (Editors) 1995 Use of
      Superconductivity in Energy Storage (Singapore: World Scientific
      Publ.) (in print)

[21] Huang X, et al. in [20]

[22] Ishikawa F and Higasa H 1995 VDI-Berichte Nr. 1187 177-93

Inst. Phys. Conf. Ser. No 148
Paper presented at Applied Superconductivity, Edinburgh, 3–6 July 1995

SIR JAMES DEWAR

R. J. Soulen, Jr.

Materials Science and Technology Division, Naval Research Laboratory, Washington, D. C., 20375-5000, USA.

Abstract Sir James Dewar is best known for his work in cryogenics, although he made many important contributions to other areas of chemistry and physics. This paper will outline his career and show how the two disciplines were unified in his research.

1. Introduction

Sir James Dewar is intimately associated with the discipline which we know as cryogenics. The achievements of this pioneer are many, not the least of which was the first liquefaction of hydrogen. Dewar was widely recognized for this and other work during his lifetime: He received the Rumford Medal from the Royal Society in 1894, the First Hodgkins Gold Medal from the Smithsonian Institution in 1899, the Lavoisier Medal from the French Academy of Sciences in 1904 [1], the first Matteucci Medal awarded by the Italian Society of Sciences in 1906, and the Albert Medal of the Society of Arts in 1908. He was knighted in 1904.

Even Dewar's name survives today in common usage, for he is also remembered as the inventor of the Dewar flask. This double-walled container, constructed with reflective surfaces insulated by a vacuum, is universally used to store cryogenic fluids. So ubiquitous are these vessels in cryogenic laboratories throughout the world and so frequently does the term referring to them appear in ordinary conversation, that they are simply called "dewars".

Dewar's research interests and accomplishments, however, were not restricted to cryogenics. He made significant contributions to such disparate fields as spectroscopy, organic chemistry, magnetism, and dielectric phenomena. Indeed, a curious fact is that the Dewar flask, so well known to the cryogenics community, actually had its origin in chemistry research. This article will not only indicate how the Dewar flask evolved naturally from its originator's broad research interests, but it will also trace Dewar's career, highlighting some facets which are less well known.

2. Education

James Dewar was born in Scotland on September 20, 1842 in the village of Kincardine-on-Forth, which is roughly 15 miles upstream and on the opposite shore of the river Forth from Edinburgh. He was the youngest of seven sons born to a Scottish innkeeper, Thomas, and his wife, Ann. A near drowning at the age of 10 left him in poor health and for the next few years he learned how to make fiddles from the village joiner [2]. The manual dexterity he developed during this time was to serve him later as an experimental chemist and physicist.

Dewar was educated at the Dollar Academy (1855-1858) where he excelled in mathematics. From 1859 to 1861 he was enrolled at Edinburgh University where he was introduced to the physical sciences by physicist and geologist, David Forbes. Dewar eventually became a student of Lyon Playfair, Professor of Chemistry, and consequently he graduated as a chemist. Dewar remained at Edinburgh to serve as an assistant to Playfair and later to Playfair's successor, Crum Brown. He held a lectureship in chemistry during this period.

3. Benzene

Dewar's first scientific publication [3] appeared in 1867 and addressed the structure of benzene, one of the most important issues in organic chemistry at that time. The composition of this material had been determined, but its structure was still unknown. A scant two years before, August Kekule, then at the University of Ghent, framed his famous proposition that benzene consisted of a ring of six carbon atoms trimmed with six peripheral hydrogen atoms. In an attempt to confirm Kekule's model, Dewar conducted a series of experiments in which benzene was decomposed by oxidation. The article does not form a conclusion about the structure of benzene, but it goes on to describe a mechanical model for hydrocarbons which Dewar had developed. He used it to illustrate seven possible structures for benzene which were consistent with the chemical formula and with experiments of that time. One of these structures still appears in textbooks [4] discussing benzene and is known as the "Dewar formula". Another was the Kekule formula. By noting how many independent reactants resulted when substitutions were made for the hydrogen atoms, chemists were eventually able to eliminate all but one of the possible structures. The survivor is the formulation attributed to Kekule.

Dewar subsequently left Edinburgh University to hold positions as lecturer at the Royal Veterinary College, Edinburgh (1869-1873) and as assistant chemist to the Highland and Agricultural Society of Scotland (1873-1875). During this time he continued his research in organic chemistry, but he also collaborated with J. G. M'Kendrick on the physiological effect of light on animals and with Professor P. G. Tait (Edinburgh University) on production of vacuum and on estimates of the high temperatures encountered in the sun and in electric sparks.

4. The First Dewar Flask

During this period Dewar dealt with another topic of contemporary interest which, coincidentally, was the first step to his invention of the Dewar flask. Thomas Graham, who was by that time Master of the Mint, had conducted several experiments on diffusion, including diffusion of gases in metals. Graham noticed a peculiar phenomenon: While exposed to an atmosphere of hydrogen when heated and subsequently cooled, palladium absorbed between 500 to 600 times its volume of hydrogen. Graham argued that a new alloy, dubbed "Hydrogenium", had been formed. An alternative hypothesis was simply that the hydrogen had been absorbed in the palladium without chemical reaction. To decide the issue, Dewar measured the specific heat of hydrogen-charged palladium in a high-temperature calorimeter. He reported his conclusion in 1872 [5], and published the data and description of the apparatus in 1873 [6]. The heart of Dewar's apparatus was a double-walled calorimeter made from brass. The sample resided in the inner chamber which was surrounded by an outer chamber. *In order to improve the thermal isolation of the calorimeter, Dewar evacuated the volume between the two chambers.* Calorimeters used previous to that time had double walls which contained at most a desiccant. Dewar realized the benefit of a vacuum in providing better thermal isolation for the calorimeter. This experiment was the first in a series of studies Dewar conducted to improve calorimeters which led to the development of the Dewar flask as we know it today. Incidentally, Dewar's specific heat measurements indicated that the hydrogen was absorbed by the palladium.

5. Cambridge and the Royal Institution

In 1875 Dewar was elected Jacksonian Professor of natural experimental philosophy at Cambridge University. From all reports he was not happy there, so that two years later, when he was offered the position of Fullerian Professor of Chemistry at the Royal Institution, he accepted. He held both positions until his death in 1923. At Cambridge, Dewar did not teach, choosing to pass this responsibility onto associates, but he did maintain a collaboration with G. D. Liveling on spectroscopic studies of vaporized metals which lasted several years (1877-1891).

At the Royal Institution, Dewar took up the work of his predecessor, Michael Faraday, on the liquefaction of gases. Faraday, who had been director of the Royal Institution, set out to liquefy all the known gases. Six of them (oxygen, nitrogen, carbonic oxide, nitric oxide, "marsh gas", and hydrogen) resisted his attempts and, thinking that they would not liquefy under any conditions, he labeled them "permanent gases". Several of these gases were subsequently liquefied at cryogenic laboratories which were established by Cailletet in Paris, by Pictet in Geneva, and by Wroblewski and Olszewski in Crakow. One year after joining the Royal Institution, Dewar gave his first Friday evening discourse held at the theatre at the Royal Institution. The talk was entitled "The Liquefaction of Gases" [7] and Dewar reviewed the recent history of cryogenics. Using a Cailletet apparatus obtained from Paris, he also liquefied a few droplets of liquid oxygen for his audience. This lecture was the first demonstration of liquefaction of oxygen in England, but it also marked the inauguration of Dewar's legendary career as a lecturer at the Friday evening discourses.

It took several more years for Dewar to establish his own working cryogenics laboratory at the Royal Institution. It was unsurpassed in England and was on competitive par with the few similar institutions scattered throughout the world (including the aforementioned, which were eventually joined by Onnes in Leiden). These laboratories were engaged in an effort to liquefy the remaining "permanent" gas hydrogen, as well as the recently-discovered gas, helium. The account of this race has been eloquently told elsewhere [2], [8]. Suffice to say, Dewar was the first to liquefy hydrogen in 1898 [9], while Onnes won the race to liquefy helium in 1908.

6. Evolution of the Dewar Flask

Dewar found that he needed better containers to hold the liquids he prepared for his research laboratory and for his lectures at the Friday evening discourses. Having demonstrated the utility of vacuum insulation for calorimetry in 1872-73, Dewar turned to a series of experiments from 1893-1898 to determine the best way to prepare the vessels. By this time the dewars were blown from glass and by 1893 Dewar had demonstrated that vacuum insulation improved the holding time of liquid nitrogen by a factor of five [10]. In the same publication he refers to the efficacy of coating the glass with a thin layer of silver or mercury to reduce losses due to radiation. Experiments were carried out to test various shielding techniques which included metallic coatings such as mercury and silver [11]. Dewar also experimented with metallic foils [12] and barely missed inventing superinsulation. Improvement in holding time by a factor of 13 was obtained by good radiation shielding.

The Dewar flask, when used at the opposite extreme of temperature, i.e., to keep liquids hot, is commonly known as the *Thermos* flask. According to R. G. Scurlock [13], this application was discovered accidentally. In his words:

> Dewar had considerable difficulty in finding competent glass blowers willing to undertake the construction of his double-walled vessels, and was forced to get them made in Germany; by 1898, a ready supply became available. However, it was said that the discovery by a German glass-blower, Muller of Coburn, that a silvered vacuum flask could also be used for keeping milk hot overnight for feeding his baby, led to a major commercial development, namely the "*Thermos Flasche*" for keeping liquids hot. The manufacture of Thermos Flasks rapidly developed into an important industry, first in Germany and then in the UK and USA. Dewar never patented his silvered vacuum flask and therefore never benefited financially from its invention.

In investigating the history of the Dewar flask, I conclude that Dewar had clear and plain title to this invention [14], yet he chose not to profit from it by applying for a patent. The most probable explanation is that Dewar was an academic and it was not common practice for an individual in his position to patent the fruits of his research.

7. Low Temperature Physics

More or less simultaneous with his efforts to liquefy gases and to improve cryogenic storage vessels, Dewar carried out (1892-1897) a sweeping canvas of the low-temperature behavior of materials in collaboration with Professor J. A. Fleming, of University College, London. Dewar and Fleming studied the temperature dependence of the electrical resistance of several materials and elements, studied the thermopower of some 24 metals and alloys, studied the dielectric constant of many materials and inorganic substances, and studied the magnetic behavior of steels, of liquid oxygen and liquid air at low temperatures [15].

The study of electrical resistance was perhaps the most interesting of these studies because resolution of a theoretical controversy was at stake. There were diametrically opposed predictions about what would become of the resistance of metals when the temperature approached absolute zero; one suggested that the resistance must vanish at $T=0$, whereas another argued that it should increase indefinitely as this limit was approached. One of the metals surveyed by Dewar and Fleming was mercury. In 1896 [16] they extended the measurements below the limit of -100°C reached by previous experimenters. They noted the very large drop in resistance at a temperature of -50°C as the liquid froze and noted further that the resistance continued to drop as the temperature decreased. In a later publication [17], the measurements were extended to lower temperatures (down to 20 K) where the resistance was found to cease to depend on temperature. Dewar characterized the complete R(T) curve as:

> ...the electrical resistance of mercury, in which we were able to observe the resistance of the metal far below its melting-point, and considerably above it when in the molten state. The curve connecting the resistance of mercury with temperature, throughout this range, including the change in state, was somewhat like the disused old English ∫, the temperature being measured horizontally to the right, and the resistance vertically upwards. ..

Dewar and Fleming found that the R(T) curves for all the metals they measured were similar in shape to that of mercury described above. Since the resistance became insensitive to temperature at the lowest temperature, this general experimental finding could be used to support either theoretical view. Thus, resolution of the issue was destined to wait until lower temperatures could be attained. In fact, it was only after Kammerlingh Onnes liquefied helium, that the issue was decided in favor of a resistance which would be zero at $T=0$ in a pure metal. This achievement placed Onnes in a unique position to study properties of materials at lower temperatures than his cryogenic competitors. This position had an unexpected bonus: it led to the discovery of superconductivity and to a Nobel Prize for Onnes.

8. Cordite

During the same time that Dewar was so very actively engaged in cryogenics, he served as a member of the Ordnance Committee on Explosives (1888-91). The chairman was Sir Frederick Abel who had distinguished himself as a researcher in explosives having been professor of chemistry at the Royal Military Academy and chemist to the War Department. Together, Dewar and Abel developed the propellant-explosive cordite which consisted of 58% nitroglycerine, 37 % gun cotton (nitrocellulose), and 5% mineral jelly (vaseline). Brownish in color, less prone to accidental explosion than other explosives, and waterproof, this material could be extruded into long continuous lengths (hence the name). Cordite became the smokeless propellant used by the British army and navy and supplanted the far less appealing black powder. Whatever Dewar's total contribution was to the development of cordite, at least one feature bears his unmistakable trademark. Determination of the stability of the finished product was very important, and thus tests were developed by the committee. The so-called Abel test consisted of heating the cordite to 70°C and noting the change in color of an indicator which signaled the start of decomposition. The longer the period before decomposition, the better. Another test, labeled the "vacuum silvered vessel process" [18], consisted of heating a sample of cordite in what was clearly a Dewar flask and noting when the temperature increased (a reaction indicator). It is not difficult to believe that development of this test was uniquely Dewar's contribution.

9. Soap Bubbles

The liquefaction of helium required a handsome complement of machinists, glassblowers and technicians. Onnes succeeded in this effort because he was able to set up such an infrastructure, whereas Dewar was not able to match such stakes. Thus in the late 1890's and early 1900's cryogenics flourished at Leiden and languished at the Royal Institution. Dewar's interest gradually turned from cryogenics to soap bubbles and films. Still the showman, his Friday evening discourses were now devoted to displays of soap film patterns and their response to sound waves. In an article published in the last year of his life, Dewar reported on his study of the use of soap films as sound detectors [19].

10. Conclusion

Dewar's career spanned the disciplines of chemistry and physics and, by any standards, justly earned him the accolades he received. Much of his work can be appreciated for its historical value as it comprised an important step in the evolution of cryogenics. The remainder, however, influences our present every-day activities. Indeed, every time a dewar flask is decanted or a Thermos flask is filled, or a magnetic resonance image is taken, another iota in our debt to him is accumulated.

11. Acknowledgments

I would like to thank several individuals who helped me to bring forth this article. Alan F. Clark (NIST) and Jack. H. Colwell (NIST, ret.) loaned me several valuable reference books. Terrel Vanderah (NIST) provided information on benzene. Edward F. Hammel (LANL, ret.) and Linda Norton (NRL library) provided additional references. Alan Clark also is to be thanked for a critical reading of the manuscript.

12. References

[1]   First British subject to be so honored.

[2]   Mendelssohn K 1977 *The Quest for Absolute Zero* (London: Taylor & Francis)

[3]   Dewar J 1867 *Proc. Roy. Soc. Edin.* **VI** 82-86

[4]   See for example, Morrison R T and Boyd R N 1992 *Organic Chemistry* (New York: Prentice Hall)

[5]   Dewar J 1872 *Phil. Mag.* **XLIV** 400

[6]   Dewar J 1873 *Trans. Roy. Soc. Edin.* **XXVII** 167-173

[7]   An account of this lecture appears in Dewar J 1878 *Proc. Roy. Inst.* **VIII** 657-663

[8]   Dahl Per Fridtjof 1992 *Superconductivity* (New York: American Institute of Physics)

[9]   There is some controversy here concerning a prior claim by Olszewski, which is resolved in favor of Dewar. See K. Mendelssohn, p. 67.

[10]   Dewar J 1893 *Proc. Roy. Inst.* **XIV** 1-12

[11]   Dewar J 1894 *Proc. Roy. Inst.* **XIV** 393-404

[12]   Dewar J 1898 *Proc. Roy. Inst.* **XV** 815-829

[13]   Scurlock R G 1994 *Proc. Roy. Instit.* **65** 145-167

[14]   Weinhold in Germany and D'Arsonval in France used vacuum insulated containers and are therefore sometimes considered claimants, but Dewar's experiment in 1873 preceded either by several years. See K. Mendelssohn, p 56-57.

[15]   A broad view of this body of work is found in *The Collected Papers of Sir James Dewar*, Edited by Lady Dewar (London: Cambridge University Press) 1927.

[16]   Dewar J and Fleming J A 1896 *Proc. Roy. Soc.* **LX** 76-81

[17]   Dewar J 1904 *Proc. Roy. Soc.* **LXXIII** 244-251

[18]   *Encyclopedia Britannica*, 13th ed., 1926 (London: The Encyclopedia Britannica Co., Ltd) **7-8** 138-139

[19]   Dewar J 1923 *Proc. Roy. Instit.* **XXIV** 197-259

Inst. Phys. Conf. Ser. No 148
Paper presented at Applied Superconductivity, Edinburgh, 3–6 July 1995
© *1995 IOP Publishing Ltd*

Current limiting mechanisms in Bi-Sr-Ca-Cu-O tapes

D.C. Larbalestier, S.E Babcock, X.Y. Cai, S.E. Dorris*, H.S. Edelman, A. Gurevich, J.A. Parrell, A. Pashitski, M. Polak**, A. Polyanskii***, I-Fei Tsu, Jyh-Lih Wang

Applied Superconductivity Center, University of Wisconsin, 1500 Johnson Drive, Madison, WI 53706, USA
* Argonne National Laboratory, Argonne, IL 60439 USA
**Visitor from Electrotechnical Institute, Slovak Academy of Sciences, Bratislava, Slovakia
***Visitor from Institute of Solid State Physics, 142342 Chernogolovka, Russia.

Abstract. Three recent experiments addressing the current limiting mechanisms of polycrystalline BSCCO compounds are summarized. In the first the irreversibility field (H*) of a BSCCO-2223 tape was found to be a monotonically increasing function through 3 heat treatments and two deformation cycles. It is widely thought that H* is determined by the *intra*grain flux pinning properties but since the relevant flux pinning defects are likely to be ionic defects, it is surprising that they would create a continuous increase of H* in BSCCO-2223 since the single phase field is so small. The second experiment measured H* for a strongly coupled 8° [001] tilt boundary in a BSCCO-2212 bicrystal. H* and the *inter*grain J_c were both slightly depressed in comparison to the *intra*grain properties, suggesting that low angle grain boundaries are one source of a varying H*. The third experiment used magneto-optical imaging (MOI) to observe the current paths in BSCCO-2223 tapes under both magnetization and transport current flow. Some current, particularly that near the Ag sheath, flowed over the whole sample length in both cases but there were many short range current loops flowing over a few grain lengths ($\approx$ 50μm) in the central region of the tape in the low-electric-field (E) magnetization experiment, these being almost eliminated in the higher-E transport measurement. Extended E-J curves show that J is a strong function of E, consistent with the MOI data, which additionally show that both the local current paths and the local current density are highly variable, implying that many barriers to current flow have a strong E-dependence. The experiments show that current limiting mechanisms exist on many length scales and point to two important conclusions, first that BSCCO-2223 tapes are a long way from their full optimization and second that the path to this optimization lies through a better understanding of their defect materials science.

1. Introduction

A striking fact of the great effort to raise the J_c of BSCCO tapes is the variability of the properties obtained even by experienced groups. Moreover, experiments that explicitly look for evidence of locally variable J_c properties find it [1-3]. Since all technological prospects for HTS applications would be greatly enhanced by higher J_c values, it is of vital interest to understand and to control those factors that limit the J_c of present BSCCO tapes. The present paper summarizes some recent experiments of ours addressing this issue. Recent experiments have mostly been performed with tapes using "two powder" mixtures [4] of nominal overall

30

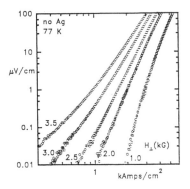

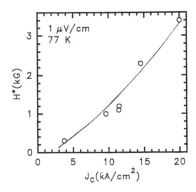

Figure 1: E-J curves of bare BSCCO-2223 tape
after the first heat treatment.

Figure 2: H* vs. Jc (0T, 77K) for tapes at
different process steps. The upper right point is
from ref. 8.

composition $Bi_{1.8}Pb_{0.4}Sr_2Ca_2Cu_3O_x$. These powders develop a high phase purity after reaction,
with small, typically 5μm or smaller, alkaline earth cuprate as the principal residual non-
superconducting phase. Such a relatively phase-pure microstructure would be expected to
develop a high J_c, but a complication of such an expectation is that a high phase purity also
implies that the liquid phase, necessary for growing the 2223 phase during heat treatment,
becomes exhausted, consequently inhibiting the healing of cracks produced in the deformation
steps which occur between heat treatments. This may explain why no tape made from such a
powder has achieved a J_c(0T, 77K) exceeding 30,000 A/cm² [4,5], unhealed cracks being a
significant and largely unconsidered barrier to current flow [5,6]. The inference that cracks are
always present is consistent with the general science of ceramics but it is unsatisfying to
postulate them without being able to be more specific about how prevalent they are, to define
where they are located, or to understand their characteristic length scales. Moreover, other
current limiting mechanisms, especially the primary ones of intra-granular flux pinning and the
electromagnetic grain-to-grain coupling strength are also very important. This diversity of
factors complicates assignment of the exact current limiting mechanisms operating in any one
composite, particularly since all may be operating in particular regions, temperatures, and field
regimes. This paper discusses three recent experiments relevant to the current limiting
mechanisms of BSCCO-2223 tapes. The experiments provide evidence for current limiting at
different length scales and by different mechanisms.

2. E-J characteristics of a BSCCO-2223 tape throughout the fabrication process

A standard process is to roll the Ag:BSCCO composite to a tape thickness of ~0.2 mm (core
thickness ~70 μm), heat treat, press, heat treat, press and then give a final heat treatment [5].
In the experiment of Edelman et al. [7], the heat treatment took place at 825° C in 7.5% O_2
and the pressure applied during the deformation between heat treatments was either 0.5 or 2
GPa. Extended E-J characteristics (~ 0.005 to 100 μV/cm) were acquired with a Keithley
1801 nanovoltmeter (Fig. 1). Measurements were made both with and without the Ag sheath;
the differences were negligible below 1 μV/cm. The very extended nature of the transition is
clear. The field at which the curvature of the characteristics changes sign, is denoted the

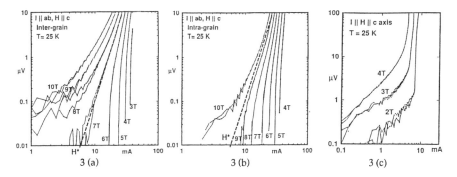

Figure 3: V-I curves for BSCCO-2212 bicrystal at 25 K. a) intergranular properties along ab plane, b) intragranular properties along ab plane, c) current passed along the c-axis of one grain. H is parallel to the c axis in all cases.

irreversibility field H*. Two interesting and surprising features are that H* increases monotonically with processing step and that it correlates to J_c(77K, 0T) (Figure 2). An additional point at higher J_c from another study [8] adds further weight to the correlation. The AC susceptibility transitions of samples beyond the first heat treatment did not show any significant evidence for 2212 intergrowths in the current path, as seen earlier in single pot powders studied by Umezawa et al. [9]. On the hypothesis that *intra*granular flux pinning is determined by the cation defect density of the 2223 phase, whose phase field is small [10], we expected that H* would rise to one characteristic value appropriate to the particular composition of 2223 phase produced by the reaction.. This hypothesis is reasonable but not borne out by the experiment. This prompts further consideration of what a macroscopically calculated J_c really measures. In fact it cannot be taken for granted that J_c defined by experimental measures of the critical current (I_c) and the total cross-sectional area (A) measures the magnitude of the flux pinning, since nothing constrains the current density to be constant within a cross-section, or even that all of the cross-section is active in carrying the current. The next two experiments show explicitly that J_c does vary spatially, first at the nanometer scale of a grain boundary and then on scales of several grains, at dimensions of order 100 μm.

3. V-I characteristics of an 8° [001] tilt 2212 BSCCO bicrystal

The naturally grown 8° [001] tilt BSCCO-2212 bicrystal was carefully cut with a laser so as to produce a uniform cross-section.. The V-I measurements of Wang et al. [11] of the *inter*- and *intra*grain ab plane J_c (Fig. 3a and 3b) and their similarity to those shown in Figure 1 for the polycrystalline 2223 tapes is clear. However, H* is not the same for both grain and boundary, H* (25K) for the grain being ~9.5 T while H* for the grain boundary is ~7 T. J_c for the boundary is ~65% of that across the grain. HRTEM images of this boundary showed it to be composed of 1/2 [010] dislocations, spaced by 1.9nm in accordance with Frank's formula [11]. Comparatively undisturbed lattice lay between the dislocation cores. Thus the electromagnetic and microstructural observations are consistent in suggesting that the supercurrent flows between the dislocation cores. H* in the strongly coupled channels may be locally depressed by the influence of the dislocation strain field on the local order parameter. After 2212 intergrowths are removed from the current path [9], the increasing H* seen in Figure 2 is then a natural consequence of reducing the distribution of angular misorientations in the current path. The increase in J_c with process step then has two aspects, one connected to the improvement of H*, which measures the *quality* of the connections in the current path, and one connected to the *density* of connections, which determines the fraction of cross-section carrying current. As liquid is exhausted by the reaction going to completion, the second factor becomes more important and the determining current-limiting mechanism switches from a nanoscale flux pinning or grain boundary one to a more macroscopic connectivity or crack-limited one.

Another aspect of the properties of polycrystalline tapes is illuminated by measurement of the c-axis E-J characteristics of one of the single crystals in the bicrystal. These c-axis characteristics (Fig. 3c) are very different from the ab-plane characteristics (Figure 3b), in that $J_c(c)$ is much lower than J_c (ab) and in particular because a low level ohmic voltage appears in magnetic fields at which none is apparent for current flow along the ab planes. We speculate that this resistance is provided by the Bi-O layers, which form an SNS stack [12]. We interpret linear resistances in polycrystalline BSCCO-2223 tapes [13] as being due to the appearance of c-axis components in the current path. They appear when the ab plane current path is interrupted, either by too low a density of low angle grain boundaries or by cracks or other progressive damage (e.g. thermal cycling damage as in ref. 13). This interpretation explicitly supports the view that the preferred current path is along low angle grain boundary connections with a large tilt component, rather than along the c-axis [14,15].

4. Magneto-optical images of current paths in 2223 BSCCO tapes.

Starting originally with microslicing as a way of determining that there is a significant variation of properties in BSCCO tapes [1], we have since shown that magneto-optical imaging (MOI) is an even more effective method of observing the variation of properties in BSCCO tapes [16-19]. The MOI technique can be used both to image the flux shielding capabilities of the tape [17,18] and to extract the current flow paths [18,19]. Figure 4 presents two reconstructions of the perpendicular component of field above the polished surface of a BSCCO-2223 tape having a transport J_c (=I_c/A) of 18,000 A/cm^2 at 77 K, 0 T. The data were taken in a field of 21 mT, which was applied after zero field cooling to 77 K. The upper data were taken in magnetization mode with only induced currents flowing, while the lower were taken with a transport current of ~0.8 I_c, where I_c is defined at 1 μV/cm. Full details of the reconstruction are given by Pashitski et al. [18,19]. As described there, the field contours are also the current stream lines. The stream lines clearly have a very different character in the two cases, being mostly granular in the magnetization case and largely laminar in the transport case. The local current density is given by the two dimensional gradient of the field. It is much lower in the magnetization than in the transport case. One cause is the strong dependence of J on E, as illustrated in Figure 1. E for the magnetization data is significantly lower than for the transport data and the effective local J would therefore be expected to be lower, just as is seen. The second characteristic of the data, the variable length scale of the current loops, was not predicted. In the magnetization data, the current percolates rather uncertainly from one end of

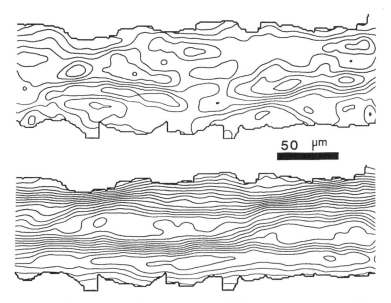

50 μm

Figure 4: Contours of constant perpendicular component of field measured above the polished surface of a BSCCO-2223 tape. The field contours correspond to current streamlines whose major axes tend to follow the ab planes. The monofilament tape was cut symmetrically in two so as to approximate a slab, the longest axis (10 mm) of which lies along the tape axis approximately parallel to the ab planes of the BSCCO-2223. This is the horizontal axis of the figure. The vertical axis of the figure is the thickness of the tape (~50 μm) which is approximately parallel to the c axis of the grains. The filament depth below the imaged surface is about 1 mm. The upper image is taken in induced current mode, while the lower is taken with transport current of 0.8 I_c applied. Both data sets were taken at 77 K with 21 mT applied perpendicular to the plane of the paper. The irregular edge of each plot represents the Ag-BSCCO interface. Fuller details are presented in refs. 18 and 19.

the field of view to the other and closed current eddies having dimensions of 50-100 μm are common. Although small in comparison to the tape length, these scales are significantly larger than the BSCCO grain size (~ 10 μm), which proves that there are many strongly coupled grain boundaries. At the higher E of the transport measurement, the granular behavior has largely disappeared and long range current flows throughout the whole cross-section, even though the local J is still strongly variable [19]. Thus some of the important barriers to current flow can be overcome by raising the electric field. An important task is now to identify the microstructural origin of these barriers and to devise strategies for mitigating them. The recent experiments of Parrell et al. [20] address one aspect of this issue. By cycling the oxygen partial pressure during heat treatment, liquid may be formed late in the heat treatment process, thus helping to control the residual crack density.

5. Summary

The three experiments reviewed above provide explicit evidence that current limiting can occur at grain boundaries and on much larger length scales. This diversity of scale is one of the important reasons why continuously improving the J_c values of long lengths of BSCCO tapes is difficult, since controlling the processing of complex materials at both nanometer and

34

macroscopic scales is inherently complex. It is the goal of studies of the type presented here to provide a better fundamental understanding of the current limiting mechanisms, so that raising the J_c will become more rapid and predictable. In any case it is clear that we are far from any intrinsic limits of performance in polycrystalline samples of BSCCO-2223.

Acknowledgments

Discussions of many points by collaborators at UW, particularly Y. Feng (now at Philips, North America), E. Hellstrom and R. Kelley, and within the Wire Development Group, particularly D. Kroeger, M. Maley, V. Maroni, G. N. Riley, and M. Ruppich are gratefully acknowledged. Support of different aspects of the work by ARPA, EPRI and NSF (Materials Research Group Program) is gratefully acknowledged.

References

[1] D.C. Larbalestier, X.Y. Cai, Y. Feng, H. Edelman, A. Umezawa, G.N. Riley Jr. and W.L. Carter, *Physica C*, **221**, 299 (1994).
[2] G. Grasso et al. *Physica C*, **241**, 45 (1994).
[3] Y. Yamada et al. *Proc. of the 7th International Workshop on Critical Current in Superconductors*, World Scientific, 76-81 (1994).
[4] S. E. Dorris, B.C. Prorock, M.T. Lanagan, S. Sinha and R.B. Poeppel, *Physica C*, **212**, 66 (1993).
[5] J.A. Parrell, S.E. Dorris and D.C. Larbalestier, *Physica C*, **231**, 137 (1994).
[6] D. C. Larbalestier et al. Proc. of the 7th International Workshop on Critical Current in Superconductors, World Scientific, 82-87 (1994).
[7] H. Edelman, J. A. Parrell and D. C. Larbalestier in preparation (1995).
[8] Q. Li, H. J. Wiesmann, M. Suenaga, L. Motowidlo, and P. Haldar, *Phys. Rev.* B, **51**, 1 (1995).
[9] A. Umezawa, Y. Feng, H.S. Edelman, T.C. Willis, J.A. Parrell, D.C. Larbalestier, G.N. Riley Jr., and W.L. Carter, *Physica C*, **219**, 378 (1994).
[10] G. N. Riley et al. *Proc. of the 1992 International Workshop on Superconductors*, 216-219 (1992).
[11] Jyh-Lih Wang, I-Fei Tsu, X. Y. Cai, R. J. Kelley, M. D. Vaudin, S. E. Babcock, and D. C. Larbalestier, submitted to *J. of Materials Research* (August 1995).
[12] R. Kleiner and P. Muller, *Phys. Rev.* B, **49**, 1327 (1994).
[13] Y. Fukumoto, Q. Li, Y. L. Wang, M. Suenaga, and P. Haldar, *Appl. Phys. Lett.* **66**, 1827 (1995).
[14] B. Hensel, J.-C. Grivel, A. Jeremie, A. Perin, A. Pollini, and R. Flukiger, *Physica C*, **205**, 329 (1993).
[15] L. N. Bulaevski, L. L. Daemen, M. P. Maley, J. Y. Coulter, *Phys. Rev.* B, **48**, 13798 (1993).
[16] A.A. Polyanskii, A.E. Pashitski, A. Gurevich, D.C. Larbalestier, V. Vlasko-Vlasov, and V.I. Nikitenko, in: ICMC-94, Honolulu, Hawaii, October 1994. To appear in *World Scientific Publ.* Singapur (1995).
[17] D.C. Larbalestier et al., *J. Metals* **46**, 20 (1994).
[18] A.E. Pashitski, A. Polyanskii, A. Gurevich, J.A. Parrell, and D.C. Larbalestier, *Physica C*, **246**, 133 (1995).
[19] A. Pashitski, A. Polyanskii, A. Gurevich, J. A. Parrell and D.C. Larbalestier, To appear *Appl. Phys. Lett.* (Fall 1995).
[20] J. A. Parrell, S. E. Dorris and D. C. Larbalestier *Subm. to J. of Materials Research* (July 1995).

Inst. Phys. Conf. Ser. No 148
Paper presented at Applied Superconductivity, Edinburgh, 3–6 July 1995
© *1995 IOP Publishing Ltd*

Superconducting Bearings and Flywheel Batteries for Power Quality Applications.

T.S. Luhman, M. Strasik, A.C. Day, D.F. Garrigus, T.D. Martin, K.E. McCrary, and H.G. Ahlstrom

Boeing Defense & Space Group, Seattle, Washington 98124, U.S.A.

1. Abstract

After many years of development, flywheels utilizing high temperature superconducting bearings may offer an economical and environmental alternative to batteries in Uninterruptable Power Supply (UPS) systems. Through the innovative concurrent use of permanent and bulk superconducting single-grain magnets, Boeing has developed a practical flywheel design with nearly 10 times the energy density of lead-acid chemical batteries.

In this paper we briefly describe a Boeing study which has leveraged the advantages of superconducting magnetic bearings into a Flywheel Energy Storage System (FESS) design suitable for replacing lead acid batteries in UPS systems. By utilizing the unique properties of bulk $YBa_2Cu_3O_{7-x}$ (YBCO) superconducting single grain magnets in the bearing, the traditional limitations of flywheel designs are circumvented. For example, Boeing's design (patent pending) eliminates the flywheel's central shaft, controls radial strains in the rotor, and minimizes bearing losses by replacing complex active electromagnetic bearings required for rotational stability with a hybrid superconducting design. A three-dimensional representation from a finite element model for a 1 MW-hr flywheel is presented in Figure 1. With bearing losses less than 0.1%/hour this flywheel system is expected to show overall efficiencies exceeding 90%.

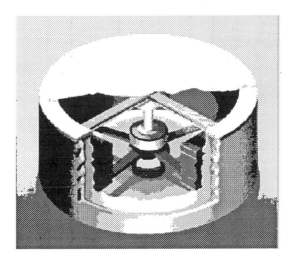

Figure 1. Finite element model representation of a 1MW-hr FESS.

2. Background

2.1. Economic opportunity

Interruption of line power to a computer system, even momentarily, can cause a loss or corruption of data and lead to severe economic loss to the user. Therefore, all critical computer installations employ UPS systems to provide AC back-up power during voltage drops and/or power outages. In most existing UPS systems, a battery (usually lead-acid) supplies the temporary power back-up power, typically for 15 minutes. The storage of energy in a rotating device for such UPS applications is appealing because the flywheel energy densities can be up to 10 times that of the lead acid batteries. For other longer term energy storage applications, such as electric utility load leveling, flywheels do not compete effectively. These applications rely on gas turbine generators for back-up power where the flywheel energy density is approximately ten times less than that of the fossil fuels used in the turbines.

Batteries in standard UPS installations must be replaced every 3-5 years with costs including not only replacement but, for environmental reasons, also disposal expenses. Continued system maintenance is required to ensure adequate battery fluid levels and contact resistances, system health, hydrogen gas removal etc. Life cycle battery installation and use costs for UPS systems are about $ 2000 per kW-hr. Therefore, a flywheel UPS system with a design life of 25-30 years, and an equivalent $/kW-hr cost, could show an early return on investment.

2.2. Enabling technologies

The flywheel energy storage system, based on superconducting bearings, can eliminate both momentary voltage and frequency changes and the longer term power interruptions. If power is interrupted, the rotating kinetic energy in a flywheel (connected in parallel with the load) is instantaneously brought on line through its motor/generator to provide back-up electrical power. In this application, as with batteries, energy is continuously stored in the flywheel. Therefore, economic considerations dictate that system losses must be low. Until the recent development of bulk superconducting, self-centering, YBCO bearings, the energy loss associated with either mechanical or electromechanical bearings has been prohibitively high. These losses range from 1-5%/ hour. We have demonstrated hybrid superconducting bearings with less than 0.1%/hr losses.

Figure 2 presents data collected in collaboration with Dr. John Hull of Argonne National Laboratory, using bulk single grain materials produced at Boeing. Details of Argonne's bearing design will be presented elsewhere [1]. The figure shows that bearing power consumption per kg of mass rotated can be six orders of magnitude below that for a typical 6-pole active magnetic bearing [2]. There are additional significant reasons for using bulk superconductors in flywheel bearings. Table 1 summarizes a few of the more compelling ones.

Table 1. Enhancements with superconducting bearings

Active electromagnetic bearings	Flux-trap superconducting bearings
Inductive coupling limits response time and increases power consumption	Instantaneous force response
High stiffness values required for active control, critical frequencies within operational ranges	Low stiffness values enable critical frequencies outside of operational ranges
Changes in component performance with temperature and time reduces reliability	No sensors, no servo systems, no electromagnetic coils adds operational simplicity and reliability

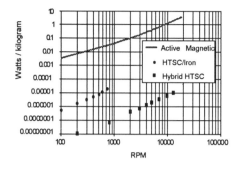

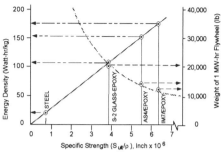

Figure 2. Bearing power consumption per kg of mass rotated. HTSC results adopted from ref. 1.

Figure 3. Comparison of energy densities and weights for candidate rim materials

As is often the case with the emergence of new technological capabilities, several technologies, often developed for alternate applications, converge to provide enabling capabilities in the new area. The second enabling technology, now in hand for FESS, is the ability to build high strength rim and spoke structures with graphite composites.

The figure of merit for flywheel energy storage is the kinetic energy stored per unit mass. For a thin-rim rotor this is $\sigma/2\rho$ where σ is the hoop stress and ρ is the rotor density. The flywheel, therefore, must be able to withstand high stresses and should be low density. The dependence of energy density on $\sigma/2\rho$ illustrates why fiber graphite composites are the materials of choice for FESS's. Figure 3 compares the energy densities achievable in thin-rim rotors using different materials in a 1 MW-hr design. After discounting 30% for matrix volume fraction and 30% for safety, usable graphite rotors can be built with energy densities exceeding 150 Wh/kg. Usable energy densities for lead-acid batteries are ~20-30 W-hr/kg.

3. Technical challenges and solutions

3.1. Thin-rim rotor concept

In a rotating disk 75% of the energy is contained in the outer 30% of the rotor. From an energy storage standpoint, there is really no reason to place rotating material in the central three quarters of the rotor. Doing so only adds material cost and produces large radial strains as noted in Figure 4. Figure 4 also illustrates that with an inner to outer radius ratio of 0.7 radial strains can be maintained below the strain allowables for the preferred IM7 graphite fiber.

Typical flywheel designs are driven by the requirement to provide lift along with radial and axial stability. In such traditional cases a central shaft is used with a bearing at either end. Our thin-rim concept, Figure 1, does not use a central shaft, thus avoiding having to deal with difficult attachment stresses between the shaft and the rotor (the central structure in the figure is the motor/generator support). Instead, our hybrid permanent magnet/superconductor bearing design utilizes a rotating set of spokes to lift the rotor. The spokes are designed to grow in length as the rotor's diameter increases during spin-up. With this design, rotor attachment stresses are minimized, excess material is removed from the central portion of the rotor and radial strains are kept within strain allowables for commercially available composite material. Key to employing this design is the superconducting bearing.

38

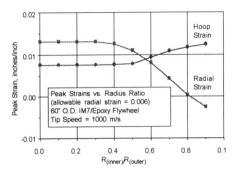
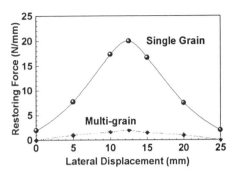

Figure 4. Peak strains as a function of radius ratios   Figure 5. Lateral restoring force comparison

3.2. Bulk single-grain YBCO superconductors

As shown in Figure 2 above, extremely low energy losses can be achieved in the superconducting bearing. In our thin-rim concept, lift forces are provided by a ring of permanent magnets, configured to enhance horizontal stability and bulk single grain superconductors, in combination with permanent magnets, to provide rotational stability. To ensure critical frequencies are outside of the normal operating ranges, soft, self-centering bearings are required. To provide this capability we developed processes to produce large single-grain YBCO material.

Oriented YBCO single-grains were grown using $SmBa_2Cu_3O_{7-x}$ (SBCO) seeds to induce crystallization during a modified melt-growth process [3]. The SBCO seeds were grown with the a-b plane parallel to the surface. YBCO disks made from melt-processed powders yielded single grains extending to the edge of the samples. No grain boundaries were observed in good samples as determined by SEM, high resolution TEM, and x-ray analysis. Grains up to 5 cm in diameter were grown using this technique, as reported previously [4].

Flux trapping was used as a measure of quality control. Magnetic fields of up to 8000 Gauss were trapped at 77K in the 2.5 cm diameter single-grain samples. The flux profiles clearly showed a presence of a circulating current in a one turn single-grain conductors. Macroscopic current densities exceeding 3×10^4 A/cm^2 at 77K were calculated. The trapped field increased to 1.3 Tesla at 65K in a 2.5 cm diameter single grain.

3.3. Lateral restoring force measurements

Lateral restoring force measurements were performed on 2.5 cm diameter single-grain and multi-grain samples as a function of horizontal displacement. Restoring force data were collected on field cooled samples 77K with a 4 mm magnet to superconductor gap. The applied field, approximately 4000 gauss, was provided by a matching 2.5 cm diameter permanent magnet covering the superconductor.

Lateral restoring forces for the single-grain and multi-grain sample are shown in Fig. 5 as a function of horizontal displacement across the sample. The restoring force achieved in a multi-grain sample was 2 N/mm and about 20 N/mm in a single-grain sample. The low restoring force in the multi-grain sample is mainly attributed to the significantly lower intergranular current density.

3.4. Dynamic stiffness

A spin-down test was carried out with an annular NdFeB rotor suspended over four YBCO disks, each 4 cm in diameter . The rotor had an O.D. of 10 cm, an I.D. of 5 cm, and a mass of 1.96 kg. The thrust load was supported by the repulsion of two small permanent magnets along the axis, maintaining a space of 2-3 mm between the rotor and the superconductors. Thus, the superconductors provided radial stiffness but were not subject to static de-pinning forces. The rotor was spun up with a small air turbine and allowed to spin down in air. The results are shown in Figure 6. The conspicuous drop in speed at about 630 rpm corresponds to the first critical speed of the rotor/bearing system, from which we calculate a radial stiffness of 8.54 N/mm. Static force measurements of the system carried out with an Instron were in good agreement, yielding a stiffness of 8.53 N/mm. We found that the sub-critical behavior could be fitted well with a velocity-proportional loss term, while above the critical speed a v^2 dependence worked much better. The most obvious loss source is air resistance, particularly since Hull measured much lower effective frictional coefficients in vacuum with Boeing's material [1]. It is also true that in going supercritical, the rotor changes from rotation about a magnetic center to rotation about the mass center. Supercritical oscillations of the magnetic axis of symmetry would contribute additional drag losses in the superconductors.

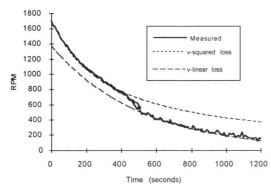

Figure 6. Dynamic stiffness spin-down test

4. Summary

An optimized flywheel design utilizing hybrid superconducting bearings has been developed. We use permanent magnets for levitation and single-grain YBCO for stabilization. Incorporation of the hybrid superconducting bearing has enabled a soft efficient bearing design which makes possible a thin-rim rotor concept. Bearing element testing has confirmed that single-grain YBCO material exhibits the necessary stiffness characteristic. We are currently pursuing the construction, testing and evaluation of a 450 kW-hr engineering prototype.

5. References

[1] Hull J.R., Hilton E.F., Mulcahy T.M., Yang Z.J., Lockwood A., and Strasik M., To be submitted to J. Appl. Phys.
[2] Genta G., *Kinetic Energy Storage*, Butterworths, Boston (1985).
[3] Salama K., Selvamanickam V., Gao L., and Sun K. 1989 Appl. Phys. Lett. 54, 2352
[4] Blohowiak K.Y., Garrigus D.F., Luhman T.S., McCrary K.E., Strasik M., Aksay I.A., Dogan F., Hicks W.B., Lui J., and Sarikaya M., 1993 IEEE Trans. Appl. Supercond. 3, 1049

Inst. Phys. Conf. Ser. No 148
Paper presented at Applied Superconductivity, Edinburgh, 3–6 July 1995
© *1995 IOP Publishing Ltd*

Superconducting materials and coils for operation at power frequencies

M. Polak

University of Wisconsin, Madison, Applied Superconductivity Center, 1500 Engineering Drive, Madison, WI 53706

M. Majoros

Institute of Electrical Engineering, Slovak Academy of Sciences, 84239 Bratislava, Slovak Republic

Abstract. A review of actual problems and new results on superconductors and superconducting coils operating in AC regime at power frequencies 50/60 Hz is presented. Materials examined include submicron multifilamentary NbTi wires with high resistivity matricies , cables made from stacks of such , wires and Ag-clad BSCCO tapes . The behavior of these superconductors under self- field conditions is also described and discussed.

1. Introduction

Low AC losses, a high critical current density, and a high degree of electromagnetic stability are the most important requirements for an AC superconductor. The "cost" of a Watt of losses dissipated in a superconducting device depends on the properties of the cooling medium and the efficiency of the liquifier. In liquid helium we need 500 W/W, while in liquid nitrogen only 10 W/W. This fact is a very big advantage of high temperature superconductors. AC applications of superconductors can be divided into 2 categories depending on factors controlling the losses:
1) superconducting devices in which the superconductor operates mainly in its self - field ;
2) superconducting devices in which the superconductor is exposed to magnetic fields much larger than its self- field.

Actually, good prospects exist for AC applications of low-Tc (LTC) as well as high-Tc (HTC) superconductors in devices operating close to the self field condition [1-6]. For applications in the second category, attention is focused on problems of cable design, stability, current distribution, AC losses, winding structure, etc. [7].

In our contribution we focus on problems of both LTC and HTC superconductors, which have an impact on their AC loss behavior. Some recommendations for improvement of AC wires and coils performance, as well as some open-ended questions, are formulated in the last part of the paper.

2. Specific problems of NbTi multifilamentary wires for AC use

In Fig.1 we show electric field versus transport current characteristics measured on a sample of wire manufactured by GEC - Alsthom having a diameter 0.12 mm, containing 14496 NbTi filaments of diameter 570 nm in CuNi matrix [8]. We can observe two interesting phenomena of these curves: they are exponential over 5 decades of electric field, and intersect in one point at I=0.5 A and E=5×10^{-6} μV/cm. An analysis of this result in [8] indicates that defects like a high degree of strong filament sausaging and filament breaking are responsible for this behavior.

Tachikawa et al. [9] stated that the mechanical hardness of a Cu-30%Ni matrix is too high for

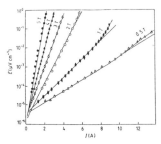

Fig.1. Electric field - transport current curves of the wire GEC Alsthom with the diameter 0.12 mm

smooth drawing of NbTi composites, and intermetallic compounds are produced on the surface of NbTi filaments. Measurements of I-V curves on a sample of NbTi multifilamentary wire with CuMn matrix showed [10] that they are much steeper and do not intersect. Another alternative material is CuSi , which allowed for the fabrication of wires with 110 nm filaments, and a critical current density close to 7000 A/mm^2 at 1 T and 4.2 K [9]. Filament breaking and sausaging also cause a variation of the critical current along the length of a wire. As a result, the electric field along the wire is very inhomogeneous. Results of an inhomogeneity study on AC wires will be presented at this conference[11]. These inhomogeneities are one of several reasons for the strong quench current degradation observed in multistrand cables [12].

Hysteresis losses in filaments decrease with decreasing filament diameter. However, the filaments may be coupled together if the matrix material between the filaments becomes superconducting due to the proximity effect. The interfilamentary coupling can be suppressed by increasing the matrix resistivity (using CuNi, CuMn, or CuSi matrix). Filaments become decoupled if the external magnetic field is increased above the so called "decoupling field". We have found that the decoupling field for filaments in a Cu matrix strongly increases with decreasing interfilamentary spacing. For interfilamentary spacings of 600 nm, 180 nm and 90 nm, the decoupling field was 5 mT, 20 mT and 1 T, respectively. The addition of 10 % Ni to the Cu lowered the decoupling field for a 90 nm spacing to 3 mT [13]. We also found that transport currents considerably lower then I_c strongly reduces the interfilamentary coupling [14].

3. Specific problems of BSCCO tapes exposed to an AC external magnetic field

Requirements to a low loss AC HTC conductor which could effectively operate in external magnetic fields with an arbitrary orientation will probably lead to a conductor structure similar to that used for AC wires based on LTC materials: many small cross - section superconducting twisted filaments embedded in a resistive matrix. The development of an AC HTC wire will profit from the experience with LTC materials, but specific properties of HTC must be understood and taken into consideration.

Let us suppose we have a multifilamentary tape with N_f filaments embedded in Ag matrix. A time varying external magnetic field induces several types of currents (see Fig. 2) :

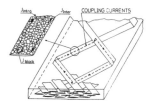

Fig.2 Different kind of currents flowing in a multifilamentary HTC tape

1) Shielding (magnetization) currents in the filaments. Due to the granular structure of the filament material (well aligned grains in BSCCO 2223 or 2212), intergranular as well as intragranular currents flow in each filament. Currents also flow inside of blocks of well-coupled grains in 2223 tapes, as shown by magnetooptical studies [15] ;

2) Coupling currents flowing from filament to filament through the normal conducting matrix.

3) Shielding currents in the silver cladding in the non - filamentary zone. The losses associated with these currents are small in comparison with the other loss components.

AC losses in single core as well as multifilamentary BSCCO tapes have been intensively studied, and many experimental data can be found. Here we refer to careful studies of AC losses published by Kwasnitza and Clerc [16]. Hysteresis losses per cycle of the external magnetic field Q_h depend not only on the amplitude ΔB , but also on the magnetic field sweep rate dB/dt. As coupling losses Q_c also depend on dB/dt , the separation of Q_h and Q_c is not simple. In [16] a method of separation is proposed. The method is based on the fact that the hysteresis losses per unit volume do not depend on the sample length or twist pitch length, while Q_c increases with the square of the twist pitch length.

An exact understanding and description of hysteresis losses must take into account another property of BSCCO superconductors - the magnetic history effect on J_c. The critical current density in BSCCO tapes measured in increasing external magnetic field differs from that measured in decreasing field, as shown in Fig. 3a. As a result, magnetization loops above the penetration field are not symmetric, as shown in Fig. 3b.

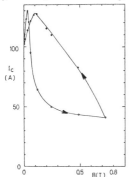
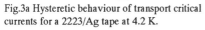
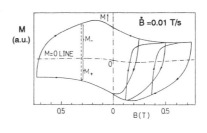

Fig.3a Hysteretic behaviour of transport critical currents for a 2223/Ag tape at 4.2 K.

Fig.3b Magnetization characteristic of a 2223/Ag tape at 4.2 K. Note the difference between the values measured at increasing and decreasing field.

Another factor affecting the hysteresis losses is the strong anisotropy of the the critical current density. As shown by Ashworth and Glowacki [17], if the current flows perpendicular to the broad plane of a BSCCO 2223 /Ag tape, (i.e. nominally parallel to the "c" direction), resistive voltages appear at current densities less than 1% of the longitudinal (a-b plane)critical current density. We observed similar behavior in 2212 tapes. The sample length may strongly influence the hysteresis losses in magnetic fields parallel to the tape, as shielding currents must flow perpendicular to the a-b plane at the edges of the sample.

Coupling losses P_c are an important component of the total losses. According to Kwasnitza and Clerc [16] the losses P_c in a multifilamentary conductor exposed to a sinusoidal external magnetic field with frequency f in the power frequency range are $Q_c/V = (1/2\mu_0)$ n ΔB^2 F($\omega\tau$), where $\omega = 2\pi f$, n is the shape factor and F is the loss parameter depending on f and the time constant of the coupling currents τ. The time constant is $\tau = k_g l_t^2 / \rho_e$, where k_g is a geometrical factor depending on the form of the conductor cross - section ($k_g = \mu_0/8\pi^2$ for a round conductor), and ρ_e is the effective transverse resistivity of the matrix between the filaments and l_t is the the twist

44

pitch length. Similarly to LTC materials, low coupling losses require a high resistivity matrix (Ag - alloys) and short twist pitch.

4. Behavior of AC wires and HTC tapes under self - field conditions

In some AC applications of LTC superconductors, wires operate in rather low external magnetic fields. The current density is very high, and the changing transport current induces a self magnetic field B_{sf} on the wire surface. The amplitude of the electric field on the wire surface E_{sf0} can attain values one to two orders of magnitude higher then the electric field commonly used to define I_c (0.1 up to 1 μV/cm). As a result, the local loss power density in the self field mode, $P_{sf} = E.J$ can be very high . The self field effect in AC wires was studied in [18, 19, 20]. As an example, in Fig.4 we show the self field voltage versus time measured on a NbTi AC wire with a diameter 0.25 mm in a DC magnetic field of 1 T, carrying 50 Hz transport current having different amplitudes [18].

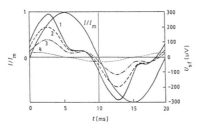

Fig.4 Self field voltage versus time of a wire with diameter 0.25 mm, containing 597 102 NbTi filaments 0.136μm in diameter, measured in an external DC magnetic field of 1 T. The transport current has a sinusoidal form and frequency f = 50 Hz. Current amplitudes are :1 - 65 A, 2 - 55 A, 3 - 47 A, 4 - 28.5 A. The temperature is 4.2 K [18].

In [18] we showed that the self field power loss per unit length is $P_{sfl} = f I_c g(i)$ where g(i) is a function strongly increasing with increasing $i = I_m / I_c$, where I_m is the transport current amplitude. Losses of similar samples measured in a 50 Hz external magnetic field with an amplitude of 1 T without transport current were of the order of 10^3 W/m^3 , while the self field losses were considerably higher, 10^4 up to 10^5 W/m^3 , depending on the current amplitude.

The measurement of losses in short samples of HTC tapes is complicated. Experiments have showed that the measured loss voltage, and hence the loss, are dependent on the position of the potential taps on the tape surface [21]. Clem [22] explained this effect and showed that the potential wires must be brought far from the tape (to a distance at least 3 times the tape width) in order to obtain correct losses. Similar conclusions were presented in [23].

In [24] we reported results of loss measurements on a small 1 layer coil made of 2223 tape. The critical current was relatively low and the losses due to the broad resistive transition of the tape dominated over the losses due to the self-field. In this case the potential wires were co - wound with the tape. Further studies are needed to investigate the influence of the position of the potential leads on losses in the case of coils.

5. Superconducting wires , cables and coils for power frequency

In Table 1 we show parameters of some multifilamentary wires and cables developed by GEC Alsthom , Showa Electric Wire & Cable (SEWC), and Furukawa Electric (a wire with artificialy introduced pins, which is under development [25]). The artificial pins of the Furukawa wire improved the critical current density of the wire by a factor of about 2.

In order to achieve higher current carrying capacity, single multifilamentary wires are assembled to form a cable. GEC Alsthom developed a cable consisting of six "C" wires [26]. Showa developed a cable of the type (6+1)x(6+1) with a diameter of 9 mm using the strand described in the Table 1.

Superconducting coils operating at 50/60 Hz can be made from a single strand or from a cable. As strands are thin, coils have high inductance and operate at voltages in kV range [27]. The

Table 1
Parameters of some NbTi multifilamentary wires for AC applications

	SEWC	GEC Alsthom	Furukawa (APC)
Wire diameter (mm)	0.209	0.3	0.16
Number of filaments	446 383	920 304	78 625
Filament diameter (μm)	0.13	0.14	0.29
Filling factor	0.2	0.191	0.24
Twist pitch (mm)	1.7		
I_c at 1 T and 4.2 K	40	120	46
J_c overall (A/cm^2)	1.17×10^5	1.7×10^5	2.29×10^5

degradation of the quench current compared with that at DC operation is quite small. However, high power applications require coils operating with high currents and have to be wound with cables. Currently, a major problem with such coils is a strong quench current degradation. The quench current of a coil made from a cable consisting of N_w wires , is $I_{QC} = k_d \, N_w \, I_{QS}$, where I_{QS} is the quench current of a similar coil wound of a single strand and k_d is the coefficient of degradation. Studies of the degradation problem have shown that various construction details of coils, as well as the structure and parameters of the cable used in the winding, influence the degradation. For example, impregnated coils are mechanically stable, but the transfer of heat due to AC losses is strongly reduced, and the coil quenches due to heating. The bonding of the cable to the coil bobbin improves the coil performance considerably, as shown by Ise et al. [5]. We used a winding structure shown in Fig.5, which enables good cooling as well as mechanical stability. Thermomechanical instabilities in AC cables and instabilities due to the longitudinal component of the magnetic field are other limitations of the quench current.

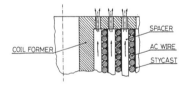

Fig. 5 The winding structure with cooling channels. Only one side of each layer is impregnated.

A relativaly large number of various AC coils were made and tested. As examples we mention the laboratory AC coil operating at 50 Hz in DC backround field [27], primary and secondary coils of a 40 kVA transformer [5], and a coil made of NbTi with APC [25].

6. Conclusions

The broad resistive transition of NbTi multifilamentary wires with CuNi matricies is due to filament breaks. The matrix material considerably influences the value of the external magnetic field above which the filaments are not coupled by the proximity effect. Hysteresis losses in fine NbTi filaments depend on frequency . More data on I - V curves, losses, and magnetization for AC strands with different matrix materials would be very helpful in search for an optimal AC strand.

Specific effects observed in BSCC , like magnetic history dependent I_c and asymetric M of HTC tapes, are due to the existence of several current systems. Intrinsic electric field - current density characteristics of BSCCO in wide interval of electric fields are needed for an exact description of the electromagnetic behavior of these materials.

Self-field losses are very important for both LTC and HTC materials, as the most realistic applications of both types of superconductors will operate very close to the self-field regime. More data are necessary on self-field losses in short samples as well as on spiral coils of HTC tapes which are basic elements of hollow conductors in superconducting cables.

Our results show that local inhomogeneities of strands may play an important role in determining the behavior of cables, in addition to other more intensively studied reasons for degradation of AC coils made from cables.

Acknowledgment

We would like to thank T. Kumano and Y. Laumond for providing us with information on technical parameters of commercially available AC superconductors.

References

[1] Von Dollen D W, Hingorani N G, Samm R W, 1993 6th Int. Symp. on Superconductivity ISTEC Hiroshima Japan SA-4

[2] Fujikami J, Saga N, Ohkura K, Sato K, Ishii H, Hara T, 1995 Workshop on ac losses , San Francisco April 17 - 18

[3] Yoneda E S, Tasaki K, Yazawa T, Maeda H, Matzuzaki J, Tsurunaga K, Tada T, Fujisawa A, Ito D, Hara T, Wakade M, Ookuma T, 1994 ICEC 15 Genoa to be published in Cryogenics

[4] Verhaege T, Cottevielle C, Estop P, Therond P G, Thomas P, Laumond Y, Berkhaled M, Bonnet P, Pham V D, 1994 Appl Superc Conf Boston to be published in IEEE Trans on Superconductivity

[5] Ise T, Marutani Y, Murakami Y, Yoneda E, Sugawara R, ibid ref 4

[6] Fevrier A, Tavergnier J P, Laumond Y, Bekhaled M, 1988 IEEE Trans on Magn 24 1477-80

[7] Lacaze A, Laumond Y, Bonnet P, Fevrier A, Verhaege T, Ansart A, 1992 IEEE Trans on Magn 28 767-770

[8] Polak M, Hlasnik I, Rakhmanov A L, Ivanov S S, 1995 Superc Sci Technol 8 112-118

[9] Tachikawa K, Koyama S, Akita S, Torii S, Kasahara H, Tanaka Y, Matsumoto K, 1993 IEEE Trans on Appl Superc 3 1374-7

[10] Ivanov S S, not published

[11] Majoros M, Suchon D, Polak M, this conference

[12] Torii S, Akita S, Ishikawa R, Uyeda K, Amemyia M, Tsukamoto O, 1993 IEEE Trans on Superc 3 126

[13] Majoros M, Zanella S, Ottoboni V, Jansak L, EUCAS'93 Appl Superc 1 27-30

[14] Polak M, Krempasky L, Majoros M, Suchon D, Kirchmayr H, 1993 IEEE Trans on Superc 3 150-152

[15] Pashitski, A E, Polyanskii A, Gurevich A, Parrell J A, Larbalestier D C, 1995 Physica C 246 133-144.

[16] Kwasnitza K, Clerc St, 1994 Physica C 233 423-35

[17] Ashworth S P, Glowacki B A, 1994 Physica C 226 159-64

[18] Polak M, Hlasnik I, Fukui S, Ikeda N, Tsukamoto O, 1994 Crygenics 34 313-24.
    Hlasnik I, Fukui S, Ito M, Tsukamoto O, Polak M, Kokavec J, Kottman P, 1995 Appl Electromagnetics in Materials Suppl to the Journal of Japan Society of Electromagnetics 3 37-48

[19] Schneider M, 1995 Diplomarbeit, TU Vienna

[20] Estop P, Cottevieille C, Poullain S, Tavergnier J P, Verhaege T, Lacaze A, Laumond Y, Le Naour S, Ansart A, Manuel P, ibid ref4

[21] Ciszek M, Campbel A M, Glowacki B A, 1994 Physica C 233 203-8

[22] Clem J, ibid ref 2

[23] Hlasnik I, Jansak L, Majoros M, Kokavec K, Chovanec F, Jergel M, Martini L, Zanella S, 1995 Mag Tech Conf Tampere

[24] Paasi J, Polak M, Kottman P, Suchon D, Lahtinen M, Kokavec J, ibid ref 4

[25] Miura O, Inoue I, Suzuki T, Matsumoto K, Tanaka Y, Iwakuma M, Yamafuji K, Matsushita T, 1995 Cryogenics 35 181-8

[26] Verhaege T, Estop P, Weber W, Lacaze A, Laumond Y, Bonnet P, Ansart A, IEEE Trans on Appl Superc 3 164-7

[27] Polak M, Pitel J, Majoros M, Kokavec J, Suchon D, Kedrova M, Kvitkovic J, Fikis H, Kirchmayr H, ibid ref 4

Inst. Phys. Conf. Ser. No 148
Paper presented at Applied Superconductivity, Edinburgh, 3–6 July 1995

Industrial impact of the applications of superconductivity in large scale research programs

R. Penco

Ansaldo Energia, Magnets Dept., Genova, 16152 Italy

Abstract. The branches of science using superconducting machines in an extensive way are High Energy Physics and Fusion research: the first requests the serial production of a big number of superconducting magnets and cavities while the second requires large volumes, high magnetic fields and huge coils.

The applications in the large scale research programs represents, nowadays and in the next future, the second market for the superconductivity after the magnetic resonance.

A general review of past and future programs involving superconducting machines is given here, paying a particular attention to the impact of these applications on industry and, in particular, on the European one.

1. Introduction

The High Energy Physics and Fusion Research have been the first large applications of s.c. technology and they had the merit to have always pushed this technology to higher limits. At the beginning, in the early 70's, the large scale applications of superconductivity were limited to the detectors magnets (like the large bubble chamber of CERN and other magnets for experiments). In the same period many of the European electrical firms were engaged in other research programs for s.c. applications for industrial scope. Researches were developed on s.c. motors and alternators, levitation train and SMES; sometimes they were financially supported by governments, sometimes financed directly by industry.

In the period from 1970 to 1980, anyway, several tenders were called for the constructions of single magnets, prototypes or models (e.g.: the European and Swiss LCT Coils). As usual, in Europe the offers were requested for fixed prices also for works connected to the research development. On the contrary in the USA, even today, most tenders related to s.c. magnets for research are made by the cost + fee method, clearly reducing the risk for industry (but finally increasing the cost for the customer). In that period Brown Boveri, Gec-Alsthom, Siemens and later Ansaldo were the major electro-mechanical industries involved in the construction of large super-conductive apparatus.

2. The TORE SUPRA

In the 80's, some large projects actually started. The construction of Tore Supra was for the industry the first relevant contract for a small series production of s.c. magnet [1]. Ansaldo got the order to build nineteen toroidal field coils. Each coil was composed of 26 double pancakes, made by winding a bare conductor. The turn insulation was given by a 0.2

48

m thick glass epoxy strip, the layer insulation related to 5 millions of G11 pellets. The Tore Supra was, and probably will also be in the future, the biggest magnet cooled by the superfluid helium directly in contact with the cable, without any insulation barrier. Later the positive results of LCT and the following experiments on cable in conduit, moved the designers to obtain the stability requirements using the cable in conduit conductor more than bath cooled magnets.

For the first time, at least in Ansaldo, it was necessary to design and set up a dedicated partially-automatic production line for s.c. magnets.

Several reasons are involved in this choice:

1) The delivery time requested by the customer were (and usually are) very tight. The time needed to calibrate the process is always long and when it is finally fixed it is necessary to guarantee a high production rate.

2) The number of workers in the magnet department during this kind of productions increases very much. The new ones are normally taken from other similar departments (conventional winding dept.) but they are less familiar with the product: an automatic tooling helps to maintain a good standard of reproducibility.

3) Some technical requirements can be obtained only using complex tooling [2].

The total area dedicated to the Tore Supra toroidal field coil was of about 3400 m², with a clean room of 1400 m², where the coils were wound (see Fig. 1). The first coil was delivered after 41 months from the order and after that the mean production rate was kept at one coil for month from Nov. '85 to May '86, when coil No. 19 was delivered.

3. The HERA project

The European industries had to wait for a second big project until the end of 1985 when the HERA accelerator ring construction started.

Fig. 1. The clean area for Tore Supra manufacturing   Fig. 2. The HERA dipoles clean room

While a Tore Supra model coil was built by an industry (Gec-Alsthom) the prototype of the HERA dipoles were made directly by the DESY laboratory. Prototypes of quadrupoles were built by French laboratories. After the order the "technology transfer" was achieved in a very simple way: workers from the industries were involved in the construction of some of the prototypes at DESY Laboratory, together with local workers. Anyway the problems to set up a real production line able to reach the request production rate had to be solved by industry.

The total number of magnets to be produced was very high: 416 dipoles, 220+36 quad. plus minor magnets up to a total number of 2000 considering also the beam pipe coils [3].

In particular dipoles were built by Ansaldo and ABB, the quadrupoles by Gec-Alsthom and Noell, most other magnets by Holec. In Ansaldo the total dedicated area for this production was of 3150 m^2 (1710 m^2 of clean room (see Fig. 2).

Ansaldo spent about 2 years to complete the first prototype of the 10 preseries magnets (Nov. '87). The series production was carried out from the end of 1988 to July 1990 with an average production of 11 magnet per month. For a certain period the production reached the level of 22 magnets per month. A total of 32 collared coils were also cold tested in a factory tests facility up to their quench current [4]. A statistical analysis of the ANSALDO dipoles production is reported in [5].

4. The SSC project

In 1990 USA starts the SSC project where about 8000 s.c. magnets were used. After the not completely satisfactory experience with the Tevatron (where all the magnets were built in the laboratory) the American DOE decided to give to industry the full responsibility for design, manufacturing and tests of SSC magnets. On the other end, in the mean time, tens of prototypes were built in different laboratories and a technology transfer program for industries was forecast.

The industries that took the orders for dipoles (General Dynamics and Westighouse) began to build huge facilities to guarantee the requested production rate. Other prototypes were built in the laboratories by personnel of industries. In the mean time new and very large buildings ,completed with automatic and partially-automatic tooling, were built.

The SSC project was finally stopped in 1994, when, perhaps, only one dipole was built using the new facilities.

5. The LEP s.c. cavities

The third and last big project known, still to be completed, is related to the construction of the LEP s.c. cavities [6]. A total amount of 218 Nb sputtered cavities (divided into 54.5 modules of 4 cavities each) are now under construction for CERN at Ansaldo, Cerca and Interatom.

Also for this project the prototype cavities were built by the customer. The order for production was given to Ansaldo in Nov. '90 and a technology transfer was needed.

The time to set up the production line for a reliable production cycle was really long. The first module with 4 cavities was delivered by Ansaldo in June '93. Actually 11 of the foreseen 17.5 modules have been delivered for a mean production rate of 1 module in 2 months. The situation for the other firm is similar (1 or 2 more delivered modules).

The CERN specification on these s.c. cavities were very high (Q > 3.4 10^9 at 6 MV/m). The number of bare cavities that passed through the CERN test at the first attempt were about 10% in 1992, 20% in 1993 up to about 68% in 1995 [7]. The cavities that did not pass the specification had to be treated to remove the Nb and sputtered again.

The total dedicated area for this production is of 2500 m², inside this area "white rooms" are installed : 40 m² class 10, 75 m² class 100 and 60 m² class 1000 (see Fig. 3). Mechanical machinery and electron beam welding (70 kV in a 3x3x6 m vacuum chamber) are located elsewhere.

6. The LHC project

The construction of LHC at present the next large scale future project. In this case the involvement of industry started from the very beginning. Ansaldo made its first LHC single dipole prototype (1 m long) in 1988 [8].

After that other short single and twin prototypes were built by Ansaldo and others different companies (Holec Gec-Alsthom) using different techniques (NbTi but also Nb_3Sn dipole were built). Ansaldo received orders from CERN and INFN for the construction of three full scale (11 m) twin magnets, complete with cryostat. Other orders were placed for full scale prototypes (2 for Noell, 1 to GEC-Alsthom, 1 for Holec). The first Ansaldo magnet was successfully tested by CERN in Oct. '93 and the success obtained gave a certain influence on the decision to build the machine.

The approach used by CERN for the LHC project was different from the one used by DESY or SSC: industry were involved since the production of prototypes. The advantage of this solution is clear. There is no need for technology transfer to start the series production, some of the tooling are ready, the production methods have already been tested., the nucleus of designers and workers for the series production has already been formed.

Today, Ansaldo is working on a new prototype 15m long while Noell and Gec-Alsthom received an order for some collared coils that will be mounted in the iron directly by CERN. The fact that CERN decided to start again making part of the work inside the laboratory shows a certain change in its politic.

The total number of s.c. magnets forecast for the LHC project is about 1500 with 1100 dipoles and 150 quadrupoles. From a preliminary estimation the total dedicated area needed to produce, for example, 500 of dipoles should be of about 7000 m² with 3300 m² of clean area. Also the construction of the detector magnets of LHC will deeply involve the industries due to the large dimensions of ATLAS and CMS.

7. The Fusion projects

After the LCT and Tore Supra no other big fusion projects involving superconductivity applications started in West Europe while in Russia the T15 machine was built. The main activity in this field during the last years have been the construction test facilities (PSI, KfK), small models (like the 12T ENEA CIC coil) as well as conductor development.

When the world-wide collaboration on the ITER task started, several local projects stopped: a project for the construction of a NET model coil (already requested in a tender) was cancelled in Europe; in the last months also the TPX project in USA had a dramatic slow down .

Anyway, in the last two years, the European and the international industry, was involved in the feasibility analysis and design of ITER. A big effort was required to industry to develop and construct the CS and TF cable. Several firms set up production lines for Nb3Sn strands (in Europe, USA and Japan), on cabling (Europe and Japan), on the cable jacket material (USA) and on the cable jacketing (Europe and Russia). Ansaldo got a contract to set up a jacketing production line for the central solenoid conductor (see Fig. 4). The line is able to jacket single lengths up to 340 m but the ends plant are dimensioned for one Kilometre cable length.

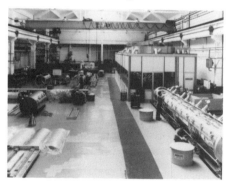

Fig. 3. Part of the LEP cavities production area

Fig. 4. A partial view of the Ansaldo ITER jacketing line

At present American and Japanese industries are involved in the construction of the CS model coil, while European industry are going to be called for tender on the toroidal model coil.

The construction of ITER magnets will be a challenge for industries involved in superconductivity. In the present design[9], 20 toroidal field coils are forecast. They are made by 14 single pancakes of thin INCOLOY 908 jacket cable in conduit put inside a grooves in seven 316LN radial shear plates . The toroidal coils are surrounded by a stainless steel case. The central solenoid will be wound using a thick jacket cable in conduit like the eight poloidal coils that will surround the system. Toroidal coils, central solenoid and two of the poloidal coils will be made of Nb$_3$Sn wind and react cable.

The building of the tokamak will be 71 m large and 108 m high . The overall dimension of s.c. coils are : Toroidal 18m x 12m x 1.2 m, Central Solenoid 5.2m dia. x 12.2 m height , Poloidal up to 24m of dia. The weight will be of 12800 tons for toroidal coils (but to realise them about 30000 tons of 316LN steel are required), CS 850 tons, PF 4500 tons , Structure 4200 tons for a total cold mass weight with miscellanies of about 23800 tons. The weight of the only superconducting cable will be about 1500 ton.

All the coils (but the large PF coils) could be transported if the construction side ward the ITER side will be placed near the sea or a navigable river.

8. Conclusions

From this general view, it is clear that to fit the requests coming from these large scale research projects the industries must have:

- a team of experts in design and manufacturing of s.c. magnets for research.

- the capability to machine, move and stock large amount of materials and big components.

- the financial power to invest in dedicated production lines

- a large number of workers to manage possible fast changes in manpower request during different phases of the production

- laboratories, tooling and qualified personnel to ensure an efficient quality control

All these characteristics are present only in large electrical companies with a specific know-how in the superconductive field. Clearly these companies do not need to built s.c. magnets to survive. Normally the amount of these kind of orders is small compared with the total turnover of a big company while the industrial risk is usually high. Furthermore, the time between a big project and another is usually very long.

We believe that the general spin off of this technology is enough to justify the effort, but it is not a general opinion: in the last 10 years three big companies like ABB, SIEMENS and ELIN got out of this particular business. Other companies seem to go in the same direction. Research foundations have to meditate on this tendency.

Clearly other companies could enter in the business, but the set up of a team of experts, like the one you can find in an industry involved in a superconducting technology for twenty years, is very difficult or even impossible.

The only way to overcome this problem is the solution partially adopted in the last year by CERN and ITER: to involve big industries since the beginning of a project, feed them with preliminary orders for feasibility, design and prototype construction in a way to keep involved in the field expert people in the industry.

On the contrary, if preliminary works or feasibility studies or prototype construction will be made directly in the laboratories or assigned to very little industries (able, perhaps, to do a good job at a lower price but not able to receive orders for large scale production) none of the historical firms could be, in the future, in the position to answer to a call for tender.

References

[1] Bessette, Hamelin, Lafon, Libeyre, Monbrun, Moutet, Portafaix, Quet, Torossian, Cannizzo, Landucci, Larini, Laurenti, Martinelli, Martini, Montanari, Pignatelli, Prini, Renzetti, 1984, Manufacturing of the Tore Supra superconducting magnet, *SOFT 13*

[2] A. Laurenti, P. Renzetti, C. Martinelli, E. Ferrari, D. Bessette, 1986, Dimensional accuracies obtained in the Tore Supra toroidal field coils, *SOFT 14*

[3] R. Meinke, March 1991, Superconducting magnet system for HERA, *IEEE Transaction on Magnetics* **27** No. 2

[4] R. Musenich, A. Bonito-Oliva, G. Gaggero, S. Parodi, S. Pepe, P. Valente, R. Penco, Southampton UK, 1988, Cryogenic tests on HERA dipoles, *ICEC 12*

[5] A. Bonito-Oliva, P. Gagliardi, R. Penco, P. Valente, 1990, A statistical analysis of the whole Ansaldo "HERA dipoles" production, *ICEC 13,*

[6] G. Cavallari, C. Benvenuti, P. Bernard, D. Bloess, E. Chiaveri, F. Genesio, E. Haebel, N. Hilleret, J. Tuckmantel, W. Weingarten, Washington D.C., 17-20 May 1993, Superconducting cavities for the LEP energy upgrade, *Particle Accelerator Conference in Washington D.C.*

[7] E. Chiaveri, W. Weingarten, Newport News, Oct. 4-8, 1993, Industrial production of superconducting niobium sputter coated copper cavities for LEP, *6th Workshop on RF Superconductivity at CEBAF*

[8] R. Perin, D. Leroy, G. Spigo, The first, industry made, model magnet for the CERN Large Hadron Collider,*ASC 88,San Francisco USA*

[9] ITER EDA,Technical Review presented by Joint Central Team and Home Teams, *NAKA, May 1995*

Inst. Phys. Conf. Ser. No 148
Paper presented at Applied Superconductivity, Edinburgh, 3–6 July 1995
© *1995 IOP Publishing Ltd*

A 1.5 kW HT superconducting synchronous machine

J.-T. Eriksson, R. Mikkonen, J. Paasi and L. Söderlund

Laboratory of Electricity and Magnetism, Tampere University of
Technology, P.O.Box 692, FIN-33101 Tampere, Finland

Abstract. A HT superconducting synchronous machine of 1.5 kW and 1500 rpm has been built as a first step in an attempt to develop electric machinery for special purposes such as ship propulsion. The four pole machine comprises Bi-2223 racetrack coils for magnetization. The working temperature of the SC coils is chosen to 20 K, liquid hydrogen will be used for cooling. The design is of the inversed type meaning that the exitation is located in the stator and the armature in the rotor. The nominal voltage of the machine is 400 V. The present paper reviews some superconductivity aspects and the impact of HTS on rotating electric machinery. Finally we report the design parameters of the NORPAS motor along with the test results of the HTS coils.

1. Superconductivity, state of art

The main challenge of HTS research is to produce long wires capable of carrying currents of densities 200 A/mm^2, overall, in fluxes of 3 - 5 T. Best results have been achieved with bismuth based materials, e.g. Bi-2223/Ag. In short sample tape wires a current density of 600 A/mm^2 at 77 K has been achieved. Longer tapes, say 100 m, can stand 80 A/mm^2. These values, however, relates to the pure superconductor, the overall density is typically reduced by a factor of five. An even more severe drawback is the low flux densities these materials can take at 77 K. Around 40 K the density B drops from several teslas to a level below 0.1 T. Flux creep causes internal losses and the breakdown of the superconducting state. At Tampere University of Technology (TUT) a group led by Jaakko Paasi attempts to model and simulate the magnetic behaviour on the grain scale, the objective being to offer agendas for more efficient flux pinning. The theoretical work has been verified by very sophisticated experimental investigations [1].

Although the goal is to use liquid nitrogen at 77 K as the final coolant, one has to remember that once 20 K was a dream for designers of SC devices as this temperature facilitated liquid hydrogen for refrigeration. Hydrogen would make an ideal coolant, which is reflected by the exellent heat capacity of water. However, hazards related to the handling raise some objections.

2. Superfields in rotating machines

Once electric machinery, next to pure magnets, was one of the most attractive applications of superconductivity. During the 70ies several d.c. motors of the homopolar type were built raging from a few kW to 2.5 MW. At Helsinki University of Technology a 100 kW motor was successfully tested in 1979 [2]. Homopolars were favoured because the reaction torque is not transferred to the SC field winding, as in conventional d.c. machines, but is taken by the stator member of the armature circuit. The economical break even for these machines was estimated to 4...6 MW.

What has changed through the discovery of HTS? First, the refrigeration will be more simple relying on a well-established technology. Both first costs and the cost of heat losses will be significantly reduced. Several technological advancements have occured since the 70ies, among them the development of high power frequency converters. For the motor drive concept a synchronous motor seems today more attractive than the d.c. motor. However, one has to deal with the alternating stray and transient flux trying to penetrate the SC field winding. The synchronous motor is, in principle, very simple. A clear-cut geometry results in a robust construction. Utilizing superconductivity leads, however, to conflicting design interests. A rotating SC field winding requires both current supply (through slip rings, also required by the conventional construction) and coolant supply.

The opposite, a static cryostat and a rotating armature winding will have the advantage of an easy access to the cryogenic part offering perhaps higher heat loss margins. The construction will also allow more freedom in designing the mechanical torque transfer from low temperature to room temperature parts. Further the magnetic flux can be more efficiently used in the central room temperature space of the cryostat. The drawback rises from the requirement to transport the main, ie. armature current to and from the rotor. No commutator is needed as for the d.c. machine, but the slip rings should be able to handle, say, 2000 V and 1500 A, equivalent to 5 MVA.

Although some 80 % of industrial energy use is consumed by electric motors, superconductivity will not revolutionize paper mills or distributed low power motor drives. SC motors will find their way to uses, where their extraordinary properties result in exclusive solutions. For instance, the small size make them attractive for ship propulsion by offering more payload or facilitating mounting in narrow twin hull spaces. For a 4 MVA water feeding pump of a power station an increase of the efficiency by 1 -1.5 percentage units results in an annual saving of $ 20,000. The pay-back time for superconductivity will be of the order 2 - 3 years.

3. The NORPAS motor

Four Nordic countries decided in 1992 to start a cooperatative research program on applied superconductivity. In Finland the main player was Tampere University of Technology, who has gained substantial support from Finnish Industry and the NEMO energy program. Activities at Tampere has been focused on energy applications and fundamental research on magnetic properties of HTS materials.

In order to demonstrate the potential, it was suggested that Tampere should design and build a model motor, however capable to produce shaft power [3]. The target was set to 1500 W, 1500 rpm. The synchronous machine type was chosen because of its relative simplicity. In order to study the performance of the SC field winding the inside-out concept was preferred. A cross sectional view is seen in Fig. 1. Iron is used in the magnetic poles both to enhance magnetic conductivity and to shield the SC windings from high flux densities. The machine has been optimized with great care accounting for maximizing total flux in the armature and minimizing flux density in the field winding. Main technical characteristics are given in table 1. The calculated efficiency of 95.5 % is extremly good for this size of motor. The efficiency does not account for refrigeration losses.

4. Cryogenics

The four SC race track coils were manufactured by American Superconductor Co. Originally the operating temperature was set to 77 K, as it was assumed that the HTS conductor would allow a sufficient current density, 12 A/mm^2 (overall), in a field of 0.5 T. Later, control measurements showed that the available flux density drops very sharply at 40 K and would be less then 0.1 T for said current density. The temperature was accordingly altered to 20 K facilitating cooling by liquid hydrogen. The cryostat was however, designed to be efficient even at liquid helium temperature, 4.2 K. According to current plans the motor will be tested up to ultimate limits of the field quantities at all available temperatures, 4.2 K, 20.4 K and 77 K.

Table 1. Technical data of the NORPAS SC synchronous motor.

Power	1500 W	Copper losses	24.2 W
Speed	1500 rpm	Eddy current losses	18.9 W (armature)
Voltage	400 V	Iron losses	8.8 W
Current (armature)	2.17 A	Additional losses	18.9 W
Field current	130 A	Efficiency	95.5 %

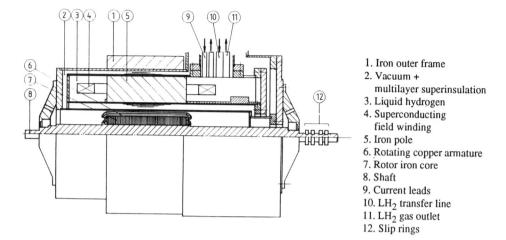

1. Iron outer frame
2. Vacuum + multilayer superinsulation
3. Liquid hydrogen
4. Superconducting field winding
5. Iron pole
6. Rotating copper armature
7. Rotor iron core
8. Shaft
9. Current leads
10. LH_2 transfer line
11. LH_2 gas outlet
12. Slip rings

Fig. 1. Longitudinal cross-section of the NORPAS motor. The 4 pole synchronous machine is characterized by the inside-out configuration. A 3-phase air-gap armature winding is attached to the laminated rotor core.

The motor outlook can be seen from Fig. 2, which specifically shows the cryostat. The inner vessel is thermally shielded by vacuum and superinsulation. Supporting is provided by 5 "horns" and the cryostat neck. The horns make it possible to center the field windings. They also take some of the tangential torque force, which, however, mainly is balanced by the stiff neck. This unconventional construction has the advantage of being easy to assemble. Further, there are no mechanical connections between the outer and the inner vessel in the active region, where the armature provides an extra heat load to the cryostat bore.

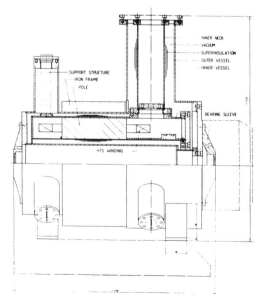

Fig. 2. The cryostat of the NORPAS motor.

56

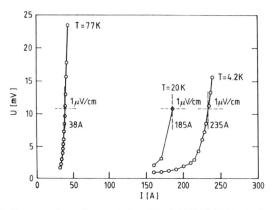

Fig. 3. Current-voltage characteristics of the NORPAS HTS exitation coils.

The field coils were delivered at the beginning of 1995. They were tested at TUT. All coils met expected requirements. Typical V(I) characteristics of one coil are shown in Fig. 3. No external field was applied. The transition range of the critical current is comparatively wide. This allows more margin for the adjustment of the exitation current. Even a current above the Ic definition value did not cause an avalanching quench, provided the coil was immersed in the liquid coolant. The critical current quantities for the complete magnet system are given in table 2.

Table 2. Characteristics of the field winding.

Temperature	I_c	Ampere-turns	Overall J_c
4.2 K	235 A	47,235 A	82 A/mm^2
20.4 K	185 A	37,185 A	64.6 A/mm^2
77 K	38 A	7,638 A	13.3 A/mm^2

5. Conclusions

The early optimism of applying superconductivity to electrical machinery was based on the fact, that electromechanical energy conversion requires high magnetic fields. On the other hand electric motors and generators have reached a very high level of development and reliability. A new technology has to offer extraordinary characteristics in form of energy and material savings in order to compete with established solutions. Rough calculations have shown that the superconducting option decreases weight to one third and size to at least one half. These reductions are strongly dependent on machine size and rotational speed. Especially for slow speed motors in the MW range there seem to be a substantial saving in both weight and volume. The simplicity of synchronous machines combined with modern power electronics underlines the potential of the field.

Acknowledgement The contributions of H. Nieminen, R. Perälä, J. Salonen, S. Tuomela and M. Lahtinen, all of the Tampere University of Technology are highly acknowledged. We also want to thank J. Kellers, G. Papst, A. Rodenbush, R. Schwall and G. Snitchler of American Superconductor Co. for their supporting attitude.

References

[1] Paasi J., Lahtinen M. and Plechácek V., 1995 *Physica C* **242** 267-276, and references therein.
[2] Arkkio A., Berglund P., Eriksson J.-T., Luomi J. and Savelainen M., 1981 *IEEE Transactions on Magnetics* **MAG-17** 900-903.
[3] Eriksson J.-T., Mikkonen R. and Söderlund L., Paper presented at *the 14th Int. Conf. on Magnet Technology*, 12-15 June, 1995, Tampere, Finland.

Inst. Phys. Conf. Ser. No 148
Paper presented at Applied Superconductivity, Edinburgh, 3–6 July 1995

Microstructure, thermodynamics and critical current densities in Bi,Pb(2223) tapes

R. Flükiger[a,b], G. Grasso[a], B. Hensel[a], M. Däumling[a], A. Jeremie[a], J.C. Grivel[a], A. Perin[a], R. Gladyshevskii[a], F. Marti[a] and C. Opagiste[b]

[a] *Dépt. de Physique de la Matière Condensée,* [b] *Groupe Appliqué de Physique, Université de Genève, 24, quai Ernest Ansermet, 1211 Genève 4, Switzerland*

Abstract. Recent developments at the University of Geneva in the field of long, mono-filamentary Bi,Pb(2223) tapes produced by rolling with j_c values of 30'000A/cm^2 at 77K,0T are reviewed. A particular attention is given to the mechanism of formation of the Bi,Pb(2223) phase. A correlation between the degree of texturing of the Bi(2212) grains before reaction and of the Bi,Pb(2223) platelets after reaction is found. The variation of j_c at 77 and 4.2K vs. applied magnetic field is presented. The lateral distribution of the transport j_c inside the filament shows a minimum at the center of the filament, reaching j_c(77K,0T) = 53'000 A/cm^2 at the sides, the zone of highest compression.

1. Introduction

Envisaging possible large scale applications of Bi(2223) tapes in transmission lines, transformers or motors, an important requirement is a high critical current density (j_c), at least 3 to 5 × 10^4 A/cm^2 and preferably even > 10^5 A/cm^2 at 77K, the temperature of liquid nitrogen. Serious difficulties have been encountered when trying to transpose the high j_c values obtained on short, pressed tapes [1] to long tapes, where industrial rolling procedures have to be applied. In spite of the metallurgical complexity of the multinary system Bi-Pb-Sr-Ca-Cu-0 in the region of the Bi,Pb(2223) phase, a substantial progress has been achieved recently. At the end of 1994, the highest reported value of j_c at 77K and 0T for 118 m long, multifilamentary Bi,Pb(2223) tapes was 28'500 A/cm^2. Meanwhile, several companies have produced tapes with lengths exceeding 1 km and critical current densities > 10'000 A/cm^2. For shorter lengths (<1m) of Bi,Pb(2223) tapes produced by industrial rolling procedures, higher values were reached: Sato et al. [2] reported 42'000 A/cm^2. The progress is confirmed by the recent achievement of an insert coil of 60 mm bore producing the very high magnetic field of 24 T at 4.2K by the same authors [2]. For comparison, j_c values of 30'000 A/cm^2 at 77K and 0T have been reached at the University of Geneva on monofilamentary Bi,Pb(2223) tapes of > 1 m length [3,4]. It is the aim of the present paper to present our data on long, monofilamentary Bi,Pb(2223) tapes with high critical current densities, giving a particular attention to the crystallographic and thermodynamic aspects of the Bi,Pb(2223) phase formation. The present review is restricted to monofilamentary tapes, which allow a better study of the thermodynamical and physical properties. The critical current densities of

Bi,Pb(2223) tapes as a function of the magnetic field are briefly described. For more details, see various posters at this conference.

2. The formation of the Bi,Pb(2223) phase

The precise knowledge of the complex Bi,Pb(2223) phase formation process is essential for further substantial enhancements of the critical current density, and various models have been proposed. The formation process starts with the Bi(2212) phase, which forms the majority of the calcined powder used for the Powder-In-Tube process, in addition to Ca2PbO4, CuO and traces of other phases. As first reported by Grivel et al.[5] and Jeremie et al.[6], the reaction from Bi(2212) to Bi,Pb(2223) passes through the intermediate phase Bi,Pb(2212). From single crystal X ray refinement [4], it follows that this new phase is in reality a solid solution of Bi(2212), with the same orthorhombic average crystal structure, but with a larger b/a ratio [5]. Note that the Bi(2212) structure is in reality described by a monoclinic superstructure [7], but this point is not of importance for the following considerations. The solid solution is formed very rapidly, and the incorporation of Pb can be observed from the very first beginning of the reaction heat treatments at 840 °C by means of X ray difraction and of EDX. Pb atoms partially substitute the Bi atoms, and the ratio of the other elements with respect to each other is changed. During the first hours of reaction, a growth of the Bi,Pb(2212) grains up to sizes > 30 mm is observed, the atoms being supplied by the surrounding liquid (30 % at the beginning of the reaction). As a consequence of the Bi,Pb(2212) grain growth, the amount of liquid is continuously decreasing.

The kinetics of the Bi,Pb(2223) formation is more difficult to study. Due to the fact that different compositions and temperatures were used, the activation energies differ from one author to another. Grivel et al. [8] have recently studied the variation of the exponent n in the Avrami equation for isothermal transformations to study the kinetics of Bi,Pb(2223) phase formation. For a series of different, a kink in the curve $\ln[-\ln(1-V)]$ vs. $\ln t$ was found after approximately 12 h of reaction [8], where V is the volume fraction of the Bi,Pb(2223) phase.

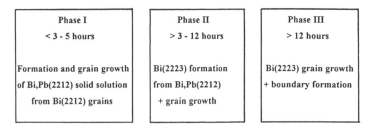

Fig. 1. Intermediate steps in the formation of the Bi,Pb(2223) phase

The value $n_1 = 1.5$ for times < 12 h indicates that at this stage of the reaction the Bi,Pb(2223) grains grow essentially from particles already nucleated at the beginning of the transformation. The value $n_2 = 0.85$ after 12 hours reflects the Bi,Pb(2223) grain growth (a value of $n = 1$ corresponds to a pure 2D growth). The whole reaction process is schematically represented in Fig. 2, showing 3 stages of reaction. These thermodynamic considerations do not reflect the behavior of the critical current density: most grain boundaries are formed after times well above 12 hours, as can be deduced from j_c measurements.

2. Deformation Correlations

Grasso et al. [9] have found that at the end of the cold rolling process, the Bi(2212) grains in unreacted tapes are already oriented in the tape plane. Above tape thicknesses of 90 μm, where the maximum of j_c was found, a drop of the degree of texturing of the Bi(2212) phase was observed, corresponding to a lower effective critical misalignment angle ϕ_e (analogous to the rocking angle) of the Bi,Pb(2223) phase. There is thus a correlation [9] between the degree of texturing in the Bi(2212) phase after deformation and in the Bi,Pb(2223) phase after reaction. This correlation is a direct consequence of the growth mechanism where the direction of the c axis for the three modifications, Bi(2212), Bi,Pb(2212) and Bi(2223) is the same.

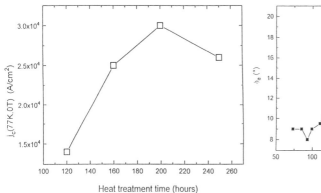

Fig. 2. j_c(77K,0T) as a function of tape thickness. (Grasso et al. [9]).

Fig. 3. Correlation between tape thickness and effective transport misalignment angle, ϕ_e [9].

3. Critical current densities of Bi,Pb(2223) tapes

The variation of the critical current density vs. B at 77K for Bi,Pb(2223) tapes is shown in Fig. 4 for a rolled tape with j_c(77K,0T) = 30'000 A/cm^2. The anisotropy, as well as the field dependence are not

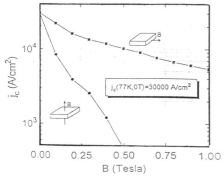

Fig. 5. j_c vs. applied magnetic field at 77K for Bi,Pb(2223) tapes obtained by rolling [4].

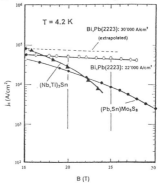

Fig. 6. j_c vs. applied magnetic field at 4.2K for various superconducting materials

are not essentially different from those of Sato et al. [3] with 40'000 A/cm^2. This indicates that the essential difference between these tapes resides in a higher density rather than in a higher degree of texturing. The distribution of the transport critical current density in tapes of 23'000 and 28'000 A/cm^2 was measured and is shown in Fig. 6. For both tapes, the distribution is very similar and shows a parabolic behavior. The local j$_c$ value in the central part is approximately 2.5 times lower than at the sides, the highest value measured so far being 53'000 A/cm^2 (Grasso et al, this conference).

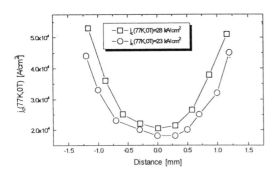

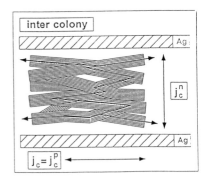

Fig. 6. Lateral j$_c$ distribution for two Bi,Pb(2223) tapes with 23'000 and 28'000 A/cm^2 at 77K and 0T.

Fig. 7. Schematic representation of the "railway switch" model.

The current transport in textured Bi,Pb(2223) tapes is described by the « railway switch model ». There is evidence that the critical current density is not limited by the intergrain j$_c$ across the dominating small-angle c axis tilt grain boundaries, but rather by the low intragrain critical current density j$_c^c$ along the c axis of the individual grains [10]. A high density of crystalline defects is induced by the mechanical deformation and persists after the heat treatment. This leads to an enhanced performance of the tapes in magnetic fields with respect to thin films.

References

[1] Yamada Y, Satou M, Murase M, Kitamura T, Kamisada Y, 1993 in *Proc.5th Int. Symp. on Superconductivity*, eds. Y. Bando and H. Yameuchi (Springer, Tokyo) 717

[2] Sato K, Ohkura K, Hayashi K, Ueyama M, Fujikami J, Kato T, 1995 *Intl. Workshop on Advanced Magnetic Fields,* February, 20-22, Tsukuba (Japan), to be published

[3] Grasso G, Hensel B, Jeremie A, Perin A, Flükiger R 1994 *Il Nuovo Cimento* **16D** 2073

[4] Flükiger R, Grasso G, Hensel B, Däumling M, Gladyshevskii R, Jeremie A, Grivel J C, Perin A, in *Bismuth-based High Temperature Superconductors,* eds. Maeda H and Togano K, to be published

[5] Grivel J C, Jeremie A, Hensel B, Flükiger R, 1993 *Supercond. Sci. Technol.* **6** 725

[6] Jeremie A, Alami-Yadri K, Grivel J C, Flükiger R 1993, *Supercond. Sci. Technol.* **6** 720

[7] Gladyshevskii R, Flükiger R, accepted for publication by *Acta Crystallographica*

[8] Grivel J C, Flükiger R, submitted to *Physica C*

[9] Grasso G, Jeremie, Flükiger R, submitted to *Physica C.*

[10] Hensel B, Grasso G, Flükiger R, 1995 *Phys. Rev.B* **51** 15456.

Inst. Phys. Conf. Ser. No 148
Paper presented at Applied Superconductivity, Edinburgh, 3–6 July 1995
© 1995 IOP Publishing Ltd

The Current State of Maglev Development in Japan

H Nakashima

Maglev System Development Division, Railway Technical Research Institute,
1 - 24 - 1 Nishishinjuku, Shinjuku - ku Tokyo

Abstract — Magnetically levitatied (Maglev) vehicles using on - board super - conducting magnets have been under development for the past 25 years in Japan.
These vehicles can be levitated magnetically as much as 100 mm above the guideway surface, and propelled by synchronous motors using the superconducting magnets.
The target speed in the commercial service of this system is 500 km/h, which will be able to connect Tokyo and Osaka in one hour.
The current state of the development of Maglev system, especially on the development of the superconducting magnets will be reviewed.

1. Introduction

The Maglev system using the superconducting magnets was proposed by Dr.Gorden Danby and Dr.James Powell in 1966. The development of this system in Japan was started as early as in 1969 to study its feasibility as the high speed ground transportation system.

After several years of basic tests and developments, a 7 km test track was constructed in Miyazaki prefecture in 1977. Using the 1st test vehicle ML - 500 a world speed record of 517 km/h was attained in 1979. Also, this test track has served to yield important data for the development of the Maglev system these 17 years.

In 1990 the project of a new test line was decided in a steady step toward realization of the revenue service using the the Maglev system. Now, this new test line (Yamanashi test line) is being constructed in Yamanashi prefecture.

The running test will be started soon after the completion of the new test line for a maximum target speed of 550km/h. Also, each part of the system will be improved and confirmed.

In the whole plan, the final decision on the construction of a new revenue service Malev line between Tokyo and Osaka will be made in 1999.

Here, the development of the Maglev system in Japan is traced.

2. Outline of Maglev development

2.1. Miyazaki test track

Table 1 shows a brief history of the development of the Maglev vehicle in Japan.

In 1979, an unmanned vehicle ML - 500 attained a record maximum of 517 km/h surpassing the target speed. This demonstrated the feasibility of high speed running.

Subsequently, the inverted - T shaped guideway was transformed into U - shape one to secure the passenger room on the vehicle. After this tranformation the test vehicle MLU001 registered 400.8 km/h in a manned run.

The test vehicle MLU002 made its debut in 1989. Unfortunately this vehicle had many troubles in its superconducting magnets and finally was destroyed in a fire accident in 1991.

The latest car MLU002N (Fig.1) was unveiled in January 1993. This vehicle attained the maximum speed of 431 km/h last year. The test results contributed to the ultimate design of a new test vehicle for Yamanashi test line.

Table 1   Maglev development in Japan

1970	START OF DEVELOPMENT		
	BASIC TEST	(1970-)	
	ML-100	(1972-1974)	60 km/h
	ML-100A	(1975-1977)	60 km/h
1977	7 km MIYAZAKI TEST TRACK		
	ML-500	(1977-1980)	517 km/h
	MLU001	(1981-1987)	400 km/h
	MLU002	(1987-1991)	
	MLU002N	(1993-)	431 km/h
1990	DECISION ON YAMANASHI TEST LINE		

Fig. 1 Test vehicle MLU002N

2.2. Yamanashi test line

Fig.2 shows the planned route of Yamanashi Test Line. The new test line is 42.8 km long including about 35 km of tunnel sections, partially with double tracks. The maximum radius of curvature on main line will be 8,000 m and the maximum gradient 40 permill (4 percent).

Two sets of trains (3 cars and 5 cars) are designed and will be used for the running tests. The first set with 3 cars is already completed and will be delivered to the car depot this July. The maximum test speed is expected to be 550km/h. All kinds of tests including riding comfort at high speed and confirmation of durability are planned.

Fig.3 shows the configuration of the new Maglev vehicle. Fig.4 shows the nose-shapes of the test vehicle. These nose-shapes were selected through the wind tunnel test and the computer simulations of the aerodynamic characteristics. They will finally be compared in the real high speed running test, especially for aerodynamic drag and noise characteristics.

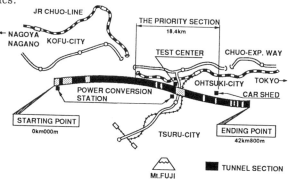

Fig. 2 The route of Yamanashi test line

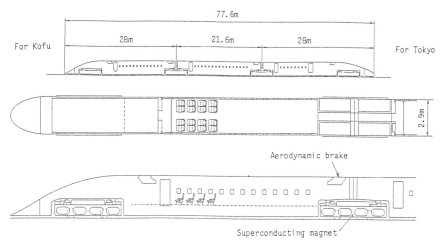

Fig. 3 The configuration of the test vehicle for Yamanashi test line

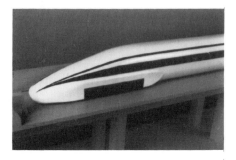 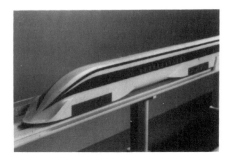

Aerowedge style   Double cusp style

Fig.4 Nose shapes of Yamanashi test vehicle

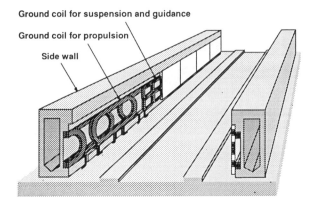

Ground coil for suspension and guidance

Ground coil for propulsion

Side wall

Fig.5 The configuration of ground coils in the guideway

Fig.5 shows the configuration of the ground coils in the guideway of Yamanashi test line.

3. Superconducting magnet

The superconducting magnet is one of the most important parts of Maglev system. They have to stand strong forces levitating, propelling or guiding the vehicle. Also, they have to work stably in severe circumstances such as flux flactuation from the ground coils and mechanical vibration.

Fig.6 shows an image of the superconducting magnet and the super conducting coil. Fig.7 shows an other view of the superconducting magnet for Yamanashi test vehicle.

Table 2 gives the basic characteristics of the superconducting magnet in one of the models for the Yamanashi test vehicle.

Table 2 Basic characteristics of the superconducting magnet

Items	Specifications
Dimensions of SCM	5.32 m(L) × 1.17 m(H)
Weight of SCM	1500 kg
Magnetomotive force	700 kA
Number of coils	4
Pole pitch	1350 mm
Maximum empirical magnetic field	4.23 T
Heat leakage to inner vessel	3 W
Levitative force per magnet	98 kN
Refrigeration capacity	8W at 4.3K

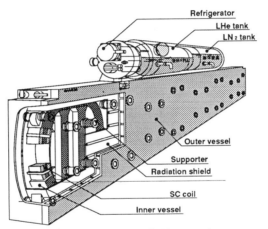

Fig.6 Image of the superconducting magnet

Fig.7 The superconducting magnet for Yamanashi test vehicle

About 4 km length of the superconducting wire is used for one superconducting coil. The wound coil is impregnated with epoxy resin to suppress the micro - movement of the superconducting wire.

The impregnated superconducting coil is fixed to the stainless steel inner vessel. Between this inner vessel and the superconducting coil come spacers. Liquid helium (LHe) supplied from the upper tank is delivered to this narrow space keeping the coil temperature at 4.3 to 4.4K. The on - board refrigerator is mounted on this LHe tank. Liquid nitrogen is also stored in the LN_2 tank, which is delivered to the thermal radiation shield plate in the superconducting magnet.

The superconducting coils installed in the test vehicle MLU002 suffered so many quenching troubles. This fact promoted an improvement on the stability characteristics of the superconducting coils.

The main cause of the quenching trouble in the superconducting coils was the mechanical friction between the coil surface and the spacers. It is clear that the superconducting coils using aluminum alloy spacers instead of FRP has obviously an improved stability. This means that the increased thermal conductivity of the aluminium alloy spacers contributes to releasing the heat generated between the coil surface and the spacers.

Now, we believe that the superconducting magnets will serve on Yamanashi test line with good stability.

4. Concluding remarks

With the test vehicle MLU002 we experienced many coil quenching troubles. Then much improvement has been done to increase the stability of the superconducting coils.

Inprovement is practically implemented in the designing of the superconducting magnets for Yamanashi new test line under construction.

One of the remaining problems to be solved is the confirmation of the reliability of the total system. This will be one of the important jobs to be under taken on the new test line.

The development of the Maglev system will enter a new stage with the start of the new Yamanashi test line. The running test will begin in 1997. It will prove the superiority of this system and give many important data for the realization of the revenue service.

References

[1] H.Nakashima 1994 IEEE Trans. on Magnetics 4 1572 - 1578
[2] H.Tsuchishima, T.Herai 1991 IEEE Trans. on Magnetics 2 2272 - 2275

Inst. Phys. Conf. Ser. No 148

Paper presented at Applied Superconductivity, Edinburgh, 3–6 July 1995

Centrifugal Casting of BSCCO 2212 Form Parts and their First Applications

J. Bock, S. Elschner Hoechst AG, Central Research, D-65926 Frankfurt/Main, Germany

P.F. Herrmann Alcatel Alsthom Recherche, Dept.Electrotechnique, F-91460 Marcoussis, France

B. Rudolf Siemens AG, Accelerator and Magnet Technology, D-51425 Bergisch-Gladbach, Germany

Abstract - The Melt Cast Process (MCP) and the centrifugal casting technique for BSCCO 2212 bulk parts could be upscaled to technical dimensions (Ø up to 400 mm for rings and L up to 500 mm for tubes). The electrical properties of the parts have reached a perfomance needed for power applications. Computer calculations of the expected critical current under self field conditions were confirmed by DC transport measurements. Tubes with integrated silver contacts are able to carry transport currents of up to 7.5 kA DC ($1\mu V/cm$, 77K) which marks a new record value for HTS bulk parts. Ic is significantly improving with decreasing temperature (gain by a factor of 3- 4 from 77K to 64K), an effect which is even more pronounced under external magnetic fields. First results show, that the losses are not prohibitively high also for operation under AC conditions. The applicability of MCP bulk parts for power applications has already been shown in several prototypes and now also in first commercial high-Tc current leads. Shaping possibilities offer additional options, e.g. manufacturing of coil shaped samples out of centrifugally cast tubes.

1 Introduction

The technical application of high-temperature superconductors (HTS) in magnet technolgy and power engineering requires superconducting half-stuff (wire or bulk parts) with operating currents of several hundred A´s or even better in the kA-range. The intermediate parts are processed to cables or magnets in the case of wires or can be directly integrated into electrotechnical devices in the case of bulk parts. A commercial exploitation of the ceramic superconductors will depend on the development of manufacturing methods that are appropriate for industrial scale up and that simultaneously yield reproducible electromagnetical properties according to the specific applications.

Common to all the different material options and manufacturing methods for bulk parts is that only processes involving at least a partial melting during the growing of the superconducting phase are leading to sufficient current transport properties. The Melt Cast Process (MCP) of Hoechst is especially straightforward and a now well established method for the manufacture of BSCCO 2212 bulk parts [1]. Available by the basic process are solid parts e.g. in the form of rods and plates. An extension of the process applies the centrifugal casting technique to yield hollow parts such as rings or tubes. Low resistance current contacts in the form of silver sheets can be integrated right in the process, mechanical treatment offers additional options for shaping. This paper resumes the current status of the basic MCP together with the centrifugal casting technique. Discussed are the results of electromagnetical characterisation of little slabs, which reflect the intrinsic properties of the MCP material, in comparison to the transport properties of the larger bulk parts. Already achieved is the technical application of MCP BSCCO 2212, which is the most advanced HTS bulk material, in hybride current leads. Examples for functional models, demonstrators and a first commercial application are presented.

2 Experimental

In general shaping techniques for ceramic form parts comprise the manufacturing of the corresponding powder on a solid state route (with different single steps) as a first main step, forming of the powders into the desired shapes often with the aid of e.g. organic binders [2] in a pressing step and annealing to achieve sintered ceramic bodies in the third step. The entire process is limited by the pressing step which makes the manufacturing of large sizes and complex shapes very costly. The electrical properties of the so prepared ceramic parts are determined on the one hand through the addition of organic aids which may lead to grain boundary impurities, on the other hand the sintered ceramics behave granular with large angle grain boundaries. In contrast to this are zone melting processes which indeed allow to achieve the so far best properties for HTS bulk parts, but only with high expenditures of time and money.

2.1 Basic MCP for solid parts

In contrast to the classical ceramic route the Melt Cast Process needs only two main steps because the shape forming is managed by melt casting instead of pressing. Consequently no superconducting or other especially prepared precursor material are required and also no organic aids, which might generate carbon impurities, are needed. The starting materials, oxides of the metals Bi, Sr, Ca, and Cu, are mixed in a ratio near 2:2:1:2 together with small amounts of $SrSO_4$ [3]. The mixture is molten under release of oxygen in a furnace operating at 1000-1200°C. The homogeneous melt is poured into a mould where solidification and slow cooling to room temperature occurs.

The obtained cast form parts are not superconducting but are already in the final shape of bars, plates, rods, tubes or rings (hollow cylinders). They consist of a heterogeneous phase mixture which is oxygen deficient compared to the starting mixture. Their microstrucure is, depending on the cooling conditions after melting, more or less coarse grained. In order to achieve the desired superconducting BSCCO 2-layer phase and to adjust the oxygen content the cast parts are subjected to an annealing around 800°C. The complex diffusion reactions in the solid bodies are favoured by the formation of a partial melt which also accelerates oxygen uptake [3]. The concrete heating conditions such as temperature, duration and gas atmosphere depend on the dimensions of the individual parts.

2.2 Centrifugal casting of hollow parts

While melt casting into static moulds leads to solid form parts, hollow forms can be obtained by casting the melt into rotating moulds (centrifgual casting). Different issues had to be adressed for a successful adaption of this technique known for metal casting to the fabrication of ceramic shapes. Due to the centrifugal force the cast melt is evenly distributed on the inner side walls of the mould. During solidification the melt shrinks and leaves a gap between the mould wall and the ceramic cylinder which begins to clatter in the rotating metal mould. To avoid the breaking of the cylinder the rotation speed is carefully monitored and adapted during the different stages of the solidifcation process.

Cracks and tensions in the ceramics in the as cast state, especially in the parts of larger diameters, are prevented by a controlled cooling of the melt after casting. Together with this centrifugal casting technique the MCP has proven to be a very powerful and flexible method to fabricate rings and tubes of arbitrary sizes. Typical dimensions achieved so far are given in tab.1 and listed in fig.1. The present limitation on parts up to 2kg due to the handling of the hot melt will be eased in the near future and larger parts will become available.

tab. 1: Sizes (in mm) of centrifugally cast rings and tubes

	Ø	max. length	wall thickn.
min. achieved	13	100	4
	26	200	4-7
	35	500	5-8
	43	300	5-8
	70	200	
	150	130	5
max. achieved	400	50	

fig. 1: MCP tubes with integrated contacts

2.3 Mechancial Treatment

High-Tc superconducting material is ceramic and brittle in nature and this holds also for the solidified melt of BSCCO in the as cast state. It becomes surprisingly maleable after annealing and has excellent machining properties making it easy to cut, grind, or drill. Mechanical turning e.g allows to obtain hollow cylinders with thin walls between 1.5 and 2 mm. The turned parts are particularly homogeneous and dense (shrink holes due to the casting procedure are removed) with a smooth surface. They are exhibiting better electrical properties compared to original castings with comparable thin side walls. An example for the excellent machining properties of the MCP material is the helical sample in fig.2. It was cut by a small high rotating diamond tool out of a BSCCO tube which was supported with an inserted metal tube.

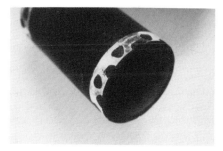

fig.2: Helical sample (turn width of ~4 mm, 2 mm gap) cut from a MCP-tube (Ø 26 mm, L 75 mm)

fig.3: contact area of a Ø70 mm tube

3 Electrical contacts

While the application of contacts to superconducting bulk materials normally requires an additional manufacturing step, MCP allows a coprocessing of silver sheets right in the fabrication process of rods and tubes. Recently improved was the contact design for tubes (see fig.3). The now used punched silver foils are providing a better integrity of the ceramic tubes in contrast to the earlier applied folded sheets. The measurement of the contact resistance is not straightforward and reason for several misinterpretations [4]. The designing engineer is not so much interested in the specific transfer resistance silver /BSCCO but mainly in the overall value achieved for the current transfer copper/solder (or even solder1/ solder2)/ silver/ BSCCO. For an experimental determination of this value a silver tube (Ø26 mm) was soldered (for details of soldering see [5]) to the normally applied copper holder. The position of the voltage taps inside the copper holder near to the silver surface can be taken from fig.4. The resistance of both contacts was 1.1 $\mu\Omega$ for the silver tube. For Ø26 mm BSCCO tubes with 2cm contact width a contact resistance in the range of 1.5 $\mu\Omega$ is measured which indicates that the major part of the contact resistance is caused by the solder. Experience from series of bulk parts with different dimensions show that reliable low loss contacts $\leq 2\mu\Omega$ cm^2 (77 K), 40 nΩ cm^2 (4 K) and good mechanical properties are achieved.

4 Electrical characterization

The critical current density of the MCP material is influenced by several parameters. Among them are the corresponding microstructure of the particular part and the self field effect which in turn is determined through the sample size and the operating current. The microstructure of MCP rods, in detail described elsewhere [3], is consisting of a chilled zone (about 200 μm) in the surface regions, an inevitable shrinking zone in the center of the rod and a coarse grain structure in the main part in between.

tab. 2: Transport currents in small slabs and large parts of the MCP material at differrent temperatures

		jc [A/cm²]	Ic [A]	jc [A/cm²]	Ic [A]
sample type	size [mm]	*(77 K, self field, 1µV/cm)**			
rodlet	1x1x40	2000-4250		*(64 K, self field,1µV/cm)***	
rod	Ø 5, L 100	1100-1200	200-250	4100	800
	Ø 8, L 150	900-1000	450- 500	*(65.5 K, self field,1µV/cm)***	
tube	Ø 26, L 75	600-680	1700-2200	1820	6000
	Ø 35, L 400	600-650	2600- 3100		
	Ø 43, L 100	560-620	3500 -3900		
	Ø 70, L 200	450-550	6000-7500		

* numerous measurements, different wall thicknesses ** single measurements

4.1 Critical current and temperature dependence in small samples and rods

Not surprisingly, the current flows essentially in the coarse grained region in between with 2500 A/cm² as typical value. For one sample (1x1x40 mm³), cut off from this zone, we obtained 4250 A/cm² at 77K. The critical current of an entire rod is an average value over the whole cross section [1] further reduced by self field effects. Typical and well reproducible values that are now achieved for rods are 1100 A/cm² (Ø 5mm) and 950A/cm² (Ø 8mm), see tab.2. For a rod (Ø 5mm, L 50mm) the field and temperature dependence of jc was also determined in liq.N2 at different pressures. With decreasing temperature a strong increase of jc can be observed, an effect which is much more pronounced in external fields [1] and which significantly enhances the applicability of the MCP BSCCO 2212 bulk parts.

4.2 High transport currents and temperature dependence in large tubes

A DC current up to 7.5 kA (77K, 1µV/cm) was measured under self field conditions (approx 43mT) in a tube conductor of Ø70/54 mm (outer/inner) and L200 mm which marks a new record value for bulk parts of these dimensions. Different experiments show, that the MCP parts can also safely be operated at overcurrents (i.e. criterions > 1µV/ cm). The tolerance of these conductors to overcurrents is a consequence of the low n-value (n=9) of the superconducting transition.

In order to test the current carrying capability potential of a tube also below 77K, a measurement was performed under pumped liquid nitrogen with a tube Ø26 mm. The critical current increased from 1970A (600A/cm²) at 77K to 6000A (1820 A/cm²) at 65.5K by a factor of about three which is in the same range as for the Ø5 mm rods (see 4.1).

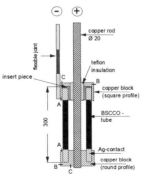

fig.4: Set up for measurement of critical current and
contact resistance of BSCCO tubes

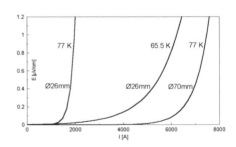

fig..5: U(I) curves of a Ø26 mm tube (65.5K, 77K)
and a Ø70 mm tube (77K)

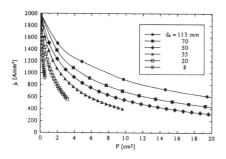

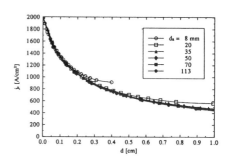

fi.g 6 : Calculated jc-values of MCP tubes with
different outer diameters

fi.g 7: Calculated jc-values of MCP tubes
with different wall thicknesses

4.3 Impacts of the magnetic self field

A limiting factor for the current carrying capability of BSCCO bulk parts is the magnetic self field effect which at large absolute currents (kA range) is important enough to substantially reduce the critical current density. One way to reduce this impact is the use of tubes instead of solid rods leading to a substantially increased critical current for a given cross section. In order to fully exploit the potential of the material and to be able to tailor the bulk parts for certain rated currents, a quantitative analysis of this effect by a computer code was developed. The calculation bases on the intrinsic critical current density and its field dependence [6] and provides the achievable overall critical current for tubes with different outer diameters and wall thicknesses. For a given cross section, jc is improving with growing tube diameter i.e. decreasing wall thickness. For a given tube jc, decreases with growing wall thickness and increasing operating current (fig.6). Tubes of different outer diameters but same wall thickness have the same critical current density (fig.7). The calculated results have been verified in DC transport measurements performed on MCP-tubes of various dimensions.

The impact of the magnetic self field effect was also directly demonstrated: In a MCP-tube, Ø35 mm and L200 mm, a critical current Ic= 3.1kA (77K, 1μV/cm), corresponding to jc= 620 A/cm², was measured under self field (about 35 mT). After this the tube was longitudinally cut into ten segments of approx. 10x5x200 mm³. The jc-values of the single segments (800-850 A/cm²) are significantly improved compared to the complete tube and are in the range of MCP rods (tab.2).

4.4 Operation under AC conditions

A Ø70/58 mm tube was tested with 50Hz AC current up to 10 kÂ (peak value). The full current was ramped up several times without damaging the tube, in one case even 11.5 kÂ were achieved. This also demonstrates the quality of the integrated silver contacts. AC loss measurements of the MCP BSCCO 2212 conductors at and below the critical current are indicating losses that are sufficiently low for self field applications at technical frequencies. For all samples losses inferior to 8.5 10^{-3} W/Am were observed. This value is already taking into account the refrigeration for the dissipated heat at 77K and is thus indicating an economical interest for AC applications with this material at liq.N2 temperature [7].

5 HTS hybrid current leads

Most advanced among the different possible applications of HTS form parts such as current leads, current limiters and magnetic shielding are undoubtly hybrid current leads with incorporated HTS bulk parts. They are developed as electrical links between superconducting magnets or power devices operating at low temperatures (about 1.8K to 30K) and an intermediate temperature of 20K to 77K. They are the first real electrotechnical application for HTS conductors which already has been achieved with

MCP BSCCO parts. They are not only leading to a significant loss reduction but as a consequence of this, enabling new innovative cooling concepts (no-boil-off, cryogen-free™) for low-Tc magnet systems.

5.1 Demonstrators and prototypes

In the frame of an EC-project [8] 2 kA/ 20kV DC leads based on MCP tubes were developed which are reducing the heat input to the 4K level by a factor of 10 and the total refrigeration load by a factor of 5 compared to an optimized all metal lead. Goal of the actual project between Alcatel, Hoechst and Siemens is the development of a 5 kA$_{rms}$ /50 kV AC lead. A second liquid-He-free (cryogen-free™) Nb$_3$Sn magnet system, now for 4-5T, was built by Oxford Instruments with leads also made of MCP rods [9].

5.2 First real application

At Siemens an innovative current lead design has successfully been introduced. The hybrid current lead consists of a normal metal part and a HTS module. In principal this design can easily be adapted to a variety of different cooling environments (i.e. to 1-, 2-, or 3- stage refrigerators). The present design is optimized for cryostats operating with zero-Helium-boil-off and thus no helium vapor for cooling of the current leads. It has a refrigerator cooled and current optimized metallic part, thermally anchored at about 70K and 20K. The thermal contact to the cryocooler and the electrical insulation is extremely good. The HTS module design is current independent up to currents of more than 300 A. It consists of a MCP BSCCO 2212 rod encapsulated in a thin walled stainless steel tube. The ss-tube takes all the mechanical forces and serves as current bypass in emergency situations.

The complete current lead is available in a retractable version. It is designed as a pure heat conduction cooled system ooperating in helium atmosphere, but gas cooling is also possible for emergency (cryocooler not working) or overcurrent situations. The most striking benefit of this lead is the low thermal heat load at 4K (15 mW/ pole @ 120A, retractable 31 mW/pole) without any cooling gas needs. This type of hybrid current lead (Ø8 mm for 120A, Ø5 mm for 25A) is already successfully used in commercial sc magnet systems (as e.g. a 4/5T Wiggler at ESRF, Grenoble, F) built by Siemens.

6 Conclusions

MCP offers simple route to form BSCCO bulk parts in a wide variety of shapes and dimensions. Rings with a diameter up to Ø 400mm have been fabricated. The critical currents (DC 7.5 kA, 1µV/cm, 77K, selffield) are record values for HTS bulk parts. A lowering of the operation temperature leads to a significant further enhancement of the transport currents (e.g from 2kA at 77K to 6kA at 65.5K) in the rods and tubes. In comparison to other HTS bulk materials, MCP parts with integrated current contacts are most advanced with regard to electrotechnical applications. They have been successfully used in demonstrators and prototypes and have now been transferred to first real applications.

This work was partly funded from the EU in the frame of Brite-EuRam under the projects no. BE 4071 and BE 7856. We thank C. Albrecht (Siemens AG , Erlangen) for stimulating discussions and an excellent project management.

7 References

1. J.Bock, S.Elschner, P.F. Herrmann, Proc. of ASC 1994 (Boston), IEEE Trans. Appl. Supercond. in press

2. D.R.Watson , M.Chen, J.Evetts, Supercond. Sci. Technol. **8** (1995) 311

3. J.Bock, H.Bestgen, S.Elschner, E.Preisler, IEEE Trans. Appl. Supercond. **3,** 1 (1993) 1659

4. S.Elschner, J.Bock, Adv. Mat. **4** (1992) 242

5. E.Preisler, J.Bayersdörfer, M.Brunner, J.Bock , S.Elschner, Supercond. Sci. Technol. 7 (1994) 389

6. S.Elschner, J.Bock, G.Brommer, P.F.Herrmann, submitted to proc. of MT14 (June 12-16, 1995), Tampere (SF)

7. E. Béghin, G.Duperray, D.Legat, P.F.Herrmann and J.Bock , submitted to Appl. Superconductivity

8. C.Albrecht, J.Bock, P.F.Herrmann, Supraleitung in der Energietechnik II, VDI-Bericht 1187, München, Mai 1995

9. L. Cowey et al., this conference

Inst. Phys. Conf. Ser. No 148
Paper presented at Applied Superconductivity, Edinburgh, 3–6 July 1995
© *1995 IOP Publishing Ltd*

Fault Current Limiters Based on High Temperature Superconductors

W. Paul*, J. Rhyner*, Th. Baumann* and F. Platter[#]

*ABB Corporate Research, CH-5405 Baden, Switzerland,
[#]ABB High Voltage Technologies, CH-8050 Zürich, Switzerland.

Abstract. Various types of fault current limiters based on High Temperature Superconductors are currently discussed. One is based on the so called "shielded iron core concept" where the superconductor is coupled magnetically to the network. ABB has built and tested several prototypes of this limiter. It mainly consists of a copper coil, a superconducting tube, and an iron core which are concentrically arranged. The superconductors used for the prototypes were bulk Bi2212 ceramics. They were fabricated by a partial melting process, where the precursor powder is molten in a cylindrical Ag-mould which is rotating inside a furnace. The material has the voltage-current characteristic $V \sim I^\alpha$ with $\alpha \approx 5$. Its critical current density, defined by the 1 μV/cm criterion, is about 1400 A/cm^2.

Tests of a 100 kW prototype were performed in a 480 V circuit, with a prospective fault current of 8 kA. The superconducting tube of the prototype has a diameter of 20 cm and a length of 35 cm. Depending on the number of turns of the copper coil the device has a nominal current between 130 A and 250 A. In normal operation the AC-losses in the superconducting tube are of the order of 10 W. In short circuit tests performed at different phase angles of the voltage source, the current limiter was loaded for 50 msec, then the circuit was opened by a breaker. The peak fault current was limited to less than 5 times the nominal current. The limitation was always smooth without any overvoltages. The superconductor did not show any degradation after about 20 shorts.

Simulations of both the normal operation and the behaviour under fault conditions have been performed taking into account the nonlinearity and temperature dependence of the I(V) characteristics of the superconductor and the saturation of the iron core. The agreement between experiments and calculations is excellent

Typical applications for upscaled devices are networks with high prospective short circuit currents and low nominal currents. In power stations such conditions can be found in the excitation, auxiliary, and start up branch.

1. Introduction

The worldwide increasing demand for electrical power leads to a growing interconnection of electrical networks. As a consequence, the short-circuit currents (I_{SC}) increase. They can be orders of magnitudes higher than the nominal current (I_N) and lead to high mechanical and thermal stresses, which both are proportional to I_{SC}^2. All electrical equipment has to be designed to withstand these stresses. It is obvious, from both technical and economical points of view, that a device that reduces I_{SC} is needed.

Common measures to limit I_{sc} are increased transformer reactances or additional reactors. These, however, substantially influence the network under normal conditions. Conventional and triggered fuses [1] are also used, but they have to be replaced after every fault event. Recently, circuit breakers with current limiting properties have also been proposed [2].

All the solutions realized up to now have substantial disadvantages. They fulfill only a few of the required characteristics. The characteristics of an ideal fault current limiter, are:

- Negligible influence on the network under normal conditions.
- Instantaneous limitation of all types of fault currents.
- Repetitive operation without any replacements, and short recovery time.
- No over voltages during a fault
- No external trigger or external energy source.
- High reliability and fail safety.

2. Main concepts of superconducting fault current limiters

2.1. Saturated iron core reactor

The device uses a superconducting (SC) secondary winding to keep the iron core of a reactor in saturation during normal operation [3]. The normal conducting coil of the reactor is in series to the line which has to be protected. During a fault the iron core departs from saturation and thus increases the impedance of the reactor.

The SC-winding sees only a DC-current and always stays in the SC-state so that it needs no recovery time after a fault. The main disadvantage of the device is its mass, which is about two times that of a transformer with the same rated power.

Based on Low Temperature Superconductors (LTS) a 5 MVA prototype of this device has been built in 1997 (NEI, U.K.) [3]. Among the available High Temperature Superconductors (HTS) the Bi2223 tapes seem to be the most suitable to realize this device. However, an application of the tapes at 77 K is prohibited by the strong magnetic field dependence of the critical current density (j_c) [4].

2.2. Resistive Limiter

In this concept the SC is connected directly with the line, which has to be protected [5,6]. During a fault, the critical current of the SC is surpassed, its resistance increases and thus limits the current.

This superconducting fault current limiter (SCFCL) is very compact, but it needs long lengths of SC-wire, and current-leads which connect the SC to the normal conducting line and lead to additional thermal losses.

Based on LTS a 5 MVA (GEC Alsthom, France) [5] and a 10 MVA (Toshiba, Japan) [6] prototype have been built. Small prototypes with a rated power of about 100 VA have been realized using thin films of YBCO [7]. A realization of larger devices based on HTS has up to now been prohibited by the lack of long SC with sufficient high normal resistivity. The available Bi2223 tapes are not suitable because of their low resistivity due to the high silver content.

2.3. Transformer with shorted secondary winding / Shielded iron core

This device is essentially a transformer with the primary winding connected in the line and a superconducting secondary winding, which is shorted. Because of the inductive coupling between the line and the SC, the device is often referred to as "inductive" SCFCL.

This SCFCL needs no current-leads. Since the number of turns of the secondary windings can be much smaller than the primary turns, only short SC are needed and the voltage drop in the cryogenic part of the device is reduced. The number of secondary turns can even be 1 and the winding may be a tube [8,9].

3. ABB Concept

At ABB we pursue the concept of the "shielded iron core" SCFCL based on HTS. It mainly consists of a normal conducting coil, a superconducting tube, a cryostat, and an iron core, all of which are concentrically arranged (Fig. 1). The device is essentially a transformer in which the secondary winding is the superconducting tube. The primary winding (coil) is connected in series with the line which has to be protected. Only the superconducting tube is cooled with liquid nitrogen (77 K).

Fig. 2 shows the equivalent circuit of the device. The primary (normal) and secondary (superconducting) side are labeled by the numbers 1 and 2, respectively. R_1,R_2 are the resistances of the coil and SC-tube, L_S is the stray inductance, L_{11}, L_{22}, L_{12} are the inductances of and between the primary and secondary sides, and I_1, I_2 are the currents in the primary and secondary sides. The equivalent circuit is exact, even in the presence of nonlinearities and temperature dependencies [9].

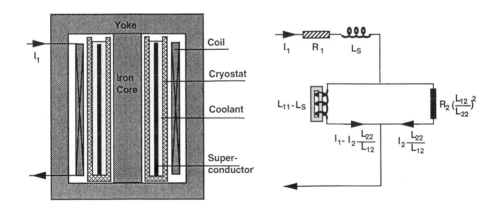

Fig.1 Scheme of our SCFCL　　　　　　　　Fig.2 Exact equivalent circuit

4. 100 kVA Prototype

4.1. Coils

In order to vary the nominal current of the device various copper coils with n = 45, 60, 74, and 86 turns, have been used. Their DC-resistances are 26, 29, 36 and 53 mΩ, respectively. The height and radius of the coils are h $\approx$ 35 cm and $r_{pr} \approx$ 14 cm.

4.2. Iron-core and Yoke

Since the iron is not magnetized in normal operation, core and yoke can be made of rather thick and cheap steel. We used 2 mm thick construction steel. The height and radius of the core are $h_{co} = 48$ cm and $r_{co} = 7.2$ cm, respectively. The effective permeability of the closed core is about $\mu = 60$. The total mass including the yoke is about 200 kg.

4.3. Cryostat

Since the cryostat sees a magnetic AC-field even in normal operation, it should be made of non-magnetic and electrically insulating material. For our cryostat we used glass-fibre reinforced epoxy and vacuum insulation.

4.4. Superconductor

We produced SC-rings with 8 cm height by processing Bi2212 powder in cylindrical Ag-moulds which were rotating in a furnace. The rings have radius $r_{sc} = 10$ cm and wall thickness d = 1.9 mm. The tube finally used in the SCFCL was built from a stack of 4 of these rings. The total height of the stack was 35 cm. The material has the voltage-current characteristic $V \sim I^{\alpha}$ with $\alpha \approx 5$ [9,10]. The critical current density, as defined by the 1μV/cm criterion, is about 1400 A/cm^2.

5. Tests and Discussion

5.1. Nominal Current

The critical current of the tube as defined by the 1μV/cm criterion was $I_2(1\mu V) = 8500$ A. However, it turned out that, due to the smooth I(V) curve of the Bi2212 the superconductor stayed thermally stable up to about 2.5 times that value. For the maximum operation current we chose $I_{2max} = 16000$ A, which corresponds to a voltage drop of about 30 μV/cm. The nominal current of our device thus was $I_N = I_{2max}/(n\sqrt{2}) = 251$, 189, 153, or 132 A, depending on the number of turns of the coils. Below I_N the impedance of the SCFCL is mainly given by $\omega L_S = 0.07, 0.12, 0.19, 0.25$ Ω, depending on the different coils.

5.2. Losses

For low currents the measured AC-losses increase with $I_1{}^3$, reaching about 12 W at 80 % of I_N. This is in nice agreement with the theoretical value of 10 W [9]. At I_N the losses are about 35 W. This is 15 W above the theoretical value and is obviously due to the increased "DC-part" of the losses.

5.3. Short Circuit Tests

Short circuit tests were performed in a 480 V, 50 Hz circuit comprising the SCFCL, an inductive load of 13 mH and a circuit breaker in series. A short was simulated by bypassing the load. After 50 ms the circuit was opened by the breaker. The prospective fault current was 8 kA. About 20 tests were performed with different coils and at different phase angles of the source voltage The limitation was always smooth without any over voltages (Fig. 3). The Bi2212 first limits the current by its I(V) characteristic at 77 K, then it warms up and thus further reduces the current. The SC-rings needed about 30 s to become superconducting again and did not show any degradation during the tests.

Fig. 3 shows the time evolution of the circuit current and the voltage drop over the SCFCL ($n = 86$, $I_N = 132$ A$_{rms}$) for two extreme phase angles at which the short was triggered, namely (a) at peak source voltage, and (b) at zero source voltage. In both cases the fault current is limited to a peak value of about 900 A, i.e. 5 times I_N. In Fig. 3b we find a distinct double peak. This occurs because the iron core saturates after the first current maximum, so that the inductance is reduced and thus the current rises again to a second maximum. For coils with less turns this second peak is higher. For $n = 86, 74, 60, 45$ it was about 700, 1100, 1600 and 2200 A, respectively. Fig. 4 shows a simulations of the tests, taking into account all nonlinearities and temperature dependencies of the device.

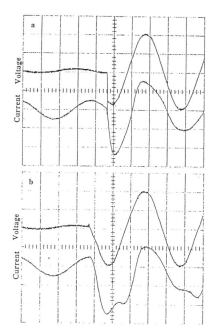

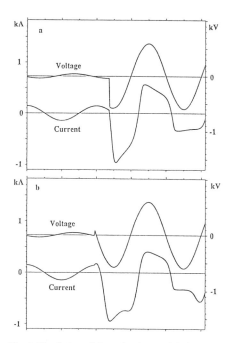

Fig. 3. Test oscillograms: (a) short triggered at peak and (b) at zero crossing of source voltage. Upper traces : voltage drop at SCFCL (337 V/ div). Lower traces: current in test circuit (365 A/div).

Fig. 4. Simulation of short circuit test: (a) short triggered at peak and (b) at zero crossing of source voltage. Upper traces : voltage drop at SCFCL. Lower traces: current in test circuit

78

6. Conclusion

Because of the lower cooling costs, SCFCL based on HTS are more likely to be applied in power electricity than LTS. Up to now, however, it is not possible to realize at 77 K any of the SCFCL-concepts based on long-length conductors, because the required HTS are not available.

A prototype based on HTS-tubes with a rated power in the 100 kVA range has been built and successfully tested. The device limits fault currents to about 5 times the nominal value without showing any over voltages or degradation of the HTS. A simulation program has been established which yields very good results both for the normal operation of the SCFCL and for its behavior under fault conditions.

The device can be scaled up to the MVA-region where we envisage first applications in networks with high prospective fault currents and rather small nominal currents, e.g. in the excitation, auxiliary, and startup branches of power stations.

Acknowledgements

We thank J. Meier, P. Unternährer, and R. Weder for technical assistance. Financial support from the National Energy Research Foundation "NEFF" and the Swiss National Fund "NFP30" is gratefully acknowledged.

References

[1] T. Keders, A.Leibold, "A current limiting device for service voltages up to 36 kV", IEEE PES Summer Meeting, 1976,

[2] N. Engelman, E. Schreurs and B. Drugge, "Field Test Results for a Multi-Shot.12.47 kV Fault Current Limiter", IEEE PES Summer Meeting, 1990,

[3] B.P. Raju, K.C. Parton and T.,C. Bertram, "Fault Current Limiter Reactor with Superconducting DC Bias Winding", paper no. 23-03 CICRE (1982)

[4] J.X. Jin, S. X. Dou, H.K. Liu and C. Grantham, "Preparation of High Tc Superconducting Coils for Consideration of their Use in a Prototype Fault Current Limiter", Presented at the ASC 94 in Boston, to appear in IEEE Trans. Appl. Supercond.

[5] T. Verhaege et al., "Experimental 7.2 kV$_{rms}$/1 kV$_{rms}$/3kApeak Current Limiter Systems", IEEE Trans. Appl. Supercond., vol. 3, no.1 , 574-577 (1993)

[6] T. Hara et al. "Development of a new 6.6 kV / 1.5 kA-class superconducting fault current limiter for electric power systems", IEEE Trans. Power Delivery, vol. 8, 182-192, (1993)

[7] Y. Terashima et al., "Application of Superconducting Films to Fault Current Limiter", Jpn. J. Appl. Phys., vol 33, 1592-1594 (1994)

[8] D.W.A. Willen and J.R. Cave, "Short Circuit Test Performance of Inductive High Tc Superconducting Fault Current Limiter" Presented at the ASC '94 in Boston, to appear in IEEE Trans. Appl. Supercond.

[9] W. Paul, Th. Baumann, J. Rhyner, and F. Platter "Test of 100 kW High-Tc Superconducting Fault Current Limiter" Presented at the ASC '94 in Boston, to appear in IEEE Trans. Appl. Supercond.

[10] W. Paul and J.P.Meier, "Inductive mesurements of voltage-currentcharacteristics between 10^{-12}V/cm and 10^{-2}V/cm in rings of Bi2212 ceramic", Physica C, vol. 205, 240-246 (1993)

Inst. Phys. Conf. Ser. No 148
Paper presented at Applied Superconductivity, Edinburgh, 3–6 July 1995
© *1995 IOP Publishing Ltd*

Constructing HTS Coils for Practical Applications

R G Jenkins and H Jones

University of Oxford, Clarendon Laboratory, Oxford OX1 3PU, UK

Abstract. Practical HTS conductor, combining high Jc and long length, at present tends to be in tape form. We describe how its highly anisotropic Jc(B) dependence, together with the brittle nature of the HTS material affects coil design and construction. The two main coil fabrication techniques, React-and-Wind and Wind-and-React, are reviewed and notable results from the worldwide research effort are presented. In addition we discuss HTS coil application areas and fabrication technique potential. The Oxford University conductor/coil development programme is briefly reviewed, and lastly we outline a project investigating the controllability of HTS coils for electromagnetic maglev and discuss the design of HTS coils for a large-scale demonstrator.

1. Introduction

Since their discovery in 1986 the development of HTS materials has progressed to the point where practical conductor (combining high Jc and long length) is being produced at various locations around the world, and indeed is becoming commercially available. Large-scale HTS coils are now being produced and we begin by discussing their main application areas. The various features of an ideal conductor from the viewpoint of a magnet engineer are described and then we discuss how present day practical HTS conductors differ. The implications of these differences on coil design and construction are discussed, and we review the two main coil fabrication techniques, React-and-Wind and Wind-and-React. Notable results in HTS coil fabrication from the worldwide research effort are reviewed, and we comment on the prospects for the near future.

The Oxford University conductor/coil fabrication programme which includes some novel approaches, is briefly described, and we end with a description of HTS coil design for an electromagnetic maglev demonstrator.

2. HTS coil applications

In general terms HTS coil applications fall into 2 areas: 1) where they do something unique (i.e. for which there is no alternative) e.g. superconduct in very high fields or at high temperatures, and 2) where they offer an improvement on balance over alternatives, be they resistive or LTS windings. Considerations include power consumption, refrigeration costs, current densities, stability, cryogen-free operation and size/weight of the device.

More specifically HTS coil developments fall into 3 main application areas:
1) very high field inserts (>20T) for LTS magnets operating at 1.8-4.2K, exploiting the extremely high B_{c2} of HTS materials at low temperatures. At present there are no viable alternatives although one must bear in mind possible advances in the PMS materials.
2) medium field magnets (2T$<$B<10T) operating at intermediate temperatures (20-30K)

using cryocoolers. This application is attractive for the compactness and ease of operation of cryogen-free systems.

3) high-temperature low-field ($<2T$) applications. The coils operate in liquid nitrogen (63-77K) and often provide ampere-turns ·for magnetic circuits. Examples include rotating electrical machines, maglev, SMES and transformers. All involve quasi-ac operating conditions in that the coils see variations in B and current with time. Advantages over copper coils are that resistive losses are eliminated and higher current densities can be achieved, leading to smaller, lighter machines. Compared with LTS windings, refrigeration costs are substantially reduced and stability is vastly increased due to much higher specific heats at the higher temperatures.

3. Practical HTS conductor: properties and implications for coil construction

In very general terms a magnet engineer would like the conductor from which he/she winds a coil to be homogeneous (so that a short sample test is representative of the whole) with high Jc and isotropic Jc(B) dependence. It should be available in infinitely long lengths, be superconducting (i.e. pre-reacted if necessary), flexible (without Jc degradation), strong (to withstand Lorentz forces), cheap and readily available. In addition it should be round wire for easy layer winding, have a metallic matrix (for stability and ease of contacting), and be multifilamentary (for stability).

A practical HTS conductor is one which goes some of the way towards meeting these requirements, especially the combination of high Jc and long length. It is being produced by various companies/research organisations around the world, for example 1Km long tapes with Jc >12000 Acm^{-2} at 77K (compared with 32600 Acm^{-2} for 1cm samples) have recently been reported [1]. However present practical HTS conductors differ from the ideal in two important ways: they tend to be in the form of tapes with highly anisotropic Jc(B) dependence (either internal HTS composites produced by the PIT process, or external HTS e.g. dip-coated tapes), and the HTS materials are brittle.

The tape geometry enables high Jc s to be obtained in the BSCCO materials by producing a microstructure of well-aligned grains. A narrow tape can be layer wound, but a wide tape necessitates a pancake coil construction, with many joints. A tape geometry incorporating well-aligned HTS grains also exhibits the intrinsic Jc(B) anisotropy of the BSCCO compounds, and this has implications for coil design.

3.1 Design implications of anisotropic Jc(B)

When designing a superconducting magnet incorporating isotropic conductor one is normally concerned only with $|B|$ which is a maximum at the inner windings on the coil midplane. However, when the coil is to be wound from HTS tape with anisotropic Jc(B) one must also take into account the radial component of field (which is perpendicular to the tape surface and therefore in the weak pinning direction) which for a simple solenoid is a maximum at the middle turns at the ends of the coil. Maximum operating current is determined by plotting load lines for peak radial field (Br), and peak axial field (Bp) together with the tape critical current as a function of parallel and perpendicular applied fields. An example is shown in fig. 1. In this case, as for many others, it is the Br load line and Ic(B∥c) which intersect at the lower current, hence it is the radial component of field which determines the maximum operating current and field generated by the coil.

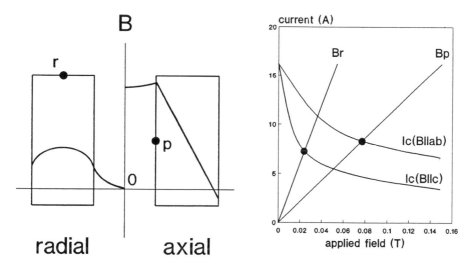

Figure 1. axial and radial field profiles in the windings of a simple solenoid: peak values occur at points p and r respectively. Load lines plotted with anisotropic Ic(B) data show that for this example it is the radial component which limits the maximum operating current of the magnet.

We note that the above example is a simplification: strictly, in designing a coil we need to know the complete Ic, B, θ surface for the HTS tape, and the B, θ profile over the coil windings, although this leads to a more complicated analysis.

3.2 Implications of brittle nature of HTS materials

HTS materials are brittle oxides and straining leads to reduction in their critical currents. This has implications for coil construction: essentially there are two possible techniques. The first is Wind-and-React (W&R), in which the conductor is wound in the "green" state (i.e. the necessary elemental ingredients are present but the sc phase has not yet been formed) and then given the appropriate heat treatment. The second is React-and-Wind (R&W), in which coils are wound from pre-reacted tape. Clearly for this second method strain must be kept within tolerable limits. This is achieved in two ways, which are not necessarily mutually exclusive, namely by producing strain-tolerant conductor comprising many fine HTS layers/filaments in a metallic matrix, or by winding the coil in such a way that strain is small e.g. by winding on a large diameter or by pre-reacting the tape on a mandrel of similar diameter to the final winding diameter. In the next section the merits and limitations of these two techniques are discussed.

4. A comparison of the Wind-and-React and React-and-Wind coil fabrication techniques

Wind-and-React: The merits of this technique are that: as the conductor is wound in the green state it can be handled more freely than a brittle pre-reacted tape and can be wound over small diameters; once the superconducting phase has formed during reaction it encounters no further strain; problems associated with reacting very long lengths of separated conductor are overcome as the whole coil is reacted as one batch; tapes can be cowound without insulation so that they fuse during heat treatment to form a multilayer composite.

Limitations include: the technique cannot be used for external HTS composites as the superconductor would react with the inter-turn insulation (ITI); there is a limited choice of material for the coil former, which must withstand the high reaction temperatures and then low temperature thermal cycling during the coil's use; choice of material for ITI is again limited, and tends to be relatively thick leading to low packing factors; the size of the coil and presence of ITI may make reaction optimisation for the whole magnet a problem, for example it may affect local oxygen partial pressures, or lead to unwanted temperature gradients.

React-and-Wind: The HTS conductor for this method can be reacted either straight, in a continuous process, or loosely coiled in a batch process. If this second approach is taken then the final stage is one of tightening rather than winding and we shall call this variation "Loosely-Wind, React-and-Tighten" (LW R&T). The comments below are relevant to the general technique. Its merits are: that it is suitable for all conductor geometries; reaction optimisation is the same as for short samples; there is a wide choice of materials for the coil former, and we can choose one with the desired low-temperature properties; there is a wide choice of materials for the ITI which can be very thin and lead to high packing factors; the conductor can be tested before winding to identify any bad sections; the LW R&T variation enables very small bore coils to be made, so long as the reaction diameter is similar to the winding diameter.

Limitations are that: careful handling of the brittle conductor is required; Jc may degrade on winding / tightening; if the conductor has been produced via the continuous reaction process then the minimum permissible winding diameter may be prohibitively large.

5. Notable world HTS coils

A wide variety of combinations of conductor geometry and coil fabrication techniques have been employed to produce HTS coils. In this section some of the best results are presented.

Early Demonstration of LW R&T Technique: In 1992 Shimoyama et al. [2] reported the construction of a single pancake coil from dip-coated tape (BSCCO 2212 on Ag) using the LW R&T technique, which generated 0.60T at 4.2K in a 12T background. Coil Ic was approximately 25% that of short sample, which the authors attributed to microcracking during tightening.

Cryogen-free 1T magnet wound from strain tolerant conductor: Using the R&W technique Ohkura et al. [3] fabricated a 40mm i.d. coil from 12 double-pancakes wound from 600m of 61 filament BSCCO 2223/Ag tape. Strain tolerance of this conductor was such that the 0.6% strain associated with a 40mm winding diameter produced a Jc degradation of only 5%. The coil was conduction cooled to 20K using a GM refrigerator and generated 1T.

World SC Record: The highest flux density generated by a purely superconducting magnet to date was reported by Tomita et al. [4]. They produced a double-pancake coil of i.d. 13mm and o.d. 46.5mm, from dip-coated BSCCO 2212/Ag tape (total length 14.6m) using the LW R&T technique. The coil was inserted in a high-field LTS magnet and generated a field of 0.9T in a 20.9T background at 1.8K, leading to a record centroid field of 21.8T.

Demonstration of Layer Winding: Haldar et al. (IGC) [5] wound a 70-layer coil from 200m of pre-reacted 37-filament BSCCO 2223/Ag tape which generated 0.1T at 77K and demonstrated the feasibility of this technique for the production of larger scale coils.

W&R Cryocooled Magnet: In the same paper [5] Haldar et al. reported the construction of a magnet for cryocooler operation comprising 4 modules, each a double pancake coil wound from 2223/Ag tape. Two tapes were cowound without insulation and fused during reaction to form quasimultifilamentary conductor. At 20K the coil generated 0.82T in a 3/4" room-temperature bore.

Highest B_HTS to date: At the 14th Magnet Technology conference K. Togano reported
[6] that Sumitomo Electrical Industries Ltd. and collaborators had produced a large R&W coil
of 60mm i.d. from multifilamentary BSCCO 2223/Ag tape which generated 3T at 21K (using
a cryocooler) and 3.9T at 4.2K. This is the highest field to date produced by a magnet
comprising HTS windings alone.

6. Oxford University conductor / coil Programme

Our inter-departmental research programme has for 6 years been producing increasingly long
lengths of practical HTS conductor using a number of techniques, including powder-in-tube,
electrophoretic deposition and dip-coating. Coils have been wound from these conductors,
initially as a means of characterising long samples and determining homogeneity, but the
production of HTS magnets for demonstrator devices and high-field inserts is now a goal in
its own right.

In addition to conventional coil fabrication techniques we have constructed an all-
ceramic magnet using a novel approach [7] in which BSCCO 2212 was melted on MgO-disc
substrates. Spiral grooves were machined in the BSCCO to produce a continuous track of
length $\sim$1m, and high-temperature annealing produced superconducting coils with critical
currents of around 10A. Four such coils were integrated in a stack to form the small magnet.

Our best coils to date have been constructed using the LW R&T technique from
BSCCO 2212/ Ag alloy dip-coated tapes. Loosely wound coils containing up to 7m of tape
have been successfully reacted, and tightened to form pancake coils with i.d.s as small as
10mm. Once impregnated with paraffin wax they are extremely robust, and after 1 year of
thermal cycling and use in a variety of experiments show no Ic degradation. A magnet
assembled from two such pancake coils generated a peak field of 0.39T at 4.2K, when
carrying its 1μV/cm critical current of 117A. Similar coils are under construction with
o.d. < 40mm for use as high-field inserts in our 18T mobile LTS research magnets.

7. HTS coils for electromagnetic maglev

In collaboration with Loughborough University of Technology and the IRC in
Superconductivity, Cambridge we are investigating the controllability of HTS coils for
electromagnetic (attractive) maglev systems [8]. Initial experiments, using Oxford University
BSSCO 2212 coils, have measured the response of coils to variations in current and the
reluctance of the magnetic circuit. Closed loop control of the excitation of the superconducting
magnet has been achieved, and we are now in a position to construct a large-scale
demonstrator. The proposed experimental arrangement is shown in fig. 2. Design targets are
that the HTS coils should generate a field of 1T in a 5mm gap between 1" diameter poles and
the iron rail, thereby generating a lift force of $\sim$400N. Two major factors influence the
HTS coil design: firstly the anisotropic Jc(B) dependence of HTS tape; and secondly the
presence of ferromagnetic material greatly alters the flux distribution in the windings
compared with that for an isolated air-cored coil. Finite element modelling has been used to
calculate accurate field profiles over proposed designs, which have been used in conjunction
with short sample Ic(B) data to check that nowhere does operating current exceed local critical
current. The final design, to be constructed from pre-reacted IGC 2223/Ag multifilamentary
tape, is shown in fig. 3. By locating the HTS windings on the back of the yoke they are kept
away from high fringe fields at the poles, and a single simple annular cryostat can be used.
Coil 1 is long and thin to reduce radial field components, which are further reduced by coils
2 and 3, which operate at half coil 1's current density.

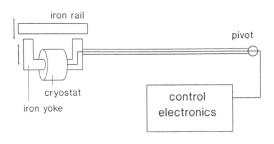

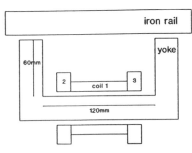

Fig. 2. The maglev demonstrator incorporating HTS coils   Fig. 3. Final design for the maglev HTS coils

8. Conclusions

The tape geometry of present practical HTS conductors has important implications for coil design and construction. In particular, in light of their anisotropic Jc(B) dependence, one must take into account radial field components.

Various combinations of conductor geometry and coil fabrication techniques have been used to produce respectable-performance high-field inserts and cryocooled coils. There is no strong reason to drop any particular approach, however on balance it does seem that the R&W technique is preferable, especially as strain tolerance of conductors continues to improve, and the LW R&T variation can be used for very small winding diameters.

The wide tape geometry of the present practical conductors in most cases necessitates a modular coil construction from pancakes, with many joints. This construction is satisfactory for driven research magnets at 4.2K, and for higher temperature operation coils but is unsuitable for NMR applications where coils operate in persistent mode. This important potential market would benefit greatly from the development of an isotropic HTS composite conductor, the most likely candidate being BSCCO 2212/Ag multifilamentary round wire. Such a material could also be used in high-field low-temperature research magnet inserts and medium field coils at temperatures up to 20K. For higher temperature operation it seems likely that one would still have to use anisotropic BSCCO 2223 tape.

Acknowledgment

Many thanks to I. Belenli, J. Burgoyne, K. Davies, D. Dew-Hughes, D. East, C. Eastell, M. J. Goringe, C. R. M. Grovenor and C. Morgan for their work in the HTS conductor/coil programme, in particular for doing the hard work of producing high quality material from which we could construct magnets. Also many thanks to our collaborators on the "Maglev" programme: R. M. Goodall, C. MacLeod, A. A. El-Abbar and A. M. Campbell. This work was supported by the EPSRC under grant numbers GRF83693 and GRJ40089.

References

[1] Kellers J and Papst G *paper E3, procs. MT-14, June 1995, Tampere, Finland*
[2] Shimoyama J et al. 1992 *Jpn. J. Appl. Phys.* **31** L163-L165
[3] Ohkura K et al. 1993 *Jpn. J. Appl. Phys.* **32** L1606-L1608
[4] Tomita N et al. *"Development of Superconducting Magnets using Bi-2212/Ag Tapes", procs. ASC'94, October 1994, Boston, USA*
[5] Haldar P et al. *"Development of Bi-2223 HTS High Field Coils and Magnets", procs. ASC'94, October 1994, Boston, USA*
[6] Togano K *"Recent Developments of HTS Tapes and Coils", paper E1, procs. MT-14, June 1995, Tampere, Finland*
[7] Davies K et al. 1994 *Applied Superconductivity* **2(1)** 61-66
[8] El-Abbar A A et al. *this conference*

Inst. Phys. Conf. Ser. No 148
Paper presented at Applied Superconductivity, Edinburgh, 3–6 July 1995

A reversible rise in the critical current of a Nb_3Sn-bronze tape due to a transverse pressure

B. ten Haken, A. Godeke and H.H.J. ten Kate

University of Twente, Low Temperature Division, P.O. Box 217, 7500AE Enschede, The Netherlands.

Abstract. The critical current of a layered Nb_3Sn tape is investigated as a function of the thermal contraction and the transverse pressure. A rise in the critical current occurs when a transverse pressure is applied to a Nb_3Sn-bronze tape. This enhancement in the critical current is reversible and the dependence on the magnetic field indicates that the upper-critical field shows a similar rise. This peculiar result is described with by dependence of the critical properties on the second invariant of the strain tensor. This so-called deviatoric strain dependence is previously proposed for the description of the critical properties in axially strained Nb_3Sn conductors. The result of the transverse pressure experiment described here, validates the deviatoric strain model in poly-crystalline Nb_3Sn.

1. Introduction

The critical properties of Nb_3Sn and other A15 conductors are sensitive to mechanical deformation. Based on a comparison of the experimental results on axially strained wires and hydrostatic-pressure experiments on single crystals, it is concluded that the influence of the non-hydrostatic strain component is large compared to the hydrostatic component [1]. A scaling law is proposed for the description of the critical properties in axially strained conductors [2]. This scaling provides an accurate description of the critical properties in the strain range that is accessible with an axial-pull set-up. Recently it appeared that the critical properties of a uni-axially compressed Nb_3Sn layer show a different behaviour. An almost linear dependence occurs in the upper-critical field versus strain, for a large compression in the same direction as the current [3]. In this uni-axial strain experiment an additional strain component can be varied by applying substrate materials with different thermal contraction coefficients. The influence of both these (in plane) strain components on the critical field (and temperature) of the Nb_3Sn layer can be described accurately by a linear dependence on the so-called "deviatoric strain" [4]:

$$B_{c2} = B_0 - C_d \varepsilon_{dev} \quad \text{with} \quad \varepsilon_{dev} = \frac{2}{3}\sqrt{\left(\varepsilon_x - \varepsilon_y\right)^2 + \left(\varepsilon_y - \varepsilon_z\right)^2 + \left(\varepsilon_z - \varepsilon_x\right)^2}. \tag{1}$$

It appears that the deviatoric-strain description can also be applied to axially deformed wires. This deviatoric-strain description predicts a rise in the upper-critical field (B_{c2}) that is proportional to the pressure applied in the transverse direction on a tape conductor. So far a reversible rise in B_{c2} has never been observed in any Nb_3Sn conductor [4].

2. Nb₃Sn bronze tape

A transverse pressure experiment on a Nb₃Sn layer requires a special shape of the sample. In a previous experiment some commercially available tape conductors consisting of a Nb₃Sn layer on a Nb ribbon and clad between two Cu tapes, are investigated. In these tapes only reversible reductions in the critical current (I_c) and extrapolated B_{c2} are observed, and the data scatter over a wide range [3]. A probable cause for non-uniformities in the strains inside these tapes is the irregular contact layer between the Nb₃Sn and the Cu layers. In this paper a different type of tape is investigated where the Nb₃Sn layer is formed on the interface of two metallic layers. This Nb₃Sn-bronze tape is especially developed for these transverse pressure experiments. A Nb layer of 25 μm is pressed between two layers of bronze (Cu 13.5% Sn) of 100 μm thick. This system is reacted following the scheme of commercial bronze wires (48hr at 570°C and 48hr at 700°C). After this heat treatment two layers of 0.7 μm with polycrystalline Nb₃Sn are formed at the interface. After reaction the tape is cut in pieces with a width of 2.5 mm and a length of 60 mm and inserted into the press apparatus [3].

3. A thermally contracted Nb₃Sn layer

After cooling down to 4.2 K the bronze layers, connected to the Nb₃Sn layer, cause a considerable thermal strain in the in-plane directions. Following the mechanical model that is proposed in ref. 3, the thermal strain in three dimensions is: $\varepsilon_x = \varepsilon_z = \delta$ and $\varepsilon_y = -\delta \cdot 2\upsilon/(1-\upsilon)$, with the y-direction perpendicular to the plane and υ is the Poisson's ratio of Nb₃Sn. The thermal-contraction difference between bronze and Nb₃Sn is large. Values of about 0.4-0.6% can occur in bronze route Nb₃Sn wires. The critical current density in the Nb₃Sn layer that is contracted by the bronze, reduces to almost zero for magnetic fields beyond 15 tesla. The critical current as a function of the applied field is presented in a so-called Kramer plot (Fig. 1).

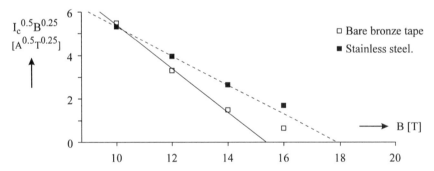

Figure 1: The field dependence of the critical current after thermal contraction at 4.2 K. The vertical axis is scaled as a "Kramer-plot". The continues line represents the bare Nb₃Sn bronze tape. The dotted line denotes the same bronze tape sample soldered on stainless steel at 475 K.

The role of the thermal contraction is investigated by soldering the same tape sample to a stainless-steel substrate. The difference in thermal contraction from soldering temperature (475 K) to 4.2 K causes a reduction of the thermal contraction of the Nb₃Sn layer by approximately 0.12%. This reduction of the thermal contraction has a strong influence on the critical current, as indicated in figure 1. Based on the difference in the thermal contraction ($\Delta\delta = 0.12\%$) the deviatoric strain difference $\Delta\varepsilon_{dev}$ equals 37% for a Poisson's ratio of $\upsilon = 0.45$. The observed shift in the $I_c(B)$ relation is $\Delta B_{c2^*} = 2.3$ T, which corresponds to $C_d = 750$ T in eq. (1). This value corresponds to a thermal contraction of -0.5% in the Nb₃Sn layer of a bare (bronze) tape and a B_{c2} for the un-strained material of: $B_0 = 25$ T.

4. Transverse pressure

A transverse pressure is applied to the tape with a press that is developed for experiments inside a 16 T solenoid with a 60 mm bore at 4.2 K, that is already described before [3]. A maximum force of about 6 kN enables a maximum pressure of approximately 200 MPa at the central 10 mm of the investigated tape sample. The critical current is determined at a level of 10^{-4} V/m and a small correction is made for the current flowing through the matrix material. The pressure is applied in small steps of 10 to 20 MPa and a complete voltage current characteristic is determined at various magnetic fields (10, 12, 14 and 16 T) at each pressure step. Additional determinations are made at zero pressure to check the reversibility of a pressure change.

The results of a complete $I_c(B)$ determination, as a function of the transverse pressure are presented in figure 2. A clear increase of the normalised critical current at a high magnetic field is visible. This (relative) critical-current rise has a strong field dependence. At 10 T the critical-current change is negligible (1%) but at 12 and 14 T it is very clear (>10%). A comparison with the critical currents at a small pressure (<10 MPa) is made after applying 85 and 130 MPa. In these cases the I_c reproduces to a value within 2% of the original value.

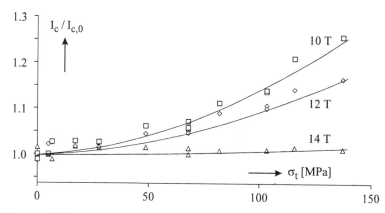

Figure 2: The normalised critical current of a Nb₃Sn tape that is pressed in the transverse direction, for three different magnetic field values ($B = B_x$, $\sigma = \sigma_y$, $I = I_z$,)

5. Upper-critical field and the deviatoric strain

The $I_c(B)$ data depicted in figure 1 do not exactly represent the linear relation that is predicted by Kramer. In this case the curvature is attributed to a certain distribution of the critical properties along the sample. First of all the stochiometry of the Nb₃Sn might be different along the height of the layer, caused by the slow diffusion of the Sn. A second possible cause for non-uniformities in the critical properties are the edges of the tape where the strain-state is different. An approximation of the mean value of the upper-critical field can be made by omitting the latter $I_c(B)$ values at 16 T. Then the upper-critical field of this tape is approximated at about 16 tesla, as can be seen in figure 1. The upper-critical field that is estimated by this approximation is presented in figure 3. A clear rise in the B_{c2*} is observed if a transverse pressure is applied.

The strain in the compressed layer depends not only on the transverse pressure but also on the properties of the sheath materials (here bronze) and the press (stainless steel). The strain in the Nb₃Sn layer is approximated in the elastic limit for an infinitely stiff press and a large friction coefficient between the bronze and the press: $\varepsilon_x = \varepsilon_z = 0$ [3]. The strain is calculated with a Young's modules of 80 GPa and the Poisson's ratio of 0.45 for Nb₃Sn. With these conditions a linear dependence for the $B_{c2}(\sigma_t)$ relation is obtained. In figure 3 a line is drawn with a slope of 2.4 T/GPa, which is in good agreement with the measured data. This slope corresponds to a deviatoric strain dependence of $C_d = 750$ T in the limit for a stiff press that does not allow the layer to be deformed in the directions parallel to the surface. A much larger slope is calculated when the friction between the press and the tape is negligible. In that case, with $E = 100$ GPa and $\upsilon = 0.25$ as the overall elastic properties for the tape, a much lager slope of 7.1 T/GPa is predicted [3].

The observed change in B_{c2*} in the compressed tape is in good agreement with the rise of B_{c2} when the thermal contraction of the Nb₃Sn layer is reduced. Both the influence of the thermal contraction and the transverse-pressure dependence can be described by a deviatoric-strain dependence of 750 T. This is comparable to the value of $C_d = 850$ T that is determined for a different type of Nb₃Sn layer, deformed by an entirely different strain device [4]. Taking into account the uncertainties in the determination of the strain in the various experimental set-ups the observed differences in the deviatoric strain constant (C_d) are small. Especially the strain in a tape compressed in the transverse direction is not very accurate.

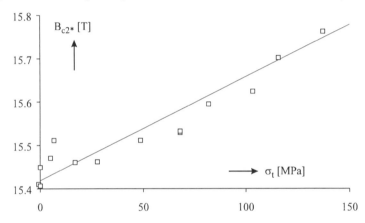

Figure 3: The extrapolated upper-critical field of a Nb₃Sn bronze tape as a function of the transverse pressure.

Conclusions

1. A reversible enhancement as a function of the transverse pressure, is observed in the critical current and the extrapolated upper-critical field of a Nb₃Sn layer, for pressures below 150 MPa.
2. The changes in the upper-critical field, due to the thermal compression and the rise in B_{c2*} after compression, can be described very well by the deviatoric strain model that is derived from other strain experiments on layers of polycrystalline Nb₃Sn.
3. The determination of the exact value of the deviatoric strain constant from a transverse pressure experiment is complicated, due to the many uncertainties in the mechanical model for the tape. Nevertheless the value deduced from this deformation experiment is in reasonable agreement with other strain experiments.

References

[1]   D.O. Welch, *Adv. in Cry. Eng. Vol.* **26**, pp 48-65, 1980.
[2]   J.W. Ekin, *Cryogenics Vol* **20**, pp. 611, 1980
[3]   B. ten Haken, A. Godeke and H.H.J. ten Kate, *Adv. in Cryo. Eng. Vol* **40**, pp 875-882, 1993.
[4]   B. ten Haken, A. Godeke and H.H.J. ten Kate, *IEEE trans. on Superconductivity* (to be published, 1995).

Inst. Phys. Conf. Ser. No 148
Paper presented at Applied Superconductivity, Edinburgh, 3–6 July 1995
© *1995 IOP Publishing Ltd*

Transport Critical Current Measurements in Very High Pulsed Magnetic Fields

H Jones, C R J Hole, D T Ryan and M van der Burgt

Clarendon Laboratory, University of Oxford, Oxford, OX1 3PU, UK

Abstract

Practical conductors incorporating high temperature superconductors are now available commercially in tape form. These are capable of sustaining supercurrents in magnetic fields in excess of 20 tesla - with all that implies for the superconducting magnet industry. We outline our attempts to measure transport critical currents in pulsed magnetic fields, up to the 50T region, on practical artefacts such as coils. We also discuss pulsed magnet technology and describe promising attempts to model the effects seen in our measurements and draw optimistic conclusions about the viability of the technique for routine, quality assurance measurements well in excess of 20T - the present day-to-day limit for continuous fields.

1. The generation of high magnetic fields

Figure 1, which is adapted from Herlach and Jones [1], shows what magnetic fields are achievable using different methods. As a generalisation it is apparent that in order to access continuous fields in excess of 20 tesla it is necessary to go to the handful of major magnet installations in the world, the big national magnet laboratories, to make measurements. Whilst this is the purpose of these centres it is difficult and expensive to use them for ongoing routine transport critical current measurements necessary for quality assurance purposes during the manufacture of superconducting magnets. At the Clarendon Laboratory we offer a $I_c(B)$ measurement service for Industry and we know that rapid turn round in response to unpredictable demand is a primary requirement. In any case, even at these world centres continuous fields in excess of 45 T (at the most) cannot be achieved. HTS and Chevrel phase materials have B_{c2}'s well in excess of this. As they become developed for technological conductors, fields >20T must be accessed for routine $I_c(B)$ measurements. Reference to fig. 1 shows that there is another way to access these high fields, namely pulsed magnets. Pulsed magnet installations can be surprisingly small and inexpensive and yet fields of the order of 50 T or more are relatively easily generated. At Oxford we have such a facility which can produce peak fields of up to 60T giving half sinusoidal pulses of length 10-100ms depending on the size of the magnet [2]. This paper addresses the use of pulsed fields for $I_c(B)$ work.

2. Pulsed magnets

The main limitation of continuous field magnets is that of heat removal as for fields >20 T classical copper windings have to be used with many MW dissipated in small volumes and therefore need to be water-cooled. By restricting the time for which the field is generated the rise in temperature can be kept within reason.

The problem is one of stresses in these magnets. Taking the magnetic pressure approximation as a guide,

$$P_m = \frac{B^2}{2\mu_0}$$ [1]

This gives at 60 T P_m=1.4 GPa(14 kbar); ie, at 60 T, the bursting pressure is some 100 times greater than that of a standard high pressure gas cylinder. There are various ways to contain these forces and space precludes in-depth discussions here but much has been published on the topic (see for instance [1][3][4]). As an example, one way is to wind the solenoid from strong wire such as the copper/stainless steel macrocomposite pioneered in Oxford [5][6]. This optimises strength and conductivity.

3. Critical current measurements

A full discussion of transport $I_c(B)$ measurements in continuous fields is given by Jones and Jenkins [7] but, briefly, a sample, preferably a small coil, is placed in a given magnetic field and current is increased until a superconducting/normal transition is observed on the IV characteristic. The critical current is arbitrarily defined as that giving a voltage appropriate to some electric field along the sample; usually 10 or 100 μVm^{-1}

4. Measurements in pulsed fields

For pulsed field work, the technique needs to be somewhat different. Here the field is rapidly changing and only at the peak does dB/dt $\rightarrow$ 0 for less than 1 ms, otherwise dB/dt can be up to 7×10^3 Ts^{-1}. In this case a steady transport current is established in the superconducting sample and the field is pulsed. This is the basic technique we use in our work in Oxford. Our objective is to establish whether critical currents measured in steady fields and pulsed fields match up, or if not, by how much and why. Ideally we would like to make the measurements in as similar a manner as possible to routine, steady field Ic(B) characterisations, ie, in small coil format. There are however a number of foreseeable difficulties. Firstly a coil by definition is highly inductive so a simple four terminal arrangement in a rapidly changing field would give rise to an inductive voltage which would swamp that from any other source such as a superconducting/normal transition. Secondly, the bores of pulsed magnets are usually smaller than those of typical laboratory scale superconducting magnets. For example, our superconducting magnets have bore sizes of 40 mm at 4.2 K whereas the largest bore pulsed coil is only 20 mm at 77 K reducing to a working bore of ~ 15 mm at 4.2 K. This has several consequences; (a) the length of the sample is shorter and therefore voltage resolution is less, (b) the smaller bore geometry tends to give lower homogeneity over the measurement region and this has particular relevance when radial field components are considered as discussed below, (c) it can make the introduction of high current leads difficult. Another problem is that the forces generated on the sample can be huge and percussive in nature and particular care must be taken to minimise these effects. Lastly a high dB/dt can generate eddy currents and other flux penetration effects with concomitant heating which then yields false results.

A useful alternative sample format possible with HTS is a meander pattern in a thick layer of BSCCO on a ceramic substrate, "ceramic on ceramic" [8]. These do not represent conductors in magnets except in one regard, the thick film, as opposed to the classical thin film used for superconducting electronics, can have similar current carrying capacities as technological conductors such as powder-in-Ag-tube without the worries of eddy currents associated with a high conductivity metal matrix so we decided to start with these to assess the viability of the method. The basic set-up is shown in fig 2 and is described in [9]. Note the counter geometry compensation circuit, placed in close proximity to the sample. This gives first order cancellation of inductive effects; further cancellation is achieved by electronic manipulation; also discussed in [9].

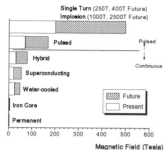

Figure 1 Methods of generating high fields [1]. Predictions for the future are speculative but realistic.

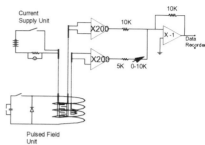

Figure 2 Schematic diagram of the 4-point measurement apparatus for pulsed fields.

5. Patterned meander samples

It is well known that $I_c(B)$ for HTS materials is highly anisotropic, as a result the radial component of the applied magnetic field has a significant effect on the measured $I_c(B)$ characteristic. Samples fabricated on ceramic substrates are frequently of low or negligible texture and thus sensitive to mod. B only. As a result of this it is possible to directly compare transport current measurements made in pulsed and continuous fields using the present pulsed field solenoids while newer, high homogeneity solenoids are under development. The coincidence of data determined using the two separate methods strongly indicates that millisecond pulsed fields attain their peak field for time periods long in comparison to the characteristic relaxation times pertinent to these measurements.

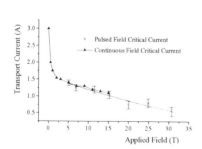

Figure 3 Critical current as a function of field as determined by the peak field signal method.

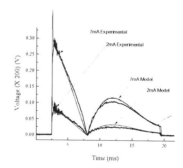

Figure 4 Voltage-time signal showing the accurate simulation of the model developed by the authors. Thick lines experimental, thin lines model.

Results presented in figs. 3 and 4 are from samples of Bi-2212 and Tl-1223 samples deposited on polycrystalline MgO and yttria stabilised zirconia respectively. Inspection of fig. 3, which shows $I_c(B)$ determined using the signal from the sample at the peak of the field pulse, shows no significant disagreement throughout the field range in which the two techniques can be compared and fig. 4 demonstrates the ability of the proposed analysis (briefly discussed later) of the observed effect to model the experimental results.

6. Prototype conductor characterisation

Of more technological relevance are composites of Ag & HTS fabricated in a variety of ways, one example of which is electrophoretic deposition [10]. This enables the fabrication of arbitrary lengths of conductor which are then deformed into pseudo-coil form as seen in fig. 5. It can be seen that both the current and voltage wires are co-wound with the sample in order to form a non inductive hairpin arrangement, one arm of which is the sample, the other is formed by conventional copper wires. An indium solder joint forms the end of the hairpin. This arrangement allows the circuit to be made almost non inductive while at the same time holding the conductor under test in a manner close to that of a conventional magnet winding. Importantly there are no joints or local bends in the wire at any point, the strain is evenly distributed throughout the length of the sample.

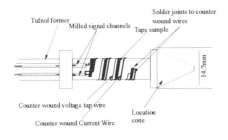

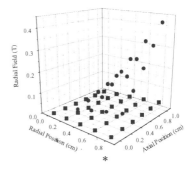

Figure 5   Pseudo-coil configuration for the transport current characterisation of prototype wire in pulsed magnetic fields.

Figure 6   Radial component of magnetic field for a conventional superconducting magnet (squares) and a high field pulsed solenoid (circles)

At present, measurements on samples of prototype conductor show a substantial disparity between critical current results from pulsed and continuous field experiments [11]. This is now thought to be the result of the highly textured nature of these samples and the large radial component of magnetic field inherent in current pulsed field solenoids (see fig. 6). As discussed above $I_c(B)$ is highly directional and comparisons of in-field performance using dissimilar magnetic field profiles should be made with caution. Eddy current heating effects are not thought to be large under these conditions as a result of the low sample thickness [12] and it is proposed that heating due to flux motion into the sample is also small.

Depressions in critical currents have been seen in the only similar experiment carried out to our knowledge [13]. This work was carried out by Benz, who did transport current work on NbZr wires. This work shows a similar, but reduced discrepancy in transport current results and we believe the origin of the reduced discrepancy lies in the directional field homogeneity of the samples under investigation. Measurements undertaken in this laboratory on NbTi wires (also field homogenous) indicate that eddy current heating can be a problem for the higher fields and shorter magnet pulses (~10ms) used here, Benz had access to 100ms pulse lengths and used a peak field of just 7T. The result of this is a substantially lower rate of change of field than that characteristic of the experiments detailed here, and the correspondingly lower eddy current induction, combined with significant sample cooling by the surrounding cryogen. Despite the higher rate of change of field in the current experiments we suggest that the primary $I_c(B)$ depression mechanism is the divergence of the applied field combined with the anisotropy of textured high Tc sample, an effect not present in the samples investigated by Benz. Magnets are being developed for longer pulses (~100ms) and these will enable clarification of the mechanisms actually responsible for the disparity.

7. Proposed model and numerical simulation

The characteristic twin maxima seen in the measurements of all superconducting samples investigated by us conform to the following characteristics:

- Such maxima/minima traces do not exist in non-superconducting samples, magnetoresistance is seen instead.
- Such signals are absent in samples above their transition temperature
- The level of signal at the voltage maxima (fig. 4) is a linear function of the applied transport current.
- The deep minima correspond exactly to peak field, and thus the region of zero dB/dt, for the rest of the pulse the voltage is closely related to the modulus of dB/dt.
- The voltage at the peak field minimum can be plotted as a function of applied transport current and gives the characteristic I-V curve seen for superconductors in continuous fields as the current is ramped
- The applied current for which this occurs is dependent on the critical current of the particular sample being measured and changes with temperature, following $I_c(T)$.

We have attempted to model these features and the evolution of the model can be followed in references [9],[14],[15] wherein it is discussed fully. Briefly, it makes use of the concept of shift in "electric centre", as discussed by Wilson and others [16][17] and thus describes a change in the inductive loop depending on whether field is increasing or decreasing in the presence of applied transport current. The signal is thus double peaked with no voltage at dB/dt = 0 apart from where I_c is exceeded. Numerically the voltage is expressed as follows:

$$V = \frac{DlI_t}{2} d\left(\frac{B}{I_c(B)}\right)\Big/ dt \qquad [2]$$

where D is the thickness of the sample, l, the length, I_t the applied transport current and B, the applied field.

It can be seen from fig. 4 that the agreement with experiment can be excellent for many samples, a discussion of the situations in which discrepancies arise is to be published shortly. Inspection of equation 2 suggests further interpretation of this data is possible, namely a complete determination of the $I_c(B)$ characteristic can be made using a single magnet "shot" if the equation is rearranged to leave $I_c(B)$ as the sole unknown. In its stated form an analytical rearrangement is difficult, though no significant error occurs when the numerator of the differential is separated thus rendering rearrangement a trivial exercise. Using the analysis described in references [14],[15] an example of the results generated by the simulation compared to experimental results is shown in fig. 4.

8. Conclusions

We have presented the results of a technique to perform transport critical current measurements on technologically relevant samples of high Tc superconductor at field intensities unattainable by any other technique and at the highest available fields for a greatly reduced cost burden. We justify our interpretation of the critical current data presented by modelling other aspects of the results and demonstrating excellent agreement between simulation and experiment. We briefly outline other ways in which we have manipulated our data although space precluded an in depth discussion of many of the issues central to its proper analysis but further work in this area is to be reported in the near future.

Acknowledgements

The authors gratefully acknowledge the contribution of Mr. A Hickman, Dr. M Van Cleemput, Mr. R Cripps, Dr. D. Dew-Hughes, Dr. M.J. Goringe and Dr. C.R.M Grovenor to this programme. This work was supported by EPSRC grant number GR/K15398 and CASE studentships from Oxford Instruments Ltd.

References

1. Herlach F, and Jones H, *"Magnets"*, *Encylopedia of Applied Physics* Ed. G.L. Trigg VCH publications **9** (1994) pp 245-259.
2. Siertsema W.J, Jones H, *IEEE Trans. Magn.* **30** (1994) pp1809-1812
3. Jones H, *Europhys. News*, **22** (1991) pp154-155
4. Van Bockstal L, Heremans G, Liang L, Herlach F, *IEEE Trans. Magn.* **30** (1994) pp1657-1662
5. Van Cleemput M, Jones H, Lee J.A, Barrau J.R, Eyssa Y, Schneider-Muntau H.J, Proc. "AHMF '95" Tsukuba, Japan *Physica B* (1995), in press
6. Van Cleemput M, Jones H, van der Burgt M, Eyssa Y, Schneider-Muntau H.J Proc. "MT-14", June 1995, Tampere, Finland *IEEE Trans. Magn.* in Press
7. Jones H, and Jenkins R.G, Ch. 6 in *High Temperature Superconducting Materials - New Concepts & Technology*, Ed. D. Shi, Pergamon Press 1995 pp259 - 304
8. Davies K, Jenkins R, Danjoy C, Grovenor C.R.M, Jones H, *Applied Superconductivity* **2** (1994) pp61-66
9. Hole C.R.J, Jones H, Goringe M.J, *Meas. Sci. Technol.* **5** (1994) pp1173-1176
10. Yang M, Goringe M.J, Grovenor C.R.M, Jenkins R.G, Jones H, *Supercond. Sci. Technol.* **7** (1994), p378
11. Hole C.R.J, Jones H, Burgoyne J.W, Dew-Hughes D, Grovenor C.R.M, Goringe M.J, *IEEE Trans. Appl. Supercond.* **5** (1995) p1313
12. Okuda K, Noguchi S, Honda M, Date M, *Jnl. Phys. Soc. Jpn.* **54** (1985) p1560
13. Benz H, Fasel R, Fischer E, *Rev. Sci. I.* (1965) p562
14. Hole C.R.J, Ryan D.T, Dew-Hughes D, Jones H, Goringe M.J, Grovenor C.R.M, "A Possible Mechanism for the Behaviour of High Temperature Superconductors Carrying Transport Currents in Pulsed Magnetic Fields" Proc. "AHMF '95", Tsukuba, Japan Feb. 1995 Submitted for publication, *Physica B*
15. Ryan D.T, Hole C.R.J, van der Burgt M, Jones H, Davies C.M, Grovenor C.R.M, Goringe M.J, Dew-Hughes D, "D.C. Transport Critical Current Measurements on High Temperature Superconductors in Pulsed Fields up to 50T" Proc. "MT-14", June 1995, Tampere, Finland. *IEEE Trans. Magn.* in Press
16. Wilson M. *"Superconducting Magnets"* Oxford University Press 1970
17. Ogasawara T, Yasukochi K, Nose S, Sekizawa H, *Cryogenics* **16** (1976) p33

Inst. Phys. Conf. Ser. No 148
Paper presented at Applied Superconductivity, Edinburgh, 3–6 July 1995
© *1995 IOP Publishing Ltd*

80/20 $DyBa_2Cu_3O_{7-x}/Dy_2BaCuO_5$ bulk textured materials as a tool for high-J_c superconducting ceramics

R Cloots (# *) and A Rulmont

SUPRAS, Institut de Chimie, B6, Université de Liège, B-4000 Liège, Belgium
(#) E-mail: cloots@gw.unipc.ulg.ac.be

H Bougrine (*) and M Ausloos (+)

SUPRAS, Institut de Physique, B5, Université de Liège, B-4000 Liège, Belgium
(*) also at Institut d'Electricité Montefiore, B28, Université de Liège, B-4000 Liege
(+) E-mail: ausloos@gw.unipc.ulg.ac.be

Abstract. Liquid phase processing methods for 123 bulk materials above the peritectic decomposition of the 123 phase, lead to an increase of intragrain critical current density J_c due to 211 precipitates acting as pinning centers within the 123 superconducting grains. The addition of the so-called 211 phase in excess in 123 oxide improves further the critical current density by consuming the excess liquid phase which segregates at the grain boundaries during the peritectic recombination process. We have probed the *intergranular current of magnetically textured composite materials* by electrical and magnetic susceptibility measurements as a function of magnetic field. At the percolation threshold in zero field, bulk shielding paths are such that the intergrain critical current density is above 10^5 A/cm^2. The extrapolated J_c value is huge. Moreover the field dependence of J_c is understood through an analytical form indicating a distribution of currents similar to the law of clusters at fracture /percolation thresholds.

1. Introduction

The existence of grain boundary current barriers are responsible for low J_c values in *unoriented* polycrystalline materials. It is thus necessary to texture the materials in order to favor high-J_c properties. To reach such a goal we sintered a 80/20% weight Dy-123/211 composite in the presence of its own liquid phase in a 0.6T magnetic induction. We chose to sinter Dy-123 because this compound is magnetically anisotropic and has a susceptibility larger than that of Y-123 based high critical temperature superconductors (HTS). The synthesis process led to very clean "weak links" at grain boundaries. We measured the AC susceptibility and the electrical resistivity. At the percolation threshold bulk shielding paths were such that the intergrain critical current density J_c was above 10^5 A/cm^2. The field dependence of J_c was understood through an analytical form indicating a distribution of currents similar to the law of clusters at fracture/percolation thresholds.

2. Experimental

The synthesis of 123 and 211 powders started respectively from a corresponding stoichiometric mixing of Dy_2O_3, $BaCO_3$, and $CuCO_3.Cu(OH)_2$ pretreated at 920°C for 48hours, including two intermediate grindings. The Dy-123 and 211 powders were mixed together in the appropriate ratio, compacted into a strip (ca. 15x10x3 mm^3), and transfered into an alumina crucible vertically inserted in a specially built furnace placed between the polar heads of a magnet. A horizontally generated magnetic induction of 0.6T was applied perpendicularly to the largest face of the sample during the whole process. The thermal cycle started at room temperature with heating to 1035°C at a rate of 150°C/h. A slow decrease of 2°C/h over several hours to 980°C is followed by a cooling process at 50°C/h to room temperature under oxygen atmosphere.

Microstructure analysis was performed with a HITACHI S2500 scanning electron microscope. Optical polarized light microscopy was also used in order to emphasize both (i) the 211 particle distribution and (ii) the oxygenation stage by looking at the twin patterns. It is remarkable that very clean grain boundaries exist as compared to Dy-123 systems without 211 excess particles. X-ray diffraction analysis showed that only (00l) peaks were recorded, whence the c-axis of most grains was perpendicular to the largest face of the sample, i.e. parallel to the applied field during the synthesis.

The classical four-point resistivity measurement method was used. Magnetic susceptibility measurements were made in a home-made susceptometer based on a CTI-21 model cryocooler. An AC field was applied parallel to the sample long axis from a wound coil. Thus the AC field was perpendicular to the c-axis of Dy-123 grains in such a way that the shielding currents were located in the a-c and/or b-c planes at the surface of most grains. A DC field could be superimposed perpendicular to the sample long axis, i.e. parallel to the c-axis of magnetically textured Dy-123 grains. The field cooling method from room temperature was used. The critical currents were deduced by the AC Campbell method. By measuring the AC flux entering the sample, it was possible to obtain the field flux profile as a function of the AC field amplitude for a fixed DC field, assuming a generalized Bean model, whence allowing intragrain and/or intergrain currents to be deduced in granular materials[1].

3. Results and discussion

The temperature dependence of the AC susceptibility $(\chi' - i\chi'')$ in zero and finite DC magnetic field is shown in Fig. 1. The resistivity measurement in zero field as a function of temperature is shown above the χ' and χ'' data in order to emphasize the critical temperatures and their role. *Only one smooth step* in χ' and one maximun in χ'' are visible for the zero DC field case. *They both appear at the temperature for which there is zero resistivity.* This is markedly the *intergrain coupling transition* which is only determined by the quality of the intergrain coupling when a bulk shielding path and a percolation path are both established through out the sample[2]. One should emphasize that there is no intrinsic susceptibility peak in χ'', peak which should coincide with the

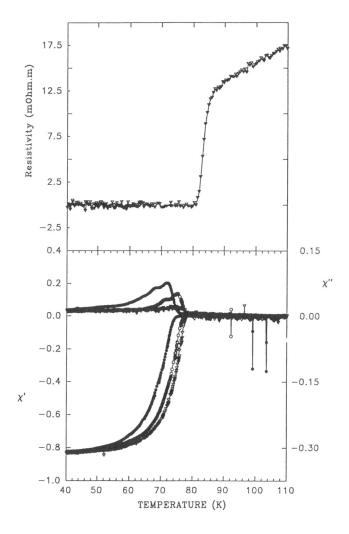

Fig.1   Resistivity and AC susceptibility measurements as a function of temperature for 80/20% Dy-123/211 composite material.

initial drop in resistivity at the intragrain superconductivity transition. At lower temperature, when there is wave function phase coherence, a complete shielding path can exist, whence the sharp transition in χ'. Closely spaced maxima in χ'' may be due to different current shielding paths or values due to various grain angle boundaries.

The critical current densities as deduced from the AC Campbell method, in granular materials, based on the flux profile acquisition technique are easily deduced. For example, J_c at 0.1T is quite high and equal to 5×10^4 A/cm^2. The numerical data when analyzed show a power law behavior : $J_c = J_{c0} \; B^{-q}$ with $q \simeq 0.3$ and $J_{c0} \simeq 4 . 10^5$ A/cm^2. This is

98

very close to the record high critical current value (10^6 A/cm^2 at 77K) in neutron irradiated YBCO single crystals[3], much above that of epitaxial thin films[4], of grain aligned YBCO and other zone melt YBCO with also some 211 excess particles[5]. The q≃0.3 value indicates that the single vortex pinning regime is not occuring [6] but interfacial effects are likely [7]. In so doing, effects based on a percolation mechanism for various pinning sites, including field percolation mechanisms can be thought of. In fact, this form of $J_c(B)$ is similar to that of the distribution of paths at dielectric breakdown and of clusters at fracture threshold[8].

In order to be more complete, let us mention that the evolution of the critical current density at different DC applied magnetic fields as a function of temperature (not shown) is very smooth and only weakly decaying. Extrapolation to low temperature leads to huge values for J_c.

This study indicates that the addition of 211 particles to the 123 system affects the microstructure[9] but also strongly *modifies the nature of the intergrain boundaries and the current carrying properties of the materials*. This results from the fact that *the 123 grains are strongly coupled* through clean interfaces. *A high value bulk shielding intergranular current can thus be established at sufficiently high temperature*. The processing route seems thus quite optimized. Secondly, the effective pinning is large, due to large interfacial pinning force density. The role of oxygen vacancies inside the grains does not seem to be the predominant mechanism at low field. The variation law of J_c with field is claimed to result from the percolative structure of the materials.

Acknowledgements

This work is part of the SSTC (Brussels) SU/02/013 and the ARC (94-99/174) contracts. MA received a British Council travel grant for attending EUCAS95. We thank Prof. H.W. Vanderschueren for allowing us to use the Measurement and Instrumentation in Electronics Laboratory (MIEL)

References

[1] Godelaine P A and Ausloos M 1990 *Solid State Commun.* **73** 759-61; 1990 *Solid State Commun.* **76** 785- 88.
[2] Brandt E H 1995 *Phys. Rev. Lett.* **74** 3025-28 .
[3] Van Dover R B, Gyorgy E M, Schneemeyer L F , Mitchell J W, Rao K V, Puzniak R and Waszczak J V 1989 *Nature* **342** 55-57.
[4] Chaudhari P, Koch R H , Laibowitz R B , McGuire T R , and Gambino R J 1987 *Phys. Rev. Lett.* **58** 2684- 86.
[5] Brand M, Elschner S, Gauss S and Assmus W 1994 *Appl. Phys. Lett.* **64** 2022- 25.
[6] Martinez B, Gomis V, Pinol S, Fontcuberta G and Obradors X 1994 *Physica C* **235-240** 3007-8.
[7] Murakami M 1992 *Melt Processed High Temperature Superconductors*, ed. M. Murakami (Singapore: World Scientific).
[8] Herrmann H J 1986 *Phys. Rep.* **136** 153-227.
[9] Sandiumenge F, Pinol S, Obradors X, Snoeck E and Roucau Ch 1994 *Phys. Rev. B* **50** 7032- 45.

Inst. Phys. Conf. Ser. No 148
Paper presented at Applied Superconductivity, Edinburgh, 3–6 July 1995
© 1995 IOP Publishing Ltd

Some properties of Tl-1223, prepared in a "quasi" open system

Chr. L. Teske and Hk. Müller-Buschbaum

Institut für Anorganische Chemie, Christian-Albrechts-Universität zu Kiel, Otto-Hahn-Platz 6/7, D-24098 Kiel;

Chr. Lang and S. Elschner

Hoechst AG, Corporate research, D-65926 Frankfurt a. M.

Abstract. The preparation of Tl-1223 in flowing gas atmosphere on a 100g scale is briefly described. The properties of the obtained material are discussed and compared with those of the other relevant HTC-ceramics. The critical current density (j_c) in terms of magnetic field dependency is superior to B(P)SCCO and even to sintered YBCO-123.

1. Introduction

The so called HTC-ceramic with nominal composition $Tl_1Ca_2Ba_2Cu_3O_{8.5+x}$ (Tl-1223) is an interesting material for technical applications with respect to its high critical temperature ($112 < T_c < 120$ K) and to current density (j_c) in terms of magnetic field dependency at 77 K. Therefore it's required to have sufficient amounts of material for further technical developments e.g. the PIT-method. The preparation of Tl-compounds including Tl-1223 [1-3] at elevated temperatures (T > 500°C) is basically problematic because of the volatility and toxicity of Tl_2O. The common lab method [4-6], carried out in hermetically closed containers (made of quartz or noble metal) is not appropriate for commercial use. Since we found out, that preparation of appropriate samples of Tl-1223 can be done in flowing oxygen without the use of such containers [7], we scaled up this process for bulk and powder samples to a 100g scale.

2. Experimental

As illustrated by DTA investigations, the solid state reaction between the precalcinated (960°C / 12h / air) precursor material (nominal $Ca_2(Sr/Ba)_2Cu_3O_7$) and Tl_2O3 takes place in the range of 900°C < T < 920°C [7], the loss of weight is maintained at a level of 3-5%. For the multistage solid state reactions we used a vertically mounted movable tube furnace. The alumina reaction chamber is closed on one side having a gas in- and outlet system on the bottom. The heat treatment schedule for each nominal composition was to be optimised due to the data from DTA investigations (STA 429 NETZSCH). The reaction times at T ≈ 900°C were relatively short (10 < t < 12 min; heating rate 10 - 30 K/min). So far samples with the nominal composition of $Tl_{1-x}Ca_{2+x}Ba_{2-y}Sr_yO_{8.5+z}$, ($0.21 < x < 0.26$; $0 < y < 0.5$ and $y = 1$) were prepared *in 3 - 5 steps*, always beginning *the first step with a surplus of thalliumoxide*. The load of the reaction chamber consisted each time of three or four pellets (20mm < ∅ < 40mm; 15 - 35g) which were reground and pressed again between each annealing step. Further precursor material was added to obtain phase pure Tl-1223 material. Some of these pellets were cut to rectangular bars (ca. 3 x 4 x 38 mm).

100

3. Results and discussion

After the first reaction step always the thallium rich phase Tl-2223 was predominant as indicated by XRD-analysis. By the subsequent addition of more and more precursor, Tl-1223 formed during the following annealing. Figure 1. shows an almost phase pure specimen with traces of Ca_2CuO_3 and $TlBa_2CaCu_2O_{7+x}$ (Tl-1212).

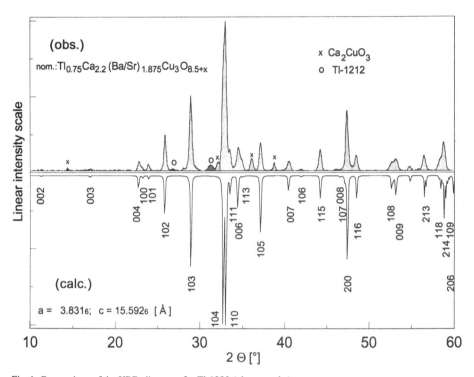

Fig. 1. Comparison of the XRD-diagrams for Tl-1223 (obs. vs. calc.)

Strontium rich samples are generally "overdoped" if prepared in pure oxygen [8] confirmed by T_c-values lower than 100K. Adjusting the oxygen content by an additional treatment in pure argon at 500°C or in Ar/H$_2$ (90:10) at 300°C reveals T_c-values of 112K. This control of the oxygen content and therefore T_c by convenient thermal treatment is nearly reversible. The susceptibility of an optimised powder sample is shown in figure 2. Hysteresis curves confirm relatively strong flux flow pinning properties of the material. So far the best measured transport current density (1µV/cm criterion) of a sintered specimen at zero field is $j_c(0) = 800$ A/cm^2. The j_c vs. B remains almost constant after decreasing sharply at applied magnetic field less than 2000 G. Comparison with the other relevant HTC-materials Bi-2212 (B(P)SCCO) and YBCO is illustrated in figure.3. In the region above 2000 G Tl- 1223 material is superior to MCP (melt cast processed) Bi-2212 also with respect to absolute values. The limited weak link behaviour is illustrated by the less pronounced degradation of j_c vs. B in comparison to sinterd YBCO-123 material.

Susceptibility of Tl-1223

- nom. comp.: Tl$_{0.75}$ Ca$_{2.25}$ Ba$_{0.85}$Sr$_{1.03}$ Cu$_3$O$_{8.5+x}$ -

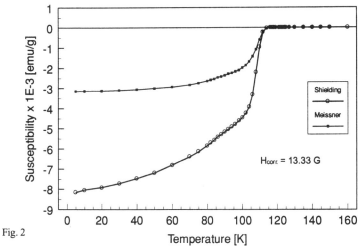

Fig. 2

Comparison Tl-1223 vs. Bi-2212, resp. YBCO-123

- j$_c$ as a funktion of applied magnetic field (in sintered bars) -

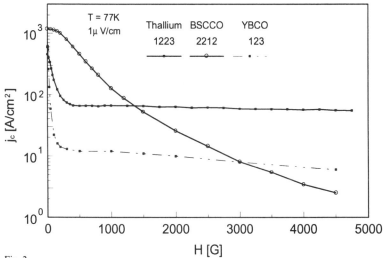

Fig. 3

4. Conclusion

We demonstrated that preparation of HTC-Tl-1223 on a relative large scale is possible without a complicated handling in a sealed environment. The properties achieved are comparable to conventional prepared Tl-1223. Especially the measured current densities are superior to MCP BSCCO -2212 and sintered YBCO -1233 with respect to the magnetic field dependency.

References

[1] M. Subramanian, C. Tonardi, J. Gopalakrishnan, P. Gai, J. Calabrese, T. Askew, R. Filippen, A. Sleight

   Science **242**, 249 (1988)

[2] S. Parkin, V.Lee, A.Nazzal, R.Savoy, R.Beyers *Phys. Rev. Lett.* **60**, 2539 (1988)

[3] R. Sugise, M. Hirabayashi, N.Tereda, M. Jo, T. Shimomura, H. Ihara *Jap. J. Appl. Phys.* **24/9**, 1709 (1988)

[4] K. Goretta, N.Chen., M. Lanagan, S.Dorris *Supercond. Sci. Technol.* **5**, 534 (1992)

[5] R. Liu, S. Wu, D. Shy, S. Hu, D. Jefferson *Physica C* **222**, 278 (1994)

[6] K.A. Richardson, S. Wu, D. Bracanovic, P. A. J. de Groot, M.K. Al-Mosawi, D. M. Ogborne and M.T. Weller
      Supercond. Sci. Technol. **8**, 238 (1995)

[7] Chr. L. Teske, Hk. Müller-Buschbaum, M. Neubacher and S. Elschner. *Supraleitung und Tieftemperatur
      -Technik , Statusseminar 13.-14. Juni 1994* (VDI-Verlag) **ISBN 3-18-401395-2**

[8] C. Martin, A. Maignan, M. Huve, Ph. Labbe, M. Ledeset, H. Leligny and B. Raveau. *Physica C 217, 106
      (1993)*

Inst. Phys. Conf. Ser. No 148
Paper presented at Applied Superconductivity, Edinburgh, 3–6 July 1995
© 1995 IOP Publishing Ltd

Crystallization of vitreous high-T_C superconducting oxide through laser zone melting method.

M   Ausloos and H   Bougrine

SUPRAS, Institut de Physique, B5, Université de Liège, B-4000 Liège, Belgium.
E-mail: ausloos@gw.unipc.ulg.ac.be

R   Cloots and A   Rulmont

SUPRAS, Institut de Chimie, B6, Université de Liège, B-4000 Liège, Belgium.
E-mail: cloots@gw.unipc.ulg.ac.be

A   Gilabert

Laboratoire de Physique de la Matière Condensée, Université de Nice-Sophia Antipolis, Parc Valrose, F-06108 Nice, Cedex 02, France.

J Y   Laval

Laboratoire de Physique des Solides, ESPCI, rue Vauquelin 10, F-75231 Paris Cedex 05, France.

Abstract. We synthesized $Bi_2Sr_2CaCu_2O_{8-y}$ ceramics through the vitreous route. After insertion in epoxy resin, we exposed the materials to a CO_2 laser beam. Various intensities and sweeping conditions were examined. We observed the samples in high resolution polarized light microscopy and with electron scanning microscopy. Crystallization zones could be detected. A systematic analysis gave the correlation between the behavior of the material and its synthesis conditions. Electrical properties were measured through micro-electrode deposited on the surface at various locations. No superconductivity was found. The optimization of such a process and potential applications will be discussed.

I. Introduction

Large scale applications of high-T_c superconductors require improvements of their transport properties in order to obtain suitable devices. Bi-based compounds are extremely promising materials for applications. Their intrinsic anisotropy can be useful for the achievement of a strong alignment of the grains necessary for high-J_c properties. Texture has been recently induced in bulk -calcium substituted by dysprosium- Bi-based 2212 materials by a melting growth process under a magnetic field.[1] The microstructure obtained is nevertheless quite unusual but leads to interesting chemical considerations. Thick layers of different chemical compositions have been observed stacked up along the c-direction with irregular intergrowth periodicities.[2] The optimization of this procedure for potential applications of these materials is not yet solved at this time, essentially due to limiting factors like a lack of chemical homogeneity and mechanical stability. On the other hand, zone melting method gives the possibility to reach such a goal.[3] Preliminary results obtained with a laser zone melting process are presented here. The precursor is a Bi-based 2212 material in the glassy state inserted in epoxy resin at room temperature, and polished before exposure to the laser beam. The produced material has been studied by high resolution optically polarized light microscopy and scanning electron microscopy with EDX analysis. Various intensities of the laser beam and sweeping conditions have been examined. A systematic analysis has given the correlation between the behavior of the material and its synthesis conditions.

II. Experimental

The synthesis of the precursor $Bi_2Sr_2CaCu_2O_{8-y}$ starts from a mixture of stoichiometric amounts of Bi_2O_3, $SrCO_3$, $CaCO_3$, and $CuCO_3.Cu(OH)_2$ pretreated at 820°C for 48 hours, including two intermediate grindings. The powder is then melted in an alumina crucible at relatively high temperature (1050°C), i.e. well above the melting point of the initial composition, during a short time (30min) in order to prevent crucible contamination through the melt. A glass precursor sample is obtained by splat quenching the melt between two room temperature copper blocks.[4] The glassy precursor is then inserted in epoxy resin at room temperature and polished with SiC papers and diamond paste. Preliminary characterizations of the precursor sample have been performed by high resolution optically polarized light microscopy. A lot of very small dendritic crystalline phases are dispersed in the bulk.[4] A moving local CO_2 laser beam was then used to texture the Bi-based 2212 material. Figure 1 gives the experimental set-up used for the zone melting process. Various intensities and sweeping conditions have been examined. They are reported in Table 1 for the samples examined here.

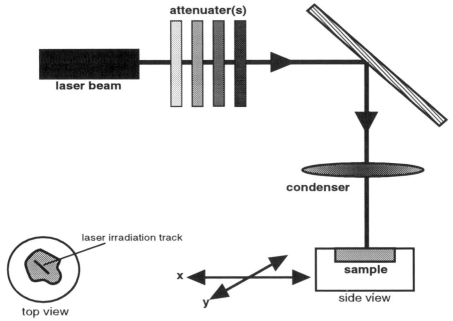

Figure 1   Schematic diagram of the experimental set-up of the laser zone melting process

sample number	number of irradiation	travelling rate (mm/min)	number of attenuator(s)
4(7)	1	8	3
4(8)	5	8	3
4(9)	10	8	3

Table 1 Experimental conditions for the three samples examined here

III. Results and discussion:
Influence of the processing conditions: analysis by optically polarized light microscopy.

Figure 2 shows the microstructure of the three different zones exposed to the laser beam adjusted with variable travelling rates as the controlling parameter (see Table 1). In some cases, crystallization zones can be detected (see the central part of figure 2 and figure 3) giving a correlation between the microstructure of the material and its synthesis conditions. The optimum parameters deduced from the observations are thus the following: the number of attenuators is 3, the travelling rate is around 8mm/min, and the number of simultaneous irradiations is 5.

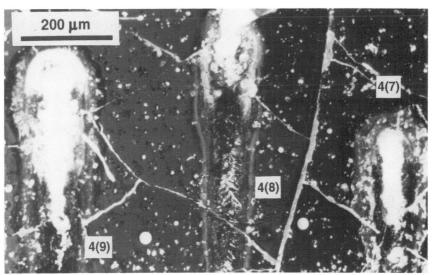

Figure 2　　　High resolution optically polarized light micrograph of three different laser tracks performed with a variable number of simultaneous irradiations (see Table 1).

Figure 3　　　High resolution optically polarized light micrograph showing a zoom of the central part of figure 2. Crystallites are clearly observed.

106

Scanning electron microscopy and Energy Dispersive X-ray analysis have been performed on the top of the chosen system. From figure 4 (back-scattered electron micrograph) it is clear that *a preferential orientation* of the crystallites is induced during the zone melting process. A systematic analysis of the chemical composition gives for the crystallites of figure 4, a chemical composition closed to $(Sr,Ca)_{14}Cu_{24}O_{41}$ (the grey phase) and for the so-called matrix, a chemical composition closed to $Bi_2Sr_2Ca_1Cu_2O_8$.

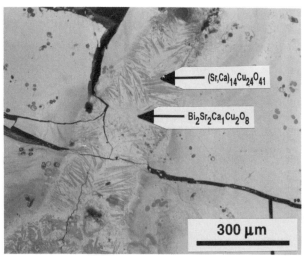

Figure 4   Back-scattered electron micrograph of the apparently best deduced recrystallized sample. The different phases are emphasized.

Micro-electrodes have been deposited on the top surface at various locations in order to probe its electrical properties. No superconductivity has been reported at all. From these observations we can conclude that no superconducting Bi-based 2212 crystallites have been produced by laser beam recrystallization in these conditions. Annealing under the laser beam with maximum attenuation should induce crystallization of the Bi-based 2212 phase by reaction between the $(Sr,Ca)_{14}Cu_{24}O_{41}$ crystallites and the Bi-rich surrounding phase. Moreover, the zone melting process used here seems to improve the deformation of the produced ceramic by reducing the cracks propagation through the surrounding glassy matrix (figure 4). From these observations, it is thus of interest to investigate the influence of all the parameters simultaneously with different cycles of exposure to the laser beam.

Acknowledgments

This work is part of the SSTC (Brussels) SU/02/013 contract. MA received a British Council travel grant for attending EUCAS 95. RC is particularly grateful to the Robert Schuman fundation for a grant.

References

[1]   Stassen S, Rulmont A, Vanderbemden Ph, Vanderschueren A, Gabelica Z, Cloots R and Ausloos M, submitted to *Appl. Phys. Lett.* (1995)
[2]   Stassen S, Rulmont A, Krekels T, Ausloos M and Cloots R, submitted to *J. Appl. Cryst.* (1995)
[3]   Pena J I, Angurel L A, Sotelo A, Martinez E, Ruiz T, de la Fuente G, Lera F and Navarro R, *Proc. of the 8th CIMTEC World Ceramic Congress and Forum on New Mater.*, Florence, Italy, June 29-July 04, 1994, P. Vincenzini, Ed., Adv. Sci. Technol. vol. 8 (Techna Faenza, 1995) p. 481-488.
[4]   Cloots R, Stassen S, Rulmont A, Godelaine P A, Diko P, Duvigneaud P H and Ausloos M, 1994 J. Cryst. Growth **135**, 496-504.

Inst. Phys. Conf. Ser. No 148
Paper presented at Applied Superconductivity, Edinburgh, 3–6 July 1995
© *1995 IOP Publishing Ltd*

(Bi,Pb)-1212 and (Bi,Pb)-0212, Superconducting Cuprates with Reduced Anisotropy

P.Zoller, B.Seling, A.Bauer, A.Ehmann, J.Glaser, W.Wischert and S.Kemmler-Sack

Institut für Anorganische Chemie der Universität, Auf der Morgenstelle 18, D-72076 Tübingen, Germany.

Abstract. The superconducting performance of (Bi,Pb)-1212 and (Bi.Pb)-0212 is strongly influenced by the method of material processing. It is shown that the application of melt processing results in bulk materials with improved intragrain and intergrain properties. The best pellets of (Bi,Pb)-1212 dispose on a transition temperature of 94K with critical current densities of 7.8 x 10^5 Acm^{-2} (5K,0T) and 2.4 x 10^3 Acm^{-2} (77K, 0T). For pellets of (Bi,Pb)-0212 a maximal transition temperature of 86K is obtained with superconducting volume fractions around 15%.

1. Introduction

In the system $(AO)_mB_2M_{n-1}Cu_nO_{2n+2+z}$ various series of superconducting cuprates are existing for m = 1 and 2. An important difference between both families consists in the reduced anisotropy of the crystal structure for the monolayer cuprates (m = 1) resulting in a stronger coupling of the CuO_2 layers of adjacent perovskite blocks. A model of Kim et al. [1] predicts a displacement of the irreversibility line towards higher temperature and magnetic field, when the thickness of the rocksalt block is decreased and the properties of several Tl cuprates strongly support this suggestion. A further reduction of the structural anisotropy is obtained by the total elimination (m = 0) of the rocksalt-like AO layers: $B_2M_{n-1}Cu_nO_{2n+2+z}$. Simultaneously, complex charge reservoir layers are missing.

Recently, we prepared several members of the series Bi-1212, Bi-0212 as well as partially Pb substituted materials for the first time [2-5]. The highest transition temperature is situated at 102K for Bi-1212, thus making this material very attractive in applied superconductivity. For Bi-0212 we demonstrated that the disadvantage of a missing charge reservoir can be overcome by introducing cations with various oxidation states.

In both structure types the superconducting properties of ceramic materials are influenced by weak links. To overcome this problem the method of material processing must be optimized. In this study we demonstrate for several examples that via a suitable melt process well textured materials with enforced grain connectivity are obtained.

108

2. Experimental

Various samples of (Bi,Pb)-1212 and (Bi,Pb)-0212 were prepared from mixtures of high purity starting materials (Bi_2O_3, PbO_2, $SrCO_3$, $CaCO_3$, La_2O_3 (prefired at 850°C/20h) and CuO). The samples were heated in form of powders or pressed pellets (diameter: 13 mm; height:1-2 mm) in corundum crucibles in various gas atmospheres at different temperatures as indicated in the text. All materials were characterized by XRD (Philips powder diffractometer, CuKα radiation, Au standard), resistance (standard four-probe DC method) and susceptibility measurement (SQUID magnetometer, Quantum Design).

3. Results and discussion

3.1. Melt texturing of (Bi,Pb)-1212

The superconducting performance of (Bi,Pb)-1212 was optimized by adjusting the (i) material composition, (ii) temperature/time schedule and (iii) gas atmosphere. It should be noted that the formation of larger grains with good connectivity is enforced by the presence of a liquid phase. For studying this process we employed different melting temperatures and times in the range between 1010°C/10 min (partial melting) and 1060°C/70 min (total melting) with several subsequent cooling rates and temperatures in different gas atmospheres.

From widespread experiments it follows that the application of higher melting temperatures gives the better results. Fig. 1a shows the χ(T) dependence of a 1212 pellet after application of a melting temperature of 1040°C (10min) and two different cooling ramps to 990°C within 35°C/h and subsequently to 983°C with 2°C/h; finally a post-treatment in flowing oxygen at 500°C (6h) was employed. This final O_2 treatment improves

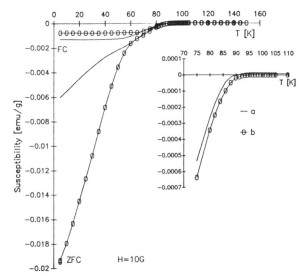

Fig. 1. χ vs. T for pellets with Bi:Pb:Sr:Y:Ca:Cu = 0.4:0.45:1.9:0.7:0.4:2.15 (a) post-treated in flowing O_2 (500°C/6h), (b) additional post-treatment in flowing Ar (700°C/6h); the inset gives an enlargement of the section of the region near T_c. FC: field cooled, ZFC: zero field cooled.

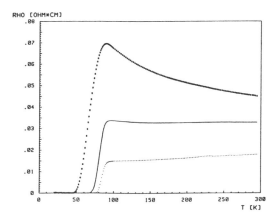

Fig. 2. ρ vs. T for pellets with Bi:Pb:Sr:Y:Ca:Cu = 0.4:0.45:1.9:0.7:0.4:2.15 after melting at 1035°C (10 min), cooling to 1010°C with 35°C/h and subsequently to 985°C with 2°C/h. (x) as prepared; (-) with additional post-treatment in flowing O_2 (500°C/6h) and (·) a pellet with Bi:Pb:Sr:Y:Ca:Cu = 0.4:0.45:1.9:0.7:0.4:2 after 1060°C/70 min; 1030°C/10 min; cooling with 35°C/h to 984°C (12h) and post-treatment in flowing O_2 (500°C/6h).

considerably the superconducting as well as the normal conducting properties as one can deduce from the $\rho(T)$ dependence in Fig. 2 of two pellets of identical composition before (a) and after O_2 post-treatment (b). In the latter case the transition to superconductivity is shifted considerably to higher temperatures, simultaneously the width of the transition is reduced and in the normal conducting region the pronounced semiconducting behaviour is reduced. As third example Fig. 2 (c) includes the $\rho(T)$ data of a pellet after melting at a still higher temperature of 1060°C for 70 min. A further improvement of the conducting properties gets obvious from the still steeper transition to superconductivity and a metallic behaviour in the normal conducting regime.

A further progress is obtained by annexing a post-treatment in flowing Ar. The $\chi(T)$ dependence in Fig. 1(b) indicates an increase of the transition temperature from 88 to 94K. The critical current density j_{cm} (calculated with the Bean model [6]) increases from 6.4 to 7.8 x $10^5 Acm^{-2}$ (5K, 0T). At 77K j_{cm} is reduced to 2.4 x $10^3 Acm^{-2}$ (0T).

3.2. Melt-texturing of (Bi,Pb)-0212

(Bi,Pb)-0212 was firstly prepared by partially substituting Fe for Cu [2,4,7]. Subsequently, superconducting Fe free materials were obtained in the systems $Bi_{0.5}La_{1.5-x}Y_xSrCu_2O_y$ and $(Bi,Pb,La,Sr)_2Sr_{1-x}Ca_xCu_2O_y$ with T_c up to 75K [4,5] as well as in the pure Pb system $(Pb,La,Sr)_2SrCu_2O_y$ with inferior T_c values up to 45K. A strong disadvantage of all as prepared powdered materials was the low superconducting volume fraction due to the formation of non-superconducting grain boundaries acting as weak links. Similar to the case of (Bi,Pb)-1212 these properties are a consequence of their crystal structure [4,5,7].

In order to improve the superconducting properties the application of melt texturing via the employment of several fluxes was studied for some (Bi,Pb)-0212 materials. Fig. 3 shows as example the increase of the superconducting transition temperature and superconducting volume fraction from 35K ($\approx 2\%$) via 62K ($\approx 6\%$) after addition of 10% Ca_2PbO_4 as flux to 86K ($\approx 17\%$) with 10% $CaBi_2O_4$ as flux. A further improvement is expected by the application of a more sophisticated preparation process.

110

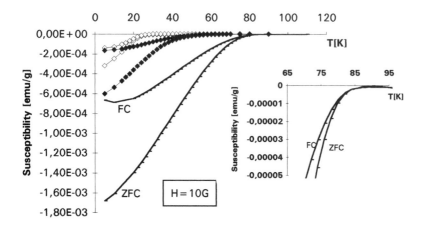

Fig. 3. χ vs. T for (□) as-prepared powder (992°C/16h) and pellets with Bi:Pb:La:Sr:Cu = 0.4:0.1:1.2:1.3:2 with (◆) 10% Ca_2PbO_4 as flux (995°C/16h) and (-) 10% $CaBi_2O_4$ as flux (994°C/3h). The inset gives an enlargement of the section near T_c for case (-).

Acknowledgements

This work was supported by the Bundesministerium für Bildung, Wissenschaft, Forschung und Technologie (FKZ 13N5842, 13N6481 and 02-KS 3TUE) and the Verband der Chemischen Industrie. We wish to thank the Rhône Poulenc GmbH for supplying the rare earth oxides.

References

[1]   D.H.Kim, K.E.Gray, R.T.Kampwirth, J.C.Smith, D.S.Richeson, T.J.Marks, J.H.Kang, J.Telvacchio and M.Eddy, Physica **C177** (1991) 431.

[2]   P.Zoller, J.Glaser, B.Seling, A.Ehmann, W.Wischert and S.Kemmler-Sack, Physica **C235-240** (1994) 955.

[3]   P.Zoller, J.Glaser, A.Ehmann, C.Schulz, W.Wischert, S.Kemmler-Sack, T.Nissel and R.P.Huebener, Z.Phys. **B96** (1995) 505

[4]   B.Seling, C.Schinzer, A.Ehmann, S.Kemmler-Sack, G.Filoti, M.Rosenberg, J.Linhart and W.Reimers, Physica C, in press.

[5]   B.Seling, A.Ehmann, C.Schinzer, J.Linhart and W.Reimers, Z.Phys., in press.

[6]   S.Elschner and S.Gauss, Supercond.Sci.Technol. **5** (1992) 300.

[7]   B.Seling, C.Schinzer, A.Ehmann, S.Kemmler-Sack, G.Filoti, M.Rosenberg, J.Linhart and W.Reimers, phys.stat.sol., submitted.

Inst. Phys. Conf. Ser. No 148
Paper presented at Applied Superconductivity, Edinburgh, 3–6 July 1995
© 1995 IOP Publishing Ltd

Decomposition and reformation of Bi-2212 during the partial melt processing in oxygen

Th. Lang, D. Buhl, M. Cantoni and L.J. Gauckler

Nichtmetallische Werkstoffe, ETH Zürich, Sonneggstrasse 5, CH-8092 Zürich

Abstract. Thick films of $Bi_{2.2}Sr_{2.05}Ca_{0.95}Cu_2O_x$ heated above 890°C on a silver substrate at an oxygen partial pressure of 1 atm decompose peritectically, forming the Bi-free phase $Sr_8Ca_6Cu_{24}O_x$, the Cu-free phase $Bi_9Sr_{11}Ca_5O_y$, a Bi-rich liquid, and oxygen, which is released to the atmosphere. The solid phases occupy a volume fraction of 17% each.
During slow cooling at a rate of 5 K/h from 893°C Bi-2212 grains start growing at a temperature around 875°C directly from the melt, without forming the one-layer compound (Bi-11905) first. With decreasing temperature the amount of Bi-2212 increases. At the same time the volume fractions of the secondary phases 014x24 and 91150 and the liquid decrease.
The conditions for the direct 2212 formation and growth are: (1) dissolution the secondary phases in the melt in order to deliver necessary cations, and, (2) the uptake of oxygen from the atmosphere because of the oxygen deficiency of the melt compared to solid 2212.
The highly aligned microstructure observed in Bi-2212 thick films is due to the combination of the extreme two-dimensional growth of the 2212 platelets with the limited macroscopic thickness of the films.

1. Introduction

Bi-2212 has a great potential for technical applications because macroscopic devices with good superconducting properties can be obtained by the partial melting process. To reach a high current carrying capacity, the microstructure has to be single phase, dense and well textured. Because of the peritectical decomposition of the superconductor during the melting process, phase purity is difficult to obtain at the end of the heat treatment. Films with thicknesses up to 20 µm show highly aligned grains with their c-axis perpendicular to the substrate [2]. By increasing the sample thickness, the texture is lost and the critical current density significantly decreased. Thus, understanding the mechanisms that lead to the texture and phase formation of Bi-2212 during cooling are important to optimize the processing conditions in order to get high critical current densities.

2. Experimental

Powder with the stoichiometry $Bi_{2.2}Sr_{2.05}Ca_{0.95}Cu_2O_x$ was prepared by a standard calcination process [3]. It was mixed with organic additives and tape-cast onto glass [4]. Samples with a diameter of 11 mm were cut from the dried tape and the organics were burnt out at 550°C. These green tapes were put on silver substrates (d=100 µm), fixed on a sample holder, and suspended in a vertical furnace one at a time. In a first set of experiments, the thick films were heated to temperatures of 885, 890, 900, and 911°C with 1 K/min and immediately quenched in vacuum-pump oil [5]. A second set of samples was submitted to a partial melting process consisting of melting at 893°C, slow cooling with 5 K/h to 850°C, and isothermal annealing at 850°C for 48 h. All heat treatments were done in flowing oxygen (pO_2=1 atm). A thermocouple was placed 2 mm above the samples to control the temperature. The samples were quenched from different temperatures during the heat treatment and analyzed by means of light microscopy, XRD, SEM, EDS, and HRTEM. Phase contents and grain sizes were measured using the line method on polished cross sections.

3. Results and Discussion

3.1. Decomposition of Bi-2212

Since the sample heated to 885°C and immediately quenched showed no signs of melting, it was not further analyzed. The sample heated to 890°C was decomposed into the Bi-free phase $Sr_8Ca_6Cu_{24}O_x$, which belongs to the 0$\underline{14}$x$\underline{24}$ family, the Cu-free phase $Bi_9Sr_{11}Ca_5O_y$ [6], and a Bi-rich liquid. Furthermore, Bi-2212 is known to release oxygen to the atmosphere during melting [3]. The decomposition reaction of Bi-2212 in oxygen (pO_2=1 atm) can be written as:

$$Bi_2Sr_2CaCu_2O_x \rightarrow Sr_8Ca_6Cu_{24}O_y + Bi_9Sr_{11}Ca_5O_z + liquid + O_2$$

The Bi-free grains were needle-shaped with a length not exceeding a few tens of micrometers, whereas the Cu-free grains grew less one-dimensional and were usually smaller than 20 μm. The solid grains were distributed homogeneously over the whole cross section of the sample and occupied a volume fraction of about 17% each. No significant differences in phase distribution and composition from the 890°C sample were found for the samples heated up to 900 and 911°C.

3.2. Reformation of Bi-2212

Figure 1 shows the amount of the secondary phases, 2212, and liquid as a function of the quenching temperature. During slow cooling from the maximum temperature (893°C), the phase composition remained unchanged until 883°C. At 875°C first Bi-2212 grains were

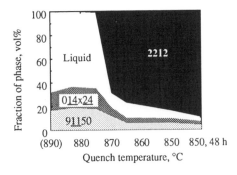

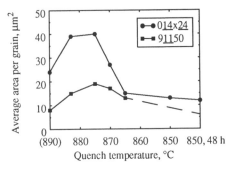

Fig. 1: Amount of phases present in a Bi-2212 thick film during slow cooling at 5 K/h from 893 to 850°C. The sample at 890°C was quenched during heating.

Fig. 2: Average area per grain in the cross sections of the quenched samples. The values were obtained by dividing the total area occupied by one of the solid phases by the number of grains.

detected, the amount of the Bi-free and the Cu-free phases was still about 17 vol% each. At 870°C almost 70% of the sample consisted of 2212 grains while most of the liquid was consumed. The amount of Bi- and Cu-free phase was reduced to approximately 10 vol% each. At 865°C most of the sample was transformed to 2212. Although the x-ray diffraction measurement indicated a single phase microstructure, light microscopy revealed the presence of 10 vol% liquid and 10 vol% secondary phases between the superconducting grains. Even after an isothermal annealing step of 48 h at 850°C these phases could not be completely removed.

At the grain boundaries of the growing 2212 platelets appeared small silver precipitates with an average size below 1 µm. At the end of the slow cooling ramp, these precipitates had grown to large silver grains. This observation can be explained by the different silver solubilities of the peritectic melt and the solid 2212 superconductor. In contrast to the 2212 phase, which shows a very low silver solubility, the Bi-rich liquid can dissolve approximately 4 wt%. This silver has to be precipitated during the solidification of the 2212 grains from the melt.

The sample quenched from 875°C was analyzed by HRTEM. Isolated 2212 grains with no single layer intergrowths were found in direct contact with the amorphous liquid. Therefore, we conclude, that the two layer compound grew directly from the melt - without forming the one layer compound first - under these conditions. No indications were found, that either the Bi- or the Cu-free grains, the film surface or the interface with the silver acted as preferred nucleation sites. However, the secondary phases had to dissolve in the liquid during cooling in order to deliver further cations for the crystallization of the 2212. Furthermore, the thick films had to take up oxygen from the atmosphere, because the melt is oxygen deficient compared to the solid superconducting phase.

The average grain size of the solid secondary phases in the Bi-rich melt as a function of the quenching temperature is plotted in figure 2. At 890°C the grain size of both phases was rather small, but it increased during the cooling period until the formation of the 2212 grains started at 875°C. A strong drop of the grain size was observed between 875 and 865°C, then the dissolution of the phases, which was responsible for the size change, slowed down. Since the grains of the Cu-free phase are difficult to detect with increasing amount of 2212 platelets, the grain size could only be estimated for the samples quenched from 850°C.

The results obtained in this work are in contrast to our previous experiences with bulk samples [1], where the one layer compound nucleated at the Bi-free grains and crystallized first. The transformation to Bi-2212 followed during a prolonged annealing at 850°C. We attribute this difference to the oxygen uptake of the samples, which is faster in thick films than in bulk samples, due to the reduced thickness.

The decomposition of Bi-2212 takes place at a defined temperature or at least in a very narrow temperature range. When a decomposed sample is cooled below this temperature, 2212 is the thermodynamically stable phase and therefore, the different solid, liquid, and gaseous phases are supposed to form the superconducting phase. Obviously, the reaction of four partners at a time is kinetically difficult. Since one of the phases is liquid, the beginning of the reaction is fast because the different ions can be transported by liquid phase diffusion. But as soon as the 2212 platelets grow to a large size, they isolate the randomly distributed secondary phases and represent a solid obstacle, which has to be overcome by solid phase diffusion, which is slower than transport in the liquid phase. This model explains the reduction of the reaction kinetics with increasing amount of 2212 present in the thick film and the necessity to anneal the samples for a long period in order to get an (almost) single phase microstructure. Of course, other effects, like Bi-loss and loss of liquid during melting, can lead to the presence of Bi-free phases [7], even after long annealing periods, because the overall sample stoichiometry is changed.

114

3.3. Texture formation

The orientation of the 2212 grains was random in the initial stage, but only grains oriented parallel to the substrate could grow to a size exceeding the film thickness, because either the substrate or the film surface (surface-tension) acted as growth barriers. Some of the initially misoriented grains were turned into a parallel position during growing, while others remained misoriented forming obstacles for the superconducting current flow. Some grains, which were oriented almost perpendicular to the substrate, grew inside the silver substrate.

The reason for the highly aligned microstructure in Bi-2212 thick films can, therefore, be found in the extreme two-dimensional growth of the Bi-2212 grains in combination with the limited thickness of these samples. This alliance forces the grains to grow with their c-axis perpendicular to the substrate.

4. Acknowledgments

The authors gratefully acknowledge the financial support of the Schweizer Nationalfonds (NFP 30).

5. References

[1] B. Heeb et al., *J. Mater Res.*, **7** (11) 2948 (1992).
[2] J. Kase et al., *IEEE Trans. Mag.*, **27** 1254 (1991).
[3] Th. Lang et al., *Proc. for the 4th World Conf. on Supercond.*, Orlando FL, USA, June 27-July 1, (1994), in press.
[4] D. Buhl et al., *Proc. for the 4th World Conf. on Supercond.*, Orlando FL, USA, June 27-July 1, (1994), in press.
[5] R.D. Ray II, *Physica C*, **175** 255 (1991).
[6] R. Müller et al., *Physica C*, **243** 103 (1995).
[7] J. Shimoyama et al., *Adv. in Superconductivity V*, Eds. Y. Bando and H. Yamauchi, Springer Verlag Tokyo, 697 (1993).

Inst. Phys. Conf. Ser. No 148
Paper presented at Applied Superconductivity, Edinburgh, 3–6 July 1995
© *1995 IOP Publishing Ltd*

Texturing of (2223) superconducting Bi-Pb-Sr-Ca-Cu-O ceramics by different processes using hot pressing and melting in a magnetic field

J.G. Noudem[1], **J. Beille**[2], **D. Bourgault**[1], **E. Beaugnon**[1], **R. Tournier**[1], **D. Chateigner**[3], **P. Germi**[3], **M. Pernet**[3].

(1) EPM-MATFORMAG, (2) Laboratoire Louis Néel, (3) Laboratoire de Cristallographie, C.N.R.S., B.P. 166, 38042 Grenoble, cedex 9, France;

Abstract : We compare several ways of texturation of superconducting (2223) Bi-Pb-Sr-Ca-Cu-O, namely hot pressing (HP), magnetic melt texturation (MMT) and an original method combining the two processes (MMHPT). Different degrees of texturation are compared from scanning electron microscopy observations and X-Ray diffraction spectra including pole figures, as well as physical characterizations based on electrical transport measurements. We analyze the relative effects of several parameters of the process : temperature, stress and magnetic field.

1. Introduction

Bi-Pb-Sr-Ca-Cu-O ceramics are interesting in view of their high critical temperature and the platelet shape of their constituting grains, which makes them easy to orientate. Their weak point is their sensitivity to magnetic field, due to a weak intrinsic pinning of the vortices in the (a,b) planes in line with their 2 D character, and to the importance of parasitic phases, especially for the (2223) phase. For industrial applications, it is important to process ceramics with oriented grains, in order to insure a better connexion between them, and to improve the intergranular medium.

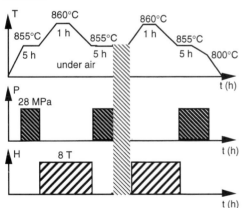

Figure. 1. Cycles of temperature, pressure and magnetic field.

In previous works [1,2], we have applied several ways of texturation of superconducting (2223) Bi-Pb-Sr-Ca-Cu-O ceramics : hot pressing (HP), magnetic melt texturation (MMT) and an original method combining the two last processes (MMHPT). In this paper, we compare the superconducting characteristics of the samples processed in the different techniques and discuss the effects of the various parameters : temperature, magnetic field, pressure.

2. The different processes of texturing the (2223) Bi-Pb-Sr-Ca-Cu-O ceramics

We used magnetic susceptibility measurements at high temperature to determine the begining and the end of melting [2].

A high pressure of 28 MPa is applied for 5 hours at a maximum temperature of 855°C, just above a magnetic susceptibility minimum ascribed to the beginning of melting. In the magnetic melt texturation, a magnetic field of 8 tesla is applied from the beginning until the end

116

of the processing. A maximum temperature of 860°C is established for 1 hour and then slowly decreased down to 700°C at a cooling rate of 1°C/hour. The parameters of the hot pressing technique have been previously optimized [1]. In our original (MMHPT) process, a pressure of of 28 MPa is applied for 5 hours while the temperature is kept constant at 855°C.

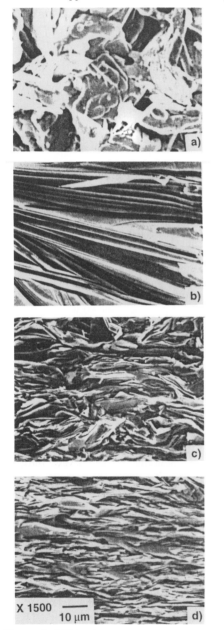

Figure. 2. SEM micrograph pictures of a face broken parallely to the applied stress and magnetic field direction for the different texturing processes.

Pressure is then released, the magnetic field is turned on and the temperature is increased up to 860°C, maintained 1 hour and then decreased at 1°C/hour down to 855°C where the magnetic field is turned off and pressure reapplied for 5 hours. The proposed cycle thus includes an initial hot pressing step, followed by melting step under magnetic field, and ending with a hot pressing step. The three cycles of temperature, pressure and magnetic field of this process may be repeated several times, as illustrated in Figure. 1,which should increase the performances of the sample.

3. Physical characterizations

3.1. Scanning electron microscopy (SEM):

SEM pictures of faces broken parallely to the axis of the texturing parameter (Figure. 2) cleary show different effects induced by these processes : a) classical sintering, b) magnetic melt texturation, c) hot pressing, d) hot pressing combined with magnetic melt texturation.

Figure 2a shows no preferential orientation of the constituting platelets $\approx 20\ \mu m$ in size, whereas figure 2b indicates a preferential orientation of the platelets perpendicularly to the magnetic texturing field with i) a larger size of the platelets ($> 50\ \mu m$), ii) a sizeable dispersion in the c axis directions, iii) quite large voids between grains. Figure 2c shows a higher density, but the platelets have a wavy shape and a reduced size probably due to mechanical defects: Figure 2d shows there is a good alignment of the platelets which appear very well compacted, their wavy aspect being largely reduced.

From these observations, we conclude that melting in a magnetic field is favorable to the growth of large, oriented cristallites, but with loose connections and voids in between. We suggest that the first hot pressing step of our MMHPT process prealigns the platelets, magnetic field favors the growth of orientated grains. Finally, while the stay in the liquid state in a the consecutive second hot pressing step confers a high density to the material.

3.2. Pole figures

Since classical Θ-2Θ X ray partterns revealed intense 00l reflections for a MMHPT Bi-2223 surface perpendicular to the applied magnetic field and uniaxial stress, we operated pole figure measurements on this type material. We used the Schulz reflection geometry [3] with monochromatic Cu(Kα) radiation.

The pole figures were corrected for background and defocusing [4] and then normalized [5] in order to calculate the distribution density of the crystallites. This value is normalized to 1 m.r.d. (multiple of random distribution) for a perfectly untextured sample. In the following χ is the tilt angle of the Schulz geometry.

According to our elaboration process, we do not expect any in-plane orientation. We first verified this fact on the $\{115\}$ pole figures (not shown here) which present a ring of strong intensities centered on $\chi = 60°$, the theoretical value of a single Bi-2223 crystal.

Figure. 3 is the $\{00\underline{1}4\}$ pole figure of the sample parially melted at 860°C which exhibits the strongest critical current density of Jc = 3800A/cm^2 obtained. This figure shows that $\{001\}$ normals are not dispersed by more than 25° from the fiber axis (magnetic field and uniaxial stress direction), taking reference at 15 % of the maximal intensity on the figure. The maximal density for this reflection is 4.5 to 5 m.r.d. No variation of the texture degree has been observed between the core of this sample and the interface with the rod. We also observe on this figure a ring beginning at $\chi = 65°$ which has not been measured further than $\chi = 72°$. This ring corresponds to the $\{020\}/\{200\}$ contribution located at a sufficiently proach Θ value, observable by defocusing at the Θ_{0014} position. Taking into account misalignments occuring from the positionning of the sample on the goniometer and from the magnetic field and uniaxial stress sample axis positionning, this $\{020\}/\{200\}$ contribution proves the a and b random orientation around c axis.

We analyzed the $\{00\underline{1}4\}$ pole figure of the sample partially melted at 865°C on which we measured Jc = 2000 A/cm^2. This pole figure indicates a dispersion of the $\{001\}$ plane normals up to 30°, with the same reference as above and consequently the orientation density is lowered to 3.5-4 m.r.d. at the maximum. This texture loss compared to with the previous sample, and the presence of parasitic phases both contribute to decrease the Jc.

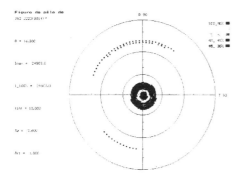

Figure. 3. $\{0014\}$ X-ray pole figure of the sample partially melted at 860°C.

3.3. Electrical transport measurements

These sample processes can be classified based on increasing transport critical current densities Jc (77 K, 0 T) as : a) classical sintering (Jc $\approx$ 800 A/cm^2), b) MMT-process (1500 A/cm^2), c) HP-process (Jc $\approx$ 2500 A/cm^2) and d) MMHPT-process (3800 A/cm^2)

The effect of the magnetic field applied during the process is highlighted by considering samples prepared by MMHPT without magnetic field. but with the melting step at 860 °C. which gives a Jc of only 2500 A/cm^2.

Figure. 4a shows an increase of critical current density Jc versus critical temperature Tc for samples elaborated according to the MMHPT process at various partial melting temperatures as usually observed.

118

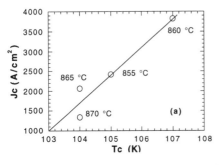
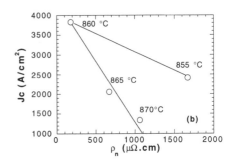

Figure. 4. Plots of Jc a) versus Tc b) versus

ρ_n.

In Figure. 4b the critical current density Jc of the same samples is plotted versus the normal state resistivity ρ_n, and shows two different regimes. For partial melting temperatures higher than 860 °C, a decrease in Jc correlates with and increase in ρ_n, attributed to a larger proportion of parasitic phases. Below 860 °C, a decrease in the temperature of the melting dwell correlates with a decrease in Jc and with an increase in ρ_n. In that case, the proportion of the liquid phase is too small to allow rotation and growth of the platelets as well as their densification. Nevertheless, ρ_n increases due to a lowering of the temperature of the upper dwell, which little affects the Jc. The defects induced by hot pressing alone may act as pinning centers for the vortices. Consequently, it is better to be a little below the optimum partial melting temperature than above.

4. Conclusion

The process of partial melting in a magnetic field combined with hot pressing (MMHPT) allows to produce very dense ceramics of Bi-Pb-Sr-Ca-Cu-O with a better degree of texture than those obtained by hot pressing or melting in magnetic field. However, the temperature window for melting under field is very narrow and has to be carefully controlled. If this temperature is too low, there is no extra effect compared to hot pressing only, if it is too high, melting induces parasitic phases which strongly depress the superconducting performances. Once the difficulty of temperature control is mastered, the process is interesting for the production of large samples with critical current densities above 3800 A/cm^2.

Acknowledgments

J.G.N acknowledges his followship from ADEME and Alcatel Alsthom. M. Bonvalot is warmly thanked for useful discussions and very careful reading of the manuscript. We are grateful to P. Amiot for her help in carrying out the SEM studies.

references

[1] J.G. Noudem, J. Beille, D. Bourgault, A. Sulpice and R. Tournier : Physica C. **230** (1994) 42.
[2] J.G. Noudem, J. Beille, D. Bourgault, E. Beaugnon, A. Sulpice, R. Tournier, D. Chateigner, P. Germi and M. Pernet : to appear to Supercond. Sci. Technol. **8** (1995) 558..
[3] L.G. Schulz : J. Appl. Phys. **20** (1949) 1030.
[4] J.C. couterne and G. Cizeron : J. Appl. Cryst. **4** (1971) 461.
[5] J.R. Holland : Adv. X-Ray anal. **7** (1964) 86.

Inst. Phys. Conf. Ser. No 148
Paper presented at Applied Superconductivity, Edinburgh, 3–6 July 1995
© 1995 IOP Publishing Ltd

Influence of Y$_2$BaCuO$_5$ Particles on the Microstructure of YBa$_2$Cu$_3$O$_{7-x}$ (123) - Y$_2$BaCuO$_5$(211) Melt-Textured Superconductors

P. Diko
Institute of Experimental Physics, SAS, Watsonova 47, 04353 Kosice, Slovakia.
W. Gawalek, T. Habisreuther, T. Klupsch, P. Görnert
Institut für Physikalische Hochtechnologie, Helmholtzweg 4, 100239 Jena, D-07702.

ABSTRACT The influence of 211 particles in 123-211 melt-textured bulk superconductors on the microstructural features such as microcracking, subgrains, twins and twin complexes was studied by polarized light microscopy. The observed linear dependence between a-b microcrack spacing, l_c, nd $d_{211}.V_{123}/V_{211}$ (d_{211} - mean 211 particle size, V_{123} - 123 volume fraction, V_{211} - 211 volume fraction) proves that the microcracking process can be described under the framework of a model devised for the multiple failure of a unidirectional composite under uniaxial tensile loading. Shorter mean free distance between 211 particles depresses the subgrain thickness, the twin complex size and twin spacing. The level of residual tensile stress in the 123 a-b plane was estimated from detwinning observed around 211 particles.

1. Introduction

Although the relatively large J_c values indicate the existence of effective flux pinning mechanisms in the melt-textured 123, the exact pinning mechanisms are still unclear. It seems that superconducting properties can be optimized mainly by controlling the volume fraction and particle size of the 211 phase. The aim of this paper is to describe the relationship between 211 particles and defects such as microcracks, twins and subgrains.

2. Experimental

In the present study, the well textured bulk YBa$_2$Cu$_3$O$_{7-x}$ samples, with different 211 volume fraction and size, were prepared by a melt-texturing process at IPHT Jena. The microstructure was analyzed mainly by optical microscopy after polishing and etching in a solution of bromine in ethanol (2 wt % Br). Observation in normal and polarized light was performed. The orientation of 123 grains was determined from their optical twin patterns (Verhoeven & Gibson 1988).). The Qantitative microstructure characteristic were measured by linear intercept and point counting methods (Richardson 1971).

3. Results and discussion

3.1. Microcracking

The first aim of this investigation was to distinguish between the a-b microcracks formed in solid state and the a-b planar defects formed during the crystallization process according to the models proposed by Alexander at al. (1992) and Goyal at al. (1993).

In the samples that were studied, it was found that some growth related a-b planar defects exist (Fig. 1). They are filled with rejected liquid (CuO, Ba-Cu-O) and are clearly distinguishable from the a-b cracks. Such growth related a-b planar defects were observed in the samples with large 211 particles but they account for less than 1% of all a-b planar defects. The remaining a-b planar defects are microcracks formed mainly at the transition from the tetragonal to orthorhombic phase (Diko at al. 1995). They are formed due to the higher thermal expansion of the 123 phase compared to the 211 phase and also because the fracture toughness of the orthorhombic 123 $K_c(001) < K_c(010) \gg K_c(100)$ (Raynes at al., 1991). It is often emphasized that in 123-211 melt textured samples with a small 211 particle size, a-b planar defects (in fact microcracks) do not exist (Meng at al., 1991). Analysis of samples in the as-polished condition, with mean particle size close to

120

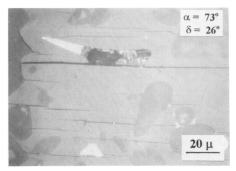

Fig. 1. Growth related a-b planar defect and a-b microcracks in the sample with larger 211 particles.

Fig. 2. The sample with small 211 particles (polished).

2 μm shows litle evidence of any a-b microcracks (Fig. 2). After etching, the a-b microcracks appear to be associated with very fine 211 particles (Fig. 3). Visualization of a-b microcracks, also in the samples with very small 211 particles, allows the measurement of the crack spacing l_c in a broad range of mean 211 particle size d_{211} and 211 volume fraction V_{211}. The true microcrack spacing was obtained knowing crystal orientation determined from its optical twin pattern. When l_c is plotted versus $(d_{211}.V_{123})/V_{211}$ the linear dependence appears with some declination for the smallest particles (Fig. 4). The observed linear dependence between l_c and $(d_{211}.V_{123})/V_{211}$ proved that the microcracking process in 123-211 composites can be described in the framework of the model devised for the multiple failure of a unidirectional composite under uniaxial tensile loading (Diko et al., 1995, Zhang, 1992).

3.2 Subgrains

If a 123 grain is in a suitable orientation, subgrains can be visible in polarized light (Diko et al., 1995) (Fig. 5). The subgrains have an approximately rectangular shape with segments of boundaries parallel or perpendicular to the a-b plane. It should be noted that the subgrain boundaries are not cracked and no secondary phase appears in them (Fig. 5).. The mean free distance between 211 particles (MFD_{211}.) does not have a great influence on the subgrain thickness l_s. The subgrain thicknees is aproximately 25 μm for the sample with MFD_{211} = 50 μm, and 12 μm for the sample with MFD_{211} = 3 μm.

Taking in to account the dimensions of the observed subgrains, we can relate their origin to the crystalization process rather than to some polygonization processes in the solid state.

3.3. Twins and twin complexes

3.3. Twins and twin complexes

The twin spacing l_t depends on the local microstructure. Generally, l_t is higher in regions with a lower

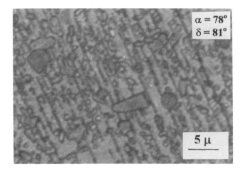

Fig. 3. A-b microcracks in the sample with small 211 particles revealed by etching.

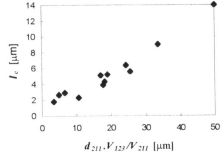

Fig. 4. Dependence between the crack spacing l_c and $(d_{211}.V_{123})/V_{211}$.

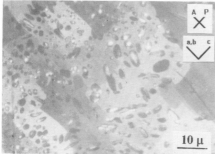

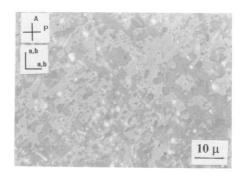

Fig.5. Subgrains visualized by polarized light.

Fig. 6. Larger twin spacing in the 211 free region (etched).

concentration of 211 particles (Figs. 6). The relationship between l_t and the crystal dimension can be expressed as (Zhu at al., 1991): $l_t = (g\gamma/CMF2)^{1/2}$ (1), where $F = 2(b - a)/(a + b)$ is the orthorhombicity and g and M are the grain size and the shear modulus, respectively. C is a constant nearly equal to 1 and γ is the twin boundary energy. In the case of MT YBaCuO the mean free distance between 211 particles (MFD_{211}) can be taken as a characteristic dimension, that determines the twin spacing l_t. According to Eq. (1) it is therefore reasonable to expect that the twin spacing l_t will have a square root dependence on MFD_{211}.. This is difficult to test from the micrographs taken by optical microscopy because, at lower MFD_{211}, the twin spacing l_t is well under the resolution limit of optical microscopy.

An interesting feature observed in melt-textured samples was detwinning around 211 particles and secondary phases. In Fig. 7 the detwinning areas, or areas with predominant one twin domain variant, developed around CuO particles and around 211 particles. These areas extend in the <100> directions forming a rosette. It is believed that they were formed by twin boundary motion. Twin boundary motion partially relaxes the tensile stresses around CuO and 211 particles created at cooling due to lower thermal expansion of 211 phase and CuO phase than 123 matrix. Stresses perpendicular and parallel to the twin boundaries do not relax by the twin boundary motion (Diko et al., 1995) . So in the <110> directions the twinning structure does not change by thermal stresses around 211 and CuO particles. The motion of twin boundaries indicates that residual tensile stresses (50 - 100 MPa) exist in the a-b plane at a level that is necessary for the detwinning process (Diko et al., 1995).

A lamellar assembley of twins with parallel (110) or (110) walls (twin complexes) can be visualized under slightly uncrossed polarizers when polarizer and analyser are nearly parallel or perpendicular to the "a" and "b" directions (Fig.8) (Rabe et al.,1990). The twin complex boundaries prefer the orientation parallel to the (110) and (110) planes. The mean linear size of the twin complexes l_{tc} in the samples with different mean free distance between 211 particles (MFD_{211}) was measured. The results, summarized in Table 1., show that l_{tc} is essentially depressed by lowering MFD_{211}.

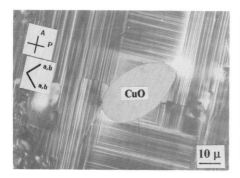

Fig. 7. Detwinned areas around CuO and 211 particles.

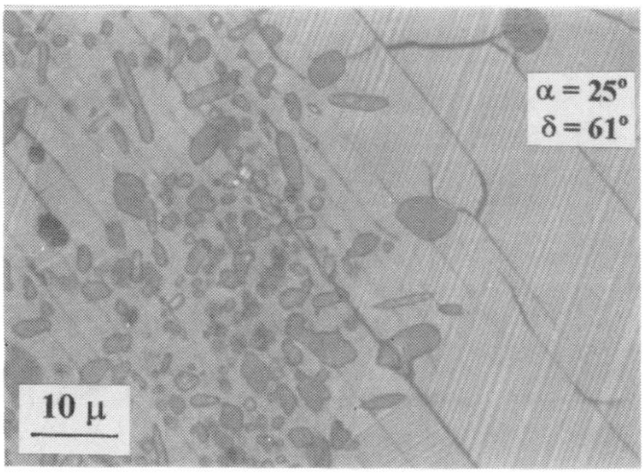

Fig. 8. Twin complexes.

122

Table 1. Masured values of 211 volume fraction V_{211}, mean free distance between 211 particles MFD_{211} and mean twin complex size l_{tc}.

Sample	V_{211}	MFD_{211} [μm]	l_{tc} [μm]
1	0.36	50	25
2	0.32	12	8
3	0.29	7.3	4.5

4. Conclusions

i) It has been shown that the a-b microcracks can be clearly visible after etching the 123-211 samples with the 211 particle size close to 1 μm. The microcracks are easily distinguishable from a-b growth related planar defects. The growth related a-b planar defects account for less than 1% of all a-b planar defects. A-b planar defects in the melt-processed YBaCuO superconductors are mainly microcracks formed preferentially in the orthorhombic 123 phase due to the lower thermal expansion of 211 than that of 123 phase.

ii) The observed linear dependence between microcrack spacing l_c and $d_{211}.V_{123}/V_{211}$ (d_{211} - mean 211 particle size, V_{123} - 123 volume fraction, V_{211} - 211 volume fraction) proves that the microcracking process in melt textured 123-211 can be described under the framework of a model devised for the multiple failure of a unidirectional composite under uniaxial tensile loading.

iii) Other observed types of crystal defect are subgrain boundaries. The subgrain boundaries are not cracked and no secondary phases appear on them. The subgrain thickness was not found to be strongly influenced by the mean free distance between 211 particles (MDF_{211}).

iv) The twin structure is refined by 211 particles. A square root dependence of twin spacing on the mean free distance between 211 particles MFD_{211} was deduced.

v) Areas, with one predominant twin type variant, extended around 211 particles in <110> directions, are formed by the twin motion under stresses arisen due to lower thermal expansion of 123 than 211 phase. The twin boundary motion proves that residual tensile stresses (of around 50-100 MPa) exist in the a-b plane.

vi) The microcracks, subgrain boundaries, twin complex boundaries , twins and inhomogenous tensile stresses around 211 particles can contribute to the flux lines· pinning in YBaCuO melt textured superconductors. Their density can be increased by higher volume fraction and refinement of 211 particles.

Acknowledgment

This work has been supported by the Grant Agency of Slovak Academy of Sciences (project: 2/1323/94) and by the Deutsche Forschung Gemeinschaft (project: YBCO - HRPLM).

References

Alexander, K.B., Goyal, A., Kroeger, D.M. Selvamanickam, V., and Salama, K. 1992 Phys. Rev. B **45** 5622.
Diko, P., Pelerin, N., Odier, P. 1995 Physica C **243** 169.
Goyal, A., Alexander, K.B., Kroeger, D.M., Funkenbusch, P.D., Burns, S.J. 1993 Physica C **210** 197.
Meng, R.L., Sun, Y.Y., Hor, P.H., Chu, C.W. 1991Physica C **179** 149.
Rabe, H., Rivera, J.P., Schmid, H., Chaminade, J.P., Nganga, L. 1990 Mat. Sci. Engn. **B5** 243.
Raynes, A.S., Freiman, S.W., Gayle, F.W., and Kaiser, D.L. 1991. J. Appl. Phys. **70** 5254.
Richardson J.H. 1971 *Optical microscopy for the Materials Science*. (Marcel Dekker, INC. New York) p 599.
Verhoeven, J.D., Gibson, E.D., 1988. Appl. Phys. Lett., **52** 1190.
Zhang, S.Y. 1992 *Handbook of Ceramic and composites, Vol.2*. (Edited by Nicholas P.Cheremisonof, Marcel Dekker, New York-Basel-Hong Kong) p 56.
Zhu, Y., Tafto, J., Suenaga, M., 1991 MRS Bulletin **15** 54.

Inst. Phys. Conf. Ser. No 148
Paper presented at Applied Superconductivity, Edinburgh, 3–6 July 1995
© 1995 IOP Publishing Ltd

Microstructural Analysis of the Growth Front in Melt-Textured YBa$_2$Cu$_3$O$_{7-x}$ (123) - Y$_2$BaCuO$_5$ (211) Composite.

P. Diko

Institute of Experimental Physics, SAS, Watsonova 47, 04353 Kosice, Slovakia

W. Gawalek, T. Habisreuther, P. Görnert

Institut für Physikalische Hochtechnologie, Helmholtzweg 4, 100239 Jena, D-07702.

Abstract The growth front in an isothermally melt-textured sample of YBaCuO (Pt) was studied by high resolution polarized light microscopy. The planar growth front was observed after quenching, and was formed by the combination of surfaces parallel and perpendicular to the "c" axis. The interlayer, consisting of 123 plates, CuO and solidified melt, was revealed between the front of 123 domains and the solidified liquid. This interlayer was formed during the quenching process and its microstructure can not be related to the growth of 123 domains. It is also shown the essential influence of gas bubbles, and uniformity of 211 particles in the melt, on the microstructure of 123 domains.

1. Introduction

Often, the cellular morphology observed at growth front during melt-texturing (MT) of YBaCuO is used as evidence confirming the formation of a-b planar defects during crystallization (Goyal et al., 1993). This observation is, however, not consistent with a recent investigation by Diko et al. (1995). It is necessary to study the growth front in more details.

2. Experimental.

An YBaCuO (Pt) MT sample, 20 mm in diameter, was prepared by isothermal crystallization at 950 °C with a SmBa$_2$Cu$_3$O$_x$ nucleation centre on the top of the sample (Habisreuther 1995). The crystallization process was interrupted by taking the covered crucible, in which the sample was placed, out of the furnace. So, the quenching was rather soft. The quenched sample was cut in to two and one half was transformed into the orthorhombic state by oxygenation. The microstructure was analyzed by optical microscopy after polishing and etching in a solution of bromine in ethanol (2 wt % Br). Observation in normal and polarized light was performed. The orientation of 123 grains was determined from their optical twin patterns (Verhoeven & Gibson 1988).

124

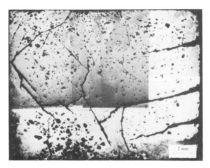

Fig. 1. Low porosity layer near the sample surface and macrocracks.

Fig. 3. Interlayer formed on the 123 domain surface during quenching.

3. Results and discussion

3.1. Macrostructure

The macrocracks parallel to the a-b plane, extending through each domain, show that the sample was composed of more domains. The maximal domain size is up to 15 mm (Fig.1). Irregular macrocracks developed at the domain boundaries. Macrocracks are created in multi-domain bulk 123-211 composites under the influence of stresses arising at cooling due to the anisotropy of the 123 phase thermal expansion (Goyal at al., 1991).

The porosity of the sample was not homogeneous. A low porosity layer is present near the surface. Its thickness increases monotonically - being zero at the point were crystallization started (Fig.1). The observed behavior is associated with the release of gasses by diffusion from the 211-melt mixture during isothermal crystallization. The evolution of gas in the melt can be connected either with the formation of oxygen duo to the $4CuO = 2Cu_2O + O_2$ reaction or with the decomposition of residual carbonates producing CO_2. Larger pores and increased porosity were observed in the part of the sample were remained melt solidified during quenching (Figs.1, 2). Here, bubbles were pushed by the growth front.

The growth front is composed of domain surfaces parallel or perpendicular to the "c" axis, with some predominance of the surfaces perpendicular to the "c" axis (Fig. 2). This behavior can be

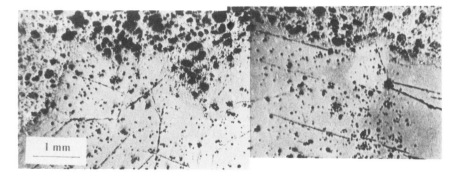

Fig. 2. Crystallization front composed of the surfaces parallel and perpendicular to the "c" axis.

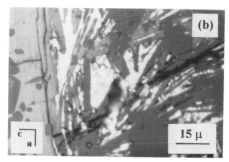

Fig. 4. Plate-like 123 crystals in the interlayer retaining the orientation of mother's 123 domain (a), and changing their orientation due to the heat transport effect (b).

explained in the framework of Müller and Freyhardt's growth model for MT YBaCuO (Müller and Freyhardt, 1995) when the processing rate is close to the growth speed in the "c" direction.

3.2. Microstructure of growth front

The interlayer, 250 μm thick, is observed between the solidified melt and planar growth front. (Fig.3). This layer is composed by 123 plate-like crystals, 211 phase, CuO and remains of solidified melt in pores (Figs.4 (a), (b)). The orientation of 123 platelets is essentially different for the surfaces parallel and perpendicular to the "c" axis of 123 domain. In the case of the surface parallel to the "c" axis, the 123 platelets keep the orientation of mother's 123 domain (Fig. 4 (a)). The 123 platelets, grown on the 123 domain surface perpendicular to the "c" axis, behave completely differently. After epitaxial nucleation, they bend into the direction perpendicular to the 123 domain surface (Fig. 4. (b)). The orientation of the 123 platelets is clearly controlled by the heat transport effect. Such behavior confirms without any doubt that the interlayer was formed during quenching from the temperature of isothermal growth and cannot be related to the mechanism of stable 123 domain growth. This result is significant because such a plate-like 123 crystal morphology, with trapped solidified liquid, is often considered as a cellular growth front of 123 domains and has been used as an argument supporting the idea that the a-b planar defects in the MT 123 are growth related (Goyal et al., 1993, Kim et al. 1994).

The morphology of 211 particles in the layer is influenced by much faster crystallization process

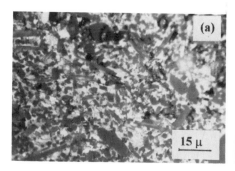

Fig. 5. Fine 211 particles in the solidified melt (b) and large particles near the bubbles (b)

126

that in the case of 123 domains. The 211 particles, whose original shape was blocky in the liquid (Figs. 5. (a), (b)), are changed to an irregular form due to the dissolution of the particles (Fig.4. (a), (b)). Longer 211 particles are divided into several parts along the growth direction of the 123 platelets.

The inhomogeneity in the 211-melt mixture has a direct influence on the homogeneity of the 123 domains. Very large 211 crystals grow around bubbles (Fig. 5. (a)). Such inhomogeneity is than copied in to 123 domain as the region with a low density of 211 particles and porosity (Fig. 3).

4. Conclusions

1. The growth front during isothermal growth of 123 domains is planar and is formed by domain surfaces parallel or perpendicular to the "c" axis. This behavior can be explained in the framework of Müller and Freyhardt's growth model for MT YBaCuO from which it follows that the processing rate is close to the growth speed in the "c" direction.

2. An interlayer between solidified melt and crystallized 123 domains is formed during quenching from the temperature of isothermal growth. The layer is composed of plate-like 123 crystals growing in the direction of the fastest heat transport which is typical for dendritic growth. This layer has no relationship to the stable isothermal growth process and therefore must also be unrelated to a-b planar defect formation in the MT YBaCuO.

3. The large 211 particles, grown around gas bubbles in the melt, are responsible for the inhomogeneity of 211 particles and porosity in the 123 domains.

4. The pore free layer is formed at the isothermal crystallization due to a diffusion degassing process in the melt.

Acknowledgment

This work has been supported by the Grant Agency of Slovak Academy of Sciences (project: 2/1323/94) and by the Deutsche Forschung Gemeinschaft (project: YBCO - HRPLM).

References

Diko, P., Pellerin, N., Odier, P. 1995 Physica C **243** 169.
Goyal, A., Funkenbusch, P.D., Kroeger, D.M., Burns, S.J., 1991 Physica C **182** 203.
Goyal, A., Alexander, K.B., Kroeger, D.M., Funkenbusch, P.D., Burns, S.J. Physica C 1993 **210** 197.
Habisreuther (1995) private communication.
Kim, C.J., Kim, K.B., Won, D.Y., Hong, G.W.1994 Physica C **228** 351.
Müller, D., FReynhardt, H.C. 1985 Physica C **242** 286.
Verhoeven, J.D., Gibson, E.D. 1988 Appl. Phys. Lett. **52** 1190.

Inst. Phys. Conf. Ser. No 148
Paper presented at Applied Superconductivity, Edinburgh, 3–6 July 1995
© 1995 IOP Publishing Ltd

Effect of Carbon Pollution on the Liquid Phase Stimulating Sintering of $YBa_2Cu_3O_{7-x}$

Y.B. Huang and P.G. Régnier

Sec. Rech. Métall. Phys. , CEREM , CE Saclay, 91.191 Gif sur Yvette cedex FRANCE

Abstract It is shown, via DTA/TG experiments + X-ray diffraction + EMPA, that decomposition of $YBa_2Cu_3O_{7-x}$ powders by residual carbon dioxide coming from the treatment atmosphere may start at 200°C and is completed at 500°C. This reaction generates copper oxide and barium carbonate or $YBa_2Cu_3O_{7-x}$ oxycarbonate. At the sintering dwell temperature, decomposition of the carbonaceous species gives additional copper oxide which, as the primary ones, reacts with YBaCuO to give a small amount of liquid phase stimulating sintering of the ceramic. Volunteer additions of CuO increase the amount of the liquid phase, but are detrimental for the critical current

1. Introduction

The critical current density, J_c, of polycristalline $YBa_2Cu_3O_{7-x}$ (123 hereafter) is known to be severely limited by the weak link nature of its grain boundaries [1]. Since pollution weakens further these boundaries, it is a major concern to understand when and how it occurs. Frozen in low-melting secondary phases and residual carbonates have been proved to be two major contamination sources. The presence of the former is usually attributed to uncomplete reaction of the precursors, whereas that of the latter is due to reaction of the powder with CO_2. Since secondary phases and carbonates are simultaneously present, we have suggested that they are related [2]. In this paper, we present a detailed study of the relationship between carbon content and liquid phase; moreover we report on efficient procedures to reduce the carbon content of 123 ceramics.

2. Experimental

Four different 123 commercial powders have been used: a very fine powder from Rhône Poulenc (labelled RP), 2 grades of coarser powders from Hoechst (HNG and HBG) and a very pure and fine powder from HTC (LC1). Their main characteristics are listed in table 1. Carbon content values of the powders were determined by infrared absorption (LEICO IR212),

Table 1 Main characteristics of the studied powders

Powder	Phase content	Carbon Content (wt%)	Average Partical Size (mm)
RP	123, 211, Y_2O_3, CuO, Ba_2CuO^*, $BaCO_3^*$	0.33	1
HNG	123, 211^*, CuO^*	0.33	5
HBG	123, 211^*, CuO^*	0.15	20
LC1	123, 211^*	0.044	2

* minute amount

and secondary phases were identified by XRD and also by EMPA when possible. The powder grains were observed by OPM and SEM which allowed a raw estimation of their average size. Thermal analysis (DTA/TG) was performed on a SETARAM TAG 24 equipment using heating and cooling rates of 10°C/min under three different atmospheres : flowing Ar or O_2, and static air. To facilitate comparisons, the amount of powder used in each DTA/TG run was always around 200 mg.

3. Results and disccussion

Despite the relatively high purity of the four powders used, the DTA heating curves recorded under flowing Ar atmosphere, shown in figure 1, reveal several pre-melting events. According to our analysis, as well as to the literature [3], the observed endothermic peaks may be identified as following:

$$BaCO_3 (\alpha) \rightarrow BaCO_3 (\beta) \text{ or } BaCuO_2 + CuO \rightarrow \text{liquid, at about } 810°C \quad (1)$$

$$YBa_2Cu_3O_{7-x} + BaCuO_2 + CuO \rightarrow \text{liquid, around } 860°C \quad (2)$$

$$YBa_2Cu_3O_{7-x} + CuO \rightarrow Y_2BaCuO_5 + \text{liquid, in the vicinity of } 870°C \quad (3)$$

As shown in figure 1, the higher the carbon content of the powder the bigger the last peak. This was confirmed by the fact that this peak was considerably magnified when the carbon content was raised from 0.05 up to 0.55 wt % by a 30 mn ball milling of the powder in heptane. Since ball milling simultaneously reduced the average particle size from about 50 to 5 μm, it may be argued that the effect is due to a higher reactivity of the finer grains [4]. However, it is not the case, since the intensity of the 870°C peak is higher for HNG than for LC1, despite the fact that the former powder is richer in C and has bigger average grain size than the latter. The contribution of CuO to this pre-melting event under Ar is further confirmed by the results plotted in figure 2, which indicate that the peak grows considerely with the addition of CuO. However, this effect is complicated by overlapping with reaction of CuO with 211 which occurs at slightly higher temperatures.

$$Y_2BaCuO_5 + CuO \rightarrow Y_2Cu_2O_5 + \text{liquid} \quad (4)$$

(1), (2) and (3) pre-melting events were also observed in air and oxygen, but shifted up to higher temperatures. In particular, reaction (3) starts at 940°C in air and 970°C in O_2, as clearly visible in figure 3. Moreover, the magnitude of this peak was higher in static air than in pure Ar or O_2. In order to understand how CO_2 can affect this pre-melting event, a series of treatments in a CO_2 atmosphere at temperatures ranging from 200 to 600°C for 1h was executed on HNG powder. It was found that even at 200°C, some $BaCO_3$, Y_2BaCuO_5, CuO and Y_2O_3 phases

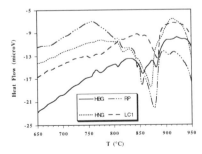

Fig. 1 Differential thermal analysis of four different commercial powders under flowing Ar.

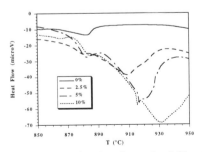

Fig. 2 DTA under Ar of powder LC1 with different amount of CuO addition.

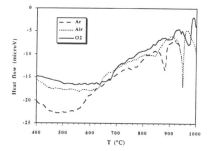

Fig. 3 DTA of powder LC1 under Ar , air and oxigen

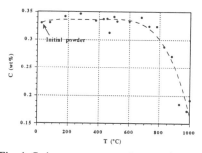

Fig. 4 Carbon content versus heat-treating temperature for 20 minutes duration in air.

appear. Decomposition of the 123 is substantial at 440°C and nearly complete at 500°C. Considering the various compounds identified by X-ray diffraction and/or EMPA, it may be infered that in CO_2, 123 decomposes via one of the 2 reactions:

$$YBa_2Cu_3O_{7-x} + 2CO_2 \rightarrow 2BaCO_3 + 3CuO + 1/2\ Y_2O_3 + \{(2\delta-1)/4\}O_2 \qquad (5)$$

$$2YBa_2Cu_3O_{7-x} + 3CO_2 \rightarrow 3BaCO_3 + 5CuO + Y_2BaCuO_5 + \{(2\delta-1)/2\}O_2 \qquad (6)$$

There are two major sources of CO_2 during our DTA/TG runs. First, in standard air there are 3000ppm CO_2+CO against 5 ppm only in commercial pure oxygen. Second, the decomposition of residual carbonates may also introduce CO_2. As shown in figure 4, when the powder is treated in air in an open-end furnace, reduction of the powder carbon content starts above 700°C. Therefore, the importance of CO_2 effect depends upon the exact values of the temperature and of the CO_2 partial pressure.

Since the pre-melting event appearing near 870, 940 and 960°C under Ar, air and O_2, respectively, strongly affects material preparation, this reaction has been studied by several groups and many different results have been published. Although it is widely believed that CuO gives the most important contribution to this event [5], not enough attention has been paid to the indirect contribution of carbonatation which generates some free CuO according to reactions (5) and (6). Probably because carbon enrichment of the powder by wet ball milling was not taken into account, it has been concluded that initial particle size has a strong effect on pre-melting events [4]. But, it is fair saying the analysis is often very delicate, since other reactions like the eutectic melting of BaO + CuO for example occur around the temperature range of reaction (3). Hence the corresponding peak may be convoluted with others and look shifted. In one of our previous papers dealing with the sinter forging of 123 [2], it was found that the carbon content of the specimens prepared in Ar was lower than that of the starting powder, but that surprisigly it increased with the temperature of sinter forging. According to the outlined ideas above, this phenomenon may be explained in terms of a production of additional CO_2 due, for example, to the use of fast glue to facilitate specimen installation or to excessive heating of greased rubber O rings. Moreover, since the gas circulating inside the sinter forging furnace is not pure, some 123 is certainly decomposed by reaction with CO_2.

A set of high temperature treatments has been performed in flowing oxygen to study the influence of temperature on the C content reduction of the various powders. Some of the results are presented in table 2. They clearly demonstrate the efficiency of the procedure. Two important points have to be emphasized: (1) 920°C appears to be the optimum temperature for treating the specimens. Above and under that point, the C content of the material was slightly higher, and (2) the final value of the C concentration is strongly related to its initial one. The first point has also been observed by Odier et al [6] who have argued that a small amount of

Table 2 Some results on decarbonatation treatments

Powder	Initial C content (wt%)	Treatment	Final C content (wt%)
RP	0.330	690°C x 10 h	0.280
		915 °C x 50 h	0.062
		1010°C x 0.1 h	0.110
HNG	0.330	920°C x 10 h	0.063
		940 °C x 10 h	0.083
		1020°C x 0.1 h	0.100
HBG	0.150	1020 °C x 0.1 h	0.062
LC1	0.044	800 °C x 100 h	0.040

liquid phase appears at this temperatureand the solubility of C in it is higher than in the parent solid. Hence according to them, C enrichment above 920°C would be related to solution of carbon of the surrounding atmosphere into this liquid phase. As for the second effect, it may be due to the fact that the stronger the initial C content the higher the CO_2 partial pressure .

These new observations also allow us to understand some puzzling results we had previously got with RP and LC1 powders [2,7]. First, to optimise the superconductiviy of our dense sinter forged specimens prepared from the highly carbonated RP powder of table 1, it was found necessarry to start reoxygenating them at a temperature as high as 900°C. According to the equilibrim phase diagramme of 123 in O_2 this seems unnecessarily high, but since carbonatation of 123 gives rise to replacement of a few Cu atoms by CO_3 ions, it is clear that when we started the reoxygenation at 700°C, the decarbonatation was just beginning and this have lead to a much lower J_c. On the contrary with the C poor LC1 powder, J_c was higher when oxygenation was started at 700°C rather than at 900°C, because contamination by carbon dioxide of the surrounding atmosphere was more important at a higher temperature.

4. Conclusion

We have shown that decomposition of 123 by CO_2 of the surrounding atmosphere starts at 200°C and is completed at 500°C. this decomposition simultaneously produces some CuO and $BaCO_3$ (and may be also of some oxicarbonate). At higher temperatures, in reduced CO_2 partial pressure atmospheres, carbonaceous species decompose and generate new CuO. Hence, at the sintering dwell there is a non negligible amount of CuO which reacts with 123 and gives rise to a small amount of liquid phase stimulating sintering. Heat treatment at a temperature of about 920°C (just before a pre-meltiing event of significant importance) is very efficient to reduce the C content of the 123 material.

Acknowledgements The EEC is sincerely acknowledged for supporting this work in the frame of the BRITE / EURAM contract n° BRE2 - CT93- 0455

References

[1] Lynn J W 1990 *High Temperature Superconductivity* (New York: Spring-Verlag)
[2] Régnier P, Huang Y and Schmirgeld L 1995 *Applied Superconductivity* **2** (in press)
[3] Aselage T and Keefer K 1988 *J. Mater. Res.* **3** 1279-91
[4] Barus A M M and Taylor J A T 1994 *Physica C* **225** 374-80
[5] Aselage T L 1994 *Physica C* **233** 292-300
[6] Gotor F J et al. 1995 *Physica C* (submitted)
[7] Régnier P, Deschanels X and Schmirgeld L 1993 *J. of Alloys and Cpds* **195** 161-64

Inst. Phys. Conf. Ser. No 148
Paper presented at Applied Superconductivity, Edinburgh, 3–6 July 1995
© 1995· IOP Publishing Ltd

The microstructure of melt-textured YBa$_2$Cu$_3$O$_x$/HfO$_2$

Y. Yan & J.E. Evetts, IRC in Superconductivity, Cambridge University

J. L. Zhang & W.M. Stobbs, Dept. of Materials Science & Metallurgy, Cambridge University

Abstract The characteristic spherulitic microstructure of melt-textured YBCO/HfO$_2$ has been studied by transmission electron microscopy. A growth mechanism model has been proposed to account for the observed structural features. Heterogeneous nucleation of the 123 phase appears to occur with a specific orientation relation from CuO particles in the melt. The growth of 123 platelets from the nucleus region is repeatedly interrupted by the orientation related nucleation of Hf rich crystalline phases in the Hf rich liquid at the solid/liquid interface followed by the formation of microchannels containing Ba-Hf-Cu-O phases. This finally leads to the characteristic divergence of the spherulitic morphology.

1. Introduction

It is well known that the critical current density of high T$_c$ superconductors is strongly dependent on microstructure. Hence microstructure and morphology control are very important for practical applications. Recently it has been found that Zr and Hf additions can change the YBCO morphology substantially [1,2]. In the present study, the microstructure of high-T$_c$ superconducting YBaCuO/HfO$_2$ has been characterized using transmission electron microscopy (TEM). The identification of the various phases and internal defects present in the materials, and the analysis of the local composition and structure of the grain boundaries in the spherules enabled us to clarify the mechanism for spherulitic structure development.

2. Experimental procedure

Specimens were prepared by a partial melting process [1,2]. YBa$_2$Cu$_3$O$_x$ (123) and HfO$_2$ powders were mixed and pressed into pellets and heat treated at about 1020°C for 2 hours, prior to ageing at 450°C for oxygenation. The TEM specimens were prepared by standard techniques, involving mechanical grinding and argon ion-milling at 4 keV at 77K.

3. Results and discussion

Microstructure of the material. As viewed optically using polarized light the microstructure suggests the radial growth of the spherules (hereafter described as a spherulitic structure for simplicity) within which there are platelets of the 123 phase (see region B in Fig. 1a) radiating from a centre (see region C in Fig. 1a). The 123 platelets in region B (hereafter termed the branch region) can be several microns in thickness and up to several hundred microns in length. A closer look at the detailed structure and composition, using SEM and EDX, revealed the presence of the BaHfO$_3$ and Ba-Hf-Cu-O phases between the 123 platelets. Some CuO particles of random shape were found both within and outside the spherulitic structure. However in the nucleus region, C, at the spherule centre, quasi-two dimensional CuO plates abutted the 123 platelets. The length of the CuO plates can be 10-100 times larger than their thickness (Fig. 1a). Fig. 2a shows an YBCO/CuO interface in this region. High resolution electron microscopy (HREM) observations indicated that there is no second phase at the YBCO/CuO interface. Several different orientation relationships between the CuO/YBCO interfaces were observed in our system. The two most common orientation relations found are shown in Figs. 2b and c. TEM observations indicate that the 123 platelets in the region B

132

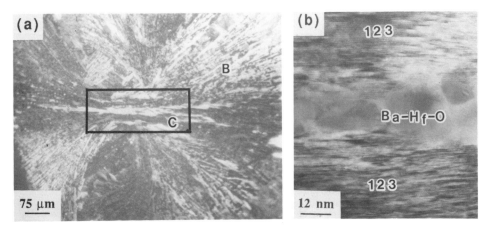

Fig. 1 a) A typical spherulitic structure in the system. b) A decorated 123/123 platelet boundary in region B

of Fig. 1 are generally disordered (see Figs. 1b and 3) and contain numerous inclusions of Y_2BaCuO_5 (the 211 phase) of varying sizes. These platelets are roughly aligned with near to common c axis orientations although there are often small orientation changes between them. Detailed analyses reveal that the 123/123 platelet boundaries are of varied form, most being decorated by impurity phases as in Fig. 1b. EDX examination of numerous such plate boundaries indicated that the impurity phases generally consist of fine grained (10~40 nm) $BaHfO_3$. The Y content in these "microchannels" is very low (normally <1%). Structural analysis revealed that the orientations of the final $BaHfO_3$ grains are totally random. However at the bifurcation point of the 123 platelet boundaries we find a crystalline phase which is probably HfO_2 whose $(0\bar{1}3)$ planar spacing matches well with the (110) planar spacing of the 123 phase (see Fig. 3a). This presumably promotes the orientation relation, as shown in Fig. 3b, which itself initiates splay of the 123 c orientation at the start of a microchannel.

Growth model for formation of the spherulitic structure A growth mechanism is outlined schematically in Fig. 4. Our previous studies have indicated that the melting point of YBCO in our system is progressively lowered by increasing HfO_2 (or ZrO_2) additions [1,2]. When samples consisting of a mixture of $YBa_2Cu_3O_x$ and HfO_2 are heated, a solid state reaction between 123 and HfO_2 occurs at about 800°C resulting in the formation of 211, $BaHfO_3$ and CuO phases [1,2]. If the temperature is increased to about 980°C (the peritectic temperature), the sample partially melts into a mixture of 211+CuO+ an yttrium deficient liquid. On cooling, the 123 phase nucleates from the liquid phase. There is a significant nucleation barrier to solidification in a pure YBCO system when melt processed [3,4]. In our system however, heterogeneous nucleation of the 123 phase occurs in a well oriented manner on the low index crystallographic planes of the pre-existing CuO particle in the liquid (see Fig. 2). It is well known that the preferred growth direction in the 123 phase is along the *ab*- planes. This results in the formation of 123 platelets thin in the *c* direction, which correspondingly forces the CuO phase that acted as the nucleus to grow along the YBCO platelet interface (see the arrow G in Fig. 4a). As a consequence a platelike structure of CuO is gradually formed at the centre of the spherulite. While several such nuclei would tend to link naturally with a common 123 *c* axis, any new CuO particles formed (see III in Fig. 4a) would also tend to grow in a like orientation from the other side of the initial 123 platelet (see II in Fig. 4a). In either way, a sandwiched 123/CuO/123/... platelet structure will be formed at the centre of the spherulite. The orientation relations, such as in Figs. 2b and c, are approximate and misorientations of the

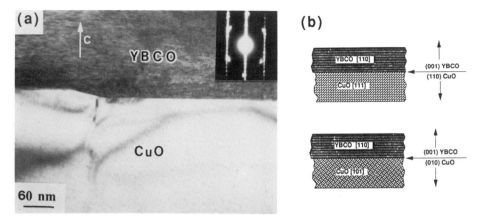

Fig. 2 a) TEM micrograph showing an YBCO/CuO interface. b) The two most common orientation relations for the YBCO/CuO interface: [$\overline{1}$01]CuO‖[110]YBCO+(020)CuO/(001)YBCO and [002]CuO‖[010]YBCO +(010)CuO/(001)YBCO.

common c direction for the various parts of the 123 platelets tend to develop in the nucleus as synergistically aided by local curvature of the growth front.

The initiation of the branch structure (B regions in Fig. 1a) only occurs well after the establishment of a well developed nucleus and can be explained as due to a cellular reaction at the advancing solid/liquid interface. The concentration of both Y and Hf vary strongly, although the trend for each element is quite different. The Y concentration is high in the solid (the 123 phase) and low, of the order 1-2%, in the equilibrium liquid with a steep concentration gradient to a very low Y concentration at the interface. As the solidification proceeds Y diffuses towards the interface down the concentration gradient and is at the same time replenished in the melt by dissolution of the 211 particles. As a consequence, even in the undoped 123 system, there is a tendency to instability in the growth front with regions of trapped or entrained liquid which can become completely depleted of Y and it is this which

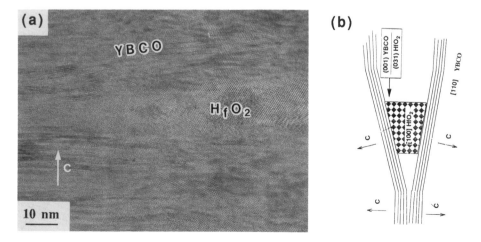

Fig. 3 a) HREM image and b) Schematic reconstruction showing the structure at the bifurcation point of the 123 platelet boundaries.

134

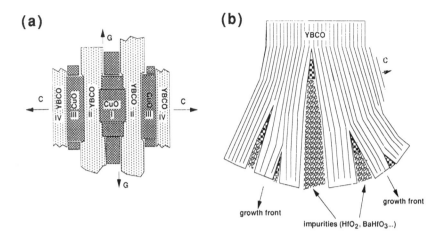

Fig. 4   a) Formation of the sandwiched YBCO/CuO/YBCO... structure at the nucleus of the spherulite by orientation related growth YBCO on CuO. b) Schematic illustration of the formation of the branch structure.

leads to nucleation of the Ba-Cu-O phases [3-5]. Because of rapid growth on *ab-* planes the regions of Ba-Cu-O tend to form paths between the 123 platelets, termed microchannels. However, in our case, with Hf doping there are further important concentration variations. Although Hf dissolves in the melt the solubility in the solid YBCO is very low, less than 1%. It is this low level of Hf within the 123 phase which leads to the characteristically irregular distorted 123 crystal structure, (Fig. 1b), in that Hf is incorporated in the structure much less easily than a Cu excess which is accommodated as extra CuO planes [5]. At the advancing solid/liquid interface most of the Hf is rejected from the solid. As a consequence the Hf concentration progressively increases in the liquid at the interface. This redistribution of Hf has consequences for the development of microstructure and leads to the heterogeneous nucleation of the HfO_2 phase at the bifurcation points for the microchannels. While subsequent Ba-Cu-O and $BaHfO_3$ precipitation in such a channel occurs with a random texture, the HfO_2 tends to be formed in a specific orientation with respect to the bifurcating local 123 platelets growing around it. The choice of the habit of the preferred orientation is presumably weighted by the orientation of the local growth front . This aids the tendency to splay which is the characteristic of the tending to radial growth from the initial more planar nucleus within the spherule. Continual renucleation of new microchannels will occur when the Hf concentration again rises sufficiently at the interface and this further develops the characteristic divergent spherulitic morphology observed in Fig. 1. As expected on this growth model (see Fig. 4b), the spherulitic structures show greater divergence and more frequent nucleation of Ba-Hf-Cu-O microchannels as the doping level of HfO_2 is increased. Equally, the spherulitic structure is more splayed for higher cooling rates because the excess Hf has less opportunity to diffuse away from the solid/liquid interface.

References

[1] Zhang J L and Evetts J E 1994 Journal of Materials Science **29** 778.

[2] Zhang J L and Evetts J E 1993 Chinese Journal of Physics **31** 1139.

[3] Goyal A, Alexander K B, Kroeger D M, Funkenbusch P D and Burns S J 1993 Physica C **210** 197.

[4] Alexander K B, Goyal A, Kroeger D M, Selvamanickam V and Salama K 1992 Phy. Rew. B **45** 5622.

[5] Yan Y, Cardwell D A, Campbell A M and Stobbs W M 1994 submitted to J. Materials Research.

Inst. Phys. Conf. Ser. No 148
Paper presented at Applied Superconductivity, Edinburgh, 3–6 July 1995
© 1995 IOP Publishing Ltd

Texturing of 123 compounds by use of a composite reaction texturing technique

B. Soylu, J. Christiansen*, D.M. Astill, R.P. Baranowski, J. Engel* and J.E. Evetts

IRC in Superconductivity and Department of Materials Science and Metallurgy,
University of Cambridge, Pembroke Street, Cambridge, CB2 3QZ, UK
*Institute of Mineral Industry, Technical University of Denmark, DK-2800 Lyngby, Denmark

Abstract. Composite Reaction Texturing (CRT) is a technique which uses a fine distribution of pre-aligned seeds as nucleating sites for texturing oxide superconductors. It has succesfully been applied to the texturing of Bi-2212 compounds. A further application of CRT is reported in which Y-123 is biaxially textured using seeds of other Rare Earth-123 compounds with higher melting points as nucleating sites. The resultant textured microstructure exhibits mainly low angle grain boundaries (up to 5° misorientation). Results will be presented on the seed alignment techniques, the development of microstructure during reaction of the composite preform and preliminary measurements of electromagnetic properties.

1. Introduction

Melt texturing techniques which depend on a single growth front moving very slowly during the solidification are known to result in small sample sizes. We have shown, for Bi-2212 compounds, that Composite Reaction Texturing (CRT) is a very versatile technique producing long textured artefacts with thick cross-sections [1,2]. CRT does not depend on a single growth front as it employs multiple nucleating sites in a preform obtained by seeding the superconducting phase with a dense aligned distribution of second phase particles of suitable geometry (e.g. fibers, whiskers or platelets) and reacting to nucleate and grow the superconducting phase on these seeds with the desired texture. In what follows the first results of applying the CRT technique to produce biaxially textured Y-123 compounds will be reported. We will describe the seed alignment techniques, the possible ways of producing preforms, report the microstructures produced during reaction and give the results of some preliminary electromagnetic measurements.

2. Experimental

2.1 Choice of seeds and the seed production

After a detailed study of many compounds including $BaZrO_3$, $BaHfO_3$ and MgO it was found that high melting point 123 compounds, such as Nd-123, Sm-123 and Eu-123 are the most suitable candidates for seeding. Fig.1 shows a back scattered electron image (BSEI) micrograph of nucleation and growth of Y-123 around a Sm-123 seed (the bright centre). The chemically diffuse area around this centre indicates that during the melt-cool cycle some Sm-123 from the seed dissolved and later solidified with the surrounding Y-123 forming an alloy of (Sm,Y)-123. The crack-like features, which form parallel to a-b planes, are typical of all melt-grown Y-123 materials. These are continuous across the seed and the growth region confirming epitaxial nucleation and growth resulting in a single crystal.

Nd-123, Sm-123 and Eu-123 seeds were grown using a self fluxing process. By careful control of composition, reaction procedure and appropriate choice of crucible material, it is possible to harvest large number of single crystal platelets (thickness>50μm, size>300μm) whose near-square shapes relate directly to sample crystallography. Due to the importance of being able to produce these seeds in bulk quantities, our experimental objectives have been in the direction of reducing the reaction time for seed growth (which takes 4-6 days for a typical experiment at a cooling rate of 1°C/hour) and the yield. We can now use cooling rates of 6°C/min to obtain a yield of 55 wt% (12.3g per batch).

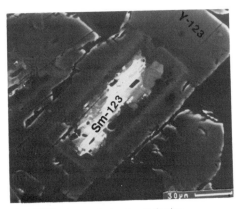

Fig.1 BSEI micrograph showing nucleation and growth of Y-123 around Sm-123 platelets (bright centre).

Fig 2 Cross-sectional microscopy of the 20wt% Nd-123 platelet seeded pellets with a unidirectional seed alignment. The nucleation and growth of the Y-123 on the seeds is unidirectional.

2.2 Seed alignment and preform fabrication

A biaxially aligned distribution of seed platelets in a preform is essential so that growth on the seeds is biaxially textured. We have employed three techniques; mechanical, magnetic or the combination of magnetic and mechanical alignments. Mechanical alignment by pressing the mixtures of seed platelet and Y-123 powder only give a unidirectional alignment as platelets align their a-b planes perpendicular to the direction of pressure. Magnetic alignment of RE-123 particles results from anisotropy in paramagnetic susceptibility that differs in sign for different rare earths. Nd-123 and Sm-123 platelets are known to align with their c-axis parallel to the magnetic field while Eu-123 aligns with the [010] axis in the field. Magnetic alignment alone only gives unidirectional alignment. The combination of magnetic alignment ([010]) during the settling of Eu-123 platelets in a slurry and mechanical alignment ([001]) during subsequent pressing can induce a biaxiallly aligned distribution of seeds in a preform [3]. For magnetic alignment experiments, slurries containing the Eu-123 seeds and superconducting powder were prepared by a method suggested by Wang et. al. [3] and were exposed to a 0.9T field.

In an alternative alignment approach, biaxial alignment of a raft of platelets may be achieved by magnetic alignment of seed platelets distributed on a glass slide. These layers of seeds may then be incorporated into a preform by sandwiching them between rubbery polymer processed sheets consisting of precursor powders (Y-123) mixed with binder, solvent and plasticizer in a rolling mill. Magnetic alignment is at present only partially successful with about 60% of the seeds aligning within a few degrees of the field direction. Other seeds are poorly aligned or fail to align for reasons that are not at present clear.

2.3 Composite reaction and sample characterisation

Two experiments are reported: (1) pellet samples consisting of 20%wt Nd-123 seed platelets (up to 150μm in size with an aspect ratio of 3-5) uniaxially aligned by mechanical pressing and (2) bilayer samples consisting of biaxially aligned Nd-123 seeds (1-3mm in size with an aspect ratio of 3-5) sandwiched between two polymer processed sheets of Y-123 and Y-211 respectively. Microstructures were studied using polarised optical microscopy, and scanning electron microscopy with the EDS and BSEI attachments. Vibrating Sample Magnetometry (VSM) was used to study the quality of textured grains and their grain boundaries (intergranularity of the current).

The development of an optimum composite microstructure depended on precise control of the melting stage, the nucleation stage, phase growth and finally low temperature oxidation. During the melting stage, the seeds tend to dissolve in the Ba and Cu rich liquid if temperatures over $1025°C$ are used for long periods (>10 mins in air). In order to minimize sample porosity it is important to go to the highest possible temperatures, if the seeds are larger higher temperatures can be used without fully melting the seeds. There is no nucleation barrier for growth on the seeds, heterogeneous nucleation can also occur at the sample surface and at porosity although this requires some undercooling, if there is strong undercooling homogenous nucleation can occur throughout the remaining liquid. There is also some uncertain evidence for "residual seeds" remaining in a liquid that has not been raised sufficiently above the melting temperature. Extraneous nucleation is minimised by increasing the melting temperature (thereby reducing porosity) and reducing undercooling during solidification. This in turn requires maximisation of the supply of yttrium to the advancing solid/liquid interface. The maximum growth rate of the Y-123 phase is limited by supply of yttrium to the growth front. One approach to increasing yttrium supply is to include up to 30%vol fine Y-123 particles into the composite.

A series of pellet samples were partially melted at temperatures between 1015 and $1035°C$ in pure O_2 followed by cooling at $0.6-360°C/hr$. Fig.2 shows the unidirectionally textured microstructure of a pellet sample. The extent of epitaxial growth from the seeds was studied by measuring the normalized volume increase of the large 123 grains as a function of maximum temperature (T_{max}), soak time at T_{max} and cooling rate. Epitaxial growth increased strongly with increasing T_{max} as long as the seeds did not dissolve, the maximum normalised volume at $T_{max}=1035°C$ was 360% (>70% of the sample uniaxially textured). The volume of epitaxial material also increased with soaking time and with decreasing cooling rate. All these trends are consistent with minimising secondary heterogenous nucleation. The quality of the texture was however poor because the alignment of low aspect ratio platelets (3:1) by pressing is difficult. It is estimated that only 23% of the grains were aligned within $10°$ of the pressing direction. The critical current density at 77K was $420Acm^{-2}$ at 0.3T falling to $130Acm^{-2}$ at 1T, very much improved on granular sintered samples but poor compared to melt textured material.

The bilayer samples were prepared using a variation on a method to grow large single crystals [4]. A raft of biaxially aligned seed platelets was sandwiched between polymer processed preforms of Y-123 (3mm thick) and Y-211 (2mm thick). It is considered that Y-211 additions to Y-123 are not a sufficient source of yttrium. During melting of Y-123+211, dissolution of smaller Y-211 particles and the preferential nucleation and growth of new Y-211 on the larger Y-211 result in large Y-211 particles. In this respect, more yttrium supply through a higher Gibbs-Thompson effect (very small Y-211 particles with a high curvature) was created using a layer of Y-211 with a fine powder size under the Y-123 layer. After slow binder burn-off to $500°C$, samples were heated rapidly to T_{max} ($1025-1030°C$), soaked for 30 mins and cooled ($1-6°C/hr$) to $900°C$. Large grains grew from each seed with biaxial alignment determined by the initial seed alignment. The grains were 5mm thick with a size (up to 15mm) determined by the seed spacing, there was no secondary heterogenous nucleation. Samples up to $40x40mm^2$ were prepared. Fig. 3 shows a typical very low angle grain boundary ($>1°$), the particles where the grains join are 211 or BaCuO phases and they do not cover more than 50% of the joint area. Typically for this reaction cycle, $100\mu m$ of the seed was dissolved from each surface indicating the delicate balance between seed size and optimum processing cycle.

138

The large samples produced have almost single crystal quality with very low angle grain boundaries (0-5°C) and have correspondingly so far proved exceedingly difficult to oxygenate, particularly since the large planar surface area is normal to the c-axis with diffusion rates some 10^6 times less than for a-b plane diffusion. After 100hrs at 550°C in O_2 the diffusion from the planar surface is negligible while oxygenation from the sides of the sample is estimated as less than 0.1mm, various techniques for increasing oxygenation are being investigated. At present much of the bulk sample has Tc below 50K. The critical current of a rectangular sample (8mmx8mmx5mm thick) with a single central grain boundary has been investigated by VSM. The sample shows no evidence of granularity with a characteristic length equal to the sample size. The critical current is shown in Fig.4 and is in excess of $2.10^5 Acm^{-2}$ at 4T.

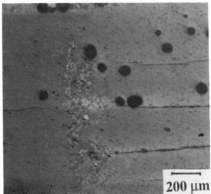

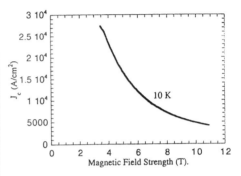

Fig.3 Grain boundary of two biaxially textured crystals joining to form a single crystal. The angle between the crystals is controlled by the seed angles. Residual phases are Y-211 and Ba-Cu-O phases which cover up to 50% of the boundary area.

Fig.4. Jc-B measurement of the biaxially textured bicrystals (Tc~50K) at 10K. A weak-link free current transport across the grain boundaries support the microstructural observation shown in Fig. 3.

4. Conclusions

The application of Composite Reaction Texturing to YBCO composites is reported, demonstrating for the first time the achievement of full biaxial texture without growth from a single growth front. For appropriate processing conditions each seed, acting as a single nucleating site, joins with a rather clean low angle boundary to the neighbouring grain independent of the temperature gradiant. The technique is therefore very suitable for the production of long, textured artefacts with large cross sections. The high quality of the samples produced has brought with it very considerable problems of oxygenation. The large near single crystal samples produced by the CRT process will require special design features and renewed attention to oxygenation procedures.

Acknowledgements

Research supported by Brite EuRam Contract BRE2-0208. The authors are grateful to Drs. A.P. Baker and W.J. Clegg of Cambridge University.

References

[1]   Soylu B., *et al.*, (*1992*), *Appl. Phy. Lett.* **60**, *3183-3185*
[2]   Watson D.R. et. Al., (1995), Super. Sci. Tech., 8, 311-316
[3]   Wang S. *et al. presented in ASC 94, Oct. 16-21, 1994, Boston, USA*
[4]   Chen Y. L. *et. al., 1994, Physica C*, **294**, *232-236*

Inst. Phys. Conf. Ser. No 148
Paper presented at Applied Superconductivity, Edinburgh, 3–6 July 1995
© *1995 IOP Publishing Ltd*

The Growth of Large Grain YBCO by Seeded Peritectic Solidification

Wai Lo and D. A. Cardwell

IRC in Superconductivity, University of Cambridge, Madingley Road, Cambridge, CB3 0HE, U.K.

Abstract. The peritectic growth of large grain YBCO by a seeding technique has been studied in detail. The melting point of the growth interface at the $Sm_{1-x}Y_xBa_2Cu_3O_{7-\delta}$ seed is dependent on the concentration ratio of Sm:Y, which varies between ~1030°C and ~1015°C with distance from the seed. Preferential nucleation and growth of YBCO occurs at the seed as a result of epitaxial crystallisation of the interfacial layer as the temperature is decreased. Solidification of YBCO occurs simultaneously along the [100] and [010] lattice directions, which gives rise to the formation of a rectangular facet plane in the vicinity of the seed. A four-fold planar growth morphology is observed on a macroscopic scale with growth planes parallel to the orthogonal facets. The greatest rate of growth is observed subsequently at the intersection between two growth planes where the local conditions critically determine the extent of the grain growth.

1. Introduction

Large grain superconducting $YBa_2Cu_3O_{7-\delta}$ (YBCO) grown by peritectic solidification from a Sm-Y-Ba-Cu-O (SmYBCO) single crystal seed yields material with a high critical current density at 77K[1-3]. This material is able to trap magnetic fields greater than those associated with iron-based permanent magnets and hence offers significant potential for a variety of engineering applications including magnetic bearings and fault current limiters[4-6]. It is essential, therefore, that a detailed understanding of the seeded melt growth process is developed if the properties of large grain YBCO are to be optimised and put to practical use.

Bulk YBCO samples fabricated using $Sm_{1-x}Y_xBa_2Cu_3O_{7-\delta}$ seeds of irregular geometry are characterised by the presence of orthogonal facet planes in the vicinity of the seed. These planes give rise to further continuous growth (Fig. 1) which results in a rectangular grain morphology. Thus seeded peritectic solidification processing of YBCO can be described by three distinct processes; seeding, facet plane development and continuous grain growth. A full understanding of the complete seeded peritectic solidification process can only be gained by detailed studies of these related sub-processes.

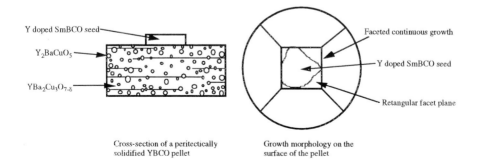

Y doped SmBCO seed

Y_2BaCuO_5

$YBa_2Cu_3O_{7-\delta}$

Faceted continuous growth

Y doped SmBCO seed

Retangular facet plane

Cross-section of a peritectically solidified YBCO pellet

Growth morphology on the surface of the pellet

Fig. 1 Schematic diagram showing the general features and different growth stages of a peritectically solidified YBCO pellet using a SmYBCO seed.

2. Experimental

Sm$_{1.02}$Y$_{0.48}$Ba$_{2.3}$Cu$_{3.3}$O$_{8.5-\delta}$ crystals prepared by a melt process have been used to seed peritectic solidification of YBCO. These seeds, which are typically grown in the form of irregularly shaped platelets of dimensions ~ 1 x 1 x 0.5 mm^3, exhibit a microstructure similar to that of peritectically solidified YBCO but with coarse (~10μm) inclusions of (Sm,Y)$_2$BaCuO$_5$. The peritectic melting temperature of the seeds was determined by differential thermal analysis to be between 1030°C to 1050°C, which is higher than the peritectic temperature of YBa$_2$Cu$_3$O$_{7-\delta}$ (typically between 1000°C and 1020°C in air[7]). As a result the maximum temperature of the YBCO specimens during peritectic solidification was limited to ~1025°C to avoid significant melting of the seeds. The YBCO precursor green bodies were prepared by uniaxial die pressing and sintering[7] of chemically prepared precursor powders containing sub-micron Y$_2$BaCuO$_5$ particles[8]. Peritectic solidification of the YBCO specimens was performed isothermally at 980°C using a purpose-built box furnace with controllable horizontal and vertical temperature gradients[9]. The microstructure of the resulting samples was examined using optical and scanning electron microscopy and the cation distribution at the seed/YBCO interface was determined by energy dispersive X-ray (EDX) and electron probe micro-analyses.

3. Results and Discussion

The SEM micrograph in Fig. 2a shows the microstructure of the seed/YBCO interface. The coarse (Sm,Y)$_2$BaCuO$_5$ inclusions are characteristic of the SmYBCO seed, whereas the finer Y$_2$BaCuO$_5$ inclusions (the lower part of the micrograph) are present as a consequence of the sub-micron Y$_2$BaCuO$_5$ inclusions in the green body. EDX analysis at positions A and B (indicated in the figure) and another displaced from the interface reveals a variation in the relative concentrations of Sm and Y (Fig. 2b), suggesting that significant diffusion of these elements occurs across the interface at high temperatures. This implies that the seed partially melts at temperatures below 1030°C, given that the peritectic decomposition temperature of SmYBCO depends on the Sm:Y cation ratio which was established by thermal analysis.

The solidified seed/YBCO interface significantly influences further solidification of the material as the temperature is lowered. A rectangular solidification symmetry is observed around a seed with c-axis oriented perpendicular to the surface of the YBCO specimen (Figs. 3a and 3b), which reflects the pseudo-tetragonal crystal structure of YBa$_2$Cu$_3$O$_{7-\delta}$ at high temperature. The growth directions apparent in Fig. 3b correspond presumably to the [100] and [010] lattice vectors which are the directions of slowest growth[10]. Furthermore, the solidification rate is enhanced when the tangent to the seed edge is parallel to either [100] or [010]. Conversely, the growth rate along the normal direction to the seed edge is suppressed under these conditions (Fig. 3c).

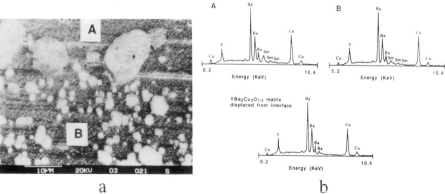

a   b

Fig. 2 (a) A SEM micrograph showing the structure of a seed/YBCO interface after peritectic solidification. (b) EDX results of locations A and B marked in (a) and a third location distant from the interface.

These growth conditions result in the rapid development of tangential growth fronts with sides parallel to [100] and [010] until they impinge on one another to define a rectangle which encompasses the seed. The facet plane for irregular seeds with more than four tangents parallel to the [100] and [010] directions of enhanced growth (i.e. with more than four enhanced growth fronts) still forms but with additional growth fronts impingent on the sides of the rectangle (Fig. 4a). An example of the development of a rectangular facet plane around a seed with six enhanced growth fronts is given in Fig. 4b (the points of impingement are indicated on this figure). The facet plane is established when the space inside the rectangle becomes fully occupied by the other growth fronts such as those illustrated in Fig. 3b.

A planar growth morphology of the YBCO grain is observed on a macroscopic scale following the initial formation of the faceted area in the vicinity of the seed. A total of four

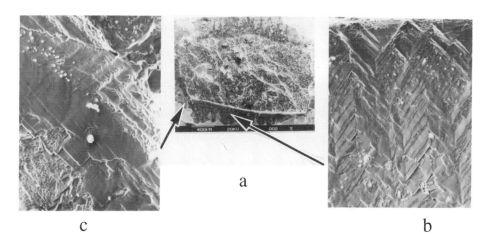

Figure 3 SEM micrographs showing solidification at different positions of YBCO around a SmYBCO seed. Note the rectangular solidification symmetry in (b) and the enhanced growth rate when the tangent to the seed is parallel to one of the growth directions, as shown in (c).

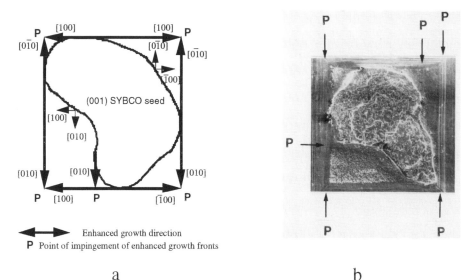

Fig. 4 (a) A schematic diagram illustrating the formation process of a facet plane under the influence of an irregularly shaped seed with more than four enhanced growth fronts. (b) An example of the development of a facet plane around a seed with six enhanced growth fronts.

142

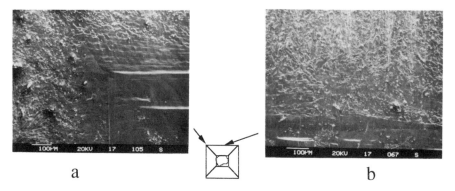

Fig. 5 SEM micrographs showing the morphology at different locations of the faceted continuous growth front on the surface of an YBCO pellet.

growth planes characterise each sample, each parallel to a side of the rectangular facet. Fig. 5 illustrates (a) the central region of growth front and (b) the intersection of two perpendicular growth fronts. It can be seen that the growth rate at the intersection is substantially greater than that along the growth plane. As a result the growth front at the intersection leads that elsewhere by hundreds of microns with growth at the centre of the front lagging furthest behind (Fig. 5a). The deviation from true planar growth was observed to increase with increased solidification temperature.

4. Conclusions

The growth of seeded YBCO by peritectic solidification has been studied in detail. Partial melting of the seed has been observed at temperatures higher than the peritectic melting point of YBCO which results in diffusion of Y and Sm across the seed/YBCO interface. A rectangular solidification symmetry is observed around a c-axis oriented seed due to an enhanced growth rate when the tangent to the seed edge is parallel to either the [100] or [010] lattice directions. A four-fold planar growth morphology is observed on a macroscopic scale following the initial facet formation, with each growth front parallel to a different side of the rectangle. The rate of grain growth is greatest at the intersection of two growth planes and typically leads the growth front at the centre of the growth plane by hundreds of microns. It may be concluded from this study that seed crystals of regular geometry are preferred for the growth of uniform bulk material and that the extent of grain growth is determined critically by the conditions at the intersection between macroscopic planar growth fronts.

References

[1] M. Murakami, Supercond. Sci. Technol. 5(1992)185.
[2] K. Salama, V. Selvamanickam and D. F. Lee, in "Processing and Properties of High Tc Superconductors, 1: Bulk Materials", edited by S. Jin (World Scientific, Singapore, 1993)pp. 155-212.
[3] M. Morita, S. Takebayashi, M. Tanaka, K. Kimura, K. Miyamoto and K. Sawano, in "Advances in Superconductivity III", Proceedings of the Third International Symposium Supplement, Sendai, November(1991)pp. 733-6.
[4] A.M. Campbell, D. A. Cardwell, S. P. Ashworth and T. A. Coombs, "IRC Research Review" (University of Cambridge, UK, 1994)p.174.
[5] H. Fukuyama, K. Seki, T. Takizawa, S. Endou, M. Murakami, H. Takaichi and N. Koshizuka, Adv. Supercond. 5(1993)1313.
[6] R. Takahata, H. Ueyama and A. Kubo, Adv. Supercond. 5(1993)1309.
[7] Wai Lo, D. A. Cardwell, S.-L. Dung and R. G. Barter, J. Mater. Sci. accepted.
[8] Wai Lo, D. A. Cardwell, S.-L. Dung and R. G. Barter, J. Mater. Res. accepted.
[9] D. A. Cardwell, Wai Lo, A. Roberts and H. Thorpe, J. Mater. Sci. Lett. accepted.
[10] Y. Nakamura, K. Furuya, T. Izumi and Y. Shiohara, J. Mater. Res. 9(1994)1350.

Inst. Phys. Conf. Ser. No 148
Paper presented at Applied Superconductivity, Edinburgh, 3–6 July 1995
© *1995 IOP Publishing Ltd*

On the mechanism of the melt texturing growth of YBCO bulk samples with $YBa_2Cu_3O_{7-\delta}$ + n Y_2O_3 $(0 < n < 1.1)$ initial composition

M. Wu[1], P. Schätzle[2], W. Bieger[2], K. Fischer[1], W. Gawalek[1], P. Görnert[1], A. Köhler[1], G. Krabbes[2], D. Litzkendorf[1], G. Stöver[2], T. Strasser[1] and U. Wiesner[2]

[1]Institut für Physikalische Hochtechnologie e. V., Postfach 100 239, D - 07702 Jena, Germany
[2]Institut für Festkörper- und Werkstofforschung e. V. Postfach 270016, D - 01171 Dresden, Germany

Abstract. On the base of thermodynamic and balance considerations, the propagation of melt processing of YBCO starting from 123 / Y_2O_3 mixtures at $p(O_2) = 0.21 \times 10^5$ Pa is discussed in detail. The experimentally observed continuous increase of 211 content in direction of the solidification front is explained by a model for isothermal growth which is characterized by an effective distribution coefficient of 211 between solid and suspension.

1. Introduction

Numerous modified techniques of melt texturing of YBCO have been developed to improve the superconducting properties of bulk materials, especially to increase its current density (see e. g. [1]). Usually, up to 20 wt% Y_2BaCuO_5 (211) is initially admixed to $YBa_2Cu_3O_{7-\delta}$ (123) to prevent a considerable melt loss during the high temperature step. Moreover, small amounts of further metals or oxides (e. g. Ag, CeO_2 or $BaTiO_3$) seem to inforce the flux pinning of the melt textured material. Especially Pt or PtO_2 are accepted as promising admixtures to stabilize the microstructure even in the presence of liquid phases.

It was shown by thermodynamic considerations that a substitution of 211 by Y_2O_3 in the initial powder mixture leads to a completely different behaviour of melt processing [2]. The process window of stable growth of YBCO is significantly enlarged which results from the fact that over a wide range of temperature (940°C < T < T_{max}, T_{max} depends on the amount of Y_2O_3 addition and is limited to 1020°C) the crystallizing 123 phase can coexist with a suspension consisting of 211 and melt. Therefore, the system is more resistant against temperature fluctuations and the material may be processed at lower temperatures. The model derived in ref. [2] which presumes in a first approximation the solidification of pure 123 (without 211 inclusions) proves that the suspension will continuously be enriched by 211 during the propagating solidification process. This enrichment is reflected by an enhanced incorporation of 211 in real samples. Furthermore, the effect of melt creeping which leads to a further increase of 211 is discussed in detail.

The present paper will explicitly consider the influence of the inclusion of 211 particles into the growing YBCO sample on the process stability as well as the properties of the resulting bulk material.

144

2. Experiments

Large cylindrical bulk samples ($\varnothing$ = 20 to 30 mm, h = 10 to 20 mm) of the initial composition $YBa_2Cu_3O_{7-\delta}$ + n Y_2O_3 (0.15 $\leq n \leq$ 0.4) with an addition of 0.5 to 1.0 wt% PtO_2 were prepared by mixing, grinding and processing of prereacted 123, Y_2O_3 and PtO_2. The samples were processed by directional solidification using a modified MTG technique as described in [2]. The local chemical compositions of the processed samples were determined by EDX.

3. Results and discussion

3.1. Influence of 211 insertion

Cylindrical specimen of the initial composition $YBa_2Cu_3O_{7-\delta}$ + 0.25 Y_2O_3 are shown to reveal a continuous change of metal ion concentrations in the direction of the propagating solidification front which can be attributed to a decreasing 123 / (123+211) ratio [2]. The extent of the considered decrease can be restrained by changing the process conditions (Fig. 1). Whereas a cooling rate of 6 K/h leads to a slightly decreased ratio, a rather slow rate of 1 K/h causes a gradual fall.

The observed behaviour can be described by the following consideration which is illustrated in Fig. 2. A homogeneously mixed suspension consisting of melt (L) and 211 coexists with the surface of the growing YBCO sample at a temperature T. The mole fraction x^P_{211} of 211 in the suspension as well as the composition of the corresponding melt are uniquely defined by the temperature T. The surface (region) of the solid consists of 123 and 211. The mole fraction of 211 in the textured solid at a position d is termed with $x^s_{211}(d)$.

Based on this general presumption the model considers two additional suppositions which are closely related to the processing parameters:

(i) The mole fraction of 211 at the top of the stable growing YBCO sample $x^s_{211}(d=0)$ is known and is a border value of the mathematical problem. Obviously, this composition is mainly determined by the undercooling at which the crystallization starts (compare [2]).

(ii) At fixed processing conditions (especially a fixed cooling rate) a constant effective distribution ratio K_{eff} of 211 between the textured solid and the suspension is assumed which is valid at every position d of the sample

$$K_{eff} = x^s_{211}(d) / x^P_{211}(d) \tag{2}$$

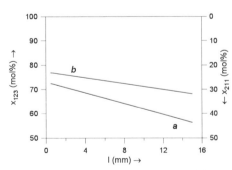

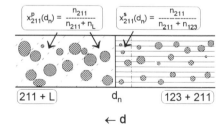

Fig. 1. Change of the 123 mole fraction in differently processed YBCO samples along the propagating crystallization front, (a) cooling rate 1 K/h, (b) cooling rate 6 K/h

Fig. 2. The main supposition of the proposed model: A homogeneously distributed suspension consisting of 211 and melt coexists with the surface of the growing YBCO sample at a temperature T

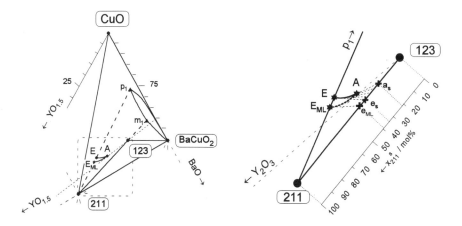

Fig. 3a. Change of the total suspension composition during the texuring process taking into account 211 insertion (A → E) and melt loss (A → E_{ML}), schematically

Fig. 3b. Change of the composition of the crystallized solid during the texturing process taking into account 211 insertion (a_s → e_s) and melt loss (a_s → e_{ML}), schematically

Now, the change of the composition within the textured sample can be calculated step by step.

The generalization of the model is illustrated in Fig. 3. The total composition of the suspension at the starting point of the crystallization shall be A, and the process ends by touching the 211 - p_1 line (e. g. at point E) [2]. The corresponding bulk compositions result from the prolongation of the tangents along the A - E path on the 123 - 211 coexistence line leading to a change from a_s to e_s.

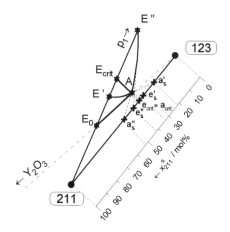

Fig. 4. The effect of the effective distribution coefficient on the resulting YBCO sample
(a) A → E_0, $K_{eff} = 0$, no 211 insertion, (b) A → E_{crit}, $K_{eff} = K_{crit}$, homogeneous 211 insertion,
(c) A → E', $K_{eff} < K_{crit}$, 211 enrichment, (d) A → E'', $K_{eff} > K_{crit}$, 211 depletion

146

The effect of several distribution coefficients is illustrated in Fig. 4:

(a) If K_{eff} is zero then no 211 is incorporated into the growing matrix. The suspension is continuously enriched by 211 and the overall composition moves along the 123 - Y_2O_3 line. The process is finished by touching the 211 - p_1 line at point E_0 [2].

(b) If an inclusion of 211 into the growing sample is taken into account, the composition of the suspension is generally depleted of Y_2O_3. There is a critical distribution coefficient K_{crit} for every initial mixture A which causes a homogeneous sample with a constant 211 concentration, $e_{crit} = a_{crit}$.

(c) If K_{eff} is smaller than K_{crit}, the composition of the suspension is changing towards E'. Consequently, the resulting sample is enriched of 211 in direction of the solidification. This case usually corresponds to experimental observations and will be the subject of a forthcoming paper.

(d) Hypothetically, a 211 depletion in the textured sample can be reached when the effective distribution coefficient becomes larger than K_{crit}. Then, the suspension composition moves towards E''. Such a case could not be found experimentally so far.

3.2. Influence of melt loss

The influence of melt loss on the processing mechanism of $YBa_2Cu_3O_{7-\delta}$ + n Y_2O_3 initial mixtures was shown in [2] to cause an enhanced content of 211 in the quasiperitectic suspension. According to the proposed model it leads to a shift of the curve A - E in Fig. 3 towards the 211 phase (E_{ML}) resulting in an enlarged 211 incorporation into the growing material. Also in this case the process is finished when the suspension composition reaches the 211 - p_1 line. The final suspension (E_{ML}) as well as the final bulk material (e_{ML}) are enriched by 211 as indicated in Fig. 3b.

Conclusions

The influence of 211 insertion as well as melt loss on the melt processing starting from $YBa_2Cu_3O_{7-\delta}$ + n Y_2O_3 ($n < 1.1$) precursors is discussed in detail. A model is derived which is based on two constants, the composition of the textured sample at the top and the effective distribution coefficient of 211, both are being closely related to processing conditions. The development of the model as well as the correlation to experimental results will be the topic of a forthcoming paper.

Acknowledgements

This work was supported by the German Federal Minister of Research and Education under contract No. 13 N 5897A and 13 N 6100. One of the authors (G. K.) thanks the Fonds der Chemischen Industrie.

References

[1] M. Murakami, Melt Processed High Temperature Superconductors (World Scientific, Singapore, 1992)
[2] G. Krabbes, P. Schätzle, W. Bieger, U. Wiesner, G. Stöver, M. Wu, T. Strasser, A. Köhler, D. Litzkendorf, K. Fischer and P. Görnert, Physica C 244 (1995) 145-152

Inst. Phys. Conf. Ser. No 148
Paper presented at Applied Superconductivity, Edinburgh, 3–6 July 1995
© 1995 IOP Publishing Ltd

Optimized melt processing conditions of NdBaCuO materials

W. Bieger, P. Schätzle, G. Krabbes, L. Zelenina[1], U. Wiesner, G. Stöver, P. Verges and J. Klosowski

Institut für Festkörper- und Werkstofforschung e. V., Postfach 270016, D-01171 Dresden, Germany

Abstract. The influence of several admixtures and of oxygen partial pressure on the melt texturing process of NdBaCuO samples is discussed. By using Nd_2BaO_4 as initial mixture instead of Nd-422 or Nd_2O_3, a stabilization of Nd depleted Nd-123 leads to satisfactory bulk samples. Texturing at reduced oxygen partial pressure causes further improvements ($T_c = 91$ K, $j_c(77$ K, 0 T, B $\parallel$ c$) = 21.8$ kA/cm^2).

1. Introduction

Melt textured NdBaCuO is reported to possess a higher irreversibility field for B $\parallel$ c and a higher J_c at B > 0.5 T and T = 77 K with B $\parallel$ c in comparison to YBCO materials [1, 2]. Due to the comparable ion radii of Nd^{3+} and Ba^{2+}, the ions of Nd^{3+} can substitute Ba^{2+} on its lattice sites forming a solid solution $Nd_{1+x}Ba_{2-x}Cu_3O_y$ (Nd-123). The homogeneity region was approximately determined to be $0.04 < x < 0.6$ [3]. The superconducting properties deteriorate with increasing x. Therefore, it is necessary to stabilize Nd-123 with a small neodymium content during the melt texturing process.

The present paper deals with a systematic study of the primary melt reactions involving the Nd-123 phase. Based on these results, promising conditions for melt processing of NdBaCuO materials are derived and testet.

2. Experiments

Besides Nd-123, the quaternary phases $Nd_4Ba_2Cu_2O_{10}$ (Nd-422) and $NdBa_6Cu_3O_{10.5}$ (Nd-163) and the ternary phases $BaCuO_2$, Ba_2CuO_3, Nd_2CuO_4 and Nd_2BaO_4 are described to be involved in the Nd-Ba-Cu-O system [4]. Like Nd-123, Nd-422 is a solid solution the structure of which is completely different from that of Y-211.

Nd-123 with x = 0, 0.25 and 0.5, Nd-422, Nd_2CuO_4 and Nd_2BaO_4 have been synthesized as follows: Stoichiometric amounts of Nd_2O_3, $BaCO_3$ and finely ground CuO were intimately mixed, two times heat treated at 920 to 930°C in dry and CO_2 free air for 24 h and ground, followed by two times heat treatment at 1000°C (Nd-123) or 1050°C (Nd-422, Nd_2BaO_4, Nd_2CuO_4) also in dry and CO_2 free air for 24 h. The purity was proved by XRD

---

[1] Permanent address: Sibirian Branch of Russian Academy of Sciences, Institute of Inorganic Chemistry, 630090 Novosibrisk, Ac. Laventyev Prospect 3, Russia

and DTA. The synthesis of Nd-163 has not succeeded at temperatures between 930 and 950°C and, possibly, this compound is not a stable one.[2] $BaCuO_2$ was prepared in the usual way as described elsewhere.

3. Results and discussion

3.1. Primary melt reactions at $p(O_2) = 0.21 \times 10^5$ Pa

A tentative subsolidus phase diagram of the Nd-Ba-Cu-O system at 890°C and $p(O_2) = 0.21 \times 10^5$ Pa was proposed by Yoo et al. [4], see Fig. 1. Nd-123 with a low Nd content coexists with $BaCuO_2$ and Nd-163, whereas a Nd rich Nd-123 coexists with Nd_2CuO_4. Both, Nd-422 and CuO are forming an equilibrium with Nd-123 within the whole region of homogeneity.

The primary melt reactions at $p(O_2) = 0.21 \times 10^5$ Pa were determined by a combination of DTA/TG with different heating rates and soaking experiments. Pure Nd-123 was usually mixed with 10 mol% secondary phase. The studied mixtures are marked in Fig. 1 and the results are summerized in Table 1.

Like Y-123, Nd-123 decomposes in a peritectic like reaction according to

(m_1) $\quad Nd_{1+x}Ba_{2-x}Cu_3O_y \rightarrow a\,Nd_{1.5+y}BaCu_{0.75+y/2}O_{z2} + b\,L + c\,O_2$

The decomposition temperature decreases with increasing x from 1086°C (x = 0) to about 1067°C (x = 0.5). The residue of soaking experiments consists of only Nd-422.

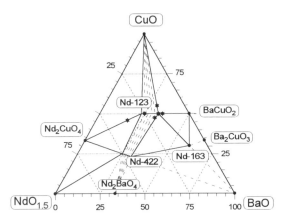

Fig. 1. Subsolidus phase diagram of the Nd-Ba-Cu-O system at 890°C and $p(O_2) = 0.21 \times 10^5$ Pa [4] and compositions investigated to determine the primary melt reactions

Table 1. Primary melt temperatures in the system Nd-Ba-Cu-O at $p(O_2) = 0.21 \times 10^5$ Pa

Investigated composition	Nd-123 (x=0)	Nd-123 (x=0.5)	Nd-123 (x=0) + Nd-422	Nd-123 (x=0) + CuO	Nd-123 (x=0) + $BaCuO_2$	Nd-123 (x=0.5) + Nd_2CuO_4
Primary melt temperature	1086°C	1067°C	1071°C	944°C	995°C	1067°C

---

[2] However, the discussion in this paper starts from the existence of Nd-163.

3.2. Melt texturing experiments

Influence of admixtures. The influence of the admixtures Nd_2O_3, Nd-422 and Nd_2BaO_4, respectively, on the microstructure formation and superconducting properties of melt textured NdBaCuO bulk material is illustrated in Fig. 2. Using individually optimized temperature - time profiles, well textured NdBaCuO samples with homogeneously distributed small Nd-422 insertions were obtained for all admixtures (Fig. 2). However, despite of its very similar microstructure the oxydized samples (48 h in O_2 at 480°C) reveal a completely different superconducting behaviour. T_c raises from 50 K to 65 K and 88 K by using Nd_2O_3, Nd-422 and Nd_2BaO_4 additions, respectively, (Fig. 3). These transition temperatures cannot be improved by a long time annealing in O_2 at low temperatures. This observation becomes clear by comparing the EDX intensities of Nd and Ba in the textured matrix. The material processed with Nd_2O_3 additions consists of Nd rich Nd-123 whereas the application of Nd_2BaO_4 leads to a significant decrease of the Nd/Ba ratio in the Nd-123 matrix. These ratios are reproducible at each position of the considered sample and do not change considerably by slightly modifying the processing conditions.

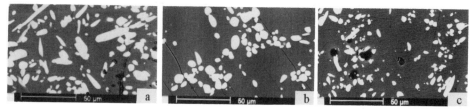

Fig. 2.
Microstructures of NdBaCuO melt textured samples processed with different initial admixtures at $p(O_2) = 0.21 \times 10^5$ Pa, (a) Nd_2O_3, (b) Nd-422, (c) Nd_2BaO_4

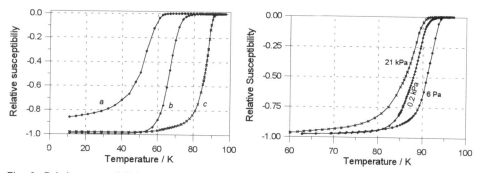

Fig. 3. Relative susceptibilities of NdBaCuO melt textured samples processed with different initial admixtures at $p(O_2) = 0.21 \times 10^5$ Pa, (a) Nd_2O_3, (b) Nd-422, (c) Nd_2BaO_4

Fig. 4. Relative susceptibilities of NdBaCuO melt textured samples in dependence on environmental oxygen partial pressure during melt texturing

The experimental observation can be explained by regarding the phase relations. According to Fig. 1, the solid state prereactions using Nd_2O_3 or Nd_2BaO_4, respectively, lead to different initial materials:

(ssr1) $NdBa_2Cu_3O_{7-\delta} + n\ Nd_2O_3 \rightarrow \alpha\ Nd_{1+x}Ba_{2-x}Cu_3O_{z1} + \beta\ Nd_{1.5+y}BaCu_{0.75+y/2}O_{z2}$

(ssr2) $NdBa_2Cu_3O_{7-\delta} + n\ Nd_2BaO_4 \rightarrow (1 - 0.45\ n)\ NdBa_2Cu_3O_{7-\delta} + 1.6\ n\ Nd_{1.5}BaCu_{0.75}O_z$
$+ 0.05\ n\ NdBa_6Cu_3O_{10.5}$

x and y in *(ssr1)* depend on the amount n of Nd_2O_3 in the initial mixture. Consequently, a Nd rich Nd-123 is stabilized which reveals bad superconducting properties. On the other hand, the conversion *(ssr2)* leads to a mixture of Nd-123, Nd-422 and Nd-163. As shown in Fig. 1 such produced Nd-123 possesses a minimum Nd/Ba ratio thus providing promising properties. Therefore, the following melting process starts from completely different positions. According to the reverse reaction of *(m₁)* a Nd rich Nd-123 ($x > 0$) is crystallized when Nd_2O_3 is initially added:

$$L + Nd_{1.5+y}BaCu_{0.75+y/2}O_{z2} \ (+ \ O_2) \ \rightarrow \ Nd_{1+x}Ba_{2-x}Cu_3O_{z1}$$

On the other hand, by using Nd_2BaO_4 additions, the melt formation and Nd-123 crystallization has to be attributed to a partial melt reaction between Nd-123 and (probably) Nd-163 which principially stabilizes Nd poor Nd-123.

Influence of oxygen partial pressure. It was shown by Murakami et al. [1, 2] that the superconducting properties of melt textured NdBaCuO can be improved by processing at lower oxygen partial pressures. Therefore, samples of the optimum initial composition ($NdBa_2Cu_3O_{7-\delta}$ + n Nd_2BaO_4) have been processed at various oxygen partial pressures using an identical temperature - time profile. The transition temperature T_c increases with decreasing oxygen partial pressure as demonstrated in Fig. 4. However, whereas processing at $p(O_2)$ = 6 Pa leads to samples with T_c > 91 K and j_c(77 K, 0 T, B $\parallel$ c) = 21.8 kA/cm^2, samples treated at $p(O_2)$ = 1 Pa did not become superconducting which may be caused by a change of the primary reformation reaction of Nd-123 between 6 Pa and 1 Pa, or an insufficient oxygen reservoir for recrystallization.

4. Conclusions

Starting from thermodynamic considerations the influences of system inherent admixtures and of the environmental oxygen partial pressure on the microstructure and the superconducting properties of melt textured NdBaCuO bulk samples have been discussed. It is shown that the choice of a suitable admixture is a necessary precondition to get high quality materials. An improvement can be reached by processing at reduced oxygen partial pressure. Further progress may be attained by adding small amounts of metals or oxides as well as by optimizing the temperature - time profiles. Experiments concerning the last two aspects are in progress.

Acknowledgements

This work was supported by the German Federal Minister of Research and Education under contract No. 13 N 5897A. The authors want to thank Mr. F. Hansch for careful experimental work. One of the authors (L. Z.) thanks the Saxon Minister of Science and Art as well as the Deutsche Akademische Austauschdienst (DAAD).

References

[1]   M. Murakami, S.-I. Yoo, T. Higuchi, N. Sakai, J. Weltz, N. Koshizuka and Sh. Tanaka, Jpn. J. Appl. Phys. 33 (1994) L 715
[2]   S.-I. Yoo, M. Murakami, N. Sakai, T. Higuchi and Sh. Tanaka, Jpn. J. Appl. Phys. 33 (1994) L 1000
[3]   W. Wong-Ng, B. Paretzkin and E. R. Fuller, Jr., J. Solid State Chem. 85 (1990) 117
[4]   S. I. Yoo and R. W. Mc Callum, Physica C 210 (1993) 147

Inst. Phys. Conf. Ser. No 148
Paper presented at Applied Superconductivity, Edinburgh, 3–6 July 1995
© 1995 IOP Publishing Ltd

Growth of melt-textured $Y_1Ba_2Cu_3O_{7-\delta}$ [1]

D.Müller‡, M.Ullrich†, K.Heinemann‡ and H.C.Freyhardt‡†

‡ Institut für Metallphysik, Universität Göttingen, Hospitalstraße 3/7, 37073 Göttingen (Germany)

† Zentrum für Funktionswerkstoffe, Windausweg 2, 37073 Göttingen (Germany)

Abstract. A growth model is proposed to explain the growth of highly textured melt-processed Y-123. These considerations take into account the anisotropic growth and distinguish between the growth rate and the processing rate. Due to the anisotropic growth of Y-123, differently oriented grains grow with different speeds parallel to the direction of the temparature gradient, $\vec{\nabla}T$. During the growth process grains with fast growing crystal directions surround grains with slow growing crystal directions and stop the latter. Therefore, a selection between differently oriented grains occur during the growth process. While other models only consider the growth of the fast growing ab-planes parallel to the temperature gradient, this modified growth model could explain the observed growth of Y-123 textured with the c axis parallel to the $\vec{\nabla}T$.

1. Introduction

Melt-textured $Y_1Ba_2Cu_3O_{7-\delta}$ (Y-123) is the most promising superconducting material for the development of levitation applications, e.g. magnetic bearings. For this application it is necessary to grow large, single grained Y-123 monoliths. Therefore, one has to understand the growth mechanism and the formation of the texture in this material. Y-123 is grown via a peritectic reaction of the solid Y_2BaCuO_5 (Y-211) phase and the BaCu-rich liquid. The Y-211 particles stongly control the growth process during melt-texturing. They provide the Y for the growing Y-123 phase [1, 2]. Therefore, the dissolution of the Y-211 particles and the subsequent Y diffusion through the BaCu-rich liquid, is the internal rate limiting factor for the growth of the melt-textured Y-123 phase. Different types of furnaces and processing methods are used for melt-texturing Y-123. Common to all of these procedures is that either the sample is moved through a stationary tempearture gradient, $\vec{\nabla}T$, or $\vec{\nabla}T$ is moved through a stationary sample. Thus, the general measure for all these methods is the relative velocity, R_P, between the $\vec{\nabla}T$ and the sample. This processing rate, R_P, is directly correlated to the velocity,

[1] Supported by the BMBF under grant number 13N5493A and 13N6566

152

with which the sample is cooled below the peritectic temperture. Because Y-123 forms
only below the peritectic temperature, the growth speed of Y-123 phase cannot be faster
than the rate with which the temperature is lowered below the peritectic temperature.
Therefore, the processing rate, R_P, is the external rate limiting factor for the growth of
melt-textured Y-123. In this paper the rate limiting factors as well as the anisotropy
of growth are taken into account to explain the high degree of texture in melt-textured
Y-123 with $\vec{c} \parallel \vec{\nabla}\mathrm{T}$ [3].

2. Growth model

The anisotropic crystal structure results in a crystallographically anisotropic growth
speed of the Y-123 phase. The anisotropy of growth is about $R_G^{a,b}/R_G^c \approx 2$ (eqn.I) in the
volume and about $R_G^{a,b}/R_G^c \approx 12$ (eqn.II) at the Y-211 interface [4], where $R_G^{a,b}$ is the
growth rate in a,b direction and R_G^c in c direction. The enhancement of the growth ratio
near the Y-211 particles can be caused by an increasing gradient of the Y concentration
at the Y-211 interface [5] and by undercooling effects due to the Gibbs-Thompson relation leading to an additional driving force for the Y diffusion [1]. This interface effect is
important for the texturing process as is discussed below.

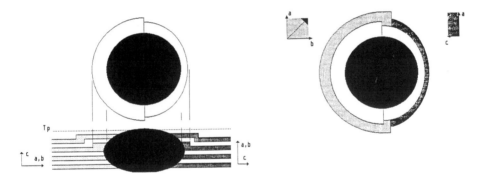

Figure 1) *a) Near the Y-211 particles the growth is enhanced due to additional undercooling. b) This leads to a proceeding growth front naer the particle which accelaretes the
growth of grains with the $\vec{c} \parallel \vec{\nabla} T$ in the plane perpendicular to the tempearture gradient.
The sequence of grey (dark grey: $\vec{a}, \vec{b} \parallel \vec{\nabla} T$, light grey: $\vec{c} \parallel \vec{\nabla} T$) and white regions represents the growth sequence.*

The Y-123 phase can be formed only below T_P. The processing rate, R_P, is the rate
limiting factor for the growth of melt-textured $Y_1Ba_2Cu_3O_{7-\delta}$ in the direction parallel
to $\vec{\nabla}T$. Perpendicular to the $\vec{\nabla}T$ the growth speed is limited by the internal rate limiting
factor, i.e. the Y diffusion and the dissolution of the Y-211 particles. This is expressed by
the maximum achievable growth rate, R_G^{max}. It will be assumed that the fast growing a,b
directions grow with the maximum achievable growth rate $R_G^{max} \approx R_G^{a,b}$ (eqn.III). The
size of R_P and R_G^{max} with respect to R_G^c and $R_G^{a,b}$, respectivly, determine which of the
randomly nucleated grains can grow faster and can surround the slower growing grains.

If the processing rate, R_P, is approximatly equal to R_G^c, equation (I) leads to $R_P \approx R_G^c \approx R_G^{a,b}/2$. Grains randomly nucleated with their fast growing $\vec{a}, \vec{b} \parallel \vec{\nabla}T$ can not grow faster than the velocity with which the sample is cooled below T_P. This speed is given by R_P. Grains with their a,b direction parallel to the $\vec{\nabla}T$ grow as fast as grains with their c direction parallel to the $\vec{\nabla}T$ in this direction, because in this direction growth is limited by $R_P \approx R_G^c$. Thus, adjacent grains, which are oriented perpendicular to each other, form a planar growth front, which moves with $R_P \approx R_G^c$ in a direction parallel to the $\vec{\nabla}T$. These grains grow, however, with different rates perpendicular to the $\vec{\nabla}T$. Grains with $\vec{c} \parallel \vec{\nabla}T$ grow with their fast growing a,b direction perpendicular to the $\vec{\nabla}T$. In this direction these grains can easily surround Y-211 particles. Considering the growth in three dimensions one has to take into account the enhanced growth near the Y-211 particles. Thus, the growth front proceeds at the Y-211 particles (fig.1a). In plane perpendicular to $\vec{\nabla}T$ the grains with $\vec{c} \parallel \vec{\nabla}T$ grow faster as differently oriented grains because the grains with $\vec{c} \parallel \vec{\nabla}T$ contain the fast growing [110] direction in the plane perpendicular to the $\vec{\nabla}T$ (fig.1b).

 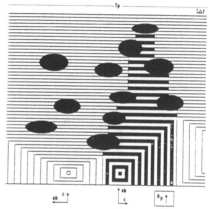

Figure 2) *a) Grains with $\vec{c} \perp \vec{\nabla}T$ will be surrounded by grains with $\vec{c} \parallel \vec{\nabla}T$ for processing rates, R_P, equal to the growth rate in the c direction, R_G^c. The vector R_P gives the direction of $\vec{\nabla}T$ and the direction in which the temperature is lowerd below T_P through the crystal. The length of the arrows $\vec{c}$ and $\vec{a}, \vec{b}$ indicate the growth speed of the c and a,b crystal direction. b) Optical micrograph of a longitudinal cross section of a VGF processed melt-textured $Y_1Ba_2Cu_3O_{7-\delta}$ monolith Polarized light was used to distinguish between different oriented melt-textured grains. The lower part of the sample is not heated above the peritectic temperature and stays polycrystalline. It can be easily seen that several melt-textured domains nucleate heterogenously at the polycrystalline sinter material.*

Grains with $\vec{c} \perp \vec{\nabla}T$ surround Y-211 particles slower than the grains with $\vec{a}, \vec{b} \perp \vec{\nabla}T$. In the areas below the Y-211 particles perpendicular oriented grains grow in a ratio of $R_G^c : R_G^{a,b} = 1 : 12$ due to the interface effect mentioned above. Y-123 grains with the c axis parallel to the $\vec{\nabla}T$ make perpendicular oriented grains to stop growing below every Y-211 particle, after surrounding only one twelfth of the Y-211 size. Thus, at each Y-211 particle Y-123 grains with their c axis parallel to the $\vec{\nabla}T$ surround not

154

only most part of the Y-211 particle but also parts of adjacent grains with perpendicular orientation. Therefore, grains with $\vec{c} \parallel \vec{\nabla} T$ dominate in the crystal, after having surrounded a few Y-211 particles. This leads to a texture with the c axis parallel to the $\vec{\nabla} T$ (fig.2a).

3. Conclusions

Melt-textured $Y_1 Ba_2 Cu_3 O_{7-\delta}$, which is processed by the VGF (vertical gradient freeze) method, is highly textured [3]. In this process the lower part of the sintered $Y_1 Ba_2 Cu_3 O_{7-\delta}$ material is not heated above the peritectic temperature, while the upper part of the sample is heated above T_P and undergoes the transformation from Y-123 phase to Ba-Cu-rich liquid and Y-211 phase. Therefore, the lower part of the polycrystalline sinter material does not melt and stays polycrystalline (fig.2b). During the cooling process heterogeneous nucleation of the Y-123 phase takes place at the interface of the polycrystalline sinter material and the liquid phase (fig.2b). Thus, randomly oriented grains form by the peritectic reaction. Differently oriented grains grow with different growth rates because of the anisotropic growth of the various crystal planes of the Y-123 phase. Thus, a few grains grow faster than other grains and dominate the texture of the sample. Samples prepared with a processing rate of $R_P \approx 0.4$mm/h exhibit a texture where the c axis lies parallel to the temperature gradient [3]. According to the proposed growth model the relation $R_P \approx R_G^c \approx R_G^{a,b}/2$ should be valid. By using equation (III) the maximum achievable growth rate of the Y-123 phase in a,b direction should be twice of the processing rate ($R_P \approx R_G^{max}/2$). Izumi et al. [1] were able to estimate the maximum achievable growth rate, R_G^{max}, of the Y-123 phase taking into account the size distribution of the Y-211 particles. They calculated R_G^{max} of about 1mm/h for Y-211 particles of average radius of 5μm and a narrow size distribution. The average radius of the Y-211 particles in the VGF melt-textured $Y_1 Ba_2 Cu_3 O_{7-\delta}$ is close to this value [6]. Therefore, it is reasonable to assume that the maximum achievable growth rate of these VGF grown $Y_1 Ba_2 Cu_3 O_{7-\delta}$ samples is $R_G^{max} \approx 1$mm/h. This is about twice of the processing rate $R_P \approx 0.4$mm/h. Therefore, the observed texture with the c axis parallel to the $\vec{\nabla} T$ is in good agreement with the prediction of the proposed growth model.

References

[1] T. Izumi, Y. Nakamura and Y. Shiohara, J. Mater. Res. **7** (1992) 1621

[2] M.J. Cima, M.C. Flemmings, A.M. Anacleto, M. Figueredo, M. Nakade, H. Ishii, H.D. Brody and J.S. Haggerty, J. Appl. Phys. **72** (1992) 179

[3] M. Ullrich and H.C. Freyhardt, *Superconducting Materials*, eds.: J. Etourneau, J.B. Torrence and H. Yamauchi, Proc. ICMAS (1993) 203

[4] D. Müller and H.C. Freyhardt, Physica C **242** (1995) 283

[5] G.J. Schmitz, J. Laakmann, Ch. Wolters, S. Rex, W. Gawalek, T. Habisreuther, G. Bruchlos, and P. Görnert, J. Mater. Res. **8,** (1993) 2774

[6] D. Müller, M. Ullrich, K. Heinemann, and H.C. Freyhardt, *Applied Superconductivity 1993*, ed. H.C. Freyhardt, Proc. EUCAS (1993) 329

Inst. Phys. Conf. Ser. No 148
Paper presented at Applied Superconductivity, Edinburgh, 3–6 July 1995
© 1995 IOP Publishing Ltd

155

Enhanced grain growth in melt textured YBCO materials with seeds

P. Schätzle, W. Bieger, G. Krabbes, J. Klosowski and G. Fuchs

Institut für Festkörper- und Werkstofforschung e. V., Postfach 270016, D-01171 Dresden, Germany

Abstract. Enhanced oriented grain growth in YBCO HTSC bulk material is established by seeding with a Sm-123 seed. The growth will not be supported by additionally applied temperature gradients. Magnetic fields up to 130 mT and $j_c = 8.5 \times 10^3$ A/cm^2 were trapped in Y -123 samples ($17 \times 17 \times 3$ mm^3) melt textured with Sm-123 seeds.

1. Introduction

Melt texturing is the most promising method utilised to prepare bulk superconducting YBCO material with superior transport and magnetic properties. Numerous modified methods have been developed in order to optimize the processing conditions, such as maximum heating temperature, temperature gradient, cooling rate and ambient atmosphere [1].

Because the grain boundaries are responsible for the weak link behaviour the main work was directed to enlarge the grain size. Usually, pressed YBCO samples were solidified in a high temperature gradient to ensure the growth of only few grains. Zone melt experiments of sintered rods show that only few grains were growing by a natural seed selection [2].

Due to the anisotropy of the transport and magnetic properties of the Y-123 superconducting material it is desirable to solidify the 123 samples with respect to the a, b orientation. As melt texturing of YBCO in a thermal gradient produces large grains in the samples a preferred orientation can be introduced by using oriented seeds. Different seeds have been proposed, e. g. Al$_2$O$_3$, MgO, SrTiO$_3$, CaAlO$_3$. The most promising seed materials are SmBa$_2$Cu$_3$O$_{7-\delta}$ (Sm-123) and NdBa$_2$Cu$_3$O$_{7-\delta}$ (Nd-123) because their lattice parameters are close to that of Y-123 and their peritectic decomposition temperatures are higher (1060°C and 1080°C, respectively). Usually, Sm-123 seeds were obtained by melt texturing [3] and Nd-123 seeds were grown as single crystals [4].

The present paper deals with several modified methods of melt texturing of YBCO using Sm-123 seeds in different thermal gradients (2°C/cm - 15°C/cm).

2. Experiments

The YBCO precursor samples ($18 \times 18 \times 5$ mm^3) were prepared by thoroughly mixing and compacting (2000 bar) powders consisting of YBa$_2$Cu$_3$O$_{7-\delta}$ and various amounts of Y$_2$O$_3$ together with 0.1 - 0.2 wt% Pt. The Sm-123 seeds which were achieved by a melt texturing process similar to YBCO were cleaved and sliced with respect to the orientation.

For investigations of the seeding behaviour in different thermal gradients the seeds were pressed on the top of the samples during compacting. The samples were placed in a two zone tube furnace with temperature gradients from ≈ 2°C/cm to 25°C/cm. The solidification proceeded by cooling the temperature of the two zones (gradient freezing).

Alternatively, a chamber furnace was applied to generate small gradients of up to 2°C/cm which permits the seeds to be placed on the top surface of the heated sample surface during the melt texturing process at a temperature of 1050°C. After the seeding the samples were cooled down with 1°C/h. Finally the samples were oxidised by annealing at 380°C for 72 h in flowing oxygen.

The microstructure and the chemical composition were examined by SEM and energy dispersive analysis of X-ray (EDAX). The trapped magnetic field was measured by scanning with a miniature hall probe.

3. Results and discussion

3.1 Seed crystals

Seed crystals of Sm-123 were grown by a melt texturing process. Compacted bars (5 × 5 × 20 mm³) were solidified in a tube furnace with a temperature gradient of 15°C/cm by positioning the bar in the gradient and ramp down the temperature. To prevent a considerable melt loss and to improve the microstructure, 0.25 to 0.4 mol% Sm_2O_3 and small amounts of Pt (0.1 to 0.2 wt%) were added. Well oriented Sm-123 grains of 3 to 4 mm³ with small Sm-211 inclusions and some Pt insertions were obtained (Fig. 1).

3.2 Influence of the thermal gradient

YBCO samples with pressed Sm-123 seeds which were located at the centre or at the corner of the sample top surface were processed in a temperature gradient of 25°C/cm. The corner with the seed was at the lowest temperature.

If the seed was located at a corner of the sample several grains were grown starting at this corner. Both, temperature gradient and seed, are responsible for the growth of the grains. The seed does not determine a preferential growth direction. A seed located in the centre of the sample top surface is completely dissolved and Sm was detected by EDAX at the corner which was at lower temperature during the solidification.

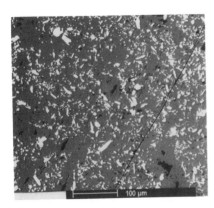

Fig. 1. Microstructure of a melt textured Sm-123 bar with Sm-211 inclusions

Fig. 2. Melt textured Y-123 sample prepared with a Sm-123 seed

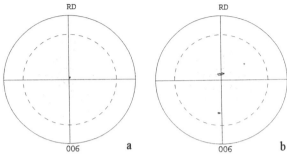

Fig. 3. Y-211 inclusions with different Sm content (near the seed region)

Fig. 4. Polfigures of seed (a) and sample (b)

Further experiments with samples where the seed was placed in the middle of the surface were performed with lower temperature gradients (2°C/cm). Two large single grains around the seed were grown (Fig. 2). The dimension was $10 \times 10 \ mm^2$ but the grains did not grow through the whole sample. EDAX measurements of the region around the seed show a variable content of Sm in the Y-123 phase. In contrast, the 211 inclusions are observed to consist of two different regions (Fig. 3), a dark Y-211 rich region in the middle surrounded by a bright Sm enriched region. Obviously, there is an exchange of Sm and Y in the Y-211 phase leading to a mixed compositions.

3.3 Seeding methods

Two different methods were used to fix the seed on the sample surface. In the first case the Sm-123 seed was pressed on the surface of the YBCO sample during compacting before melt texturing. The second method was to place the seed on the surface during the melt texturing process.

For both methods the temperature cycle in the melt texturing process is different. If the seed was initially pressed on the surface (before melt texturing) the maximum temperature is limited by the peritectic temperature of the seed material. In the applied process this temperature was 1050°C. This restriction is not necessary if the seed is placed onto the surface of the sample during the melt texturing process but the placing must be performed after cooling below the peritectic temperature of the seed material.

The influence of the seeding method was studied in a melt texturing process with low temperature gradients (2°C/cm) and cooling rates of 1°C/h. There was no significant difference in the seeding behaviour. The geometrical form of the seeds was not changed during the process.

The orientation of the seed is directly transfered into the grain which grows starting from the seed. Polfigures of such a sample (Fig. 4) show a sharp (001) orientation of the seed and also a (001) orientation within the YBCO sample (halfwidth of the (006) peak: 1°), where the a, b plane is 15° tilted corresponding to the sample corner and the (001) plane is 5° inclined to the surface.

If the seeds are very small ($2 \times 2 \times 0.5 \ mm^3$) EDAX measurements of the chemical composition show that the Sm-123 seed was completely dissolved in the YBCO sample and replaced by yttrium in the seed but the geometrical form of which still remains.

158

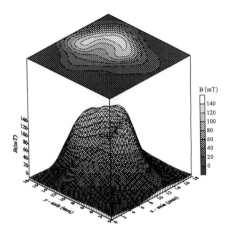

Fig. 5. Mapping of the trapped magnetic field at 77 K of a YBCO sample grown with a seed

4. Magnetic field trapping

The most important parameter for bulk HTSC applications is the effective critical current parameter of the samples. For large dimensions of the YBCO samples a mapping of the trapped magnetic field [5] by scanning the surface with a hall probe is therefore suited. B_t is generated by field cooling the YBCO sample in an external field (0.4 T, 77K). Figure 5 shows the field mapping of a sample ($17 \times 17 \times 3$ mm^3) grown with a Sm-123 seed in a low temperature gradient. There is only a single peak detected with B_t of 130 mT in the surrounding of the seed. According to Beans model j_c was estimated to 8.5×10^3 A/cm^2. For samples with a size of $22 \times 22 \times 8$ mm^3 processed on the same way B_t was more than 300 mT.

5. Conclusions

Oriented seeds of Sm-123 are suitable to prefer oriented grain growth of melt textured YBCO samples. The seeds have been prepared by a melt texturing process which produces large grains with small Sm-211 inclusions. Successful seeding can be performed by pressing or placing the seeds on the surface of the YBCO samples. During the melt texturing process low temperature gradients and cooling rates are preferred.

Acknowledgements

The authors wish to thank Mrs. G. Stöver for technical assistance. This work was supported by the German Federal Minister of Research and Technology the contracts No. 13 N 5897A. One of the authors (G. K.) wish to thank the Fonds der Chemischen Industrie.

References

[1]   M. Murakami, „Melt Processed High Temperature Superconductors", World Scientific Publishing Co. Pte. Ltd. 1992
[2]   G. Krabbes, P. Schätzle, U. Wiesner, W. Bieger and G. Stöver, Electroceramics IV, Aachen, Germany, 1994, p. 957
[3]   C. Varanasi, P. J. Mc Ginn, V. Pavate and E. P. Kvam, to be published in J. Mater. Res.
[4]   R. L. Meng, L. Gao, P. Gautier-Picard, D. Rameirez, Y. Y. Sun and C. W. Chu, Physica C 232 (1994) 337
[5]   I. G. Chen, J. Lin, R. Weinstein and K. Lau, J. Appl. Phys. 72 (1992) 1013

Inst. Phys. Conf. Ser. No 148
Paper presented at Applied Superconductivity, Edinburgh, 3–6 July 1995

LARGE GRAINS AND HIGH PERFORMANCE BULK YBa2Cu3Ox SUPERCONDUCTORS FOR LEVITATION FORCES APPLICATIONS

X. Chaud, E. Beaugnon, R. Tournier

CNRS EPM-MATFORMAG, BP166, F 38042 Grenoble Cedex 09, France;

P. Hiebel, E. Hotier

CNRS CRTBT/LEG, BP166, F 38042 Grenoble Cedex 09, France.

Abstract. A systematic use of magnetic field and gradient has been made to increase orientation and grain size of superconducting pellets. The availability of a continuous monitoring of the process through susceptibility measurement has allowed its optimization in an original way. Large levitation forces have been reached in a reproducible manner and can be compared with that of seeded samples.

1. Introduction

Although the processing of YBaCuO superconducting wires is still difficult, processing of bulk YBaCuO pellets is now widely developed. Because of the significant levitation and pinning forces that can be obtained, the most direct application is magnetic bearings. The use of such superconducting pellets introduces passivity and stability as compared to conventional magnetic bearings [1]. The forces are proportional to the product of the field gradient by the superconductor magnetization. This latter is related to the size of the current loops and to the critical current of the superconducting pellets. Since highest Jc are obtained when the current flows in the a-b planes, the superconducting pellets must exhibit large grains oriented with their c-axis along the direction of the magnetic field in order to achieve high forces.

Solidification in a strong static magnetic field has been shown to induce an orientation of the a-b planes perpendicular to the applied field [2]. However, this method has to be combined with a suitable process such as Melt Textured Growth to enhance the grain size. In addition to the orientation effect, the magnetic field can be used to monitor the sample growth since the magnetic susceptibility of the sample can be derived continuously during the process from the force exerted on the sample placed in a magnetic field gradient [3]. This paper reports the optimization (with regard to levitation forces) of superconducting pellets processing by a systematic use of a magnetic field and gradient.

Work supported by Framatome and the Brite Euram II contract N° BRE2-CT92-0274

160

2. Experimental

2.1. Apparatus

Apparatuses have been especially designed to combine high temperature and high magnetic field. They consist in a vertical tubular furnace fitting inside the 120 mm diameter room temperature bore of an Oxford superconducting magnet. Since the free space is small, a technical effort has been made to reduce the size of the insulation and heating elements, leading to a 70 mm inner diameter furnace. Temperatures up to 1100 °C can be reached in 7 Tesla. Samples with a diameter up to 50 mm and a weight up to 200 g can be processed.

An electronic balance, placed far below the magnetic field center, is associated to the furnace so that the complete apparatus has also the function of a high temperature magnetometer. A schematic view of the setup is presented in figure 1. The magnetic susceptibility is traced down from the force ($F_z = m.\chi.H.dH/dz$) exerted on the sample in a magnetic field gradient. The accuracy of the balance, combined with the mass of the samples (up to 200 g) and the high value of the product H.dH/dz (2.10^8 Oe^2/cm), leads to a high value of the sensitivity of the apparatus, below 10^{-9} emu/g.

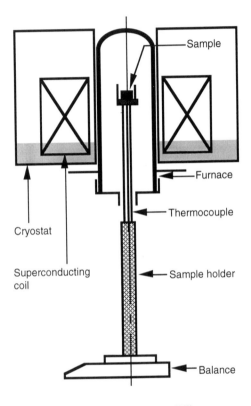

Fig. 1 : Diagram of the in-situ susceptibility apparatus.

2.2. Samples

The starting composition is 123 Rhône-Poulenc or Hoechst precursor with addition of 12.5 wt% Ag_2O, 12.5 at% Y_2O_3 and 0.3 wt% PtO_2 in excess. The mixing is performed manually by a thorough grinding in an agate mortar and pestle. No differences are seen in the performances of the materials prepared with either commercially available precursors.

Small pellets of 25 g are uniaxially pressed at 100 MPa in a Ø 20 mm die. They are processed as pressed with no prior sintering into ZrO_2 crucibles. A diameter of 17 mm and a height of 17 mm is obtained after melt processing.

Large pellets of 125 g are pressed in the same condition in a Ø 50 mm splitted die. Due to their size, pellets have to be sintered to avoid cracking while heating. These pellets are then processed on a Y_2O_3 buffer layer into Al_2O_3 crucibles. They have a diameter of 40 mm and a height of 17 mm after melt-processing.

3. Results

The magnetic susceptibility and temperature of the sample are continuously recorded by a computer that controls the temperature of the furnace. The susceptibility curves obtained that way [4] provide information about the liquid state (oxygen loss and uptake related to the oxidation state of the copper) and the solidification. Figure 2 shows the susceptibility curve of a pellet during a classical MTG, i.e. a rapid heating (240°/h) up to 1100°C, a 30 mn superheating dwell, then a slow cooling ramp (1°/h) from 1020°C to 890°C. The solidification temperature range for a pellet with a given composition, shape and mass is known precisely. This range is narrow and depends on materials composition, cooling conditions as well as on thermal history (oxygen loss and superheating). Although the cooling rate was very low (1°/h), the solidification is associated with an important undercooling.

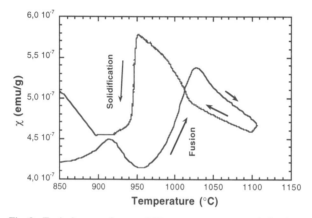

Fig. 2 : Typical curve of susceptibility versus temperature during heat treatment.

Modifications of the heat treatment were investigated to control and enhance the grain growth. By allowing time for YBaCuO grains to grow in the right solidification temperature range, a single grain was obtained on the whole surface of a Ø 17 mm pellet. The levitation force reaches 15 N using a NdFeB magnet (Ø 22 mm, h 20 mm). This represents a threefold increase of the force measured on a similar pellet textured using a conventional MTG process. A sample processed in the very same way but without any field applied shows the same macrostructure (grain size). The force is still important (11 N) but lower due to an uncontrolled grain orientation.

The same kind of optimization has been successfully applied to larger pellets (Ø40 mm). Results show that only two or three grains can be obtained without seeding or thermal gradient. An example is given on figure 3. The macrography is taken under polarized light to distinguish grains. The surface observation is correlated to flux mapping that evidences the actual current loops in the bulk sample. Vertical cuts on some samples also confirm that the grain growth is not occurring only at the surface but also in the bulk.

The levitation forces of such pellets are reproducible and range between 35 and 40 N using the Ø 22 mm magnet mentioned above (fig. 4). The measured force depends strongly on the magnet used and forces up to 73 N have been measured using a Ø 30 mm h 30 mm magnet. In both case, the magnetic pressure, as reported to the surface of the magnet, reaches 10 N/cm^2. Magnetization measurements at 77 K were performed on the whole Ø40 mm samples and confirmed the overall good quality of the samples. The maximum diamagnetism during the first magnetization is around 200 emu/cm^3 including bad quality parts at the bottom

162

of the pellet due to reaction with the crucible or the buffer layer.

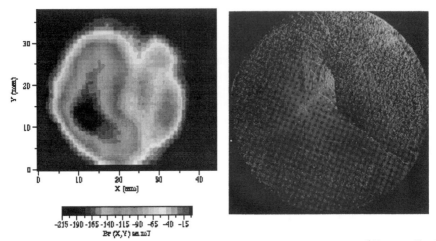

Fig. 3 : Macrography (right) showing a very large grain and two smaller ones on a Ø40 mm pellet. The corresponding flux map (left) confirms that the currents are flowing on the grains scale.

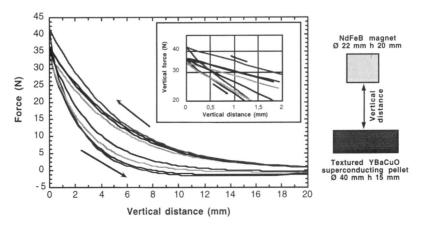

Fig. 4: Large reproducible levitation forces of large pellets.

These performances are obtained on samples which are not single grain and with a composition that should be optimized. The levitation forces reach 70%-80% of the ones measured on single grain seeded samples (35-40 N compared to 50 N with the same Ø 22 mm magnet) as a result of a good grain orientation and controlled solidification.

References

[1] H.J. Bornemann et al., Proceedings of the Second International Symposium on Magnetic Suspension Technology, Seattle, Aug. 1993.
[2] M.R. Lees et al., Superconducting. Sci. Technol. 5, 362 (1992)
[3] P de Rango et al., Proceeding of Applied Superconductivity, Göttingen, Oct. 1993, 305.
[4] X. Chaud et al., M2SHTS IV, Physica C 235-240, 433 (1994)

Inst. Phys. Conf. Ser. No 148
Paper presented at Applied Superconductivity, Edinburgh, 3–6 July 1995
© 1995 IOP Publishing Ltd

Unidirectional solidification of $Y_1Ba_2Cu_3O_{7-\delta}$, effect of a magnetic field applied during solidification

L. Durand*, D. Dierickx¨, D. Chateigner° and P. Régnier*

* CEA/C.E. Saclay/CEREM/SRMP, 91 191 Gif sur Yvette, France.
¨ KUL-MTM de croylaan 2 B-3001 Heverlee Belgium
° Laboratoire de Cristallographie associé à l' Université Joseph
  Fourier, CNRS, 38 042 Grenoble cedex 09, France.

Abstract. The growth of large 123 grains containing homogeneously distributed ultra fine 211 precipitates can be achieved by acting on the grain size of the 123 powder, adding properitectic 211 particles and Pt powder to the precursor.
Application of a vertical magnetic field during horizontal solidification tends to align the a-b planes parallel to the main axis of the sample. It was found that 211 precipitates can be oriented by a magnetic field ($b_{211}//H$).

1. Introduction

Looking for 123 grain of size as large as possible and containing homogeneously distributed ultra fine 211 precipitates, has led us to study 123 peritectic melting and 211 coarsening in the peritectic liquid. We have investigated the effects of grain size of 123 starting powder, properitectic 211 and platinum additions on the final distribution of 211 particles trapped in the crystal. The consequences of the 211 distribution size variation along the growth direction [1] on the super cooling at the solidification front level are discussed.

Unfortunately, application of a thermal gradient alone is not sufficient to align the 123 a-b planes of the grains parallel to the main axis of the sample. P. de Rango et al. [2] have discovered that a magnetic field induces an alignment of the c axis of 123 grains parallel to it and hence controls their a-b planes orientation. A set-up has been built allowing to apply a 4T vertical magnetic field during horizontal solidification. The effect of magnetic field application on c_{123} orientation during the growth is reported. Moreover, looking for a possible effect of the magnetic field on the 211 particles in the melt, EPR and X-ray pole figures of 211 particles trapped in 123 crystals have been performed.

2. Experimental

In order to investigate the effect of grain boundaries on the peritectic 211 nucleation, a 123 HOECHST powder of mean grain size d=3μm has been used. Two different routes have been used to prepare precursor samples containing 123 of two different grain size. Precursor powder A: has been uniaxially cold pressed (100MPa) and then sintered at 930°C during 12 hours in air. Precursor powder B: has been isostatically cold pressed at 13000 bar (the grain are broken during pressing) without sintering. Precursor sample A contains large 123 sintered grains and precursor sample B contains small 123 grains unsintered. The influence of additions to the 123 precursor powder on the size distribution of the 211 precipitates has been studied on the following samples: a) pure 123, b) 123 + 20% wt 211 sol-gel powder [3] and c) 123 + 20% wt 211 + 0,5% wt Pt wich has been prepared by mixing 123 + 20% wt 211 and Pt powder in an agate mortar. Those powders have been isostatically cold pressed (not sintered to prevent any coarsening of the grain before melting).
Details of experimental procedure used for unidirectional solidification have been published elsewhere [4]. Microstructural characterization has been performed on longitudinal polished sections of the samples by optical microscopy. Observation of the 211 morphology in quenched samples were made with EMPA Cameca SX50.

164

The super cooling temperatures have been estimated from measurements of the length of the solidified part of the bar knowing the thermal gradient in the furnace.

3. Results

3.1. Unidirectional solidification

Using exactly the same experimental procedure, precursor A and B have been melted and slowly solidified (3°/h) under a thermal gradient [4].
It is very difficult to quantitatively compare the size distribution of the 211 particles trapped in the 123 solidified crystal for each precursor because samples A exhibit large areas without any precipitate. Table 1 qualitatively gives the main trends.

Table 1 : Main 211 distribution tendency for precursor A and B

Precursor	A	B
211 volume fraction	low	high
211 morphology	acicular	more spherical
211 distribution	non homogeneous	homogeneous

Figure 1 is an X-ray image of Yttrium on 123 and 123+20% wt 211precursor quickly melted (10⁴ °/h) at 1150°C during 5 mn and then air quenched. We can see the morphology of the 211 precipitates and the volume fraction changes.

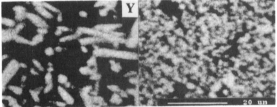

Fig. 1 : EMPA image of Yttrium on 123 and 123 + 20% wt 211 precursor melted 5 mn at 1150°C and then air quenched.

Figure 2 shows 211 mean diameter and volume fraction along the growth direction, from the cold end (firstly solidified) to the hot end for precursor b) and c). Note that the Pt doped sample has a 211 mean diameter smaller and a higher volume fraction than those of the undoped sample.

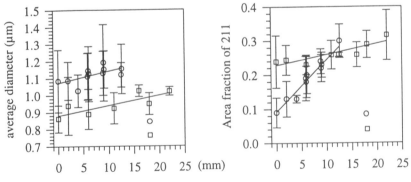

Fig. 2a and 2b : 211 size distribution along the bar from cold end to hot end. Respectively mean diameter and area % 211

Pt addition to 123 + 20% wt 211 powder has allowed us to grow a single 123 grain 20mm x 2mm x 2mm and to homogenise the 211 distribution along the bar.

Table 2 shows the temperature T_s of the solidification front for different lengths of solidified crystal. The super cooling ΔT is calculated using the melting temperature which is Tp = 1030°C under PO_2=1 atm of flowing oxygen. Super cooling increases with the length of the solidified part of the sample.

Table 2: Super cooling versus the length of the solidified part of the sample

sample	T_s (°C)	ΔT	length of the solidified part (mm)
D4	1022	8	6
D2	1016	14	15

3.2. Unidirectional solidification under 4T magnetic field

The c axis orientation of the grains has been determined on the face perpendicular to the direction of the magnetic field and ploted on a stereographic projection (Fig. 3). Applying a vertical magnetic field during the growth tends to align the c_{123} axis of the grain parallel to the direction of the magnetic field.

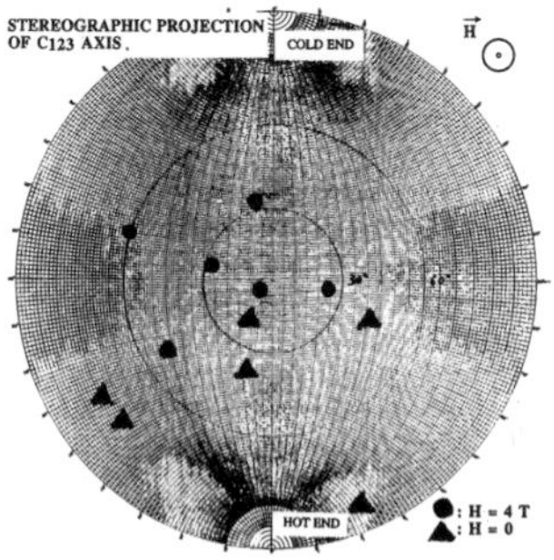

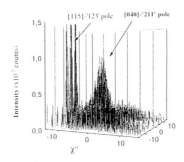

Fig.3 : Stereographic projection of the c_{123} axis on the face perpendicular to the magnetic field direction of samples processed with and without applied magnetic field.

A careful study has been made on one sample particularly well textured; the c_{123} axis of the grain is at less than 8° from the direction of the magnetic field [4]. Fig.4 presents the central zone of the {040}211 pole performed on a face perpendicular to the applied magnetic field. While it subsists a proportion of 211 phase randomly oriented (constant diffracted signal over the entire pole figure), there is a preferential orientation of 211 crystallites.

Fig. 4 : {040}211 pole figure on the face perpendicular to the magnetic field and the c_{123} direction on a particularly well textured sample processed under magnetic field.

166

4. Discussion

4.1. Unidirectional solidification

The difference in size distribution between samples A and B confirms that 211 precipitates nucleate at grain boundaries. Besides, as Griffith et al [5] have shown, when equiaxed properitectic 211 are added to 123, the resulting 211 precipitates are still equiaxed.

The fact that the mean radius and volume fraction of the 211particles increase along the growth direction (fig. 2a and 2b) is believed to be due to several phenomena: 211 coarsening and agglomeration in the melt, precipitates pushed by the solidification front motion and 211 dissolution in the super cooled liquid in front of the solidification front [1].
The increase in super cooling with the solidified length seems correlated with that of the mean diameter and volume fraction of the 211 particles (figure 2a and 2b). Indeed 211 dissolution ahead of the solidification front is more and more difficult as the 211 precipitates coarsen and consequently a stronger driving force is needed. This is in disagreement with P. de Rango et al [6] who explain the increase in super cooling with increasing cooling rate by a decreasing in the oxygen exchange rate.

Pt addition modifies the particle distribution along the bar. Indeed Pt particles should provide nucleation sites for peritectic 211 precipitates but also, as shown in figures 2a and 2b, they should inhibit their coarsening in the melt [7].

4.2. Unidirectional solidification under a 4T magnetic field

The essential result shown in figure 4 is a strong orientation effect of the b_{211} axis parallel to the magnetic field direction. Anisotropy of the susceptibility of the 211 precipitates has been confirmed [8]. The fact that we have here c_{123} // H // b_{211} lead us to wonder about the mechanisms involved during solidification under magnetic field. Does this 211 orientation induced by the magnetic field impose the 123 growth direction?

5. Conclusion

Peritectic 211 precipitates nucleate on the 123 grain boundaries and on introduced nucleation sites like 211 or Pt particles. Moreover, Pt inhibits 211 particles coarsening in the melt. A single grain: 20mm x 2mm x 2mm was grown in a sample prepared with isostatically cold pressed (13000 bar) 123 + 20% wt 211 + 0,5% wt Pt powder.
Application of a 4T magnetic field during solidification tends to align the c_{123} axis of the grains parallel to the magnetic field direction. On a particularly well textured sample, we have found that a substantial proportion of 211 particles had their b_{211} axis parallel to the magnetic field direction. It has been confirmed that 211 phase is magneticaly anisotropic [8].

References

[1] Durand L and Pastol J L 1994 *Materials Letters* **19** 291
[2] de Rango P, Lees M , Lejay P, Suplice A, Tournier R, Ingold M, Germi P and Pernet M 1991 *Nature* **349** 770
[3] Van der Biest O and Kwarciak J 1992 *J. Austr. Ceram. Soc.* **27** 51.
[4] Durand L, Kircher F, Régnier P, Chateigner D, Pellerin N, Gotor F J, Simon P and Odier P 1995 *Supercond. Sci. Technol.* **8** 214
[5] Griffith M L, Hauffman R T and Holloran J W 1994 *J.Mat.Res.* **9** 1633.
[6] de Rango P, Chaud X, Gautier-Picard P, Beaugnon E and Tournier R 1993 Eucas-93 305
[7] Varanasi C, Black M A and McGinn P J 1994 *Supercond. Sci. Technol.* **7** 10
[8] Pellerin N, Gotor F J, Simon P, Durand L and Odier P 1995 *Solid. State Commun.* submitted

Inst. Phys. Conf. Ser. No 148
Paper presented at Applied Superconductivity, Edinburgh, 3–6 July 1995
© *1995 IOP Publishing Ltd*

Numerical simulation of YBaCuO-growth phenomena using the phase field method

G.J. Schmitz, B. Nestler, H.J. Diepers, F. Pezzolla, R. Prieler, M. Seeßelberg, I. Steinbach

ACCESS e. V., D-52072 Aachen, Intzestraße 5, Federal Republic of Germany

Abstract. Numerical modelling of the growth of superconducting materials is beneficial with respect to both process control and the understanding of growth phenomena. The phase field method can be used to treat the evolution of a solid-liquid interface and has been extended to describe aspects of the multiphase system YBaCuO. The method is appropriate to simulate the faceted growth of both the properitectic and peritectic phases and its dependence upon Y-diffusion in the liquid phase. Numerical results show qualitative agreement with microstructures being experimentally observed in a transparent organic model system of YBaCuO and the YBaCuO system itself.

1. Introduction

Modelling of microstructural formation is a relative new topic in materials' sciences. Three different approaches can be distinguished: Monte-Carlo-simulations [1-3], cellular automata models [4-7] and the phase field method [8].

Recently the phase field concept has been extended to multiphase systems [9]. It allows the simulation of the peritectic solidification process of the YBaCuO high T_c superconductor from the melt. This process is driven by the local concentration of Y. Within the present report, preliminary numerical calculations qualitatively reveal several features of YBaCuO growth.

2. Phase field equations

For the sake of brevity, in this section we discuss the phase field equation of a single solid phase. This equation is a partial differential equation governing the phase field $p(\bar{x}, t)$ whose values lie in the unit interval $[0,1]$. The values $p(\bar{x}, t) = 0$ and $p(\bar{x}, t) = 1$ mean that at the point $\bar{x}$ and at time t the system is in the liquid or solid phase respectively. The boundary between the liquid and solid phase is characterized by values of $0 < p(\bar{x}, t) < 1$. The phase field equation can be derived by minimizing the free energy density of the system and reads [10]

$$\tau(\theta, \varphi)\dot{p}(\bar{x}, t) = \varepsilon \nabla^2 p(\bar{x}, t) - f'(p) + d(p, T, c). \tag{1}$$

ε is the gradient energy coefficient. $\tau(\theta, \varphi)$ determines the growth velocity of the phase front. It is an angular dependent function taking into account anisotropic and faceted growth of the solid phase. The derivative $f'(p)$ of a double well potential $f(p)$ ensures the stability of the phase boundary in equilibrium. Minima of $f'(p)$ are at $p = 0$ and $p = 1$; the derivative $f'(p)$ obeys $f'(p) > 0$ for $0 < p < 0.5$ and $f'(p) < 0$ for $0.5 < p < 1$. The driving force $d(p, T, c)$ is proportional to the deviation from thermodynamical equilibrium, it depends on temperature $T(\bar{x}, t)$ as well as on the concentration field $c(\bar{x}, t)$ of certain chemical species, e.g. Y in the YBaCuO system. The solutal field $c(\bar{x}, t)$ is described by a diffusion equation with diffusion constant D and a source term $s(\dot{p})$, i. e.

$$\dot{c}(\bar{x}, t) = D \nabla^2 c(\bar{x}, t) + s(\dot{p}) \tag{2}$$

The source term $s(\dot{p})$ depends on growth and dissolution rate of the solid phase.

These equations can be generalized to describe multicomponent systems with several solid phases, i. e. the 123 and 211 phases of YBaCuO. This yields differential equations similar to (1) for each of the phases. However, the description of the phase field equations for multiphase systems goes beyond the scope of this article and will be described elsewhere [9].

3. Numerical results and experimental observations

3.1 Faceted growth of a single grain

As initial condition $p(\bar{x},0)$ for the phase field equation (1) we considered a single nucleus in a slightly undercooled melt. The anisotropy function $\tau(\theta,\varphi)$ was chosen to reproduce a faceted symmetry. We investigated the phase field equation (1) numerically. The evolution of the phase field $p(\bar{x},t)$ in time reveals a growing faceted grain in three dimensions (Fig. 1).

3.2 Single phase systems with more than one grain

Also growth processes involving more than one grain (e. g., several growing 123 YBaCuO platelets) can be treated (Fig. 2). The growth process is competitive, i. e., grains with certain crystallographic orientations in relation to the temperature gradient tend to overgrow neighbouring grains [11].

3.3 Multiphase systems

A multiphase system is described by different equilibrium concentrations $c(\bar{x},t)$ at the solid-liquid interfaces of the individual phases. The local undercooling of each interface is treated with respect to the metastable extension of the equilibrium phase diagram. In case of the YBaCuO system, the different phases interact via the diffusion of Y between the interfaces. Numerical results are in qualitative agreement with properties of the 211-123-YBaCuO system and its transparent organic analogue [12,13].

A 2-dimensional simulation of a dissolving properitectic needle and a faceted growing peritectic grain (Fig. 3.a) shows a similar situation as observed in the transparent salicylic acid/acetamide model system (Fig. 3b). The observation of a large properitectic 211 needle being pierced by several small 123 grains (Fig. 3.c) is also in qualitative agreement with both the simulation and the organic model system.

The phase field calculations also describe effects of bridge formation and engulfment (Fig. 4).

4. Conclusions

The phase field model has been shown to describe qualitatively several features of the growth of YBaCuO:
- faceted growth of the individual phases
- competitive growth between different growth orientations
- dissolution of the properitectic 211 phase in the vicinity of the peritectic 123 phase
- bridge formation
- engulfment of 211 particles.

The model can also be used to simulate aspects of other microstructures, e. g. emerging from solidification of silicon, steel or any other alloy.

Future work will focus on quantitative simulations using thermodynamical data sets of the various phases in the YBaCuO system. The simulation will then become a powerful tool to investigate the formation of microstructures. An extension of the existing software is required to consider more phase fields and the solutal diffusion of more species. The simulation of larger specimen requires a further improvement of the algorithms to enhance the software performance. However, it seems that no mayor obstacles need to be overcome.

5.Acknowledgement

The presented work is funded by the German Ministry of Education and Research (BMBF) under grand No. 13N5565 A.

References

[1] JA Spittle, SGR Brown 1989 Acta metall. 37 No. 7 p. 1803
[2] A Pekalski, M Ausloos 1994 Physica C 226 188-198
[3] ZX Cai, DO Welch 1994 Philosophical Magazine B 70 141-150
[4] HW Hesselbarth, IR Göbel 1991 Acta metall. mater. 39 No. 9 p. 2135
[5] CH Wolters, J Laakmann, S Rex, GJ Schmitz 1993 Proceedings of EUCAS 93, p. 353
[6] EA Gudilin, NN Oleinikov, AN Branov, YD Tretyakov 1993 Inorganic Materials 29 1285-1290
[7] N Vandewalle, R Cloots, M Ausloos 1995 J. Mat. Res. 10 p. 268
[8] AA Wheeler, WS Boettinger, GB Mc Fadden1992 Phys. Rev. A 45 No. 10 p. 7424
[9] I Steinbach, B Nestler, F Pezzolla, J Rezende, M Seeßelberg, GJ Schmitz 1995, A phase field concept for multiphase systems, in preparation
[10] R Kobayashi 1993 Physica D 63 410-423
[11] I Steinbach, F Pezzolla, R Prieler 1995, in: Proceedings of Modelling of Casting, Welding and Advanced Solidification processes VII (London, ed. by M. Cross)
[12] HO Yasuda, I Ohnaka, Y Matsunaga, Y. Shiohara 1995, In situ observation of peritectic growth with faceted interface, submitted to: J. Cryst. growth
[13] GJ Schmitz, R Terborg: Modeling of peritectic $YBa_2Cu_3O_{7-x}$ growth using transparent organic analogues, in preparation.
[14] GJ Schmitz, J Laakmann, CH Wolters, S Rex, W Gawalek, T Habisreuther, G Bruchlos, P. Görnert 1993 J. Mater. Research 8 2774-2779

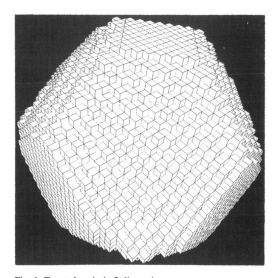

Fig. 1: Faceted grain in 3 dimensions

170

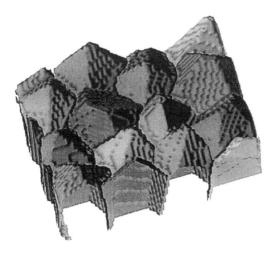

Fig. 2: Several grains of the same phase

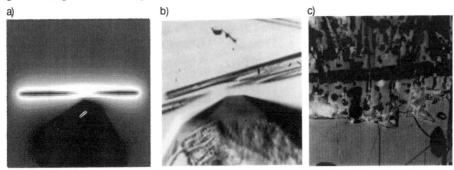

Fig. 3: Dissolving properitectic needle and faceted growing peritectic grain in a) numerical simulation, b) transparent organic model system, c) real YBaCuO (from W. Gawalek)

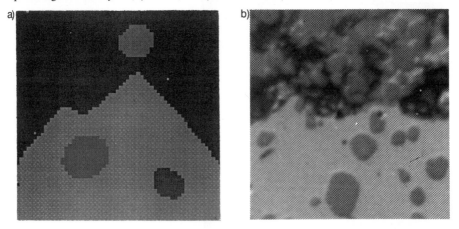

Fig. 4: Bridge formation and engulfment in a peritectic system. Diagram a) shows a simulation with two enclosed properitectic particles (engulfment). Due to the solutal concentration profile the peritectic phase grows rapidly towards the properitectic particle showing the effect of bridge formation [14]. The same phenomena can be observed experimentally in the YBaCuO system (diagram b). Dark (medium, light) grey corresponds to the liquid (properitectic, peritectic) phase.

Inst. Phys. Conf. Ser. No 148
Paper presented at Applied Superconductivity, Edinburgh, 3–6 July 1995
© 1995 IOP Publishing Ltd

Non-Equilibrium Processing of YBCO

G.J. Schmitz, G. Eßer, O. Kugeler

ACCESS e.V., Intzestr. 5, D-52072 Aachen, Germany

Abstract. 123-phase formation in non-equilibrium conditions was investigated with respect to higher growth rates. Non-equilibrium was achieved by rapidly heating a $BaCuO_2$-CuO-Y_2O_3 powder-mixture to temperatures slightly below the peritectic temperature T_p of the superconducting phase. In situ hot-stage SEM revealed the formation of a liquid-phase at temperatures exceeding 900°C, in which nucleation phenomena were observed. Samples heated to temperatures below T_p exhibited a microstructure similar to samples rapidly cooled to these temperatures from the 211+liquid regime. A homogeneous distribution of micron sized 211 inclusions inside 123 grains, which were in the millimeter range, could be achieved without any dopant. The degree of undercooling determined both the final size and the number of 123 grains. The powder-mixture can be assumed to represent an undercooled melt.

1. Introduction

Melt processing of bulk high temperature ceramic superconductors is widely accepted to yield the best superconducting properties within these materials. All processes proposed up to now (for a review see e.g.[1]) include at least one processing step - slow equilibrium cooling through the peritectic temperature -, which limits the growth rate of this material to the order of mm/h or requires cooling rates in the range of only a few K/h. For the economical application of these materials, however, significantly higher growth rates are very desirable.

Non-equilibrium processes, in general, allow higher growth rates and such processes accordingly represent an interesting alternative processing route. Deviations from equilibrium may be achieved either thermally [2,3,4] or constitutionally [5]. Processes involving mixtures of powdered precursors (e.g. PMP [6], SLMG [7]) can be considered as constitutional non equilibrium processes.

2. Objectives

Scope of the presented work is the investigation of non-equilibrium phase formation in an undercooled regime slightly below the peritectic temperature T_p of the superconducting 123 phase. In this temperature regime, the 211-phase is highly undercooled, which corresponds to a high nucleation rate and a small critical radius for stable nuclei.

By exploiting this, it should be possible to yield a homogeneous dispersion of 211-particles as fine as possible, which is beneficial to the diffusion controlled growth of the 123-phase [8,9,10].

Aiming for higher growth rates we applied a low temperature process, that exploits the high 211-nucleation rate slightly below T_p in order to achieve optimum growth conditions for the 123-phase.

In addition to the potential enhancement of growth rate this method allows:

- control of precusor shape [11,12,13]
- low temperature processing and accordingly less corrosion
- avoidance of complex precursor manufacture.

3. Experimental

Powder mixtures with overall stoichiometries corresponding to a 5:1 mixture of the final 123 / 211 phases were prepared from commercial or freeze dried precursors of $BaCuO_2$, CuO, Y_2O_3 with mean diameters in the micron range.

The precursors were individually ball milled in aceton to equalize their size distributions. Subsequent mixing was carried out by additional ball milling for approximately 24 h. After evaporating the aceton the powders were dry mixed in an achat motar and uniaxially pressed into small pellets (10 mm in diameter, 3-5 mm long). These pellets were put onto a MgO single crystal substrate to avoid the loss of liquid phase and inserted into a tubular furnace at room temperature.

Temperature measurement at high temperatures was calibrated using a sintered 123 reference sample.

The samples were heated to various temperatures slightly below the peritectic temperature with heating rates of up to 40 K/min. In addition to changing the power input into the furnace, the heating rate could be slightly varied by putting the sample into an alumina tube which acted as thermal buffer.

The samples were kept isothermally in the undercooled region for some hours, rapidly cooled and analyzed by SEM/EDX and optical microscopy. Hot-stage in situ SEM was carried out in order to clarify the reaction sequence in the powder-mixture.

4. Results

As revealed by hot-stage in situ SEM, a strongly wetting liquid phase forms at temperatures exceeding 900°C. The amount of liquid phase increases with increasing temperature and nucleation is observed on the surface of the wetted powder particles, Fig. 1.

Cross sections of the polished samples reveal a fine dispersion of 211 particles, which has mean diameters in the micron range, embedded into grains of 123 which are in the millimeter range, Fig.2. This fine dispersion was achieved without any doping.

The number and size of the 123 grains strongly depends on the heating rate, which when only slightly decreased, results in a drastic change of the microstructure, Fig. 3. A comparison of samples quenched to the undercooled region from higher temperatures (involving liquid phases) to samples produced by heating powdered precursors, reveals a similar microstructure, Fig. 4.

5. Conclusions

Samples produced by heating powder precursors to a temperature region just below the peritectic temperature reveals a microstructure similar to samples which were quenched to this temperature region from higher temperatures. Therefore, the powder mixture can be assumed to represent an undercooled liquid.

Fig. 1

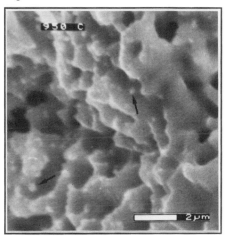

Hot stage SEM shows the formation of a liquid phase upon heating the reactive powder-mixture to temperatures exceeding 900°C. Nucleation of probably 211 and 123 is observed on the surface of the wetted powder particles.

Fig. 2

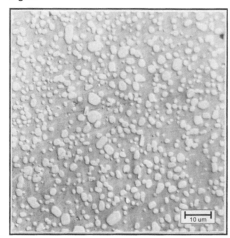

Microstructure of a 123 domain grown from the reactive powder-mixture at temperatures below the peritectic temperature of the 123-phase. Fine 211 inclusions are achieved without any dopant.

Fig. 3

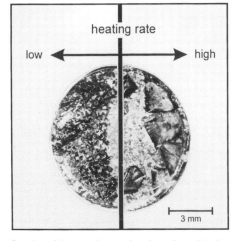

Samples of the reactive powder-mixture heated to the same final temperature below Tp and held for the same time reveal a strong dependence of the microstructure on the initial heating conditions. Relatively large 123 grains are achieved with high heating rates.

Fig. 4

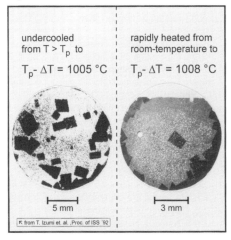

Samples of the reactive powder-mixture heated from room-temperature to a temperature below Tp reveal a microstructure similiar to samples cooled to the same temperature from the 211+Liquid regime. Millimeter sized 123 grains are embedded in quenched 211+Liquid.

174

Although higher growth rates, which were one of the objectives of this work, could not be confirmed so far, the fact that not a single step within the presented process exceeds the peritectic temperature of this material makes it a potential joining method for large domains of YBCO, even transferring the domain orientation to regions grown reactively from the powder mixture.

6. Acknowledgment

The presented work is funded by the German Ministry of Education and Research (BMBF) under grant No. 13N5565A.

Thanks are due to Ch. Wolters from the NHMFL Tallahassee / Florida for the in situ hot stage SEM investigations.

References

[1]   K. Salama, D.F. Lee: "Progress in melt texturing of $YBa_2Cu_3O_x$" Supercon. Sci. and Techn. 7 (4) (1994) 177-193

[2]   V.R. Todt, G.J. Schmitz: "Containerless Solidification of YBaCuO (123) in a Drop Tube", J. Mater. Res. 8 (3) (1993) 411

[3]   J.R. Olive, W.H. Hofmeister, R.J. Bayuzick, G. Carro, J.P. McHugh, R.H. Hopkins, M. Vlasse, J.K.R. Weber, P.C. Nordine, M. McElfresh: "Formation of tetragonal YBaCuO (123) from an undercooled melt" J. Mat. Res. 9 (1) (1994), 1

[4]   B.J. Chen, M.A. Rodriguez, S.T. Misture, R.L. Snyder: "Effect of undercooling temperature on the solidification kinetics and morphology of YBaCuO during melt texturing". Physica C 217 (1993), 367-375

[5]   V. Milonopoulou, K.M. Forster, J.P. Formica, J. Kulik, J.T. Richardson, D. Luss: "Infuence of Oxygen Partial Pressure on the Kinetics of YBACuO (123) Formation". J. Mat. Res. 9 (2) (1994), 275

[6]   Z. Lian, Z. Pingxiang, J. Pin, W. Keguang, W. Jinguong, W. Xiaozu: "The properties of YBCO superconductors prepared by the powder melting process". Supercon. Sci. Technol. 3 (1990), 490

[7]   C. Varanasi, S. Sengupta, P.J. McGinn: "An alternative Method to Introduce Fine YBaCuO (211) Precipitates into YBaCuO (123) with Enhanced Flux Pinning" Appl. Supercond. 2 2 (1994) 117-121

[8]   T. Izumi, Y. Nakamura, Y. Shiohara, J. Mater. Res. 7 (1992), 1621

[9]   M.J. Cima, M.C. Flemings, A.M. Figueredo, M. Nakade, H. Ishii, H.D. Brody, J.S. Haggerty, J. Appl. Phys. 72 (1) (1992) 179

[10] G.J. Schmitz, J. Laakmann, Ch. Wolters, S. Rex, W. Gawalek, T. Habisreuther, G. Bruchlos, P. Görnert, J. Mater. Res. 8 (11) (1993), 2774

[11] Ch. Wolters, J. Laakmann, S. Rex, G.J. Schmitz "Numerical Simulation of the influence of 211 particles on the growth morphology of peritectically solidifying $YBa_2Cu_3O_7$" Proc. EUCAS 93, ed. H.C. Freyhardt, DGM, Oberursel (1993)

[12] G.J. Schmitz, B. Nestler, H.J. Diepers, F. Pezzolla, R. Prieler, M. Seeßelberg, I. Steinbach: "Numerical Simulation of YBaCuO-Growth Phenomena Using the Phase Field Method" this conference

[13] F. Frangi, T. Higuchi, M. Deguchi and M. Murakami: "Optimization of Y_2BaCuO_5 morphology for the growth of large YBCO grains" preprint

Inst. Phys. Conf. Ser. No 148
Paper presented at Applied Superconductivity, Edinburgh, 3–6 July 1995
© *1995 IOP Publishing Ltd*

INFLUENCE OF PRECURSOR PREPARATION ON THE Y₂BACUO₅ SIZE DISTRIBUTION THROUGH OUT MELT PROCESSED BULK YBA₂CU₃O₇₋ₓ SUPERCONDUCTORS.

D. Dierickx[1], K. Rosseel[2], W. Boon[2], V.V. Moshchalkov[2],Y. Bruynseraede[2], O. Van der Biest[1].

1) K.U. Leuven, dept. of Metallurgy and Materials Science, de Croylaan 2, B3001 Heverlee, Belgium.
2)K.U. Leuven Laboratory of solid state physics and Magnetism, Celestijnenlaan 200D, B3001 Heverlee, Belgium

Abstract Precursor production routes have an important influence on the melt processing behaviour of the superconductor compacts. Two one step chemical processing procedures for $YBa_2Cu_3O_{6+x}$ + Y_2BaCuO_5 stoichiometries are compared on the basis of Y_2BaCuO_5 grain size distributions measured during, and after fast melt processing. Freeze drying is used to prepare a synthetic $Y_2O_3+Ba_2Cu_3O_x$ mixture, while sol-gel EDTA leads to a Y_2BaCuO_5 + $YBa_2Cu_3O_{6+x}$ precursor. Influence of the properitectic Y_2BaCuO_5 and $YBa_2Cu_3O_{6+x}$ grain size distribution on the final microstructure is discussed. Optimisation of the average Y_2BaCuO_5 grain size population and spatial distribution induces an enhancement of magnetic hysteresis at 77K. Superconducting behaviour is prolonged up to 15 T in pulsed magnetic fields at liquid nitrogen temperature.

1. Introduction

Melt processing is known to have a beneficial influence on the magnetic properties of $YBa_2Cu_3O_{7-x}$ bulk material. The exact role of the Y_2BaCuO_5 precipitates on flux pinning is still under discussion, but generally is recognised as being of high importance, either directly or indirectly. Additionally, during growth of the superconducting $YBa_2Cu_3O_{7-x}$ phase the diffusion of yttrium through the peritectic liquid will be influenced by the precipitate grain size and spatial distribution [1]. As properitectic $YBa_2Cu_3O_{7-x}$ grain size has been reported to influence final Y_2BaCuO_5 precipitate size, freeze drying and EDTA sol-gel were chosen as precursor preparation routes. The precursor phase mixture consists of Y_2O_3 + $Ba_2Cu_3O_x$ in the freeze drying case, thus minimising the $YBa_2Cu_3O_{7-x}$ grain size effect. Sol-gel prepared precursor material consists of $YBa_2Cu_3O_{7-x}$ + Y_2BaCuO_5 which gives rise to both a properitectic Y_2BaCuO_5 size distribution combined with a peritectic size distribution influenced by the pre-existing $YBa_2Cu_3O_{7-x}$ grains. The differences in view of high temperature Y_2BaCuO_5 precipitate growth for both precursor materials will be documented and discussed in view of further application in melt processing. At this point detailed physical measurements have only been performed on fast melt processed freeze dried material.

2. Experimental procedure

For both methods, three starting solutions were prepared by dissolving stoichiometric amounts (80wt% $YBa_2Cu_3O_{7-x}$ + 20 wt% Y_2BaCuO_5) of Y_2O_3 (Janssen chimica 99.999%) and Cu (Merck 99.7%) using necessary amounts of HNO_3 (Merck 65%) and $Ba(NO_3)_2$ (Aldrich >99%) in distilled water upon heating. In the freeze drying case no further additions were made to the solutions. Details of the freeze drying procedure have been reported in [2]. For the sol-gel material, separate solutions of EDTA (ethylene di-amine tetra acetic acid) were prepared and added to the respective cation solutions for complexation. The solutions were mixed together and left to gelate and decompose, followed by an oxygen calcination. Details of the processing can be found in [3]. Precursor material was cold isostatically pressed at 1000 MPa and subjected to the heat treatments as summarised in table 1. Resistivity measurements were performed with a four point set-up using a lock in technique. Transport I-V measurements were made in a four point set-up using a pulsed current with a pulse duration of +/- 20 ms. Magnetisation in fields up to 0.15T was measured with a 50 Hz magnetic field. Pulsed field measurements reaching applied fields up to 50Twere done using the set-up of [4], while Squid measurements were performed using a Quantum design squid magnetometer with a scan length of 3 cm, corresponding to a field homogeneity better than 0.05%.

Table 1. Heat treatments performed on both precursors.

Code	Precursor	ramp °C/min	temp °C	hold hr.	ramp °C/min
FQ1	Freeze Dried	3	1030	0.5	quench
FQ2	Freeze Dried	3	1030	1.0	quench
FQ3	Freeze Dried	3	1030	3.0	quench
F1	Freeze Dried	3	1030	2.0	1
SQ1	sol-gel	3	1045	1.0	quench
SQ2	sol-gel	3	1045	1.5	quench
SQ3	sol-gel	3	1045	2.0	quench
S1	sol-gel	3	1045	2.0	1

3. Results and discussion.

Figure 1 shows the DSC measurements on both the freeze dried and the sol-gel prepared precursors. From these curves the freeze dried powder seems tó be molten completely at a lower temperature compared to the sol-gel precursor. This was used in determining the processing temperatures for the quench and fast melt processing experiments. As the heating rate was set at 3 °C per minute only part of the freeze dried precursor will be transformed to $YBa_2Cu_3O_{7-x}$ before the melting sets in while the original composition is quite similar to a liquid composition and therefore melts very fast. While the freeze dried material was already completely molten after a hold of 30 minutes, the sol-gel required at least 60 minutes. The different behaviour of both precursors is further evidenced by Y_2BaCuO_5 formation during the quench experiments. From figure 2 the freeze dried precursor clearly displays a nucleation and growth behaviour and can be fitted by an Avramy type of curve. The sol-gel precursor shows a sharp increase in Y_2BaCuO_5 content from an initial value determined by the original composition in combination with the decomposition reaction of $YBa_2Cu_3O_{7-x}$.

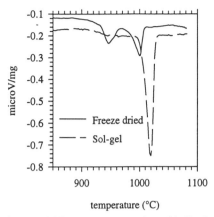

Figure 1. DSC measurements performed in flowing oxygen. Heating rate 3°C/min.

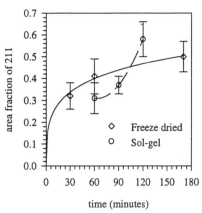

Figure 2. Area fraction of Y_2BaCuO_5 in as quenched microstructures, error bars are 95% confidence intervals

The F1 and S1 samples were cut and polished for optical microscopy. The precipitate size distributions were measured and are compared in figure 3. The size distribution for the F1 sample seems to be broader then expected for a log normal fit. This is interpreted as being a result of the recombination reaction of Y_2BaCuO_5 and liquid in order to form $YBa_2Cu_3O_{7-x}$. The smallest precipitates are preferentially removed from the distribution, but the process is incomplete due to the speed of cool down. The S1 sample contains a higher amount of larger sized precipitates as is evidenced by the more important tail in the size distribution. Area measurements showed the Y_2BaCuO_5 precipitates to be more homogeneously spaced throughout the F1 sample (0.4 mean area fraction) while differences in concentration were clearly visible in the S1 sample (from 0.2 to 0.5 area fraction). The nature of the Y_2BaCuO_5 growth from the freeze dried precursor necessitates a holding time at low (1030°C) temperature for the required precipitate size distribution to form prior to further heating. This

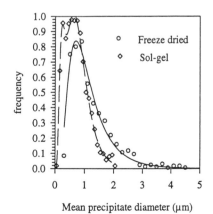

Figure 3. Precipitate size distribution for samples F1 (freeze dried), and S1 (sol-gel) measured from crossed polarised optical micrographs.

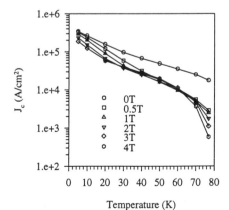

Figure 4 Squid measurement results of sample F1. A large field insensitivity can be observed in the range of 20 to 60K, indicating the high intrinsic flux pinning capacity of the material.

way a size distribution with smaller centre is formed and a sufficient amount of precipitates is grown in order retain the samples shape. Due to the properitectic precipitate distribution present in the sol-gel precursor, this effect is less pronounced in the sol-gel case.

The critical temperature T_c of the samples was determined from four point resistivity measurements using a lock-in method. The transition to the superconducting state occurs typically at $T_c \approx 92K$ with a transition width $\Delta T_c < 1K$. Magnetic measurements were performed in low magnetic fields up to 0.15T using an AC -setup and in high pulsed magnetic fields of up to 50T in the temperature range 5K-77K. From the low field measurements, no sign of a weak link decoupling could be resolved. This seems to indicate that a very weak josephson coupling exists between the $YBa_2Cu_3O_{7-x}$ grains. To check this, transport measurements were performed at 77.3 K using a pulsed current setup. The very small transport critical current density J_{ct} derived ($\approx 10A/cm^2$ at 77K and 0T), unambiguously showed that the samples can be seen as consisting of rather large, but very weakly coupled grains, with a Josephson decoupling field $\mu_{0*}H^J_{c2} < 0.001T$. To study the temperature and field dependence of the intragrain critical current density J_{cm} , pulsed field measurements were performed. As a first estimate, the Bean critical state model was used : $J_{cm} \sim \Delta M$, with ΔM the width of the hysterisloop. A weak field dependence of J_{cm} over a large field range (~10T at 77K) could clearly be observed. An estimate of the irreversibility field $\mu_{0*}H^{irr}$ could only be made for temperatures above 60K and was found to be $\mu_{0*}H^{irr} \approx 35T$ at 60K and 20T at 77K. The relatively large field insensitivity can also be observed in the squid results summarised in figure 4. Squid and pulsed field derived J_c are of the same order of magnitude, indicating the strong flux pinning inside the $YBa_2Cu_3O_{7-x}$ grains. Concerning the exact pinning mechanisms involved and the role of the 211 inclusions, no definite conclusions can be drawn at this point.

5. Conclusion

Precursor preparation clearly influences both precursor melting and precipitate growth behaviour. Control of precipitate size follows normal nucleation and growth considerations for the freeze dried precursor. Fast melt processing of a freeze dried precursor produces weakly coupled superconductor grains with a high intrinsic flux pinning capacity.

6. Acknowledgements

D. Dierickx is supported by the Belgian High Temperature Impulse (HTSI) Program.
K. Rosseel is a Research Fellow of the Belgian I.W.T.
W. Boon is a Research Associate of the Belgian N.F.W.O.

References

1) Izumi T, Nakamura Y and Shiora Y 1992 J. Mater. Res. **7** 1621
2) Dierickx D and Van der Biest O Eur. J. Sol. St. Inorg. Chem. to be published
3) Dierickx D and Van der Biest O 1993 Third Euro Ceramics Spain **2** 547
4) Herlach F, Bogaerts R, Vanacken J and Bockstall L 1994 Physica B **201** 5

Inst. Phys. Conf. Ser. No 148
Paper presented at Applied Superconductivity, Edinburgh, 3–6 July 1995
© 1995 IOP Publishing Ltd

Quenched flake precursors by a special flame fusion process used for the melt texturing of YBCO bulk material

M. Ueltzen, Ch. Seega, H. Altenburg

FH Münster, FB Chemieingenieurwesen, Supraleitertechnologie und Kristalltechnik Stegerwaldstr. 39, D-48565 Steinfurt, Germany

D. Litzkendorf, K. Fischer, G. Bruchlos, P. Görnert

IPHT Jena e.V., Helmholtzweg 4, D-07702 Jena, Germany

Abstract. A crucible-free melting and quenching process for the preparation of YBCO is described. The use of precursors prepared by means of this newly developed process results in higher levitation forces of the superconductors.

1. Introduction

To achieve a compact textured material, directional solidification from a partially melted state is the most successful method. A high temperature step at temperatures obvious above the peritectic melting of the superconducting phase seems to be necessary in order to attain high jc material. So temperatures of about 1100 °C (modified MTG) or 1400 °C (QMG) are applied. With this high temperature step a homogeneous distribution of all components is possible and the pore formation is minimized. An effective quenching prevents segregation and coarsening of the different phases. The controlling of the 211 particle size is very important for optimizing the critical current density. Doping materials, e.g. platinum influence the 211 particle size. By working with platinum crucibles, an uncontrollable but essential platinum doping is achieved. [1]

This paper deals with a new melting process for the high temperature step in the 123 preparation route. The material is melted in the oxyhydrogen flame of a Low-temperature Flame Fusion (LTFF) burner. The process is crucible free and enables the exact doping of the YBCO material.

2. Precursor preparation

Fig. 1 shows the principle of the precursor preparation. The used YBCO rods were made by cutting pieces from pre-sintered cold isostatic pressed bodies and had a cross-section of a few millimeters and a length of about 100 mm. In the upper part, the special oxyhydrogen burner can be seen. The LTFF burner [2] was modified in order to realize the material support in form of thin ceramic rods. A ring nozzle around the central opening for material support enables the pressure-less operation of the central tube without any non-return valve. The burner is designed to operate in the diffusion mode of the flame. There are cold oxygen-rich and hydrogen-rich zones in the flame as well as hot combustion zones. Furthermore, comparatively low temperatures can be realized with this

180

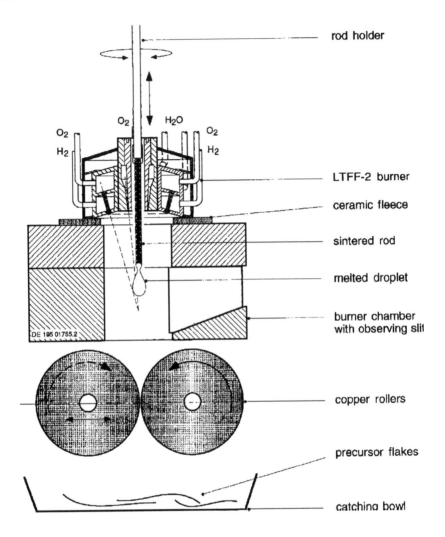

rod holder

O_2 H_2O

O_2 O_2

H_2 H_2

LTFF-2 burner

ceramic fleece

sintered rod

melted droplet

burner chamber
with observing slit

DE 195 01755.2

copper rollers

precursor flakes

catching bowl

Fig. 1: Principle of the precursor flakes preparation process

oxyhydrogen flame. In this way, evaporation of copper is prevented [2]. The oxygen inlet through the inner ring nozzle protects the burner against a backkick of the oxyhydrogen flame into the material support system. Useful gas flows are 12.5 l/min hydrogen and 5.5 l/min oxygen through the 24 nozzles arranged on a circle and 2.8 l/min oxygen through the ring nozzle. The oxyhydrogen flame burns into a ceramic chamber. Through the central opening of the burner, a ceramic rod is supplied. From this rod, droplets of 250-500 mg fall off with a rate of up to one droplet per second. Below the ceramic chamber, a twin roller made of copper was positioned. Between the rollers, the droplets are pressed to flakes. In this way, a very effective quenching of the material with cooling rates between 10^3 and 10^4 K/s is realized. The flakes are sheets, which reach dimensions of $(50x30x0.1)$ mm^3.

3. Preparation of melt textured YBCO bulk materials

The flakes are powdered in a ball mill. With the resulting powder, ceramics are
processed using the modified MTG route. It is possible to add dopants like
platinum to the ceramic rods before the melting process or after the quenching
during the milling of the flakes. Cylindrical samples of 20 to 30 mm diameter
and a height of 10 to 20 mm of the initial composition $Y_{1.5}Ba_2Cu_3O_{7-x}$ with an
addition of 1.0 wt-% Pt were prepared by mixing, grinding and pressing of the
molten and quenched 123, Y_2O_3, and PtO_2. Afterwards, the melt texturing
growth by directional solidification was used [3].

4. Results

The quenched material consists of a mixture of barium cuprates $BaCuO_2$ and
$BaCu_2O_2$, the green phase Y_2BaCuO_5 and yttrium oxide (cf. fig. 2). This was
confirmed by X-ray powder analysis of powdered flakes as well as by SEM
investigations. Powdered flakes have the total composition $Y_1Ba_2Cu_3O_x$.

The superconducting pellets were confirmed by measuring the integral
magnetic levitation forces [4]. The new developed preparation method results in
higher levitation forces at comparable superconductors (see fig. 3).

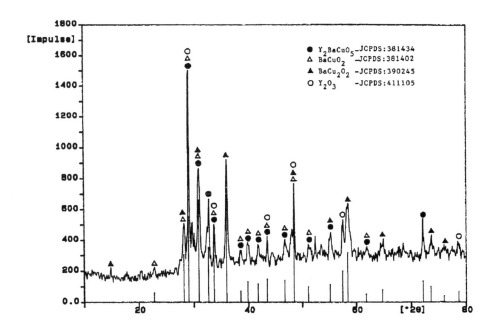

Fig. 2: X-ray diffraction pattern of powdered flakes

182

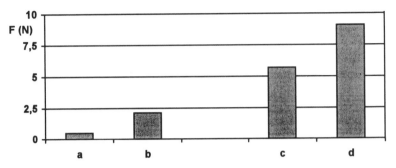

Fig. 3: Comparison of levitation forces, D = 20 mm
a $Y_{1.5}Ba_2Cu_3O_x$ MTG   b $Y_{1.5}Ba_2Cu_3O_x$ flake processed + MTG
c $Y_{1.5}Ba_2Cu_3O_x$ + 1 wt-% Pt MTG   d $Y_{1.5}Ba_2Cu_3O_x$ + 1 wt-% Pt flake processed +MTG

5. Conclusions

The crucible-free overheating of a Y-Ba-Cu-O melt in the oxyhydrogen flame of
a special flame fusion burner increases the homogeneity of the reactive mixture
and the rapid quenching diminishes phase segregation during the solidification
step. So, yttrium oxide and 211 phase exist in the precursor material. This step
is to be optimized in order to diminish the coarsening of the 211 phase in the
precursor material. The formation of pinning centers is influenced positively.
The homogeneity of the material results in good conditions for the regular
growth of the superconducting phase and with that in improved integral
levitation forces of the superconductor.
    The advantages of the process in detail are, that pollution by crucible
material is avoided, as well as pollution during the milling is minimized
because of the special shape of the material. Furthermore, the quenching of the
melt is very effective.
    For further developments, the process has to be scaled up from 100 g
batches to the kg scale.

References

[1]   McGinn P 1995: *Bulk Processing and Characterization of YBa₂Cu₃O₆₊ₓ* in
     High-Temperature Superconducting Materials Science and Engineering
     (ed. by Donglu Shi, Oxford: Pergamon)
[2]   Ueltzen M, Brüggenkamp T, Franke M, and Altenburg H 1993 *Rev. Sci.*
     Instrum. **64** 1089-1090
[3]   Krabbes G, Schätzle P, Bieger W, Wiesner U, Stöver G, Wu M, Straßer T,
     Köhler A, Litzkendorf D, Fischer K, and Görnert P 1995 *Physica C* **244**,
     145-152
[4]   Gawalek W, Habisreuther T, Straßer T, Wu M, Litzkendorf D, Fischer K,
     Görnert P, Gladun A, Stoye P, Verges P, Ilushin KV, Kovalev LK 1994
     Appl. Supercond. **2**, 465-478

This work was supported by the German ministry of education and research by
grant BMFT 13N5555 and 13N6100.

Inst. Phys. Conf. Ser. No 148
Paper presented at Applied Superconductivity, Edinburgh, 3–6 July 1995
© *1995 IOP Publishing Ltd*

Growth of large $REBa_2Cu_3O_{7-y}$ and $Bi_2Sr_2CaCu_2O_{8-y}$ crystals

Kunihiko Oka and Toshimitsu Ito

Electrotechnical Laboratory, 1-1-4 Umezono, Tsukuba, Ibaraki, 305 Japan.

Single crystal of $REBa_2Cu_3O_{7-y}$(RE= Y, La, Pr, Nd and Sm) have been grown by the travelling-solvent floating-zone (TSFZ) method with low oxygen pressure atmosphere and twice-scanning technique. Single crystal of $Bi_2Sr_2CaCu_2O_{8-y}$ have been grown by the top-seeded solution growth (TSSG) method with $Bi_{2.4}Sr_{1.5}Ca_{1.0}Cu_{1.8}Ox$ as the solvent.

Introduction

It is very important to obtain large single crystals of $REBa_2Cu_3O_{7-y}$ (RE= rare earth elements and Y) and $Bi_2Sr_2CaCu_2O_{8-y}$ for the measurement of physical properties and the development of the superconducting devices. $REBa_2Cu_3O_{7-y}$ and $Bi_2Sr_2CaCu_2O_{8-y}$ are known to be incongruently melting substances. We had to grow single crystals of these substance only from solution, eg., slow-cooling, travelling-solvent floating-zone (TSFZ) and top-seeded solution-growth (TSSG) method. it is very difficult to grow them from solution containing CuO. We have been succeeded crystal growth of $REBa_2Cu_3O_{7-y}$(RE= Y, La, Pr, Nd and Sm) by the TSFZ method (1) and $Bi_2Sr_2CaCu_2O_{8-y}$ by the TSSG method (2). In this note we describe the procedure of growing large single crystals of $REBa_2Cu_3O_{7-y}$ and $Bi_2Sr_2CaCu_2O_{8-y}$, and present some problems on crystal growth.

Crystal growth of $REBa_2Cu_3O_{7-y}$ by the TSFZ method

The TSFZ method is a valuable growth technique to produce large single crystals without contamination from crucible materials. $REBa_2Cu_3O_{7-y}$ has not been grown successfully by the TSFZ method, because the solvent falls easily down onto the seed and penetrates deeply into the feed rod. We found it is possible to keep the stable molten zone with the $REBa_2Cu_3O_{7-y}$ system by using a low oxygen pressure atmosphere and a twice-scanning procedure. A low oxygen pressure atmosphere stabilizes the molten zone. The twice-scanning technique has been applied in order to avoid the penetration into the feed rod.
Single crystals were grown in infrared radiation furnaces with a 750W halogen lamp as radiation source as shown in Fig.1. The starting feed and solvent materials were prepared from Y_2O_3, La_2O_3, Pr_6O_{11}, Nd_2O_3, Sm_2O_3, $BaCO_3$ and CuO of 99.9% purity. The feed rod and solvent zone consist of RE, Ba and Cu mixture at 1 : 2 : 3 and 1 : 3 : 6 - 1 : 16 : 22, respectively. The well-mixed powder in the required molar ratios were first calcined at 880°C for 12h, then after grinding and milling they were formed into cylindrical shape of 6-8 mm in diameter and 10 cm in length by pressing at a hydrostatic pressure of about 3.5 k bar. The $REBa_2Cu_3O_{7-y}$ ceramic polycrystalline feed rod and the solvent material was then sintered at 950-1030°C and 900-920°C for 15h in air, respectively. About 1.0-1.5 g of solvent was connected to the sintered feed rod. A "twice-scanning" technique has been applied in order to avoid the penetration into feed rod. First, the feed rod was suspended at the bottom of the upper shaft ; the seed, made from a sintered rod, was held at the top of the lower shaft. The feed and the growing crystal were rotated at about 20 rpm in opposite directions. The lamp power was elevated gradually until the solvent melted. The molten zone was attached to the top of the seed. In order to densify the feed rod the molten zone was passed through the feed rod at a relatively high rate of about 5.0 mm/h in 0.1% oxygen mixed argon gas.
Secondly, the regular growing procedure was carried out, using once-scanned rod as the feed rod to which the solvent of about 0.3-0.5 g was connected. The growth rate was in the

range of 0.5 mm/h in 0.1% oxygen mixed argon gas. The twice scanning technique has been achieve stable growing processes at low growth rate.

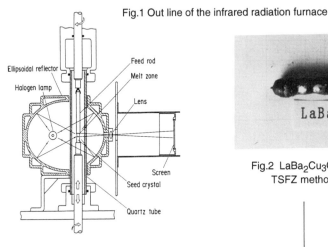

Fig.1 Out line of the infrared radiation furnace

Fig.2 LaBa₂Cu₃O₇-y crystal grown by the TSFZ method

A crystal grown by the TSFZ method is shown in fig.2. The X-ray diffraction pattern of twice scanning technique process for $LaBa_2Cu_3O_{7-y}$ system were observed using Cu Kα line as incident radiation. The first scanned crystallized feed bar for rate of 5 mm/h showed $La_4Ba_2Cu_2O_x$ phase and $LaBa_2Cu_3O_{7-y}$ phase as shown in fig.3 (a). The as grown crystal had only tetragonal $LaBa_2Cu_3O_{7-y}$ phase as shown in fig.3 (b) and showed no superconductivity. The as-grown crystal was annealed at 500 °C for 25 h, cooled at 250°C/h, held at 400°C for 60 h in an oxygen atmosphere, and cooled down slowly to room temperature. The oxygen annealed crystal showed orthorohmbic symmetry as shown in fig.3 (c).Some results of characterization of crystals are summarized in Table 1.

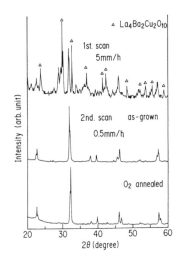

Fig.3 X-ray diffraction pattern of 1st. scanned feed rod and grown crystals of $LaBa_2Cu_3O_{7-y}$

Table 1 Characterization of the $REBa_2Cu_3O_{7-y}$ crystals

RE element involved	Y	La	Pr	Nd	Sm
size of crystal boule (mm x mm)	4 x 15	5 x 45	4.5 x 16	5 x 15	5 x 35
onset of Tc (K)	91.5	90	none	91	94
secondary phase	weak	very weak	none	very weak	very weak
grain area (mm x mm)	1 x 1	3 x 3	1.5 x 1.5	1.5 x 1.5	3 x 2

Crystal growth of $Bi_2Sr_2CaCu_2O_{8-y}$ by the TSSG method

Large single crystal of $Bi_2Sr_2CaCu_2O_{8-y}$ has been grown by TSFZ method (3,4). We have made attempt to grow the $Bi_2Sr_2CaCu_2O_{8-y}$ crystal by the TSSG method with seed crystal which, in principle, is equivalent with the TSFZ method. The crystals were grown in a 8 kW resistively heated furnace (Fig.4) and an rf heated ADL-MP furnace (Fig. 5). The 50 x 35 mm platinum crucible is heated by SiC heating elements and the furnace temperature is regulated by a SCR program controller, and also by induction from rf power source working at 400kHz and 30kW.

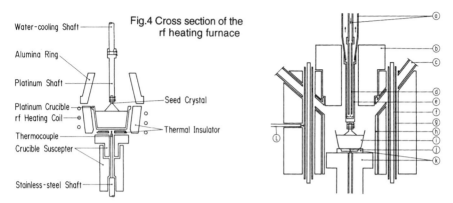

Fig.4 Cross section of the rf heating furnace

Water-cooling Shaft

Alumina Ring

Platinum Shaft

Platinum Crucible

rf Heating Coil

Thermocouple

Crucible Suscepter

Stainless-steel Shaft

Seed Crystal

Thermal Insulator

Fig.5 Cross section of the resistance heating furnace.
(a) Air-cooled shaft (stainless steel). (b) Plug. (c) Viewing port. (d,e) Air cooled shaft (platinum, platinum/rhodium). (f) Heating element guard tube. (h) Alumina tube. (i) Crucible (alumina or platinum). (j,l) Thermocouple. (k) Crucible support.

The starting material in a platinum crucible was heated to complete the melting Then the seed was dipped into the melt, and which was slowly cooled. Pulling was then started after the diameter of the crystal was recognized to be increased. At the end of the run, the crystal was removed from solution and kept suspended above the surface while cooling to room temperature. In case of the resistance-heated furnace, the temperature gradient above melt was 5°C/cm and the drawing double tube shaft was cooled by air at a flow rate of 20 liter/ min. The temperature gradient above melt of rf-heated furnace was 200°C/cm.The growth conditions were as the following (Table 2).

Fig.6 $Bi_2Sr_2CaCu_2O_{8-y}$ crystal

Table 2 The growth condition of TSSG method for $Bi_2Sr_2CaCu_2O_{8-y}$

Starting composition	Bi : Sr : Ca : Cu
	2.4 : 1.5 : 1.0 : 1.8
growth rate (mm/h)	0.3-0.5
crystal rotation (rpm)	30-40
melt cooling rate (°C/h)	0.5-1.0
atmosphere	air
cooling rate (°C/h)	20-50

A crystal boule grown by the TSSG method is shown in Fig.6. The crystal has dimensions of 10 mm in diameter, 6 mm in length and is 4 g in weight. The concentration ratio of grown crystal were examined on X-ray fluorescent analysis.

We examined different compositions of the starting solution to find optimum growth conditions, and found that the growth from a solution with composition near $Bi_{2.4}Sr_{1.5}Ca_{1.0}Cu_{1.8}O_x$ yields crystals of good quality. Figure 7 is a triangular coordinate diagram showing the composition of $Bi_2Sr_2CaCu_2O_{8-y}$ crystals grown from $BiO_{1.5}$-(Sr,Ca)O-CuO solution by the TSSG method.

Measurement of the temperature dependence of the magnetic susceptibility at 8.6 Oe was performed on the as-grown crystal. The $Bi_{2.2}Sr_{1.8}Ca_{1.0}Cu_{1.9}O_x$ crystal exhibited superconductivity at about 86 K as shown in Fig.8.

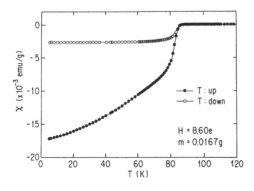

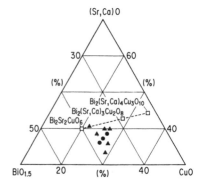

Fig.7 Temperature dependence of the magnetic susceptibility at 8.6 Oe for an as-grown shown $Bi_{2.2}Sr_{1.8}Ca_{1.0}Cu_{1.9}O_x$ crystal .

Fig.8 The composition range for the growth of $Bi_2Sr_2CaCu_2O_{8-y}$ crystals is in area (l),and both $Bi_2Sr_2CaCu_2O_8$ and $Bi_2Sr_2CuO_8$ crystal in area (s), as grown by the TSSG method from $BiO_{1.5}$-(Sr,Ca)O-CuO solution.

Conclusion

Single crystals of $REBa_2Cu_3O_{7-y}$ have been grown successfully by the TSFZ method by using a high-density feed bar and a low oxygen pressure atmosphere. The $Bi_2Sr_2CaCu_2O_{8-y}$ crystals growth was successfully done only narrow range from 34-37% $BiO_{1.5}$, 35-39% (Sr,Ca)O and 26-27% CuO solution by the TSSG method.

References

(1) K.Oka and T.Ito : Physica C 227 (1994) 77.
(2) K.Oka, T.S.Han, D.H.Ha, F.Iga and H.Unoki : Physica C 215 (1993) 407
(3) S.Takekawa, H.Nozaki, A.Umezono, K.Kosuda and M.Kobayashi: J.Cryst. Growth 92 (1988) 687.
(4) I.Shigaki,K.Kitahama,K.Shibutani,S.Hayashi, R.Ogawa, Y.Kawate, T.Kawai, M.Matsumoto and J.Shirafuji, Jpn.j.Appl.Phys. 29 (1990) L2013.

Inst. Phys. Conf. Ser. No 148
Paper presented at Applied Superconductivity, Edinburgh, 3–6 July 1995
© 1995 IOP Publishing Ltd

THERMAL STABILITY OF Bi CUPRATES

J. Hauck, K. Bickmann, S. Chernyaev and K. Mika
Institut für Festkörperforschung, KFA Forschungszentrum Jülich, D–52425 Jülich

Abstract:
The pseudobinary phase diagrams of Bi 2 2 0 1, 2 2 1 2 and 2 2 2 3 are closely related to the CuO_x and $SrCuO_x$ systems. $Bi_{2.1}Sr_{1.9}Ca_2Cu_3O_x$ with $10.0 \leq x \lesssim 10.3$ is a stable phase at 800 – 950 °C and $10^{-3} \leq p(O_2) \lesssim 100$ bar. Tapes of the 2 2 2 3 phase can be obtained by annealing of a 2 1.7 1.3 2, 0 0.2 1.8 1, CuO phase assemblage at 830 – 870 °C in air, rapid cooling to room temperature and rolling or pressing before the next heat treatment. The Bi 2 2 1 2 compound is stable above ∼ 700 °C at $8.04 \leq x \leq 8.3$ and $10^{-4.5} \leq p(O_2) \lesssim 10$ bar. It decomposes to 2 1.25 1.75 2 and various other decomposition products at long annealing below 700 °C. The decomposition of 2 2 1 2 to melt, $Sr_8Ca_6Cu_{24}O_{41}$ and $Bi_2Sr_2CaO_6$ ($x \approx 7.9$) above 891 °C and $p(O_2) \approx 0.5 - 1$ bar is superior to the usual melt texture process with a decomposition to melt, $Sr_{0.5}Ca_{0.5}CuO_2$ and $Bi_2Sr_2CaO_6$ ($x \approx 7.6$) at 883 – 891 °C in air.

(1) Introduction
The binary or pseudobinary phase diagrams of CuO_x, $SrCuO_x$, $Bi_2Sr_2CuO_x$, Bi_2Sr_2Ca Cu_2O_x and $Bi_2Sr_2Ca_2Cu_3O_x$ show the variation of oxygen content x with temperature T and oxygen partial pressure $p(O_2)$ (fig. 1–5). The five systems are related by the successive addition of components C = Sr, Bi, Ca to the binary system CuO_x. The parameters $p(O_2)$, T and x of the binary system CuO_x (C = 2) can not be varied (F = 0) according to the Gibbs phase rule F + P = C + 1 (at constant total pressure), if P = 3 phases are in equilibrium. Cu_2O, CuO and the liquid L_2 with the composition $CuO_{0.65}$ coexist e.g. at 1091 °C and $p(O_2) \approx 0.6$ bar (figure 1). The number of phases P = C + 1 at F = 0 increases at increased numbers of components in $SrCuO_x$ (C = 3), $Bi_2Sr_2CuO_x$ (C = 4), $Bi_2Sr_2CaCu_2O_x$ and $Bi_2Sr_2Ca_2Cu_3O_x$ (C = 5). The number of phases of the latter two systems however is usually reduced to P = 5, because of the similarity of Ca and Sr and the formation of solid solutions in the decomposition products (1). Each vertical line of the binary or pseudobinary phase diagrams contains one phase for CuO_x (C = 2), one or two phases for $SrCuO_x$ (C = 3) and up to three phases in the other systems. The decomposition products of the latter systems are characterized by the ratio BiSrCaCu and can be projected in a Bi, Cu, (Ca,Sr) triangle (fig. 6).

(2) Experimental
The pseudobinary phase diagrams were obtained from log $p(O_2)/T^{-1}$ data for CuO_x (2), $SrCuO_x$ (3) 2 2 1 2 and 2 2 2 3 (4) various other phase diagrams, microprobe measurements (5), thermogravimetric analyses at constant $p(O_2)$ and x–ray analyses. The thermogravimetric analyses show the variation of x with temperature, the x–ray analyses of quenched samples the decomposition products, if equilibrium conditions were obtained at higher temperatures.

(3) Results
$Bi_2Sr_{3-y}Ca_yCu_2O_x$ can be obtained at $1 \leq y \leq 1.75$ and $7.9 \leq x \leq 8.3$. The y = 1 compound (2 2 1 2) is stable above ∼ 700 °C at $8.04 \leq x \lesssim 8.3$ and $10^{-4.5} \leq p(O_2) \lesssim 10$ bar. It decomposes to 2 1.25 1.75 2 and various other decomposition products at long

annealing below 700 °C. Metastable phases with x = 7.9 were obtained by reaction with H_2 at 210 – 310 °C. The lattice parameters a, c and the T_c values depend on the O and Ca content x and y by the approximations a(pm) = $-y$ – 3x + 563.2, c(pm) = $-30y$ + 19x + 2964, T_c^{on}(K) = $-19y$ + 12x + 14 at x < 8.22.

$Bi_{2.1}Sr_{1.9}Ca_2Cu_3O_x$ with $10.0 \leq x \lesssim 10.3$ can be obtained at 800 – 950 °C and $10^{-3} \leq p(O_2) \lesssim 100$ bar.

$Bi_{2.1}Sr_{1.9}CuO_x$ with $6.07 \leq x \leq 6.17$ melts at ~ 910 °C and $p(O_2) \sim 10$ bar. This phase is frequently observed as a decomposition product of Bi 2 2 1 2 and 2 2 2 3.

The melt–texture process of 2212 can be performed by

(A) decomposition of the 2 2 1 2 compound with x $\approx$ 8.04 at $\sim$798 – 865 °C and $p(O_2) \approx 10^{-3}$ bar to the phase assemblage melt, $Sr_{0.2}Ca_{1.8}CuO_3$ and Cu_2O with a total oxygen content x $\approx$ 7.25.

(B) decomposition of 2 2 1 2 with x $\approx$ 8.1 to melt, $Sr_{0.5}Ca_{0.5}CuO_2$ and $Sr_{0.2}Ca_{1.8}CuO_3$ (x $\approx$ 7.54) at $\sim$865 – 883 °C and $p(O_2) \approx 10^{-1.5}$ bar.

(C) decomposition to melt, $Sr_{0.5}Ca_{0.5}\dot{C}uO_2$ and $Bi_2Sr_2CaO_6$ (x $\approx$ 7.6) at $\sim$883 – 891 °C and $p(O_2) \approx 10^{-1}$ bar or air.

(D) decomposition to melt, $Sr_8Ca_6Cu_{24}O_{41}$ and $Bi_2Sr_2CaO_6$ (x $\approx$ 7.9) above 891 °C and $p(O_2) \approx 0.5$ – 1 bar.

(E) A combination of process (A) and (D) with a prereaction of 2 2 1 2 at $\sim$ 800 °C and 10^{-3} bar to x $\approx$ 8.04 before the decomposition to x $\approx$ 7.9 to minimize the variation of oxygen content to Δx $\approx$ 0.14.

Oxygen bubbles are formed in the silver tube in process (A) because of the large difference of oxygen content Δx $\approx$ 0.8. The process (B) and (C) are successful, if the temperature variation is performed in such a way that small crystals of $Sr_{0.5}Ca_{0.5}CuO_2$ are grown. The $Sr_8Ca_6Cu_{24}O_{41}$ phase formed in process (D) and (E) can react faster to 2 2 1 2 than $Sr_{0.5}Ca_{0.5}CuO_2$. The 2 2 1 2 samples obtained in the melt–texture process should be cooled quickly with $\sim 1000°$/hr to prevent decomposition at decreased temperatures.

Tapes of the 2223 phase can be obtained by annealing of a 2 1.7 1.3 2, 0 0.2 1.8 1, CuO phase assemblage at 830 – 870 °C in air, rapid cooling to room temperature and rolling or pressing before the next heat treatment.

References

(1) S. Chernyaev, J. Hauck, A. Mozhaev, K. Bickmann and H. Altenburg, Physica C **244**, 139 (1995)
(2) Phase diagrams for high T_c Superconductors, J.D. Whitler and R.S. Roth, eds., The Am. Ceram. Soc. (1993)
(3) R.O. Suzuki, P. Bohac and L.J. Gauckler, J. Am. Ceram. Soc. **75**, 2833 (1992)
(4) J.L. MacManus–Driscoll, J. Bravman, R.J. Savoy, G. Gorman and R.B. Beyers, J. Am. Ceram. Soc. **77**, 2305 (1994)
(5) P. Majewski, Adv. Mater. **6**, 460 (1994)

189

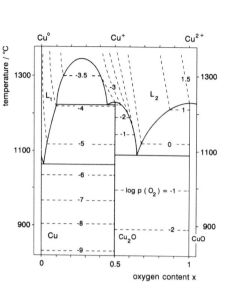

Fig. 1: CuO$_x$ phase diagram

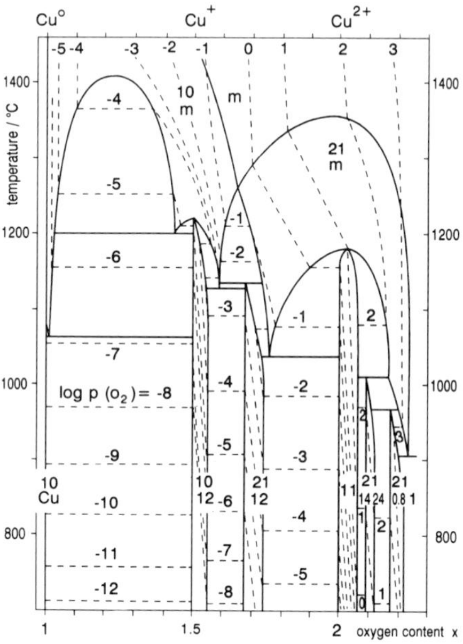

Fig. 2: SrCuO$_x$ phase diagram

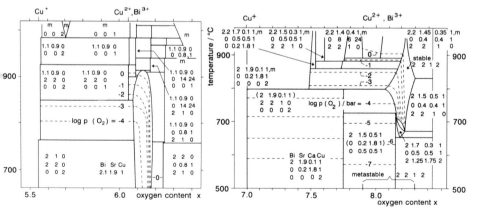

Fig. 3: Bi$_{2.1}$Sr$_{1.9}$CuO$_x$ phase diagram

Fig. 4: Bi$_2$Sr$_2$CaCu$_2$O$_x$ phase diagram

190

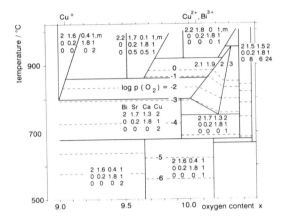

Fig. 5: $Bi_{2.1}Sr_{1.9}Ca_2Cu_3O_x$ phase diagram

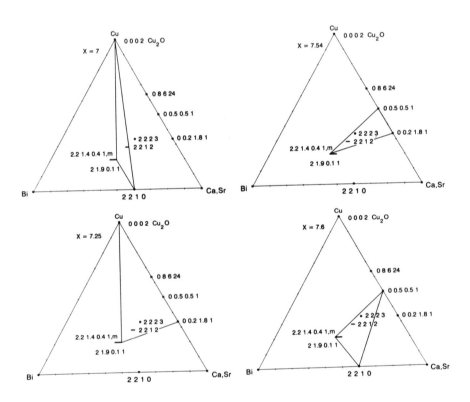

Fig. 6: Decomposition products of 2 2 1 2 in the Bi–Cu–Ca,Sr system

Inst. Phys. Conf. Ser. No 148
Paper presented at Applied Superconductivity, Edinburgh, 3–6 July 1995
© 1995 IOP Publishing Ltd

YBCO-Film formation on ceramic substrates by dip-coating and post-annealing

G. Risse, K. Fischer, B. Schlobach, D. Schläfer

Institut für Festkörper- und Werkstofforschung Dresden e. V., Helmholtzstraße 20, 01069 Dresden, Germany

Abstract

YBCO films were grown using precursor films, which were prepared by dip-coating on single crystalline ($SrTiO_3$ and $LaAlO_3$) and polycrystalline substrates (ZrO_2). Annealing conditions were varied in the temperature range between 740 and 925 °C in N_2-0.5 %O_2-atmosphere and between 900 and 960 °C in pure O_2. The crystallization behaviour of the 123-phase and the orientation of these films were observed. In the range between 840 and 850 °C the randomly oriented 123-phase fraction disappears and only 001 and (on single crystalline substrates) 100 textured $YBa_2Cu_3O_x$ remains. These films can be subsequently densified and smoothed by heating up to 915 - 925 °C for a short period of time. In-plane aligned films with FWHM up to 0.4 ° were obtained at temperatures between 920 and 930 °C in pure O_2. No a-oriented $YBa_2Cu_3O_x$ was formed on polycrystalline ZrO_2. The 123-crystals formed on this material above 850 °C are similar to these grown from a partial molten state.

Introduction

There have been many reports about the formation of YBCO-films on ceramic substrates, mostly on single crystalline $SrTiO_3$ and $LaAlO_3$, from precursor films that were made from a solution of metal organic compounds [1] [2] [3]. It was pointed out that precise control of annealing parameters, such as temperature and pO_2, is essential for producing high quality YBCO-films. The objective of this work was to determine annealing conditions of YBCO-films on single crystalline $SrTiO_3$ and $LaAlO_3$ substrates and to investigate the formation of YBCO-films on polycrystalline ZrO_2-substrates.

Experimental

Precursor films were prepared by dip-coating on single crystalline $SrTiO_3$ and $LaAlO_3$ and polycristalline ZrO_2 substrates. The coating solution was prepared from Y, Ba and Cu-acetylacetonates as described in [1] but using trichlorethylene instead of methanol suggested there. With trichlorethylene as the solvent, the concentration of the coating solution could be increased that allows to prepare smooth and faultless films up to 0.5 μm thickness after calcination in one step. The film thickness was controlled by varying the lifting speed and viscosity. The films then were dried and pyrolysed at 600 °C in air. The coating procedure can optionally be repeated so as to increase the film thickness. The results described in the following were obtained for films of 0.5 μm thickness. The prefired films were annealed at temperatures of 740 up to 925 °C in a N_2-0.5%O_2-atmosphere and in pure O_2 between 900 and 960 °C. Thereafter the films were slowly cooled down to room temperature in pure O_2-atmosphere. The film morphology was studied by scanning electronmicroscopy and the $YBa_2Cu_3O_x$-orientation was investigated by x-ray diffractometry. A standard four-point probe technique was used to measure the superconducting transition temperature. A profilometer was used to measure surface roughness and the thickness of the films.

Results and Discussion

Figure 1 shows x-ray diffraction patterns of films that were heat-treated at 750, 840 and 850 °C under a N_2-0.5%O_2-atmosphere. Diffraction patterns of films obtained below 850 °C exhibit c-axis orientation, concluded from the presence of 004 the reflection at 35.73 °. The 004 reflection of films deposited onto ZrO_2 is covered by the substrate signal and hence for these films the 003 reflection at 26.60 ° indicates the texture. For all films prepared below 850 °C a randomly oriented 123-phase was found as indicated by the 110 and 103 reflection at 38.33 °. It was not possible to enhance the texture using longer annealing times in this temperature range. Only grain coarsening was obtained. From the maximum intensities of the 110/103 (r) and the 004 (for ZrO_2 003) reflection (t) the relation $r/(r + t)$ was determined to estimate the change of the randomly oriented fraction of the 123-phase with temperature (Table 1).

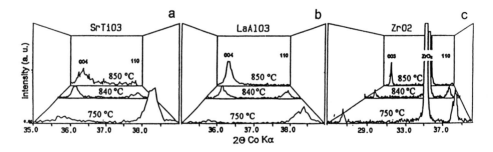

Figure 1
X-ray diffraction patterns of $YBa_2Cu_3O_x$-films on $SrTiO_3$ (100) (a), $LaAlO_3$ (100) (b) and polycrystalline ZrO_2 (c) annealed at different temperatures showing the increase of texture of the 123-phase with temperature.

Table 1
Relation $r/r + t$ (maximum intensities of the 004 (003 for ZrO_2) (t) and 110/103 reflection (r) corresponding to figure 1

Substrate	$SrTiO_3$			$LaAlO_3$			ZrO_2		
annealing temperature	750 °C	840 °C	850 °C	750 °C	840 °C	850 °C	750 °C	840 °C	850 °C
relation $r/(r + t)$	0.84	0.26	0	0.82	0.39	0	0.73	0.60	0

In the range between 840 and 850 °C the $YBa_2Cu_3O_x$ crystallization behaviour was found to change drastically with increasing temperature. Annealing temperatures of 850 °C and higher cause the 110 and 103 peak to disappear. This behaviour is found for films on all substrates examined in this study. The SEM micrographs in Figures 2 to 7 show the effect of different annealing temperatures on the film morphology. YBCO-films annealed below 850 °C consist mostly of small platelets which are clearly visible as individual crystals with different orientations (Figure 2). At temperatures above 850 °C films grown on single crystalline substrates consist of very small crystals with a very large number of extremely small pores after short time annealing (Figure 3). Films on polycrystalline ZrO_2-substrates consist of platelike crystals that are similar to crystals grown from a partial molten state (Figure 4) with a relatively strong c-axis orientation which was proved by narrow rocking not contain a-oriented YBa$_2$Cu$_3$O$_x$. Marella et al. [4] described, that during sintering of sol-gel YBCO powder at 850 °C in reduced pO$_2$ an incongruent melting of initial grains occurs and YBa$_2$Cu$_3$O$_x$ growth becomes subsequently similar to that of partial melted bulk samples. The sudden change in crystallization behaviour of YBCO-films at the same temperature observed in our study confirms these results given in [4]. The density and smoothness of the films obtained on single crystalline substrates above 850 °C can considerably be improved by a short heating up to 925 °C (Figure 5).This additional short heating was already emphasized by Chu and Buchanan [5] [6] in order to improve the film quality. Long time annealing at 850 to 900 °C results in grain coarsening. C-axis oriented as well as a-axis oriented YBa$_2$Cu$_3$O$_x$ are simultaneously growing (Figure 6). Increasing the sintering temperature above 900 °C films consisting of in-plane aligned platelike crystals were obtained (Figure 7). The strong in-plane alignment of these films was assessed by pole figure analysis using the 102 YBCO reflection. With increasing temperature an enhanced c-axis orientation of the films was found. The c-axis orientation was detected with FWHM up to 0.4 ° in films annealed at temperatures up to 930 °C for several hours in pure O$_2$-atmosphere. In agreement with [7] a considerable reaction between YBCO film and substrate was found that rises with increasing temperature. A layer of reaction products is formed immediately on the substrate surface with longer annealing times at temperatures above 850 °C.

Figures 2 to 7
Scanning electron micrographs of YBCO-films heat-treated at 840 °C for 1 hour (2), at 860 °C for 20 minutes (3) and 955 °C for 20 minutes (4), at 860 °C for 20 minutes and 920 °C for 2 minutes (5), at 860 °C for 30 minutes (6), at 920 °C for 30 minutes (7).

194

The films that were obtained at annealing temperatures above 900 °C on SrTiO$_3$ are characterised by semiconductive behaviour in the normal state and show low T$_c$ and broad transition (Figure 8 SrTiO$_3$ - curve c). Obviously, the interaction between YBCO and substrate, inparticular at very high temperatures, reduces the superconducting properties of the YBCO-films. Three reasons may explain this behaviour (1) diffusion of substrate material into the films, (2) reducing the film thickness and (3) changing the starting composition of the YBCO-material due to a different reactivity of the YBCO-components with the substrate. This leads to traces of unreacted YBCO components at grain boundaries. After annealing at 850 up to 860 °C the films are metallic in the normal state. Their T$_{c,onset}$ is relatively high but the transition is broad (Figure 8 curves a). Films short heated at temperatures between 915 to 925 °C after annealing at 850 to 860 °C show a more narrow transition width (Figure 8 curves b).

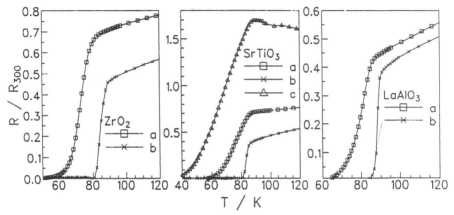

Figure 8
Normalized resistance versus temperature curves for YBa$_2$Cu$_3$O$_x$-films on SrTiO$_3$, LaAlO$_3$ and polycrystalline ZrO$_2$ annealed at different temperatures. Annealing conditions: curves a - at 850 °C for 20 minutes, curves b - at 850 °C for 20 minutes and at 920 °C for 5 minutes, curve c - at 920 °C for 30 minutes.

Conclusions

It has been shown that YBCO-films can be made from a metal organic deposition process onto single crystalline as well as on polycrystalline substrates. Although the c-axis texture is dominating during film growth superconducting properties still have to be improved. Thus an improvement requires to compromise between the overall film growth orientation and the interfacial reaction between film and substrate and will be the focus on future work.

References

[1] Kumagai, T. et al 1990 *Jpn. J. Appl. Phys.*, Vol. 30, No 1A, L28
[2] McIntyre, P.C. et al. 1992 *J. Appl. Phys.*, Vol. 71, No. 4, 1868
[3] Bauer, W., Tomandl, G. 1991 *J. Mat. Sc. Letters* 10 190
[4] Marella M. et al. 1994 *J. Mat. Sc. Letters* 13 1108
[5] Chu, P.-Y. and Buchanan, R.C. 1993 *J. Mater. Res.* Vol. 8, No 9, 2134
[6] Chu, P.-Y. and Buchanan, R.C. 1994 *J. Mater. Res.* Vol. 9, No 4, 844
[7] Manabe, T. et al. 1994 *J. Mater. Res.* Vol. 9, No 4, 85
The work was supported by the Bundesminister für Bildung, Wissenschaft und Forschung under contract 13 N 6101

Inst. Phys. Conf. Ser. No 148
Paper presented at Applied Superconductivity, Edinburgh, 3–6 July 1995

Critical Current Densities in Bi-2212 Thick Films

D. Buhl, Th. Lang, M. Cantoni and L.J. Gauckler

Nichtmetallische Werkstoffe, ETH Zürich, Sonneggstrasse 5, CH-8092 Zürich

Abstract $Bi_2Sr_2CaCu_2O_{8+\delta}$ (Bi-2212) films of 130 μm thickness with reproducible critical current densities of 5000 A/cm^2 (best value 6200 A/cm^2) at 77K/0T were produced via tape casting and partial melting. The critical current densities (j_c) are strongly depending on the conditions of the partial melting process (temperature, pO$_2$) and can significantly be increased with appropriate post partial melting heat treatments. Controlling the maximum temperature at 5 - 8°C above the solidus enables to achieve high critical current densities, but only slow cooling with 5 K/h from the maximum temperature to 850°C and isothermal annealing for 70 hours at 850°C allows to exhaust the potential for high j_c's. These improvements of the critical current densities are accompanied by microstructural changes caused by the different heat treatment steps.

1. Introduction

Partial melting has been established as an appropriate processing method for the production of Bi-2212 material with high critical current densities [1,2]. Nevertheless, the processing of Bi-2212 via a liquid phase is seriously complicated by the incongruent melting. As a consequence, different phases occur during solidification leading to a complex microstructure instead of the desired single-phase Bi-2212 [3].

The critical current densities of Bi-2212 are increasing by orders of magnitudes by decreasing the thickness of the material to films of less than 20 μm thickness and thus force the strongly two-dimensional grains, that are randomly distributed in the bulk material, into an aligned microstructure parallel to the substrate. This dependence of j_c and the texture on the film thickness is described in detail elsewhere [4].

However, concerning components for electrical devices, the more important absolute current carrying capacity is drastically decreased for thinner films even in case they have high j_c's [5]. For applications in electrical engineering, devices with macroscopic dimensions are required. Therefore, the production of Bi-2212 material with dimensions between those of thin films with high j_c and those of bulk material with low j_c becomes important.

In this work we focus on films with a thickness > 100 μm. The purpose of this study is to find appropriate heat treatment steps in order to optimize the microstructure for high critical current densities as well as a high current carrying capacity at 77K. Therefore, special emphasis is laid on the development of single-phase material with perfect Bi-2212 grains and a good grain connectivity.

2. Experiments

The Bi-2212 powder was prepared by the standard calcination process, leading to material of the stoichiometry $Bi_{2.2}Sr_{2.05}Ca_{0.95}Cu_2O_{8+\delta}$. The powder was then mixed with organic additives into a slurry and cast to 1,7m long, 150 mm wide and 350 μm thick tapes by doctor blade tape casting onto glass. Samples were then cut from the green tape, put on a 50 μm thick silver foil and subjected to the heat treatment. The details of this fabrication procedure are reported elsewhere [6].

The heat treatment consisted in removal of the organic additives, partial melting, slow cooling and annealing. The entire process was done in pure oxygen (pO_2 = 1 atm.), except an additional annealing at 500°C for 30 hours in flowing nitrogen (pO_2 < 0.001 atm) to adjust the oxygen content of the $Bi_2Sr_2Ca_1Cu_2O_{8+\delta}$ to $\delta = 0.16$, leading to an optimum superconducting transition temperature. The different steps of the heat treatment were varied as follows:

T_{max} and holding time	cooling rate from T_{max} to 850°C	holding time at 850°C
880°C 0h	5, 10, 20 and 40°C/h	2 and 70h
880°C 1h	40°C/h	70h
885°C 0h	5°C/h	5, 10, 30 and 50 h
880°C 0h	5 °C/h	2, 40 and 70 h
875, 880, 883, 885, 890°C	5 °C/h	> 50 h

In order to minimize the loss of Bismuth, the samples were processed together with a silver crucible containing powder of $Bi_2Al_4O_9$. Thus, the atmosphere was saturated with Bismuth and the Bismuth loss due to evaporization from the film was supressed.

The critical current density of the samples was determined via magnetization measurements in a VSM in liquid nitrogen. The criterion was transformed to the common 1 μV/cm criterion proposing a power law for the I-V characteristics with an exponent $\alpha = 6$ [7].

The microstructures of the samples were analyzed by means of X-ray diffraction analysis with CuK_α radiation, scanning electron microscopy and transmission electron microscopy in combination with energy dispersive X-ray analysis. The onset of melting was determined with DTA measurements on thick films on silver substrate using Na_2SO_4 as an internal standard.

3. Results and Discussion

Properties

Each processing step was perfomed using a batch of at least 10 samples. This way, two different points of view could be taken into account: the trend of the average critical current density of each batch ($j_c^{average}$) and the scattering of j_c for equally processed samples inside one batch as a function of maximum temperature, cooling rate and annealing time.

As indicated in figure 1, $j_c^{average}$ is significantly improved by about a factor 2 during a prolonged annealing from 2'200 A/cm^2 after 2 hours annealing to 5'000 A/cm^2 after 70 hours annealing. The enhancement is most pronounced in the first 10-15 hours annealing, where 80% of the increase in j_c takes place. This trend was found to be the same for batches processed with different maximum temperatures. The shape of the curve j_c versus annealing time in figure 1 remains the same for different maximum temperatures, but the j_c's for the samples heated up to 885°C are about 40% lower than for those heated up to 880°C.

The wide scattering of j_c is not reduced by annealing the samples. Investigating one batch of samples before and after annealing shows, that although $j_c^{average}$ is increased from 2'200 A/cm^2 to about 4'700 A/cm^2, the scattering of about a factor 2 is not reduced.

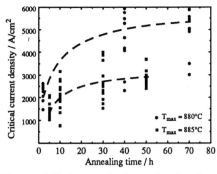
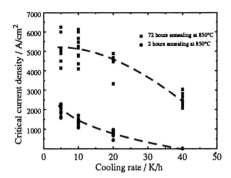

Figure 1: Critical current density as a function of annealing time for two different maximum temperatures.

Figure 2: Critical current density as a function of cooling rate before and after annealing.

The dotted lines are guides to the eyes and illustrate the tendency of $j_c^{average}$.

Figure 2 shows the critical current density as a function of the cooling rate from the maximum temperature 880°C to 850°C . The slowly cooled samples result in higher j_c's compared to the faster cooled samples. This tendency for $j_c^{average}$ remains the same before and after 70 hours annealing but is more pronounced before annealing, where the fast cooled samples (40 K/h) show no critical current density at all. This might be due to the fact, that the slower cooled samples already experienced an annealing around 850°C, whereas the faster cooled samples did not. However, even after 70 hours annealing the drop of $j_c^{average}$ is still significant: from 5'100 A/cm^2 for 5K/h cooling rate to 2'400 A/cm^2 for 40K/h cooling rate.

$J_c^{average}$ for a batch of samples, that was hold at the maximum temperature for 1 hour and fast cooled with 40K/h is slightly higher ($j_c^{average}$ = 3'200 A/cm^2) than for the batch directly cooled with the same rate, but the detrimental effect of the fast cooling rate is still significant. The trend of $j_c^{average}$ to decrease for faster cooling rates is evident, but the cooling rate has no influence on the scattering of j_c inside the different batches. The scattering remains in the region of ±25% around $j_c^{average}$

In figure 3, the influence of the maximum temperature on the critical current density is shown. A maximum temperature of 875°C (which is the onset of melting determined by DTA measurements) leads to no critical current densities. Shifting the maximum temperature 5°C above the solidus results in the highest $j_c^{average}$ = 5'100 A/cm^2. In comparison, the samples heated up 8°C above the solidus show already a slightly decreased $j_c^{average}$. Processing 10 and more degrees above the solidus temperature further decreases the critical current densities to $j_c^{average}$ = 1'800 A/cm^2.

Altough the maximum temperature strongly influences $j_c^{average}$, allowing only high values in a narrow range of 5 - 8°C above the solidus, the scattering of j_c inside one batch of samples is not affected. It remains in the region of ±25% around $j_c^{average}$ for every maximum temperature.

Every processing step significantly influences the critical current density and can not be compensated by the other steps. The potential for high j_c's is given by controlling the maximum temperature inside a narrow gap and can be exhausted only by subsequent slow cooling and long-time annealing. Following these optimized processing steps leads to a $j_c^{average}$ of about 5'000 A/cm^2. However, the scattering of the values could not be eliminated by the processing steps. Thes suggests, that there exists still a basic intrinsic parameter that is not affected by the heat treatment steps but strongly influences j_c. Correlating this intrinsic feature with the degree of texture is a reasonable assumption.

198

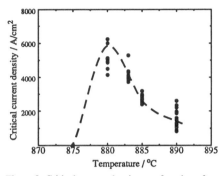

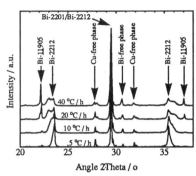

Figure 3: Critical current density as a function of maximum temperature. The dotted line is a guide to th eye and shows the tendency of $j_c^{average}$.

Figure 4: XRD patterns of samples with different cooling rates. The four phases visible are Bi-2212, Bi-$\underline{1}$1905, Cu-free 9$\underline{11}$50 and Bi-free 068$\underline{24}$.

Microstructure

The XRD patterns of the samples cooled at different rates are shown in figure 4. Four different phases are clearly visible: a Cu-free phase of the stoichiometry 9$\underline{11}$50 (determined by EDX), a Bi-free phase with the cation ratio 068$\underline{24}$ and the two superconducting phases Bi-$\underline{1}$1905 and Bi-2212. The volume fraction of the phases is changing for different cooling rates. Fast cooling with 40K/h promotes the crystallization of Bi-$\underline{1}$1905 out of the liquid or at least Bi-2212 grains with a high amount of Bi-$\underline{1}$1905 intergrowths (asymmetric widening of the Bi-2212 reflections towards the Bi-$\underline{1}$1905 reflections). This can be explained by the reduced oxygen uptake for faster cooling. The actually stable Bi-2212 phase only crystallizes directly from the melt, if sufficient oxygen can enter the sample. This is the case only for slow cooling at 5-10K/h. The appearance of the Bi- and Cu-free secondary phases cannot be avoided by changing the cooling rate. However, slow cooling enables them to resolve in the liquid upon crossing the solidus temperature, providing the cations for the crystallization of the superconducting phase. .

During annealing at 850°C the liquid still present after slow cooling at the grain boundaries is transformed to Bi-2212 , hence the grain connectivity is improved. At the same time the transformation of Bi-$\underline{1}$1905 to Bi-2212 takes place and the Bi-2212 grains are perfectionned. A study going more into details of the microstructural changes during annealing is published elswhere [8].

4. REFERENCES

[1]   J. Kase, N. Irisawa, T. Morimoto, K. Togano, H. Kumakura, D.R. Dietderich, H. Maeda, Appl. Phys. Lett. **56** [10] (1990) 970-972.
[2]   W. Paul, Th. Baumann, Physica C **175** (1991) 102-110.
[3]   B. Heeb, S. Oesch, P. Bohac, L.J. Gauckler, J. Mater. Res. **7** [11] (1992) 2948-55.
[4]   T. Lang, D. Buhl, M. Cantoni, Z. Wu, L.J. Gauckler, this issue.
[5]   B. Heeb, L.J. Gauckler, Proc. of the 4th World Congress on Superconductivity, Orlando Fl, July 1-4, (1994), in press.
[6]   D. Buhl, T. Lang, B. Heeb, L.J. Gauckler, Physica C **235-240** (1994) 3399-3400.
[7]   W. Paul, J.P. Meier, Physica C **205** (1993) 240-246.
[8]   B. Heeb, L.J. Gauckler, J. Mater. Res. **8** [9] (1993) 2170-2176.

Inst. Phys. Conf. Ser. No 148
Paper presented at Applied Superconductivity, Edinburgh, 3–6 July 1995

Enhanced Properties in Screen Printed $YBa_2Cu_3O_{7-\delta}$ Thick Films with Additions of Platinum and / or $Ba_4Cu_{1+x}Pt_{2-x}O_{9-z}$

J. Langhorn, Y.J. Bi, J.S. Abell

School of Metallurgy and Materials, University of Birmingham, Edgbaston, Birmingham.
B15 2TT

Abstract. Screen printed thick film technology is recognised as an inexpensive and effective means for the production of superconductors with potential applications to the electronics and microwave device industries. Platinum and $Ba_4Cu_{1+x}Pt_{2-x}O_{9-z}$ (0412) additions have been made to thick films of YBCO, and have shown significant improvements in the physical properties of the processed films, in particular critical current densities (Jc). Controlled concentration additions have yielded Jc values in excess of 5×10^3 Acm^{-2} at 77K and zero applied field, an increase by a factor of three over undoped specimens. Optical and transmission electron microscopy of thick film sections show a highly refined distribution of sub-micron sized 211, the larger precipitates being highly anisotropic in nature. TEM analysis shows an increased density of dislocations associated with the 123 / 211 interface with this refinement of 211 morphology, suggesting that smaller 211 precipitates may act as heterogeneous nucleation sites for flux pinning defects. Doping additions have also yielded increased homogeneity, and optimised Tc's in the system by aiding oxygen diffusion through the system.

1. Introduction

Screen printing technology may offer important advantages in the production of superconducting oxide coatings, especially in the deposition of large areas and / or complex shapes; for example in microwave components and magnetic shielding applications [1]. It has now been well established that significant improvements in superconducting properties can be achieved in thick films of YBCO on yttria-stabilised zirconia (YSZ) substrates by employing a processing regime above the peritectic temperature [2,3].

An inherent, and very prominent microstructural feature of melt processed bulk 123 YBCO is the formation of non-superconducting Y_2BaCuO_5 (211) precipitates. These precipitates are formed as 123 YBCO decomposes incongruently ($123 \rightarrow 211$ + liquid) upon heating above the peritectic temperature (Tp) (which is approximately 1030°C in a pure oxygen atmosphere), and have been reported to aid in flux pinning, increasing Jc [4,5]. The enhancement in superconducting properties can be attributed to microstructural improvements achieved as a result of partial melting, *i.e.* increased densification, grain growth, and the presence of relatively large 'globular' shaped 211 particles non-uniformly distributed throughout the film.

Additions of Pt or PtO_2 appear to have a dramatic effect on the size, morphology and distribution of 211 precipitates in melt textured bulk YBCO specimens and have subsequently been shown to increase Jc's in processed material [6-10]. It has been shown that platinum metal preferentially reacts within the YBCO system on heating to form $Ba_4Cu_{1+x}Pt_{2-x}O_{9-z}$ (0412) [9], which it has been reported may act as heterogeneous nucleation sites for the deposition of Y_2BaCuO_5 (211) precipitates.

Here we report the results of incorporating Pt and Pt containing compounds into YBCO thick films, and discuss the microstructural and superconducting properties of melt processed films on YSZ substrates.

2. Experimental

Material inks for screen printing of the required viscosity were prepared by intimately mixing commercially available YBCO powder (Rhone Poulenc 'superamic' Y123) and Pt / $Ba_4Cu_{1+x}Pt_{2-x}O_{9-z}$ doped powders with a suitable organic binder. Platinum powder (particle size 0.5-1.2μm Johnson Matthey), and 0412 (average size 2.5μm), prepared by calcining and grinding appropriate amounts of BaO, CuO, and Pt sponge, were added in varying concentrations to the precursors. Tracks of 25mm x 5mm were screen printed onto YSZ substrates to a green thickness of 90-100μm, and processed in a flowing oxygen atmosphere to a temperature of 1050°C for 6 minutes with heating rates of 5°Cmin^{-1}, and cooling rates of 4°Cmin^{-1} to 900°C, and 2°Cmin^{-1} to room temperature.

The superconducting properties of the films were examined using a dc transport critical current (77K) method, and vibrating sample magnetometry (VSM). The microstructural characteristics of the films were evaluated by XRD, optical analysis of film surfaces and polished cross sections, together with transmission electron microscopy (TEM).

3. Results

DC transport critical current characteristics, undertaken at 77K and zero applied field, of the processed films are summarised in figure 1 and show a considerable effect on intergranular Jc characteristics on addition of Pt / 0412, increasing on doping up to an optimal concentration. Results obtained from the VSM (figure 2) show a directly comparable intragranular increase in Jc in the material, from $1.8 \times 10^3 Acm^{-2}$ in control specimens to values in excess of $5 \times 10^3 Acm^{-2}$ in samples doped with 0.1 and 0.4wt% Pt and 0412 respectively. It is concluded that improved properties can be attributable to enhanced flux pinning in the films.

Optical microscopy of thick film cross sections are shown in figure 3. Control specimens (figure 3a) were observed to contain a random distribution of relatively large 211 precipitates throughout the films. Additions of Pt and 0412 (figures 3b, and c) were observed to dramatically refine the morphology and distribution of 211 throughout the film sections, and optimised doped specimens shown are observed to exhibit a high density of spherical sub-micron sized particles and a smaller volume of larger acicular precipitates.

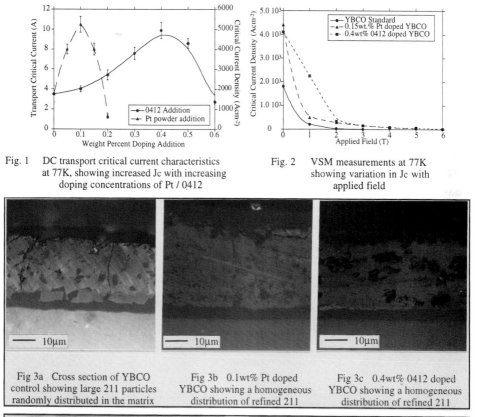

Fig. 1   DC transport critical current characteristics at 77K, showing increased Jc with increasing doping concentrations of Pt / 0412

Fig. 2   VSM measurements at 77K showing variation in Jc with applied field

Fig 3a   Cross section of YBCO control showing large 211 particles randomly distributed in the matrix

Fig 3b   0.1wt% Pt doped YBCO showing a homogeneous distribution of refined 211

Fig 3c   0.4wt% 0412 doped YBCO showing a homogeneous distribution of refined 211

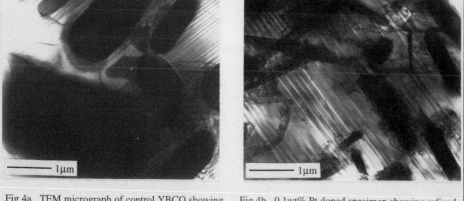

Fig 4a   TEM micrograph of control YBCO showing large 211 precipitates with relatively 'clean' interfaces

Fig 4b   0.1wt% Pt doped specimen showing refined 211 particles and increased dislocation density

202

Transmission electron microscopy (figure 4) revealed the presence of extremely fine precipitates in doped specimens together with highly anisotropic larger precipitates. The presence of these refined particles was also observed to be associated with an increased volume of dislocations at the 123 / 211 interfacial regions (figure 4b). Networks of dislocations appeared to be nucleated from high surface curvature areas of the 211 particles, being concentrated around smaller precipitates and the ends of larger acicular ones.

4. Conclusions

The addition of controlled concentrations of Pt and / or $Ba_4Cu_{1+x}Pt_{2-x}O_{9-z}$ particulates to precursor powders of YBCO for the subsequent deposition of thick film material can effectively improve the critical current characteristics of the processed films. Microstructural observations of processed films have shown that the size, morphology, and distribution of 211 precipitates within the matrix material are highly refined on doping with Pt and 0412. Smaller precipitates in the doped films are observed to be relatively spherical in nature, and the larger ones to be acicular. An increased volume of defects is observed at the 123 / 211 interface which are associated with the increased surface curvature of these precipitates. It is proposed that platinum based additions act as nucleation sites for the deposition of refined 211 particles, which in turn act as nucleation sites for dislocation networks and aid flux pinning in the films.

5. Acknowledgements

The authors wish to acknowledge the support of the School of Metallurgy and Materials at the University of Birmingham, and the financial support of the Engineering and Physical Sciences Research Council and Johnson Matthey Plc.

6. References

1] Alford, N. McN., Button, T.W., Peterson, G.E., Smith, P.A., Davis, L.E., Penn, S.L., Lancaster, M.J., Wu, Z. and Gallop, J.C., 1991 *IEEE Trans. Magn.* **27** 1510

2] Button, T.W., Alford, N. McN., Wellhofer, F., Shields, T.C., Abell, J.S. and Day, M.J., 1991 *IEEE Trans. Magn.* **27** 1434

3] Bi, Y.J., Wellhofer, F., Day, M.J. and Abell, J.S., 1993 *Mat. Sci. and Eng.* **B21** 19-25

4] Murakami, M., Gotoh, S., Fujimoto, H., Yamaguchi, K., Koshizuka, N. and Tanaka, S., 1991 *Supercond. Sci. Technol.* **4** 543

5] Lee, D.F., Selvamanickam, V. and Salama, K., 1992 *Physica C* **202** 83

6] Pavate, V., Williams, L.B., Kvam, E.P., Varanasi, C. and McGinn, P.J., 1994 *J. of Electronic Materials* **23** 1131

7] Park, J.H., Kim and H.W., Song, J.T., 1993 *J. of Mat. Sci.: Mat in Electronics* **4** 77

8] Ogawa, N., Hirabayashi, I. and Tanaka, S., 1991 *Physica C* **177** 101

9] Kim,W., Shim, G., Jang, D., Suh, C., Shin, W. and No, K., 1994 *Jpn. J. Appl. Phys.* **33** 999

10] Pavate, V., Kvam, E.P., Varanasi, C. and McGinn, P.J., 1994 *Proc. ICEM 13-Paris 17-22 July* 939

Inst. Phys. Conf. Ser. No 148
Paper presented at Applied Superconductivity, Edinburgh, 3–6 July 1995
© *1995 IOP Publishing Ltd*

Melt processing of Bi-2212 thick films and bulk components

Th. Lang, D. Buhl, M. Cantoni, Z. Wu and L.J. Gauckler

Nichtmetallische Werkstoffe, ETH Zürich, Sonneggstrasse 5, CH-8092 Zürich

Abstract. Critical current densities of melt processed Bi-2212 on Ag-substrates are remarkably depending on the component thickness in the range from 1000 µm to 10 µm. They increase from several 10^3 A/cm^2 to 10^4 with decreasing thickness.

The most obvious microstructural feature in thin samples with high critical current densities is the pronounced alignment of the platelet like grains with their c-axis perpendicular to the substrate. This texture gets lost with increasing sample thickness. For a fixed sample thickness, the melting temperature determines the amount of liquid and solid (second) phases in equilibrium. At the optimum melting temperature just enough liquid is formed to densify the sample without coarsening of the second phases which otherwise lead to inhomogeneous microstructures in the final product and suppressed critical current densities.

There exists an optimum melting temperature window of ±2°C for thin and ±8°C for thick samples. Processing below this windows leads to porous microstructures, processing above produces coarse secondary phases, Bi-loss and loss of liquid. These conditions lead to suppressed critical current densities.

Silver additions lower the solidus temperature of Bi-2212 as much as 25°C and improve the critical current density in thicker components processed under oxygen. Reduced oxygen partial pressures (e.g. 0.21 atm) lower the solidus temperature of Bi-2212 also, but lead to an increased frequency of amorphous grain boundary layers and, therefore, much lower critical current densities compared to samples processed in oxygen (pO$_2$=1 atm).

1. Introduction

Partial melting of Bi-2212 is an elegant way to produce large dense HTC components with high critical current densities [1]. The superconducting properties depend strongly on the maximum melting temperature [2, 3] and atmosphere as well as on Ag-additions. Different sets of processing parameters are reported in literature for Bi-2212 materials with optimum critical current densities. The aim of this study is to elucidate the influence of the melting temperature on the microstructure and critical current densities of Bi-2212 of various thicknesses processed with and without Ag under different oxygen partial pressures.

2. Experiments

Melting of Bi-2212 was studied using simultaneously differential thermal and gravimetrical analysis (STA). 100 mg Bi$_2$Sr$_2$Ca$_1$Cu$_2$O$_x$ powder prepared by a standard calcination process [4] were used with various amounts of Ag$_2$O additions. Heating was performed at 5 K/h to 1000°C at 1 atm total pressure in the range of 1 to 10^{-3} atm oxygen partial pressure.

Films and bulk samples in the thickness range from 10 to 1000 µm were prepared using Bi$_{2.2}$Sr$_{2.05}$ Ca$_{0.95}$Cu$_2$O$_x$ powders [4] with and without Ag$_2$O additions. Thin samples (10 µm) with 0 and 5 wt% Ag were prepared by drying butanol suspensions of the powders on Ag foil.

The thicker samples with 0 and 2.7 wt% Ag additions were prepared by axial pressing the powders at 150 MPa in tablets, subsequently they were processed on Ag-foil. The different processing parameters, e.g. maximum melting temperatures, cooling rates and isothermal annealings are summarized in table 1.

Table 1: Processing parameters

Sample thickness [μm]	Melting temperature T_{max} [°C]	Holding time at T_{max} [h]	Cooling rate to 850°C [K/h]	Annealing time at 850°C [h]
10	870-900	0.001	5	0
100	870-900	0.001	5	0
500	870-920	0.25	40	60
1000	870-920	2	40	84

To control the oxygen stoichiometry in order to ensure maximum T_c, the atmosphere was switched to $pO_2 = 0.001$ atm at 700°C during final cooling. The 10 and 100 μm samples, which were not annealed at 850°C were reduced at temperatures below 600°C for 20 h at $pO_2 = 0.001$ atm. The critical current densities were measured in an AC magnetometer at 77 K in self field. T_c was measured resistively by the four probe technique in self field.

3. Results and Discussion

Figure 1 shows the influence of the silver content and the oxygen partial pressure on the solidus temperature of $Bi_2Sr_2CaCu_2O_x$. The onset of melting was shifted to lower temperatures under reduced pO_2. Changing the atmosphere from oxygen ($pO_2 = 1$ atm) to air ($pO_2 = 0.21$ atm) led to a decrease of $T_{solidus}$ of 13°C. In nitrogen ($pO_2 = 0.001$ atm) melting started already at 835°C, e.g. 58°C earlier than in oxygen. The solidus temperature was further decreased by the silver additions. In oxygen as well as in air, $T_{solidus}$ reached a minimum at approximately 2 wt% Ag addition. The maximum reduction of the solidus temperatures due to silver was smaller under reducing conditions and leveled to a ΔT of 25°C in oxygen, 20°C in air and 10°C in nitrogen.

During melting, Bi-2212 releases oxygen leading to a weight loss of the sample. This loss was smaller at higher pO_2 and in silver containing specimens. The influence of the silver was found to be pO_2-dependent (most pronounced in oxygen, least in nitrogen). It is concluded from TG measurements, that the oxygen solubility of the liquid phase is increased when silver

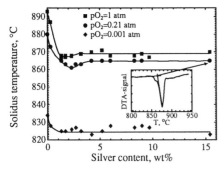

Fig. 1: Influence of silver addition and oxygen partial pressure on the solidus temperature of $Bi_2Sr_2CaCu_2O_x$. The inset shows how the solidus temperatures were determined from the DTA data.

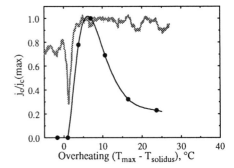

Fig. 2: Relative j_c of Bi-2212 thick films (d=10 μm) on silver without further silver addition. The background curve shows the DTA signal of the melting of such a sample in oxygen.

is present in the melt.

Processing the samples in the STA allowed a precise temperature control during the partial melt processing. The difference between the onset of the melting signal and the maximum melt process temperature is defined as *overheating* ΔT_O. This ΔT_O is less affected by absolute measuring errors of temperature and guarantees reproducible processing conditions. In figure 2, the relative critical current density of Bi-2212 thick films without silver additions - but processed on a silver substrate - is plotted against the overheating ΔT_0. The background curve shows the DTA signal of the melting reaction. The measured solidus temperature was 875°C, the maximum critical current density 8000 A/cm^2 (77 K, 0 T). If the overheating was too small ($\Delta T_0 < 5$°C), not enough liquid was formed to densify the thick film, resulting in a reduced critical current density. Overheating the samples above $\Delta T_0 \geq 10$°C led to a microstructure with a large amount of secondary phases such as $Sr_{14-x}Ca_xCu_{24}O_y$ ($0\underline{14}x\underline{24}$) and $Bi_9Sr_{11}Ca_5O_z$ ($9\underline{11}50$), resulting in a disturbed connectivity of the 2212 microstructure with a suppressed critical current density, too. In the 24°C overheated sample an increasing amount of one-layer-intergrowths in the 2212 grains was detected by XRD, whereby intergrowth free 2212 grains were obtained when cooling slowly from the optimum processing temperature. In the overheated samples the grain size of the secondary phases was larger (≥ 100 μm). Due to their size and slow kinetics they could not redesolve in the liquid during slow cooling parallel to the 2212 formation and remained in the final microstructure. In addition, overheating led to melt loss through the grain boundaries of the Ag-substrate and possible Bi-loss by evaporation. The Bi-content in the 24°C overheated sample was 15% lower compared to the one with the highest critical current density. In summary, for thick films a temperature control better than ±2°C is needed to reproducibly process in the optimal window achieving high critical current densities.

Increasing the samples thickness from 10 to 1000 μm led to a significant decrease of the critical current density from 8000 A/cm^2 for thin samples to 1800 A/cm^2 for bulk samples (fig. 3). These values correlate with the loss of grain alignment in thicker samples. In thin samples the microstructure was highly textured and the platelet like grains were aligned with their c-axis perpendicular to the substrate. By increasing the film thickness the almost perfect alignment got lost and randomly oriented bundles of 2212 grains occurred in the microstructure [5]. Surprisingly the absolute current carrying capacity increased with the sample thickness. This means that the reduction of the sample thickness led to higher j_c but this gain could not compensate the loss of cross section area. For processing bulk samples with high critical current densities the processing window is ±8°C.

The influence of silver additions on the maximum critical current density for each sample set is shown in figure 3 also. The critical current densities of Ag containing samples were equal

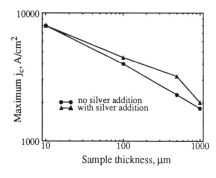

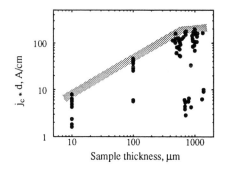

Fig. 3: Influence of the sample thickness and the addition of silver on the maximum j_c of melt processed Bi-2212.

Fig. 4: Absolute current carrying capacity of melt processed 2212 as a function of the sample thickness.

206

or higher compared to those without silver. The silver addition promoted the homogeneous melting of the entire sample cross section as soon as the solidus temperature was passed in contrast to silver free samples. In those the interface between the 2212 powder grains and the silver substrate melted at lower temperatures than the Ag-free bulk material. These samples, especially when heated slightly below the solidus temperature of the silver-free bulk material, showed large amounts of secondary phases at the interface to the silver substrate. The reaction of the substrate silver with the 2212 powder led to melting in the vicinity of the substrate only. Immediately after this melt is formed it was sucked into the Ag-free bulk by capillary forces. Without this liquid the secondary phases formed during melting at the interface could not be redesolved any more and used for the 2212 formation upon cooling. This behavior was pronounced in thicker samples and therefore Ag additions to the starting powders of those was beneficial and led to enhancement of the critical current density. In case of the thin samples (10 μm) the silver from the substrate reacting with the powder formed enough liquid to densify the microstructure at low overheatings and additional silver in the starting powder had no further influence.

The processing of the samples with a thickness of 500 μm in synthetic air instead of oxygen resulted in suppressed critical current densities by one order of magnitude. TEM investigations revealed amorphous phases in most of the grain boundaries. Such decorated grain boundaries were found in the oxygen processed samples too, but at reduced frequencies.

4. Acknowledgments

The authors gratefully acknowledge the financial support of the Schweizer Nationalfonds (NFP 30).

5. References

[1] J. Kase et al., *IEEE Trans. Mag.*, 27 (2) (1991) 1254-1257.
[2] J. Shimoyama et al., *Jpn. J. Appl. Phys.*, 31 (2) (1992) L1326-L1331.
[3] T. Kanai et al., *J. Mater. Res.*, 9 (6) (1994) 1363-1368.
[4] T. Lang et al., *Proc. for the 4th World Conf. on Supercond.*, Orlando FL, USA, June 27-July 1, (1991), in press.
[5] B. Heeb et al., *Proc. for the 4th World Conf. on Supercond.*, Orlando FL, USA, June 27-July 1, (1991), in press.

Inst. Phys. Conf. Ser. No 148
Paper presented at Applied Superconductivity, Edinburgh, 3–6 July 1995

Transport critical currents in $(Y,Gd)Ba_2Cu_3O_7$ long bars for current lead applications

X.Granados , V.Gomis , S.Piñol , M.Carrera , B.Martínez , F.Sandiumenge , N.Vilalta , J.Fontcuberta X.Obradors

Institut de Ciència de Materials de Barcelona , CSIC
Campus Universitat Autònoma de Barcelona
08193 Bellaterra , Catalunya , Spain

J.Iglesias , S.Portillo

Servicio de Técnicas Físicas , CEDEX
Alfonso XII , 3 , 28014 Madrid , Spain

Abstract: Long single domain bars (100mm , ϕ=6mm) of $YBa_2Cu_3O_7$ superconductors have been fabricated by a directional solidification technique . The transport critical current density at 77 K is $2 \ 10^4$ A/cm^2 in samples having the ab planes at 60° from the long axis and currents above 1.000 A have been injected without dissipation . It's demonstrated that the performances of the bars are good enough for current lead applications .

1. Introduction

The potential of melt procesed YBCO superconductors for the development of current leads is very high [1] but progress has been hindered by the difficulties in the fabrication of long single domain bars , where the critical current densities are much higher than the melt cast BSCCO counterparts [2] and its field dependence is also strongly reduced [3] . Directional solidification is the best alternative to growth long lengths of YBCO . A vertical Bridgman geometry has been demonstrated to be a very simple and convenient arrangement for this purposes [4,5] and a continous progress has been demonstrated in the performances of materials grown by this technique [6] or a modified version including seeding [7] .

We have recently shown that small additions of CeO_2 have very beneficial effects in this technology : it increases the capilarity sustentation in the semisolid state and refines the final size of 211 precipitates . In this work we report further progress in the fabrication of long single domain bars of YBCO superconductors with characteristics very useful for current lead applications .

2. Directional solidification

In this work we report results concerning the preparation and testing of bars having final dimensions of L=120 mm and ϕ=6 mm .Other geometries and dimensions are also being investigated for other application such as magnetic levitation and current limiters .

Ceramic preforms were prepared by sintering commercial powders of 123 , 211 and CeO_2 [7] . The size of 211 powders was reduced to d≈0.5 µm after ball milling and the content of 211 was around 25 % wt . The cross section of the bars after the melting process was 6 mm but samples having ϕ≈18 mm have also already been fabricated . The growth rate in our temperature gradient of 20 °C/cm was fixed at 1mm/h to preserve the conditions for plane front growth [9] . In Figure 1 we show a picture of one of these bars .

208

Microstructural investigations were carried out by optical microscopy , SEM and TEM . A single domain is observed after a small region (few mm) where the nucleation and competition growth processes occur . The size of 211 precipitates has been strongly reduced by our combined procedure of CeO_2 additions and ball milling of 211 precursors . In Figure 2 we show a SEM picture where 211 precipitates are identified , together with the results of an image analysis which show that the maximum of the distribution curve has been shifted to 0.3 μm [5] .

The directional solidification process has also been extended succesfully to Gd-123 . In both cases, Y-123 and Gd-123, we have found that the final orientation of a-b planes, as compared to the temperature gradient, is completely dominated by the cross section of the bars. While in thin fibers (∅ = 1 mm) the a-b plane are oriented parallel to the axis of the fibers [10], in the bars with ∅ = 6 mm the a-b plane present, typically, an angle of 60° with the axis of the bar. Microstructural observation by TEM has shown, however, that microcraks are greatly reduced in size by the small 211 precipitates [11] and this has very beneficial effects when external current is injected through the bars and there's a component parallel to the c-axis.

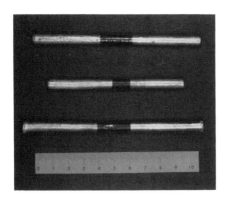

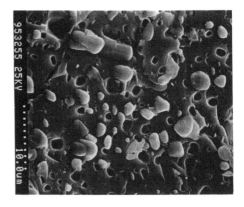

Figure 1 : Directionally solidified bar

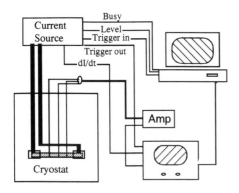

Figure 3: General schema of the pulsed critical currents experimental set up.

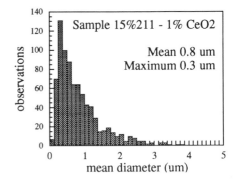

Figure 2: (a) SEM image of a fractured bar where 211 particles are observed , (b) Distribution of the size of 211 particles .

3. Superconducting properties

The inductive critical currents of small pieces of the directionally solidified bars have been investigated by SQUID magnetometry and the results have been reported elsewhere [6]. Critical currents up to $J_c^{ab} = 10^5$ A/cm^2 have been measured at 77 K when H//c and $J_c^c = 10^4$ A/cm^2 when H//ab . Electrical contacts for transport critical current measurements and current lead applications have been developed. Large area gold evaporation around the whole perimeter of the bars was applied (Figure 1), followed by a short heat treatment at 450 °C under O_2 atmosphere . Wood metal was directly applied to these Au pads and contact resistances at 77 K of 30 $\mu\Omega$ were measured (i.e. specific surface resistances of about 10^{-5} Ω/cm^2). With these electrical contacts we measured at 77 K a dissipative rate of 8 mW/A under a DC current of 100 A. This value should still be reduced to have a good performance as current lead [1], but it's very likely that after some improvement the contact resistance of the present large bars will be reduced to those observed in smaller melt textured YBCO samples [1,3].

Transport critical currents were measured at 77K in self field by applying pulses of current during a time as short as 200 μs. This technique allows to rise to very high current values avoiding heat dissipation which introduces an important indetermination in the temperature of the sample.

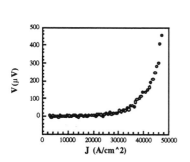

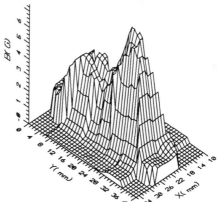

Figure 4 : Voltage vs current measured at 77 K in self field conditions .

Figure 5 : Magnetic remanence profile after field cooling under 200 Oe .

For that purpose we have designed and build the experimental set-up schematized in Fig.3 . The source of current produces current pulses by discharging a capacitor system with a whole capacity of $5 \cdot 10^{-3}$ F and a maximum voltage value of 80V. This system is previously charged at a voltage determined by an external analog control tension generated by the computer. When the computer sends the trigger signal to the current source three processes are started: the capacitor system is isolated from its charge system, the data collection process is triggered and the discharge of the capacitor is initiated . The first process is begun 1ms before the data colection is triggered and the discharge starts 50 μs later . The discharge process is driven by a MOSFET transitor bridg allowing to discharge intensities up to 3 KA and can be turned off when necessary . This allows to diminish the total energy dissipated . A digital oscilloscope collects two curves: the voltage drop along two points of the sample and the voltage induced in a small solenoid in series with the high current circuitry which provides dI/dt. This last signal allows to determine the current flow by integration . In order to avoid the inductive contribution, we read the voltage drop when the derivative dI/dt crosses the zero line . This process which provides a point in the V-I diagram is repeated increasing the voltage level of the capacitor charge and so the maximum current .

A typical experimental curve obtained with a bar having the a-b planes oriented at about 60° from the current is represented in Figure 3 where it's seen that $J_c = 2 \times 10^4$ A/cm^2 . This means that the critical current of our bars having Ø=6mm should be around 6700 A , which is much higher than the values we can generate with our experimental facility. We have, however, indeed verified that no dissipation is observed up to I = 1000 A which is already within the limits required for most current lead applications.

The last parameter which is of major concern when the performance of current leads is considered is thermal conductivity. In the present directionally solidified bars it is very useful that ab planes are oriented out of the long axis because of the anisotropic thermal conductivity in YBa$_2$Cu$_3$O$_7$ single crystals [12] , being about a factor 4 smaller along the c-axis. We hope then that values near to $\kappa_c = 2$ W/mK will be reached in these samples and this is already very similar to those observed in BSCCO melt cast processed [1].

A final concern in the fabrication of bars for current leads is the homogeneity achieved during the crystallization process. In order to test non-destructively for any possible inhomogeneity we have used the Hall probe magnetic profile technique which allows to detect grain boundaries. The potentiality of the technique is demonstrated in Figure 4 where the magnetic remanence profile of a bar is shown , after a field cooling process under 200 Oe. Transport critical current measurements showed an anomalous low Jc value in this sample and after observation of Figure 4 it may be clearly stated that some kind of microstructural defect exists which induces a discontinuity in the field profile. The origin of these microstructural defects is still unknown , but it is very likely that a macrocrack has developed during the oxygenation process. As we have previously mentioned the formation of cracks is strongly limited by the refinement of 211 precipitates, so any inhomogeneity in their distribution can induce the formation of bigger cracks which can be the main source of nuisance.

In conclusion we have demonstrated that the directional solidification process can be considered as a very promising technology for the fabrication of high performance current leads for electrotechnical applications.

Acknowledgements: We are grateful to: CICYT (MAT91-0742), Programa MIDAS (93-2331) and EC-EURAM (BRE2CT94-1011).

References

[1] P.F.Herrman et al.. IEEE Trans. on Appl. Supercon. 3, 876 (1993)
[2] J. Bock et al. IEEE Trans. on Appl. Supercon. 3, 1659 (1993)
[3] P.F. Herrman et al. Cryogenics 33, 555 (1993) ; F. Grivan et al. Physica C. 235-240, 34 (1994)
[4] K. Salama et al. in "Processing and properties of high T$_c$ superconductors". Ed. Sungho Jin. World Scientific. Singapore (1993)
[5] S. Piñol et al. Appl. Phys. Lett. 65, 1448 (1994); IEEE Trans. on Appl. Superc.(in press)
[6] B. Martinez et al. Phys. Rev B (in press)
[7] K. Salama et al. Physica C 235-240, 213 (1994)
[8] Seattle Speciality Ceramics Inc.
[9] M. Cima et al. J. Appl. Phys. 72, 179 (1992)
[10] S.Piñol et al. in these Proceedings
[11] F. Sandiumenge et al. Phys. Rev. B 50, 70 (1994)
[12] C.Uher , in "Physical properties of High Temperature superconductors" , D.M.Ginsberg ed. , Vol.3 , p.159 , (1993)

Inst. Phys. Conf. Ser. No 148
Paper presented at Applied Superconductivity, Edinburgh, 3–6 July 1995

211

AC loss of High-Tc superconductors in self-field by a conctactless method

A Díaz, G Domarco, J Maza and F Vidal

LAFIMAS, Departamento de Física de la Materia Condensada, Universidad de Santiago de Compostela, 15706 Santiago de Compostela, Spain

Abstract. A transformer method well suited for measurement of AC resistance, specially AC loss, in which a ring-shaped superconducting sample acts as a terciary winding, is presented. With this device, more than two orders of magnitude (according to the numbers of turns at the secondary) in precision may be easily gained over the conventional four-lead arrangement. Very neat signals from AC loss are in fact obtained as shown here, quite similar to those found in the voltage-tap configuration.

1. Introduction

AC loss in superconductors is a topic closely related to applications, having for instance a direct bearing on multifilamentary design, etc [1]. There are two experimental limit cases. In the first, the sample, with or without a transport current, is subjected to an alternating magnetic field and induced currents are measured by means of a pick-up coil embracing the sample. All measurements of AC susceptibility fall within this class. In the second, an alternating transport current is applied and response by means of the voltage drop is measured. No external magnetic field is applied. This paper deals with the second method which may be referred to as self-field loss since it is associated with the time-varying current themselves. Currently, measurement is implemented by using voltage taps on the superconducting specimen. One of the practical difficulties in using voltage taps on copper oxides is achieving low contacts resistance on the sample, but a more fundamental one is the fact that the lossy component of the measured voltage is very dependent on the position of the taps [2]. In this work quite a distinct experimental approach is explored. It is an inductive, and so contactless, method but at the same time it is a transport mode. A brief description follows together with first results and assessment of adequacy.

2. Experimental setup

The experimental arrangement is schematically drawn in figure 1. A ring-shaped superconducting sample is used as a terciary winding in a transformer configuration. The primary winding is driven by a function generator. The changing flux so caused acts upon the ringed sample which responds inductively. Both the secondary and the primary currents are recorded by a digital oscilloscope as a funtion of time (eventually, two lock-in amplifiers were used instead). For these recordings, low frequency signals from 35Hz to 600Hz have been used. A ferrite core (Philips 3E5) is also used to lead the magnetic flux and to close the magnetic circuit. As will be seen, voltages and/or currents at the primary and secondary windings (to be denoted as V_p, I_p and V_s, I_s respectively) allow the determination of the corresponding quantities V, I on the ring-shaped sample. By using a number of turns on the hundreds (200-400 in our case) more than two orders of magnitude in resolution can be gained over the direct four-lead arrangement.

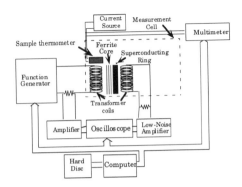

Figure 1. Experimental arrangement diagram for measuring AC resistance in self-field mode on a superconducting ring

We recall that one of the standard methods of loss determination are the electromagnetic ones, a group of which is named superconducting transformer method [1]. The impossibility of studying dissipation in the presence of transport current makes this technique quite diferent from ours. It is also worth pointing out that our transformer method has been used to measure critical currents [3] but no utilization in AC losse are known to us.

3. Results and discussion

As mentioned, the accesible quantities are the current intensity at the primary $I_p(t)$ and the secondary $I_s(t)$, both as a function of time t. Since the total load at the secondary R_s is known, the output voltage V_s is also accesible. Let us first relate these quantities with the sample's intrinsic characteristics V, I. The total magnetic flux ϕ may be expressed as

$$\phi = L\left(nI_p + I + nI_s\right) \tag{1}$$

where L is the inductance per turn and n the number of turns. The equation at the secondary is simply

$$-n\frac{d\phi}{dt} = V_s. \tag{2}$$

By combining (1) and (2) one gets on integration

$$I(t) = -\frac{1}{nL}\int_0^t V_s(t')dt' - nI_p(t) - nI_s(t) + \frac{\phi(0)}{L}. \tag{3}$$

We are interested in the "subcritical" regime $I < I_c$, I_c being the critical current, in which case the superconducting ring nearly cancels all the flux through the ferrite core and acordingly the output at the secondary is vanishingly small compared with the primary excitation. In short, from (3)

$$I(t) \cong -nI_p(t). \tag{4}$$

The sample's voltage in turn is readily obtained from the flux conservation, i.e.,

$$V = \frac{V_s}{n}. \tag{5}$$

Equations (4) and (5) map the translations into the intrinsic quantities V, I in the transformer method. Figure 2 show the outcome for a series of runs using I/I_c=0.5, 0.7, 0.9 and 0.95 respectively. Both the nonlinearity and the phase shift are as expected and quite similar to those obtained from direct four-lead measurements [4].

We have checked that no spurious effects such as flux leaks in the magnetic circuit may account for the observed behavior at the secondary. A first argument is that the nonlinearity displayed in figure 2 is incompatible with the linearity of our ferrite core permeability (in the range of very low neat fluxes as it is the case). Secondly, there is the observed fact that the secondary voltage is 90° delayed with respect to the primary current, while just the opposite is expected, as it is not difficult to see, in case of flux leakage.

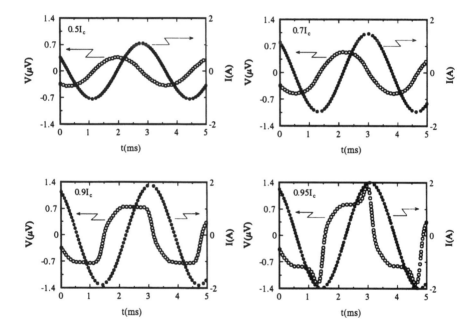

Figure 2. Time evolution of the transport current through the ring-shaped sample (solid symbols) and the total voltage (open symbols) as measured by the transformer method. Current intensity amplitude ranges from 0.5 to 0.95 of the critical current I_c as indicated in the figures. Note the increasing nonlinearity in the voltage for values near I_c.

Figure 3 illustrates the dependence of the secondary voltage on frequency for increasing driven current at the primary coil. This time both signals were fed into two phase sensitive detectors. Again this frequency behavior bears close resemblance to that found in the more conventional potential taps configuration (see, e.g. Ref. [2]). Below I_c, proportionality with frequency is characteristic of histeretic phenomena while above I_c behavior should follow that of a normal resistor, i.e., independent of frequency. Notice that the whole voltage and not only the lossy component is shown in figure 3.

214

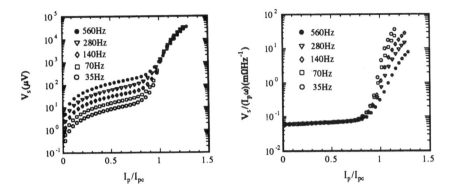

Figure 3. Frequency dependence of the secondary voltage V_s (proportional to the sample voltage) vs primary current I_p (proportional to the current through the sample) in the transformer device. Scaling shown is distinctive of hysteretic AC losses.

In sum, the transformer method presented here allow measurement of AC resistance, in particular AC loss in self-field mode, in structured superconducting samples to be made with high accuracy avoiding voltage taps. Though results agree qualitatively with those obtained from the more conventional configuration, further tests on the same samples are needed and are under way.

Acnoledgements

This work has been supported by the Spanish CYCYT (Project MAT 92-0841).

References

[1]   Kovachev V 1991 *Energy Dissipation in Superconducting Materials* (Oxford: Clarendon Press)
[2]   Ciszek M, Campbell A M and Glowacki B A 1994 *Physica C* **233** 203-8
[3]   Harris E A, Bishop J E L, Havill R L and Ward P J 1988 *J. Phys. C* **21** L673-7
       Díaz A, Pomar A, Domarco G, Maza J and Vidal F 1993 *Appl. Phys. Lett* **63** 1684-6
[4]   Basso M, Lambardi L, Marini S, Marinelli M and Morpugo A 1994 *Appl. Supercond.* **2** 71-91

Inst. Phys. Conf. Ser. No 148
Paper presented at Applied Superconductivity, Edinburgh, 3–6 July 1995
© 1995 IOP Publishing Ltd

PREPARATION, CHARACTERISATION AND TRANSPORT PROPERTIES OF EPITAXIAL $YBa_2Cu_3O_7$/$REBa_2Cu_3O_7$ SUPERLATTICES

A. Schattke, Th. Becherer, S. Eckert, G. Jakob and H. Adrian

Institut für Physik, Johannes Gutenberg Universität, 55099 Mainz, Germany

Abstract. In order to investigate the magnetic interaction between the vortex lattice and the rare earth in high temperature superconductors, measurements on $YBa_2Cu_3O_7$/$GdBa_2Cu_3O_7$ superlattices were performed. Resistive transitions in applied magnetic field up to 6 Tesla in the orientations $B \parallel c$ and $B \perp c$ were analysed in order to obtain the activation energy. The magnetic field dependence of the activation energy shows clear deviations from the typical three dimensional behaviour. The critical current density or rather the pinning force density shows scaling behaviour in agreement with predictions due to the flux creep theory.

1. Introduction and Preparation

The spatial separation of superconductivity and magnetism of the rare earth ions in the high temperature cuprates leads to a fascinating coexistence. The aim of this work is to look at the magnetic interaction between a vortex and the rare earth ions. The existence of a 'magnetic' pinning mechanism originating from the interaction between the vortex lattice and a superlattice structure consisting of alternating superconductive layers with and without magnetic ions is considered.

To this aim, superlattices of $YBa_2Cu_3O_7$/$GdBa_2Cu_3O_7$ have been prepared in situ onto $SrTiO_3$ (0 0 1) substrates by DC magnetron sputtering. The high crystalline quality of the samples was confirmed by x-ray diffraction. The superlattice structure expresses itself by additional satellites in the diffraction pattern. The position, intensity and sharpness of these satellites allow to estimate the degree of crystalline order. The patterning of the films into a suitable structure was accomplished by standard photolithographic technique. This structure includes a $2\,mm \times 200\,\mu m$ stripline for resistive measurement and a $100\,\mu m \times 10\,\mu m$ microbridge in order to perform the j_c measurement.

All voltages were measured in true four lead geometry while the dc-currents were pulsed and reversed in order to eliminate the thermovoltage. The resistive data were taken with a current density of $4 \cdot 10^3\,A/cm^2$, which is significantly below the threshold where the activation energy becomes influenced by the current itself.

2. Activation energy

The activation energy $U(B,T)$ was extracted by analysing the resistive transitions in applied magnetic field up to 6 Tesla in the orientations $B \parallel c$ and $B \perp c$. Fig. 1 depicts the resistive transitions for a superlattice of the modulation 10:40 (Y:Gd) in the orientation $B \parallel c$. The samples exhibit without exceptions metallic behaviour with $\rho(300\,\mathrm{K}) = 150\,\mu\Omega\mathrm{cm}$ and $\rho(300\,\mathrm{K})/\rho(150\,\mathrm{K}) = 2$. The resistive transitions were investigated with regard to the theoretical model of Palstra, Anderson and Kim [1]. The resistivity ρ is predicted to behave due to:

$$\rho(T,B) = \rho_0(T_c^\star, B)\exp\left(-\frac{U(T,B)}{k_B T}\right) \quad \text{with} \quad U(T,B) \approx U_0(1 - T/T_c)^q \cdot \frac{1}{B} \quad (1)$$

In fig. 2 the resistive transitions are shown as Arrhenius plots. The lines represent the theoretical model of Palstra fitted to the experimental data. The preexponential factor $\rho(T_c^\star, B)$ is chosen as the resistivity corresponding to the magnetic field dependent temperature $T_c^\star(B)$ at the inflection point. From this evaluation the activation energy $U(T,B)$ is only determined by a single parameter, the dimensionality q of the vortex. The slope of the resitive transitions in the Arrhenius plots is well described by the theoretical model with $q = 1.5$. Tinkham considered the magnetic field dependence of the activation energy, which yields $U_0 \propto B^{-\alpha}$ with $\alpha = 1$. In fig. 3 the extrapolated activation energies are plotted against applied magnetic field. The lines are fits to the data with the following exponents α:

Gd #64:	pure GdBa$_2$Cu$_3$O$_7$ film	$\alpha_\parallel = 0.4$	$\alpha_\perp = 0.1$
YGd #14:	Y:Gd = 40:10	$\alpha_\parallel = 0.6$	$\alpha_\perp = 0.4$

which are much lower than the predicted value of $\alpha = 1$.

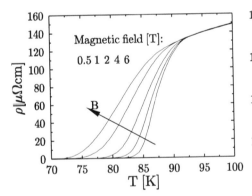

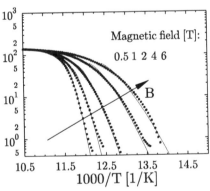

Figure 1. *Resitive transitions of a YBa$_2$Cu$_3$O$_7$/GdBa$_2$Cu$_3$O$_7$ superlattice in magnetic field up to 6 Tesla in the orientation $B \parallel c$.*

Figure 2. *Arrhenius plot of the transitions. The continuous lines are fits to the data due to the theoretical model of Palstra, Anderson and Kim [1].*

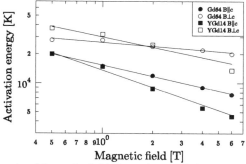

Figure 3. *Magnetic field dependence of the activation energy of a pure GdBa$_2$Cu$_3$O$_7$ film (Gd 64) and a superlattice (YGd 14) of modulation 10:40 (Y:Gd).*

3. Pinning force density

The critical current density was determined by a $1\,\mu$V criterion across the $10\,\mu$m microbridge. As proven for the classical superconductors the pinning force density $f_\mathrm{p} = |\boldsymbol{j}_c \times \boldsymbol{B}|$ obeys a scaling law:

$$f_\mathrm{p}(B,T) = (B^\star(T))^n \cdot f(b) \tag{2}$$

with $b = B/B^\star$ as reduced magnetic field. Recent investigations on the scaling behaviour of the pinning force density in high-T$_c$ superconductors were done by Hettinger *et al* [4] and Yamasaki *et al* [5] for YBa$_2$Cu$_3$O$_7$ and Bi$_2$Sr$_2$Ca$_2$Cu$_3$O$_{10}$, respectively. They report a scaling behaviour in a limited temperature range which corresponds to a theoretical scaling law based upon the flux creep theory. The crucial difference between both investigations consists in the anisotropy of the systems and the magnetic field dependence of the activation energy. The YBa$_2$Cu$_3$O$_7$/GdBa$_2$Cu$_3$O$_7$ superlattices with a typical three dimensional anisotropy parameter of $\gamma = 7$ point towards a Hettinger scaling. But the magnetic field dependence of the activation energy implies a Yamasaki scaling. From this point of view we compared our experimental data with both predictions. A convincing coincidence is found between the experimental data and the Yamasaki scaling. At the low fields regime $0 < b \ll 1$ the movement of the flux lines against the driving force is neglected. This results in the low field approximation:

$$f_\mathrm{p} = (\mu_0 K/\beta)\, B^\star \ln(E_0/E_c)\, \sqrt{b}\,(1 - (\beta^2 b)^{1/3}) \tag{3}$$

In the high field regime we cannot neglect the possibility of flux motion in the opposite direction against the driving force. By following the same procedure used to obtain the low field approximation, we get:

$$f_\mathrm{p} = (2/\pi)\mu_0\, K\, B^\star\, b\, \mathrm{asinh}\left(\frac{1}{2}\exp\left[\left(\frac{1}{\beta\sqrt{b}} - 1\right)\ln(E_0/E_c)\right]\right) \tag{4}$$

The important features of both equations are:

- linear $f_\mathrm{p}^{\max}$ vs. $B^\star$ dependence

- three independent fit parameters: K, β and $\ln(E_0/E_c)$

218

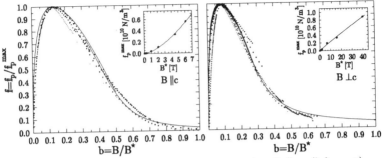

Figure 4. *Pinning force density in the orientation $B \parallel c$ (left part) and $B \perp c$ (right part). The continuous lines are fits to the data due to the Yamasaki scaling. The inserts show the $f_{\mathrm{p}}^{\mathrm{max}}$ vs. B^* dependence.*

We take the critical current density $j_{\mathrm{c}}(B)$ in applied magnetic field up to 6 Tesla at fixed temperature ranging from 60 K to 86 K. Fig. 4 shows the reduced pinning force density f_{p} as a function of reduced applied magnetic field b in the orientation $B \parallel c$ (left part) and $B \perp c$ (right part). All data for different temperatures coincide in a single curve which emphasize a scaling behaviour of the pinning force density. Fig. 4 also shows the flux creep scaling behaviour predicted by eqs. (3,4) (represented by the lines). In the inserts of the figures the dependence of the reducing parameters $f_{\mathrm{p}}^{\mathrm{max}}$ and B^* are depicted. The linear prediction is in the case of $B \perp c$ fulfilled, whereas the $B \parallel c$ orientation shows an upward curvature due to a power law $f_{\mathrm{p}}^{\mathrm{max}} \propto B^{*n}$ with exponent $n = 1.55$. The best agreement between the data and the theoretical description is reached for the following parameters:

$$B \parallel c: \quad \beta = 1.28, \ln(E_0/E_c) = 11 \quad \text{and } K_{\parallel} = 3.1 \cdot 10^8 \text{ A/m}^2$$
$$B \perp c: \quad \beta = 1.58, \ln(E_0/E_c) = 7 \quad \text{and } K_{\perp} = 2.5 \cdot 10^8 \text{ A/m}^2$$

Acknowledgments

Financially support by the Ministerium für Bildung, Wissenschaft, Forschung und Technologie der Bundesrepublik Deutschland (contract No. 13N6437) and the Deutsche Forschungsgemeinschaft through Sonderforschungsgemeinschaft 252 are greatfully acknowledged.

References

[1] Anderson P W 1962 *Phys. Rev. Lett.* **9** 309
   Anderson P W and Kim Y B 1964 *Rev. Mod. Phys.* **36** 39
   Palstra T T M, Batlogg B, van Dover R B, Schneemeyer L F and Waszczak J W 1990 *Phys. Rev. B* **41** 6621

[2] Tinkham M 1988 *Phys. Rev. Lett.* **61** 1658

[3] Geshkenbein V, Larkin A, Feigel'man M and Vinokur V 1989 *Physica C* **162–164** 239

[4] Hettinger J D, Swanson A G, Skocpol W J, Brooks J S, Graybeal J M, Mankiewich P M, Howard R E, Straughn B L and Burkhard E G 1989 *Phys. Rev. Lett.* **62** 2044

[5] Yamasaki H, Endo K, Kosaka S, Umeda M, Yoshida S and Kajiamura K 1993 *Phys. Rev. Lett.* **70** 3331

Inst. Phys. Conf. Ser. No 148
Paper presented at Applied Superconductivity, Edinburgh, 3–6 July 1995
© *1995 IOP Publishing Ltd*

The supercurrent transport and structure of grain boundaries in the $Bi_2Sr_2Ca_{n-1}Cu_nO_x$ system

Y. Yan & J.E. Evetts   IRC in Superconductivity, Cambridge University, UK
B. Soylu & W.M. Stobbs   Dept. of Materials Science & Metallurgy, Cambridge University, UK

Abstract HREM studies show that the form of the grain boundaries in the BiSrCaCuO system is unusual in exhibiting changes in the local composition as well as in the structure. Our work has demonstrated that low angle boundaries in the system exhibit compositionally modulated facetting. Although we see local regions of the low-T_c (T_c=20K) $Bi_2Sr_2CuO_x$ phase (hereafter the (2201) phase) at such boundaries which would be highly resistive at liquid nitrogen temperature, it is also clear that the form of the boundary transformations is such that, for low angles of misorientation (<~10°) there are "pathways" crossing the boundary plane made up of the high-T_c $Bi_2Sr_2CaCu_2O_x$ (T_c=80K) and $Bi_2Sr_2Ca_2Cu_3O_x$ (T_c=110K) phases (hereafter the (2212) and (2223) phases). The structural models which we have developed here for the low-angle tilt boundaries in the $Bi_2Sr_2Ca_{n-1}Cu_nO_x$ system provide a general insight into why this particular system exhibits good J_c values for low texture breadths.

1. Introduction

In the BiSrCaCuO system, the observation of higher values of the critical current density, J_c, for an aligned microstructure compared with those for randomly oriented polycrystals indicates that the form of the grain boundaries must change in a significant way in relation to their current carrying capacity as a function of the misorientation across them. However it is not clear why the critical current carrying capacity in this system can be strongly affected by the nature and form of the grain boundaries. In the present study, grain boundary structures in the "composite reaction texturing" (CRT) processed $Bi_2Sr_2Ca_{n-1}Cu_nO_x$ system [1] have been studied by transmission electron microscopy (TEM). We also discuss why high J_c values might be expected for the $Bi_2Sr_2Ca_{n-1}Cu_nO_x$ system in the presence of low tilt angle misorientations in the light of the local structure of the low angle tilt boundaries for the material.

2. Experimental procedure

The samples had a composition such that the majority phase present was 2212, and they were processed to make use of inert MgO whiskers to align the superconducting grains and control their morphology by the CRT [1]. The preparation procedure for the specimens used has been described elsewhere and involved a partial melting process to temperatures of 910°C - 920°C for short periods and then slow cooling to 840°C - 850°C for oxygenation over 20hrs [1]. For an optimized microstructure, the transport current density is high, in excess of $2 \times 10^4 Acm^{-2}$ at 4.5 K and 12 T and $4 \times 10^3 Acm^{-2}$ at 77 K in zero applied field. The high resolution electron microscopical data described here were obtained using a JEOL-4000EXII microscope with a point-to-point resolution of 0.17 nm. Thin foils were prepared by standard techniques, involving mechanical grinding and argon ion-milling at 4 keV with liquid nitrogen cooling. The precise imaging conditions for the micrographs shown here are not of importance for the characterization of the major features of the grain boundary structures of interest to us here, but the images shown were all obtained in thin areas at close to Scherzer defocus.

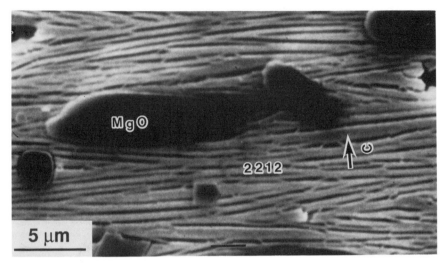

Fig. 1   SEM micrograph showing a well-c-axis-aligned cross-sectional area of a CRT 2212 specimen.

3. Results and discussion

It is relevant that the different phases exhibited by the system are closely related: they all contain BiO double layers sandwiching perovskite-type units containing Sr, Ca, Cu and O, and are characterized by n this being the number of CuO layers parallel to the (001) crystallographic plane in the half unit cell [2,3,4]. The n=1, 2, 3 phases $Bi_2Sr_2CuO_x$ ($c=2.46$ nm), $Bi_2Sr_2Ca_1Cu_2O_x$ ($c=3.09$ nm), $Bi_2Sr_2Ca_2Cu_3O_x$($c=3.7$ nm) are called (2201), (2212) and (2223) for brevity and have critical temperatures, T_c, of about 20, 80, and 110 K respectively. It should be particularly noted that the grain boundary structural data we report for the system would be expected to be of equal relevance for other $Bi_2Sr_2Ca_{n-1}Cu_nO_x$ alloys however they were processed to exhibit a tendency to c-axis alignment.

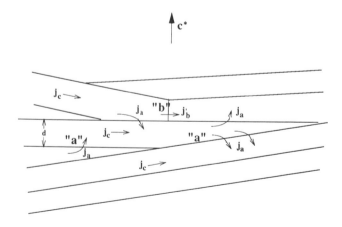

Fig. 2   A schematic of the characteristic grain structure in which two different kinds of paths, "a" across the low habit angle boundaries, and "b" across the high habit angle boundaries are labelled. The current density along the a-b plane within grains and across the grain boundaries are labeled by j_a and j_b.

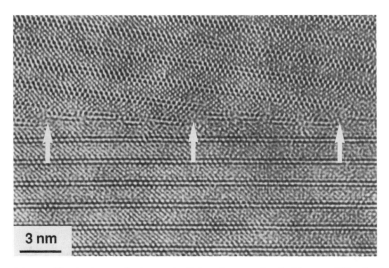

Fig. 3   High Magnification image of a low angle tilt boundary associated with a near 90° twist about the c axis, viewed along [110] and [1$\bar{1}$0] in the two respective grains. The (2201), (2212) and (2223) compounds can be seen locally at the boundary with the repeat distances for the modulated boundary structures indicated.

It has been found that the microstructure of the CRT processed (2212) system had a strong c-axis texture (perpendicular to the pellet plane). Texture goniometry studies showed that more than 50% of the diffracted intensity lay within 10° of a perfect single crystal structure. As viewed using scanning electron microscopy, the microstructure suggests that the (2212) phase tends naturally to form flat grains which can be tens of microns wide and just fractions of a micron thick in the c direction (see Fig. 1). The plate-like grains commonly have an aspect ratio of length, l, to thickness, d, ratio of greater than 10^2. A corresponding schematic of the characteristic grain structure in these materials is shown in Fig. 2. Grains are terminated by low angle wedge like intersections with habits of near (001) normal (marked "a") or less frequently in "head-on" boundaries where there can be continuity of the a-b planes from grain to grain (marked "b").

Fig. 3 shows an HREM image of a low tilt angle boundary (about 8°) though, as is common for this material, the grains are also twist misoriented by nearly exactly 90° about their c axes. The upper grain is oriented accurately at [110]. Image simulations indicate that the dots with strong dark contrast are at the positions occupied by Bi atoms and that, in general and over large areas of the image, the other black dots approximate the positions of projections of Sr, Ca and Cu columns [5]. As is discussed in more detail elsewhere [5], the characteristic of the form of such boundaries is that they are both structurally and compositionally modulated. The structure changes along the boundary, as also noted by Eibl [6], can be conveniently described using the notation (n,n'[m]), in which n and n' denote the value of n (1,2,3...) for the $Bi_2Sr_2Ca_{n-1}Cu_nO_x$ phases locally present on each side of the boundary while m denotes the number of (a/2,b/2,0) planes over which the given arrangement extends along the boundary. In this way the structure present can be seen to be periodically repeated by units of (2,3[6]), (2,2[7]), (1,2[4]) and finally (1,1[8]). It is interesting that the relative distances (the value of 'm') along the boundary over which we see the (2223), (2212) and (2201) compounds differ and this will reflect the small differences in their relative formation energies. We believe that this relates to the mechanism explaining why this type of phase reconstruction occurs for the system at low angle tilt boundaries. We have characterized boundaries at a variety of tilt angles and found that, for tilts up to about 10°, local composition modulation of the type described above is general with larger values of m the lower the tilt angle.

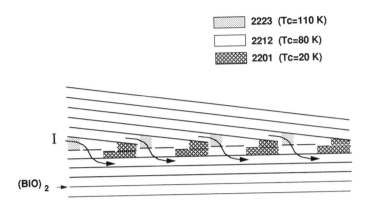

Fig. 4   A schematic diagram of a low angle boundary of the type shown in Fig. 3. High-T$_C$ "pathways" consisting of the (2212) and (2223) phases can be seen to pass from grain to grain, allowing supercurrents, as labelled, to pass between the grains on a-b planes

A schematic diagram of a typical low angle boundary (as for an 8° tilt) is shown in Fig. 4 indicating the way local units of the (2201), (2212), and (2223) phases accommodate the gross mismatch. Although we see local regions of the low-T$_c$ (20K) 2201 phase at the boundary which would be highly resistive at liquid nitrogen temperature, the form of the boundary transformations is such that there are "pathways" crossing the boundary plane made up of the high-T$_c$ (2212) and (2223) phases. "Short circuits" of this general form were found to be present for the full range of tilt angles examined up to about 10°. These pathways which, if of low height, are relatively wide would be expected to allow supercurrent passage except to the degree that the incomplete misfit relaxation associated with the incorporation of the differing phases leaves them at varying degrees of strain. Increasing the misorientation causes the length along the boundary of such pathways to decrease and the locally retained strains to be larger. It is suggested that the remarkable structure of these boundaries is the origin of the high critical currents seen in textured (2212) material at low temperatures. At higher tilt angles (>10°) of the habit of the boundaries, the interfaces still tend to exhibit (001) plane facets, but increasingly steps of height equal to a one-half a unit-cell-layer of the n=1 and 2 phases. Thus while CuO planes still tend to cross the boundary plane, the way the composition changes modulating the boundary structure now tend to occur (over shorter distances and using relatively more units of the low Tc (2201) compound) prevents pathways of the type described above from being formed. Nonetheless the current carrying characteristics of such boundaries might be expected to be intermediate between those suggested by the pathway structures present for lower angle tilts and those for boundary habits and misorientations which are still larger at which the incorporation of amorphous interlayers is more common.

References

[1]   Soylu B, Adamopoulos N, Glowacka D M and Evetts J E 1992 Appl. Phys. Lett., **60** 3183.
[2]   Akimitsu J, Yamazaki A, Sawa H and Fujiki H 1987 Jpn. J. Appl. Phys. **26** L2080.
[3]   Maeda M, Tanaka Y, Fukutomi M and T. Asano T 1988 Jpn. J. Appl. Phys., **27** L209.
[4]   Michel C, Hervieu M, Borel M M, Grandin A, Deslandes F, Provost J and Raveau B 1987 Z. Phys. **B68** 421.
[5]   Yan Y, Evetts J E, Soylu B and Stobbs W M 1994 Phil. Mag. Lett. **70**, 95.
[6]   Eibl O 1990 Physica C **168** 239.

Inst. Phys. Conf. Ser. No 148
Paper presented at Applied Superconductivity, Edinburgh, 3–6 July 1995
© 1995 IOP Publishing Ltd

Fishtail effect and current-voltage characteristics of high-T$_c$ superconductors

K I Kugel*, T Matsushita**, E Z Meilikhov*** and A L Rakhmanov *

* Scientific Center for Applied Problems in Electrodynamics, Russian Acad. Sci., Izhorskaya 13/19, 127412 Moscow, Russia
** Faculty of Computer Science and System Engineering, Kyushu Institute of Technology, 680-4 Kawazu, Iizuka 820, Japan
*** Russian Research Center "Kurchatov Institute", 123182 Moscow, Russia

Abstract. For a superconductor with fishtail-like field dependence of the critical current, the I-V characteristics are discussed in the framework of the collective creep theory. In this approach, the I-V characteristics may be presented as $V = V_0 \exp[-(j_c/j)^\mu)]$. The value of j_c can be related to the elementary pinning potential v_p and the size of moving flux line bundle L_c. The both dependencies $\mu(H)$ and $L_c(H)$ are different in the cases of superconductors with and without fishtail. In the former case, $L_c(H)$ is nonmonotone: $L_c \leq a_f$ (where a_f is flux line lattice (FLL) constant) and increases with H at low magnetic fields $H < H_p$ (where H_p is the field corresponding to the fishtail maximum of $j_c(H)$), so that $L_c > a_f$ in the intermediate field region. At high fields, $L_c(H)$ decreases and becomes smaller than a_f. Hence, at low and high fields the collective creep of individual vortices with $\mu << 1$ should be observed. At intermediate fields the collective creep of FLL with $\mu > 1$ takes place. Then, the behavior of I-V characteristics differs from that predicted for superconductors without fishtail.

1. Introduction

The fishtail or peak effect in the magnetic field dependence of the critical current j_c is commonly observed in high-T$_c$ superconductors. In [1,2] the I-V characteristics of YBCO single crystals were measured within the magnetic field range near the peak field H_p of the curve $j_c(H)$. It was found that the evolution of the I-V curves of these samples with the magnetic field differs from that of the samples without fishtail or with slight fishtail effect [3].

It was argued in [4,5] that the fishtail relates to the proximity effect in normal or weak superconducting inclusions. These inclusions are assumed to be pinning sites with the elementary pinning potential v_p depending on the magnetic induction B. The difference between the values of the order parameter in the superconductor and in the inclusion is small at low H due to the proximity This difference increases with H since superconducting state in the inclusion is more sensitive to the effect of the magnetic field than that in the superconductor. Thus, v_p increases with H, giving rise to the fishtail in j_c.

We shall analyze the I-V characteristics in the vicinity of fishtail in the framework of collective creep theory [6,7]. We demonstrate that the same mechanisms which determine the form of I-V curves in the absence of fishtail are also responsible for characteristic features of these curves in the case of peaked $v_p(H)$ dependence. The function $v_p(H)$ is assumed to have the form predicted by the theory [5]. Such an approach allows us to interpret qualitatively the available experimental data [1].

2. Correlation length and creep modes

The pinned flux line lattice (FLL) is characterized by two correlation lengths: L_c in the direction transverse to **B** and L_z in the direction along **B**. These values are related by [6]:

$$L_z = \gamma L_c, \quad \gamma = (C_{44}/C_{66})^{1/2}, \tag{1}$$

where C_{44} and C_{66} are tilt and shear moduli of FLL. A pinned FLL is correlated within the volume $V_c = \gamma L_c^3$, and flux lines move as bundles having volume V_c. The correlation length depends on H, temperature and v_p. The concept of flux bundles is valid if $L_c >> a_f$ where

$a_f = 1.075 \cdot (\phi_0/B)^{1/2}$ is the FLL constant. At $L_c \approx a_f$ the crossover occurs from three-dimensional (3D) collective FLL creep to the 1D creep of individual flux lines. The I-V curve for 3D and 1D collective creep modes can be presented in the form [6,7]

$$V = V_0 \exp\left[-(j_c/j)^\mu\right], \tag{2}$$

where V is the voltage, V_0 is some prefactor, and μ has different values for 1D and 3D modes. According to [6,7], μ decreases at the crossover from 3D ($\mu \geq 1$) to 1D ($\mu << 1$).

To calculate L_c we should minimize the FLL free energy F of the correlated volume V_c [6,7]. The free energy F can be estimated in 3D case as [6]

$$F = F_{el} + F_p$$
$$F_{el} = C_{66}(u/L_c)^2 \cdot V_c = \sqrt{C_{66}C_{44}} u^2 L_c, \quad F_p = -v_p(V_c n_p)^{1/2}, \tag{3}$$

where F_{el} is the contribution due to elastic deformation and F_p is the term related to pinning, u is the FLL displacement, and n_p is the density of the pinning sites. In the case of 3D collective pinning at $H >> H_{c1}$, the FLL will be torn from the pinning sites if its displacement is of the order of a_f. With the equation $\partial F/\partial L_c = 0$ one finds L_c by substituting $u = a_f$ and using (1) and (3). This expression, being correct to a constant factor, is written as

$$L_c = C_{66}^{3/2} C_{44}^{1/2} a_f^4 / v_p^2 n_p. \tag{4}$$

Within the usual approach [6,7], the magnetic field dependence of the elementary pinning potential is neglected. In this case, L_c increases with H at $H_{c1} << H << H_{c2}$. Indeed, when the latter conditions are met, we have [7]

$$C_{66} = B\phi_0/(8\pi\lambda)^2, \quad C_{44} = B^2/4\pi, \tag{5}$$

where λ is the London penetration depth and $B \approx H$. We get $\eta = L_c/a_f \propto H$ at $v_p(H) = const$. It is natural to assume that $L_c \leq a_f$ at low H. Thus, in the low field range one should expect 1D creep with $\mu << 1$. At higher fields, η increases and the transition from 1D to 3D creep occurs. The latter is characterized by μ of the order of unity or higher. The crossover occurs at $\eta = 1$.

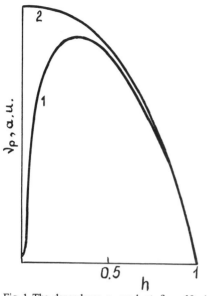

Fig. 1 The dependence v_p vs h at $\delta_n = 10$: (1) $\theta = 0.01$, (2) $\theta = 10$

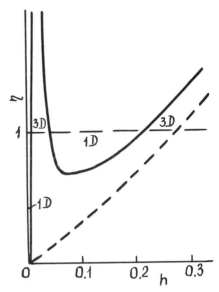

Fig. 2 The dependence η vs h at $\delta_n = 10$: $\theta = 0.01$ (solid line), $\theta = 10$ (dashed line)

3. Fishtail and I-V characteristics

The results described above are in agreement with the experiments [3] where no or small fishtail was observed. The theoretical prediction is 'self-consistent', since in the collective creep model j_c decreases with H at $v_p(H) = const$ for both 1D and 3D [6]. Then, the collective creep approach needs some modification to account for the fishtail.

Several mechanisms are suggested to explain fishtail. In [4] it was supposed that the increase of v_p with H due to suppression of the superconductivity in some 'weak points' is responsible for the fishtail. We believe the increase of $v_p(H)$ to be essential for the observed behavior of I-V curves in the case of fishtail effect. Thus, we base the following discussion on such concepts.

The elementary pinning potential is calculated in [5] in the framework of Ginzburg-Landau theory assuming that the pinning sites are the normal inclusions. The function $v_p(H)$ is approximated as

$$v_p = H_c^2 \xi_s^2 d_{nz}\left[\left(1 - R_n^2\right) - h^2\right]/8, \quad h = H/H_{c2}, \quad \theta = \xi_s^2/\xi_n^2, \quad \delta_n = d_n/\xi_s \quad (6)$$

$$R_n = \frac{\gamma\sqrt{1-h}\,M[(1+\theta/h)/4,1/2;h\delta_n^2]}{(\gamma + h\delta_n)M[(1+\theta/h)/4,1/2;h\delta_n^2] + \delta_n(\theta - h)M[(1+\theta/h)/4,3/2;h\delta_n^2]},$$

where ξ_s and ξ_n are the coherence lengths in the superconductor and in the inclusion, γ is a parameter characterizing NS interface, $M(a,b;x)$ is the confluent hypergeometric function, d_n and d_{nz} are the dimensions of the normal inclusion in the directions transverse and along $\mathbf{H}$. The $v_p(H)$ function is plotted in Fig. 1. This function is nonmotone at $d_n, \xi_n > \xi_s$ [5].

226

Nonmonotone behavior of $v_p(H)$ results both in the fishtail in $j_c(H)$ and in the nonmotone behavior of $\eta(H)$. The $\eta(H)$ curves are shown in Fig. 2. In contrast to the case of $v_p(H) = const$, Eq. (6) has now three solutions. At $H < H_1$, Fig. 2 demonstrates that $\eta < 1$ which corresponds to 1D behavior, whereas in the field range $H_1 < H < H_2$ we have $\eta > 1$ corresponding to 3D behavior. The crossover from 3D to 1D occurs at $H = H_2$ and 1D creep exists at $H > H_2$ where H_2 is in the vicinity of the peak field H_p. At high fields, $H > H_3$, the 3D creep should be expected again. However, the field H_3 is too high for any reasonable values of parameters and the last crossover could not be observed owing to approach to the irreversibility line.

4. The comparison with the experiment

Using (2) we can express index μ as follows

$$\mu = \frac{1}{\log(V/V_0)} \times \frac{\partial \log(V/V_0)}{\partial \log(j/j_0)}, \tag{7}$$

where j_0 is a constant introduced for convenience. It is difficult to estimate the factor V_0. Moreover, V_0 is different for 1D and 3D cases. Then, we can not find index μ quantitatively and only qualitative behavior of $\mu(H)$ can be revealed. Neglecting the change of $\log(V/V_0)$ at the crossover from 1D to 3D and using the experiment [1] we find that the ratio of indices μ at $H < H_p$ and at $H > H_p$ is approximately 4-5. This result is in agreement with the predictions of collective creep theory which gives $\mu = 1/7$ in 1D and $\mu = 0.5$-1.5 in 3D.

5. Conclusions

Thus, the fishtail behavior of the critical current and elementary pinning potential can produce a pronounced effect on the I-V characteristics of high-T_c superconductors. We interpret this phenomenon in terms of the 'reentrant' creep of individual vortices in the increasing magnetic field. The existing experimental data give certain indications in favor of the proposed explanation of I-V characteristics evolution near the maximum in $j_c(H)$ curves.

The work is supported in part by the Japan-FSU Scientists Collaboration Project and the International Science Foundation (Grant No 9500).

References

[1]   Gordeev S N, Jahn W, Zhukov A A, Kupfer H and Wolf T 1994 *Phys. Rev. B* **49** 15420
[2]   Kupfer H, Gordeev S N, Jahn W, et al 1994 *Phys. Rev. B* **50** 7016
[3]   Civale L, Krusin-Elbaum L, Thompson J R and Holtzberg F 1994 *Phys. Rev. B* **50** 7188
[4]   Daeumling M, Seuntjens J M and Larbalestier D C 1990 *Nature* **346** 332
[5]   Kugel K I, Matsushita T, Meilikhov E Z and Rakhmanov A L 1994 *Physica C* **228** 373
[6]   Fisher K H and Natterman T. 1991 *Phys. Rev. B* **43** 10372
[7]   Blatter G, Feigelman M V, Geshkenbein V B, Larkin A I and Vinokur V M 1994 *Rev. Mod. Phys.* **66** 1125

Inst. Phys. Conf. Ser. No 148
Paper presented at Applied Superconductivity, Edinburgh, 3–6 July 1995

Proton doped layers in YBCO bulk

S.Colombo, R.Gerbaldo, G.Ghigo, L.Gozzelino, E. Mezzetti, B. Minetti

INFN Torino; INFM Torino; Dept. of Physics, Politecnico di Torino, c.so Duca degli Abruzzi 24, 10129 Torino, Italy

R.Cherubini

L.N.L.-I.N.F.N., via Romea 4, 35020 Legnaro (Padova), Italy

F.Abbattista, R.Albanese, M.Vallino

Dept. of Science of Materials, Politecnico di Torino, c.so Duca Abruzzi 24, 10129 Torino, Italy

Abstract. Effects of monoenergetic proton irradiation on superconducting properties of good quality sintered and melt-textured YBCO samples are investigated. For sintered materials intragrain and intergrain critical current changes on different pellets of YBCO are reported for different irradiation fluences, ranging from $1.8 \cdot 10^{16}$ to $8.6 \cdot 10^{16}$ p/cm^2. Lower fluences seem to be more effective in enhancing intergranular pinning. For better quality samples, modulations of critical current density as a function of field set up. For melt-textured materials, proton induced defects do not significantly affect the critical current density for H ∥ ab, while strongly modify the critical current density in H ∥ c. The changes are field and temperature dependent. Strong enhancements are found up to a factor of about 15 at 25 K and 4 T.

1. Introduction

The aim of the paper is to outline some characteristic aspects of the flux pinning induced by proton irradiation, by comparing proton-induced effects on good quality sintered materials and melt-textured YBCO samples.

In the range of proton energies used in the experiment (3.5 and 6.5 MeV), defects are mainly produced by Coulomb scattering with target atoms along the whole proton path (Fig.1a). The number of interactions increases with depth due to the progressive decrease of the proton energy. At the end of the path protons implant. Depth and width of the implantation layer depend on the energy of protons as well as on the density of the specimen. For 6.5 MeV protons and our sample density of about 5.2 g/cm^3, the implantation peak lies at a depth of about 210 μm and is about 20 μm wide (Fig.1b). These values were estimated by a TRIM 95 Monte Carlo simulation [1]. The estimated average dimension of defects along the proton path is less than 1 nm; strain fields are mainly produced along the beam direction [2].

2. Experimental results and discussion

Irradiation experiments were performed at room temperature in vacuum at the APT scattering chamber of the 7 MV Van de Graaff CN accelerator (Laboratori Nazionali Legnaro - INFN).

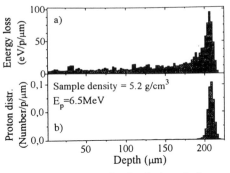

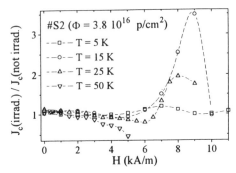

Figure 1 - Proton energy loss by elastic scattering as a function of depth (a) and implantation width distribution (b).

Figure 2 - Intergrain J_c enhancements for sample S2 (with respect to the non-irradiated sample S0) for different temperatures as a function of magnetic field. Data were obtained from low d.c. hysteresis loop.

For sintered specimens, irradiation doses range from $1.8 \cdot 10^{16}$ to $8.6 \cdot 10^{16}$ p/cm^2. For melt processed materials the fluence of $1.8 \cdot 10^{16}$ p/cm^2 was chosen. The samples were characterised by means of a Lakeshore 7225 susceptometer-magnetometer.

2.1 Sintered materials

The sintered specimens were cut from two different pellets, coming from different batches:

a) four "twin" slabs (labelled "S-samples", 0.60 mm thick) were irradiated with 6.5 MeV protons at different fluences (0, 1.8, 3.8, $8.6 \cdot 10^{16}$ p/cm^2, respectively)

b) two slabs (labelled "A-samples", 0.23 mm thick) were irradiated with 3.5 MeV protons at the doses 1.8, 3.8, $7.5 \cdot 10^{16}$ p/cm^2, and were fully characterised before and after every irradiation.

There seem to be two different types of regions at the grain boundaries in the sintered sample: "fully developed" boundary regions and "underdeveloped" boundary regions. Small portions of fully developed regions appear to be always present after liquid-phase sintering. The area ratio of this two types of regions is controlled by the amount of liquid phase [3]. The presence of strongly coupled regions is revealed under proton irradiation [4], because in these regions intergrain critical currents can be enhanced by means of proton induced pinning, although so far this process can not be fully controlled. As a general trend, the critical current enhancement is more significant at lower fluences, i.e. at the highest average defect distance.

In Table 1 enhancement factors of the critical currents due to irradiation are reported.

Table 1 - Intergrain characterisation: Jc relative enhancements at H_{dc} = 4000 A/m and T = 50 K

	$\Phi=1.8 \cdot 10^{16}$ p/cm^2	$\Phi=3.8 \cdot 10^{16}$ p/cm^2	$\Phi=7.5 \cdot 10^{16}$ p/cm^2	$\Phi=8.6 \cdot 10^{16}$ p/cm^2
S-samples (E$_p$=6.5MeV)	1.7	1.5	----	1.4
A-samples (E$_p$=3.5MeV)	1.12	1.05	0.85	----

Table 2 - Intragrain characterisation: Jc relative enhancements at H_{dc} = 2400 kA/m and T = 50 K

	$\Phi=1.8 \cdot 10^{16}$ p/cm^2	$\Phi=3.8 \cdot 10^{16}$ p/cm^2	$\Phi=7.5 \cdot 10^{16}$ p/cm^2	$\Phi=8.6 \cdot 10^{16}$ p/cm^2
S-samples (E$_p$=6.5MeV)	1.0	1.1(5)	----	1.5
A-samples (E$_p$=3.5MeV)	0.9(5)	2.0	2.2	----

These data, obtained from a.c. susceptibility measurements, are presented as a function of fluence, for a given field (4000 A/m) and a given temperature (50 K). At this temperature the chosen field is characteristic of the intergrain network and it is lower than H_{c1} for single grains. Moreover, different proton energies were used in order to study the dependence of the induced effects on defect distribution within the sample. It seems that higher energies are more effective in enhancing intergrain J_c.

The ratio among intergrain critical current densities after and before irradiation, is field and temperature dependent [5]. An example of this characteristic feature is presented in Fig.2. Modulations of the J_c vs. H response were observed (Fig.3). In some case, as in transport measurements, cross-over from weak J_c damage to J_c enhancement can be found when a magnetic field is applied [5].

In Table 2 relative intragrain critical current enhancements, obtained by means of d.c. magnetisation cycles up to 5 T, are reported as a function of fluence at a field of 3 T. By comparing results presented in Table 1 and Table 2 an opposite matching between fluence and critical current density enhancement emerges. The lowest proton dose affected intergrain critical currents, while no significant changes were observed in intragrain currents. Moreover, intergrain enhancements decrease with the fluence and intragrain enhancements increase with the proton dose.

It should be noted that the enhancement factors are underestimated because, due to proton implantation, the samples are not affected by irradiation in their whole thickness.

2.2 Melt-textured materials

In conventional sintering processes (which imply temperatures between 880 and 1000 °C), the fraction of strongly coupled regions remains rather limited. These fraction can be significantly increased when a melt growth technique is used. Melt-textured samples have been investigated [6] and irradiated. Our samples, obtained by TLD-MG process [7], consist of well aligned 123-phase domains and homogeneously distributed inclusions of 211 phase. These inclusions, as it is well known, are active in vortex pinning. The question arises if smaller pinning centres, such as those induced by protons, are also effective, as in good quality sintered materials, and in this case which are the peculiar effects of these defects. A irradiation fluence of $1.8 \cdot 10^{16}$ p/cm^2 was chosen, the same dose which had induced the highest average intergrain critical current enhancements in sintered materials. The results show that these pinning centres do not significantly affect the critical current densities flowing along the c axis J_c^c, measured with H ∥ ab (Fig.4).

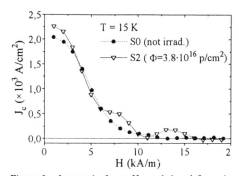

Figure 3 - Intergrain J_c vs. H, as deduced from d.c. hysteresis loops, for reference sintered sample #S0, and for sample #S2 irradiated with $3.8 \cdot 10^{16}$ p/cm^2.

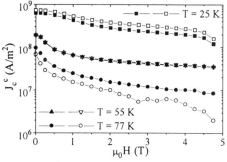

Figure 4 - J_c^c vs. μ_0H (H ∥ ab) obtained at different temperature before (solid symbols) and after (open symbols) proton irradiation (E_p = 3.5 MeV, Φ = 1.86 p/cm^2) for a melt-textured sample.

230

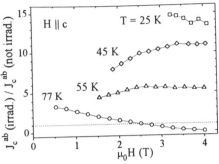
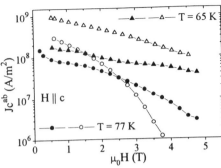

Figure 5 - Ratio of J_c^{ab} after and before irradiation for different temperature as a function of magnetic field.

Figure 6 - J_c^{ab} vs. μ_0H obtained before (solid symbols) and after (open symbols) proton irradiation ($E_p = 3.5$ MeV, $\Phi = 1.86$ p/cm^2) for a melt-textured sample.

It seems rather evident that these defects, which probably work at the same scale in which intrinsic pinning due to ab planes already dominates, only introduce some disorder or some further pins. On the contrary this kind of pinning centres strongly affects the critical current density flowing in ab planes J_c^{ab}, measured with H ∥ c. This effect depend on field and temperature: we obtained a J_c^{ab} enhancement more pronounced at lower temperatures (Fig.5). Moreover, at higher temperatures the enhancement is also strongly dependent on magnetic field. At T=77 K a crossover from enhanced to damaged current occurs at about 2.5 T (Fig.6).

3. Conclusions

The main result of the paper is that in strongly coupled regions (both grains and domains), pins such as the ones obtained by monoenergetic proton irradiation can be effective in enhancing the critical currents involving the whole specimen. Suitable fluences are needed. These pinning centres positively operate only in given field and temperature ranges.

Although the irradiation with 3.5-6.5 MeV protons can not produce continuous tracks, further measurements are in progress in order to establish if the alignment of the beam direction (i.e. the orientation of strain fields induced by protons) with the applied magnetic field is in some extent effective in enhancing pinning.

References

[1] Ziegler J F, Biersack J P, Littmark U 1985 *The stopping and range of ions in solids* vol 1 (ed. by Ziegler J, Oxford: Pergamon Press)
[2] Kirk M A 1993 *Cryogenics* **33** 235-242
[3] Shi D 1993 *Applied Superconductivity* **1** 61-70
[4] Mezzetti E, Bona F, Castagno G, Colombo S, Gerbaldo R, Ghigo G, Gozzelino L, Minetti B, Cherubini R 1994 *Cryogenics* **34** ICEC Supplement 825-828
[5] Mezzetti E, Cherubini R, Colombo S, Gerbaldo R, Ghigo G, Gozzelino L, Minetti B, Zafiropoulos D 1995 *Journal of Superconductivity* **8** 321-328.
[6] Abbattista F, Albanese R, Delprete C, Genta G, Gerbaldo R, Ghigo G, Gozzelino L, Mezzetti E, Mossolov A B, Ronchetti S, Vallino M 1994 *Il Nuovo Cimento* **16 D** 2087-2094
[7] Abbattista F, Albanese R, Vallino M, Gerbaldo R, Ghigo G, Gozzelino L, Mezzetti E, Minetti B, Mossolov A B, *this conference*

Inst. Phys. Conf. Ser. No 148
Paper presented at Applied Superconductivity, Edinburgh, 3–6 July 1995

231

Dissipation processes in transport-current-carrying Bi₂Sr₂Ca₂Cu₃O₁₀₊δ and Bi₂Sr₂CaCu₂O₈₊δ thin films[†]

L Miu,[1,2] U Frey,[1] F Hillmer,[3] Dana Miu,[1,2] G Wirth,[3] and H Adrian[1]

[1]Department of Physics, TH-Darmstadt, Hochschulstrasse 8, D-64289 Darmstadt, Germany
[2]Institute of Atomic Physics, Bucharest, P.O. Box MG-7 Bucharest-Magurele, Romania
[3]Gesellschaft für Schwerionenforschung, Planckstrasse 1, D-64291 Darmstadt, Germany

Abstract. The interlayer Josephson coupling in highly anisotropic Bi-based superconducting cuprates manifests itself in the existence of a finite critical-current density in zero applied magnetic field. By analysing the shape of the current-voltage characteristics in terms of quasi-two-dimensional vortex unbinding, it was found that the interlayer coupling vanishes above the hypothetical Kosterlitz-Thouless transition temperature, in agreement with Monte Carlo simulations. The nonvanishing coupling between the superconducting layers makes possible the occurrence of a vortex-glass state even in the case of an accentuated anisotropy. Bose-glass scaling of the $E(J)$ curves were obtained for both Bi₂Sr₂Ca₂Cu₃O₁₀₊δ and Bi₂Sr₂CaCu₂O₈₊δ films containing columnar defects oriented along the c axis. The scaling revealed wide critical regions and large values of the dynamic critical exponent.

1. Introduction

The maximum T_c values of Bi₂Sr₂Ca₂Cu₃O₁₀₊δ (Bi-2223) and Bi₂Sr₂CaCu₂O₈₊δ (Bi-2212) high temperature superconductors (HTSC's) are accompanied by a large anisotropy (with the anisotropy factor γ of the order of a few hundreds). In the two dimensional (2D) case (decoupled layers), a vortex glass cannot exist at $T > 0$ [1], whereas in zero applied magnetic field the critical-current density is essentially zero at all temperatures [2] (neglecting pinning and the size limitation [3]), due to the presence of thermally excited vortices of opposite helicity. In this work we show that, even though small, the interlayer coupling in Bi-2212 and Bi-2223 films with high T_c values leads to a finite critical-current density in zero magnetic field (J_{c0}) and makes possible the occurrence of a glass transition in the vortex system.

2. Sample preparation and characterisation

The films were prepared by an *in situ* sputtering method on (100) oriented SrTiO₃ substrates, as described in detail in Ref. [4,5]. The strongly c axis oriented growth and in plane epitaxy were confirmed by means of X-ray diffraction in a Bragg-Brentano and four circle geometry. The Bi-2212 films are single phase, whereas the Bi-2223 films contain ≈ 20 % of Bi-2212 and Bi-2234 compounds, intercalated through stacking faults, as revealed by TEM studies. The anisotropy factor of the investigated films was estimated from the predicted crossover in the vortex system at low applied field values, when the intervortex distance becomes larger than the Josephson length [1]. This is reflected in the significant change of the magnetic field dependence of the activation energy in the TAFF regime [6]. In the case of a large anisotropy, the crossover field $B_{cr} \approx 4\Phi_0/\gamma^2 s^2$, where s is the distance between the CuO₂ double and triple layers, respectively. The determined B_{cr} values (≈ 0.03 T for the Bi-2212 films, and ≈ 0.1 T in the case of the Bi-2223 films) lead to $\gamma_{2212} \approx 350$, and $\gamma_{2223} \approx 200$.

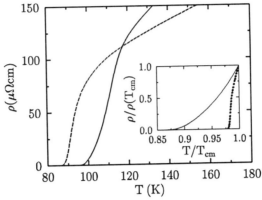

Fig. 1. Characteristic resistive transitions of the investigated films in zero magnetic field: Bi-2212 (---), Bi-2223 (—). The inset illustrates (in normalised scales, with T_{cm} - the midpoint critical temperature) the „resistive tail" of the Bi-2223 film (—) compared with that of a polycrystalline sample with preferential grain orientation (•).

The mean-field critical temperature T_{c0} (taken at the inflection point in the $\rho(T)$ curve) was $T_{c0} = 91.6$ K for Bi-2212 and $T_{c0} = 110.4$ K in the case of the Bi-2223 films (Fig. 1). The structures, measured in plane, consist of a 1.8 mm × 200 μm strip and a 0.1 mm × 10 μm bridge. Some of these films were irradiated with 2.7 GeV ^{238}U, parralel to the c axis, at doses corresponding to a dose equivalent field $B_\phi = 1$ T for the 200 μm strip line, and 0.5 T for the 10 μm bridge. The irradiation decreases T_{c0} by 2 - 3 K.

3. Results and discussion

3.1 Vortex fluctuations in zero external magnetic field

Besides some sample inhomogeneities, the resistive tail in the case of highly anisotropic Bi-based HTSC's is mainly caused by the thermal excitation of vortices and antivortices. A direct proof could be the partial dissappearance of the resistive tail for polycrystalline samples with preferential grain orientation (Fig. 1, inset). In such a case, a strong size limitation of vortex fluctuations is present due to the ribbonlike crystallites and „good" superconducting intergranular contacts with a small extension w in the (a, b) plane. This leads to a finite critical current density in zero magnetic field $J_{c0} \sim 1/w$ [3] even in the 2D case ($w < \gamma s$), for temperature values lower than the Kosterlitz-Thouless transition temperature T_{KT}. The same size limitation of vortex fluctuations will increases T_{KT}.

By analysing the shape of the I-V curves in the framework of the quasi-2D current induced vortex-antivortex unbinding model from Ref. [7], which attributes the finite J_{c0} to the interlayer Josephson coupling, it was found [8] that this intrinsic critical-current density vanishes slightly above the hypothetical T_{KT} value. This is shown in Fig. 2, for a Bi-2223 strip, and indicates the occurrence of a vortex fluctuation induced layer decoupling, in agreement with Monte Carlo simulations [9] and recent analytical studies [10]. The layer decoupling is confirmed by the rapid decrease of the activation energy U in the TAFF regime with increasing temperature at low fields (Fig. 2).

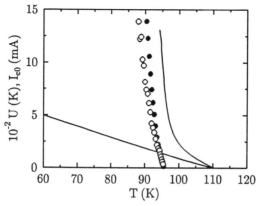

Fig. 2. Temperature dependence of the critical current in zero external magnetic field I_{c0}, resulting from the fit of the I-V curves of a Bi-2223 strip with the quasi-2D vortex unbinding model (O), and that determined at 0.5 μV/cm ($\bullet$); Temperature dependence of the activation energy U for $\mu_0 H = 2$ mT (right) and 2 T (left). The activation energy was determined from the resistive data, $U(B, T) = k_B T \ln[\rho (B, T_{c0})/\rho(B, T)]$.

The appearance of layer decoupling would be the reason why the $\rho(T)$ dependence for highly anisotropic HTSC's can be described by the Kosterlitz-Thouless picture [2] whereas the I-V curves show the occurrence of a finite J_{c0}.

3.2 Bose-glass behavior of the vortex system in the presence of columnar defects

Despite the large intrinsic anisotropy, the resistive transitions of the irradiated films in a magnetic field applied along the defects can be fitted at low resistivity levels with the „vortex-glass expression" $\rho(T) \sim (T - T_{BG})^{\nu(z-2)}$, where ν and z are the critical exponents [11,12].

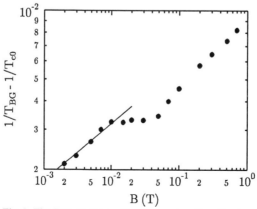

Fig. 3. The Bose-glass transition temperature T_{BG} of an irradiated Bi-2212 strip ($B_\phi = 1$ T) *vs* external magnetic field ($\bullet$). The continuous line represents the fit of the data at low field values with Equation (1) (one parameter fit).

234

At low fields, $B \leq B_0 = \Phi_0/\lambda^2$ (with λ - the in plane component of the penetration depth), the magnetic field dependence of the Bose-glass transition temperature (T_{BG}) is in good agreement with the predictions of theory of boson localization in the presence of correlated disorder [11,12], as illustrated in Fig. 3,

$$1/T_{BG} - 1/T_{c0} \sim B^{1/4}. \tag{1}$$

Just above the field value B_0, the $T_{BG}(B)$ decrease is limited to a certain extent through vortex-vortex interactions. In the case of a Bose-glass transition [11,13], for a given field value, the electric field-current density isotherms $E(J)$ measured below and above T_{BG} must collapse in two universal curves $E^*(J^*)$, valid below and above T_{BG}, respectively, with

$$E^* = E|1 - T/T_{BG}|^{-\nu(z + 1)}, \quad J^* = (J/T)|1 - T/T_{BG}|^{-3\nu}, \tag{2}$$

where ν is the static critical exponent, and z is the dynamic one. The scaling performed in the case of our highly anisotropic samples reveals wide critical regions and large values of the dynamic critical exponent (Fig. 4).

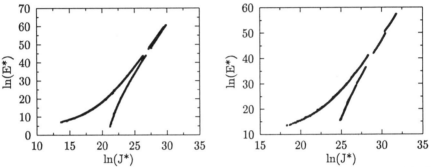

Fig. 4. The $E(J)$ curve scaling [Rel. (2)] for a Bi-2212 bridge (left) and for a Bi-2223 bridge (right) in a magnetic field $B = B_\phi = 0.5$ T. Bi-2212: T (K) = 30, 35, 40, and between 44 K and 64 K in steps of 2 K; T_{BG} = 47 K, $\nu = 0.95$, $z = 14.5$. Bi-2223: T between 53 K and 73 K in steps of 2 K; T_{BG} = 62.3 K, $\nu = 0.85$, $z = 14$.

The occurrence of the Bose-glass transition in the case of intrinsically very anisotropic HTCS's can be understood through the increase of the tilt modulus in the vortex ensemble in the presence of correlated disorder [12].

References

[1] D. S. Fisher et al., Phys. Rev. B **43**, 130 (1991); V. M. Vinokur et al., Physica C **168**, 29 (1990).
[2] J. M. Kosterlitz and D. J. Thouless, J. Phys. C **6**, 1181 (1973).
[3] L. Miu, Phys. Rev. B **50**, 13849 (1994).
[4] P. Wagner et al., Physica C **215**, 123 (1993).
[5] P. Wagner et al., Phys. Rev. B **51**, 1206 (1995).
[6] L. Miu et al., Physica C **234**, 249 (1994); J. of Superconductivity **8**, 293 (1995).
[7] H. J. Jensen and P. Minnhagen, Phys. Rev. Lett. **66**, 1630 (1991).
[8] L. Miu et al., Phys. Rev. B **52** (1 August, 1995).
[9] H. Weber and H. J. Jensen, Phys. Rev. B **44**, 454 (1991).
[10] S. W. Pierson, Phys. Rev. B **51**, 6663 (1995).
[11] D. R. Nelson and V. M. Vinokur, Phys. Rev. B **48**, 13060 (1993).
[12] L. Miu et al., Phys. Rev. B **51**, 3953 (1995).
[13] M. Wallin and S. M. Girvin, Phys. Rev. B **47**, 14642 (1993).

† Work supported by the Ministerium für Forschung und Technologie der Bundesrepublik Deutschland (Contract No. 13 N 5748 A). The kind assistance of the Alexander von Humboldt Foundation is gratefully acknowledged.

Inst. Phys. Conf. Ser. No 148
Paper presented at Applied Superconductivity, Edinburgh, 3–6 July 1995
© *1995 IOP Publishing Ltd*

Relaxation measurements in a $Bi_2Sr_2CaCu_2O_{8+\delta}$ single crystal with inhomogeneous pinning

Koblischka M R†[1], A J J van Dalen†, H Kuhn‡, Th Schuster‡ and M Konczykowski§

†Free University, Faculty of Physics and Astronomy, De Boelelaan 1081, 1081 HV Amsterdam, The Netherlands.

‡Max-Planck Institut für Metallforschung, Institut für Physik, D-70506 Stuttgart, Germany

§Laboratoire des Solides Irradiés, Ecole Polytechnique, F-91128 Palaiseau, France

Abstract. Measurements of torque hysteresis loops, of the conventional relaxation rate S and of the dynamical relaxation rate Q were performed by means of torque magnetometry on a $Bi_2Sr_2CaCu_2O_{8+\delta}$ (Bi-2212) single crystal with enhanced edge-zone pinning. Measurements were performed at two temperatures, $T = 30$ K and 50 K. At $T = 50$ K, the inner non-irradiated area is above the depinning line, whereas at $T = 30$ K the inner area shows irreversible behaviour but considerably less than in the irradiated edge zone. The torque measurements are compared to magneto-optical investigations of the flux distributions obtained on the *identical* sample.

1. Introduction

The understanding of the pinning mechanisms is the key to improve the critical current densities for future applications of the high-T_c superconductors. This entails also the study how magnetic flux penetrates and exits from a superconductor. The irreversibility line (IL) at low temperatures ($T < 30$ K) is governed by bulk pinning. If bulk pinning is weak at temperatures close to the irreversibility temperature T_{irr}, flux penetration can be governed by edge pinning.

With a framelike zone of enhanced bulk pinning an edge barrier for flux penetration can be simulated. The enhancement of the pinning within the sample is performed using heavy ion irradiation, which procedure was demonstrated to create very effective pinning centres in Bi-2212 [1]. On such samples, experiments were carried out using direct

---

[1] present address: Groupe de Physique appliquée, Université de Genève, Rue d'Ecole de Medecine 20, CH-1211 Genève 4, Switzerland.

visualization of the flux distributions by means of magneto-optics [2] or by means of Hall probes [3]. These measured flux density profiles were compared to calculated profiles using a newly developed model [4] but here with a steplike critical current density. A very good agreement between these model calculations and the experiments was found [2, 3]. Aim of this paper is to study the relaxation behaviour in a sample with an artificial pinning barrier at temperatures close to T_{irr}. We present measurements of the dynamical creep rate [5] $Q \equiv \mathrm{d} \ln j / \mathrm{d} \ln(\mathrm{d}H_a/\mathrm{d}t) = \mathrm{d} \ln j / \mathrm{d} \ln(\mathrm{d}j/\mathrm{d}t)$ and the conventional creep rate $S \equiv -\mathrm{d} \ln j / \mathrm{d} \ln t$. The experiments were performed on the *identical* sample as used in [2]. Two temperatures were chosen for the experiments; one where the central unirradiated area of the sample is above the irreversibility line (IL), and the other one below the IL so that there is irreversible behaviour also in the central area.

2. RESULTS AND DISCUSSION

The way of flux penetration in this edge-zone irradiated sample shows some pecularities [2] where the flux distributions of the identical sample were visualized using magneto-optics. On applying a magnetic field to a zero-field cooled state, the flux penetrates first into the irradiated area. When the vortices reach the unirradiated central area at the narrowest place of the irradiated belt, an interesting phenomenon occurs: Magnetic flux appears at H_1 in the centre of the unirradiated zone. Since vortices cannot nucleate at the sample centre, they have to cross the Meissner area being driven to the centre by screening currents which are much higher ("overcritical") than the critical current in this unirradiated region. Further increase of the external magnetic field leads to a completely filled inner area which occurs at $H_a = H_2$ and vortices start from there to penetrate into the irradiated area until the full penetration field H^* is reached. Torque hysteresis loops

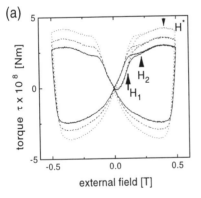 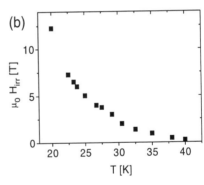

Figure 1. (a): Torque loops measured at $T = 50$ K using various sweep rates. Outer curve: $\mathrm{d}H_a/\mathrm{d}t = 20$ mT/s, intermediate curve: $\mathrm{d}H_a/\mathrm{d}t = 10$ mT/s, inner curve: $\mathrm{d}H_a/\mathrm{d}t = 5$ mT/s. The fields H_1, H_2 and H^* are defined in the text. (b): IL measured on an unirradiated $Bi_2Sr_2CaCu_2O_{8+\delta}$ single crystal of the same batch.

were measured by a capacitance technique [6] at $T = 50$ K using various sweep rates $\mathrm{d}H_a/\mathrm{d}t = 20$ mT/s, 10 mT/s and 5 mT/s (figure 1 (a), from top to bottom). It is clearly visible from this plot that the faster the sweep rate is the larger is the measured torque. The critical currents measured using a high sweep rate are less influenced by creep effects than the currents in the case of using a low sweep rate. The deviation point from the initial linear behaviour denotes H_1. H_1 depends slightly on the sweep rate used in the experiment and varies between $H_1 = 122$ mT (5 mT/s) and 155 mT (20 mT/s). These values correspond well to the magneto-optically determined value of 107 mT, where a sweep rate of 5 mT/s was used. H_2 is marked by the joining of the virgin branch with the torque curve after completing a full loop. Also this field depends on the sweep rate; the values vary from 240 mT (5 mT/s) up to 267 mT (20 mT/s). Due to creep effects during the magnetization H_1 in reversed fields is smaller than for virgin magnetization. Figure 1 (b) shows the IL measured on an unirradiated $Bi_2Sr_2CaCu_2O_{8+\delta}$ single crystal. This plot clearly demonstrates that at $T = 50$ K the central area is above the IL. At $T = 30$ K, however, also the central area shows pinning but considerably smaller than in the irradiated edge zone.

Flux creep in such an edge-irradiated sample is composed of two contributions: Starting from a zero-field cooled state, for $H_a < H_1$ the relaxation in the irradiated belt is measured where the flux relaxes towards the sample centre. For $H_1 < H_a < H_2$, there is a relaxation away from the central flux pile-up towards the edge zone. The measured creep rate, R^{obs}, is therefore a superposition of two components, R^{irr} and R^{unirr} denoting the rates in the irradiated and unirradiated parts of the sample, e. g. $R^{obs} = (m^{irr}R^{irr} + m^{unirr}R^{unirr})/(m^{irr} + m^{unirr})$; m denotes the magnetization of the sample. This relation is valid for both S and Q. First, we look at the dynamical creep rate Q.

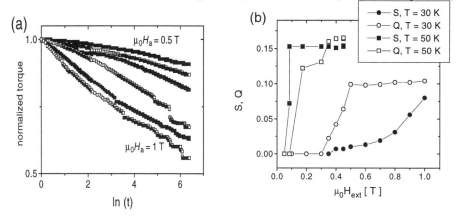

Figure 2. (a): Conventional relaxation experiment at $T = 30$ K; the curves are normalized to their initial values. The central area of the sample is below the IL. The applied external magnetic fields range from 0.5 T up to 1 T. (b): S and Q as a function of field for $T = 30$ K and 50 K.

As directly visible from figures 1 and 2, Q is minimal when the external magnetic field is increased from zero. As soon as flux leaks into the central area $(H_a > H_1)$, Q increases immediately and then stays constant above H_2, i.e., the torque curves are equidistant up to the maximum field. At $T = 30$ K, the behaviour of Q is nearly identical. In figure 2 (b), Q and S are presented as function of field at $T = 30$ K and 50 K.

Conventional relaxations at $T = 30$ K are presented in figure 2 (a). The external magnetic field is varied between 0.1 T and 1.0 T in steps of 0.1 T. The relaxation measurements below H_1 are shown in an inset for clarity. Up to 0.4 T, no relaxation is measured; these curves are omitted in figure 2 (a). Above 0.4 T, vortices have penetrated the central area and we measure relaxation rates which are a function of the applied field. At $T = 50$ K, all determinations of S yield a common value of $S = 0.153$ above H_1. At $\mu_0 H_a = 0.085$ T, which is only slightly below H_1, the relaxation does not follow a simple $M \propto \ln t$ law and two values can be determined for S. At times $t < 20$ s, $S = 0.072$, for larger times $S = 0.153$. This behaviour is due to vortices leaking into the central area *during* the relaxation.

In figure 2 (b), S and Q are compared to each other. As expected, the relaxation rates at 30 K are considerably smaller than at 50 K. For small fields both relaxation rates are equal to zero due to sensitivity limits of the torquemeter. However, as soon as vortices penetrate the central area, the relaxation rates can be measured. The behaviour of Q is very similar at both temperatures investigated here whereas the behaviour of S is clearly different when the unirradiated area shows irreversible behaviour. Here, one has to realize that close to the IL the relaxation is strongly deviating from the simple $M \propto \ln t$ behaviour. Hence, the relaxation rate S becomes time dependent. As a difference, Q is evaluated at the beginning of the relaxation process and S has been taken as the long time limit. Additionally, the initial delay time τ_i [7] causes a suppression of S compared to Q. This effect is enhanced by the non-linearity of the current dependence of the activation energy $U(j)$. The observed differences between S and Q reduce with a further increase of the external magnetic field. For long times $t \gg \tau_i$, S and Q will be approximately equal. This determination of creep rates with a part of the sample above the IL demonstrates the advantages of the dynamical method in the case of a fast relaxing sample.

Acknowledgments

The authors wish to thank E. H. Brandt (MPI Stuttgart), M. V. Indenbom (Chernogolovka) and Prof. R. Griessen (Free University of Amsterdam) for their collaboration. We wish to thank also T. W. Li (University of Leiden) and A. A. Menovsky (University of Amsterdam) for the $Bi_2Sr_2CaCu_2O_{8+\delta}$ single crystals.

References

[1] Gerhäuser W *et al* (1992) *Phys. Rev. Lett.* **68** 879

[2] Schuster Th *et al* (1994) *Phys. Rev. Lett.* **73** 1424; Schuster Th *et al* (1994) *Phys. Rev. B* **50** 16684

[3] Khaykovich B *et al* (1994) *Physica* C **235-240** .

[4] Brandt E H (1993) *Phys. Rev. Lett.* **71** 2821; (1994) *Phys. Rev. B* **49** 9024; (1994) *Phys. Rev. B* **50** 4034

[5] Pûst L *et al* *J. Low Temp. Phys.* **78** (1990) 179

[6] Qvarford M *et al* (1993) *Rev. Sci. Instrum.* **63** 5726

[7] Schnack H G *et al* (1992) *Physica* C **197** 337

Inst. Phys. Conf. Ser. No 148
Paper presented at Applied Superconductivity, Edinburgh, 3–6 July 1995

Flux creep measurements on a melt-textured YBCO sample

J Lorenz, M Reissner, W Steiner, P Diko[1], N Pellerin[2] and P Odier[2]

Institut für Angewandte und Technische Physik, T.U.Wien, A-1040 Wien, Austria
[1] Institute of Experimental Physics, SAS, Watsonova 47, 04353 Košice, Slovak Republic
[2] Centre de Recherches sur la Physique des Hautes Températures, CNRS, 45071 Orleans cedex 2, France

Abstract. A melt-textured YBCO sample, which shows a pronounced fishtail effect, was measured magnetically before and after reoxygenation. From flux creep investigations it is found, that at the field B_1, where the fishtail starts to develop, pinning behaviour changes drastically. Below this field the sample behaves in the usual way: a transition from 3D-pinning of small flux bundles to pinning of large flux bundles is obtained by increasing the temperature. At higher fields, where flux creep is dominated by the pinning centres which are responsible for the fishtail, a change from single vortex creep to collective vortex creep in 2D with temperature appears. Reoxygenation of the sample leads to an increase of T_c and a strong decrease of the fishtail effect, whereas the width of the hysteresis below the fishtail-region is nearly unchanged.

Whereas a tendency towards two dimensionality in the microstructure of melt-textured YBCO by reoxygenation was found [1], a tendency towards 2D behaviour of flux creep with increasing oxygen defect concentration in powder-melt-processed (PMP) YBCO was reported [2]. The role of oxygen in samples of such kind is very peculiar and not fully understood. Therefore we have investigated by flux creep measurements a well-textured YBCO sample, prepared using the melt-texturing procedure [3].

The volume fraction of the 211 particles was 18%. They are homogeneously distributed in the 123 grains on a macroscopic scale, but on a microscopic scale some fluctuations in the density of the 211 particles appear with a periodicity of 75 μm. The mean diameter of the 211 particles is 5 μm, the largest grain volume is 130 mm^3. Planar defects, which are found to be oriented parallel to the (a,b)-plane have a mean distance of 8 μm. It can be shown, that the oxygenation of the sample is a combination of oxygen volume diffusion, microcracking and penetration of oxygen through the cracks. Details of the microstructural analysis are given in [4].

For magnetic measurements with a vibrating sample magnetometer (0.1 mT<B_a<7 T, 1.5 K<T<300 K) a sample with dimensions of 3.11x1.16x1.75 mm^3 was separated. It consists of rectangular subgrains, the c-axis of which are tilted against the mean c-direction by about 6°. The subgrain boundaries are clean and free of cracks. All measurements were performed with $B_a \| c$. After finishing the magnetic measurements the sample was reoxygenated (10 h at 550°C, 15 h at 500°C and 20 h at 450°C) and the measurements were repeated.

T_c-onset was found to be 88 K (89.6 K) and the transition width to be 1.2 K (1.4 K) before (after) reoxygenation.

Weak links were not observable in magnetic measurements in low fields. The sample

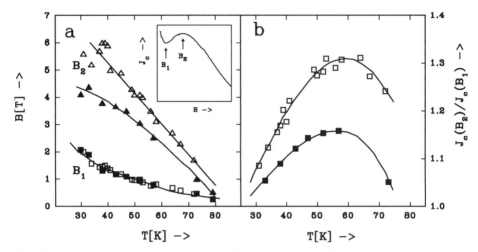

Fig.1 Temperature dependence of B_1, B_2 (a) and $J_c(B_2)/J_c(B_1)$ (b) measured before (open symbols) and after (full symbols) reoxygenation.

behaves like a single crystal. Therefore from hysteresis loops critical current densities (J_c) could be determined by $J_c(A/cm^2) = 40|M_{irr}|/(a(1-a/3b))$ [5], where $2M_{irr}$ is the width of the hysteresis in emu/cm^3, a the width (1.16 mm) and b the length (3.11 mm) of the sample. Because the main planar defects are all oriented parallel to the (a,b)-planes, they do not influence the current path for measurements with $B_a \| c$. For an external field of 1 T J_c-values of $1.5 \cdot 10^5$ and $2.0 \cdot 10^4$ A/cm^2 were obtained at 35 and 70 K, respectively. Because of a strong fishtail effect, J_c increases with field up to a maximum value (B_2) followed by a slow decrease at higher fields (inset Fig.1). B_2 decreases with increasing temperature (Fig.1a). After reoxygenation B_2 is shifted to lower fields, but the field where the fishtail starts to develop (B_1, Fig.1a) remains the same. Also the relative strength of the fishtail $J_c(B_2)/J_c(B_1)$ is strongly reduced after reoxygenation (Fig.1b). For fields below B_1 the critical current densities are nearly the same before and after reoxygenation. Thus reoxygenation strongly influences the pinning centres which are responsible for the fishtail, whereas those pinning centres which are dominant outside of the fishtail region are only weakly changed.

During the measurement of the hysteresis cycle the field sweep (3.9 mT/s) was interrupted at 1, 3 and 5 T, and the decay of magnetic moment was recorded for up to 1 h. This time dependence was assumed to be in first approximation logarithmic, although especially at high temperature a clear curvature in $M(\ln t)$ was obtained. In Fig.2a the temperature dependence of the normalized creep rate $S=(1/M_0)(dM/d\ln t)$ is shown. At low temperatures a peak (S_{max}) appears, which shifts to lower temperatures for higher fields (Inset Fig.2a). The temperature dependence of S_{max} seems to be related to the one of B_1 (Fig.1a). From S a mean effective activation energy ($<E>$) was deduced following Anderson [6]. The maximum in $<E>$ also shifts to lower temperatures for higher fields.

Following Maley et al [7] the current dependence of the effective activation energy was determined (Fig.3). To bring the relaxation measurements for all temperatures on one smooth curve two unknown parameters have to be adjusted. First a factor $C = \ln(H_a v/2\pi d)$, which is related to the flux moving velocity v and the applied field H_a (d is the sample

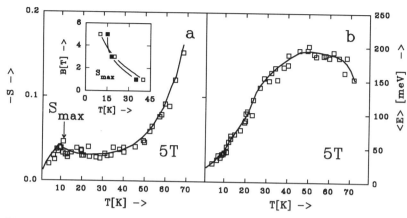

Fig.2 $S(T)$ (a) and $<E>(T)$ (b) measured before reoxygenation. **Inset:** $S_{max}(T)$ measured before (open symbols) and after (full symbols) reoxygenation.

thickness), and second the function g(T) which describes the temperature dependence of the true pinning potential. Following Tinkham [8] g(T) = 1 - $(T/T_c)^2$ was chosen. Replacement of T_c by the irreversibility temperature (before reoxygenation T_{irr} = 84, 79 and 76 K for 1, 3 and 5 T, respectively) as proposed by [9] changes the result only marginally. We have also checked the influence of a variation of C, which should be small, because it is only a logarithmic correction term, and found that it does not change the conclusions given below. Therefore in the following we only give the results for C = 20.

In the collective pinning theory for the effective pinning energy a power law $U = U_c(J/J_c)^{-\mu}$ is proposed for $J \ll J_c$ [10]. The exponent μ depends on the dimensionality and the particular flux regime. Values of 1/7, 3/2 and 7/9 are found for single vortex creep, pinning of small flux bundles (*sfb*) and pinning of large flux bundles (*lfb*) in 3 dimensions whereas μ should be 9/8 for single vortex creep (*svc*) and 1/2 for collective vortex creep (*cvc*) in 2 dimensions. μ can be directly determined from the slope of the effective pinning energy versus the irreversible magnetization in a double logarithmic representation (Insets Fig.3). For 1 T up to nearly 70 K μ = 7/9 is obtained, whereas for the higher fields μ = 9/8 fits the data for a large temperature range much better. A change from 3D pinning of large flux bundles to single vortex creep in 2D at higher fields and temperatures is unexpected in this material. One may argue that 9/8 is between 3/2 and 7/9 and therefore it might be that we are at higher fields in a situation where the change from *sfb* to *lfb* creep takes place. But if this is valid the regime of *sfb* pinning is larger at higher than at lower fields. This is also not reasonable. Therefore it seems that at higher fields pinning is dominated by different pinning centres, which are also responsible for the fishtail effect. The data can be explained in a consistent way by assuming the following scenario. Below a line defined by both $B_1(T)$ and $S_{max}(T)$, which coincide within the experimental error, the system behaves 3-dimensional. This is the regime where no fishtail appears. In this range the system changes with increasing temperature from *sfb* to *lfb* creep. Above $B_1(T)$, where the fishtail appears, flux pinning

242

seems to be 2-dimensional. In this region the system changes from *svc* to *cvc* with increasing temperature at fields which coincide roughly with the maximum of the fishtail effect ($B_2(T)$, Fig.1a)).

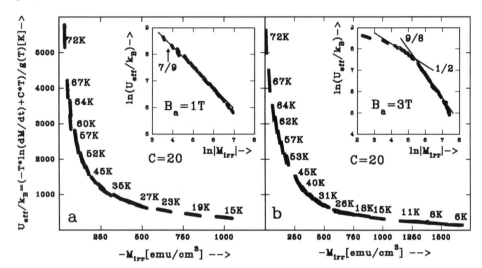

Fig.3 Effective pinning energy versus irreversible magnetization measured for the sample before reoxygenation for 1 T (a) and 3 T (b). **Insets:** Same in double logarithmic representation.

Therefore we can conclude, that there exist two different types of pinning centres. Ones which are responsible for the fishtail effect and which are strongly influenced by reoxygenation, and the others, which dominate in the field and temperature range where no fishtail appears. The pinning of the latter ones is not changed by reoxygenation. When the first kind of pinning centres are dominant, the pinning seems to change from 3D to 2D behaviour.

Work supported by the Austrian Ministry of Science under contract GZ45.262/2-46a/93.

References

[1] Sandiumenge F, Vilalta N, Piñol S, Martínez B and Obrados X 1995 *Phys.Rev.B* **51** 6645-6654
[2] Reissner M, Steiner W, Czurda C, Weber H W, Zhang P X and Zhou L 1994 *Critical Currents in Superconductors* (Singapore: World Scientific) 451-454
[3] Pellerin N, Gervais M and Odier P 1992 *Mater.Res.Soc.Symp.Proc.* **275** 537
[4] Diko P, Pellerin N and Odier P 1995 *Physica C* **247** in print
[5] Chen D X and Goldfarb R B 1989 *J.Appl.Phys.* **66** 2489-2500
[6] Anderson P W 1962 *Phys.Rev.Lett.* **9** 309
[7] Maley M P, Willis J O, Lessure H, and McHenry M E 1990 *Phys.Rev.B* **42** 2639-2642
[8] Tinkham M 1988 *Phys.Rev.Lett.* **61** 1658-1661
[9] McHenry M E, Simizu S, Lessure H, Maley M P, Coulter J Y, Tanaka I and Kojima H 1991 *Phys.Rev.B* **44** 7614-7624
[10] Feigel'man M V, Geshkenbein V B, Larkin A I and Vinokur V M 1989 *Phys.Rev.Lett.* **63** 2303-2306

Inst. Phys. Conf. Ser. No 148
Paper presented at Applied Superconductivity, Edinburgh, 3–6 July 1995
© 1995 IOP Publishing Ltd

Short time flux creep in $Bi_2Sr_2Ca_2Cu_3O_{10}$ tapes

A. Gupta, E. D. Tuset, M. G. Karkut* and K. Fossheim

Dept. of Physics, Norwegian Institute of Technology, N-7034 Trondheim, Norway
* SINTEF Applied physics, N-7034 Trondheim, Norway

Abstract. We report magnetic relaxation, measured on a $Bi_2Sr_2Ca_2Cu_3O_{10}$ tape, of a field step (50 G) superimposed on a static field of 0 - 0.2 T in a time window of 0.1 - 3600 s. The relaxation curves show a unique behaviour at the "vortex glass transition" ($T_G(B)$) determined on the same tape by transport V-I measurements. We observe a downward and an upward curvature in M(lnt) below and above $T_G(B)$, respectively. A characteristic time, where the curvature in M(lnt) starts to appear, shows a "diverging" behaviour as a function of temperature at $T_G(B)$. This characteristic time apparently corresponds to a full penetration time and a crossover from logarithmic to exponential relaxation behaviour, below and above $T_G(B)$ respectively.

1. Introduction

Using commercial squid set-ups, relaxation measurements in high-T_c cuprates start mainly from 10 -100 s onwards. As a consequence there are not many reports on relaxation around and above the irreversibility line. To probe this region of the B-T phase diagram, we have used a home-assembled squid set-up capable of measuring 0.1 s onwards [1]. In addition, we can measure the relaxation of a small field step superimposed upon a homogeneous persistent dc field. Our results highlight the correspondence between the magnetic relaxation and transport V-I characteristics in $Bi_2Sr_2Ca_2Cu_3O_{10}$ tapes. Especially the "vortex glass transition" $T_G(B)$, obtained from the V-I scaling on the same tape [2], shows characteristic signatures in relaxation measurements.

2. Sample and experimental details

The measurements were performed on a silver - sheathed $Bi_2Sr_2Ca_2Cu_3O_{10}$ tape with a T_c = 108 K and J_c = 32000 A/cm^2 (at 77 K and 0 T). The superconducting part of the sample had a size of about 2.5 x 2.1 x 0.06 mm^3. See [1] for the details of the squid set-up. A field step δB (= 50G), superimposed on a persistent field B (= 0 - 0.2 T), could be switched off / on within a few ms. The sample was first field (B + δB) cooled from above T_c to a measurement temperature, followed by a stabilisation of the temperature within a few mK. The relaxation was then measured between 0.1 - 3600 s after switching the δB off. The measured signal above T_c for each run allows us to determine the precise zero level, so that the reported quantity M (in a.u.) is proportional to the magnetisation of the sample. All the measurements were performed with both B and δB applied perpendicular to the flat surface of the tape, i.e. along the c- axis of the textured $Bi_2Sr_2Ca_2Cu_3O_{10}$.

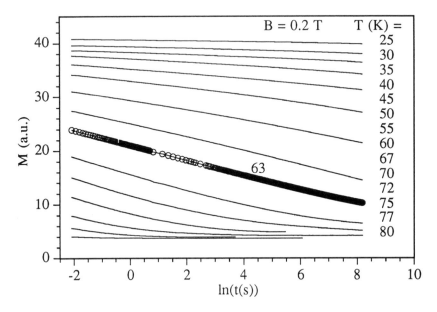

Fig.1 Magnetisation as a function of logarithm of time at different temperatures.

3. Experimental results and discussions

In fig. 1 is shown the magnetisation as a function of logarithm of the time at different temperatures with an applied field of 0.2 T. At first glance the results appear to be the same as obtained earlier by several other groups ([3] and references therein): nearly a logarithmic relaxation at low temperatures, and at high temperatures the relaxation becomes increasingly non logarithmic. At 80 K, it takes ~ 2 s for the magnetisation to relax to its equilibrium value within our experimental sensitivity. However, careful observation of the data in fig.1 shows a remarkable evolution of the curvature of the relaxation curves with temperature. For T < 63 K a clear downward curvature is observed and for T ≥ 63 K the curvature becomes upwards. Interestingly, the temperature 63 K at B = 0.2 T matches well with the T_G(0.2 T) determined from the scaled transport V-I characteristics on the same tape [2]. We observe similar behaviour at B = 0.05 T and 0.1 T. This coincidence strongly suggests that the change from negative to positive curvature in our relaxation data is a signature of the "glass transition".

To illustrate the point further, for B = 0.2 T, we plot in fig. 2 a characteristic time at which M(lnt) deviates from an initial linearity seen in fig.1. We get a unique "diverging" behaviour of this characteristic time at about 63 K (≈ T_G(0.2 T)). Since the deviation in M(lnt) does not occur suddenly, there is a large uncertainty (see below) in determining this time; however, that should not effect the "divergence" seen on a semi-log scale in fig.2.

Now, we first discuss the downward curvature in M(lnt) below 63 K as seen in fig.1. At these temperatures, the applied field step might not penetrate the sample completely at the beginning of the relaxation. In this case [4, 5], a downward curvature in M(lnt) is expected at a characteristic time t*, marking the crossover from incomplete to complete penetration of the field step (δB) in the sample. Thus the characteristic time for T < 63 K may be identified with

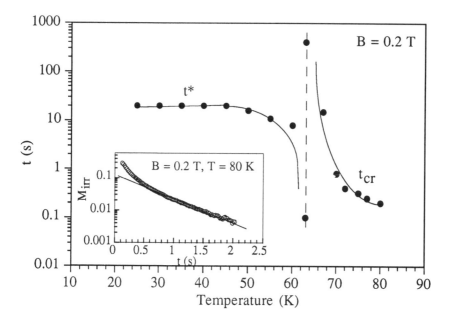

Fig.2 Characteristic time as a function of temperature. The determination of this time, and the labelling t* and t_{cr} are discussed in the text. Solid and dashed lines are only a guide to the eye. Inset: Irreversible part of the M as a function of time. The straight line in the semi-log scale shows the exponential nature of the relaxation.

t* [4], which is found to be nearly constant at low temperatures and decreasing rapidly near 63 K ($\approx T_G(0.2$ T)), see fig.2. This can be understood by a decrease of j_c near $T_G(B)$, so that the complete penetration of the field step gets quicker, i.e. t* decreases. At T = 63 K, t* was undetectable even within ~ 0.1 s of the start of the relaxation. We would like to make two comments here. Firstly, the downward curvature does not result from the transient effects reported in [6]. We can estimate the transient time (τ_0) from $\tau_0 = 0.41[d \, \mu_0 \, s \, j_c / (\delta B / \delta t)]$, where s is the normalised creep rate [6, 7]. With d = 60 x 10^{-6} m, $\delta B/\delta t$ = 1 T/s, and upper limits of s = 0.2 and $j_c = 10^{10}$ A/m^2, we still get τ_0 = 10 ms < 0.1 s, where the first relaxation data is recorded. Secondly, as argued in [5], the complete penetration of the field step occurs smoothly at t* generating a downward curvature, and no kink is observed in M(lnt) [4].

Now we discuss the upward curvature in M(lnt) above 63 K shown in fig.1. The curvature in relaxation results from a crossover of an initially logarithmic to a later exponential relaxation regime. To show this more clearly, we plot in the inset of fig. 2 log(M_{irr}) as a function of time at T = 80 K and B = 0.2 T. M_{irr} represents the irreversible part of the magnetisation, determined by subtracting the equilibrium value. The straight line indicates the exponential behaviour at longer times, and a deviation from it at shorter times indicates the logarithmic nature of the relaxation. It should be noted here that the exponential M(t) reflects linear V-I characteristics [8], whereas the nearly logarithmic M(t) should be a consequence of the non-linear V-I characteristics (see e.g. [2]) of the sample. This shows a consistency of magnetic relaxation and typical s- shaped transport V-I characteristics

246

observed above $T_G(B)$ in high-T_C materials [2, 9 - 11]. Since $M \propto I$, which in a relaxation measurement decays with time; at the beginning of the relaxation I (M) is large resulting in a non-linear V-I (logarithmic relaxation), and at longer times the I (M) decays to small values giving a linear V-I (exponential relaxation). This behaviour of flux lines above $T_G(B)$ is still controversial, and may be explained with TAFF dynamics of a 2D - plastic creep [12] or a pinned liquid [13] or a vortex slush [10]. Thus the characteristic time for T > 63 K in fig.2 may be identified with t_{cr}, where with increasing time a crossover occurs to a 2D- plastic creep regime or a pinned liquid or a slush. We would remark here that the width (the uncertainty in t_{cr}) of such a crossover can be broad depending upon the disorder in the sample [10]. The decrease of t_{cr} with temperature indicates the decrease of the effective energy barrier, determining the crossover, with increasing temperature.

4. Conclusions

Using a squid set-up, capable of measuring relaxation at 0.1s onwards, we have probed the B-T diagram of a BSSCO-2223 tape both much above and below the transport determined $T_G(B)$. Our results indicate a unique signature of the "vortex glass transition", in the relaxation measurements, where we observe a change of curvature in the relaxation curves. We find a characteristic time "diverging" at $T_G(B)$ at all the measured fields 0.05 T, 0.1 T and 0.2 T. This characteristic time below $T_G(B)$ has been identified with the complete penetration of the field step, and above $T_G(B)$ it corresponds to a crossover from non linear V-I to linear V-I characteristics. The latter crossover might be related to the TAFF - dynamics of a 2D-plastic creep or a pinned liquid or a vortex slush. The results show an accordance between the magnetic relaxation and the transport V-I characteristics in these materials.

Acknowledgements: One of the authors (A.G.) is grateful to Dr. L. Fabrega for carefully reading the manuscript. The tape is provided by NKT Research Centre A / S, Denmark.

References

[1] E D Tuset et al. 1994 *Physica C* **235 - 240** 2927
[2] M G Karkut et al. this conference
[3] C J van der Beek et al. 1992 *Physica C* **195** 307
[4] V Vinokur et al. 1991 *Phys. Rev. Lett.* **67** 915
[5] H G Schnack and R P Griessen 1992 (Comment) *Phys. Rev. Lett.* **68** 2706
[6] A Gurevich and H. Küpfer 1993 *Phys. Rev. B* **48** 6477
[7] A Gurevich and E H Brandt 1994 *Phys. Rev. Lett.* **73** 178
[8] E H Brandt 1994 *Phys. Rev. B* **49** 9024
[9] R H Koch et al. 1989 *Phys. Rev. Lett.* **63** 1511
[10] T K Worthington et al. 1992 *Phys. Rev. B* **46** 11854
[11] Qiang Li et al. 1995 *Phys. Rev. B* **51** 701
[12] C J van der Beek and P H Kes 1991 *Phys. Rev. B* **43** 13032
[13] V Vinokur et al. 1990 *Phys. Rev. Lett.* **65** 259

Inst. Phys. Conf. Ser. No 148

Paper presented at Applied Superconductivity, Edinburgh, 3–6 July 1995

© 1995 IOP Publishing Ltd

The critical current densities of some Tl-based superconductors: Influence of the various atomic compositions

A. Kilic, S. Senoussi, H. Traxler

Laboratoire de Physique des Solides (URA 2 associée au CNRS), Université Paris-Sud, 91405

Orsay Cedex, France.

J.C. Moore, A.J. Collier, M.J.Goringe and C.R.M. Grovenor,

Department of Materials, Oxford University, Parks Road Oxford, OX1 3PH, UK.

Abstract

We investigate the correlation between the critical current density and the critical temperatures of five Tl-based superconductors with T_c varying from 88 to 115 K. We find that for some of the samples J_c is as high as $10^7 A/cm^2$ at T = 4.2 K and H = 0. Perhaps, much more interestingly, for our best samples $J_c(H=0)$ exceeds $10^5 A/cm^2$ at 100 K. It is also very important to emphasize that, contrary to Bi-based materials, the critical current density is not very sensitive to the magnetic field here. For instance, at T = 77 K and H = 20 kG, J_c is still as high as $2 \times 10^4 A/cm^2$ while it is virtually zero in Bi-based systems.

Introduction

Numerous investigations of the Thallium- [1-5] and Bismuth-based high temperature superconductors (HTSC) showed that, despite their higher T_c, these compounds exhibit much lower critical current densities (J_c) than $YBa_2Cu_3O_{7-\delta}$. The difference between the pinning properties of the two HTSC families is especially large in the T-H region around 77 K and 10 kG where technical applications are most promising. The weak pinning properties of Tl and Bi (HTSC) has been ascribed [6] to their extremely high mass anisotropy ratio: $\gamma = (m_c/m_{ab})^{1/2} > 50$ for Bi and $\gamma > 500$ for Tl against 5 to 6 in the case of YBCO [7-10]. It is widely believed that the physical origin of this behavior is due to the fact that pinning barriers against flux creep decrease exponentially with γ making the critical current densities of Tl and Bi layered cuprates intrinsically much lower than that of YBCO. We shall show here that this pessimistic conclusion is perhaps not justified for some Tl-based HTSC. More generally, this paper compares the critical current densities of five Tl-based high-Tc superconductors having the following compositions: $(Tl_{0.85}Pb_{0.15})Ba_2CaCu_2O_x$ (refered to as TLP here, T_c=88 K), $Tl(Sr_{0.8}Ba_{0.2})_2Ca_2Cu_3O_x$ (called TLS, T_c=93 K), $TlBa_2CaCu_2O_x$ (called TLBA, T_c=97K), $(Tl_{0.7}Bi_{0.3})(Sr_{0.8}Ba_{0.2})_2CaCu_3O_x$ (called TLB, T_c=115K) and $(Tl_{0.78}Bi_{0.22})(Sr_{0.8}Ba_{0.2})_2Ca_2Cu_3O_x$ (TLBI, T_c= 114K). All of the present samples are made of randomly oriented fine powders with grain sizes ranging from 5 to 10 μm in the a-b planes and about 2 μm along the c-axis. This synthesis techniques used to prepare these powders are described in J.C. Moore et al (this volume). The critical current density was determined from the hysteresis cycle via a generalized Bean-model which accounts for the high anisotropy of the material and assumes that J_c is restricted to the a-b basal planes exclusively.

The hysteresis cycles were obtained by means of a home made fast vibrating sample magnetometer (VSM) with a field sweeping rate dH/dt of about 1 kG/s and a sensitivity better than 10^{-5} emu at low fields. To characterize our samples further and investigate the first stage of field penetration within the grains, we also made systematic VSM-investigation of the M(H) curves for the same temperature as above but for H, ranging from field values as low as 10 mG to about 20 G. In the VSM measurements, the temperature was sensed by means of two different platinum thermometers located just below and just above the sample. As a result, the uncertainty in the absolute temperature was less than 0.2 K over the whole interval of measurements (4.2 - 120 K) while the temperature was kept stable to about 10^{-2} K during a given fixed M versus H measurement.

Results and discussion

Because of the lack of space we only present some critical current values here. More complete description of the measurements together with detailed analysis of the data will be reported elsewhere.
Fig. 1 shows the field dependence of the critical current densities of the five specimens at 4.2 K. We note that in the low field limit J_c is very high and comprised between 10^6 and 10^7 A/cm^2.

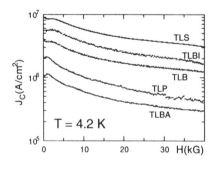

Figure1. Field dependence of the critical current density $J_c(H)$ de from magnetization measurements on the basis of an extended Bean for randomly oriented fine powders of five Tl-based superconduc 4.2 K

$TLP = (Tl_{0.85}Pb_{0.15})Ba_2CaCu_2O_x$
$TLS = Tl(Sr_{0.8}Ba_{0.2})_2Ca_2Cu_3O_x$
$TLBA = TlBa_2CaCu_2O_x$
$TLB = (Tl_{0.7}Bi_{0.3})(Sr_{0.8}Ba_{0.2})_2CaCu_3O_x$
$TLBI = (Tl_{0.78}Bi_{0.22})(Sr_{0.8}Ba_{0.2})_2Ca_2Cu_3O_x$

Moreover, the five specimens exhibit roughly the same slow variation with H. It is also interesting to note that these J_C values are comparable to other results of the literature (especially for sample TLBA [4]).

The critical current density of the same materials are presented in fig. 2 - 4 at T = 50, 80 and 100 K respectively. We note that for samples TLBI and TLB, J_C stays quite large and depends

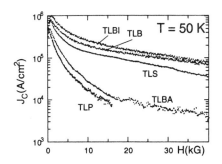

Figure 2. Field dependence of $J_c(H)$ of the same samples as in figure 1 but for T=50 K

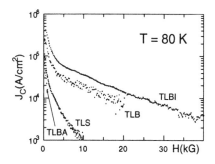

Figure 3. Field dependence of $J_c(H)$ of the same samples as in figure 1 and 2. but at 80 K

rather slowly on H for all the investigated temperatures. Fig. 5 compares the temperature dependences of J_c at H = 0 for the five specimens. We observe that for T > 30 K, the critical current densities of TLBI and TLBA vary much more slowly with T than those of TLS, TLP and TLBA. As a result, we observe that J_c of TLBI and TLB is still as high as 10^5 A/cm^2 at 100 K.

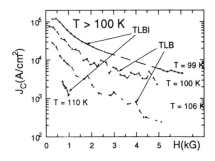

Figure 4. Field dependence of $J_c(H)$ of TLBI and TLB (see also in figure 1-3) at T>100 K. Note that the current densities of both samples are quite high at temperatures above 100 K.

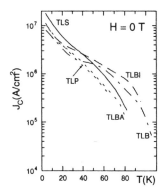

Figure 5. Temperature dependence of $J_c(T)$ of the same Tl- based superconductors as in figure 1-4 at zero applied field (H=0). Note that there is a factor of 2 as compared to figure 1-4. This difference is obtained when we use B instead of H in the Bean model. (This correction is important only when the applied field is close to H=0)

250

Finally, we would like to add that we have also measured the intragrain critical current density of a ceramic sample of composition $(Tl_{0.5}Pb_{0.5})(Sr_{1.6}Ba_x)Ca_2Cu_3O_x$ and found that the J_c is almost the same as that of the sample TLBI at T=4.2 K and $0 \leq H \leq 50$ kG. This suggests that Bi and Pb doped 1223 have very similiar pinning properties at least at low temperatures. Comparision between two materials will be reported elsewhere.

In conclusion, the 1223 samples have higher magnetisation J_c values than the 1212 samples even very far below T_c. The altering of the Bi content in the Bi-1223 samples seems to have no effect on the magnetisation properties. There is an interesting crossover in J_c performance between the Bi-doped (TLB and TLBI) and undoped 1223 (TLS) at low temperatures. TLS gives significantly higher magnetisation J_c values below about 50 K.

Acknowledgements: This work is in part supported under a Brite Euram Project, BRE2-CT93-0455. One of us (A.K.) thanks TUBITAK/TURKEY for financial support and the other one (H.T.) would like to thank the ERASMUS-fund for financial support.

References

[1] M.R. Presland, J.L. Tallon, N.E. Flower, R.G. Buckley, A. Mawdsley, M.P. Staines and M.G. Fee, Cryogenics 33 (1993) 502.

[2] D.N. Zheng, A.M. Campell, R.S. Lieu and P.P. Edwards, Cryogenics 33 (1993) 46.

[3] M. Ledésert, A. Maignan, J. Chardon, C. Martin, Ph. Labbé, M. Hervieu and B. Raveau, Physica C, 232 (1994) 387.

[4] A. Wahl, V. Hardy, A. Maignan, C. Martin, B. Raveau, Cryogenics 34 (1994) 941.

[5] Toshiya J. Dot, Toyotaka Yuasa, Takeshi Ozawa and Kazutoshi Higashiyama, Jpn. J. Appl. Phys. 33(1994) 5692.

[6] G. Blatter, V.B. Geshkenbein and A.I. Larkin, Phys. Rev. Lett., 68 (1992) 875; Zhidong Hoa and J.R. Clem, Phys. Rev. B 46 (1992) 5853.

[7] K.E. Grey, R.T. Kampwirth and D.E. Farrell, Phys. Rev. B41 (1990) 891.

[8] D.E. Farrell, R.G. Beck, M.F. Booth, C.J. Allen, A.D. Bukowski and D.M. Ginsberg, Phys. Rev. B42, (1990) 6758.

[9] U. Welp, W.K. Kwok, G.W. Crabtree, K.G. Vander-Voort and J.Z. Liu, Phys. Rev. Lett. 62, (1989) 1908.

[10] D.E. Farrell, C.M. Williams, S.A. Wolf, N.P. Bansal and V.G. Kogan, Phys. Rev. Lett. 61, (1988) 2805.

Inst. Phys. Conf. Ser. No 148
Paper presented at Applied Superconductivity, Edinburgh, 3–6 July 1995
© 1995 IOP Publishing Ltd

Transport critical current measurements of a HIP'ed and unHIP'ed bulk samples of PbMo$_6$S$_8$ in magnetic fields up to 12 Tesla

H A Hamid, D N Zheng and D P Hampshire

Superconductivity Group, Department of Physics, University of Durham, Durham, UK.

Abstract. Transport critical current measurements have been made on bulk PbMo$_6$S$_8$ (PMS) samples from 5 K up to T$_c$ in magnetic fields up to 12 T. The properties of hot isostatically pressed (HIP'ed) PMS fabricated at 1.3×10^8 N.m^{-2}(1300 bar) are compared with those of PMS fabricated at ambient pressure. We have found that HIP'ing PMS increases the transport critical current density by ≈ 7. The volume pinning force (F$_p$) of both samples has the same functional form F$_p = \alpha (B^*_{C2})^n$(T)b(1-b) where b=B/B$^*_{C2}$ and n=1.6 $\pm$ 0.06 and 1.67 $\pm$ 0.06 and α is 6.0×10^5 Am^{-2}T$^{-0.6}$ and 0.9×10^5Am^{-2}T$^{-0.67}$ for the HIP'ed and unHIP'ed sample respectively. B$^*_{C2}$ for the HIP'ed sample lies close to the irreversibility line derived from complementary magnetic measurements.

Our results demonstrate that the critical current density of HIP'ed PbMo$_6$S$_8$ in high magnetic fields is sufficiently large that this material is a potential candidate for the next generation of high field applications operating in magnetic fields up above 25 Tesla.

1. Introduction

The Chevrel phase PbMo$_6$S$_8$ (PMS) is a potential candidate for the next generation of high field applications because its upper critical field can exceed 50 T at 4.2 K [1]. Magnetic studies on bulk PMS samples have shown that the critical current density of PMS can be markedly increased by fabricating the material using a hot isostatic press (HIP) [2,3].

In this work, transport critical current measurements (J$_C$) are presented on a HIP'ed bulk sample of PbMo$_6$S$_8$ as a function of magnetic field and temperature and compared to those of bulk PMS prepared at ambient pressure (unHIP'ed). These data are used to calculate the functional form of the volume pinning force (F$_p$) for each sample from J$_C$ and the effective upper critical field B$^*_{C2}$. We determine the effect of HIP'ing PMS on J$_C$ in high magnetic fields.

2. Sample Fabrication

The HIP'ed PbMo$_6$S$_8$ sample preparation involved a two step reaction procedure in a controlled environment. The starting materials, Pb (99.9999% purity), Mo (99.95%, 4-8 μm) and S (99.999%) were mixed in the atomic ratio 1:6:8. The material was then heat treated at 450 °C for 4 hours, then at 650 °C for 8 hours with further treatment at 1000 °C for 44

252

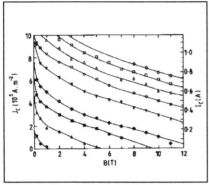

Figure 1 J_C of the unHIP′ed sample as a function of field and temperature (from top to bottom: 5, 6, 7, 8, 9, 10.5, 11, 12, 13 K)

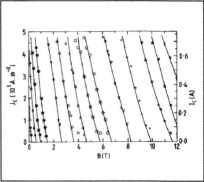

Figure 2 J_C of the HIP′ed sample as a function of field and temperature (right to left: 8.7, 8.9, 9.1, 9.9, 10.2, 10.5, 11, 11.4, 11.7, 12, 12.3, 12.5, 12.7, and 12.9 K)

hours in Argon atmosphere to form the $PbMo_6S_8$ phase. The sample was then HIP′ed at 1300 bar for 8 hours at 900 °C. A detailed account of the fabrication procedure is provided elsewhere [3].

3. J_C measurements

The samples were carefully cut into rectangular shapes with dimensions 6.0 x 2.4 x 0.5 mm³ (unHIP′ed) and 6.0 x 1.9 x 0.8 mm³ (HIP′ed). The critical current densities were measured at a $2\mu V$ (10 μVcm^{-1}) criteria. The four terminal measurements were made using a probe which can measure critical current densities as a function of temperature in high magnetic fields to an accuracy of ±70 mK at 10 K [4]. For the HIP′ed sample, data were taken from 8.7 K up to 12.9 K and for the unHIP′ed sample, from 5 K up to 13 K, in magnetic fields up to 12 T. In all measurements, the magnetic field was orthogonal to the transport current.

4. Results

Figures 1 and 2 show the J_C data for both samples. At each temperature, as the field is decreased, J_C increases much more rapidly for the HIP′ed sample than the unHIP′ed sample. Figure 3 is an example of an E-J characteristics of the HIP′ed sample at 10.5 K for three different magnetic fields. The baseline is rather flat and has a low noise level which demonstrates sufficiently low contact resistances have been achieved for good temperature control over the range of J_C investigated. Figure 4 shows the effective upper critical field B^*_{C2} as a function of temperature for both samples. The B^*_{C2} curve of the unHIP′ed sample lies above that of the HIP′ed sample in this temperature range. On the same figure are magnetic and specific heat data for a similarly prepared sample, HIP′ed at 800 °C. It can

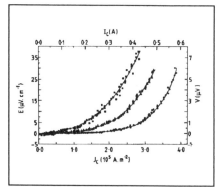

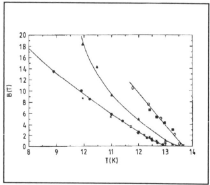

Figure 3 The E-J characteristics for the HIP′ed sample at 10.5 K in magnetic fields of (from right to left): 3.6 T, 3.8 T and 4 T.

Figure 4 The B^*_{C2} line for the HIP′ed (●) and unHIP′ed (▲) sample. ($\times$:B_{irr}, ■:B_{C2}-magnetic measurements[3]; $\bigcirc$:B_{C2}-specific heat measurements[5])

be seen that the B^*_{C2} curve of our sample lies close to the B_{irr} line from the magnetic measurements and significantly below the upper critical field data B_{C2}.

5. Analysis of the experimental data

The J_C characteristics are linear throughout most of the field range at each temperature in figures 1 and 2. The effective upper critical fields B^*_{C2} has been calculated at each temperature by linearly extrapolating J_C to zero. The pinning force F_p can then be represented by the relation

$$F_p = \alpha (B^*_{C2})^n (T) b (1-b) \qquad \text{Eqn. (1)}$$

where $b=B/B^*_{C2}$ is the reduced magnetic field and $n=1.6 \pm 0.06$ and 1.67 ± 0.06 and α is 6.0×10^5 Am^{-2}T$^{-0.6}$ and 0.9×10^5Am^{-2}T$^{-0.67}$ for the HIP′ed and unHIP′ed sample respectively.

6. Discussion

The functional form of F_p, more specifically the reduced field dependence and the values of n, are similar for both samples. This suggests that the mechanism that determines J_C in both samples is the same. However the value of α, and hence the magnitude of J_C for the HIP′ed sample is larger by a factor of ≈ 7 than that of the unHIP′ed sample. We expect that the primary reason for the improvement in the HIP′ed sample is due to an increase in the density and the contact area between the grains, although the results are also consistent with stronger pinning.

In contrast to most transport measurements on wires which show a Kramer dependence, we have found a functional form for F_p that is most commonly found in systems where core pinning dominates [6]. The transport measurements presented here are very sensitive to the intergranular current density. Whether the regions of contact between the grains in these bulk samples are off-stoichiometric or highly defective is to be addressed by detailed electron microscopy.

The comparison between these transport measurements and the magnetic and specific heat data support the conclusion that the irreversibility line of bulk HIP'ed and unHIP'ed PMS is markedly less than the upper critical field. The similarity between the B^*_{C2} curve from the transport measurements and the B_{irr} line from magnetic measurements suggests that the difference in HIP'ing temperature does not significantly affect these parameters. We attribute the lower values of B^*_{C2} and T_C found in the HIP'ed sample to some oxygen contamination during processing [7].

Most work to determine the J_C on bulk samples comes from magnetic measurements which are sensitive to intragranular J_C. We have completed transport current measurements which are particularly sensitive to intergranular transport J_C.

7. Conclusions

We have found that the critical current density increased by a factor of ≈ 7 when the sample was hot isostatically pressed.

Both samples can be described by a single functional form $F_p = \alpha B^n_{C2}(T)b(1-b)$ which suggests that the same pinning mechanism determines J_C.

The B^*_{C2} for the HIP'ed sample lies close to the irreversibility line derived from magnetic measurements on a HIP'ed sample. However it is significantly lower than the unHIP'ed sample over the same temperature range.

The high B^*_{C2} values for the HIP'ed sample extrapolate to above 30 T at 4.2 K which makes this material a promising candidate for use in high field steady state magnets operating above 25 T.

Acknowledgements

We wish to acknowledge P. Russell for her help with the production of the drawings. This work is supported by the EPSRC. GR/J39588 and The Royal Society, UK.

References

[1] Odermatt R, Fisher Ø , Jones H and Bongi G 1974 *J. Phys. C: Solid State Phys.,* 7 L13

[2] Alekseevskii N E, Dobrovol'skii N M, Eckert D and Tscbro V I 1977 *J. Low Temp. Phys.* **29** 565

[3] Zheng D N, Ramsbottom H D and Hampshire D P 1995 *to be published in Phys. Rev B*

[4] Friend C M and Hampshire D P 1994 *J. Meas. Sci. Tech.* 5 1-9.

[5] Ali S, Zheng D N and Hampshire D P 1995 *paper presented at this conf.*

[6] Dew-Hughes D 1974 *Philos. Mag.* **30** 293-305

[7] Foner S, McNiff Jr. E J and Hinks D G 1985 *Phys. Rev. B* **31** 6108-6111

Inst. Phys. Conf. Ser. No 148
Paper presented at Applied Superconductivity, Edinburgh, 3–6 July 1995
© 1995 IOP Publishing Ltd

Critical currents of the Chevrel phase $Pb_{1-x}Gd_xMo_6S_8$

D N Zheng and D P Hampshire

Department of Physics, University of Durham, South Road, Durham DH1 3LE, UK

Abstract. A series of Chevrel phase $Pb_{1-x}Gd_xMo_6S_8$ bulk samples have been made using a hot isostatic pressing (HIP) process. Dc magnetic measurements have been carried out on the samples. The critical current density J_c has been calculated from magnetic hysteresis data using the Bean model. The data show that by increasing the substitution of Gd for Pb J_c is decreased. In particular, the field sensitivity of J_c increases markedly as more Gd is added to the samples. However, it was found that although the magnitude of J_c changes significantly with Gd substitution the pinning force curves for all samples can be described by the Kramer relation $F_p = J_c \times B \propto b^{1/2}(1-b)^2$ over a wide temperature range. This indicates that J_c is limited by one dominant flux pinning mechanism in these samples. The irreversibility field B_{irr} was obtained by extrapolating linearly the Kramer plots ($J_c^{1/2}B^{1/4}$ versus B) to $J_c^{1/2}B^{1/4}=0$. The data again show that B_{irr} decreases dramatically with the addition of Gd.

1. Introduction

The high upper critical field of the Chevrel phase material $PbMo_6S_8$ offers the potential for making high field superconducting magnets. In order to generate high magnetic fields, high critical current densities are required. The reasons for the comparatively low J_c values are the subject of ongoing research because the mechanisms that control J_c are still not well understood.

In this study, a series of samples with partial Gd substitution for Pb were fabricated and the critical current density of the samples was investigated. It has been reported that adding the magnetic element Eu can increase B_{c2} in $PbMo_6S_8$ and $SnMo_6S_8$ [1]. Therefore, it is interesting to study the effect of Gd substitution.

2. Experimental

Samples used in this study were prepared using a procedure described elsewhere [2]. In brief, ceramic samples were first made using a solid state reaction process. Pure elements were mixed with nominal composition $Pb_{1-x}Gd_xMo_6S_8$ (x=0, 0.1, 0.2, and 0.3) and reacted in an evacuated and sealed quartz tube at 450 °C for 4 h. This was followed by a slow temperature ramp at 33 °C.h⁻¹ to 650 °C and held for another 8 h. The material was then ground and pressed into discs. The discs were reacted again, in an evacuated and sealed quartz tube, at 1000 °C for 44 h. The sample was then wrapped in Mo foil and sealed under vacuum in a stainless steel tube prior to the hot isostatic pressing treatment, carried out at 2000 bar and 800 °C for 8 h. For each sample, a rectangular piece was cut off and used for magnetic measurements.

During the fabrication procedure, the samples were handled in a glove box. This ensured that the samples were not exposed to oxygen which strongly suppresses the superconducting transition temperature T_c.

The magnetic hysteresis measurements were performed on a home-made vibrating sample magnetometer using a field ramp rate of 15 mT.s⁻¹.

3. Results and discussion

The curves of magnetic moment versus temperature for the $Pb_{1-x}Gd_xMo_6S_8$ (x=0, 0.1, 0.2 and 0.3) samples are shown in Fig.1. The applied field was 5 mT. The figure shows that the change of T_c is not substantial up to Gd level of x=0.3. The same effect has been observed by Fischer et al. [1] for $Sn_{1.2(1-x)}Eu_xMo_{6.35}S_8$ samples. It is believed that superconductivity in Chevrel phase compounds occurs in the Mo-clusters. The substitution of Pb can affect T_c through the change in charge transfer from the Pb-site to the Mo-cluster. Since Gd has the same valence as Pb, it is to be expected that the change in T_c is small.

In Fig.2 and Fig.3, the critical current density J_c, calculated using the Bean model, is shown as a function of applied field and temperature for samples x=0 and x=0.2 respectively. Clearly, J_c is significantly lower in the sample with Gd. More importantly, the data show that J_c decreases more quickly with increasing field for the x=0.2 sample. The same trend is observed in the sample of x=0.1 and 0.3.

Figure 4 shows the Kramer plot (i.e., $J_c^{1/2}B^{1/4}$ versus B) for the data of the $Pb_{0.8}Gd_{0.2}Mo_6S_8$ sample. The straight lines suggest that the pinning force $F_P=(J_c \times B)$ can be expressed by the Kramer formula [3],

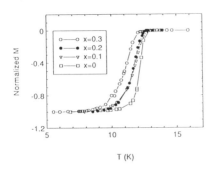

Figure 1. Temperature dependence of the reduced zero-field-cooling magnetic moment for the $Pb_{1-x}Gd_xMo_6S_8$ (x=0, 0.1, 0.2 and 0.3) samples.

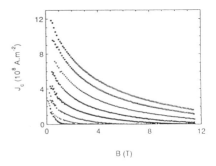

Figure 2. J_c of $PbMo_6S_8$ as a function of field and temperature. From top to bottom: 4.2, 5, 6, 7, 8, 9, 10 and 11 K.

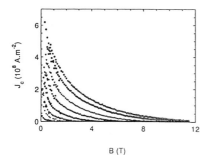

Figure 3. J_c of $Pb_{0.8}Gd_{0.2}Mo_6S_8$ as a function of field and temperature. From top to bottom: 4.2, 5, 6, 7, 8, 9, 10 and 11 K.

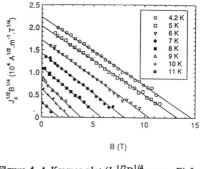

Figure 4. A Kramer plot ($J_c^{1/2}B^{1/4}$ versus B) for the $Pb_{0.8}Gd_{0.2}Mo_6S_8$ sample. Linear fitting to the data are shown by solid lines.

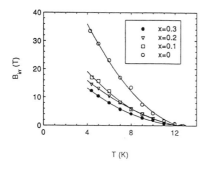

Figure 5. The irreversibility field B_{irr} data for the $Pb_{1-x}Gd_xMo_6S_8$ (x=0, 0.1, 0.2 and 0.3) samples. The solid lines are guides for the eye.

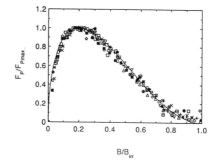

Figure 6. Normalized pinning force as a function of normalized field for the $Pb_{1-x}Gd_xMo_6S_8$ (x=0, 0.1, 0.2 and 0.3) samples. For each sample, data at 4.2, 6, 8, 10 K are shown.

$$F_P \propto b^{1/2}(1-b)^2$$

where $b=B/B_{irr}$. For the other three samples, the same linear behaviour in the Kramer plot is also observed. The linear extrapolation of the data for each temperature to $J_c^{1/2}B^{1/4}=0$ gives the irreversibility field B_{irr} value at corresponding temperatures. The B_{irr} data for the samples are shown in Fig.5. Clearly, the addition of Gd reduces B_{irr} significantly. This is consistent with the J_c data shown in Figs. 2 and 3. The addition of Gd could result in more second phase material in the samples and lead to a lower value of J_c since the superconducting volume would be smaller than the actual sample volume. However, the presence of second phase material should not affect B_{irr} if the Gd is not substituted into the

258

lattice. Thus, a more plausible reason of the reduction in J_c and B_{irr} could be that the substitution of Gd for Pb changes the superconducting parameters, such as the thermodynamic critical field B_c and the coherence length ξ, and consequently changes J_c and B_{irr}. Specific heat measurements on the x=0.3 sample have shown the presence of material with a lower B_{c2} value than the undoped sample. More work is required to understand the interaction between the magnetic Gd ion and the superconductivity and in particular why J_c, B_{irr} and B_{c2} are reduced by the substitution of Gd.

Figure 6 shows a normalized plot of F_P as a function of the applied field for the samples. F_P is normalized to the peak value F_{Pmax} for each curve. The scaling behaviour is a striking feature, considering the significant difference in J_c and B_{irr} between the samples. It suggests that the J_c in the samples is primarily determined by one flux pinning mechanism. The form of the scaling law is similar to that commonly observed in Nb_3Sn (where grain boundaries are major pinning centres), and has led to the hypothesis of a grain boundary pinning mechanism in Chevrel phase materials such as $PbMo_6S_8$ and $SnMo_6S_8$ [4].

4. Conclusions

A series of Chevrel phase $Pb_{1-x}Gd_xMo_6S_8$ bulk samples have been made using the hot isostatic pressing process and the critical current density J_c of the samples has been studied. The results show that upon the addition of Gd J_c is reduced and becomes more field dependent although the transition temperature T_c is almost unchanged. More importantly, despite the significant change in J_c, the pinning force curve for the samples can be scaled according to the Kramer relation: $F_P = J_c \times B \propto b^{1/2}(1-b)^2$. This suggests that J_c in the samples is limited by one dominant flux pinning mechanism, possibly grain boundary pinning.

Acknowledgements

The authors acknowledge: the contributions of R. Luscombe, L. Le Lay (BICC Superconductors, UK) and T.C. Willis (University of Wisconsin-Madison, USA); A. Crum of Engineered Pressure Systems for use of the Hot Isostatic Press; the support of the Engineering and Physical Sciences Research Council UK and the Royal Society.

References

[1] Fischer Ø, Decroux M, Roth S, Chevrel R and Sergent M 1975 *J. Phys. C: Solid State Phys.* **8** L474-7
[2] Zheng D N, Ramsbottom H D and Hampshire D P 1995 *to be published in Phys. Rev. B*
[3] Kramer E J 1977 *J. Appl. Phys.* **44** 1360-1370
[4] Bonney L A, Willis T C and Larbalestier 1993 *IEEE Trans. Appl. Supercond.* **3** 1582-5

Inst. Phys. Conf. Ser. No 148
Paper presented at Applied Superconductivity, Edinburgh, 3–6 July 1995
© *1995 IOP Publishing Ltd*

Critical current densities, scaling relations and flux pinning in NbTi as determined by flux penetration measurements

H D Ramsbottom and D P Hampshire

Superconductivity Group, Department of Physics, University of Durham, Durham, UK.

Abstract. Flux penetration measurements have been made on a commercial multifilamentary NbTi wire from 4.2 K up to T_C in magnetic fields up to 10 T. The results have been used to determine the magnetic field profiles and the spatial variation of the critical current density as a function of field and temperature. This is the first report of such analysis on a commercial wire in magnetic fields up to $B_{c2}(T)$. The critical current density has been calculated as a function of field and temperature from the flux penetration measurements. The values obtained compare well with complimentary transport data on the same wire. The functional form of the pinning force (F_P) obeys the Fietz-Webb scaling law throughout the entire field and temperature range. The spatial variation of $J_c(B,T)$ is small for all fields and temperatures. This confirms that bulk pinning operates in this technical superconductor.

1. Introduction

Flux penetration measurements provide a unique opportunity to study both the functional form and spatial variation of the critical current density J_C. They make it possible to quantify the inhomogeneous properties of superconducting materials and locate regions of interest. Such measurements are used to determine the functional form of the inter-granular and intra-granular J_C, the role of granularity, and the relative contributions of bulk and surface pinning within a superconducting sample.

In this paper, results are presented on a commercial, multifilamentary, superconducting NbTi wire. NbTi is chosen because it is probably the most extensively studied and industrially useful superconductor. Hence these measurements serve as a standard for assessing conductors being developed with new materials for high field applications.

Flux penetration measurements have been completed on the wire from 4.2 K up to T_C in magnetic fields up to 10 T. The results are used to calculate the magnetic field profile, the spatial variation of J_C and the average J_C as a function of field and temperature. These are analyzed within the framework of a universal scaling law and compared with transport current data on the same sample.

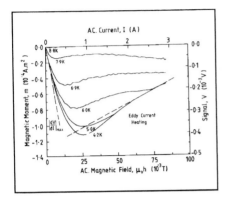

Figure 1 Magnetic moment versus ac. field for NbTi as a function of temperature at 3 T.

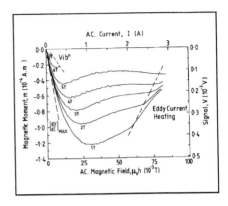

Figure 2 Magnetic moment versus ac. field for NbTi as a function of dc. field at 6 K.

2. Experimental Procedure

Flux penetration measurements are made by applying an ac. field ($\mu_0 h$) onto a superconducting sample. In agreement with Lenz's law, lossless supercurrents flow to a depth δ to expel the field. Bean's critical state model shows that the local current density that can flow is equal to J_C. At high applied ac. fields, the sample is fully penetrated ie. the depth to which the field penetrates (δ) is equal to the radius of the sample. At this point, a current density equal to J_C flows throughout the entire sample.

A primary coil was used to generate an ac. field and a set of oppositely wound secondary coils was used to measure the magnetization of the sample [1]. The sample consisted of a superconducting NbTi wire of 61 filaments, each 28 μm in diameter . The copper matrix was removed using nitric acid and the NbTi filaments stuck together with GE. varnish. 60 sections were cut, each 3 mm long and formed into a sample. The sample was positioned so that the ac. field was perpendicular to the axis of the filaments.

Measurements were made from 4.2 K up to T_C in magnetic fields up to 10 T.

3. Results and Discussion

Figures 1 and 2 show the magnetic moment versus ac. field as a function of both dc. field and temperature. In these figures the experimentally determined quantities are also included. At all dc. fields and temperatures there is a sharp fall in the magnetic moment to a minimum value when there is full penetration of the superconductor. The marked increase in magnetic moment at the lowest temperatures and highest ac. fields is attributed to eddy current heating in the copper components of the probe. Only limited data can be obtained in high dc. fields due to the coupling of the ac. and dc. fields which causes the probe to vibrate.

By following the analysis outlined by Campbell [2], and using the data in figures 1 and 2, it is possible to determine the magnetic field profile inside the NbTi sample. The differential of the magnetic moment with respect to the ac. field gives the penetration depth.

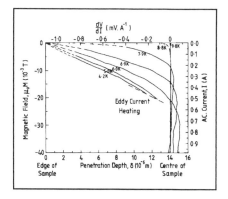

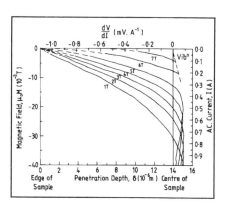

Figure 3 The magnetic field profile inside a NbTi sample as a function of temperature in a dc. field of 3 T.

Figure 4 The magnetic field profile inside a NbTi sample as a function of dc. field at a temperature of 6 K.

Figures 3 and 4 show the spatial variation in the magnetic field ($\mu_0 M$) as a function of both field and temperature. The applied dc. field ($\mu_0 H$) and ac. field ($\mu_0 h$) can be added to the magnetic field produced by the NbTi ($\mu_0 M$) to give the net field B. The gradient of the lines ($dM/d\delta$) as a function of penetration depth gives the spatial variation of J_C. It can be seen that the gradients of these lines and hence J_C is approximately constant throughout the sample. As expected, surface pinning is not a significant mechanism in this wire. The minima in the magnetic moment, (shown in figures 1 and 2), correspond to the apparent penetration of the magnetic field to a depth greater than the radius of the sample, (shown in figure 3 and 4). This is a artifact of the analysis and is expected even for a bulk pinning superconductor [3].

From the minimum value of the magnetic moment, the coil geometry and the dimensions of the sample, it is possible to calculate the average critical current density at each field and temperature. This is shown in figure 5. The values of J_C are accurate to approximately 10 %, primarily due to uncertainties in the sample dimension. At 6 T and 6 K, J_C is 6.2 x 10^8 A.m^{-2} which is in good agreement with transport current data on the same sample. This gives a value of 6.0 x 10^8 A.m^{-2} [4].

Figure 6 shows the reduced pinning force (F_P/F_{PMAX}) versus the reduced magnetic field (b=B/B_{C2}) as a function of temperature. The reduced pinning force is calculated using F_P= J_C x B while B_{C2} is taken to be the magnetic field at which the critical current density drops to zero. A single curve can be drawn through the data, consistent with Fietz-Webb scaling [5]. Alternatively, the J_C data in figure 5 can be extrapolated linearly to zero. This gives an effective upper critical field (B_{C2}^*) which is about 10 % lower than that used in figure 6. The data can then be parametrized by a scaling law of the form $F_P = \alpha B_{C2}^n b(1-b)$, where b=B/$B_{C2}^*$, the index n = 2.18±0.05 and the constant α = (3.14±0.12) x 10^8 T$^{-1.18}$A.m^{-2}.

262

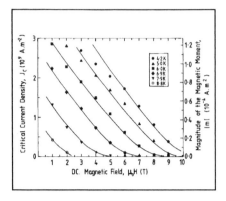

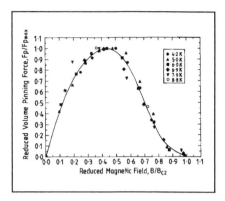

Figure 5 The critical current density of NbTi as a function of field and temperature.

Figure 6 The reduced pinning force versus the reduced magnetic field as a function of temperature.

4. Conclusions

Flux profile measurements have been made on high J_C superconductors which exhibit strong bulk pinning. The results have been used to calculate the functional form and spatial variation of J_C as a function of field and temperature.

The gradient of the field profiles suggest that the spatial variation of J_C is small. The apparent penetration of the magnetic field to a depth greater than the sample radius is a artifact of the analysis and is not indicative of sample granularity.

At 6 T and 6 K, the flux penetration measurements give a J_C of 6.2 x 10^8 A.m^{-2}. This is within 3.5 % of transport current measurements on the same sample which give a value of 6.0 x 10^8 A.m^{-2}.

The functional form of the pinning force obeys the universal Fietz-Webb scaling relation. Except at fields close to $B_{C2}(T)$, the pinning force can be parametrized in the form $F_P = \alpha B_{C2}{}^n b(1-b)$, where n = 2.18±0.05 and α = (3.14±0.12) x 10^8 T$^{-1.18}$A.m^{-2}.

Acknowledgements

The authors wish to acknowledge P. Russell for help with the production of the drawings. This work is supported by the EPSRC. GR/J39588 and The Royal Society, UK.

References

[1] Ramsbottom H D and Hampshire D P 1995 *Accepted for publication in Meas. Sci. Technol.*

[2] Campbell A M 1969 *J. Phys. C* **2** 1492-1501

[3] Ramsbottom H D and Hampshire D P 1995 *To be submitted to J. Phys. C.*

[4] Friend C M and Hampshire D P 1993 *Applied Superconductivity Vol. 1 Proc. Europ. Conf. on Appl. Supercond.* 23-26

[5] Fietz F A and Webb W W 1969 *Phys. Rev.* **178** (2) 657-667

Inst. Phys. Conf. Ser. No 148
Paper presented at Applied Superconductivity, Edinburgh, 3–6 July 1995
© 1995 IOP Publishing Ltd

The Abnormal Temperature Dependence of j_c in YBCO Textured High-T_c Superconductors

L.M. Fisher[1], A.V. Kalinov[1,2], J. Mirkovic[1,2], V.M. Soukhov[3], and I.F. Voloshin[1]

[1] All-Russian Electrical Engineering Institute, 111250 Moscow, Russia.
[2] Physics Department of Moscow State University, 117234 Moscow, Russia.
[3] Russian Ministry of Sciences and Technologies, Moscow, Russia.

The temperature dependencies of the critical current density j_c in melt textured $YBa_2Cu_3O_x$ samples have been studied. Our investigations were carried out by measuring the ac magnetic susceptibility. We have obtained $j_c(T)$ dependencies for $\mathbf{H} \parallel \mathbf{ab}$ plane as well as for $\mathbf{H} \parallel \mathbf{c}$ axis of samples. The temperature dependence of the critical current density for $\mathbf{H} \parallel \mathbf{c}$ is smooth enough at $T < 0.8T_c$ but at $T \sim 0.9T_c$ we have observed a peak in $j_c(T)$. In some samples the kink rather than the peak appears at the same practically temperature. We relate the anomaly observed to the 2D-3D transition in the vortex lattice.

1. Introduction

The investigation of temperature and magnetic field dependencies of the critical current density j_c in high temperature superconductors (HTS) is very vital for the fundamental physics of the vortex state as well as for their technical applications. The detailed study of these dependencies and their connection with the material microstructure is of great interest for the development of new HTS with a higher current capability, since it is the characteristics that restrict now their practical usage. The perfect melt textured (MT) samples prepared in different laboratories are well known to be characterized by the typical microstructure. They consist of thin (1-10 μ) crystallites with the **c** axis perpendicular to their plane. It seems to be true that the transport and electromagnetic properties of MT HTS should have some universal features owing to their similar microstructure. But the question: "Does microstructure *completely* determine the ability of MT HTS to carry the electric current ?" is now far from being settled. The present paper is devoted to the investigation of $j_c(T,H)$ in $YBa_2Cu_3O_x$ (YBCO) MT samples.

2. Experimental technique

Textured YBCO ingots were prepared by different techniques: by zone-drawing method (the sample G21) and by different modifications of MTG method (samples P50, A94 and C17). The samples A94 and G21 are characterized by the close values of average thickness of the crystallites about 2-3 μ. This thickness is greatest in sample C17 (about 10-20 μ) and has an intermediate value in sample P50 (5-7 μ). The concentration of 211-phase has a maximum value in the sample A94 (~10%) and a minimum value in the samples C17 and P50 (~1-2%).

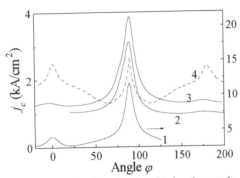

Fig.1. The angular dependencies of j_c for: the sample A94 (curve 1, the right axis), the sample C17 (2), the sample G21 (3), and the sample P50 (dotted curve 4). $H = 1$ T, $T = 77$ K.

We saw two types of the structure of the interlayer dividing the adjacent crystallites inside one textured domain: the ordered ones and the amorphous ones with thickness of about 3-5 nm. We cut the plate-like pieces from ingots with the typical sizes about $5 \times 3 \times 0.3$ mm^3 by such a way that the **c** axis was directed orthogonal to the plate or was parallel to it. The misorientation of **c** axis over the sample have been observed by SEM and was little more than 5°. All samples have $T_c \approx 90$K. The textured structure of our samples is revealed clearly in the orientation dependence of $j_c(\mathbf{H})$ (see Fig.1).

The experimental procedure is based on the measurements of the ac magnetic susceptibility $\chi = \chi' + i\chi''$ that may be easily transformed into the critical current density [1] by means of the critical state model. The sample with pick-up coil was placed into ac magnetic field $h = h_o \cos\omega t$ where the frequency $\omega / 2\pi = 130$ Hz, as well as into the dc magnetic field H. The applicability conditions of our model to obtain adequate j_c values are follows: $H_{c1} < H$, $h_o \ll H$, where H_{c1} is the lower critical field. The typical $h_o = 0.02$ T and we assert our calculation to be fairly reliable for $H > 0.1$ T. As it have been established earlier [2], the experimental results do not depend practically on the angle between the ac and dc magnetic fields over a wide range of the experimental conditions. To study the anisotropy of our samples we continuously alter the angle φ between $\mathbf{H}$ and the **c** axis. The reliability of our method for the j_c measurements have being experimentally tested by comparing the contactless data against the four-probe ones [3] and we have not revealed a significant discrepancy.

3. Results and discussion

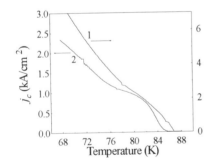

Fig.2. The dependencies of j_c on the temperature for the sample C17 at $H = 1$ T; $\mathbf{H} \parallel \mathbf{ab}$ (the curve 1, right axis) and $\mathbf{H} \parallel \mathbf{c}$ (the curve 2, left axis).

The abnormal magnetic field dependence of j_c (the fishtail effect [4]) is rather widely investigated and discussed but the anomalous $j_c(T)$ are significantly less studied. The interesting phenomenon of such a type is the existence of a local maximum of the function $j_c(T)$ that have been observed and discussed for HTS ceramics [5]. The declination (kink) of experimental data from the typical behavior described by the formula $j_c(T) \sim (1-T/T_c)^\alpha$ at $T \sim 0.9T_c$ in melt textured YBCO samples has been reported in [3]. Our further study reveals that the kink is essentially influenced by the magnetic field orientation as well as by the value of H. We have investigated $j_c(T)$ for different angles between

H and the crystallite plane. Fig.2 demonstrates such a kink at the temperature about 80 K for the sample C17, with the external magnetic field being parallel to the **c** axis (curve 2). However, we have not revealed a significant kink in the orientation **H** || **ab** (curve 1). The ultimate case of the anomaly of $j_c(T)$ have been observed in the sample G21 at **H** || **c**. With H being strong enough ($H > 0.5$ T), $j_c(T)$ demonstrates a distinct peak at $T = T_p = 80$ K $\sim 0.9T_c$ and the j_c increase is magnetic field affected (Fig.3): the increase of H is followed by the relative increase of the peak. The peak location of $j_c(T)$ is practically unaffected by H. Moreover, the kink or peak locations for different samples are very close to each other. The curves $j_c(T)$ for the different φ also confirm the stability of T_p (Fig.4). The dependence $j_c(T)$ persists practically the same shape in the angular range $0° < \varphi < 50°$ and only at $\varphi = 70°$ the peak is substituted by the plateau. The substantial effect of φ on this anomaly is observable in the vicinity of **H** || **ab** only. It is the angular region that may be characterized by the dominant influence of the planar defects (see Fig.1). Another feature of $j_c(T)$ dependencies is common for all samples investigated. With **H** being parallel to **c** axis, the j_c subsides faster when the temperature is higher than T_p (Fig.2). On the other hand, at **H** || **ab** $j_c(T)$ does not undergoes the marked transformation (Fig.2). The above mentioned common character of $j_c(T)$ curves suggests that there is the principal difference in the pinning processes at temperatures $T > T_p$ against the lower temperature region..

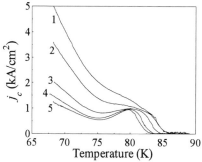

Fig.3. The dependencies of the critical current density j_c on the temperature for the sample G21 at **H** || **c**. The magnetic field $H = 0.3$ T (curve 1), $H = 0.5$ T (curve 2), $H = 1$ T (curve 3), $H = 1.5$ T (curve 4), $H = 2$ T (curve 5).

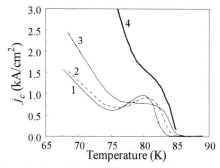

Fig.4. The dependencies of the critical current density j_c on the temperature for the sample G21 at following angles φ between **H** and the **c** axis: $\varphi = 0°$ (curve 1), $\varphi = 50°$ (dotted curve 2), $\varphi = 70°$ (curve 3) and $\varphi = 90°$ (curve 4). $H=1.5$ T.

The experimental data obtained can be explained in the frame of the following consideration [6]. The coherence length across the **ab** plane ξ_c is known to depend on the temperature by the relation $\xi_c(T) \sim \xi_{c0} (1-T/T_c)^{-1/2}$ and at the certain $T = T_D$ the length ξ_c is about the distance between Cu-O planes d. At $T < T_D$ the inequality $\xi_c < d$ is valid and the vortex lattice consists of the 'pancake' vortices which are pinned independently in neighboring Cu-O layers. It is the so-called two-dimensional (2D) regime. At $T > T_D$ another relation $\xi_c > d$ is valid and pancakes are united into the flux lines lattice. It is the three-dimensional (3D) regime. As it have been estimated for YBCO systems, $T_D \sim 0.9T_c$ and this value coincides fairly well with our $T_p = 80$ K $\sim 0.9T_c$. The critical temperature T_c of our samples is about 90 K. (Note that the temperature where j_c becomes equal to zero in Figures 2-4 is defined by the irreversibility line rather than T_c). Let us presume the low concentration of pinning centers in our samples: the mean distance between these centers L satisfies to the relation $L >> a = (\Phi_0/B)^{1/2}$ where a is the vortex lattice period and Φ_0 is the magnetic flux quantum. This presumption seems to be not failed since the sample G21, where the anomaly is most pronounced, is characterized by the least j_c, other things being equal. The condition mentioned implies that many vortices should be pinned by one pinning center and there are few vortices which are effectively pinned. At $T < T_D$ the number of the pancakes which are not pinned is large enough. We can assume that the pancakes in Cu-O planes placed near the intercrystallite plane are effectively pinned, whereas most of pancakes in other Cu-O planes are not pinned. As T closes to T_D, the interaction of the pancakes in neighboring Cu-O planes becomes stronger and the same pinning center may pin more than the single pancake owing to this interaction. As a consequence, an extra amount of pancakes proves to be pinned hence j_c may increases with the temperature growth. When T exceeds T_D there are no pancakes any more but we have the conventional flux line lattice and each a line is pinned as a whole.

The above phenomenological speculation agrees with the theoretical study [7] that states that 3D vortices are pinned more effectively than 2D ones. However, our argumentation will not be true if there exist strong pinning centers inside the crystallites since j_c provided by such defects should be dominant. For instance, the twin boundaries may be the centers that makes this phenomenon unobservable in YBCO single crystals.

Acknowledgments

We would like to thank V.A. Yampol'skii and S.E. Savel'ev for helpful discussions. This work has been done in the frame of the Programs in high-T_c superconductivity (projects 93087 and 92100) and has been supported by Russian Science Fundamental Foundation (project 93-02-14768).

References

[1] Fisher L M et al 1992 Phys. Rev. B **46** 10986-96
[2] Fisher L M et al 1995 Appl. Supercond. to be published
[3] Fisher L M, Voloshin I F, Gorbachev V S, Savel'ev S E and Kalinov A V 1994
    Proceedings of the 7th International Workshop on Critical Current in Superconductors
    (Alpbach, Austria) ed. Harald W. Weber, World Scientific 439-442
[4] Daeumling M, Seuntjens J M, Larbalestier D C 1990 Nature **346** 332-335
[5] Fisher L M, Il'in N V, Podlevskikh N A 1992 Solid. St. Commun. **81** 983-988
[6] Feinberg D 1994 J. Phys. III France **4** 169
[7] Artemenko S N and Kruglov A N 1990 Phys. Lett. A **143** 485-489

Inst. Phys. Conf. Ser. No 148
Paper presented at Applied Superconductivity, Edinburgh, 3–6 July 1995
© *1995 IOP Publishing Ltd*

Influence of point-like defects on the melting transition in YBa$_2$Cu$_3$O$_{7-\delta}$ single crystals

W. Jahn[1], S.N. Gordeev[1,2], A. A. Zhukov[1,3], H. Küpfer[1], T. Wolf[1]

[1]Forschungszentrum Karlsruhe, Institut für Technische Physik und Universität Karlsruhe, 76021 Karlsruhe, Germany
[2]Moscow Institute for Radioengeneering, Electronics and Automation, Moscow 117454, Russia
[3]Moscow State University, Physics Department, Moscow 117234, Russia

Abstract. Pure and Sr-doped YBa$_2$Cu$_3$O$_{7-\delta}$ single crystals were investigated by transport and magnetic measurements. The pure samples show low current densities ($<10^4$ A/cm^2) but high irreversibility fields (about 8T) whereas the doped samples have current densities up to 10^5 A/cm^2 but significantly smaller irreversibility fields (about 5T). By transport measurements in the superconducting transition regime three regions are found, the first related to fluctuations of the normal-superconducting transition, the second to the vortex liquid regime and the third to the transition from the vortex liquid to the vortex solid. Further it is seen that twin boundaries and point-like defects suppress the melting transition. In addition the influence of twin boundaries becomes small with high density of point-like defects.

1. Introduction

The dominating pinning centers in YBa$_2$Cu$_3$O$_{7-\delta}$ single crystals are expected to be point-like defects originating from different impurities or oxygen vacancies and twin boundaries. In pure samples without oxygen deficiency the influence of twin boundaries is expected to be dominant [1-3].

In this work we report on magnetic and transport measurements of pure and Sr doped twinned YBa$_2$Cu$_3$O$_{7-\delta}$ single crystals to investigate the influence of point-like defects on the current density and the melting transition.

2. Experimental

The starting material of the YBa$_2$Cu$_3$O$_{7-\delta}$ crystal W1 was very pure, whereas sample W2 was doped with Sr. EMP measurements have shown that Sr-atoms replaced 6% of the Ba-atoms. The resistives at 100K were 90 μΩcm for W2 and 50 μΩcm for sample W1. Both samples exhibit a quite similar twin structure.

Magnetic measurements were performed using a vibrating sample magnetometer. The shielding current densities j_s were measured with a sweep rate of 120 Oe/s which corresponds to a voltage criterion of about 10^{-2} μV/cm.

Two different kinds of transport measurements were carried out. In the conventional dc-method currents up to 100mA were used. Measurements up to 15A were performed with a pulsed current technique. The current pulse was of rectangular shape with a pulse duration of 3 ms. The voltage resolution was about 5μV.

3. Results and Discussion

3.1 Magnetic Measurements

The different behavior of the current density j_s is demonstrated in Fig.1. Crystal W2 shows large current densities ($8 \cdot 10^4$ A/cm^2 at 1.5T, 77K) and a pronounced fishtail behavior whereas the current densities in crystal W1 decrease monotonously. In the low field region the values of j_s are significantly smaller than in crystal W2. From this one may expect that also the

irreversibility field in sample W1 should be smaller than in sample W2. But just the opposite is observed, the irreversibility field of the purer sample W1 is significantly larger.

Such a behavior contradicts to a simple de-pinning model in which with increasing strength of the defect structure an increase of the current *and* the irreversibility line is expected.

To understand the influence of point-like defects on the irreversibilty line we investigated the superconducting transition region of both samples with transport measurement methods.

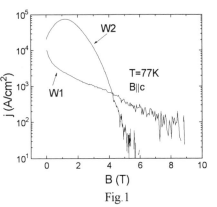

Fig.1

3.2 Transport Measurements

3.2.1 The ρ(T)-dependence

The behavior of the resistivity at an external field of 5T is shown in Fig. 2a. In both samples three temperature regions can be destinguished. In spite of the different defect density the behavior of both samples in regions I and II is quite similar.

In region I the resistance decreases slowly with decreasing temperature down to 0.6 $\rho(T_c)$. This change of the resistivity is probably determined by superconducting fluctuations. An exponential drop of the resistivity is observed for both samples in region II. This decrease of the resistivity in this regime may be explained by viscous plastic movement of the vortices in the vortex liquid regime [4] caused either by interaction with point-like or other defects or by vortex entanglement. In all cases the viscous plastic movement may be described by a current independent potential U_{pl}. Due to the linear dependence of U_{pl} on the temperature, $U_{pl} = U_{pl0}(1 - T/T_c)$, and the exponential relationship between the resistivity and the potential, $\rho = \rho_0 exp(U_{pl}/kT)$, the potential U_{pl0} can be obtained from $\partial ln\rho/\partial(1/T) = -U_{pl0}/k$. Thus region II, in which the derivative is temperature independent, can be connected with plastic deformations (Fig. 2b). Because of similar values of U_{pl0}/k in both samples, 1560K in sample

W1 and 1600K in sample W2, but different defect density, we exclude viscous movement caused by point-like defects and favour vortex entanglement as the reason for this behavior.

In region III the temperature dependence of the resistivity differs significantly for both samples. A sharp decrease of the resistivity in W1 is observed at 0.3 $\rho(T_c)$ wheras the decrease of the resistivity in W2 occurs at 0.1 $\rho(T_c)$ and is shifted to lower temperatures. This is in agreement with the observation that the irreversibility fields of purer samples are larger than the corresponding values of samples with high concentration of point-like defects (Fig.1). To investigate this region further we measured current-voltage characteristics.

3.2.2 Current-voltage characteristics

In Fig.3 current-voltage characteristics of both samples at 3T in the geometry B∥c are shown. The region III is marked by dotted lines. Both samples do not show a vortex- or bose-glass transition as it is predicted for samples with randomly distributed weak point disorder or for samples with extended correlated defects. Instead a continuous change from an ohmic to a power-like behavior is seen. To exclude the influence of twin boundaries we turned the field along the a-b-plane (Fig.4). A drastic change from linear to a negative curved current-voltage characteristics in crystal W1 is observed. Such a sharp change is attributed to a melting of the vortex lattice [2]. The different behavior in both geometries of crystal W1 can be explained by twin boundaries. In the geometry B∥c the vortices are aligned parallel to the twin boundaries. Thus pinning by twin boundaries becomes efficient. The number of free moving vortices is reduced and thus the resistivity drops down. If the field is aligned along the a-b-plane the vortices are not parallel to the twin boundaries and pinning by these planar defects becomes inefficient.

Consequently twin boundaries in the sample W1 suppress the melting transition in the geometry B∥c and a continuous change from ohmic behavior to a negative curved current-voltage characteristic is present.

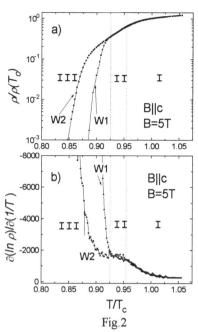

Fig.2

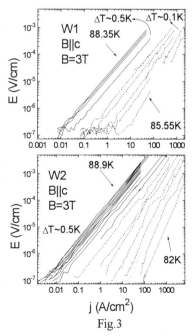

Fig.3

270

In sample W2 the behavior in the geometry B∥a,b is similar to that in the geometry B∥c which indicates that twin boundaries have negligible influence on the shape of the transition in this sample. A reason might be the competition of pinning by point-like defects with twin boundary pinning, which suppresses an alignment of vortices along the twin boundaries in the geometry B∥c. Thus in this sample point-like defects are dominant.

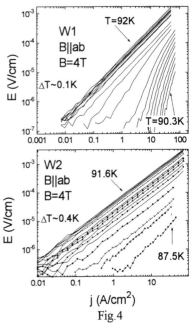

Fig.4

Unlike to sample W1, W2 shows a continuous change of current-voltage characteristics in the geometry B∥a,b (Fig.4). We refer this behavior to point-like defects because twin boundaries should not be efficient in this geometry. Thus point-like defects also suppress the melting transition.

Comparing both samples in the geometry B∥c we see a shift of region III to higher temperatures in the pure sample. This shift may be related to pinning by twin boundaries. On the other hand point-like defects broaden the transition as can be seen in the geometry B∥a,b. Thus a solidification of the vortex liquid may be shifted to lower temperatures in the sample W2 with high concentration of point-like defects. Both reasons may explain the observed behavior in Fig.1 and Fig.2.

4. Conclusions

The behavior of the resistivity in $YBa_2Cu_3O_{7-\delta}$ single crystals can be devided into three regions. The first region is related to the superconducting fluctuations. Plastic deformation due to vortex-entanglement determines the resistive behavior in region II. The transition from a vortex liquid to a vortex solid takes place in region III.

Twin boundaries influence the transition in pure samples and suppress the melting transition. In samples with high concentration of point-like defects no influence of twin boundaries in our measurements are seen probably due to the competing disorder from point-like defects. This disorder suppresses the melting transition resulting in a continuous change from a linear to a power-like behavior of current-voltage characteristics for both geometries.

References

[1] G. Blatter et. al., *Rev. Mod. Phys.* **66**, 1125 (1994)
[2] W. K. Kwok et. al., *Phys. Rev. Lett.* **69**, 3370 (1992); W. K. Kwok et. al., *Phys. Rev. Lett.* **72**, 1092 (1994)
[3] J. A. Fendrich et. al., *Phys. Rev. Lett.* **74**, 210 (1995)
[4] V. M. Vinokur et. al., *Phys. Rev. Lett.* **65**, 259 (1990)

Inst. Phys. Conf. Ser. No 148
Paper presented at Applied Superconductivity, Edinburgh, 3–6 July 1995

Pinning mechanisms in melt textured $YBa_2Cu_3O_7$-Y_2BaCuO_5 ceramics

B. Martínez, X. Obradors, A. Gou, V. Gomis, S. Piñol and J. Fontcuberta and
H. Van Tol*

Instituto de Ciencia de Materiales de Barcelona - CSIC. Campus UAB. Bellaterra 08193. SPAIN
*High Magnetic Field Laboratory, 25 Avenue des Martyrs. BP166 - F 38042 Grenoble Cedex 9.
FRANCE

Abstract. The field, temperature and composition dependence of the critical currents
of directionally solidified $YBa_2Cu_3O_7$-Y_2BaCuO_5 composites allow to identify interfacial
pinning by 211 precipitates as a very effective pinning mechanism in a narrow region below
about 1T and temperatures $40K \leq T \leq 80K$. In this region the system shows a vehabior very
akin to that described as correlated disorder. On lowering temperatures secondary weaker
pinning centres become active leading to a new single vortex pinning regime extending
up to very high magnetic fields (20T). In the high field region the characteristic footprints
of the small bundles and large bundles regimes are identified at intermediate temperatures.
Finally, amagnetic phase diagram of the mixed state of the 123/211 textured composites
is proposed.

1.-Introduction:

One of the most promising fields for immediate application of high temperature
superconductors is large scale power applications such as superconducting magnets, current
leads, energy storage systems, etc. All these applications require a high current density flowing
at tolerable low dissipation at 77 K . Even though processing techniques such as melt texturing
and thermomechanical methods [1] allow to obtain good quality samples and to solve
granularity problems intrinsic to the ceramic character of these materials, the improvement of
the flux pinning mechanism is a subject of major interest. Many efforts have been devoted to
determine the role played by twin planes, dislocations, stacking faults, oxygen deficiencies
and 211 precipitates in providing effective pinning for the flux lines. The active role played by
211 inclusions has been evidenced by some experimental results obtained in 123/211
composites that indicates that J_c at 77 K scales with the effective 123/211 interface area [2,3].

2.-Experimental and results:

The samples used in this work have been fabricated by using a directional solidification method
based in a vertical Bridgman technique that allows to obtain quasi-single crystalline
superconducting bars (up to 12 cm in length and $1cm^2$ of cross section) with a fine distribution
of precipitates of Y_2BaCuO_5 (211 phase) [4]. A complete characterization of the microstructure
of these samples by using transmission electron microscopy (TEM) may be found elsewhere
[5]. The magnetic characterization has been carried out using a SQUID magnetometer and
extraction magnetometry at Service National des Champs Magnetiques Intenses (Grenoble) up
to 22 T. Typically samples are parallelepipeds in shape with about $2.5x1.5x0.4$ mm^3 of
volume. The distribution of sizes of the 211 phase precipitates (mean sizes from 0.5 μm to 3
μm) has been deduced by means of image analysis of scanning electron microscopy (SEM)
pictures [6]. A careful oxygenation of the samples has been performed after the directional
solidification process to avoid deficient oxygenation or aging processes [7].

272

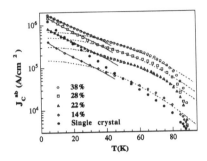

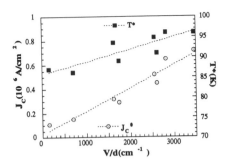

FIGURE 1: Temperature dependence of J_c^{ab} in a single crystal and textured ceramic samples with different content of 211 phase in self field. a) Solid lines correspond to vortex glass model while dashed lines correspond to correlated disorder (see text).

FIGURE 2: Dependence of J_C^0 and T^* obtained from the fitting by using Eq. [1] on the 123-211 interface specific area in the H||c geometry and in self field conditions.

We have studied the thermal dependence of the critical currents, determined by using the anisotropic Bean model $J_c^{ab} \approx 20\Delta M/[a(1-a/3b)]$, in a series of samples having different content of 211 phase ranging from 4% to approximately 38%. Samples were cooled in a maximum field of 5.5 T down to 5 K then the field is lowered to the desired value and the magnetization measured as temperature is increased up to 92 K. An exhaustive analysis of these measurements may be found in [8]. In Fig. 1 we show the thermal dependence of J_c^{ab} (H||c) in self field for several textured samples having different contents of 211 precipitates and a single crystal. First of all it is worth mentioning that a net increase of the critical currents with the increase of the 211 precipitates is clearly observed and what is even more important: the thermal dependence of the critical current in the single crystal is much higher than in the textured samples. This effect is probably a consequence of the existing 211 precipitates which also induce the observed softening of $J_c^{ab}(T)$ at intermediate temperatures. In the region of low fields, the low temperature (T≤40K) dependence of $J_c(T)$ is consistent with the single vortex behavior predicted by the vortex glass-collective pinning theory [9,10] (full line in Fig. 1) with an exponent $\mu=1/7$ (see Ref. [8] for details) that is the predicted value for the case of single vortex pinning by weak pinning centres in the framework of the vortex glass theory.

At higher temperatures (40K≤T≤80K) we have found that $J_c(T)$ may be properly described by the following expression:

$$J_C(T) = J_c^0 \exp[-3(T/T^*)^2] \qquad (1)$$

as it is also shown in Fig.1 (dashed lines). This expression has been proposed to account for the temperature dependence of the critical current in the case of linear correlated disorder [11].

In Fig. 2 we show the evolution of the parameters J_c^0 and T^*, obtained from the fitting of the experimental data by using Eq.[1], as a function of the composition of the samples given by the parameter V/d, that represents the ratio of the volume percentage of 211 phase precipitates and their mean diameter, and measures the interface area between 211 particles and 123 superconducting matrix. We observe that a linear relation does exist between the parameters J_c^0 and T^*, and V/d thus suggesting that 123/211 interfacial pinning is the dominant pinning mechanism in this high temperature-low field region of the magnetic phase diagram while at low temperatures additional pinning centres may also become active as thermal activation decreases. The differences between the single crystal and 123/211 composites become more evident in the high temperature regime (40K≤T≤80K) i.e. where 123/211 interfacial pinning plays the dominant role.

The occurrence of the interfacial pinning mechanism in our samples has also been checked by fitting our experimental $J_c(H)$ curves to the expression $J_c \approx \beta B^{-1/2}$ typical for single

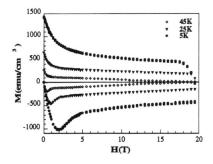
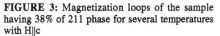

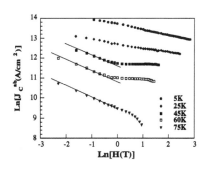

FIGURE 3: Magnetization loops of the sample having 38% of 211 phase for several temperatures with H‖c

FIGURE 4: Log-Log plot of J_c^{ab} vs. H making evident the existence of a potential relation $J_c^{ab} \propto H^{-\alpha}$

vortex pinning regime where a simple summation of the individual microscopic pinning forces may be performed [2]. Very good agreement is obtained in the range of fields below 1.5 T and temperatures above 45K (see Fig. 4) with a linear dependence of the parameter β on V/d (see [3,8] for details) corroborating the active role of 211-123 interfacial pinning.

Nevertheless, on lowering temperature a progressive departure from the above simple picture, i.e. $J_c \propto B^{-\alpha}$ with $\alpha=1/2$, is observed. After a fast decrease of the critical currents $J_c^{ab}(H)$ in which $\alpha \approx 1/2$, a regimen in which the field dependence of M(H), or $J_c^{ab}(H)$, is very smooth ($\alpha \approx 0.1$) is reached in a wide region of fields above 1.5-2 T. On lowering temperature the values of the α exponent in the low and high field regimes slowly approach each other and below about 20 K all the $J_c(H)$ curve may be described with only one value of $\alpha \approx 0.3$. The crossover field between these two regions depends on temperature but stabilizes around 1.5 - 2 T at high temperature. Above 1.5 - 2 T we observe that J_c is almost independent of the parameter V/d, thus indicating that interfacial pinning only plays a secondary role in the high field region and other pinning centers are dominant. Finally approaching the IL a much faster decrease of the critical currents is observed. In this part of the $J_c(H)$ curves it is possible to identify, for some temperatures, a region in which $J_c \propto H^{-3}$ before reaching the irreversibility line (see Fig. 5). This field dependence of the critical current has been theoretically predicted for the so-called large bundle regime [11] in the collective pinning theory.

The irreversibility line of our samples has been determined from the point where persistent currents, determined from zero-field-cooled and field-cooled (ZFC-FC) temperature-dependent curves, vanish. For fields above 5.5 T up to 22 T, the IL has been determined from isothermal hysteresis loops.It is worth mentioning here that there is no correlation between the tiny variations observed in the IL of different samples and their actual content on 211 phase. The temperature dependence of the IL is properly described by using the following expression: $H^*(T)=H_0[1-T/T_c]^{3/2}$ and no change to a more pronounced temperature dependence down to $T/T_c \approx 0.5$ has been detected

Collecting all these data we can propose the phase diagram of the mixed state depicted in Fig. 6. Region I is single vortex pinning in nature and 123-211 interfacial pinning dominates, on lowering temperature other weaker pinning centres become active and the system enters in region II that is also a regime of single vortex pinning where strong and weak pinning centres coexist. In region III, above 1-1.5T the field dependence of J_c is weaker and pinning could be understood as due to small bundles. Further increasing the field a region where $J_c(H) \propto H^{-3}$ is reached. This region, labelled as region IV in Fig. 6, has been associated to large bundles of vortices and is upper-bounded by the IL.

274

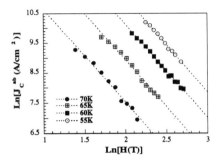

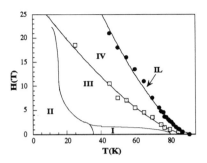

FIGURE 5: Log-log plot of J_c^{ab} vs. H showing the existence of a region with H^{-3} dependence of J_c^{ab} corresponding to the large bundle regime.

FIGURE 6: Magnetic phase diagram of the mixed state of the 211/123 composites obtained by using the sample having 38% of 211 phase. Region I: Correlated disorder single vortex pinning. Region II: Single vortex with strong and weak pinning centres. Region III: Small bundle regime. Region IV: Large bundle regime. See the text for details.

3.- Conclusions

We have investigated the field and temperature dependence of the critical currents in the H||c configuration of textured $YBa_2Cu_3O_7$-Y_2BaCuO_5 composites. The first, and more outstanding characteristic of this system, is the existence of randomly distributed precipitates which can pin vortices at the sharp interfaces. These defects can be classified as a source of correlated disorder. A single vortex pinning regime with 211/123 interfaces as dominating pinning centres has been unambiguously identified (Region I in Fig. 6). However, due to the relatively large separation between the 211 precipitates a crossover field must exist. When this crossover field is overcome the vortices will start to pin at other, weaker, pinning centres existing between these precipitates. In this new situation the interaction energy between the vortices become comparable to the pinning energy and thus we enter in the conditions of the small bundle regime (Region III). We have also identified a low temperature region extending up to very high fields (beyond 20 T at 5K) interpreted as a single vortex regime with several types of pinning centres being effective simultaneously.

Acknowledgements: We are very grateful to: CICYT (MAT91-0742), Programa MIDAS (93-2331), EC-EURAM (BRE2CT94-1011) and EC grant for the use of Large Scientific Facilities (High Magnetic Field Laboratory - Grenoble. SM-2993), for their financial support.

References:

1.- K. Salama, V. Selvamanickam and D.F. Lee, Chapter 5 in "Processing and properties of high T_c superconductors". Vol. 1 Bulk Materials. pag. 155-211, Ed. Sungho Jin. World Scientific. Singapore (1993)
2.- M. Murakami et al., Cryogenics 32, 930 (1992)
3.- B. Martínez, V. Gomis, S. Piñol, J. Fontcuberta and X. Obradors, Physica C 235-240, 3007 (1994)
4.- S. Piñol et al., in "Applied Superconductivity"; Edited by H.C. Freyhardt. (DGM-I- Verlag, Oberursel. 1993). p. 365; ibid. J. Alloys and Compounds. 195, 11 (1993)
5.- F. Sandiumenge et al., Phys. Rev. B-50, 7032 (1994)
6.- S. Piñol et al., IEEE Trans. Appl. Superconductivity. (in press).
7.- B. Martínez et al., Appl. Phys. Lett. 66, 772 (1995); F. Sandiumenge et al., Phys. Rev. B-51, 6645 (1995); S. Piñol et al., Physica C 235-240, 3045 (1994)
8.- B. Martínez et al., Phys. Rev. B (in press).
9.- D.S. Fisher, M.P.A. Fisher and D.A. Huse, Phys. Rev. B-43, 130 (1991);
10.- M.V. Feigel'man, V.B. Geshkenbein and V.M. Vinokur, Phys. Rev. B-43, 6263 (1991)
11.- D.R. Nelson and V.M. Vinokur, Phys. Rev. B-48, 13060 (1993)

Inst. Phys. Conf. Ser. No 148
Paper presented at Applied Superconductivity, Edinburgh, 3–6 July 1995
© 1995 IOP Publishing Ltd

Twin boundaries induced transformations of vortex pinning and flux creep in YBa$_2$Cu$_3$O$_y$ single crystals

A A Zhukov[a,b], H Küpfer[a], M Kläser[c], H Claus[a] and H Wühl[a]

a Forschungszentrum Karlsruhe, Institut für Technische Physik, Postfach 3640, D-76021 Karlsruhe, Germany
b Physics Department, Moscow State University, Moscow 117234, Russia
c Universität Karlsruhe, Fakultät für Physik, D-76128 Karlsruhe, Germany

Abstract. Investigating twinned and detwinned YBa$_2$Cu$_3$O$_{7-\delta}$ single crystals, we have observed differences in the shape of the peak-effect, in the field dependence of the relaxation rate, and in the values of the irreversibility field and shielding current. It is shown that twins produce an additional peak or a plateau-like behavior in the magnetization hysteresis at an intermediate temperature range. Twin boundaries increase j$_s$ in the region of small currents. However, for stronger bulk pinning they reduce the current and presumably become channels for easier flux penetration.

1. Introduction

The peak-effect in conventional superconductors was for a long time a topic of controversial and intensive discussions [1]. Experimentally, this feature corresponds to the anomalous increase of the critical current with the magnetic field, resulting in a peak significantly above the self-field region.

Recently it was found that in YBa$_2$Cu$_3$O$_y$ single crystals this single peak-behavior may be accompanied by the appearence of a second maximum [2]. The origin of this new peak was related to the matching effect of vortices with the twin boundary structure [2]. Similarly, the importance of twin boundaries for the appearence of the plateau-like behavior was pointed out recently [3]. However, the relation of twins with all these anomalies still requires a direct confirmation, which will be given in this paper.

Another important aspect of the presented study is the analysis of the influence of twins on the pinning of vortices. This problem is still not resolved and suffers from contradictory statements (e.g. [3,4]).

2. Experimental

The YBa$_2$Cu$_3$O$_y$ single crystals studied were grown in Y$_2$O$_3$-stabilized ZrO$_2$ crucibles as discussed elsewhere [5]. Two samples with different oxygen treatment were investigated. One of the crystals, sample 1, was annealed at 1 bar oxygen and 480°C for 2 weeks. This provided an oxygen content of y = 6.94. The superconducting transition temperature of this sample was 91.4 K and the transition width 0.8 K. Another crystal was annealed at 1 bar oxygen and 440°C during 2 weeks and then furnace cooled. This crystal had higher oxygen content y = 6.97, which corresponded to lower T_c = 90.5 K [5]. The transition width was 0.36 K.

In the initial state, all these samples were densely twinned, containing twin complexes of both (110)- and (1-10)-direction with characteristic size of several tenth of millimeter. Detwinning was achieved by applying a uniaxial pressure of about 10^8 N/m^2 at temperature of 400°C for about 10 min. Some twin boundaries remained, however. Samples 2, 1A and 1B were about 90 %, 60 % and 75 % monodomain, respectively. To check the reproducibility of the observed behavior, the last two crystals were twinned back by heating to 400°C for about 10 min.

The magnetization measurements were performed with a vibration sample magnetometer (VSM), in magnetic field H ≤ 120 kOe parallel to the c-axis. Measurements of the angular dependence have been performed in a split-coil VSM with magnetic fields ≤ 70 kOe. The angular resolution was ~0.1°.

3. Results

As was previously reported (e.g. [2]) most of the twinned samples with high oxygen content show three different types of magnetization hysteresis curves, occurring at different temperatures. As can be seen from Fig. 1, showing data for sample 2, detwinning produces qualitatively different effects in each of these regions.

The conventional "fishtail" behavior, observed at high temperatures T > 60 ÷ 70 K does not change qualitatively besides appearence of a minimum at low fields. However, detwinning decreases the values of the irreversibility field H_{irr}, shielding currents j_s and, to a lesser extent, influences the position of the $j_s(H)$ maximum, H_{max} (Fig. 1a). The inset in Fig. 1a shows temperature dependences of the peak position, H_{max} (circles), and the irreversibility line, H_{irr} (squares) in the initial (open symbols) and detwinned states (closed symbols). This behavior was observed for all studied samples and was reproducible after the samples were subsequently twinned and detwinned again.

The drastic changes of the magnetization hysteresis loop occur in the intermediate temperature region 20 K < T < 60 K (Fig. 1b). The double peak or plateau structure originally observed in this interval totally disappears after detwinning. Instead, the conventional "fishtail" behavior appears. The new maximum is observed roughly in the middle of the two initial peak positions or the plateau (Fig. 1b and inset of Fig. 1 a) . The current in the remanent state practically does not change with detwinning but significantly increases in the region of the fishtail peak. However, far from the fishtail peak the current in the twinned sample is higher (Fig. 1b). In correspondence with the transformations of the magnetization behavior, significant changes in the normalized relaxation rate S are observed. In the detwinned sample, the minimum of S occurs approximately at $H_{max}/2$ (Fig. 1b) which is the usual behavior for the "fishtail" peak [6]. Whereas in the twinned state the low field minimum of the relaxation rate S is close to the position of the corresponding maximum H_{max}.

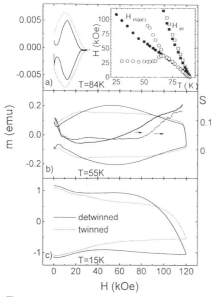

Fig. 1

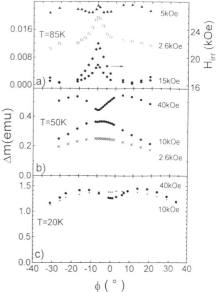

Fig. 2

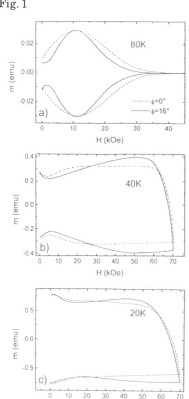

Fig. 3

In the low temperature region, T < 20 K, only a monotonous decrease of the current with field H was observed in the twinned samples. However, detwinning spreads the "fishtail" feature to this T interval, too (Fig. 1c). In the detwinned state, the values of the current increased for the whole field range.

For further elucidation of the importance of vortex trapping by the twin planes [7], we have studied the angular dependence of the magnetization hysteresis. The magnetic field was tilted away from the c-axes in the plane parallel to the sample side, thus, the tilting plane has an angle of ~ 45° with twin boundaries. As can be seen from Fig. 2, showing date for sample 1B, a sharp maximum in the dependence of the hysteresis width Δm on the tilting angle ϕ is observed for the field direction along c-axis at high temperatures (Fig. 2a). The $\Delta m(\phi)$ peak is accompanied by a similar angular behavior of the irreversibility line (Fig. 2a).

In the intermediate temperature region, for the magnetic field between the two peaks or in the region of plateau, a well pronounced dip in the hysteresis width $\Delta m(\phi)$ is found for the direction of the magnetic field along the c-axis (Fig. 2b), whereas in fields below and above this region a peak in $\Delta m(\phi)$ is observed (Fig. 2b). In the low temperature region, the angular dependence of the hysteresis width $\Delta m(\phi)$ shows a well pronounced dip, at all magnetic fields above the self-field value (Fig. 2c). For tilt angles above 15°, the influence of twins is highly suppressed. In this case, as can be seen from the comparison of Fig. 3 (sample 1B) and 1, the tilting of the magnetic field produces an influence qualitatively similar to the detwining procedure and fishtail behavior expands to the low temperatures.

4. Discussion

The data obtained, prove that the intermediate temperature anomaly is produced by twins. They also show that the "mirror-like" behavior of the relaxation rate at intermediate temperature range originates from twins. On the other hand, the absence of qualitative changes in the high temperature "fishtail" behavior despite drastic changes in the twin structure rules out their importance for the "fishtail" peak. This is in agreement with previous studies of untwinned samples (e.g. [8]. At low T, twin boundaries suppress the "fishtail" mechanism and produce a monotonous decrease of the current with H in the twinned state. The influence of twins on the current was found to change with temperature. For the twinned state, pinning is significantly increased in the high T region, pointing to twin boundaries as being strong pins. However, for the low T interval, only a decrease of the current was found. In this case twin boundaries provide easier flux penetration thus behaving as weak links. With the increase of the tilting angle ϕ, vortices should leave twin planes and align with the direction of the magnetic field. In this case the influence of twins should be drastically decreased and the observed changes in $\Delta m(H)$ are qualitatively similar to that produced by detwinning. However, this correspondence is not direct because of the change of intrinsic parameters with ϕ, which should decrease $\Delta m(\phi)$ even for isotropic pinning centres. From the analysis of the angular dependences the value of the trapping angle [7] can be estimated as $\sim 15°$.

References

[1]   Campbell A M and Evetts J E 1972 *Critical Currents in Superconductors* (London: Taylor & Francis)
[2]   Zhukov A A et al. 1994 *Physica C* **235-240** 2837; 1995 *Phys. Rev.* *B***51** 12704
[3]   Oussena M et al. 1995 *Phys. Rev.* *B***51** 1389
[4]   Fleshler S et al. 1993 *Phys. Rev.* *B***47** 14448
[5]   Hauff R et al. 1994 *Physica C* **235-240** 1953
[6]   Zhukov A A et al. 1994 *Physica B* **194-196** 1921
[7]   Blatter G et al. 1991 *Phys. Rev.* *B***43** 7826
[8]   Zhukov A A et al. 1993 *J. Alloys and Compounds* **195** 479

Inst. Phys. Conf. Ser. No 148
Paper presented at Applied Superconductivity, Edinburgh, 3–6 July 1995
© 1995 IOP Publishing Ltd

Strong pinning in melt-textured $YBa_2Cu_3O_{7-\delta}$ with non superconducting Y_2BaCuO_5 inclusions.

K Rosseel[1], D. Dierickx[2], J. Lapin[3], V.V. Metlushko[1], W. Boon[1],
L. Trappeniers[1], J. Vanacken[1], F. Herlach[1], V.V. Moshchalkov[1],
Y. Bruynseraede[1], O. Van der Biest[2], L. Reylandt[3], F. Delannay[3]

[1]Laboratorium voor Vaste-Stoffysica en Magnetisme, K.U. Leuven, Celestijnenlaan 200D, B-3001 Leuven , Belgium.
[2]Departement Metaalkunde en Toegepaste Materiaalkunde, K.U. Leuven, W. De Croylaan 2, B-3001 Leuven , Belgium
[3]Département de Sciences de Matériaux et des Procédés, Place Croix du Sud 2, B-1348 Louvain-la-Neuve, Belgium.

Abstract. We report on a detailed study of the temperature and field dependence of the critical current density J_c (T, H) in the temperqture range 5 K - 100 K and fields up to 50 tesla, in high quality melt-textured $YBa_2Cu_3O_7$ with 20wt% Y_2BaCuO_5 , fabricated by Directional Solidification. From magnetic measurements, J_c was determined as $\approx 10^6$ A/cm^2 at H = 1 tesla and T = 5 K. From the normalised relaxation rate S (T, H) and the pinning potential U (J, H), determined from detailed SQUID measurements, different pinning regimes have been identified, in accordance with 3D collective pinning theory. The kink on U (J, H) and S (T, H) can be described as a transition from a small bundle to a big bundle collective pinning regime. Furthermore, the J_c (T, H) dependence has been investigated using high pulsed magnetic fields of up to 50 Tesla. The estimated irreversibility fields show the potential of these materials for high field applications.

1. Introduction

Because of the 3D nature of the Abrikosov Flux Line Lattice in the $YBa_2Cu_3O_7$ high T_c superconductors, imperfections in these materials such as second phase precipitates, dislocations, twin planes, stacking faults,... are expected to have a large influence on the flux pinning. A substantial number of melt processing methods have been developed in order to obtain the desired microstructural characteristics. [1]. By carefully choosing the fabrication route, weak links can be reduced, thus allowing high critical currents (~10^4 A/cm^2 at 77 K and 1 T) to be obtained in these materials. However, further research is needed in order to clarify the nature of the pinning in these melt-textured $YBa_2Cu_3O_7$ materials and to extend the potential for high field applications.

2. Experimental

The $YBa_2Cu_3O_{7-\delta}$ (further denoted as 123) precursor powders were prepared by the classical method of a solid state reaction combined with several grinding and recalcination stages using appropriate amounts of Y_2O_3, $BaCO_3$ and CuO to obtain precursor powders with a nominal composition of 123 + 20wt% Y_2BaCuO_5 (further denoted as 211).

The meltprocessing was done by directional solidification in flowing oxygen, using a horizontal furnace. To avoid sample vibrations during the heat treatment, the sample is fixed while the furnace moves over the sample, with a speed of 3mm/h.

The resulting samples show large textured 123 domains (~1 mm), which is an indication that during the heat treatment a quasi stable planar 123 growth front was obtained. [2] From resistive measurements, done in a four point configuration and using a lock-in technique, the transition to the superconducting state was found to be $T_C = 91$ K with a transition width of ~1 K. For the magnetization measurements a single 123 grain with dimensions $1*1*1$ mm^3 was extracted. All the magnetisation measurements were performed in a magnetic field applied perpendicular to the (a,b) plane of the grain. SQUID measurements were done using a Quantum Design SQUID magnetometer with a scan length of 3 cm, corresponding to a field homogeneity better than 0.05 %. For the measurements in a constant applied field μ_0H, the field was swept from -5.5T to the measuring field (with a typical sweep time of 1 minute) to insure the entire sample was in the critical state. The pulsed field magnetisation measurements (PFMM) were performed in pulsed fields of up to 50 T with a typical pulse duration of ~20 ms [3], using a home made sensor with high measuring sensitivity in high applied fields [4] (better than 10^{-2} emu in fields above 30T). From the width of the obtained hysteresis loops, the critical current was extracted using the Bean model as a first approximation : $J_C \sim \Delta M$.

3. Results and discussion

Figures 1 and 2 show the critical current density J_C derived from the SQUID measurements. Clearly observable is the kink in J_C, giving a first indication of a transition between two different pinning regimes (indicated on the figures as I resp II). In the temperature interval 20 K to 50 K and field interval 0.5 T to 4 T (ie in regime II) the sample shows a very weak

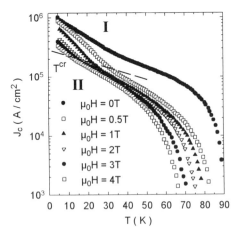
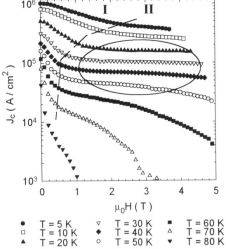

Figure 1 : Temperature dependence of the critical current density J_C of the 123 + 20 wt% 211 sample, as derived from SQUID measurements. Indicated are the different pinning regimes I and II as well as the field dependent cross-over temperature T^{cr}. The line drawn is only indicative.

Figure 2 : Field dependence of J_C as derived from SQUID measurements. The dashed line indicates the different pinning regimes I and II. The ellipse emphasises the very weak field dependence observed in regime II.

dependence on both temperature and field, which is an interesting feature especially from the point of view of applications. In order to determine the nature of the observed kink in J_c and to investigate the pinning mechanisms involved, extensive relaxation measurements have been performed in the SQUID. From the relaxation data the pinning potential U was determined using the Maley analysis [5]. Figure 3 shows the extracted pinning potential as a function of the current density J for an applied field of 1 tesla. The exponent μ can be derived as $\mu = -d(logU)/d(logJ)$ and is shown in the inset. The μ values predicted from 3D collective pinning theory [6] for small bundle and large bundle pinning are indicated.

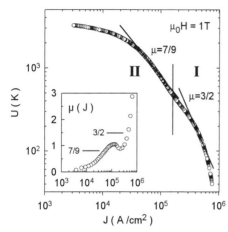

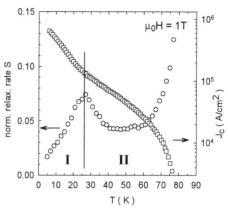

Figure 3 : Effective pinning potential U(J) as derived from extensive SQUID relaxation measurements using the Maley analysis. The inset shows $\mu = -d(logU)/d(logJ)$. The μ values predicted from the 3D collective pinning theory for small bundles (3/2) and large bundles (7/9) are indicated.

Figure 4 : Normalised relaxation rate S, calculated as $S=d(lnM)/d(lnt)$, as a function of temperature, for an applied field of 1T. For comparison the critical current density $J_c(T)$ at 1 T is shown. It is evident that the kink in J_c corresponds to a transiton to a second pinning regime where the relaxation rate is almost temperature independent.

The μ value found for the second pinning regime is ~1, ie a bit higher than the predicted value for large bundle vortex pinning. A similar μ value (9/10) was extracted by Kung et al [7], for Melt Processed Melt Growth (MPMG) materials. Because of the large fields (> 1 T , ie small lattice spacing $a_0 \sim (\phi_0/\mu_0 * H)^{1/2}$ of the vortex lines) at which the anomalies in the temperature and field dependence of J_c occur, they attribute these anomalies, which are similar to our observations, to the 211 inclusions and the related defect structure formed in these MPMG materials. The change in pinning strength is also reflected in the relaxation rate S as shown in figure 4. The kink in J_c (T, 1 tesla) clearly corresponds to a decrease in the relaxation rate ie an increase in the pinning of the vortex bundles formed in this second pinning regime. The 211 rich regions in these melttextured samples typically display a high density of twin planes with smaller twin boundary spacing than in 211 free regions [8], indicating a reduction in the 123 oxygen content around these 211 inclusions. Given the high sensitivity of T_c on the 123 oxygen content, oxygen vacancies seem to be likely candidates for the enhanced pinning in meltprocessed 123. However, at present it is not clear to which defect structure (oxygen vacancies, twinning, 211,...) the anomalies observed in J_c (T, H) in our sample correspond. This calls for further investigation.

In order to determine the irreversibility fields, pulsed field measurements were performed in fields up to 50T. The results are summarised in figure 5.

282

The insert shows a comparison with the $H^{irr} \sim (1-T_{irr}/T_c)3/2$ law. The irreversibility fields obtained show the potential of samples produced via directional solidification for high field applications.

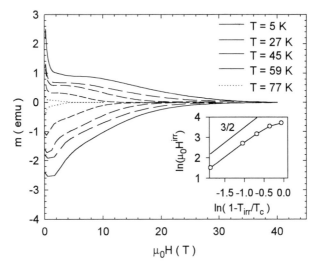

Figure 5 : Pulsed field magnetisation measurements showing high irreversibility fields obtained in the sample. The inset shows a comparison with the 3/2 power law behaviour.

4. Conclusions

Samples produced via directional solidification show a critical current density J_c which is weakly field dependent in the broad temperature and field interval of 20 K-50 K and 1 tesla to 4 tesla, demonstrating the potential of these materials for high field applications. From detailed relaxation measurements, the kink in the $J_c(T,H)$ characteristic can be attributed to a transition from a small bundle regime to a strong pinned large bundle regime. Concerning the nature of the pinning centers involved, oxygen vacancies, or other defect structures induced by the 211 inclusions are good candidates. Further research is necessary to clarify this matter.

Acknowledgements.
This work is supported by the Belgian High Temperature Superconductor, Concerted Action and Interuniversity Attraction Pole Program at the K.U.Leuven. D.D. is supported by the U.I.A.P, K.R. is a Research Fellow of the Belgian I.W.T, W.B. is a Research Associate of the Belgian N.F.W.O.

References.
[1] Salama K, Lee D F, 1994 *Supercond. Sci. Technol.* **7** 177
[2] Izumi T, Nakamura Y, Shiohara Y, 1992 *J. Mater. Res.* **7** 1621
[3] Herlach F, Bogaerts R, Vanacken J and Van Bockstal L, 1994 *Physica B* **201** 5
[4] Lagutin A S, Vanacken J, Harrison N and Herlach F, 1995 *Rev.Sci.Instrum.* **66** (8) -in press-
[5] Maley M P, Willis J O, Lessure H, McHenry M E, 1990 *Phys. Rev B* **42** 2639
[6] Blatter G, Feigel'mann M V,Geshkenbein V B, Larkin A I, Vinokur V M, 1994 *Rev. Mod. Phys.* **66** 1125.
[7] Kung P J, Maley M P, McHenry M E, Willis O J, Murakami M, Tanaka S 1993 *Phys. Rev. B* **48** 13922
[8] Miletich R, Murakami M, Preisinger A, Weber H W 1993 *Physica C* **209** 416

Inst. Phys. Conf. Ser. No 148
Paper presented at Applied Superconductivity, Edinburgh, 3–6 July 1995
© *1995 IOP Publishing Ltd*

Flux Pinning in Bi-2212 Single Crystals with Various Oxygen Concentrations and with Ti Doping

T.W. Li, P.H. Kes
Kamerlingh Onnes Laboratorium, Leiden University,
P.O.Box 9506,2300 RA Leiden, The Netherlands.
A.A. Menovsky, J.J.M. Franse
Van der Waals-Zeeman Laboratorium, University of Amsterdam,
Valckenierstraat 65, 1018 XE Amsterdam, The Netherlands.

The origin of pinning in Bi-2212 single crystals and the effect of Ti doping are reported. For an overview of pinning concepts and pinning effects in HT_cS we like to refer to previous overviews [1]. Recent developments dealing with flat geometries in small perpendicular fields are given in [2-3].

Although there are good reasons to believe that flux pinning in as-grown Bi-2212 single crystals is caused by oxygen vacancies in the CuO_2 layers [4], there is no hard experimental evidence for this belief. Such evidence can only be provided by studying the effect of a systematic change of the defect concentration on the critical current and pinning energy. Earlier work along these lines [5] gave incomplete support for the case of oxygen vacancies. We followed a slightly different scenario and started with a careful investigation of oxygen diffusion under different annealing conditions [6]. According to this work annealing in air between 500°C and 800°C reversibly removes interstitial oxygen from the BiO layers, whereas annealing in nitrogen at 500°C and 600°C irreversibly removes oxygen from the CuO_2 sheets. We therefore cut from the same single crystal four equally large ($1.5x1.5x0.02mm^3$) pieces and prepared 4 samples (Bi1-Bi4) by giving them a different heat treatment, see table 1. The difference between Bi1 and Bi2 is that oxygen was removed from the BiO layers as evidenced by a correlated increase in both T_c, the c-axis lattice parameter and the London penetration depth. The latter parameter was obtained from magnetization measurements by plotting dM/dT vs lnH [7]. The increase of $\lambda_L(0)$ upon oxygen loss shows that Bi1 is in the overdoped regime and Bi2 optimally doped. Further removal of oxygen,i.e. for Bi3 and Bi4, does not change these parameters anymore, giving strong support to the assumption that this oxygen comes out of the CuO_2 sheets creating more vacancies in these layers. If pinning is caused by oxygen vacancies we thus expect that the magnetization curves become more irreversible going from Bi2 to Bi3 to Bi4. This effect is indeed observed, as can be seen in Fig.1. We also see that this increase in pinning is preceded by a strong decrease going from Bi1 to Bi2. As will be shown below this decrease can be related to the change of $\lambda_L(0)$.

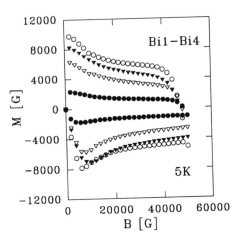

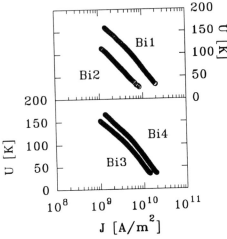

Fig.1 A typical magnetization
loop M-B for Bi1(○),Bi2(•),
Bi3(▽),Bi4(▼)

Fig.2 Magnetic relaxation
experiment at 2T on samples
Bi1-Bi4.

For a confrontation of pinning experiments with theory the data in Fig.1 are not suitable as they are influenced by flux creep. The basic pinning properties at low temperature can be deduced from magnetic relaxation measurements and a Maley analysis [8]. The results of such an analysis for our samples at a field of 2T are shown in Fig.2 where the energy barrier for flux creep U is plotted versus the current density j. The logarithmic behavior $U(j)=U_c \ln(j_c/j)$ is typical for creep in the single-vortex pinning regime [9]. The pinning energy U_c and the critical current density j_c can be determined from the linear extrapolation to $U=0$. Results for our samples are given in Table 1.

As shown in Ref.4, the relations for j_c and U_c at $T=0$, are

$$j_c = A_1 (B_c^2/\mu_0\phi_0 s) \sigma_{tr}\xi_0 n^{1/2} \qquad (1)$$

$$U_c = j_c\phi_0 sr_f = A_2 j_c\phi_0 s\xi \qquad (2)$$

Here n is the areal density of oxygen vacancies in the CuO_2 double layers, r_f the range of the pinning potential and A_1 and A_2 are constants of order 10 and 1, respectively. For Bi-2212 we use the following parameter values: $B_c(0)=0.5T$, $\xi(0)=2.1nm$, $\xi_0=2.8nm$, $s=1.54nm$ and $\sigma_{tr}=D^2/4$ with $D=2.9nm$. In order to compare our data with these predictions we plot in Fig.3 (j_c) and $(U_c)^2$ versus the change in oxygen concentration Δx. The latter is related to n via $n=6.9\times10^{18}\Delta x m^{-2}$. Within experimental accuracy the data fall on straight lines. From the slopes we determine $A_1=4.1$ and $A_2=2.0$. We thus find that $r_f=2$, which is quite reasonable. It shows that the changes in jc and U_c are consistently described by the pinning theory for single pancake vortices.

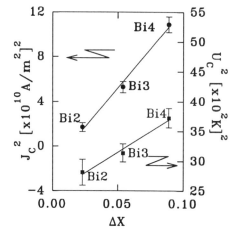
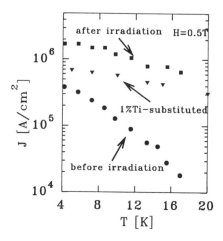

Fig.3 J_c^2 and U_c^2 vs change in oxygen concentration for Bi2-Bi4.

Fig.4 Comparison of $j_c(0)$ data 0.5T for as-grown, with columnar defects and doped with 1% Ti.

Table 1: Characteristic data of Bi-2212 samples after different heat treatments:

	T_{ann} (°C)	Atm.	T_c (K)	λ_L (nm)	ΔX	j_c (10^{10}Am^{-2})	U_c (K)	n (10^{17}m^{-2})
Bi1	500	air	83.5	180	0	3.5	55	0.7
Bi2	800	air	89.5	260	0.023	1.3	53	0.7
Bi3	500	N$_2$	89.0	260	0.054	2.3	56	2.2
Bi4	600	N$_2$	89.0	260	0.089	3.3	61	4.2

We therefore conclude to have convincingly shown that pinning in Bi-2212 is predominantly caused by oxygen vacancies in the CuO$_2$ layers.

Finally, we briefly like to report the effect of 1% Ti doping in Bi2212. This causes a high density of planar defects parallel to the c-axis and a decrease of the anisotropy. For both effects it is expected that they may enhance the critical current. In Fig.4 a comparison is shown of data we obtained after Ti doping and data obtained by Hardy et al.[10] on as-grown Bi-2212 and on

Bi-2212 with columnar defect created by heavy ion irradiation. It is well documented that these defects are optimal pinning centers for flux densities below the dose-equivalent field. It is seen in Fig.4 that the current densities in our Ti-doped sample are appreciably larger than in as-grown Bi-2212, but not as large as in irradiated samples. In view of the fact that the Ti is well dissolved, these results indicate that Ti doping provides a convenient way to enhance the current densities in Bi-2212 by at least a factor of two. Extensive further investigations are in progress

Reference:

[1]. P.H. Kes, (NATO-ASI series, E 263, Kluwer, The Netherlands,1994), ed. E. Kaldis, P.401 431; and in "Phenomenology and applications of HTS", ed. K. Bedell et al.(Addosion Wesley New York,1992), p.390-434.
[2]. E. Zeldov et al., Phys. Rev. Lett. 73, 1428(1994).
[3]. E.H. Brandt and M.V. Indenbom, Phys. Rev. B 48, 12893(1993).
[4]. C.J. van der Beek and P.H. Kes, Phys Rev. B 43, 13032(1991).
[5]. J.M. Gonzalez-Calbet et al., Physica C 203, 223(1992).
[6]. T.W. Li et al., Physica C 224, 110(1994).
[7]. P.H. Kes et al., Phys. Rev. Lett. 67, 2383(1991).
[8]. M.P. Maley et al., Phys. Rev. B 42, 2639(1990).
[9]. C.J. van der Beek et al., Physica C 195, 307(1992); V.M. Vinokuue, P.H. Kes and A.E Koshelev, Physica C 248, 179(1995).
[10]. H. Shaked et al., Phys. Rev. B 48, 1294(1993).

Inst. Phys. Conf. Ser. No 148
Paper presented at Applied Superconductivity, Edinburgh, 3–6 July 1995
© 1995 IOP Publishing Ltd

Direct correlation between TEM studies and pinning lengthscales for columnar defects in $YBa_2Cu_3O_{7-\delta}$ thin films

W.S. Seow, R.A. Doyle, Y. Yan, D. Kumar, J.D. Johnson, and A. M. Campbell

IRC in Superconductivity, University of Cambridge, West Cambridge Site,
Cambridge CB3 0DS, England.

P. Berghuis, R.E. Somekh, and J.E. Evetts

Dept. of Materials Science and Metallurgy, University of Cambridge, Pembroke St,
Cambridge CB2 3QZ, England.

G. Wirth and J. Wiesner

Gesellschaft fur Schwerionenforschung, Planckstrasse 1, D-64291, Darmstadt, Germany.

Abstract. We present HRTEM images showing the spacing between, and diameter of, columnar damage tracks produced by ^{238}U ion irradiation (parallel to the c-axis) in an epitaxial $YBa_2Cu_3O_{7-\delta}$ thin film. These are compared with the elastic limit of displacement, extracted from transport force-displacement measurements, for flux lines pinned on these defects. The transport results are consistent with both the columnar track diameter determined from the micrographs, as well as the columnar defect spacing. These results are discussed in terms of current understanding of flux pinning and the effects of heavy ion irradiation in high temperature superconductors.

1. Introduction

Ion irradiation has become a well-established method to change the random (point defects) or correlated (columnar defects) static disorder in a controlled manner, providing a powerful perturbative means to investigate the factors controlling flux pinning in high temperature superconductors[1-3]. Despite numerous reports of ion irradiation experiments on YBCO thin films, the enhancement able to be achieved for J_c is always much smaller than that of single crystals due to the different background level of disorder in the two sample systems. Angular studies[4,5] of J_c or resistivity strongly suggest that the columnar defects control the pinning. In spite of this, there is still little *direct* physical evidence relating the irradiation effects to the elastic pinning properties (pinning wells).

2. Experimental details

We use an *ac* I-V transport measurement with a static *dc* field in the maximum Lorentz force orientation (force-displacement measurement) which allows us to determine the oscillation of the pinned flux lines in the reversible (lossless) regime[3,6,7]. A high quality c-axis epitaxial YBCO film prepared by high pressure *dc* sputtering was used. The 200nm thick film was patterned into lines 20μm in width and 1mm in length and with contact pads in the four point geometry. The zero field (ZF) *dc* J_c value for the film at 77K was 1.9 x 10^6 Acm^{-2} before irradiation. An *ac* current at 3.3kHz was used. The quadrature voltage, which measures the elastic displacements of the flux lines, and the in-phase voltage, which measures the dissipation, are monitored with a dual-phase lock-in amplifier. The film was characterised at 77K before irradiation and re-characterised at a corrected reduced temperature, $t = T/T_c = 0.86$, taking into account the shift in T_c due to the effect of the irradiation. The film was irradiated at room temperature with 2.7GeV ^{238}U ions at the UNILAC heavy ion accelerator in Darmstadt with the irradiation perpendicular to the film surface (parallel to the *c*-axis). A flux of 4 x 10^7 ions cm^{-2} s^{-1} was used and beam heating effects are considered to be negligible. The matching field B_Φ is defined where the average impact site spacing equals the vortex spacing $a_o = 1.07(\Phi_0/B)^{1/2}$. The fluence of the U irradiation of 2 x 10^{10} ions/cm^2 gives $B_\Phi = 400$mT.

3. Results and discussion

Fig. 1(a) shows the field dependence of the relative enhancement of J_c, $\Delta J_c = J_c^{af} - J_c^{be}$, when the field is applied parallel to the columnar defects. Here, J_c^{be} is the critical current density before irradiation and J_c^{af}, after irradiation. The magnitude of the enhancement of J_c in the direction of irradiation is similar to other reported results at comparable fields and temperatures[1]. ΔJ_c clearly shows a maximum close to the matching field of 400mT.

Fig. 1(b) shows the field dependence of the elastic limit d_o for B//c before and after irradiation. The behaviour of d_o for B//ab was unchanged within experimental uncertainty but a clear decrease of d_o for B//c is observed. This is consistent with the recently reported results[6] showing an increase in the restoring force when the flux lines are parallel to the columnar defects. This implies an effective "stiffening" of the flux lines when they are pinned in the columnar tracks. Before irradiation, the flux lines are weakly pinned by point defects (oxygen vacancies, etc.) and hence there is considerable bending of the flux lines. This is reflected in the values of d_o which are about 8nm and 1.5nm at fields of 0.25T and 1.5T respectively. It is widely accepted that pinning of vortices for B//c in YBCO is determined by a dense array of small pinning defects of the same size as the coherence length. However, the large value of several nm we measure here cannot be reconciled with the size of oxygen vacancies which are typically of order 0.2-0.3nm. This suggests that the vortices are pinned by changes in density of oxygen defects. The large electronic energy-loss of the U ions leads to amorphisation of the 123 phase along the ion traces. The distinct surfaces between the amorphous columnar tracks and the surrounding crystalline material are expected to act as well spatially defined pinning potential wells. This is supported by the d_0 values after irradiation below and in the vicinity of the matching field. These are reduced relative to the case before irradiation to between 3.2 - 3.6nm.

TEM analysis of the film demonstrated that the YBCO material was highly oriented with the c - axis perpendicular to the surface. A plane view TEM micrograph of the epitaxial film after irradiation is shown in Fig. 2(a). Selected area electron diffractogram taken from the thin areas of the specimen contain only sharp YBCO reflections. The composition was measured to be 1:2:3 of Y:Ba:Cu from the EDX spectrum. The U ion-induced damaged regions are clearly visible in the form of circular features. The number density of these damaged tracks which was estimated from Fig. 2(a) to be around 1.8×10^{10} ions/cm^2, corresponds closely with the applied dose. The HRTEM image in Fig. 2(b) clearly shows a distinct amorphous region, surrounded by the crystalline matrix. Averaging over a large number of such images indicates that the average radius of the columnar defects is between 3 and 5nm which is very close to our measured value (converted to zero-to-peak). The coherence length, which defines the vortex core size, or the lengthscale over which the core is sensitive to variations in the superconducting order parameter is estimated as 3.6nm at this temperature. This strongly implies that the columns are responsible for the pinning and that they suppress the bending of the vortices. Before irradiation, the vortices are pinned by point defects and although the thickness of the film is only 200nm, the flux lines still bend considerably. This is supported by other measurements[3,8] which estimated the longitudinal correlation length to be a few tens of nm in YBCO thin films. After irradiation, the columnar tracks restrict the displacement of the flux lines, decreasing the d_0 of the flux lines.

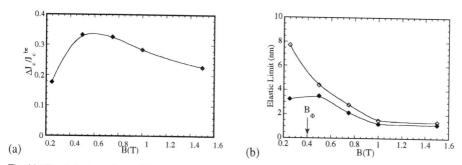

Fig. 1(a) The field dependence of the relative enhancement of J_c after irradiation at t = 0.86. ΔJ_c is the difference between the J_c values before and after irradiation at t. (b) The field dependence of the elastic limit, d_0, t = 0.86. The open diamonds and closed diamonds are for B//c before and after irradiation, respectively. The matching field, B_Φ is indicated by the arrow. Lines are guides for the eye.

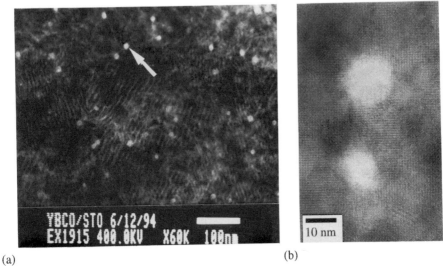

(a)   (b)

Fig. 2 (a) TEM micrograph of the columnar damage in the films showing the defect density and spacing. The arrow indicates a columnar defect. (b) HRTEM micrograph of the columnar tracks produced by the 2.7GeV U^{238} ions. The diameter of the damaged tracks are 7nm and 12nm, corresponding well to d_o values in the vicinity of the matching field.

4. Conclusions

We have carried out HRTEM studies of irradiated YBCO films and compared these with the recently reported elastic limits of flux lines *for the same film*, before and after heavy ion irradiation. The results indicate that the increase of J_c is caused by the confinement of the flux lines to the columnar tracks, causing an increased stiffness of the flux lines. The reduced elastic limits and their field dependence correspond well both with the size of the columnar tracks and their spacing, giving direct evidence of pinning by the columns along the entire vortex lengths.

References

[1] R.C. Budhani *et al.* 1992 *Appl. Phys. Lett.* **61**, 985-987

[2] B. Roas *et al.* 1990 *Europhys. Lett.* **11**, 669-674

[3] R.A. Doyle *et al.* 1995 *Phys. Rev. B* **51**, 12763-12769

[4] B. Holzapfel *et al.* 1993 *Phys. Rev. B* **48**, 600-603

[5] R. Prozorov *et al.* 1994 *Physica C* **234**, 311-317

[6] R.A. Doyle *et al.* 1993 *Phys. Rev. Lett.* **71**, 4241-4244

[7] W.S. Seow *et al.* 1995 *Physica C* **241**, 71-82

[8] O. Brunner *et al.* 1991 *Phys. Rev. Lett.* **67**, 1354

Inst. Phys. Conf. Ser. No 148
Paper presented at Applied Superconductivity, Edinburgh, 3–6 July 1995
© *1995 IOP Publishing Ltd*

'Arrowhead' magnetisation anomaly, hysteresis loop scaling properties and the irreversibility line in $Bi_2Sr_2CaCu_2O_{8+\delta}$

C. D. Dewhurst, R. A. Doyle, D. A. Cardwell, A. M. Campbell and *G. Balakrishnan

IRC in Superconductivity, University of Cambridge, West Cambridge Site, Cambridge CB3 0HE, UK.
*Department of Physics, University of Warwick, Coventry CV4 7AL, UK.

Abstract. We report the results of magnetisation measurements of $Bi_2Sr_2CaCu_2O_{8+\delta}$ single crystals. Hysteresis loops are compared using scaling analysis and the possible origin of the 'arrowhead' anomaly in the m-H behaviour is discussed. The irreversibility line (IR) is determined between 55K and 4.2K and is shown to vary with temperature exponentially over two distinct thermal ranges. These results are interpreted by considering the influence of temperature and distribution of pinning site strengths on the dimensionality of the vortex lattice. The 'arrowhead' anomaly and rapid drop in the irreversible magnetisation at about 20K are suggested to be intrinsic features of the critical state behaviour in the presence of a dimensional crossover in extremely anisotropic HTS materials.

1. Introduction

Recent magnetisation measurements performed on single crystal samples of $Bi_2Sr_2CaCu_2O_{8+\delta}$ show a double peak or 'arrowhead' structure in the m-H behaviour for B//c[1,2]. This is generally observed in a temperature window of ~20K to 35K and at an applied field of ~40mT[1,2]. The magnetic irreversibility (IR) line drops rapidly with increasing temperature[2,3] at about 20K. Further, the logarithmic flux creep rate, S(T), is reported to be almost discontinuous in this regime, rising rapidly from a value of ~0.04 to ~0.25 between 4.2K and 20K, followed by a sharp drop above ~20K[4]. Measurements of the c-axis resistivity, ρ_c, in single crystals suggest that the onset of interlayer coupling, thought to be Josephson-like in nature, is dramatically suppressed to lower temperatures in moderate applied magnetic fields[5]. The origin of the arrowhead and its relation to other anomalous magnetic and electrical properties in this system are considered in this study.

2. Experimental

Samples of $Bi_{2.2}Sr_{1.64}Ca_{1.16}Cu_2O_{8+\delta}$ were grown using a traveling floating zone technique, specific details of which are given in Ref. 6. Magnetisation measurements were carried out on two as-prepared crystals of approximate area 0.5 x 0.9mm^2 and thickness between 10 and 20μm. m-H loops were measured

292

between 4.2K and 55K with the magnetic field oriented parallel to the c-axis. The applied field was swept at rates of 10mTs^{-1} and 0.5mTs^{-1} up to a maximum of 12T, or to a value greater than the temperature dependent IR field. T_c was determined from the onset of a diamagnetic DC magnetisation and found to be 89K and 90K for the two samples respectively.

3. Results and Discussion

Hysteresis loops were scaled as described in Ref. 7. The minima in the magnetisation curves with coordinates [H*(T), m*(T)], and similarly the second peak minima on the ascending side of the magnetisation curve [H*`(T), m*`(T)], have been used as empirical linear scaling reference points. The significance of the scaling parameters is discussed below. Fig. 1a shows reasonable scaling of magnetisation curves up to about 16K. This implies that the physical processes governing sample magnetisation in this regime have the same origin. At higher temperatures, however, anomalous features begin to appear in the descending quadrant of the m-H loop. These can be seen in Fig. 1a as a reduction in the scaled low field magnetic moment at 21K, followed by the appearance of the arrowhead feature at 23K. The magnetisation behaviour deviates significantly from the scaling behaviour evident below 16K as the temperature is increased further, as shown in Fig. 1b. The arrowhead anomaly can be seen clearly at temperatures between 22K and 28K. Attempts to scale the magnetisation curves to the arrowhead feature [H*`(T), m*`(T)] were made without success, as shown in Fig. 1c.

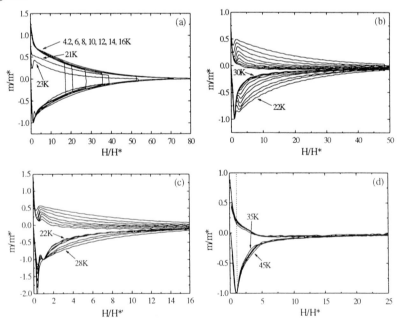

Figure 1 Scaling analysis of m-H curves for the 2212 single crystal. a), b) and d) are scaled to the parameter H*(T) and m*(T), and c) is scaled to the arrowhead peak anomaly H*`(T) and m*`(T). The subtraction of m_{rev} is invalid below the dotted line in (d).

The absence of scaling between 22K and 28K to either (H*, m*) or (H*', m*') indicates a change in the pinning mechanisms governing m-H behaviour. As can be seen in Fig. 1d, the high temperature hysteresis loops scale well after the relatively large reversible magnetisation component m_{rev}, has been subtracted from the data. The deviations from a simple scaling behaviour are seen to occur in the same broad temperature and field regime as the hysteresis width, Δm (proportional to J_c), and the IR line are observed to collapse over several orders of magnitude, as is widely observed in the literature.

We follow the analysis of Perkins et al.[7], in which, for the case of ideal scaling, the hysteresis loops retain exactly the same shape changing only in moment and field magnitude. In this case a plot of the natural logarithm of the two scaling parameters is expected to give a straight line in any region of scaling. Fig. 2a shows the scaling parameters $\ln[m*(T)]$ vs. $\ln[H*(T)]$ for the temperature range 4.2K to 45K. This shows two regions of linearity (4.2K to ~20K and 35K to 45K) which correspond to the two regimes where m-H curve scaling is observed (Fig. 1a and 1d). The self similarity of the hysteresis loops on which we base our scaling, may then be used to determine the IR line beyond our measurable range. This is shown schematically in Fig. 2b. The extrapolated IR lines can be calculated from H*, using measured reference IR fields of $H_{IR}(16K)=10.75T$ and $H_{IR}(35K)=0.1304T$.

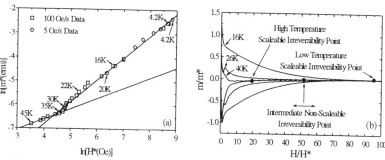

Figure 2 a) Plots $\ln[m*(T)]$ vs. $\ln[H*(T)]$ scaling parameters. The solid lines are a guides to the eye for the two observed linear regions. b) Schematic diagram showing the two universal hysteresis loop shape functions and scaleable IR points.

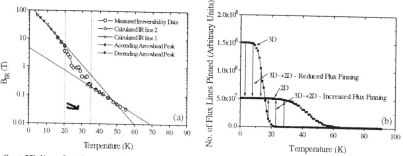

Figure 3 a) IR line for the second single crystal as measured from the point of hysteresis closure ($\Delta m<1e^{-6}emu$) for the 100 Oe/s and 5 Oe/s data, and extrapolated from scaling data. b) Calculated number of flux lines able to be pinned for the 3D and 2D cases. A transition from 3D to 2D at T <~ 18K and T >~ 18K would result in a decrease and increase in the number of pinned vortices respectively.

The results of the measured and extrapolated IR line are compared in Fig. 3a. This clearly shows the presence of two regimes each with a different exponential temperature dependence, as implied by the scaling data. The crossover temperature regime lies between ~20K and 35K. The measured IR line bounded by the two dotted lines shows the transition between the two exponential dependencies. Fig. 3a shows further that the arrowhead anomaly appears in this temperature range.

We have modelled the softening of the vortex lattice using a simple approach which assumes a crossover from 3D to 2D behaviour. This involves estimating the ratio of required pinning energies, U_{3D}/U_{2D}, as 3 (i.e. assuming 2D pancake vortices are three times easier to pin than 3D flux line segments), from which the relative number of available pinning sites can be calculated[8]. In this case the number of effective pinning sites diminishes more rapidly for the 3D flux lattice as the system approaches irreversibility. Hence the number of flux lines able to be pinned for either case can be found from the area under the pinning distribution curve, normalised to the number of pinning sites per unit flux line length. The results of this calculation are presented in Fig. 3b, which shows that the crossover of the vortex dimensionality from 3D to 2D may result in either reduced or increased flux pinning at low and high temperatures, respectively. This is observed experimentally by the rapid fall in magnetic moment below ~1000 Oe at 21K, as shown in Fig. 1a, and from the increase in flux pinning associated with the appearance of the arrowhead feature.

4. Conclusions

The m-H behaviour of $Bi_2Sr_2CaCu_2O_{8+\delta}$ has been found to exhibit scaling bahaviour in the two distinct temperature regimes of 4.2K to 16K and 35K to 45K. The IR line has been measured and extrapolated using the scaling parameters and shown to be exponential over the same temperature regimes. This enables an estimate of the IR line to be made down to 4.2K. The intermediate temperature regime of ~16K to ~35K is found to coincide with the appearance of anomalous m-H behaviour and a sharp fall in magnetic moment and the IR line with temperature. The results of an empirical model based on the relative pinning energies of 3D or 2D flux lines suggest temperature regimes where either 3D or 2D pinning may become more effective. This implies that the two exponential thermal regimes of the IR line and two scaling behaviours correspond to a 3D lattice at low temperatures and a 2D lattice at high temperatures. In the intermediate regime where a dimensional crossover may be induced by applied field, flux pinning may be enhanced which results in the appearance of the arrowhead anomaly in this temperature range.

References
[1] G. Yang et al., *Phys. Rev. B.* **48**, 4054 (1993).
[2] K. Kishio et al., *Proc. Critical Currents Conference, Alpbach* (1994).
[3] M. C. Hellerqvist et al., *Physica C* **230**, 170 (1994).
[4] V. V. Metlushko et al., *Europhys. Lett.* **26**, 371 (1994).
[5] H. Pastoriza et al., *Phys. Rev. B.* **46**, 9278 (1992).
[6] G. Balakrishnan et al., *Physica C* **206**, 148 (1993).
[7] G. K. Perkins et al., *Phys. Rev. B*, **51**, 8513 (1995).
[8] C. D. Dewhurst et al., *In Preparation* (1995).

Inst. Phys. Conf. Ser. No 148
Paper presented at Applied Superconductivity, Edinburgh, 3–6 July 1995
© 1995 IOP Publishing Ltd

Critical currents and vortex-glass melting in $YBa_2Cu_3O_7$ microstrips probed by phase-sensitive ac impedance measurements

W Lang and C Fussenegger

Ludwig Boltzmann Institut für Festkörperphysik, Kopernikusgasse 15, A-1060 Wien, Austria, and Inst. für Festkörperphysik der Universität Wien, Strudlhofgasse 4, A-1090 Wien, Austria

S Proyer, E Stangl, and D Bäuerle

Institut für Angewandte Physik, Universität Linz, A-4040 Linz, Austria.

Abstract. We report on investigations of the current-dependent ac impedance Z at 17 Hz of a $YBa_2Cu_3O_7$ thin-film microbridge in magnetic fields from 1 to 8 T. A phase-sensitive lock-in technique was applied to monitor the phase lag ϕ between current and voltage. Below the vortex-glass transition temperature T_g, ϕ saturates to finite values at low current densities j and to zero at high j. The phase lag remains below a critical value, predicted theoretically by Dorsey for low current densities, if $T > T_g$ and rapidly approaches 90° at $T < T_g$ indicating the freezing of the vortices into the dissipation-free vortex-glass state. We show that the measurement of ϕ provides a method for the direct detection of the vortex-glass transition.

1. Introduction

The determination of the critical current in the mixed state of a high temperature superconductor in magnetic fields $H_{c1} < H < H_{c2}$ is a matter of intense practical and physical interest. Commonly, the critical current is defined rather arbitrarily by introducing a voltage criterion and, thus, depends on the particular measurement method and on the sample geometry. It is of fundamental importance, however, if a genuine zero-resistance state $R(j \rightarrow 0)$ in vanishing transport current density can be observed at a finite temperature, or only at zero temperature. Several theoretical models, like flux creep [1] or thermally-assisted flux-flow [2], predict $R(j \rightarrow 0) = 0$ only for $T = 0$. Alternatively, the vortex-glass (VG) model [3], which is based on the very small coherence lengths in high temperature superconductors and on vortex pinning at randomly distributed defects, proposes a thermodynamic equilibrium phase with a glass-like arrangement of vortices. In this vortex-glass phase, $R(j \rightarrow 0) = 0$ for all temperatures below the magnetic-field dependent glass-transition temperature $T_g(B)$.

The dc resistivity vs. current-density isotherms near a VG transition obey the relation

$$\rho(j) \propto \xi_{VG}^{D-2-z} F_\pm(j\xi_{VG}^{D-1}\phi_0 / k_B T),\qquad(1)$$

where $\xi_{VG} \propto |T - T_g|^{-\nu}$ is the glass correlation length, ν and z are the static and dynamic critical exponent of the VG transition, respectively, and $F_\pm(x)$ are universal functions which characterize the VG system above and below $T_g(B)$. ϕ_0 is the flux quantum, and D, assumed to be 3, is the dimensionality of the vortex system. Dorsey [4] theoretically investigated the critical slowing-down of the vortex dynamics and found that the phase of the ac impedance $\phi = \arctan(\mathrm{Im}\,Z\,/\,\mathrm{Re}\,Z)$ is frequency-independent at T_g and has a universal value, which is related to the dynamic exponent by

$$\phi_g = \phi(T = T_g) = \frac{\pi}{2}\frac{2-D+z}{z}. \tag{2}$$

In this paper, we show that phase-sensitive measurements of the current-voltage characteristics in $YBa_2Cu_3O_7$ (YBCO) microbridges at a fixed frequency provide a convenient method for the observation of the VG transition and the critical slowing-down of the vortex dynamics.

2. Sample fabrication and experimental techniques

Thin YBCO films were grown on (100) MgO substrates by pulsed-laser deposition using a KrF-excimer laser and stoichiometric ceramic targets [5]. The film thicknesses were around 150 nm and the critical current densities $j_c(77\ \mathrm{K}) \geq 2\times10^6$ A/cm^2. The samples typically have a zero-resistance temperature of $T_{c0} \approx 89$ K and a 10–90% transition width smaller than 1 K. The films were patterned into a microstrip four-probe geometry with planar dimensions of $250\times17\ \mu m^2$ by laser ablation, using a protective layer of photoresist [6].

The investigations were performed in a superconducting solenoid at magnetic fields of 1, 2, 4, and 8 T oriented parallel to the c-axis. Temperature was measured with a platinum resistor and corrected for the magnetoresistance of the sensor. The accuracy of the temperature measurement was ± 0.1 K. The ac-impedance isotherms were measured at a fixed frequency of 17 Hz with phase-sensitive lock-in technique, eventually using an 1:100 input transformer for the lowest voltages. The particular frequency was chosen with regard to optimum performance of the transformer. Careful checking ensured that the transformer did not introduce phase errors, and amplitude errors were ruled out by overlapping the measurements with and without transformer. The total phase error was smaller than $\pm 2°$. By replacing the sample with a short-circuit it was ensured that spurious voltages, resulting from, e.g., vibration of current leads in the magnetic field, were substantially smaller than the signals reported in this analysis. In addition, a pulse technique was used and checked against the Re Z data to assure that no systematic errors in the ac measurements were evoked by sample heating or higher harmonic generation in the nonlinear region.

3. Results

The data for the real part of the ac impedance Re Z as a function of the current density j at low frequency corresponds to those obtained by dc I-V measurements [7]. The results will be presented in detail elsewhere [8]. Briefly, at low current densities and temperatures above T_g, a constant Re $Z(j)$ is observed, which can be attributed to thermally-assisted flux flow (linear regime). At larger current densities (j larger than about 10^4 A/cm^2), a highly nonlinear region is entered, where we observe a power law behavior of Re $Z(j)$. At $T < T_g$, only negative curvature of the Re $Z(j)$ isotherms in a log-log plot is observed. As predicted by the critical scaling theory in the vortex-glass model [3], represented by Eq. (1), the Re $Z(j)$ isotherms can be collapsed onto two characteristic curves. From the scaling collapse, the parameters $T_g(B)$, ν, and z can be obtained.

The (inductive) phase lag ϕ between current and voltage at 17 Hz is shown in Fig. 1 for various temperatures and current densities in a magnetic field $B = 1$ T. The phase lag is equal to zero at all experimentally accessible current densities, if the temperature is at least a few kelvin above T_g. $T_g(1T) = 82.7$ K was determined by the scaling collapse of the Re Z data. Close to the vortex-glass transition and at $T < T_g$, the phase angle changes from a finite constant value at low current densities to $\phi = 0°$ at higher current densities. This transition becomes steeper at lower temperatures, i.e., below T_g. Interestingly, the phase shift vanishes at current densities well below the transition into the flux-flow regime. Thus, ϕ provides a convenient indication of the vortex-glass to vortex-liquid transition as a function of the current density. We emphasize that, in contrast to the common definition of the critical current by an arbitrary voltage resolution, the change of the phase lag in ac impedance measurements can serve as a well-defined criterion.

According to Dorsey's theoretical prediction [4], ϕ remains smaller than a universal critical value $\phi_g = 71\pm2°$ calculated from Eq. (2), using z as obtained from the scaling collapse mentioned previously [8], for all current densities investigated here and at temperatures above the VG transition. For $T < T_g(B)$, the phase lag ϕ exceeds ϕ_g for low current densities, i.e., in the linear region, and reaches a maximum value of 90°. Thus, the dynamic vortex properties (represented by z in the critical scaling theory) can be probed not only by dc I-V characteristics, or, equivalently by Re $Z(j)$ measurements, but also by an analysis of the phase lag ϕ in ac-impedance measurements. Both approaches are linked via Eqs. (1) and (2) and are found to be in accordance. It should be noted, that Eq. (2) is valid only in the linear regime at low current densities and currently no theoretical prediction for the critical scaling of the phase lag in the nonlinear regime as a function of current density is available.

The phase angle at low current densities ($j \approx 10^3$ A/cm^2) is examined in Fig. 2 as a function of the reduced temperature T/T_g for various magnetic fields. ϕ increases rapidly when T_g is approached from higher temperatures, exceeds ϕ_g (indicated by the grey area in Fig. 2), and finally saturates to 90°. The latter observation obviously is equivalent to the dissipation-free state for dc currents in the vortex-glass phase, where the vortices are pinned

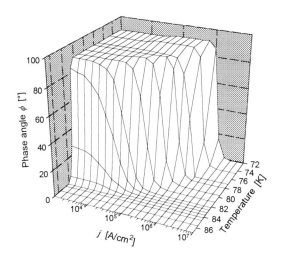

Fig. 1. Inductive phase shift ϕ between current and voltage in the ac impedance of YBCO as a function of the current density and of the temperature in a magnetic field $B = 1$ T.

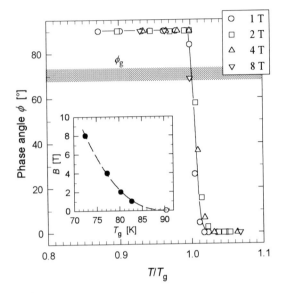

Fig. 2. Vortex-glass transition in YBCO monitored by the phase shift in ac impedance measurements at low current densities in various magnetic fields. The theoretically predicted, universal phase lag $\phi_g = 71\pm2°$ is indicated by a grey area. The insert specifies the values of the glass transition temperature $T_g(B)$.

to defects and only can oscillate around the defect potentials due to the Lorentz force generated by the ac current. The behavior found in our measurements is in good agreement with the phase angle inferred from swept-frequency measurements in the linear region [9-11].

In summary, we have presented measurements of the ac impedance in a YBCO microbridge at very low frequency as a function of the current density in various magnetic fields. We have observed the critical slowing-down of the vortex dynamics near the vortex-glass transition by measuring the phase lag between current and voltage. Our results corroborate the existence of a universal phase angle associated with the vortex-glass transition temperature for low current densities. The phase angle rapidly drops to zero, when the vortex glass melts at high current densities. The rapid change of the phase angle around the vortex-glass temperature may serve as an alternative definition for the critical current in high temperature superconductors.

Acknowledgments — This work was supported by the Fonds zur Förderung der wissenschaftlichen Forschung, Austria.

References

[1] Anderson P W 1962 *Phys. Rev. Lett.* **9** 309
[2] Kes P H, Aarts J, van den Berg J, van der Beek J C, and Mydosh J A 1989 *Supercond. Sci. Techn.* **1** 242
[3] Fisher M P A 1989 *Phys. Rev. Lett.* **62** 1415
    Fisher D S, Fisher M P A, and Huse D A 1991 *Phys. Rev. B* **43** 130
[4] Dorsey A T 1991 *Phys. Rev. B* **43** 7575
[5] Schwab P, Wang X Z, Proyer S, Kochemasov A, and Bäuerle D 1993 *Physica C* **214** 257
[6] Proyer S, Stangl E, Schwab P, Bäuerle D, Simon P, and Jordan C 1994 *Appl. Phys. A* **58** 471
[7] Wöltgens P J M, Dekker C, Swüste J, and de Wijn H W 1993 *Phys. Rev. B* **48** 16826
[8] Lang W, Fussenegger C, Proyer S, Stangl E, and Bäuerle D *to be published in Z. Phys. B*
[9] Olsson H K, Koch R H, Eidelloth W, and Robertazzi R P 1991 *Phys. Rev. Lett.* **66** 2661
[10] Reed D S, Yeh N-C, Jiang W, Kriplani U, and Holtzberg F 1993 *Phys. Rev. B* **47** 6150
[11] Wu H, Ong N P, and Li Y Q 1993 *Phys. Rev. Lett.* **71** 2642

Inst. Phys. Conf. Ser. No 148
Paper presented at Applied Superconductivity, Edinburgh, 3–6 July 1995

Experimental and analytical study of pinning mechanisms in high-temperature superconductors

C Attanasio†, C Coccorese†, V N Kushnir‡, L Maritato†, S L Prischepa‡ and M Salvato†

† Department of Physics, University of Salerno, Baronissi (SA), I-84081, Italy

‡ University of Informatics and RadioElectronics, Minsk, 220600, Belarus

Abstract. By performing magnetoresistance and critical current density measurements we have investigated pinning mechanisms in $Bi_2Sr_2CaCu_2O_x$ thin films. Analyzing simultaneously the experimental temperature, magnetic field and bias current dependencies of the pinning potential $U(T, H, J)$ we were able to obtain evidence for a crossover from a collective pinning of single vortex regime to a collective behavior of the vortex lattice. This result is in consistent agreement with the obtained magnetic field dependence of the pinning force, $F_p(H)$, including the tail usually observed at large magnetic fields.

1. Introduction

One peculiar feature of the high temperature superconductors (HTSC) is the relevant thermally assisted motion of vortices in these systems [1]. Because of the small values of the pinning potential U, a long tail in the resistive transitions, in the presence of external magnetic field, appear at small temperatures. In this case the flux creep resistivity, in the limit of small imposed stresses, can be written as [2,3]

$$\rho = \rho_0 exp\left\{ - U(T, H, J)/k_B T \right\} \qquad (1)$$

where T is the temperature, H is the applied magnetic field and J is the density of the bias current; ρ_0 is a coefficient of the order of the normal state resistivity. From equation (1) follows that analyzing the resistive transitions for different H and J it is possible to extract informations about U and its current and magnetic field dependences. Following this approach we have observed two different pinning mechanisms in our samples. At low magnetic fields, because the intervortex forces are very small, the vortices move individually in the field of randomly distributed pinning centers. Assuming an exponential type of pinning energies distribution, $U(H)$ is given by [4]

$$U(H) = U_0 + \sigma \cdot ln\left(\frac{H_0}{H} \right) \qquad (2)$$

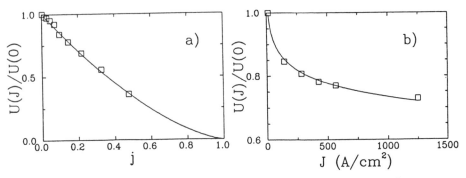

Figure 1. $U(J)$ dependencies for two values of the magnetic field for one BSCCO sample. The solid lines correspond to the equations $U(j) \sim (1-j)^{3/2}$ (1a) and to $U(J) \sim -lnJ$ (1b).

where U_0 is the activation energy corresponding to H_0, the crossover field between the single vortex and the collective behavior and σ is the variance of the distribution. For large magnetic fields, taking into account that the motion of the vortex lattice is influenced by the pinning centers, one can write, in the hypothesis that the spatial form $U(x)$ of the pinning potential is sinusoidal [4],

$$U(H) = U_0 - \gamma(a_p/2)^2 \qquad (3)$$

where the coefficient $\gamma \sim H$ and a_p is the average spatial period of the pinning potential. Equations (2) and (3) describe two different behaviors of the $U(H)$ curve: logarithmic for $H < H_0$ and linear for $H > H_0$. Moreover, at high magnetic fields the current dependence of the pinning energy becomes logarithmic. This fact, in the framework of the Anderson-Kim (A-K) model [3], has also allowed us to explain the pinning force $F_p(H)$ curve in the temperature range from $20\ K$ to $50\ K$.

2. Experimental results and discussion

$Bi_2Sr_2CaCu_2O_x$ (BSCCO) thin films (with thickness $d \simeq 0.3\ \mu m$) were prepared on MgO substrates by molecular beam epitaxy technique followed by an "ex-situ" annealing in air [5]. The critical temperatures were higher than $86\ K$ and the critical current densities J_c at $T = 4.2\ K$ were about $10^9\ A/m^2$. The samples, c-axis oriented, were patterned, by usual photolitographic procedures, in microbridges having length $l \simeq 30\ \mu m$ and width $w \simeq 25\ \mu m$. Measurements have been performed by dc four probe method. The critical current density was defined by the electric field criterion $E = E_c = 10^{-3}\ V/m$. The magnetic fields were applied parallel to the c-axis of the film. As first step we have performed the investigation of the $U(J)$ dependence in the case of weak ($\mu_0 H = 0.29\ T$) and large ($\mu_0 H = 1.83\ T$) magnetic fields. From equation (1) we get

$$U(T, H, J) = k_B T \left[ln\rho_0 - ln\rho(T, H, J) \right] \qquad (4)$$

From this relation, measuring the resistivity curves at different values of the bias current and magnetic field, we can obtain the $U(H, J)$ dependencies. The absolute $U(0)$ values

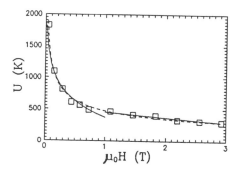

Figure 2. $U(H)$ dependence for two BSCCO samples. The solid line corresponds to the best fits obtained from equation (2) (low field) and (3) (high field). The dashed line shows the best power fit.

can be obtained in the Tinkham [6] model, in the limit $T = 0$. For $\mu_0 H = 0.29\ T$ we have found that the current dependence $U(J)$ is well described by the equation $U(j) \sim (1 - j)^{3/2}$ which follows from the assumption of washboard pinning potential spatial shape $U(x)$ [7]; here $j = J/J_{c0}$, where J_{c0} is the critical current density in the absence of flux creep. On the other hand at $\mu_0 H = 1.83\ T$ the $U(J)$ data are well described by the relation $U(J) \sim -ln(J)$. In figure 1 we show the different dependencies for $U(J)$ for the two values of the magnetic field. Assuming the two different kinds of $U(H)$ dependences, for small and large magnetic fields, we have also analyzed the $U(H)$ curves (with a bias current density of 142 A/cm^2). In figure 2 we show the experimental data of the $U(H)$ dependence for two samples, together with the best fits calculated according to equations (2) and (3), for the low field and the high field part, respectively. The crossover field is $\mu_0 H_0 \simeq 0.7\ T$, in reasonable agreement with the small bundles crossover field [1]. The $U(0)$ values have been calculated in the Tinkham model with $n = 1$ (the exponent n identifies the temperature scaling behavior of U), but we have to stress that, for our data, the $U(H)$ dependence does not depend on the specific choice of n. The dashed line in figure 2 shows the best fit $(U(H) \sim H^{-0.45})$ obtained over the entire magnetic field range not considering the presence of any crossover. The last $U(H)$ law is usually reported in literature for BSCCO films [8].

By measuring the critical current densities we also got additional informations on the pinning mechanisms below the irreversibility line. In particular, in the temperature range from 20 K to 50 K, for external magnetic fields up to 4 T, the flux pinning force $F_p = J_c B$ reveal a scaling behavior in the form $f = F_p/F_p^{max} \sim h^{0.5}(1 - h)^2$, where $h = H/H^*$, and H^* is the irreversibility field. This evidence, usually explained for conventional superconductors in the framework of the flux lines shear model (FLSM) [9], is analyzed here considering also the contribution of the thermal activation to the vortex motion which is important in HTSC [2,6,10]. We started from the classical A-K expression [3] for the $I - V$ characteristics $V \sim I exp\left\{ - \frac{U(T,H,J)}{k_B T}\right\}$. Assuming the multiplicative character of the $U(T, H, J)$ dependencies, we get the following expression for F_p

$$F_p \sim \beta^* H^* h^{1-\alpha}\left[\eta(0)\right]^{-1}\eta^{-1}\left[\eta(0)\frac{\beta}{\beta^*}h^\alpha\right] \qquad (5)$$

302

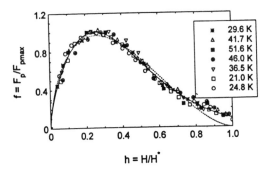

Figure 3. The normalized pinning force versus the reduced magnetic field. The solid line is the best fit curve obtained from equation 5. The dashed line is the best fit according to FLSM.

where $\beta = ln\frac{B a_p \nu_0}{E_c}$ ($\beta^* = ln\frac{B^* a_p \nu_0}{E_c}$), ν_0 is the characteristic frequency of the vortex vibration, $\eta(J)$ is the current dependence form of U and α is the exponent in the power field dependence of U. Introducing the pinning energy current dependence in the form $\eta(J) \sim -Aln(J)$ we get, after some calculations, $f(h) \sim h^{1-\alpha}exp\left\{ -\frac{h^\alpha}{A}\left(1 + \frac{lnh}{\beta^*}\right)\right\}$. In figure 3 we report the experimental data together with best fit line, obtained from equation 5 for $\alpha = 0.5$, using A and β^* as fitting parameters.

3. Conclusions

In conclusion, by performing transport measurements, we have obtained evidence for a croossover from two different vortex regimes in a region above the irreversibility line. On the other hand, in the framework of the A-K theory, assuming a current logarithmic dependence of the pinning energy, we were able to describe the all $f(h)$ curve. This seems to indicate the presence of a single pinning mechanisms below the irreversibility line.

References

[1] Blatter G et al. 1994 Rev. Mod. Phys. **66** 1125

[2] Dew-Hughes D 1988 Cryogenics **28** 674

[3] Anderson P W and Kim Y B 1964 Rev. Mod. Phys. **36** 39

[4] Inui M, Littlewood P B and Coppersmith S N 1989 Phys. Rev. Lett. **63** 2421

[5] Salvato M et al. 1994 Cryogenics **34** 859

[6] Tinkham M 1988 Phys. Rev. Lett. **61** 1658

[7] Beasley M R, Labush R and Webb W W 1969 Phys. Rev. **181** 682

[8] Kucera J T et al. 1992 Phys. Rev. **46** 11004

[9] Kramer E J 1973 J. Appl. Phys. **44** 1360

[10] Yamasaki H et al. 1993 Phys. Rev. Lett. **70** 3331

Inst. Phys. Conf. Ser. No 148
Paper presented at Applied Superconductivity, Edinburgh, 3–6 July 1995
© 1995 IOP Publishing Ltd

ANISOTROPY OF CRITICAL CURRENT DENSITY, VOLUME PINNING FORCE, AND FLUX DYNAMICS IN YBa2Cu3O7-δ SINGLE CRYSTALS

Vladimir M. Pan[1], **Vjacheslav F. Solovjov**[1], and **Herbert C. Freyhardt**[2]

[1]*Institute for Metal Physics, 36 Vernadsky Blvd., 252142 Kiev, Ukraine,* [2]*Institute for Metal Physics, University of Goettingen, Hospitalstr., 3-7, 37073 Goettingen, Germany*

In contrast to thin films and melt-textured samples the $J_c(\Theta)$ and $M_{irr}(\Theta)$ dependencies for YBa2Cu3O7-δ twinned single crystals reveal a remarkable minimum at H∥C-axis. The nonmonotonic $J_c(H)$ and $M_{irr}(H)$ functions (the "fishtail"- or peak-effect) turn out to be much more pronounced in tilted magnetic fields (and suppressed at H∥C-orientation) being strongly connected with the $J_c(H∥C)$-minimum. The effect results from the interplay between two contributions into the volume pinning force, F_p: from random point-like defects (oxygen vacancies) inducing moderately anisotropic background and from planar defects (twins) which influence on flux pinning mainly within the critical angle $\Theta_c \leq \pm 20^0$ from C-axis. We used Kramer's scaling law for the F_p (H,Θ) curves to prove the 3D collective pinning by random point defects as a major pinning source in $\Theta > \Theta_c$ range of C-axis/applied field angles and at intermediate fields. Role of twins appear to be more complicated due to interference of point-like and correlated disorders. This interference assumes to be dependent upon crossover between Larkin-Ovchinnikov transverse correlation length and twin spacing.

1. Introduction

There are several types of defects in YBa2Cu3O7-δ (YBCO) materials which are commonly thought to be the most effective pins contributing to the volume pinning force: (i) 0D point defects - oxygen vacancies, (ii) 1D line defects - dislocations and (iii) 2D planar defects - twins. The dislocation contribution to pinning is obviously dominant in thin-film [1] and melt-textured samples [2], resulting in the monotonous $J_c(H)$-dependence and $J_c(\Theta)$-curve with two maxima. It was shown in our recent works [3,4] single-crystalline YBCO exhibited unusual properties, i.e., there were observed anomalous $J_c(H,\Theta)$-dependencies. We attributed the anomalies to peculiarities of the flux pinning and motion mechanisms in perfect single crystals. Thus in "clean" case the flux behavior proved out to be quite different from that one pertinent to "dirty" or "disordered" case, e.g., the thin-films and melt-textured samples. In our previous studies we explored flux pinning and flux dynamics by means of transport current technique while in this paper we make an attempt to reconcile data (of our own and of the other groups) obtained with transport-current (pulsed and DC) as well as magnetization measurements for single-crystalline YBCO on the basis of self-consistent model for the flux pinning. The more complicated question is the flux dynamics in the "clean" case. Our previous result [3] concerns the regime of high flux velocity when the flux motion obeys predictions of the classical mean-field model [5] for elastic continuum. The resulting threshold-like approximation of current-voltage characteristics (CVC) surely fails at low dissipation rates, where thermal activation dominates over non-activated motion. Thus the problem is what kind of dynamics takes place in the range of low voltages and low currents.

304

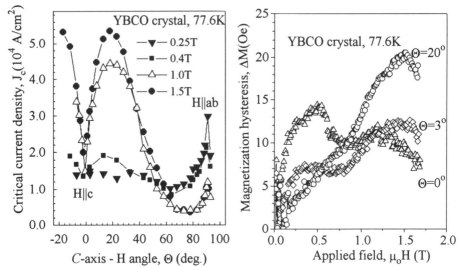

Fig.1 Angular dependencies of the transport critical current of YBCO single crystal. T=77.6K, field values are on the plot.

Fig.2 Magnetization hysteresis versus applied field for various C-axis - H angles. The angle values are on the plot, T=77.6K

2. Experiment

Direct transport current as well as magnetization measurements of the field and angular dependencies of the critical current density for single-crystalline and melt-textured samples of YBCO were performed in a wide range of temperatures (between 77K and T_c) and applied fields ($\leq$ 2T). The transport measurements were done by means of pulse technique, the experimental setup described in details elsewhere [3]. Briefly, the short (10 µs) current pulses with amplitude up to 10 Amp were used, the electric field strength criterion, E_c, was as low as 10 µV/cm. We carried out parallel measurements of the magnetic moment hystresis of the samples to cover inaccessible by transport method a range of low voltages. The miniature GaAs Hall sensor was used as a detector of the magnetic induction component normal to the surface of the crystal. The applied field has measured by an independent sensor and subtracted from the first sensor readings to obtain the magnetic moment value. The field sweep rate was 0.01T/min what corresponds to 0.01 µV/cm of the electric field strength on the surface of the sample.

3. Results and discussion

The angular dependencies of the transport J_c and field dependencies of the irreversible magnetic moment $M_{irr}(H)$ for various Θ values (Θ here means angle between the C-axis and the applied field vector) of twinned single crystal are presented in Figs.1 and 2. We argue, that major part of the volume pinning force $F_p = J_c B$ anisotropy arises from random point defect contribution (further - F_p^{0D}), while twin planes (their contribution to F_p we denote as F_p^{2D}) occur in a position to pin effectively if the external field is tilted with respect to the defect plane within some critical locking angle, $\Theta_c \leq \pm 20^o$. The point defect contribution may be analytically expressed in the form of Kramer's scaling law [6]:

$$F_P^{0D}(H,\theta) \propto B_{C2}{}^m b^p (1-b)^q \qquad (1)$$

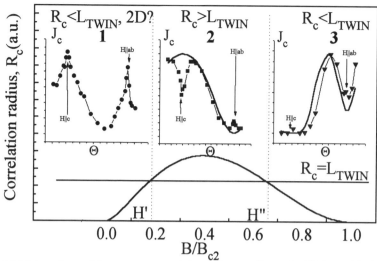

Fig.3 Model dependence of the transverse correlation radius according to the CP theory. The insets show transport critical current angular dependencies, corresponding to three field regions. Solid lines in the insets represent the approximations by expression (1).

The J_c angular dependence follows from (1) due to anisotropy of B_{c2} in the framework of the anisotropic Ginsburg-Landau theory (in (1) reduced field $b = B/B_{c2}(\Theta)$):

$$B_{C2}(\theta) = B_{C2}(H||c) / \sqrt{\cos(\theta)^2 + \Gamma^{-2}\sin(\theta)^2} \qquad (2)$$

The values of p and q indexes ($p\approx q\approx2$) [4] were found by approximations of the experimental $F_p(H,\Theta)$ curves with Kramer's law, using anisotropy parameter $\Gamma = 6$. Thus, taking for granted realization of 3D collective pinning (CP) in the higher field region we will consider existing experimental data from this point of view. There seems to be several characteristic length scales in the problem: transverse and longitudinal correlation radii R_c and L_c, twin structure spacing L_{twin}, the sample thickness d. The model R_c field dependence, following from 3D CP theory is presented in Fig.3. Let us analyze implications from possible crossover between two dimensional scales: R_c and L_{twin}, taking into account our explanation [4] for the fishtail-effect as a phenomenon, arising from the C_{44} modulus softening. The L_{twin} value is presented as a horizontal line on the plot, the insets show typical angular dependencies of the critical current for the field regions, which are separated by crossover condition: R_c>L_{twin}. Above the $R_c= L_{twin}$ line, i.e., in intermediate field region $H'<H<H''$, the interference effect is feasible, and twin planes are able to suppress effective 3D point-defect pinning. In fact the captured vortices are able to proliferate 2D state via elastic interaction on the bulk of the crystal, increasing effective values of C_{44}. The central inset demonstrates how $J_c(\Theta)$-approximation by (1) consistently (solid line) accounts for the J_c anisotropy except the Θ values less than the twin lock-in angle ~20^0. Indeed, if $\Theta<20^0$ the $J_c(\Theta)$ deviates from (1) due to the interference. Further field increase leads to R_c reduction, vacancy pinning efficiency drops down since the melting point is approaching and thermal fluctuations are smearing the pinning potential of small-sized point defects. When $R_c<L_{twin}$ point is reached the interference effects disappear and twins and point-defect contributions may be added: $F_p=F_{p0}^{0D}+F_p^{2D}$. Thus, the inset 3 implies that in the region $H>H''$ $F_{p0}^{0D}>>F_p^{2D}$, and the (1) approximation holds true over all Θ range (solid line). But the twins being defects of a greater scale are retaining their weak pinning capability up to the melting point, and at $H\approx H_{irr}$ F_p^{2D} may surpass the F_{p0}^{0D} magnitude. That is why the DC transport experiments performed in the vicinity of the irreversibility field testify in favor of strong pinning of the twins [8]. Note, that the above mentioned effects are visible only in perfect single crystals, where FLL preserves the long range order on distances comparable with the twins spacing. In the samples with strong disorder and small correlation radii the twin influence results in small J_c increase on the background of dislocation pinning.

The left inset in Fig.3 represent situation below the fishtail-effect field, where J_c is a decreasing function of the applied field. The point-defect pinning is ineffective, probably due to 2D state of the flux lattice (but we have only indirect evidences of the 2D state). Twin pinning is not affected by the vortex lattice dimensionality and F_p^{2D} contribution is a dominant one, see also Fig.2. This allows to recognize a controversy with the twin plane efficiency as pinning centers. At low fields (below the fishtail-effect) point defects pinning is weak, therefore the twins are often observable as strong pinning centers in decoration experiments [9] (when the external field amplitude is much less than = 0.5 T).

The M(H) curves have also interesting implications as for the dynamics problem. According to our results [3] the peak-effect is accompanied with non-trivial changes in flux dynamics, at least it seemed so through accessible in our transport experiment J-E window. There still question remains what happens at low rates of dissipation reduced by 2-3 orders of magnitude and how the classical dynamics with the threshold-like I-V curves, $V=V_0(1-I/I_T)^\varsigma$, transforms into thermally-activated motion below the threshold current, I_T. As we observed there exist a substantial difference of the flux behavior in the melt-textured and single-crystalline samples. The former captures the essential magnetic moment and exhibit prehistory effects, what is pertinent to glassy state. The latter captures the flux only when the twin planes are effectively interacting with vortices, i.e., within critical angle. At the strong misalignment of the C-axis and vector H ($\Theta > 20^0$ in Fig.2) M_{irr} is function of H and dH/dt and trapped flux is very small, the prehistory effects are absent. It implies at high Θ-angles flux dynamics differs dramatically from the case of melt-textured samples where formation of the glassy state is established.

4. Conclusion

Thus, we presented a model for the vortex-pins interactions in moderately anisotropic YBCO single crystals. The model is believed to reconcile data obtained by different techniques and experimental groups. Our results imply the dominance of the point-defect (oxygen vacancies) pinning in perfect YBCO single crystals in the range of intermediate fields, where the peak-effect is observed. The twin planes role proved out to be ambiguous: there is the interference effects when the crossover between the transverse correlation radius and the intertwin spacing occurs.

References

[1] J. Mannhart, D. Anselmetti, J.G. Bednorz et al. 1992 *Supercond.Sci.Technol.* 5 125-128.
[2] V. Selvamanickam, M.Mironova, S.Son, K.Salama 1993 Physica C **208** 238-244.
[3] V.F.Solovjov, V.M. Pan and H.C. Freyhardt 1994 *Phys.Rev.* **B50** 13724-13733.
[4] V.M.Pan, V.F. Solovjov et al. 1995 *IEEE Trans. on Appl. Supercond.* 5
[5] D.S.Fisher 1986 *Phys. Rev.* **B31** 1396-1428.
[6] E.J.Kramer 1973 *J.Appl. Phys.* 44 1360-1370.
[8] W.K. Kwok, G.W. Crabtree, U.Umezava, et al. 1990 *Phys.Rev. Lett.* 64 966-969.
[9] R. Bleim, M. Audier, Y.Brechet et. al. 1992 *Phil.Mag. Lett.* 65 113-120.

Inst. Phys. Conf. Ser. No 148
Paper presented at Applied Superconductivity, Edinburgh, 3–6 July 1995
© 1995 IOP Publishing Ltd

Magnetisation vector rotation in La$_{2-x}$Sr$_x$CuO$_{4-y}$ single crystals with various Sr doping and different anisotropy factors

Yu.V.Bugoslavsky,   K.V.Gamayunov,   A.L.Ivanov,   V.A.Kovalsky,   and A.A.Minakov

General Physics Institute, Vavilov st. 38, Moscow 117942, Russia;

Hironao Kojima and Isao Tanaka

Institute of Inorganic Synthesis, Faculty of Engineering, Yamanashi University, Miyamae 7, Kofu, Yamanashi 400, Japan.

Single crystal superconducting samples of La$_{2-x}$Sr$_x$CuO$_{4-y}$ (LSCO[x]) compound with Sr concentrations x = 0.07, 0.15, and 0.20 were studied by means of vector magnetisation technique at T=4.2 - 30 K, in fields up to 6.5 kOe. The effective-mass anisotropy factor is highest for LSCO[0.15] sample (which has the highest T$_c$), and decreases with both increasing and decreasing x. It was found that the structure of magnetisation curves in tilted applied field is strongly affected by anisotropic vortex penetration, when flux lines are arranges along an 'easy penetration plane'. At higher fields flux lines smoothly rotate away from the easy plane. In LSCO[0.15] sample the easy plane coincides with the direction of ab crystal plane, whereas for less anisotropic compound it takes a certain position between ab plane and H$_a$ direction. This behavior is described within the model of 3-dimensional anisotropic superconductor.

Introduction

Inclined flux line lattices (FLL) in layered, highly anisotropic high-T$_c$ superconductors (SC) are subject to vast research interest due to unusual magnetic properties of these materials. The behavior of FLL in magnetic field applied nearly parallel (within a few degrees) to CuO layers could give manifestation of quasi two-dimensional nature of HTS. A specific feature of stacked SC is the possibility of so called 'lock-in transition'[1]. At low applied tilted field magnetic vortices are locked between adjacent superconducting planes. Rotation of FLL towards the field direction starts at a certain field value, when nucleation of normal-to-plane vortex parts becomes favorable. An evidence of this transition was obtained from measurements of magnetic torque on Bi(2212) and Tl(2223) compounds [2]. It is interesting to study the properties of FLL in La$_{2-x}$Sr$_x$CuO$_{4-y}$ (LSCO[x]), where the effective mass anisotropy $\Gamma = m_c/m_{ab}$ varies in a wide range with Sr doping. On the other hand, it was observed that magnetisation (M-H) curves for LSCO crystals in tilted field have unusual structure with a broad secondary maxima, which are strongly dependent on the field direction with respect to crystal axes [3]. For simple geometry, Bean's model [4] relates high-field features of magnetisation to variation of critical current j$_c$. In tilted field, the form of M-H curve can be strongly influenced by flux-rotation effect, as argued in [3]. Indeed, the equilibrium theory [5] predicts that reversible magnetisation is double-peaked for anisotropic SC in tilted field. However, the experimental curves are evidently irreversible, j$_c$ is reasonably high, thus the applicability of flux-rotation picture requires further examination.

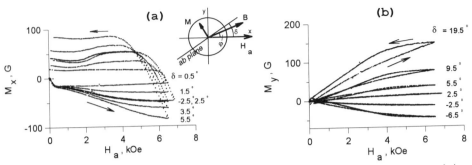

Fig. 1. Parallel-to-field (a) and normal-to-field (b) components of magnetisation for LSCO[0.15] sample in magnetic field applied at different angles δ to the ab plane, T = 4.2 K. Inset: the sketch of vectors and coordinates

Samples and experimental technique.

We report on the measurements on three crystals of LSCO[x] compound with Sr concentrations x = 0.07, 0.15, and 0.20. The respective T_c are 13, 33.5, and 23 K. The crystal LSCO[0.07] was grown by spontaneous crystallization, LSCO[0.15] and LSCO[0.20] by TSFZ technique. The samples were not annealed in oxygen, therefore T_c's have reduced values for given x. Vector magnetisation measurements were performed by means of specially equipped vibrating sample magnetometer at temperatures from 4.2 K to respective T_c, in magnetic field range 0-6500 Oe. The projections of magnetic moment on the direction of the field (M_x) and the perpendicular projection (M_y) were measured independently. To avoid angular-dependent demagnetisation effect, the samples with x=0.15 and x=0.20 were cut in the shape of disks with the c axis parallel to the disk plane. The disks are about 0.4 mm in thickness, with diameter of about 1 mm. The crystal LSCO[0.07] is a square plate of ~1x1x0.5 mm³ with c axis parallel to the smallest dimention. The dimensions, as measured with the help of optical microscope are in a good agreement with sample volume and demagnetisation factors, calculated from low-field magnetisation (H<H_{c1}) of zero-field-cooled (ZFC) sample. The magnetic field H_a was applied parallel to the disk, at a certain angle δ to the crystal ab plane. In case of plate sample, δ is the angle between H_a and the plate plane. The directions of crystal axes within disk plane were determined from a set of M_x-H curves taken at different angles δ (see Fig.1a). The narrowest curve corresponds to the field direction parallel to ab plane. The set of M_y-H curves is symmetric for angles ±δ with respect to this direction. The accuracy of sample angular alignment is 0.5 degree.

Results and discussion.

The dependencies of M_x, M_y components on H_a for LSCO[0.15] are shown in the Fig.1. The similar data with wider M_x-H curves at higher δ were obtained for the other crystals. In case of LSCO[0.07] the field scale is squizeed by a factor of 10 towards lower fields. For LSCO[0.20] the secondary maxima are wide, and therefore the bending point on M_x-H curve is less evident. The results can be presented in the form of M_x-M_y plots, with the value of H_a as a parameter (Fig.2). These curves represent traces drawn by the end of magnetisation vector while applied field is ramped. When H_a increases, $\vec{M}$ is antiparallel to $\vec{H_a}$ at low field ($H_a < H_{c1}$), and monotonically rotates away from the field direction at higher fields. This holds for the entire temperature interval studied. Upon field reversal, M_x component rapidly changes sign, while M_y stays essentially constant. At all the angles δ the tangent to M_x-M_y curve at the field reversal point is parallel to x axis, i.e. to $\vec{H_a}$. This is due to the following.

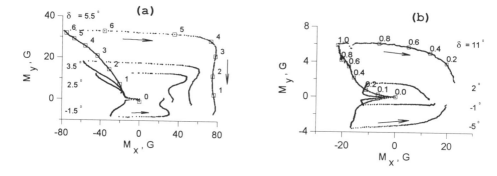

Fig.2. Mx-My plots for LSCO[0.15] (a) and LSCO[0.20] (b) samples. The values of applied field (in kOe) are shown aside the curves.

When the field is decreased from H_{max} to $H_{max} - \Delta H$, the circular critical currents in close-to-surface regions change direction of their flow in order to keep magnetic flux constant in the sample's bulk. Therefore ΔM is proportional to ΔH, and consequently $\Delta \vec{M} \parallel \Delta \vec{H}$ at small ΔH.

To understand the behavior of FLL, it is suggested to follow the variation of induction with growing field. The main obstacle is the fact that the induction is highly inhomogeneuos, it varies from point to point in the sample due to flowing critical currents. Nevertheless define formally uniform induction $\vec{B}_o$ in the following way: $\vec{B}_o = \vec{H}_a - 4\pi(1 - \hat{n})\vec{M}$, where $\vec{M}$ is measured magnetisation, $\hat{n}$ - demagnetisation tensor. At $H < H_{c1}$, B_o should be zero. This actually holds for fields applied both nearly parallel and nearly perpendicular to the ab plane (see Inset, Fig.3a). When H_a starts to decrease from its maximum value, B_o stays constant at small H_a decrements, as it is expected from the condition of constant flux. This behavior makes us convinced that the calibration, sample volume and shape are determined correctly.

The central results of this study are presented in the Fig.3. Here the induction plots are shown for different angles δ between $\vec{H}_a$ and ab plane. The respective positions of ab plane are shown by dashed lines. From the plot for LSCO[0.15] crystal it is apparent that magnetic induction penetrates the sample parallel to ab plane. When H_a increases, $\vec{B}_o$ deviates from this plane and rotates towards $\vec{H}_a$ direction. However the variation of angle ψ between $\vec{B}_o$ and $\vec{H}_a$ ($\psi = \delta - \varphi$) is monotonous, showing no plateau at low fields, which might be expected for 'locked' FLL. The dependence $\psi(H_a)$ fits well to numerical results, obtained in a

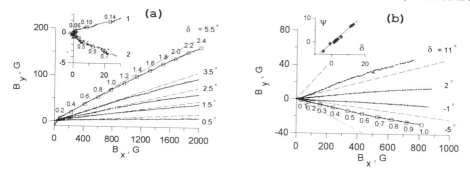

Fig. 3. Variation of induction components with increasing field for LSCO[0.15](a) and LSCO[0.20](b).. Inset (a): B at low applied fields, curve 1: $\delta = -3°$, curve 2: $\delta = 85°$. Inset (b): the dependence of flux penetration angle (measured from ab plane) on δ. Crosses: experimental; line: fit, $\Gamma = 1.9$.

framework of 3-dimensional theory.[5]. The parameters used in calculations: $\Gamma=100$ and $H_{c1}{}^c=500$ Oe agree within 10% with the anisotropy of H_{c1}, as evaluated from virgin magnetisation curves as the upper boundary of the linear part.

A similar behavior is observed for LSCO[0.20] sample, but now the flux penetration plane does not coincide with ab plane. The dependence of the angle of flux penetration on δ is presented in the Inset, Fig.3(b). For 3D anisotropic SC these angles are related via effective-mass anisotropy factor Γ: tg $\psi = (1/\Gamma)$ tg δ [5]. We evaluate $\Gamma \approx 2$. This result differs from the evaluation based on H_{c1} anisotropy. This is probably due to the fact that direct determination of H_{c1} is hampered by various factors, such as surface barrier, inhomogenety, microcracks, or penetration from sharp edges and corners. To the contrary, B_x-B_y plots reflect essentially sample bulk properties, and thus can be regarded as direct results.For LSCO[0.07] sample the evaluated factor $\Gamma = 20$, therefore it can be concluded that Γ correlates primarily with T_c, not Sr doping level.

In conclusion, vector magnetisation studies give direct evidence of anisotropic flux penetration, affected by the presence of easy plane, which coincides with ab crystal plane. The variation of FLL orientation in increasing field is an evidence in favor of 3-dimensional nature of the compound studied.

This work was made possible in part by Grant No. M41300 from the International Science Foundation and the Government of Russian Federation.

References

1. S.S.Maslov and V.I.Pokrovsky, Europhys. Lett. 14 (1991) 591
2. F.Steinmeyer, R.Kleiner, P.Muller, H.Muller, and K.Winzer, Europhys.Lett. 25 (1994) 459
3. Yu.V.Bugoslavsky, A.L.Ivanov, A.A.Minakov, and S.I.Vasyurin, Physica C 233 (1994) 67
4 C.P.Bean, Rev.Mod.Phys. 36 (1964) 31
5 A.I.Buzdin, A.Yu.Simonov Sov.Phys. JETP 71 (1990) 1165

Appendix

Consider an ellipsoid with arbitrary non-uniform magnetic induction fixed inside it. Introduce the coordinate frame of oblate ellipsoid (η, θ, ϕ), which is a generalization of sphere coordinates. Given the induction profile $\vec{B}(\vec{r})$, the following procedure can be applied to calculate sample magnetic moment. The potential of magnetostatic field in the sample exterior should be expanded into series of generalized spherical harmonics Y_j^m, in which the first term corresponds to the dipole field and thus is proportional to M. At the sample surface the normal component of induction $B_n(\theta,\phi)$ should be continuous. To obtain respective boundary conditions (BC) for each multipole component of external field, one should expand $B_n(\theta,\phi)$ into series of Y_j^m. The coefficients at first-order terms ($j=1$, $m=0,\pm1$) are proportional to components of magnetic moment. In the sample interior these terms of BC expansion yield homogeneous induction $\vec{B}_o$. From this consideration, $\vec{B}_o$ may be thought of as a uniform field giving the best overlap with $\vec{B}(r)$. If only $|B(r)|$ varies in the sample, and the direction is essentially constant, it is parallel to $\vec{B}_o$ To interpret the experiment, this consideration is to be reversed. These are components of $\vec{M}$ that are measured , and in combination with H_a, it yields formally the uniform induction $\vec{B}_o$. This value can be interpreted as a mean sample induction, which direction coincides with an average direction of FLL.

Inst. Phys. Conf. Ser. No 148
Paper presented at Applied Superconductivity, Edinburgh, 3–6 July 1995
© 1995 IOP Publishing Ltd

The peak-effect in $YBa_2Cu_3O_{7-\delta}$ single crystals with different oxygen stoichiometry

A A Zhukov[a,b], H Küpfer[a], G Perkins[c], L F Cohen[c], A D Caplin[c],
S A Klestov[b], H Claus[a], V I Voronkova[b], T Wolf[a], M Kläser[d],
H Wühl[a]

[a] Forschungszentrum Karlsruhe, Institut für Technische Physik, Postfach 3640,
D-76021 Karlsruhe, Germany
[b] Physics Department, Moscow State University, Moscow 117234, Russia
[c] Centre for HTS, Blackett Laboratory, Imperial College, London SW7 2 BZ, UK
[d] Universität Karlsruhe, Fakultät für Physik, D-76128 Karlsruhe, Germany

Abstract. The influence of oxygen stoichiometry on flux pinning in two $YBa_2Cu_3O_{7-\delta}$ single crystals, differing by pinning strength, was studied using simultaneous oxygen annealing at different temperatures. Both samples showed fishtail peak in high temperature region. In the purer sample it probably originates from oxygen vacancies, whereas in the sample with strong pinning it is induced by impurities. In the intermediate temperature region, the pure sample showed a new second peak with weak temperature and δ dependence. The probable origin of this peak is the matching effect between the vortex lattice and the twin structure in the sample.

1. Introduction

The maximum of the critical current vs magnetic field above the self-field region in high temperature superconductors (HTSC) is often called the "fishtail" or "butterfly" effect [1,2]. In contrast to the peak-effect in conventional super-conductors, the position of the maximum of the current in HTSC is significantly below the upper critical field $H_{c2}(T)$.

In this paper we report the observation of a second peak [3] arising at intermediate temperatures in addition to the conventional "fishtail" peak: we describe the influence of oxygen content on the "fishtail" effect and this new feature.

2. Experimental

The $YBa_2Cu_3O_{7-\delta}$ single crystals grown from a flux were studied. Variations of oxygen stoichiometry were produced by long time annealing (up to 500 hrs) at different temperatures with 1 bar oxygen pressure. The oxygen content was determined from a previous calibration of $T_c(\delta)$ [4]. The gradual variation of annealing temperature allowed discrimination between overdoped and underdoped states on each side of the $T_c(\delta)$ maximum.

The shielding current density j_s was determined from magnetization measurements performed with a vibration sample magnetometer in magnetic fields H ≤ 120 kOe parallel to the c-axis. The field was cycled with different constant sweep rates. The j_s values usually were taken for the sweep rate of the magnetic field of 120 Oe/s, which corresponds to the electric field E ≈ 10-7 V/cm. The superconducting transition temperature T_c was determined from onset of diamagnetic signal at H = 5 0e.

3. Results

Extensive studies of various $RBa_2Cu_3O_y$ single crystals prepared in different ways have shown large differences in the values of the shielding currents and, consequently, pinning strength. Fig. 1 represents data at 77 K for some YBa_2Cu_3O single crystals. The Sr doped sample 1 had 6 % of Ba replaced by Sr. Samples 2 and 3 were prepared in conventional way [4] and probably had different impurity contaminations. Sample 4 was prepared from high purity starting materials. All these samples were oxidized in 1 bar oxygen at ~ 400°C. The maximum value of the current decreases with decreasing impurity concentration, in contrast to this the irreversibility field increases. This shows the importance of point-like defects for pinning and indicates that point-like disorder shifts the melting transition to lower fields. The peak-effect disappeared in the high-purity sample 4 at high temperature region, where pinning by twin boundaries is probably dominating. At lower temperatures (< 70 K) this sample, too, showed the peak behavior of the current.

The similarly prepared samples, 2 and 3, were used for the study of the influence of oxygen stoichiometry. Preliminary measurements on a crystal similar to the sample 4 showed a behavior consistent with that of sample 3. The pure sample 3 showed three different types of $j_s(H)$ behavior at high oxygen content. Fig. 2 presents data for this sample which were obtained after annealing at different temperatures: 1 - 390°C, 2 - 480°C, 3 - 520°C, 4 - 550°C and 5 - 575°C. This oxygenation states were characterized by T_c: 1 - 92.1 K, 2 - 92.7 K, 3 - 92.3 K, 4 - 88.5 K, 5 - 82.9 K and δ: 1 - 6.97, 2 - 6.94, 3 - 6.91, 4 - 6.86, 5 - 6.78 ± 0.05. At high temperatures a sharp $j_s(H)$ peak corresponding to the conventional fishtail is observed. In the intermediate temperature region the increase of the current,

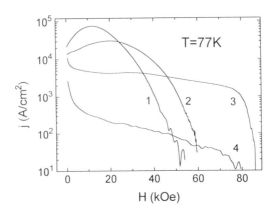

Fig. 1

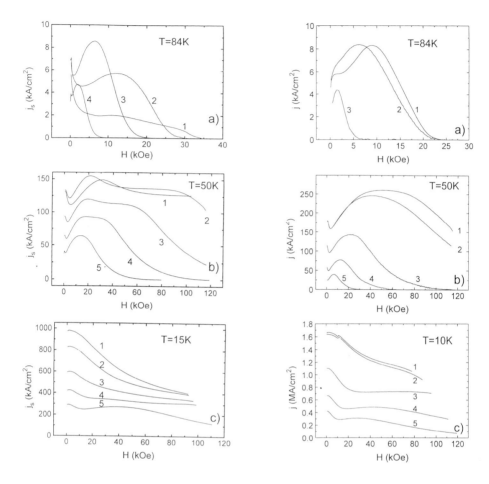

Fig. 2   Fig. 3

after the peak with nearly temperature independent position, is followed by a broad field region with nearly constant $j_s(H)$. The low temperature region shows the conventional monotonic decrease of $j_s(H)$ with magnetic field. Sample 2 with the higher current (Fig. 1) was annealed simultaneously together with sample 3 at the same temperatures, yet slightly differing by the corresponding T_c. At the high T interval it showed a pronounced fishtail (Fig. 3a). However, the decrease of the maximum current with the increase of oxygen content was almost absent. This sample did not show the second peak or plateau-like structure at intermediate temperatures (Fig. 3b). The single peak flattens and directly transforms into the monotonous decrease of $j_s(H)$ with magnetic field at low temperatures (Fig. 3c).

314

4. Discussion

The influence of oxygen content on pinning of vortices has two different reasons: (i) the intrinsic parameters changes with δ [5] leading to an increase of the elementary pinning force with the decrease of oxygen deficiency; (ii) the concentration of oxygen vacancies simultaneously decreases and consequently one should expect a decrease of the current near the fully oxidized state, if pinning by oxygen vacancies dominates. The obtained results show that at high temperatures this pinning becomes important for the pure sample. However, in sample 2 with larger concentration of background impurities, oxygen vacancies pinning is not dominant and the behavior of the current is mostly determined by the first factor - the increase of the elementary pinning force. Recently, using direct detwinning of the sample we have shown that the behavior of j_s in the intermediate and low temperature intervals is determined by twins [6]. The obtained results show that pinning by twin boundaries is dominated by the first factor (i) and, therefore, monotonously decreases with the oxygen content.

The fishtail peak shows a large sensitivity to oxygen content, but the shape of $j_s(H)$ dependence practically does not change. This contradicts to the relation of the fishtail with oxygen deficient phase [1]. This peak can originate from a synchronization effect [7] of the vortex lattice with point-like defects due to the vanishing of C_{66} at the melting transition. In sample 3 it is, probably, produced by oxygen vacancies and in sample 2 by background impurities. The position of the second peak is nearly temperature independent. Unlike to the fishtail peak (Fig. 2a) it shows almost no sensitivity to the oxygen content (Fig. 2b). This points to a matching effect, presumably, between intervortex and twin boundaries distances. According to STM studies of the YBCO single crystals the quasi-periodic twin microstructure is charaterized by the distances $d = 20 \div 50$ mm (e.g. [8]) which corresponds to the matching magnetic field of $H_m \sim \phi_0/d^2 \sim 10$ KOe in agreement with the experiment. The competing pinning by point-like disorder suppresses this matching effect. It disappeares in sample 3 at high oxygen deficiencies and is absent in the sample 2 with strong pinning by impurities.

5. Acknowledgements

This work was supported by the Bundesministerium für Forschung und Technology under Grant No. 13N6177, by the Russian Fund for Fundamental Researcher, Grant No. 93-02-14768, by the Russian Government and the International Science Foundation, Grant MSW 300, Engineering and Physical Sciences Research Council and by NATO Linkage Grant No. HT931241.

References

[1]   Daeumling M et al. 1990 *Nature* (London) **346** 332
[2]   Moshchalkov V V et al 1990 *J. Magn. Magn. Mater.* **90 & 91** 611
[3]   Zhukov A A et al. 1994 *Physica C* **235-240** 2837
[4]   Hauff R et al. 1994 *Physica C* **235-240** 1953
[5]   Ossandon J G et al. 1992 *Phys. Rev. B* **45** 12534
[6]   Zhukov A A et al. will be published in *Phys. Rev. B*
[7]   Pippard A B 1969 *Philos. Mag.* **19** 217
[8]   Rand M et al. 1993 *Cryogenics* **33** 291

Inst. Phys. Conf. Ser. No 148
Paper presented at Applied Superconductivity, Edinburgh, 3–6 July 1995
© *1995 IOP Publishing Ltd*

Field Dependence of Current Density in Strong Pinning YBCO Single Crystals with Different Microstructure

R. Hiergeist and R. Hergt

Institut für Physikalische Hochtechnologie e.V. Jena, PF 100239, D-07702 Jena, Germany;

A. Erb

Université de Genève, 24 quai Ernest-Ansermet, CH-1211 Genève 4, Switzerland.

Abstract. $YBa_2Cu_3O_{7-\delta}$ single crystals with high critical current densities and different microstructure of the twin pattern were studied by means of torque magnetometry. Pinning contributions due to random point pinning and to correlated disorder as well as the influence of thermal oscillations of the fluxlines within the pinning barriers are discussed. Different collective pinning regimes of the fluxline lattice which are characterized by different power law behaviour of the form $j(B_z) \propto B_z^{p-1}$ could be identified.

1. Introduction

High current densities are necessary suppositions for power applications of high-T_c superconducting materials. However serious restrictions arise due to the influence of thermal fluctuations on pinning properties by (i) thermally activated jumps between different metastable states of the flux line lattice which leads to significant fluxcreep effects, and (ii) thermal fluctuations of the fluxlines within the pinning barriers which smear out the elementary pinning potential [1]. Concerning the second point we report in this paper on experimental investigations of $YBa_2Cu_3O_{7-\delta}$ single crystals with different microstructure by means of torque magnetometry.

2. Experimental

The microstructure of the twin pattern of several $YBa_2Cu_3O_{7-\delta}$ single crystals which were grown under similar conditions [2] was characterized by polarization microscopy. The samples were investigated by means of torque magnetometry (e.g. [3]). The angular dependence of the torque hysteresis ΔG measured for stationary rotation of the external magnetic field H was used to derive the irreversible part of the magnetization M_{irr}, the current density j, and the pinning force volume density F_p according to

$$M_{irr} = \Delta G/(\mu_o HV\sin(\theta)) \quad (1a) \qquad j = k \cdot M_{irr}/d \quad (1b) \qquad F_p = j \cdot B_z \quad (1c)$$

(V = sample volume, θ = magnetic field direction with respect to the c-axis, d = sample width,

316

k = geometry dependent factor), which is valid under the assumption that there is only a magnetization component parallel to the c-axis [3]. From torque data measured at a temperature of 77.4K the dependence of current density on the component B_z of the flux density parallel to the c-axis direction ([001] direction) was calculated. All torque measurements were performed under a fixed voltage criterion of $E \approx 2.5 \ 10^{-8}$ V/cm.

3. Results

In Fig. 1 the dependence of the pinning force volume density on B_z is shown for a YBa$_2$Cu$_3$O$_{7-\delta}$ single crystal with only one twin domain. Here we find a transition of the

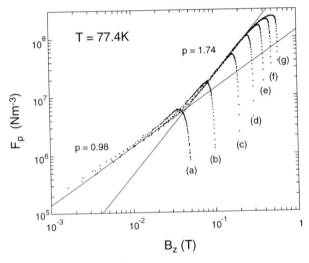

Fig.1 Pinning force volume density in dependence on B_z for different values of the external magnetic field of 40kA/m (a), 80kA/m (b), 160kA/m (c), 240kA/m (d), 320kA/m (e), 400kA/m (f), and 480kA/m (g). (Ignore the rapid drop of the F_p data for different H at high B_z which is caused by the minorloop characteristics of the torque magnetometry sweeps.)

power p in the power law $F_p(B_z) \propto B_z^p$ from $p=1$ (i.e. j = const) to $p=1.75$ (i.e. $j \propto B_z^{3/4}$) which can be interpreted in the frame of the collective pinning theory as a transition from the single vortex pinning regime to the regime of collective pinning of small flux bundles at low depinning temperatures $T_L^* << T$ and is discussed elsewhere [4,5]. In the case of very strong pinning YBa$_2$Cu$_3$O$_{7-\delta}$ single crystals with a dense twin pattern and crossing twin boundaries a transition from the single vortex pinning regime ($p=1$) towards a fishtail behaviour which is characterized by $p=2$ (i.e. $j \propto B_z$) can be observed (c.f. Fig. 2). For samples with crossing twin boundaries but only medium pinning strength a transition from a $j \propto B_z^{-1/4}$ behaviour towards a linear increase of j with B_z was found which is shown in Fig.3. The current density in this sample is more than one order of magnitude smaller than in the sample of Fig.2.

4. Conclusions

Besides the "fishtail"-effect with $p=1.75$ which is caused by collective pinning of small flux

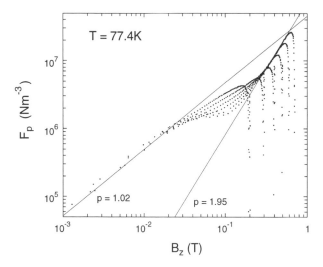

Fig.2 Transition from $p = 1$ to $p = 2$ in the pinning force volume density of a YBa$_2$Cu$_3$O$_{7-\delta}$ single crystal with high pinning strength and many crossing twin boundaries.

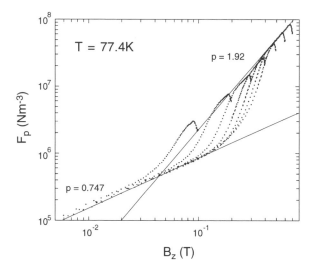

Fig.3 Transition from $p = 3/4$ to $p = 2$ in the pinning force volume density of a YBa$_2$Cu$_3$O$_{7-\delta}$ single crystal with crossing twin boundaries and only medium pinning strength. Note the bad scaling in the region of intermediate B_z values.

bundles at weak random point pinning centers, there exists a second type of "fishtail"-effect which is characterized by a linear increase of the current density with B_z. This linear increase of j can be explained by strong pinning of small flux bundles at $T \lll T_L^*$. To proof this statement the data of $\ln(j/B_z)$ of a very strong pinning heavily twinned YBa$_2$Cu$_3$O$_{7-\delta}$ single crystal are plotted in Fig.4 together with the fit data of a nonlinear fit with Equ.(10) of reference [1]. In the field range beyond the fishtail-peak, there is a good agreement of the power of B_z in the

318

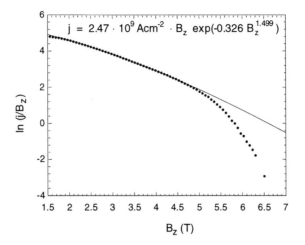

Fig.4 Nonlinear fit in the fieldrange between $B_z = 2.5T$ and $B_z = 4.5T$ of $\ln(j/B_z)$ data calculated from field sweep measurements [5] of a heavyly twinned $YBa_2Cu_3O_{7-\delta}$ single crystal. Note the good agreement of the power of B_z in the exponent (1.499) with the value 3/2 predicted by Equ.10 of reference [1]. {Ignore the rapid drop of the data at high B_z which is caused by the minorloop characteristics of the torque magnetometry sweeps.}

exponent with the value 3/2 which follows from Equ.(10) in reference [1] under the condition of $T_L^* > T$. In the case of the sample of Fig.3 the transition from the $j \propto B_z^{-1/4}$ regime at low B_z (at very low current densities) towards the "fishtail"-regime depends on the vortex orientation in the sample: For different values of H we find the inflection point of the $j(B_z)$ curve for the same angular orientation of the flux lines ($\theta \approx 52°$). This behaviour can be interpreted as a lock in of the flux lines in the linear defects which are given by the regions of high stress at crossing twin boundaries of different orientations [5]. As a consequence of this transition the pinning becomes stronger correlated and the influence of the thermal fluctuations of the flux lines within the pining centers is reduced which results in higher values of T_L^*

Acknowledgements

The authors greatly appreciate the opportunity to perform a part of the measurements in the Laboratory of Solid State Physics at the Free University of Amsterdam. This work was supported by the BMFT under contract No. 13N6100

References

[1] M.V. Feigel'man, and V.M. Vinokur 1990 *Phys. Rev. B* **41** 8986
[2] See e.g. H.Claus et al. 1992 *Physica C* **198** 42-46
[3] R.Hergt et al. 1993 *Phys. Rev. B* **47** 5405
[4] R. Hiergeist et al. 1994 ,*Proceedings of the 7th International Workshop on Critical Currents in Superconductors* 24. - 27. Januar 1994, Alpbach, Austria, World Scientific Singapore (Editor H. W. Weber) 225
[5] R. Hergt et al. *Physica C* **235–240** (1994) 2751–2752

Inst. Phys. Conf. Ser. No 148
Paper presented at Applied Superconductivity, Edinburgh, 3–6 July 1995
© 1995 IOP Publishing Ltd

Comparative Study of Anisotropy of the Critical Current Density in YBCO Melt Textured Samples and Single Crystals

L.M. Fisher[†], A.V. Kalinov[†,*], J. Mirkovic[†,*], I.F. Voloshin[†], A.V. Bondarenko[‡], M.A. Obolenskii[‡], and R.L. Snyder[♯]

[†]All-Russian Electrical Engineering Institute, 111250 Moscow, Russia
[‡]Kharkov State University, Kharkov, Ukraine
[♯]New York State College of Ceramics at Alfred University Alfred, NY, USA,
[*] Physics Department of Moscow State University, 117234 Moscow, Russia.

Magnetic field and angular dependencies of the critical current density J_c are studied for YBCO single crystals with qualitatively different arrangement of twin boundaries (TBs) and two types of bulk textured slabs differing in c-axis orientation. We observe a qualitative difference of the angular dependence of the critical current density J_c for the textured samples and single crystals. If the external dc magnetic field **H** lies within **ab** plane of a crystal, the J_c vs. angle ϑ curves exhibit a narrow peak in the vicinity of **H** ∥ TBs. The observed features of J_c are discussed.

1. Introduction

The critical current density J_c, in high-T_c superconductors is connected with the structural defects in these materials. It is commonly supposed that the anisotropy of the magnetic field dependence of $J_c(\mathbf{H})$ is appreciably determined by this structure. In melt textured (MT) samples the intercrystallite planes are well known to be the main type of pinning sites and they are parallel to the **ab** plane. The twin boundaries (TBs) which are parallel to the **c** axis are the dominant defects in the structure of single crystals and they may play a certain role in pinning processes in MT samples also. This difference in structure of defects in textured samples and single crystals stimulated us to comparative study of the anisotropy of J_c in these materials using measurements of the low frequency ac magnetic susceptibility χ of bulk superconductors.

TBs are apparently the structural defects, however, the contribution of TBs to the critical current density J_c is far from being clearly understood. There is still a certain controversy concerning the TBs-induced pinning force: some papers treat TBs as strong enough pinning centers [1, 2], the other suggests them to be channels for vortex penetration [3]. However, magneto-optical imaging studies [2] and results [1, 4] clearly demonstrate, that the pinning at TBs is efficient when the magnetic field is oriented along these defects. Usually, the experiments aimed at revealing TBs related phenomena are performed in such a way that the angle φ between **H** and **c** axis is varied [4]. For this geometry, however, the large contribution of intrinsic **ab-c** anisotropy [5] masks the effects that stem from TBs themselves. Therefore, we present here the results demonstrating clearly pronounced J_c anisotropy in the case when **H** lies in **ab** plane and can be rotated continuously at an angle ϑ with respect to TBs.

2. Experiment

The textured samples were cut from a bulk ingot ($4 \times 7 \times 60$ mm^3) and had the form of thin (0.4 mm) slabs with a preferred orientation of the **c** axis either along the length (width) of the sample (Fig 1a, 1b, and 1c) or normal to the slab (Fig. 1d and 1e). The slab contained no more than 2 macrodomains having a relative misorientation of the **c** axis of less than 5 degrees. SEM micrographs for slabs with the **c** axis directed parallel to the slab surface and to its normal **n** shows that the sample consists of extended crystallites with thickness about 1-4 μ in the **c** direction. TEM measurements showed the characteristic thickness of the interlayers between crystallites to be about 3 nm.

We studied the anisotropy of the single crystals using mainly the sample with size $0.06 \times 1.2 \times 3$ mm^3 and two high quality YBa$_2$Cu$_3$O$_x$ single crystals $2 \times 1 \times 0.06$ mm^3 and $2 \times 0.9 \times 0.04$ mm^3 in size. Further on they are referred to as Y1, Y2, and Y3, respectively. The onset temperature of the superconducting transition was 91.2 K and its width was about 0.6 K. Our single crystals are characterized by qualitatively different arrangement of TBs. Usually YBCO single crystals involve only small region where TBs are parallel to each other (we refer to as a polytwin block or domain), TBs in neighboring blocks have orthogonal orientation. The TBs pattern in our samples is clearly observed using our conventional optical microscope. At a certain direction of the incident light one can see parallel lines at an angle about 45^0 to the sample edge. The samples Y1 and Y2 have an approximately equal amount of such domains whereas Y3 is actually a single polytwin block and a volume fraction of orthogonal domains is little more than 5%.

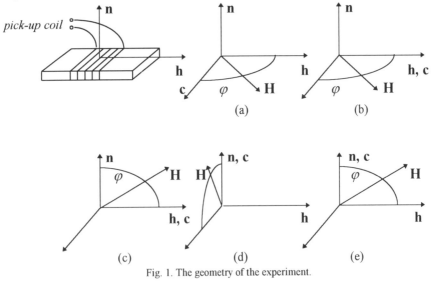

Fig. 1. The geometry of the experiment.

J_c of our samples was determined by the contactless method [6]. This method is based on the one-to-one relation between the dynamic magnetic susceptibility χ and the local value of the critical current density. According to the critical state model, the susceptibility of a hard superconductor placed into an external magnetic field **H** is inversely proportional to $J_c(H)$ if the ac field does not penetrate the whole sample bulk and is strictly proportional to $J_c(\mathbf{H})$ in the opposite case [6]. This result is valid if H is much higher than the ac magnetic field amplitude h. The value of h was chosen empirically to be about 1-10 mT at $T = 77$ K so that the critical state for the ac field was established. The frequency $\omega/2\pi$ of the ac field was in the range 10-1000 Hz.

3. Results and discussion

The method chosen to study the anisotropy of $J_c(\mathbf{H})$ turned out to be very suitable and effective. We have found that the ac response of a sample does not depend practically on the angle between the ac and dc magnetic fields over the wide range of the experimental parameters investigated. This conclusion is correct if the dc magnetic field is strong enough to penetrate the whole sample bulk. Moreover, if H completely penetrates the slab in any direction, the ac susceptibility will be defined by the relative orientation of the vector $\mathbf{H}$ and the crystallographic $\mathbf{c}$ axis only, and does not depend on the angles between $\mathbf{H}$ and $\mathbf{n}$ or $\mathbf{h}$ and $\mathbf{c}$ axis. The corresponding angular dependence of $J_c(\mathbf{H})$ presented in Fig. 2 is characteristic for bulk textured materials and epitaxial thin films. According to our results, J_c decreases with H proportional to $H^{1/2}$ at $H > 0.03$ T which is typical for the surface pinning.

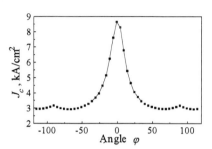

Fig.2. The anisotropy of J_c in the geometry 1(a), $T = 77$ K, $\omega / 2\pi = 130$ Hz, $h = 7$ mT

The anisotropy of J_c for the single crystal, at different temperatures and magnetic fields, is presented in Fig. 3. It differs qualitatively from the anisotropy for melt-textured samples and thin films. J_c has a maximum value at $\mathbf{H} \parallel \mathbf{c}$ for the single crystal, whereas it reaches a maximum value at $\mathbf{H} \perp \mathbf{c}$ for the melt-textured samples. J_c in the single crystal has a small local maximum at $\mathbf{H} \perp \mathbf{c}$ also. Another specific feature of the angular behavior of J_c in the single crystal is the appearance of the shoulders near the main maximum with the increase of the dc field H (the solid lines). We found that the field dependence $J_c(H)$ is

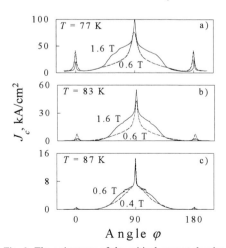

Fig. 3. The anisotropy of the critical current density for the single crystal Y1.

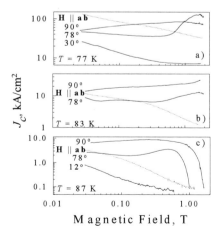

Fig. 4. The critical current density in the single crystal Y1 as a function of H at different angles between **ab** plane and **H**.

nonmonotonous in the vicinity of the **c** direction including the shoulders mentioned $|\varphi-90^{0}| \leq 50^{0}$ at 77 K, i.e. it has a 'fishtail' form (see, for example, [7]). This angle region is reduced when the temperature increases. The dependencies of the critical current density on H for different **H** directions and temperatures are shown in Fig. 4. J_c practically independent of H over a broad range of H and current decreases abruptly to zero at some magnetic field value H^* depending on the temperature. This a behavior is clearly demonstrated by the upper curve in Fig. 4c. We would like to note that the width of the angular region where the shoulders of $J_c(\varphi)$ exist correlates with the critical angles of the vortex pancake theory [8].

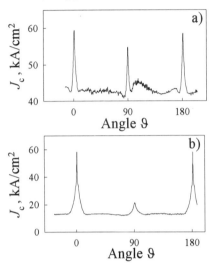

Fig. 5 The critical current density *vs.* angle ϑ between the magnetic field **H** and the twin boundary in **ab** plane at $H = 0.8$ T and T=77 K for (a) the sample Y2 and (b) the sample Y3.

To study the anisotropy of J_c in **ab** plane sample Y2 or Y3 was put into the dc magnetic field $\mathbf{H}_{dc}$ which was also strictly parallel to **ab** plane (the deviation was as small as $1/6^0$). Then the angle ϑ between **H** and the TBs direction was continuously changed to yield the required dependence $J_c(\vartheta)$, whereas **H** remained orthogonal to the **c** axis. The angle ϑ was measured from the TBs (about [110] direction).

Our $J_c(\vartheta)$ curves (Fig. 5) reveal the clearly pronounced enhancement of J_c at a certain angle ϑ, closely related to the orientation of TBs. In the sample Y1, with two orthogonal directions of TBs, $J_c(\vartheta)$ exhibits peaks of the same amplitude at intervals of 90^0 at $\vartheta = \vartheta_{max} = 0^0$, 90^0, 180^0 (Fig. 5a). Another behavior is observed for the sample Y2 (a single domain): its $J_c(\vartheta)$ peaks with equal heights appear at intervals of 180^0, the other peak at 90^0 is six times lower (Fig. 5b). The very good correlation of these peak positions with the twin arrangement suggests that TBs are the very strong pinning centers at **H** $\parallel$ TBs.

This work was supported in part by the International Science Foundation (grant N3F000) and Russian State Program on Superconductivity (grant 93027).

References

[1] Gyorgy E M *et al* 1990 *Appl. Phys. Lett.* **56** 2465-67.
[2] Turchinskaya M *et al* 1993 *Physica C* **216** 205-210.
[3] Oussena M, de Groot P A J, Porter S J, Gagnon R and Taillefer L 1995 *Phys. Rev. B* **51** 1389-92.
[4] Flippen R B and Askew T R 1994 *Physica C* **231** 352-356.
[5] Fisher L M *et al* 1995 *Appl. Supercond.* will be published.
[6] Fisher L M *et al* 1992 *Phys. Rev. B* **46** 10986-96.
[7] Daeumling M, Seuntjens J M, Larbalestier D C 1990 *Nature* **346** 332-335
[8] Blatter G, Feigel'man M V, Geshkenbein V B, Larkin A I and Vinokur V M 1993 *"Vortices in high temperature superconductors"* (Pergamon Press, London-New York)

Inst. Phys. Conf. Ser. No 148
Paper presented at Applied Superconductivity, Edinburgh, 3–6 July 1995
© *1995 IOP Publishing Ltd*

Trapped magnetic flux density in the Bi-2223 superconducting tubes

V Plecháček and J Hejtmánek

Institute of Physics , Cukrovarnická 10, 162 00 Praha 6, Czech Republic

Abstract.   Superconducting tubes were prepared from the $(Bi,Pb)_2Sr_2Ca_2Cu_3O_{10+\gamma}$ superconductor. The coaxial arrangement of five tubes of overall wall thickness of 9.4 mm traps magnetic flux density of 0.445 T at temperature of 20 K. The temperature dependence of apparent pinning potential is discussed in a frame of 3D-2D crossover and using isotropic-intrinsic pinning model.

1. Introduction

High T_c oxide superconductors, particularly $(Bi,Pb)_2Sr_2Ca_2Cu_3O_{10+g}$, provide to operate the electronic power devices and magnets at temperatures above the capability of conventional low-temperature superconductors. However, the fabrication of coils complicates the difficult processing of HTS wires or tapes. Magnetic measurements on a bulk 2223 superconducting material processed in a form of tubes, which we present here, entitles to consider Bi-2223 based tubes as a competitive electronic component with promising applications.

2. Experimental

A series of $(Bi,Pb)_2Sr_2Ca_2Cu_3O_{10+g}$ superconducting tubes with gradually increasing diameter were prepared when the fabrication procedure was oriented to enable their co-axial arrangement. The cold isostatic pressing and subsequent thermomechanical processing resulted in the priority orientation of the grains c-axis perpendicular to the tube axis [1,2] . The selected properties of the tubes are displayed in Table 1.

Table 1. Selected properties of Bi-2223 superconducting tubes at 77 K

Tube	Inner diameter (mm)	Wall thickness (mm)	Length (mm)	B_{tr} at 100 s (T)	Self field J_c (Acm^{-2})
I	6.4	1.6	31	0.0232	1130
II	9.9	1.6	39	0.0233	1190
III	15.5	2.2	42	0.0263	1030
IV	20.1	1.8	35	0.0222	1180
V	25.2	2.2	39	0.0237	1240

324

Table 2. Properties of Bi-2223 superconducting tubes in a coaxial arrangement

Arrangement	Overall wall thickness (mm)	Trapped mg.flux density at time 100 s (T) Temperature			Self field J_c (Acm^{-2}) Temperature		
		20 K	40 K	80 K	20 K	40 K	80 K
I	1.6	0.154	0.104	0.02	7910	5340	1020
I+II	3.2	0.229	0.156	0.03	5910	4010	770
I—V	9.4	0.445	0.307	0.05	4150	2870	460

The temperature dependencies of the trapped magnetic flux density B_{tr} were measured in the centre of coaxially arranged tubes using the close cycle refrigerator in the temperature range from 13 K to 100 K. After applying a pulse of an external magnetic field (up to 0.8 T) oriented parallel to the tube axis a time development of the B_{tr} of a various number of co-axially arranged tubes was detected. As a magnetic and temperature probes the Hall sensor and the silicone diode were used, respectively.

3. Results and discussion

The trapped magnetic flux densities B_{tr} (after the 100 s of relaxation) and estimated self field critical current densities for three different arrangements are given in Table 2. An increase of the trapped magnetic flux densities both with decreasing temperature and increasing number of the coaxially arranged tubes, i.e. with increasing overall wall

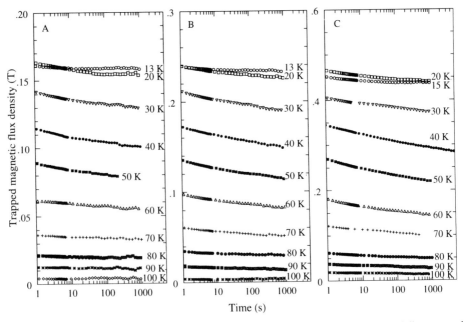

Fig1. Time decay of trapped magnetic flux density for various numbers of coaxially arranged superconducting tubes - 1 tube (A), 2 tubes (B) and 5 tubes (C).

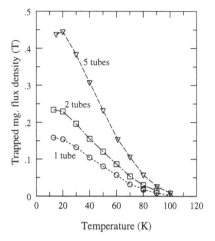

Fig.2. Trapped magnetic flux density vs temperature in the centre of three arrangements of the superconducting tubes.

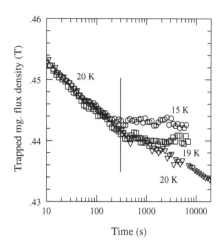

Fig.3. Time decay of trapped magnetic flux density in the centre of arrangement of five tubes at temperature of 20 K changed to 15 K (o), 19 K (□) and unchanged (∇) at the time of 300 s.

thickness, is further demonstrated in Fig.1 and Fig.2. Apparently small magnetic relaxation at temperature above 15 K (see Fig.1) together with small values of B_{tr} (see Fig.2) could be explained by magnetic flux jumps which due to a strong decrease of c_p below 20 K should lead to temperature fluctuations. This effect is further evidenced in Fig.3 where the relaxation was seemingly inhibited by the step decrease of temperature at 300 s after the switching off the external magnetic field from 20 K to 19 K and 15 K, respectively. Let us note that by the same manner nearly time independent magnetic field could be reached.

A strong logarithmic magnetic relaxation with time can be explained by a thermally activated flux creep [3]. This simple approximation enables to calculate the apparent pinning potential using the formula

$$U^* = -kTB_{tr}/(\partial B_{tr}/\partial \ln t).$$

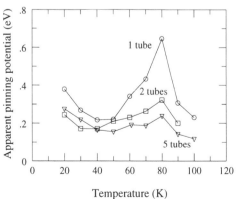

Fig.4. Temperature dependencies of apparent pinning potential for three various arrangements of the superconducting tubes.

Temperature dependencies of the U^* calculated for self magnetic flux density of the respective superconducting tubes are shown in Fig.4. All the temperature variations exhibit two extremities, a maximum at 80 K and a minimum at around T=40 K.

The unusual temperature dependence of the U^* could be explained, in spite of certain degree of priority orientation of the grains c-axis perpendicular to the magnetic field, in a frame of 3D-2D crossover [4]. Bi-based superconductors with enhanced layered crystal structure are known to exhibit 3D-2D crossover in case of vortex lattice oriented parallel to the c-axis. In the low temperature region (3D) with dominating stronger vortex-vortex interaction between individual (CuO_2) layers the flux lattice penetrates the whole sample. When the temperature is increased the thermal fluctuation energy is comparable to the magnetic interactions between the layers causing their magnetic decoupling and leading hence to the 2D behavior. It is worth to note that at temperatures when 3D-2D transition occurs a weakened interlayer links start to contribute as pinning centres to the whole pinning potential which leads to nonuniform temperature dependence of the pinning potential.

Nevertheless the complicated temperature dependence of the apparent pinning potential could be also accounted to the apparent transition of the pinning potentials as presented in the work carried out on the Bi-2212 whiskers using magnetic fields parallel to the a-axis [5]. Temperature dependence of the pinning potential at low temperatures originates mostly from the isotropic pinning with small potentials (oxygen vacancies, cationic substitutions, crystal defects, grain boundaries,.) while at temperatures above 40 K the intrinsic pinning arising from superconductive coupling between (CuO_2) planes through (Bi_2O_2) layers with high pinning potentials takes place. Taking into account the preferred orientation of the a-axis of sheet-like grains parallel to the magnetic field the latter explanation of the temperature variation of the pinning potential seems to be more probable.

4. Conclusion

The analysis of magnetic relaxation measurements was performed using Bi-2223 based superconducting tubes and the temperature dependence of the apparent pinning potential was explained using the model of isotropic-intrinsic pinning. It was shown that a system of five tubes with inner diameter of 6.4 mm, length of about 40 mm and overall wall thickness 9.4 mm can trap magnetic flux density more than 0.4 T at around 20 K. This fact indicates that superconducting magnets manufactured from the bulk Bi-based superconducting material could be in some cases a serious competitor of the HTSC superconducting coils.

References

[1] Plecháček V 1992 *Cryogenics* **32** 1010-.1013
[2] Plecháček V, Hejtmánek J, Sedmidubský D, Knížek K and Pollert E 1994 *Physica C* **235-240** 3461-3462
[3] Anderson P W 1962 *Phys.Rev.Lett.* **9** 309-.311
[4] Pradhan A K, Roy S B, Chadda P, Chen C and Wanklin B M 1994 *Phys.Rev.B* **49** 12 984-12 989
[5] Funahashi R, Matsubara I, Yamashita H, Kawai T 1994 *Appl.Phys.Lett.* **64** 646-648

*Work was supported by the Grant Agency of the Czech Republic (grant No.106/93/1152) and by the Copernicus project.

Inst. Phys. Conf. Ser. No 148
Paper presented at Applied Superconductivity, Edinburgh, 3–6 July 1995
© *1995 IOP Publishing Ltd*

AC susceptibility crossovers in composite $CuO/YBa_2(Cu_{1-x}Fe_x)_3O_{7-\delta}$.

M. Mehbod, J. Schroeder and I. Grandjean

Physique des Solides, CP 233, Université Libre de Bruxelles B-1050 Brussels Belgium.

M. Ausloos

S.U.P.R.A.S., Université de Liège,Institut de Physique B5, Sart Tilman B- 4000 Liège, Belgium.

Abstract AC susceptibility of $YBa_2(Cu_{1-x}Fe_x)_3O_{7-\delta}$ polycrystals with y% CuO (y =0 or 5) extrinsic impurity with $0 \leq x \leq 0.03$ was measured as a function of temperature and magnetic field strength. The behavior of the intergrain loss temperature T_i (secondary maximum of χ'') was analyzed through the relationship $1-t = a H^q$. Pure samples were characterized by two values for q. With increasing copper oxide concentration a crossover took place at different values of H_{ac}. Fe doped samples with or without CuO revealed three values of q, - the crossovers being at different and other H_{ac} values. Computing the pinning force densities for these samples, we found 5000 TA/m^2 for pure YBCO. For the doped sample with 5 % CuO a higher value (402 TA/m^2) than that for the iron doped one (156 TA/m^2) was found. This variation is interpreted as being due to the 'dilution' of iron into the excess CuO at grain boundaries. The results confirm the suggested model in reference 1 which was based on the effect of iron induced weak link creation inside the grains.

1. Introduction

In order to obtain informations about energy losses in high T_c superconductors like $YBa_2Cu_3O_{7-\delta}$ we measured the AC susceptibility on such materials with either intrinsic (Fe) or extrinsic (CuO) "impurities" or both simultaneously. From these measurements we determined the line: $1-t = a H^q$ with $t =T_i/T_c$, where T_i is the temperature associated with the maximum of the intergranular loss peak in the materials; a and q are parameters which gave us informations about magnetic energy losses.

The behavior of the hereby called T_i-line (T_iL) can be explained within i) the flux creep model, ii) Kim-Anderson's model and iii) Kim-Anderson's model extended to the case where the penetration length λ is larger than the characteristic grain size R_g. [1] Each model is associated to a different value of the parameter q, i.e., 2/3 for flux creep, 1 for Kim-Anderson's model and 4/3 for the extended Kim-Anderson's model.

We have observed two q values for $YBa_2Cu_3O_{7-\delta}$ + y% CuO (y = 0, 5) when going from low to high AC fields (like El-Abbar et al. [2], three for $YBa_2(Cu_{0.97}Fe_{0.03})_3O_{7-\delta}$ and three also for $YBa_2(Cu_{0.97}Fe_{0.03})_3O_{7-\delta}$ with 5% CuO. The present work was indeed performed in order to verify whether (universal) asymptotic regimes really exist also for non trivial materials, and are understandable along the theoretical lines which were proposed in ref. 1. The chemical modifications brought into classical 123 systems, on the other hand, might not be trivial perturbations, and it should be of interest to observe the predicted behavior in the (t, H) plane for non universal quantities like the crossover fields and the pinning force density.

328

2. Sample preparation and characterization

All samples were prepared by solid state reactions as described in reference 1. The structure of the different samples was controlled by X.Ray, scanning electron microscopy (S.E.M.) coupled to an energy dispersive X-ray analyser (E.D.A.X.). No secondary phase, up to the resolution inherent to the techniques, was present except that due to the excess of copper oxide in the sample containing 5% CuO. Iron has previously been shown to be substituted onto Cu sites in the YBaCuO grains.[3] On the contrary excess of CuO was localized between grains.[4] The AC susceptibility was measured with an AC Lakeshore Model 7000 susceptometer. Our measurements were made in zero DC magnetic field, for an AC magnetic field ranging between 1 and 800 A/m at 666.7 Hz.

3. Experimental results

The susceptometer allows us to obtain the real and the imaginary parts of the AC susceptibility (χ). The imaginary (χ'') represents the energy dissipation due to the motion of vortices. We define the critical temperature (T_c) as the temperature below which the imaginary part of χ is different from 0. The maximum of the intergrain peak is associated with the temperature (T_i).

This one depends on the value of the applied magnetic field according to the law: $1-t=a*H^q$ with $t=T_i/T_c$; a and q are parameters which give us informations about energy losses.

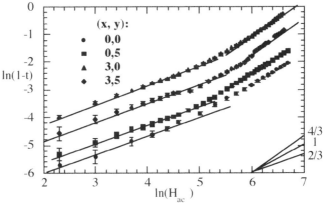

Fig.1: Logarithm of 1-t (where $t=T_i/T_c$) as a function of the logarithm of the applied magnetic field for four samples: $YBa_2Cu_3O_{7-\delta}$ with or without a 5% excess of copper oxide and 3% iron doped $YBa_2(Cu_{0.97}Fe_{0.03})_3O_{7-\delta}$ with or without a 5% excess of copper oxide. The theoretical exponents q=2/3, 1, 4/3 are shown.

The 2/3 value is connected to a flux creep behavior [5] which takes place for temperatures just below the critical temperature , i.e. for the lowest fields. The value q=1 is observed for higher AC fields, and for lower temperatures in agreement with Kim Anderson model. [6] The 4/3 value is explained by the fact that for the Fe doped samples the penetration length can be of the order of the doped grain size at high fields and low temperatures. [1]

For $YBa_2Cu_3O_{7-\delta}$ without CuO, we find the crossover field at H* =80 A/m. For $YBa_2Cu_3O_{7-\delta}$ + 5% CuO, we observe that the crossover from q=2/3 to q=1 (H*) is shifted toward higher AC fields, i.e. H* =140 A/ m as expected from arguments in ref.2. This shift is due to the fact that the excess of CuO tends to reduce the pinning force because of a diminution of the Josephson current in the superconductor grain-CuO-superconductor grain junction. Flux creep is therefore more effective at lower temperatures than in the compound with no excess of CuO.

For the doped sample without excess of CuO, H_2* for which q goes from 1 to 4/3 equals 350 A/m. In $YBa_2(Cu_{0.97}Fe_{0.03})_3O_{7-\delta}$ + 5% CuO the crossover value of H_{ac} is shifted to higher AC field, i.e. H_2* = 425 A/m. The 5% CuO has for result to dilute the iron concentration of the 123-Fe-doped grains. This leads to an enhancement of the critical temperature in fact. This dilution effect was confirmed by E.D.A.X. which showed that the concentration was reduced in the grains of $YBa_2(Cu_{0.97}Fe_{0.03})_3O_{7-\delta}$ with 5% CuO. For the Fe-doped samples with excess of CuO we also observe that the value of H_{ac} for which q moves for 2/3 to 1 (H_1*) is shifted to lower AC field, H_1* =70 A/m in comparison with that value for the doped samples for which H_1* =120 A/m.

4. Discussion

We expect that the pinning force density is higher in the doped samples with excess of CuO than in the iron doped samples. This is confirmed by using the equations of ref.1. We can determine the different parameters (a_i) in the same notations as in ref. 1.

For all cases the characteristic fields are $H_j = \phi_0/4\mu_0\lambda_g(0)R_g$. We determined the coefficients by fitting the T_jL curves with a least square method. All data are reported in Table I. From the coefficients a_i and assuming the average grain sizes and the London penetration depths given in the table we deduced the values of pinning force densities μ_j. They span the range 400-5000 TA/m^2. We found ca. 5000 TA/m^2 for pure YBCO. For the doped sample with 5 % CuO a value of 402 TA/m^2 was found to be higher than that for the iron doped one (156 TA/m^2).

	0%Fe,0%CuO	0%Fe,5%CuO	3%Fe,0%CuO	3%Fe,5%CuO
T_c (K)	91.8	91.8	76.6	88.1
a_1*10^{-4}	6.651	9.746	42.213	23.877
a_2*10^{-4}	1.566	2.354	8.185	3.811
a_3*10^{-4}	/	/	1.165	0.6205
R_g (10^{-6}m)	3	3.3	1	2
λ_g (10^{-6}m)	0.6	0.6	1	1
μ_j (TA/m^2)	5238	2684	156	402

Table I: Summary of characteristic values of parameters used to derive the pinning force density μ_j: the critical temperature T_c, the coefficients of the theoretical T_jL law a_i, defined in ref. 1, the grain radius R_g and the zero temperature penetration depth $\lambda_g(0)$.

This variation can be interpreted as being due to the 'dilution' of iron into the excess CuO at grain boundaries, whence in a reduced iron concentration with respect to the nominal one in the Fe-doped grains. This is in agreement with EDAX data.[4]

In summary, we considered whether the various regimes once observed in the behavior of the temperature T_i corresponding to the maximum of the imaginary part of the ac susceptibility of high temperature superconductors , and usually attributed to intergain losses, could hold in non trivial samples. We thus measured the AC susceptibility of polycrystalline $YBa_2(Cu_{1-x}Fe_x)_3 O_{7-\delta}$ with x equals 0 or 3 % and containing 0 or 5 % of copper oxide. We observed one crossover in the T_iL for $YBa_2Cu_3O_{7-\delta}$ + y% CuO in agreement with ref.2 and two crossovers for $YBa_2(Cu_{0.97} Fe_{0.03})_3 O_{7-\delta}$ with and without CuO just like in ref.1. The latter work and arguments given therein seem thus more general and appropriate.

A hierarchy of pinning force density was derived. We found a decrease in the YBaCuO pinning force density under addition of CuO resulting into enhancement of flux creep. For iron doped samples we found a decrease in the pinning force density compared to the pure YBaCuO and an enhancement of the pinning force density for the iron doped sample with 5% of excess of copper oxide as compared to $YBa_2(Cu_{0.97}Fe_{0.03})_3O_{7-\delta}$ without extrinsic impurity. The above work points out to quite quantitative informations of interest in such systems as derived according to the data analysis of $T_i(H)$.

5.   Acknowledgements

Thanks are due to Professor R. Deltour for his continued interest in our work. Part of this work has been financed through the Incentive Program on High Temperature Superconductors supported by the Federal Services for Scientific, Technical and Cultural (SSTC) Affairs under contracts SU/02/09 and SU/02/013. One of the authors (J.S) .is supported by an F.R.I.A. fellowship

References

[1] Mehbod M, Sergeenkov S, Ausloos M, Schroeder J and Dang A 1993 *Phys.Rev.B* **48**, 483-86.

[2] El-Abbar A A, King P J, Maxwell K J, Owers-Bradley J R and Roys WB 1992 *Physica C* **198**, 81-86.

[3] Krekels T, Van Tendeloo G, Broddin D, Amelinck S, Tanner L, Mehbod M, Vanlathem E and Deltour R 1991 *Physica C* **173**, 361-76.

[4] Grandjean I 1993 Unpublished Mémoire de fin d'études, Université Libre de Bruxelles.

[5] ShindeS L, Morril J, Goland D, Chance D A and Mc Guire T 1990 *Phys.Rev.B* **41** 8838-42.

[6] Müller K-H 1989 *Physica C* **159** 717-26.

Inst. Phys. Conf. Ser. No 148
Paper presented at Applied Superconductivity, Edinburgh, 3–6 July 1995
© 1995 IOP Publishing Ltd

Characterisation of the partial melt processing of Bi-2212

C.G. Morgan,[1] **M. Priestnall,**[2] **N.C. Hyatt**[2] **and C.R.M. Grovenor**[1]

[1]Department of Materials, University of Oxford, Parks Road, Oxford, OX1 3PH, U.K.
[2]Cookson Technology Centre, Sandy Lane, Yarnton, Oxford, OX5 1PF, U.K.

Abstract.   The melting behaviour of high phase purity, stoichiometric Bi-2212 powder on silver in air has been characterised using simultaneous thermal analysis (STA) and hot stage microscopy (HSM). Short Bi-2212/Ag tapes have been prepared via electrophoretic deposition and partial melt processing over a series of temperatures from 870°C to 900°C. The dependence of microstructure on the peak temperature has been investigated using scanning electron microscopy (SEM) and related to the STA and HSM data. The highest critical current densities of greater than $3000Acm^{-2}$ at 77K are reproducibly obtained for tapes processed in the temperature window where partial melting has occurred and the number and size of secondary phases are minimised. The critical current drops significantly when the tapes are processed at higher temperatures. It is concluded that STA and HSM can be used to predict the narrow temperature region within which the optimum partial melt temperature exists and that optimum heat treatment parameters vary for different powders and tape production routes.

1. Introduction

The Bi-2212 phase has been extensively studied for practical applications such as low temperature high-field magnets. It has a T_c of 90K, [1] supports high J_c values and irreversibility fields at temperatures below 30K [2] and can be partially melt processed on silver based substrates producing a dense and highly aligned superconductor layer on a strong and flexible tape. However, strontium calcium copper oxide needles and copper-free crystals are typically formed in the multiple phase equilibrium after melting [e.g. 3,4] and these phases are detrimental to high J_cs as they act as barriers to current flow. The processing window for achieving the best superconducting properties is narrow and the effect of powders of varying stoichiometries and processing routes on microstructure have not been fully determined.

2. Experimental

Stoichiometric, high phase purity Bi-2212 powder was made at Cookson Technology Centre using a sol-gel combustion route. Tapes were produced by electrophoretic deposition onto silver alloy substrates [5] and heated in a Carbolite furnace which was carefully calibrated to ensure that processes were repeatable to ±1°C. The heat treatment used was a ramp of 300°C/h to a peak temperature T_1 for 0.1 hours, cooling at 10°C/h to 830°C, annealing for 5 hours and cooling at 600°C/h to room temperature. The critical current at 77K, zero field was determined using a 4-point DC method. The critical temperature was measured using a 4-point AC method. The microstructure was examined using a Hitachi S530 SEM and a Philips 501 SEM with EDX.

Simultaneous thermal analysis (STA) was performed in silver crucibles using a Rheometrics 1500H analyser, allowing monitoring of both heat and weight changes. The

332

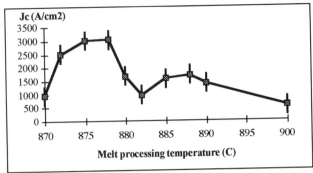

Figure 1: Variation of Jc with melt processing temperature

standard heat treatment was a ramp at 5°C/min to a peak temperature T_1 for 6 minutes and then cooling at 0.2°C/min. Hot stage microscopy was performed using a Linkam 1500H hot stage under a Nikon optical microscope with camera. Samples of powder were pressed onto a silver sheet approximately 3mm square, which was bonded to the sapphire substrate with silver electrodag to ensure good thermal contact. The hot stage was calibrated using the melting points of gold (1064°C) and silver (962°C in flowing nitrogen), and the calibration was checked by comparing with the powder melting points determined by STA.

3. Results

3.1 Tape production

Tapes made via electrophoretic deposition were melt processed over a range of peak temperatures from 870°C to 900°C. All tapes were deposited using the same conditions and were nominally identical before heating. The tapes had superconducting properties over the entire range, with critical temperatures consistently in the range from 88K to 90K. The best critical current densities (77K, 0T) of >3000Acm^{-2} were obtained for tapes processed in the temperature window between 875°C and 878°C. Values dropped significantly at temperatures above 880°C, although there was a second slightly improved region between 885°C and 890°C. (Figure 1)

3.2 STA and HSM on pure Bi-2212 powder

In the hot stage microscope, the powder was seen to peritectically decompose at a temperature of 880°C into a raft of tiny needles floating on the surface of a liquid phase. At 882°C, the surface of the liquid was completely covered with a mixture of small and large needles after a six minute dwell. At higher temperatures, the ratio of large needles to small needles increased. On cooling, most of the needles did not dissolve before recrystallisation occurs.

Small samples of powder pressed into silver crucibles were heated in the STA to 908°C in flowing air. The extrapolated onset of melting, distinguished by a large endothermic heat flow and a decrease in weight corresponding to oxygen evolution, was shown to be 880°C.

Powder samples were heated in the STA to a range of temperatures around 880°C, using a heat treatment designed to simulate that used in the production of tapes without the final annealing stage. Thus, the microstructure present after the melting and recrystallisation process can be investigated. SEM pictures of the samples clearly confirm that the powder

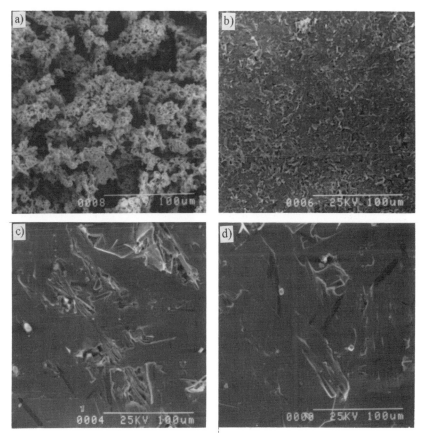

Figure 2: Microstructure of as-produced powder melt-processed at a) 877°C, b) 879°C, c) 881°C and d) 885°C

melts at a temperature between 879°C and 881°C. (Figure 2) Below the melt temperature the powder sinters together, and at higher temperatures large platelets (>100μm) of Bi-2212, $(Sr,Ca)CuO$ needles, $(Sr,Ca)_2CuO$ needles and small bismuth-free crystals are formed. The volume fraction of platelets is greatest at 885°C, approximately 5°C above the onset of melting.

3.3 Effect of residual carbon

The STA data on pure powder suggest that tapes cannot be partially melt processed at temperatures below 880°C. However, STA measurements on powder scraped from the green tape showed an extrapolated onset of melting at 873°C. By including initial burn-out periods of one hour at 835°C and five hours at 860°C, the onset of melting was increased to 878°C and 880°C respectively. This suggests that the melting onset in the tapes is significantly lowered either by residual carbon content from the coating process or its local effect on oxygen partial pressure, consistent with recent work in the literature. [6,7] A high temperature dwell must be included before partial melt processing, if all the carbon is to be burnt off.

STA and microstructural measurements were repeated on samples of the green tape coating over the temperature range 872°C to 882°C. The best microstructure, with large platelets and few needles present, was observed for the sample heated to a maximum temperature of 878°C, about 6°C above the onset of melting. (Figure 3) This corresponds exactly with the optimum processing temperature window for tapes.

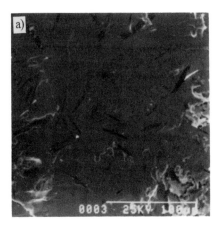 

Figure 3: Microstructure of powder from tape melt-processed at temperatures of a) 879°C and b) 887°C

4. Conclusions

From STA, the extrapolated onset of melting of pure Bi-2212 powder on silver is 880°C. By simulating typical furnace programs in the STA chamber, it is confirmed that at lower temperatures sintering occurs, and at higher temperatures the microstructure is made up of Bi-2212 platelets interspersed with SrCaCuO needles and small Cu-free phases. The optimum microstructure with large well connected platelets and a minimum of secondary phases is achieved at a processing temperature of 885°C, 5-6°C above the extrapolated onset of melting.

Bi-2212 tapes, manufactured using an electrophoretic deposition method, displayed superconducting properties over the wide range of processing temperatures from 870°C to 900°C, with the best current carrying properties over a narrow window from 875°C to 878°C. STA data for powder scraped from the green tape show a reduced onset of melting of 873°C. This value increases with initial high temperature dwells, suggesting that the reduction in temperature is due to residual carbon content. By simulating furnace programs in the STA, the best microstructure is found to occur at a temperature of 877-8°C. This is again approximately 5-6°C higher than the onset of melting, and corresponds to the temperature range where critical current densities are at a maximum in the processed tapes. Dip-coated tapes produced using a commercial powder show optimum J_c values over a temperature range from 874-6°C. [8]

STA can be used to measure the onset of melting for a given powder on silver. A maximum partial melt temperature, 5-6°C higher than the onset of melting, can be predicted where microstructure and current carrying properties are optimised. However, it is important to reproduce closely the heating profile and carbon content of the tape under investigation.

5. References

[1]   Maeda H et al 1988 Jpn. J. Appl. Phys. 27 L209
[2]   Sato K et al 1993 Cryogenics 33 243
[3]   Hellstrom E E et al 1993 Applied Superconductivity 1(10-12) 1535
[4]   Yoshida M and Endo A 1993 Jpn. J. Appl. Phys. 32 L1509
[5]   Ming Yang et al 1994 Supercond. Sci. Technol. 7 378
[6]   MacManus-Driscoll J L et al 1994 Appl. Phys. Lett. 65 22 2872
[7]   Zhang W and Hellstrom E E 1994 Physica C 234 No 1-2 137
[8]   Burgoyne J W et al to be presented at EUCAS '95

Inst. Phys. Conf. Ser. No 148
Paper presented at Applied Superconductivity, Edinburgh, 3–6 July 1995
© *1995 IOP Publishing Ltd*

A novel continuous process for the production of long lengths of BSCCO-2212/Ag dip-coated tape[*]

J.W. Burgoyne[a], C.J. Eastell[b], C.G. Morgan[b], D. East[b], R.G. Jenkins[c], R. Storey[c], M. Yang[b,], D. Dew-Hughes[a], H. Jones[c], C.R.M. Grovenor[b] and M.J. Goringe[b]**

[a]Department of Engineering Science, University of Oxford, Parks Road, Oxford OX1 3PJ, UK
[b]Department of Materials, University of Oxford, Parks Road, Oxford OX1 3PH, UK
[c]Clarendon Laboratory, University of Oxford, Parks Road, Oxford OX1 3PU, UK

Abstract   We have developed a continuous process for the production of open-surface BSCCO-2212 tape using a long multi-zone furnace which places no limits on conductor length. Critical current measurements at 77K and 4.2K, up to 10T, suggest that this is a viable method of obtaining conductor with homogeneous properties over long lengths. I_c values above 90A (4.2K, 0T) have been achieved, comparable with the same conventionally-processed material.

1. Introduction

As part of the Oxford HTS conductor programme an application of particular interest is that of high temperature superconductors used at low temperatures and high magnetic fields. This is most readily possible with the BSCCO-2212 compound which has an essentially constant critical current at high fields and remains superconducting in excess of 20T. Since high I_c values are maintained in significant fields at temperatures up to 30K there are considerable possibilities for applications mounted on closed-cycle cryocoolers. Dip-coated conductors are attractive because of their economy of processing, but problems are presented when trying to fabricate large coils by a wind-and-react (WR) method: thermal gradients across the coil in heat treatment tend to produce inhomogeneous critical currents along its length, and the total conductor length is limited by the size of the furnace and how tightly the coil can be wound during reaction. While we have successfully produced coils by a wind-react-and-tighten (WRT) method [1, 2], the above factors still apply. By continuously processing BSCCO-2212 tape through a suitable temperature profile we hope to produce much longer lengths of conductor for react-and-wind (RW) coils. Highly homogeneous properties can be achieved since each part of the tape experiences an identical heat treatment.

[*]Work supported under EPSRC/MOD Rolling Grant No. GR/H64958
[**]Present address: Redland Technologies Ltd., Crawley, West Sussex, UK

336

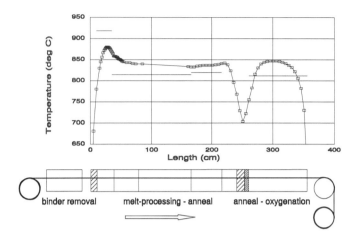

Figure 1 Schematic diagram and temperature profile of continuous process furnace

2. Experimental details

Seven controllable heating zones are used to achieve the desired temperature profile as shown in Figure 1. A 400°C burn-off removes the organic components from the tape and a five-zone furnace is programmed to give the melting/annealing profile. A single-zone furnace gives a secondary anneal and a controlled oxygenation step by determining the final cooling rate. The tape is pulled through the furnace, running over ceramic rollers on an inconel frame along the furnace bore, and is kept under slight back-tension by the feed reel.

The conductor is produced in-house by a conventional dip-coating method. Ag/0.25%Ni/0.25%Mg alloy tape, 125μm thick and 1cm wide, is pulled through a slurry of BSCCO-2212 powder (Merck Ltd.), PVB binder, PEG plasticiser and a dispersant [3]. Careful control of the slurry viscosity and coating speed allow tapes to be produced with an even coating of 70-80μm on each side which reduces to 20-30μm after reaction.

The BSCCO tapes were reacted at a partial-melting temperature of 874-876°C followed by annealing at 830-850°C. Processing speeds of 0.55, 1.1 and 1.6mh^{-1} were used.

3. Results and discussion

Predominantly phase-pure, well-aligned 2212 as shown by X-ray diffraction (Figure 2) was formed at all speeds with no significant differences in alignment. Microstructural examination by SEM/EDX showed a highly aligned BSCCO layer, 5-10μm thick, next to the Ag-alloy substrate, with much less well aligned material towards the free surface. Large flat grains with dimensions of 100-200μm were observed. EDX analysis of the open surface showed a variation in the impurity phases formed with processing speed as follows. Agglomerates (20-30μm) of $Bi(Sr,Ca)_2O_x$ (Cu-free) particles were formed at 0.55mh^{-1}, and small single particles (~5μm) at the other speeds. Alkaline-earth cuprate $(Sr,Ca)_2CuO_3$ (AEC

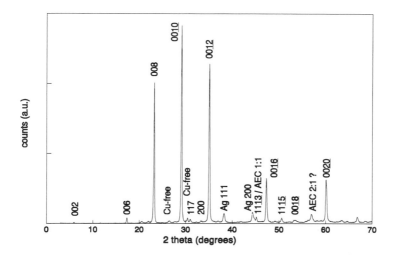

Figure 2 XRD pattern of continuously processed BSCCO-2212 tape (1.1mh⁻¹)

2:1) and $(Sr,Ca)CuO_2$ (AEC 1:1) needles (200-300μm long) were also formed. With the shortest melting time (~3 minutes) at 1.6mh⁻¹ both AEC phases were fewer than for slower speeds; the 2:1 needles were small and the more easily dissolved 1:1 needles were larger than the 2:1. At 1.1 mh⁻¹ the largest 2:1 needles were present, with some smaller 1:1 needles. The highest number of needles was found at 0.55mh⁻¹ (partial melting time ~9 minutes), although of smaller size than for faster speeds, implying some dissolution in the longer anneal. The 2:1 needles grow greater in size and number with longer partial melting. While the cooling rate from the partial melt is highly important for development of the 2212 phase, we believe that the partial-melting temperature and time are the critical parameters in determining the later phase composition [4].

To clarify the impurity phase formation a tape was run at 1.6mh⁻¹ through the partial melting zone and slowed to 0.55mh⁻¹ in the anneal; a further improved microstructure was observed due to the combination of reduced AEC needle formation and greater subsequent dissolution. Since the latter is not a realistic process for long conductor lengths we need to shorten the partial melting zone. The fact that heating/cooling rates and "hold" times are interdependent means that optimisation with respect to processing speed alone involves some degree of compromise. However, investigations of revised heating profiles are continuing.

Only small changes in T_c with speed were found (Table 1). The tapes all experience a rapid final cooling rate which is necessary to maximise T_c since slow cooling tends to over-oxygenate the 2212 phase. Despite this, post-annealing in flowing Ar increased $T_{c(0)}$ to 86.0K, indicating that the tapes are still slightly over-oxygenated.

The critical current properties of the tapes (Table 1) reflect their respective microstructures. A conventional "static" process (SP) with a partial-melting temperature of 875°C was used for comparison. With a 400°C burn-off (SP1) I_c's are lower than for the continuous process, although T_c was lower due to slower final cooling and I_c (4.2K, 10T) was

Table 1 T_c and I_c data for continuously- and statically- (SP) processed tapes

Speed (mh⁻¹)	$T_{c(on)}$ (K)	$T_{c(0)}$ (K)	I_c (A), 77K	I_c (A), 4.2K		
				0T	3T	10T
0.55	84.8	82.5	2.2	86.0	36.5	21.5†
1.1	85.7	84.0	2.7	84.7	36.0	25.5
1.6	84.9	82.7	1.9	69.5	27.5	19.6
1.6/0.55	84.7	82.9	2.2	92.5	41.5	17.5†
SP1	82.8	79.4	2.1	61.0	26.5	22.5
SP2	83.7	80.5	2.7	95.0	58.0	34.0

†possible thermal cycling damage between 3T and 10T

relatively high. A higher burn-off temperature of 830°C (SP2) gave the highest I_c at 4.2K, attributed to a lower residual carbon content, and a much improved $I_c(B)$. The intermediate continuous process I_c's may be due to faster evolution of the organic components compared to SP1 during the initial rapid heating (Figure 1). It is expected that better conductor properties will be commensurate with improvements in powders and coating techniques.

Samples taken along the length of continuously processed tapes showed variations in I_c of generally ≤10%, in contrast with SP tapes which had 50% or greater variation due to temperature gradients across the furnace, although only a few centimetres apart. These gradients are reflected in the homogeneity of SP coils [1]. Recent bending strain measurements on similar material [5] show that the continuously-processed conductor will be useful for RW coils with an inner radius of a few centimetres.

4. Conclusions

Using a novel continuous process we have obtained BSCCO-2212 tapes with homogeneous transport properties and critical current values above 90A (4.2K, 0T) which suggest that this is a viable method of conductor manufacture. By further refinement of the process it should be possible to obtain the required factor of four or five increase in critical current. While these RW tapes may be limited to use in large-bore coils, one may envisage a scenario in which they could be combined with WRT coils to build a composite HTS magnet coil, or used as high-field inserts for large-bore hybrid or conventional superconducting magnets.

References

[1] Jenkins RG, Jones H, Yang M, Belenli I, Grovenor CRM and Goringe MJ 1993 *IEEE Trans. Mag.* **30**, 1813-6
[2] Jenkins RG, Jones H, Yang M, Goringe MJ and Grovenor CRM 1994, presented at ASC'94, Boston, MA
[3] Yang M, Jenkins RG, Jones H, Goringe MJ and Grovenor CRM 1994 *Physica C* **235-240** 3435-6
[4] Morgan CG, Priestnall M, Hyatt NC and Grovenor CRM, this conference
[5] Huang SL, Schoenwaelder B, Dew-Hughes D and Grovenor CRM 1995 *Mater. Lett.*, to be published

Inst. Phys. Conf. Ser. No 148
Paper presented at Applied Superconductivity, Edinburgh, 3–6 July 1995
© 1995 IOP Publishing Ltd

Highly textured tapes prepared by sequential electrolytic deposition

F. Legendre, L. Schmirgeld-Mignot and P. Régnier

S.R.M.P., D.E.C.M., CE Saclay, 91191 Gif sur Yvette CEDEX, FRANCE

Ph. Gendre

SORAPEC, 94124 Fontenay sous Bois, FRANCE

S. Sénoussi

Lab. Phys. Sol, Université Paris Sud, 91405 Orsay, FRANCE

Abstract. We have developed a technique of preparation of superconducting tapes by oxidation of metallic precursors which makes it possible a relatively easy extension to an industrial continuous process for the production of long lengths of conductor. Constitutive elements (Bi, Sr, Ca, Cu in the case of Bi2212 compound) are deposited by a sequential electrolytic process onto silver substrates followed by thermal treatments. After process optimisation, the best sample has a zero applied field critical current density of 3.3×10^4 A/cm^2 at 77 K and a Tc of 84 K. A magnetisation Jc of 5×10^5 A/cm^2 at 4.2 K has been reached. It is to be remarked that these values have been obtained after short overall fabrication times of 4 hours.

1. Introduction

Many methods have been reported in the literature for the fabrication of $Bi_2Sr_2CaCu_2O_x$/Ag (2212/Ag) superconducting composites tapes, i.e the powder-in-tube technique [1], doctor-blade casting [2], dip-coating [3] and electrophoretic coating [4]. Although good values of critical current densities have been obtained, the main disadvantage of these techniques is the long processing time required to synthesize high-quality materials. Among them, those using oxides or carbonates include repeated mixing, grinding and thermal treatments steps during the synthesis to achieve a homogeneous product and necessitate several preparation days.

The use of an electrochemical procedure to obtain superconducting tapes has many advantages : the short times required for the deposition, the ability to form large nonplanar devices, the adaptability to large scale processes and the relatively low cost of the equipment.

We have thus developed a tape fabrication technique by oxidation of metallic precursors electrolytically deposited onto polycristalline metallic ribbons. In this paper, we describe the optimisation of some processing parameters. The results show that it is possible to synthesize reproducibly high quality superconducting tapes after a short processing time of four hours.

2. Experimental

The samples are fabricated by sequential electrolytic deposition of the constituent metals (Bi, Sr, Ca, Cu) onto 50 μm-thick, 20 mm wide and 35 mm long silver tape substrates [5]. The technique includes very short intermediate annealings and a final and relatively longer heat treatment aiming at forming the 2212 phase. The superconducting layers are formed on the two sides of the Ag foil and the final total thickness of the superconductor varies between 7 and 8 μm.

Microstructural characterisation of the specimens has been performed by light and scanning electron microscopy, X-Ray diffraction with a Philips diffractometer using Cu Kα radiation at 40 kV, and microanalysis with a CAMECA SX50 electron probe. Resistivity measurements have been performed by the standard four-probe technique using in all cases a direct current density of 500 A/cm^2; transport critical current densities were determined according to the 1 μV/cm criterion (dc). Magnetic measurements of the critical current densities have been determined in the approximation of the Bean model from hysteresis cycles measured with the field perpendicular to the surface of the sample.

3. Results and discussion

By correlating microstructural characterisations and resistivity measurements, we have investigated the effect of the annealing temperature and have selected an optimum temperature range of 830°-840°C.

We have also studied the influence of annealing time on superconducting properties. In figure 1, we have plotted the resistivity versus temperature curves of four specimens prepared at 840°C for different annealing times.

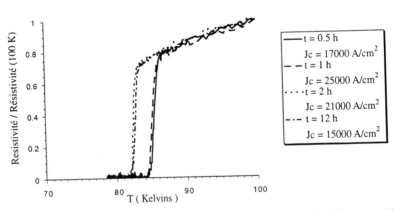

Figure 1 - Electrical resistivity normalized at 100 K plotted against temperature for different annealing times

We can see that the critical temperature (Tc) of our samples decreases with the annealing time. It is now well established that the critical temperature of Bi 2212 changes with the oxygen content, i.e the hole concentration. In particular, B. Heeb et al [6] have studied the oxygen uptake in a Bi 2212 powder during annealing in air at 850°C. The oxygen diffusion into the sample is very fast and is completed to about 99% after 10 h. Their results can provide a possible explanation of ours, i.e during the first hour of our final annealing in air at 840°C, oxygen absorption is sufficient to form the Bi 2212 phase. For longer annealing times, oxygen diffusion into the sample might result in an overdoping in oxygen and hence a lower Tc. However, our best values of Tc stay relatively low (81-84 K) suggesting that the oxygen content of our specimens is still far from the optimun. It is to be noted that we quench our samples after the final treatment and do not perform any post-annealing under reducing atmosphere.

In the inset of figure 1, we have reported the critical current densities obtained for annealing times of 30 min, 60 min, 120 min and 720 min. The size of the tapes in which the transport current density was determined was 2 cm × 0.3 cm. An important feature is that these high critical current densities have been obtained after very short final treatments. For the optimum annealig time of one hour, we have obtained reproducibly Jc's values better than 25000 A/cm^2 at 77 K in the absence of applied magnetic field. The best sample has a zero applied field critical current density of 33000 A/cm^2 for a Tc of 84 K.

It is generally accepted that in order to obtain well textured specimens of the Bi 2212 phase, it is necessary to process them by a partial melting technique, i.e to partially melt the compound and to cristallize the 2212 phase in the presence of some amount of liquid. This results in an enhancement of Jc due to improvement of the texture and densification of the material. In our case, the rapid kinetics of the reaction is inherent to the presence of an important amount of liquid phase. It enhances the cation diffusion rate and results in a rapid 2212 phase formation. Indeed, we don't start from already formed superconductor powder as in the case of other conventional techniques. After the last step of the electrolytic deposition and just before the final treatment, the specimens consist mainly of highly textured Bi 2201 phase and secondary phases. The X-ray diffraction measurements show that after only one hour of heat treatment the Bi 2201 peaks are no longer observed.

A very strong (00l) texture is obtained as a result of this type of treatment involving partial melting. This has been confirmed by texture pole analysis[5]. Besides SEM micrographs show that the 2212 thick deposits are composed of a stacking of extended, well developed plates oriented parallel to the substrate surface[5].We have thus shown that this technique provides a rapid synthesis of very good Bi 2212 superconducting tapes.

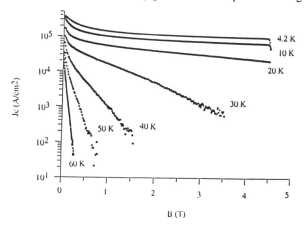

Figure 2 - Temperature dependence of the Jc - B curves obtained from magnetisation measurements for Bi 2212/Ag tape. Magnetic field was applied perpendicular to the tape surface.

The behaviour of the "magnetic" critical current density as a function of applied magnetic field is shown in figure 2 for different temperatures between 4.2 K and 60 K. Jc's values have been determined in the approximation of the Bean model from hysteresis cycles measured with the field perpendicular to the surface of the sample. We can see that at 4.2 K, 10 K and 20 K, Jc decreases slowly with increasing applied magnetic field. At and above 30 K, Jc drops precipitously with field which is consistent with the presence of the low irreversibility line in Bi 2212. At 4.2 K in self field, we have obtained critical current densities equal to $5 \ 10^5$ A/cm^2. Moreover, critical current densities greater than 10^4 A/cm^2 at 20 K under a 5 T field applied perpendicular to the tape surface have been obtained, suggesting that these textured Bi 2212 tapes could be used for practical applications at 20 K. These values are quite reproducible. On the other hand, a critical current density greater than 10^6 A/cm^2 has been reached on 2 samples processed with a different annealing procedure[7].

4. Conclusion

We have developed an electrolytic deposition technique enabling to prepare extremely textured high quality Bi 2212 tapes in 4 hours total preparation time starting from salts of the metallic constituents and not from already formed superconductor powders. By optimizing the heat treatment conditions, Jc's values better than 25000 A/cm^2 (best value : 33000 A/cm^2) at 77 K and zero applied magnetic field and of $5 \ 10^5$ A/cm^2(magnetic measurement) at 4.2 K under zero tesla have been reproducibly obtained. This result shows that the thick-film electrolytic deposition process is a competitive candidate for the production of long lenghts of superconductor.

Aknowledgments

The authors gratefully aknowledge the financial support from Electricité de France (EDF).

References

[1] K. Nomura, M. Seido, H. Kitaguchi, H. Kumakura, K. Togano and H. Maeda 1993 *Appl. Phys Lett* **62** 2131-2133
[2] W. Zhang and E.E. Hellstrom 1993 *Physica C* **218** 141-152
[3] J. Shimoyama, K. Kadowaki, H. Kitaguchi, K. Togano, H. Maeda and K. Nomura 1993 *Appl. Supercond.* **1** 43-52
[4] M. Yang and M.J. Goringe 1993 *U.K Patent* **93106637.5**
[5] Ph. Gendre, L. Schmirgeld, P. Régnier, S. Sénoussi, K. Frikach and A. Marquet 1994 *Physica C* **235-240** 953-954
[6] B. Heeb, L.J. Gauckler, H. Heinrich and G. Kostorz 1993 *J. Mat. Res* **8** 2170-2176
[7] Ph. Gendre, L. Schmirgeld, P. Régnier, F. Legendre, S. Sénoussi and A. Marquet *Appl. Supercond.* 1995 **2** 7-8

Inst. Phys. Conf. Ser. No 148
Paper presented at Applied Superconductivity, Edinburgh, 3–6 July 1995
© *1995 IOP Publishing Ltd*

343

The Effect of Lubrication on Sequentially Pressed Long Lengths of BSCCO-2223 Powder in Tube Tape.

M. P. James[‡†], **S. P. Ashworth**[‡], **B. A. Glowacki**[‡†], **R. Garré**[#] and **S. Conti**[#]

[‡] IRC in Superconductivity, University of Cambridge, Madingley Road, Cambridge CB3 OHE.

[#] Centro Ricerche Europa Metalli- LMI, Fornaci di Barga (LU), Italy.

[†] Department of Materials Science and Metallurgy, University of Cambridge, Pembroke Street, Cambridge CB2 3QZ, UK.

Abstract. This paper describes the effect of lubricating the silver surface during pressing of BSCCO-2223 superconducting tape. The effect of this lubricant is to lower the coefficient of friction at the tool-workpiece interface which lowers the average pressure at the tool-workpiece interface required for the onset of plastic flow of the tape. Thus wider tapes can be produced without increasing the pressing force. This novel method of maximising the silver to superconductor interface area has been developed due to the considerable evidence that during sintering texture development is more pronounced at the interface between the ceramic core and the silver sheath.

1. Introduction

The oxide powder in tube (OPIT) process using Bi-2223 is the most promising technique to produce long lengths of high temperature superconducting conductors for use at 77K [1]. Critical current density, J_c, values of 30,000Acm^{-2} (77 K and self field) [2] have been achieved with multiple roll-sinter single core (monofilamentary) tapes. J_c values to 69,000Acm^{-2} (77K and self field) have been reported for uniaxial pressed tapes, produced with three press-sinter cycles [3].

The microstructure of rolled and pressed tapes has revealed cracks running between blocks of grains, within the oxide core. In pressed tapes the microcracks run along the tape, parallel to the current flow direction, but in rolled tapes they run transverse to the current flow, lowering the J_c [4,5]. Pressed tapes have a higher silver to superconductor interface area per unit volume of superconductor than rolled tapes. There is considerable evidence that at this interface texture development is more pronounced and the BSCCO-2223 conversion is more complete [6]. The critical current density is highest close to the silver sheath whilst the poorly aligned core of superconductor makes little contribution to the longitudinal current [7-9]. The high forces acting on the core of the tape during pressing also increases grain alignment, grain connectivity and density. Increasing the silver to ceramic interface can be achieved with a larger pressing force, or by reducing the pressure required to induce plastic flow of the tape during pressing. The average pressure at the onset of plastic flow, P_{av}, during plane strain compression on the tool-workpiece interface without lubrication is given by Equation 1 [10]. Height h is less than width b and the maximum shear stress at the interface is the shear strength k of the work material. The average pressure at the onset of plastic flow during plane strain compression when a lubricant is added is given by Equation 2 [10], μ is the coefficient of friction and the local shear stress at the interface is now μP, where P is the local pressure.

Equation 1 $P_{av} = 2k(1 + \dfrac{b}{4h})$ Equation 2 $P_{av} = \dfrac{2kh}{\mu b}[\exp(\dfrac{\mu b}{h}) - 1]$

The effect of reducing the coefficient of friction at the tool-workpiece interface on the average pressure at the onset of plastic flow for a fully annealed pure silver tape with a width to height ratio of 10:1 is illustrated in Fig. 1. The average pressure for plastic flow without lubrication can be calculated from Equation 1 to be 140MPa.

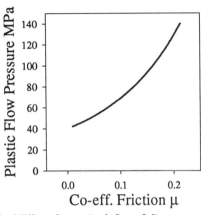

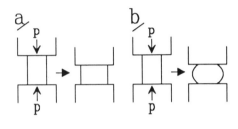

Fig. 1 Effect of μ on plastic flow of silver tape.

Fig. 2 Two dimensional block under compression. a) no friction between slab, b) friction between slab. After Han [11].

The way the tape deforms is also altered when a lubricant is added. Figure 2 shows a two dimensional block under compression. When there is no friction between the block and anvil a shorter but wider block will form. When the friction between block and anvil is higher then barrelling will occur [11]. The advantages of adding a lubricant are therefore twofold; the reduction in average pressure to initiate plastic flow enables the constant load to drive plastic flow for longer as the tape spreads. The lubricant will hopefully also allow for a smoother ceramic to silver interface due to the more uniform flow of the tape.

The production of 20cm lengths of pressed tape, longer than the few centimetres normally associated with pressing of tape, has already been described by the utilisation of sequential pressing [12]. The two aims of this work are to increase the silver to superconductor interface area per unit volume by applying a lubricant on the tape surface and to transfer this technique to the sequential pressing process.

2. Experimental

The tape was produced using the Tube-in-Tube technique, described elsewhere [13]. After rolling to 280μm the tape was sintered in a tube furnace for 25 hours at 828°C ±1°C. All sinters were under a flowing 10% oxygen in argon atmosphere and light MgO powder was used to prevent the tape from sticking to the alumina crucible. This tape was the starting material for all experiments. The effect of increasing load of a single press on tape width and J_C, whilst pressing with and without a lubricant (Clydraw 2000, Kuwait Petroleum, Leeds, UK), was investigated. These results were applied to sequential pressing in two experiments. The first involved sequentially pressing 10cm of tape. The second experiment involved sequentially pressing long lengths of tape, up to 50cm. Both experiments were performed with and without lubrication and J_C was measured along both sets of tape. For the sequential pressing the nonlubricated tape was pressed at 1600MPa, whilst the lubricated tape was pressed at 400MPa. All the samples were sintered simultaneously in a tube furnace for 100 hours at 838°C ±1°C for each section of the experiment. The critical current J_C was defined using an electric field criterion of 1µVcm⁻¹ and measured using a four point transport measurement. Samples were mounted, polished and studied by optical microscopy.

3. Results and Discussion

The theoretical behaviour for plastic flow shown in Figure 1 indicates how significant the coefficient of friction is during pressing. This correlates well with Figures 3 and 4 where a nonlubricated tape starts to spread between 100MPa and 200MPa as predicted by Equation 1, yet adding a lubricant decreases friction and flow begins well below 100MPa.

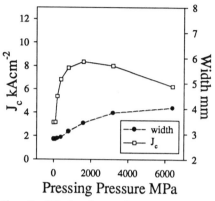

Fig. 3 Effect of pressing pressure on nonlubricated tape width and J_c.

Fig. 4 Effect of pressing pressure on lubricated tape width and J_c.

The J_c of the tape increases as the pressure and width increase, until a maximum is reached which corresponds to approximately 1 to 1.5GPa for both tapes. This maximum J_c for the nonlubricated tape is at 3.5mm wide, where both tapes have a J_c of approximately 8000Acm⁻². The ability to increase the width of the lubricated tape, without having to exceed 1 to 1.5GPa of pressure, increases the J_c by 50% to 12000Acm⁻².

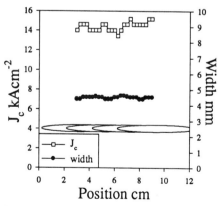 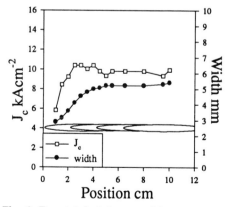

Fig. 5. Four nonlubricated sequential presses at 1600MPa with schematic of pressing footprint.

Fig. 6. Four lubricated sequential presses at 400MPa with schematic of pressing footprint.

With sequential pressing the overlap region of the next press spreads the tape wider than a single press of the same pressure, the nonlubricated tape goes from 3.5mm (single press Fig. 3) to 4.5mm (sequential press Fig. 5 and 7). The lubricated tape goes from 4.5mm

346

(single press Fig. 4) to 5.2mm (sequential press Fig. 6 and 8). This is beneficial for the nonlubricated tape, the J_c goes up above 14,000 Acm^{-2}, but no gains for the lubricated tape.

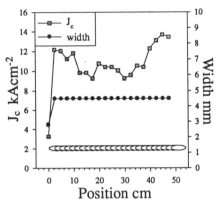

Fig. 7. Nonlubricated long sequential press, pressed at 1600MPa, with pressing footprint.

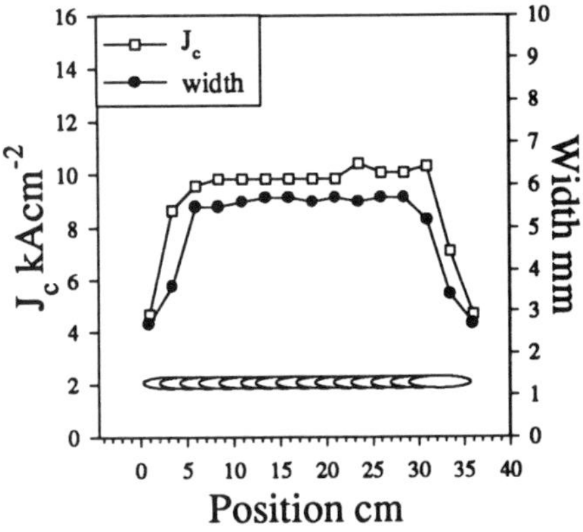

Fig. 8. Lubricated long sequential press, pressed at 400MPa, with pressing footprint.

Both the long tapes in Figures 7 and 8 carry approximately 10,000 Acm^{-2} over the full sequentially pressed section. However the nonlubricated tape has not performed to the same level as the short sequential pressed section and appears to be degraded in the central region of the tape. This highlights the problem of handling and processing tapes of this length. The lubricated tape performs in a similar manner to the short sequentially pressed section. The silver ceramic interface was studied by optical microscopy and as the tapes became wider and thinner the interface becomes more irregular for both the lubricated and the nonlubricated tapes and any benefit of a smoother interface due to lubrication was masked.

4. Conclusion

The effect of lubrication to the deformation of BSCCO-2223 tape was substantial, causing it to plastically flow under lower pressures. This produced wider and higher J_c tapes for orthodox pressing. Sequential pressing at 400MPa with a lubricant gave lower J_c values than pressing nonlubricated tapes at 1600MPa. However further work varying parameters such as initial tape dimensions and pressing pressure has potential for increasing J_c values with the sequential pressing method. As friction dominates tape width during rolling, lubrication of the deformation zone whilst rolling should also be studied.

5. References

[1] P. Haldar, L. Motowidlo, *JOM* **10**, 54 (1992).
[2] G. Grasso, B. Hensel, A Jeremie and R. Flükiger, *IEEE Transactions on Appl. Supercond.* **5**, (1994).
[3] K. Sato et al., *IEEE Trans. Magn.* 27, 1231 (1991).
[4] B. A. Glowacki and J. Jackiewicz, *J. Appl. Phys.* Vol. **75**, 2992, (1994).
[5] A. Otta et al., *IEEE Trans. Appl. Supercond.* 3, 915 (1993).
[6] N. Merchant, J.S. Lou, V.A. Maroni, G.N. Riley, W.L. Carter, *Appl. Phys. Lett.;* 65(8):1039-41, 1994.
[7] D.C. Larbalestier, et al., *Physica C*; 221:299-303, 1994.
[8] S. P. Ashworth, B. A. Glowacki, *Physica C*; 226:159-64, 1994.
[9] M. Lelovic, P. Krihnaraj, N.G. Eror, U. Balachandran, *Physica C*; 242:246-50, 1995.
[10] W. F. Hosford and R. M. Caddell, *Metal forming: mechanics and metallurgy*, PTR Prentice Hall, (1993).
[11] Z. Han and T. Freltoft, *Appl. Supercond.* Vol 2., Nos 3/4, 201, (1994).
[12] S. P. Ashworth, B. A. Glowacki, M. P. James, R. Garre and S. Conti, *IEEE Transactions on Appl. Supercond.* **5**, (1994) (in press).
[13] R. Garré, S. Conti, P. Crincoli, G. Lunardi, G. Salotti, N. Tonarelli, *Appl. Supercond.*, **1**, 205-7, 1993

Inst. Phys. Conf. Ser. No 148
Paper presented at Applied Superconductivity, Edinburgh, 3–6 July 1995
© 1995 IOP Publishing Ltd

Silver Clad Bi-2212 Tape for Low Temperature High Field Applications

M. Ionescu[1], S. X. Dou[1] E. Babic[2], I. Kusevic[2], M. Apperley[3], and E. W. Collings[4]

[1] Centre for Superconducting & Electronic Materials, University of Wollongong, Northfields Av., Wollongong, NSW 2522, Australia
[2] Dept. of Physics, University of Zagreb, Zagreb 41001, Croatia
[3] MM Cables, 1 Heathcote Rd., Liverpool, NSW 2170, Australia
[4] Dept. of Engineering, Ohio State University, Columbus, OH 43201, USA

Abstract. Single filament Bi-2212/Ag tapes were produced by the Powder-In-Tube method, and their pinning capability enhanced via an oxygen overdoping-oxygen depletion procedure. In this way, a microstructure well aligned relative to the Ag sheath, and strongly bonded to it is developed, with two predominant grain boundaries: twist and mixed. A high level of dislocations ensures a strong pinning potential.

1. Introduction

The functionality of the conventional low temperature superconductors in magnetic fields in excess of 18T is restricted by a precipitous decrease of their critical current density J_c with the applied field B.

The potential candidates to replace the conventional superconductors are the main superconducting phases in the bismuth system, Bi-2212, and Bi-2223.

Up until now, the research effort has been focused on the "high" temperature phase, 2223, which forms very slow, and under strict conditions. Thus, the manufacturing procedure could be excessively long: 250 hours in heat treatment [1], and several mechanical deformation cycles.

On the other hand, the "low" temperature phase, 2212, is more stabile and therefore much easier to synthesise, but its critical current density is lower in comparison with the 2223 phase.

In this paper we present additional evidence that the critical current density J_c of Bi-2212 phase can be enhanced via a stronger pinning, induced by oxygen depletion.

2. Experimental

Monofilament samples were produced by the PIT method, using powder prepared by co-decomposing a stoichiometric 2:2:1:2 mixture of nitrates in air and oxygen. This resulted in a homogenous 2212 powder, with particle size of between 4μm and 10μm. The powder was loaded in a silver tube, and drawn into tapes, with final dimensions 4.2mm width, and 0.15mm thickness.

Short length (~50mm) samples were heat treated in a controlled atmosphere environment, by melting the ceramic core, and then slow cooling in oxygen, followed by a nitrogen annealing. These will be called "fully treated" samples. The details of the heat treatment are presented elsewhere [2]. This preparation procedure ensures an oxygen overdoping in the first half, followed by an oxygen depletion during the subsequent annealing, which provides an increase number of defects. The duration of the heat treatment is ~30 hours, and the cross section area of the ceramic core after the heat treatment, was estimated at between 40,000μm² and 50,000μm².

Besides the fully treated samples, a reference sample was prepared under similar conditions, except the nitrogen annealing step, which was replaced by a corresponding oxygen or air annealing step.

The samples were characterised in terms of microstructure (SEM, TEM), microcomposition (WDS), and transport electric properties at 77K, and low (1T) DC field, with a 1μV/cm electric field criterion.

3. Results

The critical current I_c vs. temperature of a fully treated sample, and the reference samples is presented in Fig. 1. A notable increase in I_c up to one order of magnitude, at both low (4.2K), and high (77K) temperatures could be achieved.

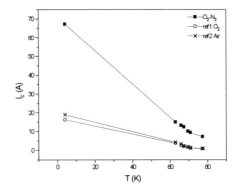

Fig. 1: Critical current vs. Temperature

The SEM microstructure of the samples was examined in two different circumstances:

a) on a bare ceramic core, for which the Ag was striped off, looking along the normal of the tape. In this case, the microstructure consists of well developed 2212 grains, having typical dimensions of (0.1-10)μm along the b direction, and (1-100)μm in the a direction. Also, a significant number of grains appears to be 90° twined.

b) on a sample polished in a plane parallel to the normal of the tape (longitudinal cross section). In this case, the c direction of the grains appears to be slightly misoriented relative to the normal of the tape, with an angle less than 4°. There appears to be no SEM difference between the microstructures of the high I_c sample and the reference sample.

The TEM microstructure, and the defects density of the 2212 tapes were studied on samples extracted in such a way as the normal to the sample would coincide with the normal to the tape, and for the Ag-ceramic core interface the normal to the sample would be perpendicular to the normal to the tape. This revealed the existence of two main types of grain boundaries: 90° twist boundaries (TB), and mixed boundaries (MB). In general, the density of dislocations for the fully treated samples is $(10^{10}-10^{11})$dislocations/cm^2, and (10^5-10^6)dislocations/cm^2 for the reference samples. The dislocations are [110] type, with Burgers vector of 1/2 [110], being concentrated preferentially at the 90 ° TB. Inside the grains with no 90° TB, the dislocation type is [001], and the dislocation density is $\sim 10^9$ dislocations/cm^2 for the fully treated sample, and $\sim 10^3$ dislocations/cm^2 for the reference sample.

The interface between the Ag and the core connects closely, with [001] direction perpendicular to it, having the coupling Bi-O layer grown on the Ag sheath.

A moderate number of stalking faults of $c/2$, 2201, and $c/2$, 2223 were found, with the distance between them ~15nm.

The secondary phases identified were $Bi_2Sr_2CuO_x$ (2201), $Sr_7Ca_7Cu_{24}O_x$ (14:24), CuO, and some amorphous structure.

The resistive dependence on temperature in low magnetic field is presented in Fig.2, and suggest a thermally activated behaviour of the flux lines at temperatures below T_c.

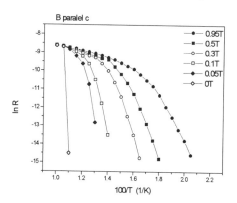

Fig2: Arrhenius plot of resistance for Bi-2212

350

The pining energy deduced from the resistive dependence in low magnetic field in the fully treated sample are compared with the pining energy in Bi-2223, Fig.3.

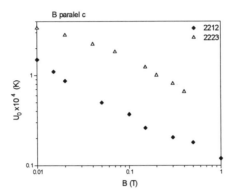

Fig.3: The pining energy U_0 vs. field for 2212 and 2223

The difference is due not only to the difference in T_c between 2223 and 2212 which would account for only ~20%, but also due to a larger anisotropy in electronic properties intrinsic to 2212. This would reduce the correlation length between the flux lines, and the flux bundle volume would be correspondingly reduced [3]. As a consequence, the coupling of the vortices between adjacent layers is weaker, which results in a weaker pining.

4. Conclusions

The increase of the defects density in Bi-2212/Ag monofilamentary tape is attributed to an oxygen overdoping-oxygen depletion procedure. This could enhance the critical current I_c by up to one order of magnitude.

Due to an intrinsic higher anisotropy, the pining energy in Bi-2212 is lower than Bi-2223, however, the behaviour in moderate magnetic field shows the potentiality of this procedure in producing silver clad Bi-2212 tapes for low temperature, high field applications.

5. References

1. Y. C. Guo, H. K. Liu, and S. X. Dou, Materials Letters, 18, 336, (1994)
2. M. Ionescu, S. X. Dou, E. Babic, I. Kusevic, and E. W. Collings, 1995 International Workshop on Superconductivity, June 18-21, 1995, Maui, Hawaii.
3. M. R. Beasley, R. Labusch, and W. W. Web, Phisical Review, 181, No. 2, 682, (1969)

Inst. Phys. Conf. Ser. No 148
Paper presented at Applied Superconductivity, Edinburgh, 3–6 July 1995
© *1995 IOP Publishing Ltd*

Novel processing of Bi-based high-T_c superconducting monofilamentary and multifilamentary wires and tapes

Y C Guo, P A Bain, H K Liu , S X Dou, E W Collings† and G Secrett‡

Centre for Superconducting and Electronic Materials, University of Wollongong, Northfields Avenue, Wollongong, NSW 2522, Australia
† Department of Materials Science and Engineering, Ohio State University, Columbus, OH43201, USA
‡ MM Cables, 1 Heathcote Road, Liverpool, NSW 2170, Australia

Abstract   Bi-based high-T_c superconducting monofilamentary and multifilamentary wires have been fabricated using a novel wire processing technique called "Continuous Tube Forming/Filling (CTFF)". In this technique loose superconductor powder is continuously fed onto a long silver strip which is then wrapped to encase the powder and form the wire. Improved superconductor core uniformity and largely reduced sausaging and cracking were found in CTFF-processed tapes due to a more uniform initial packing density and a reduction in the mechanical deformation required compared to PIT tapes. The heat treated CTFF tapes were characterised in terms of Bi2223 phase purity, microstructure, J_c and J_c behaviour in magnetic field. A J_c of 20,000 A/cm^2 at 77K and self magnetic field was achieved for 16-filament CTFF multifilamentary tapes.

1. Introduction

High-T_c $(Bi,Pb)_2Sr_2Ca_2Cu_3O_{10}$ superconducting wires are usually fabricated by the powder-in-tube (PIT) technique, in which precursor powder is packed into a silver tube and then mechanically worked into fine round wire and subsequently flat tape. The PIT process is, however, a non-continuous wire processing technique; the length of wire and tape which can be made by this method is limited by the length and diameter of the initial silver tube. "Sausaging" of the core and more seriously, microcracks, are often observed in extensively mechanically deformed wires and tapes [1].

Recently, we have used a so-called "Continuous Tube Forming/Filling (CTFF)" technique in which loose superconductor powder is continuously fed onto silver tape by a mass control unit. The silver is then formed into a tube encasing the powder. Because the length of silver tape is virtually unlimited, very long lengths of superconductor wire can be easily achieved. Moreover, the powder is fed very accurately by machine, making the initial powder packing density very uniform. As the initial diameter is small, the extent of mechanical deformation required is largely reduced, thereby reducing the sausaging effect. The CTFF procedure is, most importantly, a relatively simple and continuous wire processing technique. In this project, we investigate these aspects of CTFF processing and properties, including powder leaking, packing density, core uniformity, microstructure and transport current, firstly on monofilamentary and then for multifilamentary tapes.

2. Experimental Details

Superconductor precursor powder was prepared by the solid state reaction route. The cation ratio was Bi/Pb/Sr/Ca/Cu=1.8/0.35/1.91/2.05/3.06 and the major phase in the precursor powder was Bi2212 phase. Pure silver tape approximately of about 0.15mm thick and 6 mm wide was used as the wrapping material. Loose superconductor powder was continuously fed onto the silver tape by a mass control unit and the silver was then formed into a tube encasing the powder. The formed CTFF wire is approximately 1.5mm in diameter.

The round CTFF wires were square-rolled into thin square monofilamentary wires. For some wires an intermediate annealing was carried out between square-rolling stages to investigate the effect of annealing on the quality of the CTFF wires and tapes. The square wires were then flat-rolled into tapes. Multifilamentary tapes were fabricated by inserting 16 thin CTFF monofilamentary wires into a silver tube of 6.5mm outer and 5.5mm inner diameter and the composite silver tube was mechanically worked into thin wire and tape.

Both monofilamentary and multifilamentary CTFF tapes were subjected to a thermomechanical process consisting of several cycles of sintering and mechanical deformation. Each step of sintering was carried out at 830-836°C for a period varying from 50h to 100h in air. The heat treated CTFF tapes were characterised in terms of Bi2223 phase purity, microstructure, J_c and J_c behaviour in magnetic field.

3. Results and Discussion

3.1. CTFF-Processed Monofilamentary Wires and Tapes

About 100 meters of Bi2223/Ag superconducting wire with a diameter of 1.5mm was fabricated using the Continuous Tube Forming/Filling (CTFF) technique. Figure 1a shows the cross section of a CTFF wire. The packing density of the CTFF wires was estimated to be about 4.0 g/cm³, i.e. 65% of the theoretical value of Bi2223 compound. The packing density of CTFF wire is dependent on the feeding rate of powder during CTFF processing. Various packing densities can be easily achieved by varying the powder feeding rate.

Figures 1b and 1c are the cross section and longitudinal section of a CTFF tape of about 0.14mm thick flat-rolled from an annealed (750-800°C,15h) square wire. From Fig.1b,

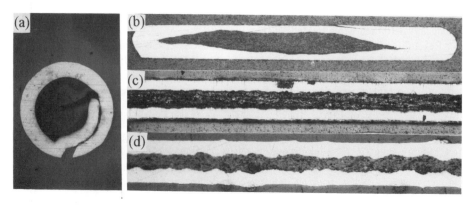

Fig.1 Transverse cross section of a CTFF wire (a) and tape (b). Longitudinal cross section of a CTFF tape (c) and PIT tape (d)

it is seen that the two silver layers have been fully diffusion bonded and hence no powder was found to leak from the gap between the two silver layers during rolling. Figure 1c shows that the superconductor core is quite uniform and the interface between core and silver sheath is very smooth. No noticeable sausaging or cracking was observed in the tape. For comparison, Figure 1d shows the longitudinal sections of a PIT tape which underwent the same mechanical deformation procedure as for the CTFF tape. It is evident that the CTFF tape has a better core uniformity and much less sausaging and cracking compared to the PIT tape. The reduction of mechanical deformation and uniform packing density is believed to be responsible for the improvement of the sausaging effect in the CTFF tapes.

Figure 2 shows the J_c (77 K) together with the high-T_c phase fraction as a function of sintering time for a CTFF tape. It is seen that the initial high-T_c phase fraction increase does not lead to a marked J_c increase because not many continuous high-T_c phase paths across the sample have been formed at this stage [2]. The J_c then increases almost linearly with increasing sintering time due to the increase of high-T_c phase fraction and improvement of grain alignment and density. During the 210-280h period the conversion of Bi2212 to Bi2223 phase is nearly completed. The J_c, however, increases substantially due to the improvement of grain alignment, grain growth and

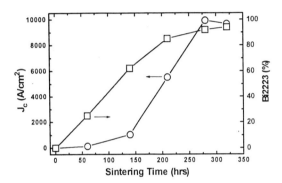

Fig.2 $J_{c,77K}$ and Bi2223 percentage vs. sintering time

connectivity between grains. It is also noted that with further sintering the J_c of the tape begins to drop, possibly due to unhealed microcracks, and Bi and Pb loss by evaporation [2,3].

3.2. CTFF-Processed Multifilamentary Wires and Tapes

Figures 3a to 3b are the cross section and longitudinal section of a multifilamentary tape of 0.14mm thickness made from the CTFF-processed wires. From the cross section, it can be seen that each filament is highly flattened and stretched, but uniformly distributed in the silver matrix. No bridging between filaments is found. From the longitudinal section of the tape, it is noted that the superconductor filaments are very straight, continuous and uniform; no noticeable sausaging or cracking is found. The uniformity of the CTFF multifilamentary tape is better than that of PIT multifilamentary tapes produced in the same experimental conditions.

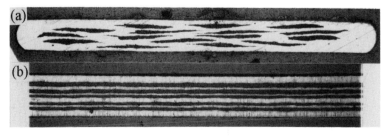

Fig.3 Transverse (a) and longitudinal (b) cross section of a CTFF multifilamentary tape

354

Figure 4 shows the dependence of the J_c on an applied magnetic field at 77K for a CTFF-processed monofilamentary tape after sintering at 834°C for 280hrs. The magnetic field was applied in two directions: parallel to the wide tape surface (B//ab) and perpendicular to the tape surface (B//C), but always perpendicular to current direction. Both J_c-B Curves (B//ab and B//C) show a rapid drop at low magnetic fields, followed by a gradual drop at high magnetic fields, which are very similar to that for PIT processed Bi2223/Ag superconducting tapes.

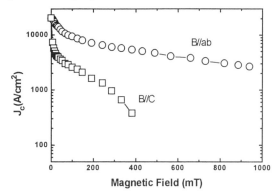

Fig.4 $J_{c,77K}$ vs. magnetic field for B//ab and B//C

A double-step characteristic in the Jc-B curves for high-T_c bulk superconductors has been observed and interpreted by Ekin. et.al. [4]. At low magnetic fields, a rapid drop in the J_c is attributed to Josephson weak links at grain boundaries. At higher fields, the Jc is thought to be governed by the intragrain pinning strength. Therefore, the rapid drop in J_c of the CTFF tape is attributed to the grain boundary weak links and the plateau of Jc-B curve at high fields is due to the strong intragrain pinning.

4. Conclusions

A novel wire processing technique, called "the Continuous Tube Forming/Filling (CTFF)" has been developed, which is able to continuously produce fine, uniform and long silver-clad Bi2223 wire directly from superconductor powder. Several aspects of CTFF processing including powder leaking, packing density and core uniformity have been investigated. The results indicate that the problem of powder leakage from CTFF wire during tape processing can be solved by annealing the samples before flat-rolling. The packing density formed during the CTFF processing produces a uniform superconductor core in the final tape. In comparison to PIT tapes the CTFF tapes show an improved superconductor core uniformity and reduced sausaging and cracking due to the uniform packing density and less mechanical deformation. The heat treated CTFF monofilamentary and multifilamentary tapes were characterised by studying the phase composition, microstructure, J_c and J_c-magnetic field behaviour. All the obtained results appear to be comparable with that of PIT tapes treated with the same conditions.

References

[1] Han Z and Treltoft 1994 *Appl. Supercond.* **2** 201-15
[2] Guo Y C, Liu H K and Dou S X 1994 *Mat. Sci. Eng.* **B23** 58-65
[3] Yamada Y, Obst B, Flükiger 1991 *Supercond. Sci. Technol.* **4** 165
[4] Ekin J W, Larson T M Hermann A M. et.al 1989 *Physica C* **160** 489

Inst. Phys. Conf. Ser. No 148
Paper presented at Applied Superconductivity, Edinburgh, 3–6 July 1995
© 1995 IOP Publishing Ltd

Influence of superconducting layer thickness on the dissipation in monofilamentary BSCCO/Ag 2223 tapes

J Everett, M D Johnston, M Dhallé , H K Liu†, S X Dou† and A D Caplin

Centre for High Temperature Superconductivity, Imperial College, London SW7 2BZ, UK
† Centre for Superconducting and Electronic Materials, University of Wollongong, NSW 2522, Australia

Abstract. We compare the critical current dependence on temperature and magnetic field of two BiSrCaCuO/Ag 2223 tapes prepared under identical conditions, but with different super-conducting layer thickness (10 μm and 100 μm). This electromagnetic characterisation was carried out using direct transport current techniques, as well as DC magnetisation measurements of both the tapes and the BSCCO powder extracted from them. Such an analysis allows us to identify the nature of the dissipation mechanisms that limit the critical current under various conditions. Differences between the two samples are discussed in terms of their microstructure.

1. Introduction

Various models of the supercurrent transport in BSCCO 2223 tapes have been proposed, attempting to identify the dissipation mechanisms as arising from commonly observed grain boundary types in these polycrystalline conductors [1, 2]. Despite the granular character of tapes, a sizeable macroscopic current survives up to relatively high magnetic fields, where *intra*granular flux motion takes over as the main current limiting mechanism [3]. Furthermore, several techniques have demonstrated the non-uniformity of the current distribution in tapes, both in the lateral [4, 5] and cross-sectional direction [6].

In this paper we present magnetic and transport measurements on two BSCCO/Ag 2223 tapes, which were prepared in a similar way, but have different superconducting layer thickness (SCLT). The techniques used allow us to distinguish between dissipation occurring on *inter* and *intra*granular levels, and show how the two mechanisms are influenced by SCLT.

2. Experimental

The two tapes measured were made using the powder in tube technique [7]; tapes D10 and D100 have a SCLT of 10 μm and 100 μm respectively. Transport measurements were performed using a pulsed current transport rig. I-V curves were corrected for the effect of the Ag sheath, and all I_c values were extracted from the I-V curves at a 5 μV criterion. The DC magnetisation measurements were carried out in vibrating sample magnetometer (VSM). Magnetic E-J characteristics were constructed using the dynamic relaxation technique [8].

The magnetisation ΔM(tape) of a tape sample can be expressed in terms of both *inter* and *intra*granular contributions [9]: ΔM(tape) = $2/3(fa J_{GR} + \Lambda J_{GB})$, where a is the average grain radius, f the filling factor, Λ the current carrying length scale, and J_{GR}, J_{GB} the *intra* and *inter*granular current densities respectively. To distinguish between *inter* and *intra*granular dissipation, we extracted the superconducting core from each tape and gently ground it to a powder. The powder was then dispersed in wax and its magnetisation ΔM(powder) measured in the VSM. By preventing the flow of any macroscopic currents, this technique effectively isolates J_{GR}, ensuring that only the *intra*granular component of the dissipation is measured. The magnetisation of the powder is then: ΔM(powder) = $2/3a J_{GR}$.

3. Results and discussion

Previous studies have shown that there are two distinct regions in the $I_c(B)$ behaviour of BSCCO 2223 tapes [10]. Fig. 1 shows the normalised field dependence $I_c(B)/I_c(B=0)$ of D10 and D100 at 75 K and 90 K. For fields less than a temperature dependent field $B_{cross}(T)$, the $I_c(B)$ data can be described by a power law dependence: $I_c \propto B^{-\nu}$. Above $B_{cross}(T)$, there is a decrease in $I_c(B)$ that is approximately exponential: $I_c \propto \exp(-B/B_g(T))$. For $B < B_{cross}(T)$ *inter*granular dissipation dominates, while $B_{cross}(T)$ represents the onset of *intra*granular dissipation [11]. At 75 K, $B_{cross}(T) \sim 0.08$ T for both D10 and D100, but a rapid decrease of $B_{cross}(T)$ with temperature is seen clearly for both samples.

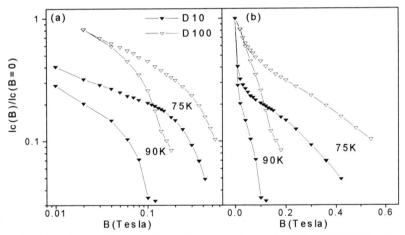

Figure 1. Field dependences of the normalised critical current for samples D10 and D100, (a) log-log plot, (b), log-linear plot (b). The magnetic field was applied perpendicular to the tape plane.

Fig. 2 shows a comparison of magnetic and transport E-J curves obtained for both tapes at fields above, below and at $B_{cross}(T)$. Transport current E-J characteristics measured for D10 were sharper than those for D100, and in a way reminiscent of the linear I-V curves commonly obtained for random polycrystalline $YBa_2Cu_3O_7$. Unlike $YBa_2Cu_3O_7$ however, the slope of the I-V curves is altered by changes in applied magnetic field [12]. Figs. 2(a) and (b) show that for $B < B_{cross}(T)$ there is no change in the slope of the E-J curves. For $B > B_{cross}(T)$ there is a clear although small decrease in slope. For both D10 and D100, the transport and magnetic E-J characteristics are in good agreement. This implies that the transport and magnetically-induced currents are flowing on a similar macroscopic scale, and are probing the same dissipation mechanisms.

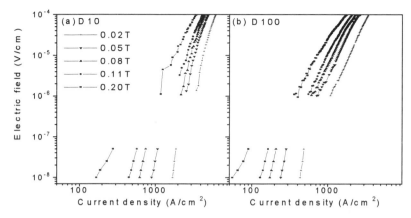

Figure 2 Magnetic *E-J* data (low *E*) compared with transport *E-J* data (high *E*) for tapes (a) D10 and (b) D100, the magnetic field was applied perpendicular to the tape plane.

Fig. 2 reveals a further important difference between D10 and D100: While the current density *J*, at E = 5 μV/cm for D10 is fairly typical for these conductors (J_{cD10} (75 K, 0 T) ~ 10^4 A/cm²), the current density in D100 is lower by a factor of ~ 5; i. e. although the SCLT of D100 is ~ 10 times that of D10, it carries only ~ 2 times higher total current.

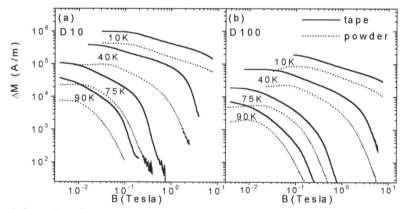

Figure 3 Comparison of tape magnetic moment with that of powder for tape samples D100 (a) and D10 (b).

Fig. 3 shows tape and powder magnetisation for samples (a) D10 and (b) D100. For D100, Δ*M*(tape) is proportional to Δ*M*(powder) with a proportionality constant that is largely independent of field temperature. For D10 on the other hand, the ratio Δ*M*(tape)/Δ*M*(powder) does show a field and temperature dependence. This may indicate that different dissipation mechanisms are dominant in the $I_c(B)$ behaviour of the tape and powder.

Measurements of the anisotropy of I_c with respect to the direction of the applied field at 75 K for D10 and D100 are shown in Fig. 4; *H* is kept perpendicular to the applied current while rotated with respect to the tape plane (see insert). Following the Kes model for dissipation in strongly 2D superconductors [13], the anisotropy curves give a measure of the *effective* average grain misorientation in the sample. Fig. 4 shows that the distribution measured for D10 is sharper than that seen for D100, so that the *effective* cross-section in D10 is better textured than that in D100.

358

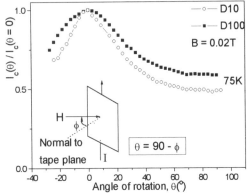

Figure 4 The normalised $I_c(\theta)$ dependence of D10 and D100., at 75 K and 0.02 T.

4. Conclusions

We have measured two BSCCO/Ag 2223 tapes of different superconducting layer thickness, using a combination of magnetic and transport techniques. The superconducting core was then extracted from each tape and its magnetisation re-measured. Comparison between these magnetic and transport measurements revealed some granular signatures of dissipation in D10. We have also compared measurements of the anisotropy of the critical current with respect to the direction of the applied magnetic field for D10 and D100. The electromagntic characterisation of the two tapes demonstrates clearly that the critical current density performance of D10 is superior to that of D100, giving a clear indication that of the two tapes, D10 has the larger *effective* fraction of cross-section. We extracted a further significant piece of information from the anisotropy measurements: The *effective* cross-section in D10 is better textured than is the *effective* cross-section in D100. This analysis highlights the important role that the BSCCO/Ag interface plays, in the formation of well textured, secondary phase free superconducting paths.

Work supported by UK EPSRC and by EC Brite Euram-Contract BRE2 CT92 0229. J. E. thanks The National Grid Company plc for financial support for this work through the CASE Award scheme.

References

[1] Bulaevski L N *et al.*1993 *Phys. Rev. B* **48** 13 798-13
[2] Hensel B *et al.* 1993 *Physica C* **205** 329-337
[3] Dhallé M *et al.*1995 *EUCAS*, submitted
[4] Grasso G *et al.* 1994 *Physica C* **241** 45
[5] M. D. Johnston *et al.* 1995 *EUCAS*, submitted
[6] Pashitski A E *et al.*1995 *Physica C* **246** 133-144
[7] Liu H K *et al.* 1993 *Physica C* **213** 95-102
[8] Zhukov A A *et al.* 1993 *Physica C* **142** 33
[9] Caplin A D *et al.*1994 to be publ. *IEEE Trans. Appl. Sup.*
[10] Cuthbert M N *et al.* 1994 *Physica C* **235-240** 3027-3028
[11] Dhallé *et al.* 1994 *Proc. IWCC7* 553-556
[12] Cassidy S M *et al.*1992 *Cryogenics* **32** 1034-1037
[13] Kes P H *et al.* 1990 *Phys. Rev. Lett.* **64** 1063-1066

358

Inst. Phys. Conf. Ser. No 148
Paper presented at Applied Superconductivity, Edinburgh, 3–6 July 1995
© *1995 IOP Publishing Ltd*

Status of development of HTS tapes for different applications

G. Papst, J. Kellers

American Superconductor Europe GmbH, 41564 Kaarst, Germany;

A.P. Malozemoff

American Superconductor Corporation, 01581 Westborough, MA, USA

Abstract. Substantial progress in the development of flexible multifilamentary high-temperature superconducting (HTS) silver and silver alloy sheathed tapes has been made. Development at American Superconductor Corporation is focused on composite tapes using Bi(Pb)SrCaCuO-2223 superconductors made by the scalable powder-in-tube technique. In this type of conductors, filament current densities exceeding 32,000 A/cm^2 (77 K, self field, 1 µV/cm) have been achieved in short length and continuous length over 1 km are produced with a pilot production basis. Recent experiments directed towards a reduction of ac losses are encouraging and thus indicate that HTS technology can be exploited for pulsed or AC applications. With the present conductor, significant application demonstrators and prototypes were enabled. We report here on the respective status as well as on the specific requirements of these applications and differentiate between low field, medium field, and high field applications.

1. Introduction

Though still quite young in years, high temperature superconductivity (HTS) is developing rapidly towards a mature technology with commercial breakthroughs in the foreseeable future. Since its advent in 1986, three branches have evolved which include HTS bulk materials, HTS thin films for electronic applications, and flexible HTS tapes for power applications.

In this paper we will review only the advancements in tape conductor development and will focus especially on progress at American Superconductor Corporation (ASC).

2. Status of HTS conductor development

The best overall results in the manufacture of multifilamentary flexible HTS tapes have been achieved using the PIT (powder in tube) technique. Here a precursor powder with the desired ratio of metal atoms is sealed into a billet of silver or silver alloy and formed into a monofilamentary rod using standard deformation processes. After rebundling, a second reduction in diameter with subsequent thermomechanical cycles gives the final multifilamentary composite. The precursor can be either a mixture of pure metals (MP = Metallic Precursor), or a mixture of oxidic ceramics (OPIT = Oxide Powder in Tube). Details on the process are reported in e.g., [1,2]; metallographic sections of OPIT conductors are shown in figure 1.

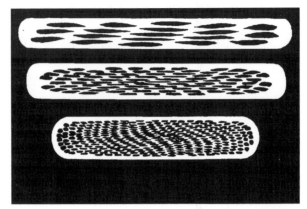

Figure 1: Filament architecture of OPIT tapes shown in optical micrographs of transverse cross-sections. From top down, 19 filament, 85 filament, and 313 filament tape conductors.

The status of ASC development of OPIT tape is shown in table 1. The table reports both th filamentary critical current density, j_c, and the respective over-all value, j_e. A general observatio that is related to the sharpness of the transition from superconductor to normal conductor, is that th critical current density measured using the 1 μV/cm criterion is just ≈20% higher than the valu determined using the more strict 10^{-11} Ω cm criterion. It is furthermore important to notice that a results have been measured end-to-end, and that even the shortest length has been manufactured b fully scalable processing steps.

Length / m	j_c, A/cm^2		j_e, A/cm^2	
	1 μV/cm	10^{-11} Ω cm	1 μV/cm	10^{-11} Ω cm
0.01	32,600	---	9,100	---
10	22,600	19,100	6,300	5,300
70	22,000	17,800	6,100	5,000
400	17,000	13,500	4,100	3,200
1,100	12,700	8,800	3,000	2,100

Table 1: Filament and engineering critical current densities of multifilamentary silver-sheathed Bi(Pb)SrCaCuO-2223 tapes at 77 K and self-field.

So far, the OPIT approach gives the better electrical performance, whereas the MP approach gives the better mechanical properties due to the potential of a higher filament count [2]. The significance of a high filament count is on the one hand the conservation of electrical properties under mechanical stress [3]. On the other hand, finer filaments contribute to the reduction of the hysteretic losses which are one of the major sources for AC losses in HTS materials.

For the reduction of the eddy current losses in the metal sheath and the filament coupling losses, it is necessary to enhance the electrical resistivity of the matrix and to twist the filaments [4]. Recent advances in the development of a twisted HTS conductor are encouraging as they show a transport critical current density of 15,300 A/cm^2 for a fully twisted tape with a twist-pitch of 7.1 mm [5]. First experimental attempts on the enhancement of matrix resistivity by alloying the silver matrix with 3 at% gold led to current leads [6] as the first HTS commercial product.

3. HTS applications

Dynamics of magnetic vortices causes degradation of electrical performance under applied or self-generated magnetic fields. This is a common feature of all type II superconductors, including all technically relevant LTS and HTS materials. However, as lateral flux motion is a thermally activated process, this motion called *flux-creep* is more prominent in the regime of higher temperatures [7,8].

Figure 3 depicts the well known dependence of critical current density on the magnitude of effective magnetic induction in a schematic way. In addition to this, figure 4 shows the inverse of Carnot's efficiency; this plot indicates the costs of refrigeration as a function temperature. Obviously, it is mandatory to balance between conflicting influences to operate a commercial application at its economic optimum[1].

A simplified rule for the optimum on the temperature scale can be deducted:

- Low-field applications (B < 0.5 T) are operated in liquid Nitrogen.

- Medium-field applications (B > 3 T) are operated between 20 K and 40 K.

- Highest-field applications (B > 15 T) are operated at T ≤ 4.2 K.

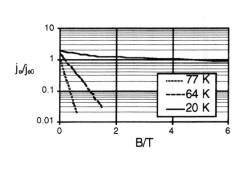

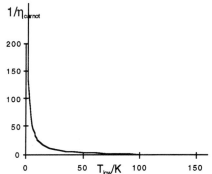

Figure 3: Schematic view on the influence of magnetic induction on critical current density of BSCCO-2223 at different temperatures.

Figure 4: Inverse of Carnot's efficiency vs. temperature. This provides the minimum cooling power needed to extract 1W out of a cryogenic envelope (T_{high}=295 K).

3.1 Low field applications

The most important HTS applications operating at a relatively low magnetic field are cables for power transmission and transformers. It is important to keep in mind that operation in liquid nitrogen does not determine the operating temperature completely. By reducing the vapor pressure above a liquid phase, temperatures as low as 64 K can be obtained. As figure 3 reveals, this moderate reduction of temperature already improves the in-field-performance of BSCCO-2223.

In a joint development effort with Pirelli Cavi S.p.A, prototype multistrand conductors have been demonstrated. In a multilayer configuration, a current carrying capacity of 4,200 A DC has been achieved with a voltage drop lower than 0.1 mV over the total length of 1 m. This result approaches the 2,000 A AC threshold required for commercial power transmission. With Pirelli, EPRI and the U.S. Department of Energy, a 30 m single-phase, machine-stranded cable is in the manufacture and will be tested late 1995.

Transformers for power transmission or power distribution applications also belong to the low-field applications as the magnetic induction in conventional transformers rarely exceeds some tenth of a Tesla. HTS transformer promise a reduces size, weight, an increased efficiency, and the environmental advantage of the oil-freedom. Based on ASC technology, ABB builds world's first HTS transformer. This prototype project is supported by Electricité de France, the Swiss utility SIG and the Swiss Department of Energy. Upon completition in late 1996, the transformer will be on-line in the power grid of Geneva.

---

[1] To shift this optimum towards higher temperatures, it is not sufficient to increase the self-field value of j_c. It is needed to substantially increase the pinning potential in which vortices move. The technical possibility of this has already been demonstrated by, e.g., heavy ion irradiation [9]. Early experimental results indicate the potential for future performance improvement in HTS wire technology.

362

3.2 Medium field applications

Medium field applications are likely to operate also in future in the temperature range between the boiling points of the most common refrigerants, Helium and Nitrogen. The most convenient and economic way of cooling high temperature superconducting devices to this temperature regime is the use of mechanical cryocoolers. In a development program supported by the Advanced Technology Program of the US Department of Commerce, ASC has already proven the feasibility of this concept by developing a conduction cooled magnet. This solenoidal magnet is cooled with a two stage Gifford-McMahon cryocooler to 27 K and generates up to 2.16 T in a warm bore [10]. This induction already exceeds the saturation of iron and is therefore consequential for, e.g., rotating machinery like AC synchronous motors and generators [11].

3.1 High field applications

In applications like Nuclear Magnetic Resonance, the spectral resolution is proportional to the magnetic field the sample experiences. Especially the complex NMR samples of biosciences create a demand for ever higher fields. Already now 1 Ghz-^{1}H-NMR-spectrometers are in the discussion, which require a magnetic flux density of 23.5 T. However, low temperature superconducting materials, which are widely used for high field magnets in scientific laboratories, are restricted by their practical current densities to inductions less than approximately 20 T. This barrier does not apply to BSCCO materials, as these are non weak-linked at low temperatures [3,12], i.e., $T \leq 4.2$ K.

For this and for other laboratory applications, HTS technology is the only practical solution for achieving uninterrupted very high magnetic fields.

4. Conclusion

The development of flexible silver and silver-alloy sheathed Bi(Pb)SrCaCuO-2223 super-conductors approaches the requirements for practical HTS wire. The robustness of the multifilamentary architecture, as well as the high and still increasing current carrying capability, permit significant prototypes like laboratory magnts, cable prototypes, transformers and many more. The development of a low-loss AC HTS conductor is very important for most of the electric utility applications. This topic is currently addressed with remarkable progress.

References

1 G.N. Riley, J.J. Gannon, P.K. Miles, D.R. Parker, Appl. Supercond. 2 (1994) 155
2 A. Otto, L.J. Masur, C. Craven, D. Daly, E.R. Podtburg, J. Schreiber *submitted to Applied Superconductivity Conference, Boston, Oct. 17-21, 1994*
3 K. Sato, T. Hikata, H. Mukai, M. Ueyama, N. Shibuta, T. Kato, T. Masuda, M. Nagata, K. Iwata, T. Mitsui *IEEE Trans. Magn., 27 (1991) 1231*
4 A.P. Malozemoff, *presented at the ICMC, Honolulu, HA, Oct. 26-26, 1994*
5 G.N. Riley, C.J. Christopherson, *submitted to 1995 International Workshop on Super-conductivity*
6 American Superconductor Corporation, *Press Release, Tampere, Finland, June 12, 1995*
7 K.A. Mueller, M. Takashige, J.G. Bednorz *Phys. Rev. Lett., 58 (1987) 1143*
8 A.P. Malozemoff *Physica C, 185-189 (1991) 264*
9 L. Civale, A.D. Marwick, R. Wheeler, M.A. Kirk, W.L. Carter, G.N. Riley, A.P. Malozemoff, *Physica C 208 (1993) 137*
10 M. Manlief, B. Bent, M. Navarro, R. Schwall, *private communications, ASC press release Oct. 13, 1994*
11 C.H. Joshi, C.B. Prum, R.F. Schiferl, D.I. Driscoll, *presented at the Applied Superconductivity Conference, Boston, Oct. 16-21, 1994*
12 J.Tenbrink et al., *Appl. Phys. Lett., 55, 2441 (1989)*

Inst. Phys. Conf. Ser. No 148
Paper presented at Applied Superconductivity, Edinburgh, 3–6 July 1995
© 1995 IOP Publishing Ltd

Processing and properties of 2223 BiPbSrCaCuO/Ag tapes

K. Fischer, U. Schläfer, Ch. Rodig, M. Schubert, W. Häßler

Institute for Solid State and Materials Research, 01069 Dresden, Germany

B. Roas, H.-W. Neumüller, A. Jenovelis

Siemens AG, Corporate Research and Development, 91050 Erlangen, Germany

B. Wolf

Technical University of Dresden, 01069 Dresden, Germany

Abstract

The densification of unreacted cores of BPSCCO/Ag composites during deformation by pressing, swaging, groove rolling, flate rolling and wire drawing has been investigated. A maximum density of about 5 g/cm^3 could be achieved independent of the deformation method employed and the core density prior to deformation. It has been shown, that further deformation beyond the maximum core density may lead to segregation of cores into hard agglomerates surrounded by material of essential lower hardness and causes strong sympathetic sausaging. Different values of microhardness of filaments could be correlated with I_c distribution over the cross section of the multifilamentary conductors. Maximum critical current densities of 7.6 kA/cm^2 (77 K, 0 T) were achieved for conductors of about 100 m in length.

1. Introduction

Best values of the critical current density j_c, calculated over the $(Bi,Pb)_2Sr_2Ca_2Cu_3O_x$ area and J_c, calculated over the full cross section of a 100 m multifilamentary BiPbSrCaCuO/Ag conductor amount to 20.5 kA/cm^2 (77 K, 0 T) and 4 kA/cm^2, respectively [1]. For commercial applications J_c of long-length conductors has to be improved up to nearly 20 kA/cm^2. This seems to be attainable by both increasing of j_c and enhancing of the fill factor, the ratio of superconductor to Ag. However, the microstructure of composites characterized by a high fill factor often reveals sympathetic sausaging of the superconductor/Ag interfaces which may contribute to the decrease of J_c. Hence the objection of this study is to determine the sausaging effect in dependence on the densification process of the precursor core (BPSCCO) during deformation and furthermore to evaluate the influence of the density of BPSCCO filaments on jc of multifilamentary conductors.

2. Experimental methods

Precursor powders with nominal chemical composition of $Bi_{1.8}Pb_{0.4}Sr_2Ca_{2.1}Cu_3O_x$ were packed and sealed into Ag-tubes 9 mm in diameter, a wall thickness of 1.25 mm and 100 mm in length.To study the role the initial powder density after tube packing plays in powder densification during deformation the powders were filled into the tubes step by step each of about 10 mm in high. Each pouring step was followed by the densification of the powder by pressing. The pressure of pressing was varied from 15 MPa to 55 MPa, which resulted in powder densities from 2.5 g/cm³ to 5.5 g/cm³. The BPSCCO/Ag composites were deformed by swaging, groove rolling, wire drawing and flat rolling. Short samples were taken at different strain rates during the deformation processes, and the core densities of these samples were determined by means of a hydrostatic weighting procedure. Details of the determination were described in [2].

To produce multifilamentary conductors the monofilamentary wires were cut, bundled into a silver tube and then drawn and rolled to tapes. The composits were subjected to alternate roll-sinter cycles to convert the precursor to the 2223 phase. Sintering was performed in air at temperatures between 830 °C and 835 °C. 19 and 49 filament tapes in length up to 100 m were fabricated. Microstructure was studied and Vickers microhardness (HV 5 and HV 10) measurements were performed on polished cross-sections with load time of 10 s was choosen for all measurments.

3. Results and discussion

Fig. 1 shows the densification behaviour of BPSCCO cores during deformation by groove rolling followed by flat rolling. It should be noticed, that cores of different densities prior to deformation were investigated. Cores of low initial densities (2.5 - 3.5 g/cm³) were rapidly densified during groove rolling up to a density value of 4.9 g/cm³ (curves 1, 2). While the core density with an initial value of 4.8 g/cm³ was only slightly increased to about 5 g/cm³ (curve 3) even a decrease of density was observed in the case, that the initial core density exceeded 5 g/cm³ essentially (curve 4). From this results it can be concluded, that independent of the initial core density a maximum value of about 5 g/cm³ can be reached by groove rolling. The curves in Fig. 1 suggest that this threshold cannot be exceeded even by use of other deformation technique such as rolling or swaging.

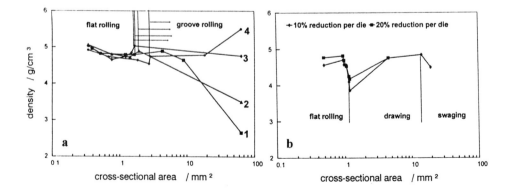

Fig. 1 Densification behaviour of BPSCCO cores during deformation

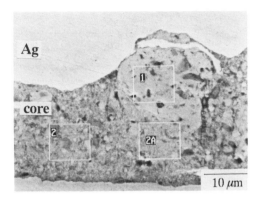

Fig. 2   Microstructure of a swaged and drawn wire

As shown in Fig. 1, the change of the deformation method from groove rolling to flat rolling results in a slight decrease of the core density during the first passes. This is probably due to material flow perpendicular to the long wire axes loosening the microstructure of the BPSCCO core. The material flow may be caused by shear stresses induced by compressing of the square shaped wire. After the wire tape has reached a nearly constant width, the deformation by flat rolling causes rather elongation than broadening of the tape and in the same time the core density increases.

Wire drawing of composites the cores of which have very high densities is accompanied by an essential decrease of the core density as illustrated in Fig. 1b. This phenomena can be understood, assuming that the wire drawing process induces tensil stresses in central regions of the composite as it was demonstrated for homogeneous metallic wires. From Fig. 2, which showes the microstructure of a swaged and drawn wire, it can be concluded that tensile stresses may lead to a break and segregation of the core into hard agglomerates surrounded by material of lower density. The microhardness of these areas marked 1 and 2 in Fig. 2 was determined to be 280 ± 83 kp/mm^2 and 120 kp/mm^2, respectively. Large fluctuations of microhardness along the core length cause fluctuations of the strain along the core during deformation as shown in Fig. 2. Hence the segregation of the core seems to be an important reason of sausaging of BPSCCO/Ag composites. This may be also true for samples which are deformed by flat rolling instead of drawing. As shown by Fig. 1b the core density at first increases by flat rolling up to a value 4.8 g/cm^2. Further deformation leads to a drop of the density which can be due to the same process as described above.

The results discussed so far show, that the parameters of deformation have to be carefully chosen so as to avoid sympathetic sausaging of the oxide/Ag interfaces. On the other hand high densities of the oxide cores should be achieved by deformation, because these correlate with high j_c-values as was shown for monofilamentary conductors [3]. This is also evident from our results of investigations on multifilamentary wire tapes, produced by drawing of BPSCCO/Ag billet to 1.8 mm in diameter and subsequent flat rolling to tapes of 0.3 - 0.4 mm thickness. Fig. 3 gives an overview of microhardness values (HV) measured at cross-sectional areas of unreacted filaments. The HV-values are significant higher for filaments located in the central region of the cross-sectional area of the composite as for filaments near the small edges of the tapes. Pieces of a tape were subjected to the thermomechanical treatment and the critical current distribution was investigated by successively grinding down the tapes from one side and measuring the transport current. As shown from Fig. 4, a maximum of the critical current density appears in the center of the tape. The j_c-maximum correlates with the maximum values of HV of filaments located in the central region of the tape.

The state of fabrication of long length conductors we have reached, is illustrated in table 1.

366

Table 1: Dates of long multifilamentary conductors

Length/m	No. of filaments	fill factor /%	j_c/kA/cm² (77 K)
17	19	28	12.0
100	49	22	7.6

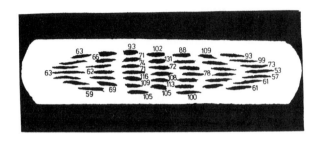

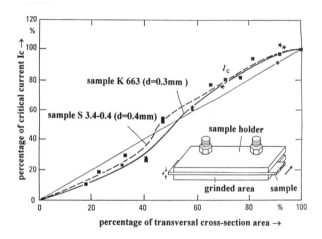

Fig. 3 Microhardness of BPSCCO filaments measured in kp/cm² and dependence of critical current of multifilamentary wire tapes on the reduction of the transversal cross-sectional area

References

[1] Fujikami J, Shibuta N, Sato K, Ishii H and Hara T 1994 *Applied Superconductivity* Vol 2 No 3/4 181 -190

[2] Wolf B 1995 *Rev. Sci. Instrum* **66** 2578 - 81

[3] Satou M, Yamada Y, Murase S, Kitamura T and Kamisada Y 1994 *Appl. Phys. Lett.* **64** 640 - 644

The work was supported by the Bundesminister für Bildung, Wissenschaft und Forschung under contract 13 N 6481.

Inst. Phys. Conf. Ser. No 148
Paper presented at Applied Superconductivity, Edinburgh, 3–6 July 1995
© *1995 IOP Publishing Ltd*

Ceramic core density and transport current density of Bi(2223)/Ag superconductors

P Kováč†, I Hušek† and W Pachla‡

† Institute of Electrical Engineering, Slovak Academy of Sciences, Dubravska cesta 9,
842 39 Bratislava, Slovakia

‡ High Pressure Research Centre, Polish Academy of Sciences, ul. Sokolowska 29,
01-142 Warszawa, Poland

Abstract. To obtain the high current density in Bi(2223)/Ag tapes the several conditions have to be fulfilled. One of them is the high BSCCO ceramics density affecting on inter-grain contacts. This paper discuss the differences in core densities of BSCCO/Ag-claded tapes manufactured by various deformation techniques, as-drawing, extrusion, rolling and uniaxial pressing, based on differences in microhardness (MH) profiles for their transverse core sections. MH profiles reflect the effect of the chosen technology on the ceramics density and on its homogeneity across the core. The effect of core density on the transport current density of press-sintered tapes at 77K and self field is shown as well.

1. Introduction

The texturization and the core density have very remarkable effect on the BSCCO/Ag transport current properties. The hardness, i.e.: the resistance to indentation, is used as the measure of the tape core density [1-3]. It was shown by Yamada et al. [1], that high transport current densities (J_c) reflect the high Vickers microhardness prior to heat treatment, and by Parrell et al. [2] that the hardness increase directly results in an increase in J_c. The mode of deformation process, is decisive for the resulting ceramics core texture and density as well [3].

2. Experimental

Four types of BSCCO/Ag tapes have been made using different modes of deformation. Precursor powder with the nominal composition $Bi_{1.7} Pb_{0.3} Sr_2 Ca_2 Cu_3 O_x$ prepared by spray pyrolisis was used. Pure Ag and Cu-Ag bimetalic tubes have been used for OPIT technique. The composition and the microhardness data are given in Table 1. For the tape 1 hard and the tape 2 soft Ag tube and only wire drawing were applied. Tapes 3 and 4 have been prepared by drawing and rolling of wires after cold (3) and hot (4) hydrostatic extrusion. The tapes were rolled with reduction per one pass smaller than 0.05mm down to 0.4mm in thickness and 0.025mm below (up to 0.1mm). The tapes (1-3) were finally uniaxialy could pressed (3 - 5 times) with pressure ranging from 1GPa to 3 GPa with intermediate annealing at 841°C in 50h intervals. The Vickers microhardness HV (10g, 20s) was measured on the transverse cross-section to monitor the core density after various mode of deformation.

Table 1. The main characteristics of BSCCO/Ag tapes after final rolling,
IF = $HV0.01_{max}/HV0.01_{min}$.

Tape number / history	BSCCO : Ag ratio	Core microhardness	Ag sheath microhardness	Inhomogeneity factor (IF)
1 /soft Ag, drawing,rolling	1 : 1	145 -155	60	1.07
2 hard Ag, drawing,rolling	1 : 2.32	150 - 175	86	1.17
3 /Ag, cold extr, drawing, rolling	1 : 1.51	100 -140	71	1.40
4 /Cu-Ag, hot extr, rolling	1 : 1.26	125 -200	76	1.60

The smoothing of microhardness profiles (HV0.01 versus x/b) by adjacent averaging was used to demonstrate the tendency of hardness (density) variation across the core, see Fig. 1. Transport critical current densities were measured by four probe method with pressed In contacts described elsewhere [4].

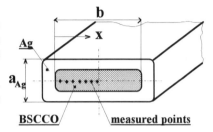

Fig. 1 Measurement of microhardness profile in BSCCO core

3. Results and discussion

The microhardness HV 0.01 in the core and in Ag sheath of tapes (1-4) are given in Table 1.

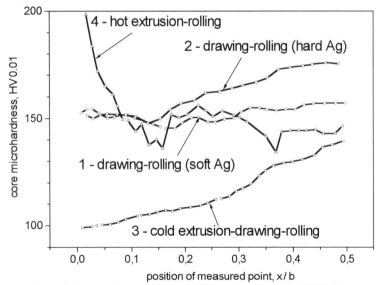

Fig. 2 Microhardness profiles of differently deformed BSCCO/Ag tapes.

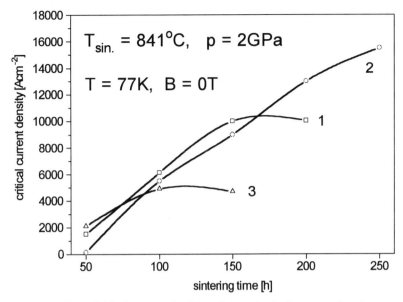

Fig. 3 Critical current densities of tapes 1 - 3 after press-sintering

Fig. 2 shows the comparison between the HV profiles in BSCCO core of tapes made by various deformation modes. Each mode of deformation is reflected by the various hardness pattern. For tapes rolled from as-drawn (1-2) and cold extruded (3) wire increase in density for ceramic closer to the tape axis is observed. This increase is higher for tape made from cold extruded wire (3). It is due to the fact, that it involves higher pattern distorting leading to higher inhomogeneity. The effect of metal sheath hardness is apparent from comparison of tapes 1 and 2. The harder the outer sheath the higher the density of the final core. In the case of hot hydrostatic extrusion tape 4, very high core density after extrusion (HV 0.01 > 300) has been destroyed by rolling as a reason of breaking of highly compacted structure and creation of the cracked and porous core in as-rolled tape [3]. Only some outer part of BSCCO core remains hard. The best core homogeneity is obtained by rolling of as-drawn wire with soft Ag sheath. To obtain dense and homogeneous BSCCO core in Ag clad tape, the small reduction ratios should be used at axisymmetrical as well as at nonaxisymmetric deformation. The core density can be increased by using of harder (Ag-alloyed) and thicker tube. As it was already shown uniaxial pressing can remarkably improve the texture. The influence of the uniaxial pressure on the homogeneity and density of the tape core has been studied alsewhere. The application of optimal pressure is especially important at the first pressing. Only pressure up to approximately 2GPa can maintain the density at high, required level. About this pressure core damage occurs resulting in significant drop in core density [3]. The critical transport current densities (J_c) at 77K and zero magnetic field were measured on the tapes (1-3) and are shown in Fig.3. The J_c values directly reflect the density of the tapes. The highest J_c is obtained for the higher core density tape (2), and vice verse the lowest J_c for the lowest core density tape (3). The various numbers of press-sinter steps is necessary to reach the J_c peak value. It may be also attributed to different densification rate between tapes 1 - 3. For less dense tape (3), lower number of press-sinter steps leads to earlier reaching of J_c maximum as a reason of higher rate of core densification and texture improvement.

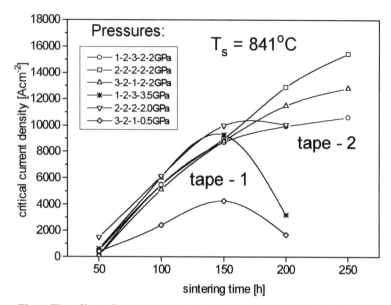

Fig. 4 The effect of pressure magnitude and its sequence on the J_c of tapes 1 and 2

Nevertheless the tape 2 has not so high core densification rate caused by pressing, J_c is still increasing with number of press-sinter steps. The reason of this difference may be in the initial core structure [5] creating the conditions for the further texture development and also in the lower diffusion rate of oxygen into BSCCO grains in the well dense core. The apparent is also difference in J_c after first press-sinter, Fig. 3. The highest J_c is reached at the tape with the lowest core density here. It may be also attributed to the differences in density influencing the grain structure and diffusion of oxygen as well. The growth of Bi(2223) grains may be quicker in low density core but the inter-grain contacts are much worse. The effect of pressing history of tapes 1 and 2 is shown in Fig. 4. It is evident that the Jc of tape 1 is much affected by the pressure magnitude itself, as well as by the pressure sequence between consecutive steps. This effect is not so apparent in the case of tape 2 differing by the more dense BSCCO core and harder and thicker Ag sheath.

Conclusion

To obtain dense and homogeneous BSCCO core in Ag clad tape, the small reduction ratios should be used at axisymmetrical (drawing) as well as at nonaxisymmetric deformation (rolling). The core density can be increased by using of harder Ag sheath. The J_c values directly corresponds to the density values of BSCCO core in tapes after final rolling. The higher the core density the lower the J_c sensitivity of tape on the press-sintering.

References
[1] Yamada et al. 1994 Critical Currents in Superconductorsedited by Weber H W 76
[2] Parrell J A Dorirs S E and Larbalestier D C 1994 Physica C 231 137
[3] Hušek I Kováč P and W. Pachla 1995 will be published in Sup. Sci. and Technology
[4] Kováč P Hušek I Pachla W Melíšek T and Kliment V 1995 Sup. Sci. and Technology 8 341
[5] Pachla W Kováč P Marciniak H and Hušek I contribution at this conference

Inst. Phys. Conf. Ser. No 148
Paper presented at Applied Superconductivity, Edinburgh, 3–6 July 1995
© *1995 IOP Publishing Ltd*

Analysis of inhomogeneities in thermomechanically processed long length Ag/BSCCO tapes

F Lera, A Villellas, A Badía, E Martínez, G F de la Fuente, A Larrea and R Navarro

Instituto de Ciencia de Materiales de Aragón, CSIC-Universidad de Zaragoza, CPS, Universidad de Zaragoza, María de Luna 3, 50015 Zaragoza, SPAIN

Abstract. Several meters long Bi(Ag/2223) tapes have been obtained from the Powder in Tube method by cold rolling-annealing procedures. Moderately good spatial homogeneity along the tape, with critical current densities $J_c(77\ \text{K})$ mainly limited by the presence of cracks, are found. The $J_c(T)$ values derived from direct transport as well as magnetic ac susceptibility techniques indicate the importance of the 2212 phase in those results.

1. Introduction

Thermomechanical processing of Ag sheathed BSCCO-2223 superconductor tapes obtained by the Powder in Tube method (PIT) has yielded very high transport critical current values (J_c). The highest J_c values have been obtained after appropriate deformation sequences (swaging, drawing, rolling and axial pressing-annealing) in short (1-5 cm long) single or multifilamentary axially pressed tapes [1-4]. However, alternative processes, such as continuous rolling thickness reduction have been envisaged in the production of long conductors.

Among the major difficulties to overcome are the spatial inhomogeneities along the tape length which would affect sensitively J_c at liquid nitrogen. This can be due to many reasons, among others: (i) inhomogeneous superconducting core cross section (sausaging) produced by inappropriate deformation or by the presence of harder secondary phase grains; (ii) development of microcracks which would depend on the followed process as well as on the degree of deformation, (iii) inhomogeneous grade of texture and/or distribution of low T_c phases (mainly 2212). To solve this problems, careful control of processing steps, as powder filling and compaction, thickness reduction rate, annealing times and temperatures, cooling down procedure, etc. is needed. In this work, results on relatively long Ag/2223 tapes processed by cold rolling are presented. A moderate critical current density is obtained, with acceptable spatial homogeneity.

2. Experimental techniques

The Bi(2223/Ag) tapes were prepared by the PIT method using precursor powders with a $Bi_{1.8}Pb_{0.35}Sr_{1.87}Ca_2Cu_3O_{10}$ nominal composition, obtained by solid state reaction [5]. The effect of powder packing was cheked by using mechanical compaction which allows to reach densities up to 80% of the theoretical value. Pure silver tubes with 3.5 mm outer and 2.5 mm inner diameters were used in all cases. The ends were sealed and the tubes slowly drawn through subsequently reducing dies to a final 1.5 mm outer diameter. Finally, the tubes were cold rolled in multiple steps to produce several meter long tapes, 3 mm wide and 100 μm thick. The superconducting core cross section mean value of the tapes was 2 mm x 40 μm. Tape sections were annealed in air at different temperatures in the range 810-840 ^oC from 40 to 100 h to determine the optimal heat treatment in a three zone furnace with a high stability region ($\pm 1^oC$) that can accomodate up to 8 (10 cm long) tapes simultaneously. The cross section and microstructure were determined by SEM and optical microscopy in selected tape segments.

Transport J_c measurements at 77 K and zero applied field were performed by the four probe method using pulsed currents with the usual 1 μV/cm criterium. The sample holder, directly immersed in liquid nitrogen, can accomodate a 10 cm long tape and nine voltage contacts, spaced 1 cm apart. For temperature dependent transport J_c measurements, a simpler four contacts sample holder was used, over 2 cm long tape segments. Furthermore, on selected samples, inductive $J_c(T)$ values were derived from the analysis of the excitation (h_0) dependence of the ac susceptibility components $\chi'(h_0,T)$ and $\chi''(h_0,T)$, measured in a Quantum Design SQUID magnetometer.

3. Results and discussion

For fixed characteristics of the BSCCO powders (granulometry and phase composition) the density of the core in the PIT precursor silver tube may change from 30% of the theoretical (crystallographic) value for hand stacked powders up to 80-85%. Although the drawing reduction rate would depend on the nature of the BSCCO core, at the end of the process the density is about 55-60%. This would imply that during drawing there is a net compaction of the core for loosely packed powders or compact breakdown for densities higher than the final. However, the best homogeneity along the wire is obtained for compacted powders with densities higher than the final drawing values, which have been used thereafter in the cold rolling procedure.

With the maximum J_c as optimizing criterium, the best results in our samples correspond to 110 h of annealing at 825 ^oC in a first step and 40 h at 815 ^oC in a second one. Important and common defects or inhomogeneities which appear after annealing are the presence of small black points (pores in the silver sheath) or bumps in the tape. Pores are originated during the mechanical deformation by perforations of the soft silver cover by the harder BSCCO or other secondary phases. Some amount of liquid phase may leak during the annealing, giving rise to small circles around a central hole, which is the characteristic fingerprint observed. On the other hand, the presence of trapped gas (air) within the tape, may give rise during heating to the above occasional bumps. A careful tube filling and compaction process helps to prevent the formation of such bumps, while a harder and thicker sheath would reduce the number of black points.

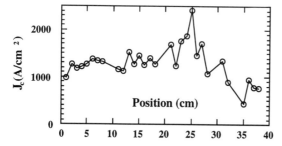

Figure 1. J_c spatial distribution of a 40 cm long Bi(Ag/2223) tape at 77 K.

The results of the spatial critical current distribution, over four contiguous 10 cm segments cut from the same tape, are presented in figure 1. The mean J_c value is 1290 A/cm^2 with a standard deviation of 370 A/cm^2. The inhomogeneities cannot be due to core cross section changes along the tape's length, which exist but show a smaller variation, below 10-15%. Other defects, visible in SEM micrographs, such as cracks, inclusions and holes, are most likely responsible for the J_c valleys. A carefull microstructure analysis (SEM,EDS) has elucidated an important additional cause for the low J_c values measured in the tapes. There is a significant amount of Ca_2PbO_4 distributed through the oxide core which is certainly responsible for a considerable reduction of the effective superconducting cross section, as well as for a deterioration of texture and grain contacts.

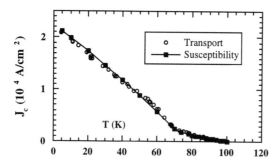

Figure 2. $J_c(T)$ determined from transport and ac magnetic measurements.

The J_c(77 K) values can be strongly influenced by the presence of the 2212 phase, which usually forms intergrowths with 2223, and thus are not enough for tape quality assessment. Transport J_c values as a function of the temperature were also determined in a short segment cut from the same tape. The results are shown in figure 2 (open symbols), and are similar to reported values for Ag/2223 tapes [6]. There is a low temperature linear regime and a high temperature quadratic regime. The somewhat high slope of the low T linear regime can be due to the presence of some remaining 2212 phase.

An alternative method to determine $J_c(T)$ is the analysis of the magnetic ac susceptibility, with advantages over transport measurements: it is contactless and more sensitive. χ' and χ'' were recorded at different temperatures, with ac fields perpendicular to the tape and frequencies below 1 Hz (to avoid any effect coming from the silver sheath). At high enough temperature, a $\chi''(h_0)$ maximum appears for a given driving field h_{max}, and following Brandt [7], $h_{max} = 0.78$ J_c d (SI units), where d is the tape thickness. The inductive J_c (T) values derived are about 20% higher than the transport values. This is satisfactory, because the sample used for magnetic measurements is a short piece cut contiguous to a longer one used in transport characterization. Further-more, the 1 μV/cm is an arbitrary criterium and 20% differences between transport and inductive values are not surprising. Below 82 K, the ac fields needed for attaining the χ'' maximum grow to unreachable values for that tape. To magnetically determine J_c we have analyzed the scaling of $\chi''(h_0,T)$ with $h_0/J_c(T)$ predicted in the framework of the critical state model. The results are shown in figure 2 (filled symbols), where the magnetic data have been scaled to the J_c transport value at 50 K, resulting in an overlap in the full temperature range. The $J_c(T)$ curve shows a crossing behaviour (inflexion) at about 70 K that is also noticeable in $\chi'(T)$ and $\chi''(T)$ measurements performed on powdered tape core samples corresponding to increases of the diamagnetic shielding. Furthermore, the somewhat high ratio $J_c(4.2$ K$)/J_c(77$ K$)\approx 16$ indicates the presence of non negligible amounts of 2212 phase.

In conclusion, this work suggests that in the fabrication of long Ag/BSCCO tapes, the axial pressing step can be substituted by cold rolling, giving a good spatial homogene-ity, but with a substantial degradation of $J_c(77$ K$)$ with values ten times lower, mainly due to the presence of cracks. Furthermore there is a complex phase equilibria with other superconducting and secondary phases [8] and the J_c values at liquid nitrogen cannot give a full assessment of the tape's quality. A more careful analysis of the $J_c(T)$ curves indi-cates that the presence of minor amounts of 2212 and other secondary phases is affecting drastically the $J_c(77$ K$)$ value. To this end, direct transport $J_c(T)$ measurements as well as inductive values have been used finding a good agreement between them. Finally, microstructure analyses also suggest the presence of Ca_2PbO_4 contributing significantly to the observed J_c lowering.

Acknowledgements

The financial support of Spanish CICYT (Projects MAT92-0896-C02-01 and -02) and MIDAS program (Project 94-2442) is acknowledged.

References

[1] Maley M P 1991 *J. Appl. Phys.* **70** 6189

[2] Sato K, Hikata T, Mukai H, Ueyama U, Shibuta N, Kato T, Matsuda T, Nagata M, Iwata K,and Mitsui T 1991 *IEEE Trans. Magn.* **MAG-27** 1231

[3] Grasso G, Hensel B, Jeremie A and Flükiger R 1995 *Physica C* **45–52** 360

[4] Huang, Y B, de la Fuente G F, Ruiz M T, Larrea A, Badía A, Lera F, Rillo C and Navarro R 1993 *Cryogenics* **33** 117

[5] Huang Y B, de la Fuente G F, Sotelo A, Ruiz M T, Larrea A, Angurel L A, Navarro R, Lera F, Ibánez R, Miao H, Primo V and Beltran D 1993 *Solid State Ionics* **63–65** 889

[6] Sun Y, Zhang F, Liu Z, Du J and Zhang Y 1995 *Phys. Rev. B* **51** 519

[7] Brandt E H 1994 *Phys. Rev. B* **49** 9024

[8] Luo J S, Merchant N, Escorcia–Aparicio E J, Maroni V A, Tani B S, Carter W L, Riley G N Jr. 1994 *J. Mater. Res.* **9** 3059

Inst. Phys. Conf. Ser. No 148
Paper presented at Applied Superconductivity, Edinburgh, 3–6 July 1995

Effects of sintering parameters on the superconducting properties of Ag-sheathed Bi(Pb)2223 tapes

A. Baldini *, E. Borchi °, S. Conti *, R. Garre' *, L. Masi ° and A. Peruzzi°

* Europa Metalli Superconductors Division, Piazzale Luigi Orlando, Fornaci di Barga (Lucca), Italy

° Dipartimento di Energetica, Universita' di Firenze, Via S. Marta, 3 50139, Firenze, Italy

Abstract. The sintering heat treatment is one of the most important and delicate steps in the fabrication process of Ag-sheathed Bi2223 tapes: the complexity of the BPSCCO phase equilibria and the fact that the final phase composition and microstructure of the superconducting core are determined by several heat treatment parameters, such as the sintering temperature and atmosphere, the heating and cooling rates and the treatment duration, render fundamental an experimental study of the effects of the parameters involved in the sintering treatment on the superconducting transport properties, in order to improve the critical current densities of the produced tapes. The investigated sintering parameters were temperature, duration, atmosphere, heating and cooling rates. The characterization of the tapes transport properties was performed by means of XRD, resistive transition and zero field I_c (77 K) measurements. A great sensitivity of I_c with respect to the sintering temperature variations was found; the cooling rate, too, was found to be a critical parameter. As expected, reduced oxygen atmosphere treatment proved more effective than air concerning critical current densities enhancement, even if also air treated tapes showed considerable values of J_c.

1. Introduction

It is now well established that Ag-sheathed BPSCCO(2223) tapes made by the OPIT technique present the most promising transport properties for technical applications among high T_c superconductors. However good performances are difficult to reproduce due to numerous parameters involved in the tape fabricating process. Heat treatment is one of the fundamental as well as the most delicate step in the tape fabrication process because it regulates the 2223 phase formation; due to the complexity of the BPSCCO phase equilibria, a strict control of the sintering parameters is difficult and at the same time essential to obtain good final transport performances. Therefore, a systematic experimental study of the influence of the sintering parameters on the cricical current density is essential for achieving satisfactory transport properties in the tapes.

In this study we investigate the dependance of the critical current density on the heat treatment temperature, atmosphere and cooling rate in Ag-sheathed BPSSCO(2223) tapes.

2. Experimental details

The samples studied in the present work were concentric conductors presenting a single superconducting layer concentric with a silver sheath and a silver core. The commercial precursor powder used for the OPIT realization had a nominal cation ratio of Bi:Pb:Sr:Ca:Cu=1.7:0.3:2:2:3. By appropriate cold working, long tapes (about 150m) 300 μm thick and 3.3 mm wide were produced. The cross-sectional area of the superconducting layer was tipically of the order of 0.158 mm². A photo of the cross section of the tape is shown in figure 1.

Fig. 1: Optical micrography of the cross-section of the tape (transverse superconducting area: 0.158 mm²).

Short straight specimens (about 6 cm) were cut from a 150 m length tape, divided in groups and each group heat treated at different conditions. The investigated sintering parameters were temperature, atmosphere and cooling rate. The sintering temperature was varied at 2 °C steps between 830 °C and 840 °C. The heat treatments were performed either in air and in atmosphere of Ar/O$_2$ (10% O$_2$) and their total duration was 80 h.

For each specimen, the critical current at 77 K in zero magnetic field was determined using a standard dc four-probe resistive measurement technique with a criterion of 1 μV/cm. The voltage was measured with a precision of about 0.02 μV and the current with a precision of 0.01 A. In some specimens four voltage taps were placed on the central zone (4 cm) and the critical current was measured for three different consecutive sections of the same specimen. The critical current density was obtained using the cross-sectional area of the superconducting core.

The resistive transition was checked for each sintering condition: the zero resistivity temperature was 105K for all the measured specimens regardless of the heat treatment parameters.

X-ray diffraction analysis radiation was used to investigate the 2223 phase purity in the ceramic core. The diffraction patterns show that the superconducting layer consists predominantly of the 2223 phase; the 2212 phase was undetectable by XRD analysis.

3. Results and discussion

The dependance of the critical current density (77 K, **B** = 0 T) on the sintering temperature and atmosphere is shown in fig. 2.

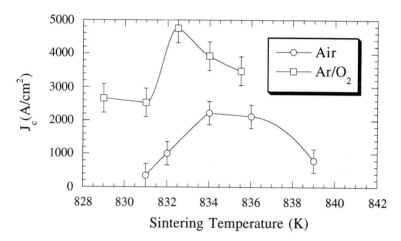

Fig. 2: Dependance of the critical current density (T = 77 K, **B** = 0 T) on the sintering temperature and atmosphere.

The reported J_c values were obtained by averaging the values corresponding to different specimens heat treated in the same conditions. As expected, the samples annealed in a reduced oxygen atmosphere yielded higher critical current densities than those sintered in air: the highest measured J_c values were 5080 A/cm^2 for the Ar/O$_2$ treatments and 2530 A/cm^2 for the air treatments. The critical current density showed a great sensitivity to temperature variations of 2 °C, confirming that the annealing temperature is a very critical parameter. The optimized sintering temperature was found to be 832.5 °C in Ar/O$_2$ and 834 °C in air.

The cooling rate was also found to be a critical parameter in determining the final transport properties of the tape: specimens cooled down at a rate of 10 °C/min showed a reduction of J_c of about 60% with respect to the specimens cooled down at the usual rate of 2 °C/min. A slow cooling rate seems to favour the oxygen take-up of the BPSSCO grains; a fast cooling rate, on the other hand, seems to freeze the optimum phase equilibrium region chosen for the annealing and yields a minimum amount of 2212 phase [1,2].

The critical current stability for thermal cycling between room temperature and 77 K was also checked: one of the specimens treated at 829 °C in Ar/O$_2$ was cooled down to 77 K and then heat treated to room temperature for ten times and after each cycle I_c was measured. Fig. 3 shows the critical current degradation with increasing the number n of thermal cycles between 77 K and room temperature. After a reduction of about 4% in the first 5 cycles, the I_c values measured for further cycling tended to stabilize around 95% of the starting value. The I_c degradation during thermal cycling is due to the fact that the coefficient of thermal expansion of the ceramic superconducting layer is lower than that of the Ag sheath and core; as a consequence, during thermal cycling tensile stresses are exerted on the ceramic layer that cause microcracks in the 2223 grains and a consequent J_c degradation [3].

378

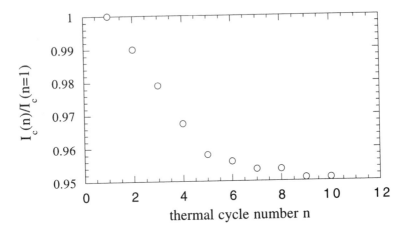

Fig. 3: Critical current degradation with increasing number of thermal cycles between 77 K and room temperature.

3. Conclusions

We investigated the dependence of the critical current density on the sintering temperature, atmosphere and cooling rate in the Ag-sheathed BPSCCO(2223) tapes.

Considerably higher J_c values were obtained in the specimens heat treated in atmosphere of Ar/O_2 (10% O_2) with respect to the air treated specimens . A great sensitivity of J_c to sintering temperature variations of 2 °C was observed, confirming the thermodynamical instability of the 2223 phase with respect to heat treatment temperature. The optimized sintering temperature was found to be 832.5 °C for the reduced oxygen atmosphere treatments (which produced a maximum J_c of 5080 A/cm^2) and 834 °C for the air treatments (which produced a maximum J_c of 2530 A/cm^2).

Acknowledgements

We would like to thank Dr. F. Del Giallo and Dr F. Pieralli of IROE-CNR (Firenze, Italy) and Prof. G. Spina of the Università di Firenze (Firenze, Italy) for their help in performing the heat treatments.

References
[1] Y.B. Huang, G.F. de la Fuente, A. Larrea and R. Navarro 1994 *Supercond. Sci. Technol.* **7** 759
[2] D.J. Brauer, D. Bush, R. Eujen, A. Glaudun and J. Hudepohl 1992 *Cryogenics* **32** 1052
[3] S. Ochiai, K. Hayashi, and K. Osamura 1991 *Cryogenics* **31** 955

Inst. Phys. Conf. Ser. No 148
Paper presented at Applied Superconductivity, Edinburgh, 3–6 July 1995

Transverse inhomogeneities of Pb/BSCCO 2223 tapes from the core-sheath interface to the centre of the core

Z Yi, L Law, S Fisher, C Beduz, Y Yang, R G Scurlock, R Riddle*

Institute of Cryogenics, University of Southampton, SO17 1BJ, UK

*Merck Ltd., Poole, Dorset, PH15 1HX, UK

Abstract. The percentage of converted 2223 phase and the Vicker's microhardness of the core have been measured across the thickness of Pb/BSCCO 2223 tapes sintered for up to 350 hours. We have found that there is a 2223 phase gradient from the sheath-core interface to the centre of the core. The rate of the phase conversion is much faster at the interface than at the centre. The microhardness is found to decrease from a maximum at the core-sheath interface to a minimum at the centre.

1. Introduction

One of the inhomogeneities in Pb/BSCCO 2223 tapes, made using an oxide powder-in-tube method, is the variation of their current carrying capacity across the thickness of the core. Recently M Lelovic et al. [1] showed that the critical current density of 1.1×10^5 Acm^{-2} at 77K in self field was sustained in a thin layer (~10µm) of a Pb/BSCCO 2223 tape at close proximity to the silver sheath. This layer appears to carry about 90% of the overall current through the tape. The reasons for the large variation of J_c may include the difference in the amount of 2223 phase , the alignment of the grains and the density across the thickness of the core. In the TEM observations [2] of the tapes kept in air at around 800°C for 72 hours, nearly pure 2223 grains were found covering the core-sheath interface while 2212 was still the major phase away from the interface. However, quantitative results of the phase conversion across the thickness of the tape have not been reported.

In the present work we have measured the percentage of 2223 phase at different positions across the thickness of tapes which were sintered for different periods up to 350 hours. We have found that the conversion from 2212 to 2223 phase is much faster at the core-sheath interface than in the centre of the core. The conversion appears to depend exponentially on sintering time after an incubation period.

2. Experimental

The precursor powder used was made by an EDS process. The nominal stoichiometry of the precursor was $Pb_{0.8}Bi_{1.34}Sr_{1.9}Ca_{2.03}Cu_{3.06}O_x$ (Merck Ltd.) with the major phase being 2212, and no 2223 phase could be detected by XRD. The precursor powder was extruded into 2.8mm diameter rods which were packed into a silver-silver alloy tube of 3mm inner diameter and 4 mm outer diameter. The wire with a final diameter of 1.3mm was rolled to a tape. The tapes were sintered in air at 832°C for up to 350 hours. Two intermediate cold rollings were applied after being sintered for 30 hours and 60 hours. The final overall thickness of the tapes was 165μm and the average core thickness was 45μm.

The evolution of 2223 phase in the tapes was measured by XRD. In order to obtain XRD information at different layers across the thickness of the core, each tape was polished down by about 10 micrometers at each step and XRD was then performed on each surface. The percentage of 2223 phase was estimated from the intensity of the 2223 peak and the 2212 peak as follows: $\%2223 = I_{2223(0010)}/(I_{2212(008)} + I_{2223(0010)})$. The microstructure of the core of the tapes was examined using SEM. The hardness of the core was measured using the Vicker's microhardness test.

3. Results and discussion

The XRD spectra of different layers of the tapes sintered for up to 200 hours are shown in Fig.1(a). There is a large gradient of 2223 phase across the thickness of the tapes sintered for a period shorter than 60 hours. For example, the percentage of 2223 phase of the tape sintered for 30 hours is ~90% at the core-sheath interface whereas it is ~25% in the core centre. In the literature the percentage of 2223 phase is usually evaluated from the XRD taken on the peeled surface of the core. The result in Fig.1(a) shows that there may be a large error in the estimation of the overall amount of 2223 phase in the core using the XRD of peeled surfaces, especially for short sintering times.

The gradient of the 2223 phase decreases with increasing sintering time as shown in Fig.1(a). After 100 hours the difference in the 2223 phase percentage between the interface and the core centre is less than 10%. The evolution of 2223 phase with sintering time is shown in Fig1(b). It appears that there is an incubation period in which the amount of 2223 phase increases very slowly. Only after a certain time does a fast, major conversion take place. The incubation period varies significantly from a very short time at the interface to about 20 hours at the centre. Once the major 2223 phase conversion has begun it appears to depend exponentially on the sintering time. The conversion rate of 2223 phase near the interface is much larger than that in the centre. The 2223 phase at the interface has already reached the equilibrium amount(~90%) after being sintered for 30 hours, whereas, it took about 100 hours for the 2223 phase at the centre to reach the equilibrium amount(~80%).

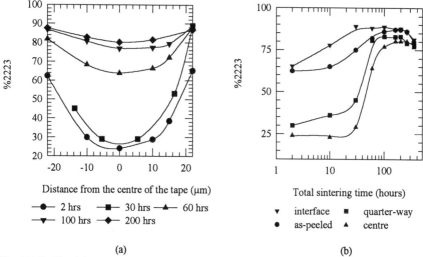

Distance from the centre of the tape (μm)

—•— 2 hrs —■— 30 hrs —▲— 60 hrs
—▼— 100 hrs —♦— 200 hrs

Total sintering time (hours)

▼ interface ■ quarter-way
• as-peeled ▲ centre

(a)   (b)

Fig. 1(a) Profile of the percentage of 2223 phase across the thickness of the core in tapes sintered for up to 200 hours, (b) the 2223 phase evolution of different layers within the core.

Along with the phase conversion gradient, there is also a gradient of the grain size and the hardness across the thickness of the core. Fig. 2(a) & (b) show the grain size near the core-sheath interface and at the centre of the core, respectively. The dimension of the well- aligned grains at the interface is about 20μm, whereas the smaller grains at the centre of the core are 2 to 15μm across. The profile of the microhardness of the core is shown in Fig.3. The hardness of the as-rolled core is nearly homogeneous across its thickness, whereas there is a gradient in the tape sintered for 160 hours. The higher values of microhardness at the interface may be due to the better alignment of the grains which would result in a higher density.

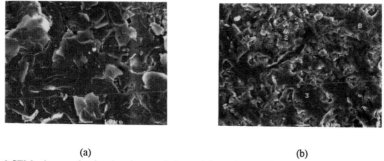

(a)   (b)

Fig.2 SEM micrographs showing the morphology of the grain size of a tape sintered for 30 hours taken (a) close to the silver-core interface and (b) close to the centre of the core.

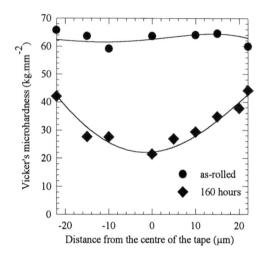

Fig.3 Vicker's microhardness as a function of distance from the centre of the tape.

Our experiment suggests that the silver sheath enhances the growth of the 2223 grains at the interface. The 2223 grains grown at the interface may promote further growth of 2223 grains towards the centre of the core. The mechanism by which the silver sheath enhances 2223 grain growth is not yet fully understood. At first glance, the gradient of 2223 phase suggests a simple diffusion process but the time dependence of the phase conversion does not appear to agree with this suggestion. The presence of the incubation period and the difference in the asymptotic percentage of 2223 phase at different positions away from the interface do not agree with a simple diffusion model.

4. Conclusions

The major conversion of 2212 to 2223 phase appears to depend exponentially on sintering time after an incubation period. Both the incubation period and the time constant for reaching an asymptotic 2223 phase percentage vary with position within the core. Consequently, there is usually a gradient of the 2223 phase percentage and the grain size as well as the microhardness across the thickness of the core during the thermal processing of Pb/BSCCO 2223 tapes. This gradient in 2223 phase percentage is still apparent even after 350 hours sintering.

References

[1] Lelovic M , Krichnaraj P, Eror N G and Balachandran U 1995 *Physica C* **242** 246
[2] Feng Y, High Y E, Larbalestier D C, Sung Y S and Hellstrom E E 1993 *Appl. Phys. Lett.* **62** 1553

Inst. Phys. Conf. Ser. No 148
Paper presented at Applied Superconductivity, Edinburgh, 3–6 July 1995
© *1995 IOP Publishing Ltd*

EFFECTS OF THE TIME SEQUENCE OF COLD WORKING ON CRITICAL CURRENT DENSITIES OF AG-ALLOY SHEATHED BiPb2223 TAPES

M Penny[*] , C Beduz, Y Yang, M Al-Mosawi, R Scurlock and R Wroe[+]

Institute of Cryogenics, University of Southampton, Southampton, SO17 1BJ. U.K.
[+]E.A. Technology, Capenhurst, Chester, CH1 6ES, U.K.

Abstract. The effect of the time sequence of cold working/sintering, on transport Jc has been studied. The results on the variation of Jc with sintering time prior to the second cold rolling, t_{S2}, are presented. Ag-Alloy sheathed, BSCCO-2223 tapes produced by PIT method were used in this study. The tapes were cold rolled to 240-260μm and processed at 825-840°C with two intermediate cold rollings. The sintering time prior to the first intermediate cold rolling was kept constant at 20h, however, prior to the second cold rolling, the tape was sintered for a time, t_{S2}, from 20 to 35h. The tapes were then sintered to give a total sintering time of 140h. Tapes produced with t_{S2}= 27h had the highest Jc of 20 x10^3 Acm^{-2} (77K, self field). This tape had an Ic of 23.3A in a monofilamentry geometry. The processing was monitored at the stages of intermediate mechanical work by SEM and XRD, in order to reveal any significant mechanism of enhancement in Jc. It seems that the percentage of the 2223 phase and the grain size before mechanical work is performed, has a very sensitive influence on the enhancement mechanism related to Jc capabilities.

1. Introduction

Critical current densities as high as 6x10^4Acm^{-2} have been achieved[1] in BSCCO tapes. These high current carrying capabilities have initiated an enormous effort to optimise the material processing conditions. It is generally believed that grain alignment, densification, and phase formation are the important parameters effecting the Jc of the tapes[2]. Total sintering time and mechanical deformation are some of the variables which govern these parameters.

Uniaxial pressed (26,000 Acm^{-2}) tapes have been proven to enhance the alignment and density more so than rolled (18,500 Acm^{-2}) tapes[3]. For engineering applications, long lengths are needed and therefore, rolling has become the main method of manufacture. The different effect produced by the mechanical work is reflected in the powder motion within the tape. Rolled tapes tend to produce ridges parallel to the direction of the rolling whereas, pressed tapes produce ridges parallel to the c-axis[1]. For each method of mechanical deformation, the time (in relation to sintering time) to apply the second deformation has been investigated in relation to the percentage of 2223 phase or grain size before the deformation.

Using these tools, optimisation of the thermomechanical processing parameters has become easier, with the Jc values increasing over the last five years to greater highs but the

[*] CASE studentship supported by E.A. Technology

progress has not accomplished the optimistic values hoped for, at the discovery of the BSCCO system. More work is necessary for the continual progress and understanding of the mechanisms involved.

2. Experimental procedure

Ag-Alloy sheathed Bi-Pb-Sr-Ca-Cu-O superconducting tapes were prepared using the conventional powder in tube method. The precursor powder was 'partially calcined' $Pb_{0.34}Bi_{1.84}Sr_{1.91}Ca_{2.03}Cu_{3.06}O_x$ which comprises of BSCCO-2212 as the major phase. The packing density of the wire before drawing was $\approx 60\%$ of the theoretical density. The tapes were drawn to ≈ 1.2mm and cold rolled to a thickness of 240μm-260μm and processed at 825-840°C (in Air) with two intermediate cold rollings. The sintering time prior to the first cold rolling was kept constant at 20 hours, however, prior to the second cold rolling (140-165μm), the tape was sintered for a time, t_{s2}, from 20 to 35 hours. The tapes were then further sintered to give a total sintering time of 140 hours.

Each tape was characterised by transport Jc (B=0 to 0.5T) measurements at 77K, XRD and S.E.M. The corresponding XRD and S.E.M. were taken on the peeled core/interface of the tape before intermediate mechanical work. Therefore, the conversion of 2212 to 2223 phase and the grain size could be monitored during the thermomechanical process. Any resulting influences on Jc could be related to the monitored parameters.

3. Results and discussion

The Jc measurements carried out on the final tapes are presented in Figure 1. These results shows a optimum Jc for tape processed with 27 hours sintering prior to the second cold rolling. Also the sensitivity of the time prior to second rolling is reflected in the fact that there is a 50% drop in Jc within ±7 hours of the peak value. This peak value being Jc = 20×10^3 Acm^{-2} at 77K (self field) using a voltage criterion of 1 μVcm^{-1}. The final core area for all the tapes were the same when polished cross sections were studied under the microscope which suggests the same mechanical deformation in the rolling process.

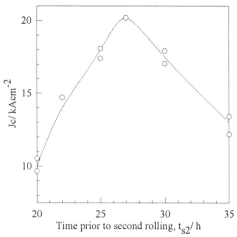

Fig. 1 Transport critical current density as a function of second rolling time.

From the XRD traces of each tape prior to the intermediate deformation, it can be shown, in Figure 2, that the percentage of 2223 phase present, changes monotonically from 50% to 80%. The final conversion at 140 hours for all the tapes was between 85% to 90%.

For t_{s2} =27 h, the percentage of 2223 phase was 77%, whereas for t_{s2}=20h was 50%. Any degree of mechanical deformation which produces microcracks or voids within the microstructure requires the remaining liquid to 'heal' and reconnect the grain boundaries. Therefore, the any optimisation in Jc could be influenced by the amount of conversion of 2212+liquid phase to 2223 yet to be completed. The enhanced Jc shown in Figure 1, at t_{s2} =27h, could be result of an optimum amount of 2223 phase or conversely, the amount of liquid and unreacted phase remaining, before final deformation was performed.

The conversion rate in Figure 2, is unique to some degree on the packing density, precursor size, sintering temperature/time and the degree of intermediate deformation. All of these parameters influence the density of the core which in turns changes the conversion rate. The line in Figure 2 represents the average conversion rate of the BSCCO tape, at the interface. This rate has shown to be different depending on the position within the core[4]. Therefore, it is important to note that the value of t_{s2} would be different for other BSCCO wires and tapes.

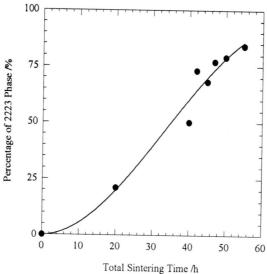

Fig. 2 Graph showing the percentage of 2223 phase with sintering time.

Although the phase content is important mechanism on the enhancement of Jc in this study, grain size may be another factor. For instance, if the grain grows too large before mechanical work, the remaining liquid may only be able to reconnect the boundaries and not the microcracks across the grain. Whereas if the grain is broken during mechanical deformation, when the grain is small the liquid remaining could 'heal' microcracks more easily. Therefore, if this is the case in the BSCCO tapes, there could be an optimum size. From Figure 3a,b,c,d, the grain size of the peeled tapes prior to second cold rolling are shown for t_{s2} =20,25,27 and 30 respectively. For t_{s2}=27h(Fig 3c) has larger grain size when compared to lesser times (Fig. 3a and Fig. 3b). However, the grain size difference between Fig.3c and Fig. 3d is harder to appreciate. Perhaps, grain size of 10-15μm (t_{s2}=27h) could be an influencing factor on the enhancement of Jc or just an equilibrium grain size which has been reached at this total sintering time and has no influence on Jc.

386

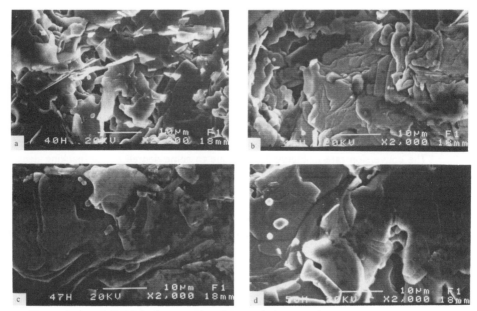

Figure 3a,b,c,d: S.E.M. photographs of the peeled tapes prior to second cold rolling, t_{s2}=20,25,27,30h respectively.

4. Conclusion

For these tapes, sintering time prior to second rolling of 27 hours, produced an enhanced Jc of 20×10^3 Acm^{-2}. Earlier mechanical work does not seem to enhance Jc due to the absence of sufficient 2223 phase. Thus the final alignment is not improved compared to later mechanical rolling times. Conversely, late mechanical work may produce a decrease in Jc shown in Fig.1, due to the insufficient liquid phase to 'heal' the microcracks produced by the degree of mechanical deformation.

From this study there seems to be an optimum time for the second rolling/final deformation. This time is dependent on the degree of mechanical work performed and the relative density and processing parameters imposed on the tape. Therefore the time, t_{s2}, can not be superimposed directly onto other BSCCO wires and tapes but this study does show a the sensitivity of just one parameter that influences the Jc. If the first rolling time is adjusted, then the conversion rate could be affected, thus, shifting the optimum t_{s2} sintering time.

Therefore, provided the first mechanical work is performed at 21%(peeled core) then an optimum t_{s2} can be found in the range of 2223 formation (68% to 79%) which enhances the overall Jc.

References

[1] Q Li et al 1993 Physica C: **217** 360-366
[2] S Dou & H Liu 1993 Supercond. Sci. Tech. **6** 297-314
[3] G Grasso, A Perin, B Hensel and R Flükiger 1993 Physica C: **217** 335-341
[4] Z Yi, L Law, C Beduz & R Scurlock EUCAS 1995 Conference Proceedings, Edinburgh

Inst. Phys. Conf. Ser. No 148
Paper presented at Applied Superconductivity, Edinburgh, 3–6 July 1995
© 1995 IOP Publishing Ltd

The Effect Of Heating Rate On The 2223 Phase formation and Core Morphology Of (PB,BI)2223 Superconducting Tapes.

D.M. Spiller, M.K. Al-Mosawi, Y. Yang, C. Beduz and R. Riddle[++]

Institute of Cryogenics, University of Southampton, Southampton, SO17 1BJ. UK.
++ Merck Ltd. Poole, Dorset, PH15 1HX. UK.

Abstract. The heating rate is particularly important in sintering ceramic systems as it is likely to affect the homogenisation of the liquid phase and the grain size. We present the results of an investigation in to the effect of the heating rate on the 2223 phase formation at the silver interface and core morphology of (Pb,Bi)2223 superconducting tapes. The precursor powder used in the investigation was partially calcined (Pb,Bi)2223. Tapes were made by the OPIT technique with two intermediate cold rolling stages and processed at 820°C ≤ Ts ≤ 850°C for sintering times up to 20h. The 2223 phase formation and core morphology was studied using XRD, optical microscopy and SEM analysis techniques. The grain size, core composition and morphology were studied for heating rates of $2°h^{-1} \leq T' \leq 40°h^{-1}$, prior to the first intermediate cold rolling. The heating rates have been seen to affect the speed of conversion of the 2223 phase in the tapes core. Using XRD analysis the phase formation as a function of time was obtained.

1. Introduction.

Since the discovery of the BSCCO system [1] there have been extensive investigations in to the optimisation of the processing parameters in an attempt to produce superconducting tapes with high transport Jc suitable for practical applications. The oxide powder in tube method has become a popular and successful route in the production of Ag sheathed (Pb,Bi)2223 superconducting tapes. Up to this point there seems to have been no reported work regarding the optimisation of the heating rate used in the production of high Jc (Pb,Bi)2223 tapes. The rate at which the tape is heated to it final sintering temperature is important as it affects the amount of liquid phase present at the sintering temperature. The heating rate has also been shown [2] to influence the grain size after sintering at the same sintering temperature T_s. In an attempt to optimise the processing parameters of the (Pb,Bi)2223 system it may be necessary to control the phase formation and the grain size of the superconducting core prior to the intermediate cold working procedures [3]. The influence of the heating rate may be an ideal way to control these parameters in an optimised sintering procedure.

2. Experimental.

The precursor powder used in this investigation was commercially available Merck partially calcined (Pb,Bi)2223, the major phase present was Bi(2212). Wires were prepared by the oxide powder in tube method (OPIT) [4]. The packed silver tubes were drawn to ~1mm and then rolled to $200 \leq d \leq 300\mu m$. Samples of the tapes were then thermally treated at T_s= 830°C in air with heating rates of 2, 20 and 40°Ch^{-1}. The tapes were processed for sintering times up to 40h, a similar time scale for the tapes prior to the initial intermediate cold rolling during a full sintering procedure. The phase composition of the tapes were then characterised using XRD of the tapes with the top silver sheath layer 'peeled' away. The core morphology and the grain size of the 2223/2212 was examined using SEM.

3. Results and Discussion.

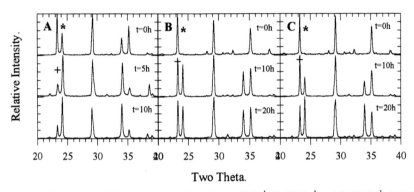

Figure 1: XRD traces of (Pb,Bi)2223 tapes heated at A) $2°Ch^{-1}$, B) $20°Ch^{-1}$ and C) $40°Ch^{-1}$ t0 830°C.
* 00$\underline{10}$ (2223), + 008 (2212).

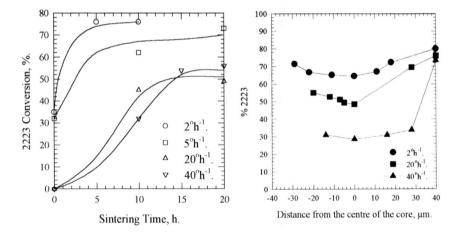

Figure 2: % 2223 Formation for various heating rates to 832°C.

Figure 3: % 2223 Formation throughout the core thickness for different heating rates after 20h sintering at 832°C.

In figure (1) and (2) t=0h represents the tape which was heated to 830°C and then immediately cooled to room temperature at $60°Ch^{-1}$. Figure (1) shows the XRD traces of the peeled sections of the tapes treated with heating rates of 2, 20 and $40°Ch^{-1}$ to 830°C in air. From figure (2) it can be seen that the initial phase conversion of the tapes with the slow heating rate ($2°Ch^{-1}$) seems to be quicker than the samples heated at 20 or 40 $°Ch^{-1}$. Samples with heating rates of 20 and 40 $°Ch^{-1}$ were found to have no appreciable 2223 phase present in the core after the initial heating up to 830°C as shown by figure(1, (B & C)), t=0h.

At first it may appear that the increased conversion for a slow heating rate may be due to the increase in the length of time the tape spends at the vicinity of the sintering temperature. The tapes heated at $2°Ch^{-1}$ spend the equivalent of 5h in the region between 820°C and 830°C where it may be possible for the 2212 to convert to the (Pb,Bi)2223 phase. However, by comparing the XRD traces of the $20°Ch^{-1}$ sample treated for 10h and the $2°Ch^{-1}$ sample treated for 5h, figure (1, (A & B)), it is evident that the sample with the slower heating rate, $2°Ch^{-1}$, has a higher proportion of (Pb,Bi)2223. Thus the time that has elapsed as the tape was heated from 820°C to 830°C does not seem to be responsible for the increase in the (Pb,Bi)2223 phase formation.

The lack of conversion in the tapes heated at 20 and $40°Ch^{-1}$ for t = 0h may be due to a change in composition or characteristics of the liquid phases present during heating to the sintering temperature. The heating rate is known to affect the viscosity, homogeneity and the amount of the liquid phase formed during heating. The heating rate of $2°Ch^{-1}$ seems to be slow enough for the liquid phases to remain in an equilibrium state during heating, thus allowing 2223 formation to take place simultaneously. The heating rates of 20 and $40°Ch^{-1}$ may heat the core 'too quickly' inhibiting the liquid phases from remaining in an equilibrium type state during heating. The non equilibrium state of the liquid phases during heating may be caused by the absorption of some of the more soluble cations of the 2212 phase in to the liquid, the change in the effective composition of the liquid phase may then inhibit 2223 formation. A change in the physical properties of the liquid phase may also be responsible for detrimental effects on the phase formation. Further investigations are necessary to determine the effect of the heating rate on the liquid phase composition etc.

Figure (4) shows the SEM micrograph of peeled sections of the tapes heated with heating rates of 2 and $40°Ch^{-1}$ and treated for 0h and 10h. The tapes heated at $2°Ch^{-1}$ for 0h (A) can be seen to have larger grains than the tape heated with a heating rate of $40°Ch^{-1}$ (C). From the XRD traces in figure(1) the grains present in the tape heated at $2°Ch^{-1}$ and treated for 0h are mainly (Pb,Bi)2223, the phase conversion was found to be ~76% 2223. The XRD trace of the tape heated at $40°Ch^{-1}$ and treated for 0h shows the phase composition of the peeled section of tape to be almost pure 2212. However, the SEM micrographs of the tapes after 10h sintering (B & D) shows a similar grain size for both heating rates. Figure (1) also shows an increase in the amount of 2223 phase present in the tape heated at $40°Ch^{-1}$ after 10h. The difference in the XRD traces and the SEM micrographs suggests that the grain size of the 2212 phase dauring the initial sintering may also be affected by the heating rate of the system.

It has been reported [5, 6] that there is an enhanced rate of 2223 formation at the interface with the silver sheath in (Pb,Bi)2223 superconducting tapes. The variation in the speed of formation of the 2223 phase leads to a distribution in the 2223 content throughout the thickness of the core. There is also a variation of the thickness of the core present on the peeled sections, thus the XRD trace of the peeled section representing an average value of the peeled surface. The silver sheath alters the phase diagram of the system in such a way that the speed of 2223 formation can been seen to increase. Figure (3) shows the phase composition through the core of tapes heated at 2, 20 and $40°Ch^{-1}$ for 20h. It is clear that at the silver interface the amount of the 2223 phase is similar for all the heating rates. However, further away from the silver interface, towards the centre of the core, the difference in the phase composition is more evident. Thus the effect of the heating rate on the phase formation is more relevant to the region further from the Ag interface.

390

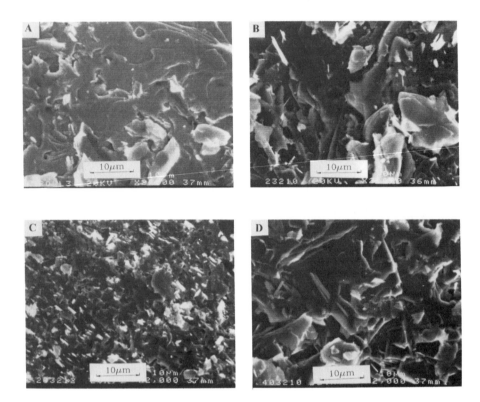

Figure 4: SEM micrographs of peeled sections (Pb,Bi)2223 tapes: A) 2°Ch^{-1}, 0h, B) 2°Ch^{-1}, 10h, C) 40°Ch^{-1}, 0h, D) 40 °Ch^{-1}, 10h.

4. Conclusions.

It is evident that the heating rate is an essential parameter that seems to have been neglected in the control of the composition and morphology during the preparation of superconducting (Pb,Bi)2223 tapes. It has been shown [3] that the core composition and the morphology at the time of the intermediate cold rolling is essential for obtaining high transport Jc. The heating rate may be a way of further controling these parameters prior to cold working. Further investigations are required to determine an optimum heating rate to produce high transport Jc tapes for practical applications.

References

[1] Koyama K, Endo U & Kawai T *Jpn. J. Appl. Phys.* 27 L1476 (1988)
[2] Rand T.J., Yi Z, Yang Y, Beduz, Riddle R & Pham K. *Applied Superconductivity* ed by Freyhardt HC, DGM Informationsgesellschaft mbH, **vol. 1**, 121 (1993).
[3] Penny M, Beduz C, Yang Y, Al-Mosawi M, Scurlock & Wroe R. *Eucas '95* Edinburgh (1995)
[4] Yamada Y, Oberst B & Flükiger R. *Supercond. Sci. Technol* 4 165 (1991).
[5] Feng Y, High YE, Larbalester DC, Sung YS & Hellstrom EE. *Appl. Phys. Lett.* 62 1553 (1993)
[6] Yi Z, Law L, Fisher S, Beduz C, Yang Y, Scurlock RG & Riddle R. *Eucas '95* Edinburgh (1995)

Inst. Phys. Conf. Ser. No 148
Paper presented at Applied Superconductivity, Edinburgh, 3–6 July 1995
© *1995 IOP Publishing Ltd*

Densification and texturing of long Bi,Pb(2223) bars by hot rolling[*]

A Perin, E Walker and R Flükiger

Université de Genève, Département de Physique de la Matière Condensée,
24 quai Ernest-Ansermet, CH-1211 Genève 4, Switzerland.

Abstract. Long bulk Bi,Pb(2223) bars of 0.4mm thickness, without silver sheath were produced by a rolling process with rolls heated at >800°C. A newly developed rolling machine allowed the precise control of the deformation of Bi(2223) bars with no limitation in length. Already reacted Bi,Pb(2223) samples, which can not be deformed at room temperature, could be deformed at 824°C without cracking. The resulting bars reached a density of 5.5 g/cm³. Microstructural observations and X-ray diffraction measurements were performed, showing strong texture in the deformed bars. A critical current density of 1800 A/cm² was measured for samples deformed at 824°C. This process opens a new possibility in view of the fabrication of long (>1m) bars for high current leads.

1. Introduction

Among the parameters influencing the current transport properties of Bi,Pb(2223) bulk superconductors, texture and density are of crucial importance. Bi,Pb(2223) is a brittle material and therefore room temperature deformation cannot be employed to improve these properties. Hot deformation at temperatures exceeding 800°C proved to be one of the most promising methods investigated in order to enhance density and grain orientation of Bi,Pb(2223) bulk samples [1-4]. However the great majority of high temperature deformation experiments were performed by applying an uniaxial force on small samples placed in a furnace. This technique has the major disadvantage that it cannot be applied to the production of bars longer than a few centimeters. Most applications, like for example currents leads, require longer lengths. Hot rolling, which is a continuous production process, allows the fabrication of bars with no limits in length. We present in this article the effects of high quality controlled hot rolling on density and texture characteristics of Bi,Pb(2223) bulk bars.

2. Experimental

Some attempts to deform preheated bulk Bi,Pb(2223) samples with rolls kept at room temperature have already been made [5]. This technique has some important limitations: the samples must be preheated at a temperature which is higher than the desired deformation

[*] Work supported by the Swiss National Foundation (PNR 30)

temperature and moreover, the rolling speed has to be very high in order to keep the sample temperature at the required value during deformation.

In order to precisely control the conditions of deformation we developed a new hot rolling machine with computer controlled rolling force, speed and temperature, thus improving the characteristics of a previously presented device [6]. In the new machine the deformation temperature is achieved by working with rolls heated at temperatures up to 920°C. The precision of temperature control is ensured by placing the rolls in a resistively heated furnace. Control of the rolling force is provided by measuring the rolling separating force. The rolling temperature can be controlled up to 920°C by varying the temperature of the furnace with a reproducibility of ±2°C. The rolling speed can be varied continuously from 10^{-1} m/min. down to 10^{-3} m/min. The rolling force can be determined with a precision better than 10N and controlled by varying automatically the distance between the rolls. In order to avoid contamination a silver foil of 75μm was placed around the bars before hot rolling.

Bi,Pb(2223) bars were prepared by solid state reaction. Coprecipitated powders with cation ratio $B_{1.72}Pb_{0.34}Sr_{1.83}Ca_{1.97}Cu_{3.13}$ were first heated at 300°C to remove the organic compounds and water. The powder was then pressed into pellets and calcined twice at 820°C for 24 hours with intermediate manual grinding. The resulting pellets were ball milled, pressed again into pellets and heat treated at 845°C for 50 hours to form the Bi,Pb(2223) phase. Finally, in order to enhance the homogeneity of the final samples, the pellets were ground again. To obtain a bar-like shape the resulting powder was pressed with a pressure of 36 kN into bars 1.5mm thick , 5 mm wide and 25mm long. The bar shaped samples were then sintered at 845°C for 50 hours. X-ray analysis showed that the samples contained almost only Bi,Pb(2223) . The resulting bulk density was measured by volume and weight measurements and was 4.45 g/cm^3.

The samples were then hot rolled at 824°C in small steps down to 0.4mm thickness. The rolling force was limited to 850N by using a speed of 4mm/min and by limiting the reduction ratio per pass. The thickness was measured before and after hot rolling.

The microstructure was observed by means of scanning electron microscopy (SEM) on longitudinal fracture sections parallel to the rolling force. Grain alignment was analyzed by X-ray diffraction on a surface perpendicular to rolling force; the samples were ground with SiC abrasive paper to reach the center of the bars and were then cleaned with isopropylic alcohol. Critical current measurement was performed by standard DC four probes technique with a criterion of 1μV/cm

3. Results and discussion

The bars were rolled up to 16 times and the thickness was measured before and after each rolling step. The reduction is defined by the ratio between the thickness variation and the sample thickness before the rolling step. The reduction per pass varied from 24% of the at the initial thickness of 1.55mm to 5% at the final thickness of 0.4mm. The samples deformed at high temperature show no cracking or macroscopic damage due to the deformation. Table 1 shows the rolling force and reduction ratio for various rolling steps. At the beginning a huge reduction of the thickness of 24% was achieved with a rolling force of only 200N. The deformation in this case consists essentially in a rapid densification, as can be seen in figure 1, together with a progressive alignment of the c-axis of Bi,Pb(2223) platelets parallel to the applied force This process is greatly facilitated by the presence of a liquid phase which "softens" the mechanical behavior of the ceramic [7].

Thickness before HR [mm]	Reduction ratio [%]	Rolling force [N]
1.55	24	200
0.70	7	550
0.42	5	850

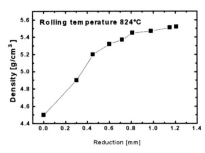

Table 1: rolling force and reduction ratio for various rolling steps. Rolling temperature. :824°C, rolling speed 4mm/min. The reduction ratio is defined by the ratio between the thickness reduction and the sample thickness before hot rolling.

Fig 1: Density of a bulk sample rolled at 824°C as a function of the thickness. Initial thickness was 1.55 mm. Speed 4mm/min

After further deformation to 0.42 mm thickness a force of 850N yielded a thickness reduction of only 5%, in this case the grains are already well aligned and deformation involves sliding of the platelets as well as a slight densification. Figure 1 shows this rapid initial densification followed by a much slower increase in density as a function of the thickness. The final density was 5.5 g/cm^3 corresponding to 86% of the theoretical density (6.4 g/cm^3). Figure 2 shows an X-ray diffraction pattern of the center of the samples. before and after a 16 steps hot rolling process. The initial and final thickness of the bars were 1.5mm and 0.4mm respectively. The sample before hot rolling consists mainly of slightly textured Bi,Pb(2223) phase, the grain orientation is due in this case to the cold uniaxial pressure applied for the preparation of the bar. After hot rolling the (00l) lines of the 2223 phase are strongly enhanced indicating strong texturing. In order to evaluate the grain orientation we define $f = I(115)/[(I(115)+I(00\underline{10})]$. The ratios given in table 2 change from 0.3 down to 0.017 indicating a strong texture induced by the hot rolling process.

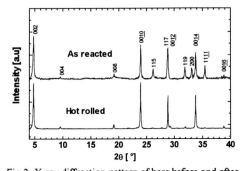

	Before HR [a.u]	After hot rolling [a.u.]
I(115)	22.90	12
I(00$\underline{10}$)	53.13	693.65
f	0.3	0.017

Fig 2: X-ray diffraction pattern of bars before and after hot rolling at 824°C. Peaks of the Bi,Pb(2223) are showed on the graph. X-ray analysis was performed in the center of the samples.

Table 2: the texture factor defined by : $f=I(115)/[(I(115)+I(00\underline{10})]$ for a bar before and after the hot rolling process.

394

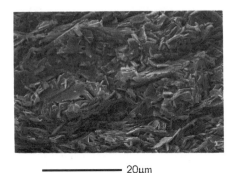

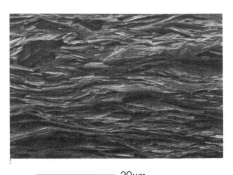

———————— 20μm   ———————— 20μm

Fig 3. SEM Micrograph of a fractured section of a   Fig 4. SEM Micrograph of a fractured
bar in the as reacted state. Thickness 1.6 mm   section of a bar after hot rolling at 824°C.
                                                         Thickness

Microstructures before and after the hot rolling process are shown in figures 3 and 4 . The platelet shaped Bi,Pb(2223) grains in the undeformed sample have no preferred orientation and the structure appears to be not very dense. The deformed sample shows an important change in the grain orientation and density of the bar. The Bi,Pb(2223) grain are aligned, their c-axis being almost parallel to the applied force while many grains are bent. The porosity of the sample is greatly reduced, resulting in dense oriented microstructure.

The bars before deformation carried no current at 77K. After the hot rolling process the critical current was 50A (77K, 0T) with a critical current density of 1800 A/cm^2. A substantial enhancement is expected after further optimization of the process.

4. Conclusion

Bulk Bi,Pb(2223) bars were successfully deformed by hot rolling at 824°C. The deformed bars showed a strong densification and grain alignment. An important deformation induced texture was observed by X-ray diffraction and confirmed by SEM microstructural observations. This densification and texturing resulted in a critical current density of 1800 A/cm^2 . This process is far from being optimized and opens a new way for the fabrication of long (>1m) bars for high current leads.

References

[1] Goretta K C et al., *Appl. Supercond.*, Vol. **2**, No. 6, (1994), 411-415
[2] Pachla W, Kovac P and Husek I, *Supercond. Sci. Technol.*, **7** (1994) 820-82
[3] Tampieri A, Masini R, Dimesso L, Gucciardo S and Malpezzi M C, *Jpn. J. Appl. Phys.* Vol. **32** (1993) 4490-4495
[4] Ichinose I and Kimihiko S, *Physica C* **190** (1991) 177-179
[5] Yang X and Chaki T, *Supercond. Sci.Technol.*, **6** (1993) 343-348
[6] Perin A, Grasso G, Däumling M, Hensel B, Walker E and Flükiger R, *Physica C* **216** (1993) 339-344.
[7] Goretta K C, Zamirowski E J, Calderoñ-Moreno J M, Miller D J, Nan Chen, Holesinger T G and Routbort J L , *J. Mater. Res.*, vol. **9**, No. 3, (1994), 541-547

Inst. Phys. Conf. Ser. No 148
Paper presented at Applied Superconductivity, Edinburgh, 3–6 July 1995
© *1995 IOP Publishing Ltd*

Enhancement of j_c in long monofilamentary Bi(2223) Ag-sheathed tapes by optimization of the fabrication process

G Grasso*, **F Marti, A Jeremie, A Perin, B Hensel, and R Flükiger**

Université de Genève, DPMC, 24, quai Ernest Ansermet, CH-1211 Genève 4, Switzerland

Abstract.   The transport properties of cold rolled Ag sheathed Bi(2223) tapes have been improved by an optimization of some crucial parameters of the tape fabrication process. Critical current densities up to 30 kA/cm^2 at the liquid nitrogen temperature in self field have been achieved. The superconducting properties of these tapes have been measured in a wide range of temperatures (4.2-77 K) and applied magnetic fields (up to 14 T). A critical current density of 70 kA/cm^2 has been measured at 4.2 K in a field of 14 T. The anisotropy of the critical current density has also been studied.

1. Introduction

In the last years it has been demonstrated that the Bi(2223) phase is suitable for the development of industrial applications both at the liquid nitrogen and at the liquid helium temperature [1]. At the beginning of the research on the improvement of the transport properties of Bi(2223) tapes, uniaxial pressing was used for the preparation of very short tapes (a few cm long). In this way, very high critical currents (up to 66 kA/cm^2 at 77 K [2]) have been achieved on very short, thin, and often fragile tapes. Recently, encouraging results have been published on the transport properties of longer tapes prepared in a reproducible way by cold rolling, their critical current densities are slowly but constantly approaching those of shorter pressed tapes ($j_c(77K,0T) \geq 30$ kA/cm^2) [3,4]. Therefore, small HTSC coils producing magnetic fields of several Tesla at 4.2 K and in excess of 0.4 Tesla at 77 K have been already successfully fabricated; the first superconducting only magnet able to produce a field of 24 T at 4.2 K has been built by adding an insert coil made by a Bi(2223) tape in the bore of a conventional Nb$_3$Sn magnet [3].

Moreover, several groups have shown that the upper limit to the critical current density of the Bi(2223) phase is far from being reached, as the local critical current density inside the filament can reach up to 110 kA/cm^2 at 77K and self field [5,6].
In this work, new results about the correlations between the critical current density and the powder heat treatment temperature and time, the cold deformation steps, as well as the tape heat treatment temperature and time are presented.

---

* on leave from Consorzio INFM, Università di Genova, Italy.
Work supported by the Swiss National Science Foundation (PNR 30), the Swiss Priority Program "Material Research and Engineering" (PPM), and the Brite Euram II Project No. BRE2 CT92 0229 (OFES BR060).

2. Results

Coprecipitated powders with cation ratio Bi:Pb:Sr:Ca:Cu=1.72:0.34:1.83:1.97:3.13 have been used as precursors. It has been already demonstrated that with this composition very high critical current densities can be achieved [2]. The coprecipitated powders require a first heat treatment at low temperatures (in the range between 300°C and 500°C) for a short time (typically about 2 hours) in order to remove the organics and the water which are retained in the powders and represent about 40% of the total powder weight. The residual powders are pressed with 40 MPa into round pellets of 25 mm in diameter and of about 5 g in weight. These pellets are put in alumina crucibles and treated in ambient air at temperatures between 780-830°C for about 20 h. Depending on the temperature, the content of lead free Bi(2212), Bi(2201) and Ca_2PbO_4 phases in the powder can be varied. A temperature of 780°C gives as dominant phases Bi(2201) and Ca_2PbO_4, while in the powders treated at higher temperatures the dominant phase becomes Bi(2212). The powders are pressed a second time into pellets and heat treated with the same parameters of the first treatment, in order to have a better mixture of the different phases, to reduce the carbon content and the size of the non-superconducting particles [7]. For this purpose, the pellets are ground by hand for a few minutes between the heat treatments without using ball milling which adds carbon and water impurities in the powders [7].

Different Ag tubes have been used for the preparation of tapes, depending on the total length which is needed, and on the ratio between superconducting core and silver sheath which has been chosen. Generally, the optimization of the deformation parameters has been carried out on pure Ag tubes of 6 mm of outer diameter and 4 mm of inner diameter which permits us to prepare typically 6 m long tapes. Longer tapes (up to 25 m long) have been prepared starting from tubes of 12 mm of outer diameter and 8 mm of inner diameter. The powder which has been prepared as described in the previous section, is loaded inside the Ag tubes and densified by means of a steel piston. The powder density inside the tubes before cold deformation results of the order of 5 g/cm^2. The tubes are properly closed using silver plugs, and they are deformed using a four hammer swaging machine, down to a diameter of about 4 mm. These wires are then drawn to a final wire diameter of about 1-2 mm. The tape-like shape is given by cold rolling in several steps. Cylinders of 105 mm in diameter and a rolling speed of about 1 m/min have been used. Sausaging of the Bi(2223) filament has been drastically reduced by limiting to about 10% the section reduction of each deformation step.

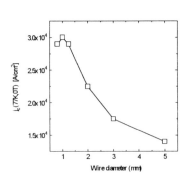

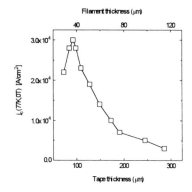

Fig.1 j_c(77K,0T) as a function of the wire diameter before rolling.

Fig.2 j_c(77K,0T) as a function of the final tape thickness

In fig. 1 and 2 the influences of the wire diameter before rolling and of the final tape thickness on the critical current at 77 K are presented: j_c is maximized for a wire diameter of 1 mm and for a tape thickness of 90 μm.

The tapes are heat treated up to four times at a temperature in the range between 830-840°C for a total time of about 160-240 hours. After each treatment, the samples have been furnace cooled down to room temperature. Between each heat treatment, the tape thickness is reduced with a cold rolling step, in order to densify the filament. Tapes of up to 1 m in length are heat treated in straight form between two alumina plates; longer tapes are treated in wound form around alumina cylinders of 90 mm in diameter. The temperature homogeneity inside the furnace is guaranteed by the use of heat pipes. The critical current reaches a maximum after a tape treatment time of about 200 hours (fig. 3). The heat treatment temperature is strongly correlated with the powder calcination temperature: j_c is maximized for a tape treatment temperature of about 838°C, provided that the powders have been calcined at 820°C (fig. 4).

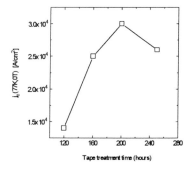

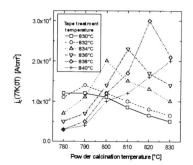

Fig.3 j_c(77K,0T) as a function of the tape treatment time.

Fig.4 j_c(77K,0T) as a function of the tape treatment time and of the powder calcination temperature

Superconducting properties of short straight pieces (about 2 cm long) representative of longer cold rolled tapes have been measured at different temperatures (4.2-77 K) and applied magnetic fields (up to 14 T). For the transport critical current density measurements, two different orientations of the magnetic field with respect to the tape plane (B//c and B//ab) have been studied. When high magnetic fields are applied, the current direction has been chosen in order to have a Lorentz force which presses the sample on the sample holder. Temperature and magnetic field dependences of j_c are shown in fig. 5 and 6 for both B//ab and B//c orientations for a typical tape with j_c(77K,0T)=30 kA/cm^2 (I_c=25 A). At 77 K, the critical current reduces by a factor of about 5 in a magnetic field of 1 T applied in the B//ab direction, in accordance with previous measurements on tapes with different j_c values. At 4.2 K, the critical current measured in self field is generally 5 to 8 times higher than at 77 K (j_c(4.2K,0T)=180 kA/cm^2), and the field dependence is much less pronounced than at higher temperatures, a reduction by a factor of 2 to 3 of j_c is usually observed when a field of 14 T is applied in the B//ab direction. Due to the high degree of texture of the Bi(2223) grains and to the high anisotropy of their superconductive properties with respect to the crystallographic orientation, the critical current density of the tapes drops much faster when a magnetic field is applied along the normal to the tape plane (B//c). In fact, at 77 K there is no detectable critical current when a field of 0.5 T is applied in that direction.

398

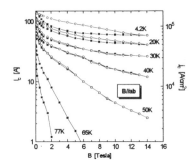

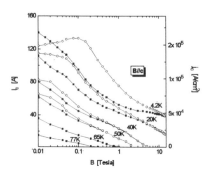

Fig. 5 Transport j_c as a function of the temperature and of the applied field for the B//ab orientation. Full symbols are measured for increasing field, open ones for decreasing field

Fig. 6 Transport j_c as a function of the temperature and of the applied field for the B//c orientation. Full symbols are measured for increasing field, open ones for decreasing field

3. Conclusions

In this work we report on the optimization of the transport properties of cold rolled Ag sheathed Bi(2223) tapes at liquid nitrogen temperature. Almost all the parameters of the fabrication process which have been investigated strongly influence the final tape properties, confirming that their reproducibility can only be obtained by carefully controlling all the steps of the preparation.

The critical current density has been measured in a wide range of temperatures (4.2-77K) and fields (up to 14 T) and for both B//ab and B//c orientations. The critical current density shows some well-known features: at 77 K j_c drops by factors of 3 and 5 in magnetic fields of 0.5 and 1 T for the B//ab orientation, respectively. For the same orientation at 4.2 K, j_c reduces by a factor of 2 to 3 when a field higher than 5 T is applied.

References

[1] Ohkura K, Mukai H, Hikata T, Ueyama M, Kato T, Fujikami J, Muranaka K, and Sato K 1994 presented at ISS'93, Hiroshima (Japan), published in *Adv. in Supercond.* **VI** ed. ISTEC 735
[2] Yamada Y, Satou M, Murase M, Kitamura T, and Kamisada Y 1993 in *Proc. 5th Int. Symp. on Superconductivity*, eds. Y. Bando and H. Yameuchi (Springer, Tokyo) 717
[3] Sato K, Ohkura K, Hayashi K, Ueyama M, Fujikami J, and Kato T 1995 to be published
[4] Grasso G, Hensel B, Jeremie A, Perin A, Flükiger R 1994 *Il Nuovo Cimento* **16D** 2073
[5] Grasso G, Hensel B, Jeremie A, and Flükiger R 1995 *Physica C* **241** 45
[6] Lelovic M, Krishnaraj P, Eror N G, and Balachandran U 1995 *Physica C* **242** 246
[7] Jeremie A, Flükiger R, and Seibt E W 1994 *IEEE Trans. Magn.* **30** 1883

Inst. Phys. Conf. Ser. No 148
Paper presented at Applied Superconductivity, Edinburgh, 3–6 July 1995
© 1995 IOP Publishing Ltd

Alternative methods for the fabrication of Bi,Pb(2223) silver sheathed tapes

A Jeremie, G Grasso* and R Flükiger

Université de Genève, DPMC, 24, quai Ernest Ansermet, CH-1211 Genève 4, Switzerland

Abstract.   We present alternative methods for the fabrication of Bi,Pb(2223)/Ag tapes with high critical current densities using different precursor powders. The Powder In Tube method currently used for the fabrication of Bi,Pb(2223)/Ag tapes has been efficient in obtaining high critical current densities j_c at 77 K and 0 T of 30 kA/cm^2 for long rolled samples. This method consists of introducing precursor powder into a Ag tube which is subsequently deformed and annealed to obtain Bi,Pb(2223)/Ag tapes with high j_c. The precursor powder is prepared by calcining oxides and carbonates. The phases present after calcination are Bi(2212), Ca$_2$PbO$_4$, CuO and others. Unfortunately, at the end of all the heat treatments, there remain secondary phases inside the Ag sheath and their amount is not always well controlled. We have studied alternative methods for the fabrication of the precursor powder before introduction into the Ag sheath.
We present two of the most promising procedures we have explored:
1. the *"Two Powder"* method. The critical current densities obtained are 26 kA/cm^2 at 77 K and 0 T.
2. the *"Pb-free calcination"* method. We have obtained critical current densities of 28 kA/cm^2 at 77 K and 0 T.

1. Introduction

Our previous studies on Pb in the Bi-Pb-Sr-Ca-Cu-O system have been used for the fabrication of Bi,Pb(2223)/Ag tapes in view of controlling the Pb content and the size of the phases present in the precursor powders. The final objective is to obtain high critical current densities j_c. We have explored two different routes for the fabrication of Bi,Pb(2223)/Ag tapes. We have compared them with the *standard* way with nominal composition Bi$_{1.72}$Pb$_{0.34}$Sr$_{1.83}$Ca$_{1.97}$Cu$_{3.13}$O$_{10+\delta}$ of producing the tapes. The latter consists of precalcining coprecipitated powder between 300°C and 500°C, then calcining in air at 820°C twice with intermediate grinding. This powder is then introduced in a Ag tube which is then deformed and annealed to obtain high current carrying Bi,Pb(2223)/Ag tapes. This method has been efficient in obtaining high critical current densities j_c at 77 K and 0 T of more than 30 kA/cm^2 for long rolled samples[1].

We have also tried the *Two-Powder* process proposed by Dorris et al [2] from Bi,Pb(2212) (the Pb doped Bi(2212) phase) and Ca-Cu-O. The last method consists of calcining the powder

---

* on leave from Consorzio INFM, Università di Genova, Italy.
Work supported by the Swiss National Science Foundation (PNR 30), the Swiss Priority Program "Material Research and Engineering" (PPM) and the European Community (Brite Euram II Project n° BRE2 CT92 0229 (OFES BR060)).

with almost the same overall stoichiometry except Pb-free. The Pb was added as PbO just before introduction into the Ag tube.

2. Results

Instead of using calcined powder composed of Bi(2212), Ca_2PbO_4, CuO and many other secondary phases with proportions that are not always well defined, we have used directly a mixture with nominal composition $Bi_{1.72}Pb_{0.34}Sr_{1.83}Ca_{1.97}Cu_{3.13}O_{10+\delta}$ of Bi,Pb(2212), Ca_2CuO_3 and CuO prepared separately and X-ray pure. This method seems attractive to better control the Pb content in the Bi,Pb(2212) phase since Pb is a very volatile element that is often lost during the heat treatments. With a limited number of phases, the transformation to Bi,Pb(2223) will probably take place in a more effective way. In addition, the grain size of the different phases can be chosen by pregrinding the seperate powders. This is an advantage since in *standard* calcined powder, the Ca_2PbO_4 grains are very big and remain almost the same size even after the deformation process. Ueyama et al [3] have found that the critical current densities are higher if the secondary phases are small and distributed homogeneously inside the tape.

Dorris et al [2] have prepared tapes by the *Two-Powder* process. One powder was of nominal composition Bi,Pb(2212) and the other CaCuO. Both powders were not single phased but the phase distribution was better controlled than in the *standard* process. These authors have obtained critical current densities of 15 to 20 kA/cm^2 at 77K and 0T.

We have used a mixture with nominal composition $Bi_{1.72}Pb_{0.34}Sr_{1.83}Ca_{1.97}Cu_{3.13}O_{10+\delta}$ of separately prepared single phased Bi,Pb(2212), Ca_2CuO_3 and CuO that was introduced into the Ag sheath.

We have obtained j_c values of 10 to 30 kA/cm^2 at 77K and 0T. These values are very close to the ones obtained by the *standard* route of a little more than 30 kA/cm^2.

In the next section, we will describe a novel preparation of precursor powders using *Pb free* calcined powder. In order to avoid the formation of Pb containing secondary phases like Ca_2PbO_4 often present in the precursor powder or $Pb(Bi)_3Sr_3Ca_2CuO_y$ noted *3321* [4] sometimes formed upon heating of samples, we have proceeded with the calcination of Pb free powder mixture of oxides and carbonates of nominal composition $Bi_{1.72}Sr_{1.83}Ca_{1.97}Cu_{3.13}O_{10+\delta}$. Before introducing the powder in the Ag tube, we added PbO in different proportions to the powder.

Figure 1 shows a DTA measurement of two tapes of nominal composition $Bi_{1.72}Pb_{0.34}Sr_{1.83}Ca_{1.97}Cu_{3.13}O_{10+\delta}$, the first one calcined in the *standard* way, and the second calcined without Pb. The measurement shows two endothermic peaks for each sample. In the figure, only the nominal composition $Bi_{1.72}Pb_{0.34}Sr_{1.83}Ca_{1.97}Cu_{3.13}O_{10+\delta}$ is shown but we have prepared different powders with Pb ratios x ranging from 0.25 to 0.38 in the nominal composition $Bi_{1.72}Pb_xSr_{1.83}Ca_{1.97}Cu_{3.13}O_{10+\delta}$. The onset temperature of the first peak of 832°C is the same for both samples. We can thus proceed with the heat treatment of the tapes at the same temperature, thus comparing tapes prepared in identical conditions. The shape of the endothermic peaks is however different indicating that the phase mixtures are not identical.

Figure 2 gives the j_c dependence of these tapes as a function of Pb content. There is strong j_c dependence on the Pb content even if the difference in proportion is small between two different compositions. It is thus extremely important to effectively control the amount of Pb inside the precursor powder.

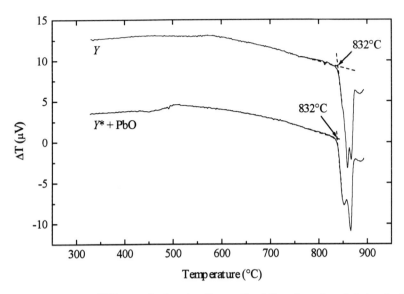

Fig. 1 *DTA measurement at 2°C/min in flowing air of tapes after deformation where Y designates the nominal composition $Bi_{1.72}Pb_{0.34}Sr_{1.83}Ca_{1.97}Cu_{3.13}O_{10+\delta}$ and Y* $Bi_{1.72}Sr_{1.83}Ca_{1.97}Cu_{3.13}O_{10+\delta}$ where the missing 0.34Pb has been added in the form of PbO.*

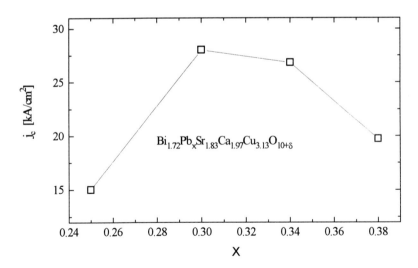

Fig. 2 *j_c at 77K and 0T as a function of Pb content in tapes prepared by first calcining the precursor powder without Pb, and adding PbO to the powder in the right proportions just before introduction into the Ag tube with nominal composition $Bi_{1.72}Pb_xSr_{1.83}Ca_{1.97}Cu_{3.13}O_{10+\delta}$*

One can also see in figure 2 that the highest j_c values of 28 kA/cm^2 are obtained with the nominal composition of 0.3<x<0.35 corresponding to the Pb ratio x of the nominal composition giving the highest j_c values at 77 K and 0 T prepared by the *standard* way. The j_c value is almost the same value as obtained by the *standard* way.

3. Discussion

All three preparation routes give the same critical current density values close to 30 kA/cm^2 at 77K and 0T. This is probably due to the fact that at the reaction temperature of the Ag sheathed tape between 830°C and 840°C, the phases present are about the same in all three cases. In a previous publication, Grivel et al [5] have shown that inside the tape, Bi,Pb(2212) forms above 835°C. In addition, Jeremie et al [6] have shown that in bulk samples, Bi,Pb(2212) decomposes into Pb free Bi(2212) and a Pb rich secondary phase *3321* at low temperature when in presence of extra Ca and Cu. Bi,Pb(2212) forms again above 835°C. We have also noticed that above 835°C, in the case of the powder prepared from Pb-free calcined powder with PbO added, Bi,Pb(2212) is also formed following a similar process. We are thus in the same situation in all three preparation routes when the reaction temperature of the tape is reached. Consequently, the tapes show similar j_c values.

4. Conclusions

In this work we have explored the use of different precursor powders for the preparation of Bi,Pb(2223)/Ag tapes. The principal goal was to better control the amount and size of the secondary phases still present in the final tape. The *standard* way of preparing rolled tapes have critical current densities of 30 kA/cm^2 at 77 K and 0 T.

We have first studied a precursor mixture of single-phased Bi,Pb(2212), Ca$_2$CuO$_3$ and CuO prepared seprately. Upon heating, the Bi,Pb(2212) decomposes into Bi(2212) and a Pb-rich phase *3321* at low temperature and Bi,Pb(2212) forms again above 835°C. These tapes have critical current densities ranging from 10 to 26 kA/cm^2 at 77 K and 0 T.

In the hope of avoiding the formation of Pb rich secondary phases, we have also explored a novel Pb-free calcination method. Just before introducing the powder in the Ag tube, PbO is added in different proportions. Again, the *3321* phase is formed upon heating, and above 835°C, the Bi,Pb(2212) phase is formed with the disapearance of *3321*. The critical current densities for these tapes range between 16 and 28 kA/cm^2 at 77 K and 0 T depending on the Pb content. The highest j_c value is obtained for the nominal composition $Bi_{1.72}Pb_{0.34}Sr_{1.83}Ca_{1.97}Cu_{3.13}O_{10+\delta}$.

All three methods give similar critical current densities close to 30 kA/cm^2 at 77 K and 0 T because at the reaction temperature between 830 and 840°C, the same phase mixture is present.

References

[1] Grasso G, Hensel B, Jeremie A, Perin A and Flükiger R, 1994 *Il Nuovo Cimento* **16D** 2073.
[2] Dorris S E, Prorok B C, Lanagan M T, Sinha S and Poeppel R B 1993 *Physica C* **212** 66.
[3] Ueyama M, Hikata T, Kato T and Sato K, 1991 *Japn. J. Appl. Phys.* **30** L1384.
[4] Kusano Y, Nanba T, Takada J, Egi T, Ikeda Y and Takano M 1994 *Physica C* **219** 366.
[5] Grivel J C, Jeremie A, Hensel B and Flükiger R 1993 *Proc. ICMAS, Paris, France* 359.
[6] Jeremie A, Grivel J C and Flükiger R 1994 *Physica C* **235-240** 943

Inst. Phys. Conf. Ser. No 148
Paper presented at Applied Superconductivity, Edinburgh, 3–6 July 1995

Formation and superconducting properties of the $(Bi,Pb)_2Sr_2Ca_2Cu_3O_{10-y}$ phase in bulk samples with oxide additions and in alloyed Ag-sheathed tapes

J-C Grivel, G Grasso, A Perin and R Flükiger

Université de Genève, DPMC, CH-1211 Genève 4, Switzerland

Abstract. In view of the mechanical reinforcement of the Ag sheath of $(Bi,Pb)_2Sr_2Ca_2Cu_3O_{10.y}$ (Bi,Pb(2223)) tapes by dispertion hardening, the effect of TiO_2, V_2O_5, Cr_2O_3 and MnO on the formation of the Bi,Pb(2223) phase has been studied. By means of DTA measurements, it was found that the onset temperature of partial melting in the precursor powders is affected by the presence of these oxides, consequently influencing the optimum temperature for the synthesis of the Bi,Pb(2223) phase. XRD and EDX analysis showed that small oxide additions result in the formation of non superconducting phases containing the transition metal elements. Lattice parameter calculations from XRD patterns, EDX measurements as well as T_c determinations by use of ac-susceptibility show that the substitution of Ti, V, Cr and Mn in the structure of the Bi,Pb(2223) phase is limited to low values. The change in the mechanical properties of the Ag-sheath induced by alloying with Ti and Mn is investigated. The influence of the sheath on the formation of the Bi,Pb(2223) phase during the heat treatment of the tapes is studied.

1.Introduction

The use of High-T_c superconductors in the form of wires or tapes for current transport at liquid nitrogen temperature require the use of a metallic sheath. For the Bi,Pb(2223) phase, Ag has been proven to be a suitable material since it does not have any detrimental effect on the superconducting properties of the material. Nevertheless, the high ductility of Ag results in inhomogeneities of the superconducting oxide layer thickness along the tape. Furthermore, during use of these tapes in high magnetic field applications the mechanical strength of Ag may not be sufficient to withstand large Lorentz forces. The reinforcement of Ag by means of dispersion hardening or alloying has been proposed to solve these problems. However, the addition of one or more elements in the sheath can result in a reaction between the alloying material and the precursor powders during the synthesis of the Bi,Pb(2223) phase inside the sheath. It is thus important to study the effects of selected element additions on the formation and some superconducting properties of the Bi,Pb(2223) phase. Recently, the use of Ag-Mg and Ag-Cu-M (M = Ti, Zr or Hf) alloys has been reported with promising results [1,2]. In a previous work [3], we reported the effect of TiO_2 additions on the synthesis of the Bi,Pb(2223) phase. Here we extend the investigation to some other 3d elements in view of the preparation of Bi,Pb(2223) tapes. The selected elements (Ti, V, Cr and Mn) were added to Bi,Pb(2223) calcined precursor powders in the form of oxides since the handling of the Ag-sheathed tapes at high temperatures in an O_2 containing atmosphere during the thermal treatments of the tapes easily results in the formation of oxide precipitates of the alloying elements.

2.Results

The DTA measurements performed on calcined precursor powders of the Bi,Pb(2223) phase without element additions reveal a partial melting reaction occurring at about 855°C [3]. The presence of TiO_2, V_2O_5, Cr_2O_3 and MnO in the mixture systematically results in a lowering of this partial melting temperature. Figure 1 shows this variation as a function of the equivalent at. % element addition in the nominal composition $Bi_{1.72}Pb_{0.34}Sr_{1.83}Ca_{1.97}Cu_{3.13}M_xO_{10-y}$ (M=Ti, V, Cr or Mn) used in this work. It appears that both V and Cr strongly influence the partial melting temperature of the precursor powders and hence the sintering temperature for the synthesis of the Bi,Pb(2223) phase. On the contrary, Ti and Mn have a comparatively low influence on this parameter. According to these results, pellets of precursor powders with various amounts of oxide additions were sintered during 130-150h in air at selected temperatures.

Figure 2 presents the X-ray patterns obtained on some sintered samples. Cr_2O_3 additions result in the formation of a foreign phase containing Cr, Bi, Sr and Ca even for very low amounts of Cr_2O_3. With V_2O_5, $Sr_3V_2O_8$ was detected by EDX in the sample with 0.03 at % V. In the case of TiO_2 and MnO, the additional elements form compounds with Sr, Ca and/or Cu. These foreign phases are also detected for 0.1 at. % Ti and Mn in the mixture. However, in presence of Ti and Mn, the Bi,Pb(2223) phase was found to be formed easily up to 0.3 at.% element additions, whereas it was not obtained above 0.1 at. % V and Cr additions. This shows that the ability of the Bi,Pb(2223) phase to form in presence of the presently investigated additions is higher for Ti and Mn.

The solubility of V and Cr in Ag being rather low, these elements are likely to be segregated from the sheath during the thermomechanical treatment of the tapes. The probability that the precursor powder reacts with them during the formation of the Bi(2223) phase is then high. On the contrary, Ti and Mn, whose solubility in Ag amounts to about 2 and 33 at. % respectively at 800°C, are expected to be better retained in the sheath. For these reasons, only Ti and Mn additions were further investigated.

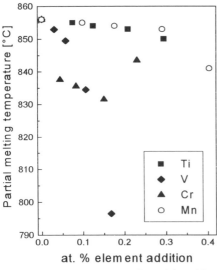

Fig 1 : onset temperature of partial melting recorded by DTA against at % Ti, V, Cr and Mn.

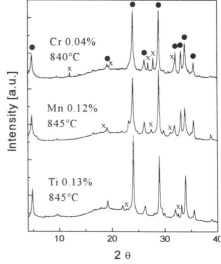

Fig 2 : XRD patterns obtained on sintered samples. Legend : ● Bi,Pb(2223), x additional phases containing Cr, Ti or Mn

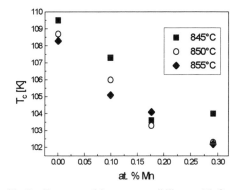

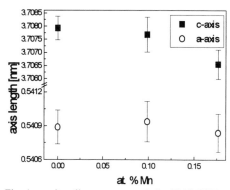

Fig 3 : T_c measured by ac-susceptibility on MnO added samples sintered for 130h at various temperatures versus at. % Mn nominal content

Fig 4 : unit cell parameters of the Bi,Pb(2223) phase formed in MnO added samples after sintering at 855°C for 130h.

The T_c of the Bi,Pb(2223) phase was determined by ac-susceptibility measurements. As shown in fig 3, the addition of MnO to the precursor powders resulted in a slight decrease of T_c. TiO_2 additions resulted in a similar decrease [3].

The values of the a and c unit cell parameters of the Bi,Pb(2223) phase are plotted against the nominal at % Mn addition in figure 4. The variation of the c parameter could be explained by a substitution of Mn^{3+} on the Cu sites but in this case the a parameter should decrease as well. However, within the accuracy of our measurements, the a parameter value does not vary. This observation together with the fact that Mn containing foreign phases were detected by EDX even for the lowest Mn content indicate that the T_c and c parameter are not primarily due to an extensive Mn substitution in the crystal lattice of the Bi,Pb(2223) phase. This conclusion is supported by the observation that no Mn could be detected by EDX measurements performed on Bi,Pb(2223) crystallites. A variation of the Sr/Ca ratio in the Bi,Pb(2223) phase could also be responsible for the decrease of T_c and the change in the c parameter value. Furthermore, the Cu-O planes being not strongly affected by such a variation, it would not be surprising that the a parameter remains unchanged.

The same argument can explain the T_c decrease in the Ti added samples that did not show any significant variation of the unit cell parameters of the Bi,Pb(2223) phase up to a proportion of 0.3 at % Ti in the nominal composition.

Ag-Ti and Ag-Mn alloys were prepared by arc-melting and annealed at 930°C in an Ar atmosphere. Slices of the ingot were cut, polished and Ar-annealed again. The Vickers microhardness of the resulting samples as well as after a further cold deformation by rolling are reported in table 1. It appears that for equivalent proportions of alloying, the microhardness of Ag-Mn alloys is higher after annealing as well as after deformation. After this deformation, the microhardness is found to be systematically slightly higher when measured perpendicularly to the rolling direction.

Figure 5 shows the onset of the diamagnetic transition of the superconducting core of $Ag_{0.991}Ti_{0.009}$ and $Ag_{0.998}Mn_{0.002}$ sheathed superconducting tapes. It appears that the T_c of the Bi,Pb(2223) phase formed in a Ti alloyed sheath is lower than that of the phase formed in the Mn alloyed Ag. The T_c value of about 108.2 K for the Bi,Pb(2223) phase formed inside the $Ag_{0.998}Mn_{0.002}$ sheath is similar to the values obtained in pure Ag sheaths.

alloy	Ar annealed	cold rolled (mid)	⊥ rolling direction	// rolling direction
Ag	30	92	92	92
Ag-Ti 0.2%	35	94	95	93
Ag-Ti 2 %	69	107	108	106
Ag-Mn 0.2%	50	99	101	97
Ag-Mn 2 %	106	141	143	139

Table 1 : microhardness after annealing and subsequent rolling for pure Ag, Ag-Ti and Ag-Mn alloys. All values are given in kg/mm^2

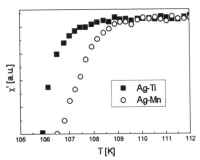

Fig 5 : ac-susceptibility measurements performed on pieces of Bi,Pb(2223) tapes with a Ag-Ti or Ag-Mn sheath.

3.Conclusions

It has been shown that TiO_2, V_2O_5, Cr_2O_3 and MnO additions to the precursor powders of the Bi,Pb(2223) phase result in a lowering of the partial melting temperature recorded by DTA. XRD and EDX measurements performed on reacted samples showed that Ti, V, Cr or Mn containing foreign phases were formed even for the lowest amounts of additions. Slight variations in the T_c value and the unit cell parameters of the Bi,Pb(2223) phase were observed for Ti and Mn additions. These variations are thought to be due to a change in the Sr/Ca ratio in the Bi,Pb(2223) phase rather than to an extended substitution of Ti or Mn on the Cu sites of the Bi,Pb(2223) phase.

The microhardness of Ag-Ti and Ag-Mn alloys was shown to be larger than that of pure Ag.

The T_c of the Bi,Pb(2223) phase formed inside a $Ag_{0.991}Ti_{0.009}$ sheathed tape was found to be lower than when formed inside a pure Ag sheath. On the contrary, a $Ag_{0.998}Mn_{0.002}$ sheath has no detrimental influence on the T_c of the Bi,Pb(2223) phase.

References

[1] Kamisada Y, Koizumi T, Satou M and Yamada Y 1994 *IEEE Trans. Magn.* **30** 1675

[2] Tanaka Y, Ishizuka M, He L L, Horiuchi S and Maeda H 1995 *to be published*

[3] Grivel J-C, Jeremie A and Flükiger R 1995 *Supercond. Sci. Technol.* **8** 41-7

The authors acknowledge the support of the Swiss National Foundation (PNR 30), of the Priority Program « Materials Research and Engineering » (PPM), and of the Brite Euram II Project N° BRE2 CT92 0229 (OFES BR060).

Inst. Phys. Conf. Ser. No 148
Paper presented at Applied Superconductivity, Edinburgh, 3–6 July 1995
© *1995 IOP Publishing Ltd*

Texture formatiom in silver-sheathed bismuth-based superconducting tapes

J Jiang and J S Abell

School of Metallurgy and Materials, University of Birmingham, Birmingham, B15 2TT, UK

Abstract. The texture formation in silver-sheathed Bi-based superconducting tapes was studied in detail. The texture measurement was made by XRD and SEM. It was found that the mechanical deformation during fabricating processes induced significant texture of BPSSCO grains in the as-pressed tapes, and that this texture degree depends on the characteristic of the precursor powder. The texture degree in the as-pressed tapes is not the determining factor for the texturing of the Bi-2223 grains in the final tapes with high Jc. The texture degree was improved significantly during the phase formation, both in the process from 2201 to 2212, and from 2212 to 2223. Intermediate mechanical deformation enchanced the texture degree. Thus, both the intermediate mechnical deformation and preferential nucleation of the 2223 grains contribute to the texturing in the Bi-2223 tapes.

1. Introduction

Despite enormous advance in high temperature superconductors, practical applications of these materials have been hampered by two major difficulties: low critical current density and poor mechanical properties. The powder-in-tube technique appears to be useful for making silver-sheathed Bi-2223 composite[1-7]. Within the Bi-2223 tapes the grains are highly textured, and the c axes are aligned almost perfectly and are perpendicular to the plane of the tape. The a and b axes are usually orientated at random from platelet to platelet. It is well known that the grain alignment and connectvity have dramatic effects on the Jc improvement. Larbalestier et al [8] showed that variations of local critical current densities occur within individual filaments cut from the tape. Within the tape, the critical current density increases gradually from the centre of the tape to the sides[9]. Proximity to the silver sheath leads to a uniform, well aligned microstructure. Lelovic et al [6] have reported a minimum critical current density of $10^5 A/cm^2$ at 77K in the thin layer of the Bi-2223 near the Ag in the Ag-sheathed tapes. Due to the complexity of phase formation in the Bi-based superconductors, the mechanism of the texture is still unclear. Recently several possible mechanisms have been reported[10-13]; for example, the reaction induced texturing, anisotropic crystalline growth, and mechanical deformation texture. In this contribution, the texture formation was investigated in detail and possible mechanisms are discussed.

2. Experimental

BPSCCO-2223 powder made by co-decomposition method with nominal composition $Bi_{1.8}Pb_{0.4}Sr_2Ca_{2.2}Cu_3O_x$ was packed into the Ag tubes, swaged, and pressed into a tape. Short lengths of the tape were cut and treated at different temperatures in air. Three kinds of precursor powders were used.

The heat treatment for powder A was 790°C/6hours, 810°C/12hours, and 835°C/10hours. The major phase was 2212 with a little amount of 2223 phase and impurity phase (Ca, Sr)$_2$PbO$_4$ (alkaline-earth plumbate, AEP), (Sr,Ca)$_{14}$Cu$_{24}$O$_x$ (alkaline-earth cuprate,AEC) and CuO.

The powder B was obtained by milling powder A into amorphouse phase and annealing it at 500°C for 2.5 hours in air. The major phase in powder B was 2201 with impurity phase (Ca, Sr)$_2$PbO$_4$(AEP), (Sr,Ca)$_{14}$Cu$_{24}$O$_x$(AEC) and CuO.

For powder C, the heat treatment was 790°C/6hours, 810°C/9hours and 842°C/10 hours. About two thirds of the major phase was 2212 and one third was 2223. The impurity phases were (Ca, Sr)$_2$PbO$_4$(AEP), (Sr,Ca)$_{14}$Cu$_{24}$O$_x$(AEC) and CuO.

The prefered orientation of the Bi-based superconducting phases can be measured from the XRD patterns[12]. The ratio of the intensity of (008) or (0010) reflection to that of the corresponding (115) reflection provides a measure of the c-axis texture of the Bi-2201, 2212 or 2223, respectively.

Microstructure of the selected samples were examined by Scanning electron microscopy(JEOL6300) in both secondary electron(SE) and backscatter(BE) imaging modes. The samples were etched with 1% percholoric acid in 2-butoxyethanol.

3. Results and discussion

For the precursor powders the texture factor F(2212) is about 0.4. Wire80, Wire81 and Wire34 were fabricated from powder B, A, and C, respectively. The texture factors of the tapes made from the three wires are listed in Table 1. The as-pressed tapes exhibited an obvious degree of texturing. This means the mechanical deformation induced texturing. The highest texture degree was observed in the as-pressed tape made from Wire34, the XRD pattern of which is shown in Fig.1. The as-pressed tape made from Wire80 exhibited the lowest degree of texturing. As we know, owing to the large separation between the two Bi-O

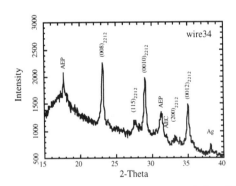

Fig.1. The XRD pattern of the as-pressed tape made from Wire 80

layers (0.3nm), the Bi-based materials exhibit a micaceous morphology and are easily cleaved. Mechanical deformation, in particular rolling and pressing, takes advantage of the plate-like morphology for grain alignment in the a-b plane direction. The main difference between Wire80, Wire81 and Wire34 is the phase content. So we deduce that the degree of texturing induced by mechanical deformation during fabricating depends on the characteristic of the precursor powder. The Bi-2201 phase in the tape made from Wire80 turned into 2212 after annealing at 823°C for 12 hours, and simultaneously the texture factor was improved from 0.72 to 1.5.

Table 1 The texture factor [F=I(008)/I(115)] of 2212 for the tapes made from different precursor powders

name	as pressed	823°C/12h	823°C/24h
wire 80	0.72*	1.5	1.4
wire81	1.4	3.7	3.8
wire34	4.1	5.1	4.9

*: This value is for 2201 and F=I(008)/I(115)

Table 2. Texture factor (F) for 2223 and 2212 and the ratio (f) of 2223 to 2212+2223 as a fuction of time at 834°C in air for the tapes made from Wire81

Times(hours)	F(2212)[a]	F_p(2212)[a]	F(2223)[b]	F_p(2223)[b]	f(2223)[c]
1	4.3	6.2			
5	3.4	5.4			
9	3.5	6.3			
18	4.4	6.5			14.7
40	4.2	9.0	34.8	42	53.2
55	7.9	11	45.5	70	70.2
70	7.7	12	50.4	76	80.3
90			56.5	80	87.6

a: F(2212)=I(008)/I(115),

b: F(2223)=I(0010)/I(115)

c: f(2223)=I(0010)$_{2223}$/[I(0010)$_{2223}$+I(008)$_{2212}$]

p: The sample was pressed at 25 kbar after sintering

Table 1 also indicates that the texture factors of 2212 were not improved obviously by extending the annealing time from 12 hours to 24 hours.

Further results of the texturing formation in the tapes made from Wire81 are listed in Table 2. Each tape was cut into two parts after annealing. From one half the Ag layer was peeled off and the core was examined by XRD. Another half was pressed at 25kbar before peeling off the Ag layer. F(2212) and F_p(2212) are the texture factor of Bi-2212 phase for the as-annealed tape and the pressed tape, respectively. From Table 2, we know that the value of F_p(2212) is obviously larger than that of F(2212), i.e. the pressing enchanced the texturing degree. So repetitive deformation and annealing are necessary to achieve a high degree of grain alignment and high Jc. With the extension of the annealing time, the Bi-2223 phase formed and the texture degree of 2212 increased. F(2223) and F_p(2223) are the texture factor of Bi-2223 phase for the as-annealed tape and the pressed tape, respectively. Table 2 provides a further evidence that the pressing enchanced the texture degree significantly.

Xi et al[10] suggested that the texture of the 2223 phase originated from the texture of the 2212 phase. But, the values of F(2223) are much larger than that of F(2212). As reported, the Bi-2223 forms via an intercalation mechanism[14]. The intercalation causes the Bi-2223 to inherit the crystallographic orientation of the Bi-2212 phase. Since the value of F(2212) is less than 10 and the value of F(2223) is larger than 50, the intercalation only at most accounts for about 20% of the texture of the Bi-2223. This result is consistent with that of Merchant et al[12]. Thus the intercalation may not be the main mechanism of the Bi-2223 nuleation.

The XRD pattern of the final tape made from from Wire80 is shown in Fig.2. Although the mechanical deformation induced different degree of texturing in the as-pressed tapes made from Wire80, Wire81 and Wire34, the final tapes exhibited almost the same texture degree after three press-annealing cycles. This means that the texture degree in the as-pressed tape did not determine the texture degree in the final tape. The texture degree can be improved significantly during the process of the Bi-2223 formation and enhanced by the intermediate pressing. A

410

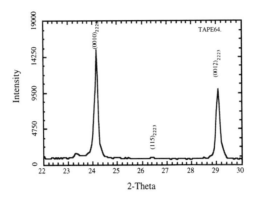

Fig.2 The XRD pattern of the tape made from Wire80

transient liquid has been proposed to accelerate the 2223 formation[7], so we think that the most of the Bi-2223 grains grow through nucleation with better preferential orientataion from the liquid route, rather than by inheriting the orientation of the 2212. thus, both the preferential nuleation of the 2223 grains and the intermediate deformation contribute to the texturing in the Bi-2223/Ag tapes.

4. Conclusion

Three kinds of precursor powders were used to investigate the formation of texture in the Bi-based superconducting tapes. It was found that the mechanical deformation during the fabricating process induced significant texture of BPSCCO grains in the as-pressed tapes and that this texture degree depends on the phase content of the precursor powder. The texture degree in the as-pressed tapes is not the determining factor for the texturing of the Bi-2223 grains in the final tapes. Both the preferential nucleation of the 2223 grains and the intermediate mechanical deformation contribute to the texturing in the Bi-2223/Ag tapes.

References

[1] Dou S X and Liu H K 1993 *Supercond. Sci. Technol.* **6** 297-314
[2] Sato K, Hikata T, Mukai H, Ueyama M, Iwata K, Kato T, Masuda T, Nagata M, Iwata K, and Mitsui T 1991 *IEEE Trans on Magnetics* **27** 1231-1238
[3] Haldar P and Motowidlo L 1992 *JOM* **44** 54-58
[4] Balachandran U, Iyer A N, Haldar P, and Motowidlo L R 1993 *JOM* **45** 54
[5] Li Q, Brodersen K, Hjuler H A and Freltoft T 1993 *Physica C* **217** 360-366
[6] Lelovic M, Krishnaraj P, Eror N G, Balachandran U 1995 *Physica C* **242** 246-250
[7] Yamada Y, Obst B and Flükiger R 1991 *Supercond. Sci. Technol.* **4** 165-171
[8] Larbalestier D C, Cai X Y, Feng Y, Edelman H, Umezwa A, Riley G N, and Carter W L 1994 *Physica C* **221** 299-303
[9] Grasso G, Aensel B, Jeremìe A, Flükiger R 1995 *Physica C* **241** 45-52
[10] Xi Z P and Zhou L 1994 *Supercond. Sci. Technol.* **7** 908-912
[11] Aksenova T D, Bratukhin P V, Shavkin S V, Melnikov V L, Antipova E V, Khlebova N E and Shikov A K 1993 *Physica C* **205** 271-279
[12] Merchant N, Luo J S, Maroni U A, Riley G N and Carter W L 1994 *Appl. Phys. Lett.* **65** 1039-1041
[13] Aksenova T D, Shavkin S V, Bratukhin P V, Akimov I I, Antipova E V, Khlebova N E and Shikov A K submitted to *Physica C*
[14] Luo J S, Merchant N, Maroni V A, Gruen D M, Tani B S, Carter W L, and Riley G N Jr 1993 *Appl. Supercond.* **1** 101

Inst. Phys. Conf. Ser. No 148
Paper presented at Applied Superconductivity, Edinburgh, 3–6 July 1995
© *1995 IOP Publishing Ltd*

Current-voltage characteristics of BSCCO-2212 wires and BSCCO-2223 tapes

B. Lehndorff [1,2]**, B. Fischer**[1a]**, M. Hortig**[2]**, R. Theisejans**[2] **and H. Piel**[1,2]
[1]Institute of Material Science, Dept. Physics, University of Wuppertal, Gauß-Str. 20,
D-42097 Wuppertal, Germany
[2]Cryoelectra GmbH, Wettinerstr. 6h, D-42287 Wuppertal, Germany

Abstract. BSCCO-2212 wires and BSCCO-2223 tapes have been prepared using the conventional powder in tube method but with an alternative rolling prozess instead of drawing. As a standard characterization J_c was measured at 77 K . Critical current densities of 25 000 A/cm^2 for the BSCCO-2223 tapes and 5 000 A/cm^2 for the BSCCO-2212 wires were obtained at 77 K. In order to examine the pinning properties current-voltage characteristics have been measured at temperatures between 4.2 K and 100 K in magnetic fields up to 8 T. At temperatures below 40 K respectively 20 K for the 2212-phase the magnetic field dependence of the critical current density is weak after an initial drop. At higher temperatures the critical current density decreases exponentially at magnetic fields above 1 T. In this region the I-V-characteristics show power law behavior as ecpected for the thermally activated flux flow (TAFF) mechanism. The anisotropy of J_c for different orientations of the magnetic field is high for the tape material and nearly zero for the wires as expected from the 2-d structure of the material and the geometry of the samples.

1. Introduction

Superconducting wires and tapes fabricated from the BSCCO materials have potential application in different fields. Tapes or wires made from the 2212-phase are ment for the fabrication of insert coils for very high field hybrid magnets (B > 20 T) at 4.2 K [1], while 2223-phase conductors are interesting for the application in refrigerator cooled magnets for intermediate fields (~ 2 T and more) at about 30 K [2] as well as for cable conductors at 77 K [3]. In addition applications like fault current limiters [4] and superconducting magnetic energy storage (SMES) devices [5] are considered and developed. In our group both conductor types are prepared, BSCCO-2223 tapes [6] and BSCCO-2212 wires using the powder-in-tube process. Furthermore a new deformation technique -profil rolling instead of drawing- is applicated and compared to the drawing technique.

2. Experimental details

The tapes and wires are synthesized using the conventional powder-in-tube process. The precursor powders were prepared by the mixed oxide/ carbonate route. The obtained overall stochiometrie was $Bi_{1.8}Pb_{0.4}Sr_{2.0}Ca_{2.2}Cu_{3.0}O_x$ and $Bi_2Sr_2CaCu_2O_x$. The powders were pressed cold isostatically and filled into a silver tube with outer diameter 10 mm, inner diameter 7mm and a filling factor of 55 %. Instead of drawing a profile rolling technique was used to achieve the final diameter of approximatly 1 mm. The BSCCO-2223 wires subsequently were rolled to tapes with 15 % reduction ratio per pass. Fig. 1 shows the cross section of a profile rolled wire

---

[a]   present address: Siemens AG, Erlangen

which differs slightly from a drawn one due to the special force distribution of this deformation technique. The critical current densities, however, of the short samples did not show a significant difference between the two methods. In general this process provides the possibility of a fast preparation of long wires in a technically simple way.

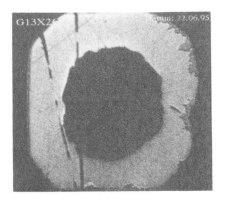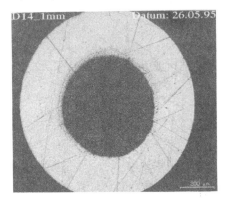

Fig. 1: Cross section of profile rolled (left) and drawn (right) BSCCO-2223 wires for comparison.

After the mechanical deformation the BSCCO-2212 wires were melt-textured at a temperature of 910 - 920°C and finaly sintered at 840°C in air. The BSCCO-2223 tapes were heat treated in a 3 step press and sintering prozess at 840°C with a total sintering time of 150 - 300 h. Critical current densities of 25 000 A/cm^2 were achieved for the BSCCO-2223 tapes and 5 000 A/cm^2 for the BSCCO-2212 wires at 77 K in self field. Besides optical and electron microscopy (SEM) combined with energy dispersive x-ray diffraction (EDX), the samples were characterized with AC-susceptibility measurements. The setup used for these experiments is in detail described in a previous paper [7]. This gives us a fast information about the critical temperature of the samples and the amount of secondary phases.

In order to study the pinning properties of the tapes and wires transport measurements were made at temperatures (between 4.2 K and 100 K) in magnetic fields up to 8 T, with field orientation parallel and perpendicular to the tape surface. The BSCCO-2212 wires were also rotated with respect to the field direction to get an information about the isotropy of the deformation process.

3. Current-voltage characteristics

The current-voltage characteristics of both types of conductors were measured between 4.2 K and 100 K in magnetic fields up to 8 T. The two materials show a quite similar behavior. At low temperatures the critical current density has a very weak magnetic field dependence after an initial steep decrease. With rising temperature the behavior changes at 20 K for the low T_c-phase BSCCO-2212 (T_c = 85 K) and at 40 K for the higher T_c-phase BSCCO-2223 (T_c = 110 K). At these temperatures the thermally assisted flux flow (TAFF) mechanism [8] starts to limit the critical current density which exhibits itself in an exponential decrease of J_c as a function of the magnetic field. Consequently the I-V-curves, which are depicted as a log-log plot in fig. 3, show a power law behavior with changing slope.

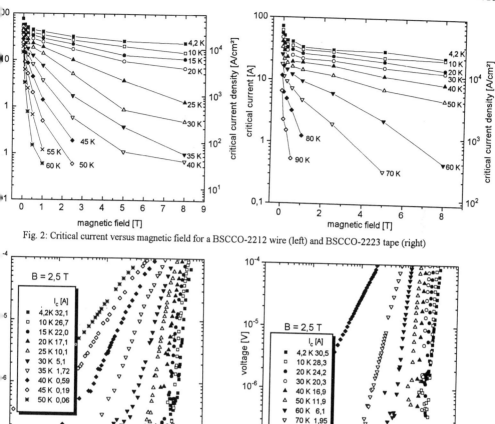

Fig. 2: Critical current versus magnetic field for a BSCCO-2212 wire (left) and BSCCO-2223 tape (right)

Fig. 3: Current-voltage characteristics for a BSCCO-2212 wire (left) and a BSCCO-2223 tape (right).

4. Anisotropy

As a further important physical property the anisotropy of the critical current density with respect to the applied magnetic field was examined. For the tape material this should provide information about the degree of texture whereas for the wire material information about the homogeneity of the deformation process should be obtained. For this purpose the critical current density was measured as a function of magnetic field with the field perpendicular to the tape plane and parallel to it respectively. The wire sample was merely rotated by 90° with respect to the magnetic field. Fig. 4 shows the normalized critical current density for the BCSSO-2223 tape as a function of the applied field.

For the 2223 tape the anisotropy is dominated by two regimes. Below 40 K the anisotropy is weak and above 40 K it is high and increasing with increasing field as expected for the 2-dimensional nature of the material in this temperature regime. The 2212 wire should exhibit no anisotropy due to its geometry. Indeed the measured anisotrpy is zero within the experimental accuracy giving evidence for the homogeneity of the texture of the BSCCO-2212 crystallites within the wire.

414

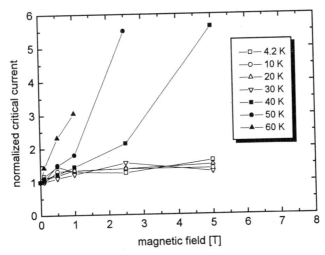

Fig. 4: Normalized critical current density as a function of magnetic field for BSCCO-2223 tapes as a measure for the anisotropy.

5. Conclusions

The new deformation process called profile rolling was presented as a technically simple method to prepare silver sheathed wires and tapes of the BSCCO materials. The obtained critical current densities are comparable to those achieved with the conventional process namely 25 kA/cm^2 for the BSCCO-2223 tapes and 5 kA/cm^2 for the BSCCO-2212 wires at 77 K in self field.

The I-V-characteristics are measured at temperatures between 4.2 K and 100 in magnetic fields up to 8 T. Thermal activation limits the critical current in BSCCO-2212 above 20 K and in BSCCO-2223 above 40 K which is in agreement with the TAFF-mechanism. At this temparatur a change in the anisotropy behavior of the critical current with respect to field orientation is found in BSCCO-2223 tapes, while BSCCO-2212 wires show no anisotropy.

Acknowledgment

We are grateful to A. Kuhn for the SEM micrographs and to J. Müller and R. Wilberg for help with the experiments. This work was supported by the Bundesminister für Bildung, Technologie und Forschung (BMBF) under contract no.: 13 N 6634

References
[1]   Tomita N et al. 1994 *Appl. Phys. Lett.* **65** 898
[2]   Ohkura K., Mukai H, Hikata T, Ueyama M, Kato T, Fujikami J and Sato K 1993 *Jpn. J. Appl. Phys.* **32** L1606
[3]   Fujikami J, Shibuta N, Sato K, Ishii H and Hara T 1993 *Adv. Supercond. VI* (Tokyo: Springer)
[4]   Jin J X, Liu H K, Grantham C and Dou S X 1994 *IEEE Trans. Appl. Supercond.* **5** to appear
[5]   Shoenung S M, Bierl R L, Schafer W J and Bickel T C 1994 *IEEE Trans. Appl. Supercond.* **5** to appear
[6]   Lehndorff B, Busch D, Eujen R, Fischer B, Piel H and Theisejans R 1994 *IEEE Trans. Appl. Supercond.* **5** to appear
[7]   Piel H, Busch D, Fischer B, Lehndorff B and Theisejans R 1994 *Adv. Supercond. VII* (Tokyo: Springer)
[8]   Paalstra T T M, Batlogg B, van Dover R B, Schneemeyer L F and Wasczak J V 1990 *Phys. Rev. B* **41** 6621

Inst. Phys. Conf. Ser. No 148
Paper presented at Applied Superconductivity, Edinburgh, 3–6 July 1995
© *1995 IOP Publishing Ltd*

Non-invasive Hall probe measurements of the lateral current distribution in BSCCO/Ag conductors.

M D Johnston[1], J Everett[1], M Dhallé[1], G Grasso[2], R Flükiger[2], M Yang[3], C R M Grovenor[3] and A D Caplin[1]

[1] Centre for High Temperature Superconductivity, Imperial College, London SW7 2BZUK

[2] Département de la Matière Condensée, Université de Genève, CH -211 Genève 4, Switzerland

[3] Department of Materials, University of Oxford, Oxford OX1 3PH, UK

Abstract. Using a 100μm resolution scanning Hall probe, we have measured the self-field profile due to a transport current in BSCCO/Ag 2223 and 2212 conductors. The field component perpendicular to the plane of the tape after analysis yields the lateral current distribution. In the 2212 samples this distribution becomes uniform (on the experimental length scale) at high currents. At lower currents the distribution shifts towards the sample edges as predicted theoretically for homogenous type II superconducting strips. The expected remanent current profiles after the external current has been switched off are also measured. In the 2223 tape, the current is found to be concentrated at the sample edges.

1. Introduction

In order to compete successfully with conventional low T_c wires, conductors made with the new ceramic superconductors must offer superior performance in terms of their response to high current and high magnetic fields. Tapes made by the powder in tube process have shown notable success, but there are clear limitations to their current handling capabilities.

The correlation of magnetic and transport current measurements has shown the tapes to be electrically inhomogenous, with the result that only a fraction of the superconducting material appears to be carrying current [1]. It has been shown that in 2223 tapes such non-uniform current distributions occur both in the plane of the tape ('lateral' distribution) [2,3] and across the thickness of the superconducting core [4,5]. BSCCO 2212 ribbons [6] however have shown to be composed of well connected grains with excellent texturing, where the current is limited by intragranular flux pinning [7].

For improved materials processing and the manufacture of commercially viable conductors it is therefore important to know where exactly the current is flowing within the tapes and which features of the microstructure are responsible. In this paper we present an analysis of the lateral current distribution in both 2212 and 2223 conductors using a non-destructive scanning Hall probe technique.

416

2. Experimental Details

The 2212 ribbon discussed was produced by electrophoretic deposition [6] of a precursor powder on a Ag carrier tape. This 'green' tape was melt-textured at 885°C and slow cooled to 850°C, where it was annealed for 50h. The thickness of the BSCCO layer is ~ 15 μm on either side and the conductor is 6 mm wide with $J_c(77K, 0T) = 1.35$ kA/cm² (E ~ 10^{-6} V/cm). The 2223 tapes were made by the standard PIT method, having a core size of ~ 30 μm and a width of 4 mm with $J_c(77K, 0T) = 10$ kA/cm² (E ~ 10^{-8} V/cm).

The micro Hall probe used was a GaAs/AlGaAs heterostructure [8] mesa-etched in a standard van der Pauw configuration. This contains a 2-dimensional electron gas (2DEG) at a depth of 100 nm below the surface which acts as a high mobility Hall plate. The active area of the probe is 150 μm square and it has a field sensitivity of ~10 μT at 77K.

All measurements were made in zero applied field. The probe was scanned across the broad face of the tape at a distance of 50 μm in steps of 100 μm, using a home-built scanning rig. A transport current was passed through the conductor and measurements made of the normal component of the self-field across its width, the full range being 7 mm (see sketch).

Scans were made for a number of current levels, which were subsequently scaled to give the spatial variation. The current distribution was extracted from the self field profiles $B_Z(x,z)$ using a 2D finite element method in which the RMS difference between $B_Z(x,z)$ and the field due to the currents I_i in N infinitely thin wires is minimised.

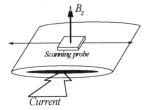

$$B_Z(x,z) \propto \sum_i^N I_i \frac{(x-x_i)}{(x-x_i)^2 + z^2}$$

3. Results and Discussion

3.1 2223 tapes

The self-field profiles in the 2223 tape are shown for increasing current levels (Fig. 1a). For clarity of comparison, all profiles have been scaled by the total current and shifted (Fig. 1b).

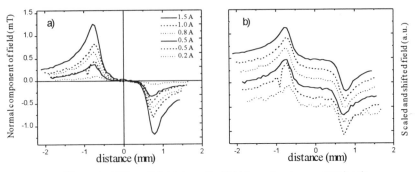

Figure 1. a) The self-field above the 2223 tape at various current levels.
b) shows the same data; this time scaled by the total current and shifted for clarity.

The extracted current distribution shows the majority of the current to be carried at the edges of the tape. Note that this profile does not appear to change with increasing current.

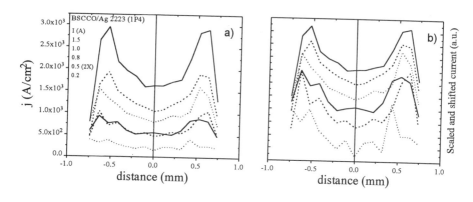

Figure 2. The extracted current distribution of the 2223 tape at various current levels:-
a) unscaled and b) scaled and shifted.

3.2 2212 ribbons

In contrast to the 2223 tape, the self field profiles in the 2212 ribbon do show a qualitative dependence on the total current amplitude. This can be seen most clearly in the remanent field profiles (Fig. 3a) which are measured after the transport current has been switched off. Clearly the higher current excursions lead to an increased remanence, the profile of which shifts towards the centre of the tape with increasing excursion amplitude.

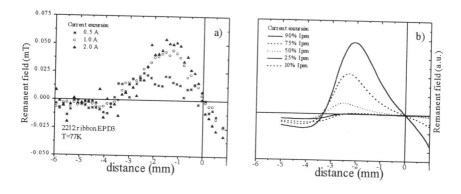

Figure 3. The remanent field a) above a 2212 ribbon after current excursion of various amplitudes and b) predicted for a homogenous type II strip.

3.3 Discussion

The current distribution in a homogeneous type II superconducting strip has been predicted theoretically [9]. In this reference it is shown how the current profile can be expected to peak at the strip edges at low currents, gradually shifting towards the centre as the current increases and becoming uniform at a penetration current I_{pen} . The calculation also shows how a current excursion sets up two remanent current loops at the edges of the strip, the amplitude and distribution of which vary for $0 \leq I < I_{pen}$ and become constant for $I \geq I_{pen}$. Figure 3b) shows the prediction for the remanent field above a strip with similar dimensions to our 2212 ribbon. Comparison shows a good qualitative agreement between this sample and the theoretical prediction, with $I_{pen} \approx 1A$. The independence of the shape of the current distribution on the total current in the 2223 tape on the other hand is in stark contrast with the predictions. This leads us to conclude that in this sample the current distribution is not governed by the intrinsic superconducting electrodynamics, but rather is dictated by inhomogeneities in the microstructure resulting from the preparation process.

4. Conclusions

The normal component of the self-field in BSCCO conductors was measured as a function of increasing transport current. The scaled fields and their corresponding current distributions were compared with those predicted for a homogenous strip.

The 2212 ribbon samples showed the profile expected of well-connected material [9], where the intragranular J_c dominates the behaviour [7]. This we attribute to the melt-texturing of the 2212, which reduces porosity without introducing fractures or cracks.

The stationary current profile found in 2223 tapes suggests that regions at the edge of the tape carry the majority of the current, as previously reported [2,3].

Acknowledgements
We would like to thank J. Harris of the Semiconductor IRC for providing us with the Hall sensors. MDJ would like to thank BICC Cables PLC for financial support through the CASE award scheme.
This work was supported by EPSRC and Brite Euram Contract No. CT920229.

References

[1] Cuthbert M N *et al.* 1994 *conference proceedings, Appl. Sup. Conf., Boston*
[2] Grasso G Hensel B Jeremie A and Flükiger R 1994 *Physica C* **241** 45
[3] Larbalestier D C *et al.* 1994 *Physica C* **221** 299
[4] Pashitski A E *et al.* 1995 *Physica C* **246** 133-144
[5] Everett J 1995 *conference proceedings EUCAS '95, submitted*
[6] Huang S *et al.* 1994 *Supercond. Sci. Technol.* **8** 32-40
[7] Dhallé M *et al.* 1994 *conference proceedings, Appl. Sup. Conf., Boston*
[8] Bending S J von Klitzing K and Ploog K 1990 *Phys. Rev. B* **42** 9859-9864
[9] Brandt E H 1993 *Physica C* **48** 12893-12906

Inst. Phys. Conf. Ser. No 148
Paper presented at Applied Superconductivity, Edinburgh, 3–6 July 1995
© *1995 IOP Publishing Ltd*

Limits on the critical current of BSCCO/Ag conductors

**M Dhallé[1], M N Cuthbert[1], J Thomas[1], J Everett[1], M D Johnston[1],
H K Liu[2], S X Dou[2], G Grasso[3], R Flükiger[3], J Kessler[4], W Goldacker[4],
M Yang[5], C Grovenor[5] and A D Caplin[1].**

[1] Centre for High Temperature Superconductivity, Blackett Laboratory, Imperial College,
London SW7 2BZ, UK;
[2] Centre for Superconducting and Electronic Materials, University of Wollongong,
NSW 2522, Australia;
[3] Département de Physique de la Matière Condensée, Université de Genève,
CH 1211, Genève 4, Switzerland;
[4] Inst. f. Techn. Physik, Kernforschungszentrum, D-76021 Karlsruhe, Germany;
[5] Department of Materials, University of Oxford, Oxford OX1 3PH, UK.

Abstract. We present an overview of the current limiting factors in BSCCO/Ag 2212 and
2223 conductors, based on a variety of transport and magnetisation experiments and consistent
over a wide range of samples of different quality and origin We show how the dominant
dissipation process determining the overall critical current density depends on the temperature
and the applied magnetic field. At low fields and temperatures weak links limit the current,
while at higher field or temperature the useful current is carried only by strong paths and is
limited by intragranular flux motion. While the effective cross section of these strong paths
- and therefore the critical current - is strongly sample-dependent and determined by
microstructural features, the flux motion in these conductors is remarkably similar from sample
to sample.

1. Introduction

In this paper we address the factors which determine the current carrying capacity of BSCCO
based large scale conductors. These factors can be either of an inter- or intragranular nature.
Material homogeneity and bad connectivity can limit current flow to a fraction of the total
cross section. However, in this fraction currents generally survive up to high magnetic fields
where flux dynamics within the grains might take over as the current limiting process.

A first section will briefly review some of the experimental evidence for the non-uniform
current flow and show how the connectivity depends on magnetic field and temperature.

In the second section we will focus on the high field behaviour. We will demonstrate how
here indeed thermally activated flux motion becomes the main dissipation mechanism and
how this process can be described in a model independent, quantitative fashion.

The conclusions of this paper are consistent with the results obtained over a wide variety
of 2223 and 2212 conductors with varying critical currents.

2. Experimental

Measurement of the magnetisation M(B,T), complementary to direct transport characterisation of $J_c(B,T)$ offers several advantages [1]. Firstly, magnetometers often allow one to probe a larger range of magnetic fields and temperaturese than standard transport set-ups. Secondly, and more fundamental, measuring both irreversible moment of the tapes and of the ground powder extracted from the sample provides a direct comparison between the inter- and intragranular current components [2].

Although the results shown in this paper were measured using a vibrating sample magnetometer, it should be stressed that the analysis presented in section 3.2. can be carried out equally well on transport data and leads to essentially the same conclusions [3].

In magnetisation experiments the current density is exteracted from the irreversible component of the magnetic moment using the Bean model. The corresponding electric field is directly related to the ramp rate of the external magnetic field and the typical length scale of coherent current flow can be determined from the shape of the M(H) loop upon reversal of the sweep direction [1].

3. Results and discussion

The techniques described allow to address both connectivity and flux dynamics. Previously it has been shown that $J_c(B,T)$ has two regions of qualitatively different behaviour. At low fields $J_c(B) \propto B^{-v}$, while at higher fields $J_c(B)$ becomes approximately exponential. Both regions are observed quite generally in BSCCO based conductors, although the cross-over field between the regimes is strongly sample dependent [4]. We first review briefly the connectivity issue, related to the first region, and then proceed to illustrate an illuminating way of assessing the flux dynamics which controls the second region..

3.1.Connectivity

Various experiments have demonstrated the non uniform character of current flow in BSCCO/Ag conductors. Mechanical slicing [5], magneto-optics [6] and miniature Hall probe measurements [7] all show how the current density in 2223 tapes tends to be larger at the edges of the sample. Comparison between magnetisation data on tapes and ex tape powder, together with length scale measurements has allowed us to demonstrate how this non uniformity increases in low magnetic fields, the region which corresponds to $J_c(B) \propto B^{-v}$ [8]. Current flow across the thickness of these tapes likewise appears to be non uniform, as can be demonstrated measuring J_c in similar tapes with different thickness [9]. Here also field induced granularity has recently been observed magneto-optically [10].

This magnetic decoupling is strongly sample dependent. In some samples it leads to a sizeable current reduction in low fields, while the effect is generally much smaller in tapes with a high zero field J_c [3]. In most tapes, however, a sizeable current survives at higher fields, where the signature of the $J_c(B)$ dependence changes to exponential [4].

3.2. Flux dynamics

Again, it is a comparison between tape and powder magnetisation which shows this exponential region to be related to flux motion. In order to address the flux dynamics in these

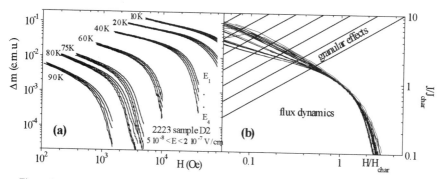

Figure 1: (a) Irreversible magnetic moment $\Delta m \propto J$ for a typical 2223 tape at various temperatures and electric fields E. The closely spaced curves correspond to E ranging from 5e-8 to 2e-7 V/cm. (b) shows the same data with both axes normalised for each E and T to the current and field where $d\ln J/d\ln B = -1$. The hatched low field region indicates the granularity discussed in section 3.1.

materials we can use the commonly observed scaling of the J(B) dependence with temperature and electric field.

Such scaling is demonstrated for a typical 2223 tape in Fig.1(a). The double log plot shows how an increase in temperature or electric field shifts the J(B) curves towards lower J and B, but leaves their general shape unchanged. This can be made more clear by plotting all curves on a reduced J/J_{char} and B/B_{char} scale, where J_{char} and B_{char} can be chosen at any fixed value of $\chi_{\ln} = d\ln J/d\ln B$, as in Fig.1(b) for $\chi_{\ln} = -1$. The 'unfolding' of the curves at low B is due to field induced changes in the connectivity discussed above.

This scaling has important implications for the nature of the flux dynamics. Generally we can write for thermally activated flux motion:

$$E = Bv \exp\{-U_{eff}(B, T, J) / kT\}.$$

Following ref.[11], we can separate the current dependence of the effective pinning potential $U_{eff}(B,T,J) = U_0(B,T) V[J/J_0(B,T)]$, where the function V expresses how the pinning barrier decreases with current. Scaling then implies that both U_0 and J_0 have a power law dependence on the magnetic field: $U_0(B,T) \propto B^n$ and $J_0(B,T) \propto B^m$.

The exponents n and m can be extracted by plotting $\chi_{\ln}$ as a function of $S = d\ln J/d\ln E$:

$$\chi_{\ln} = [1 - n \ln(Bv / E)]S + m.$$

Such a plot has been made for the previous data in Figure 2(a), together with the powder extracted from this tape. Taking into account that E in the powder is ~100 times lower due to the smaller size of the screening current loops, both tape and powder yield $n = -2.6 \pm 0.3$ and $m = -0.1 \pm 0.1$, i.e. $U_0(B,T) \propto B^{-2.6}$ and $J_0(B,T) \approx J_0(T)$. The independence on temperature of the $\chi_{\ln}$-S relation implies both exponents to be temperature independent.

Having extracted these exponents it is natural to plot J versus B^{-n}, as in Figure 2(b) for T=40K. Such plot shows V(J/Jo) to be approximately logarithmic so that we can finally write

$$J_c(B, T) = J_0(T) \exp\{-[B / B_0(T)]^{-n}\}.$$

Note how this logarithmic V(J) behaviour corresponds with the power law character of the IV curves commonly observed at higher magnetic fields.

422

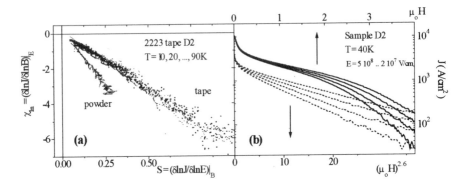

Figure 2: (a) The relation between $S = \partial\ln J / \partial\ln E|_B$ and $\chi_{\ln} = \partial\ln J / \ln B|_E$ for the data shown in fig.1.(a).
This relation enables us to extract the field dependence of the pinning potential.
(b) shows $J(B)$ for this sample at 40K and various electric fields plotted against a
linear B scale (solid lines, top axis) and against $B^{2.6}$ (dashed lines, bottom axis).

Figure 2(b) also demonstrates how the downward curvature of $J_c(B)$ at the highest fields is not related to some new dissipation mechanism setting in, but rather follows from the same pinning mechanism governing most of the H,T range.

4. Conclusions

The nature of the dissipation process limiting the critical current density J_c in BSCCO/Ag conductors depends on the temperature and magnetic field.

Below a temperature dependent crossover field magnetically induced changes in the connnectivity lead to a low field drop off in the overall current density. This field induced granularity relates to the non uniform microstructure of these tapes and is therefore strongly sample dependent.

Above this crossover field flux motion takes over as the dominant dissipation mechanism. The flux dynamics can be analysed model independently using the commonly observed scaling of $J_c(B)$ with temperature and electric field and turns out to remain qualitatively unchanged over a wide range of temperature and magnetic field. This pinning related behaviour appears to be relatively sample independent.

Acknowledgements

This work was supported by EPSRC and Brite Euram Contract No. CT920229.

References

[1] Caplin A D et al 1994 *Supercond. Sci. Technol.* **7** 412
[2] Caplin A D et al 1994 *Conference Proceedings, Appl. Sup. Conf., Boston*
[3] Dhallé M et al *to be published*
[4] Cuthbert M N et al 1994 *Conference Proceedings, Appl. Sup. Conf., Boston*
[5] Grasso G et al 1994 *Physica C* **241** 45
[6] Koblischka M R 1994 *Conference Proceedings, IWCC 7, Alpbach, Austria* 399
[7] Johnston M D et al 1995 *EUCAS '95, submitted*
[8] Dhallé M et al 1994 *Conference Proceedings, Appl. Sup. Conf., Boston*
[9] Everett J et al 1995 *EUCAS '95, submitted*
[10] Pashitski A E et al 1995 *Physica C* **246** 133
[11] Perkins K G et al 1995 *Phys. Rev. B* **51** 8513

Inst. Phys. Conf. Ser. No 148
Paper presented at Applied Superconductivity, Edinburgh, 3–6 July 1995

Low Field Magnetization Nondestructive Evaluation of HTS Tapes within the Bean Critical State Model

K.L. Telschow, L.S. Koo and K.K. Haulenbeek

Idaho National Engineering Laboratory, Idaho Falls, ID, 83415-2209, USA

Abstract. A new approach, that accounts for demagnetization, is described for calculating the flux front boundary within the Bean critical state model for a tape geometry. The method combines an analytical description in the form of an integral equation for the flux front boundary with a numerical method of resolving that equation. Results are given for a source coil above the tape and measuring coils above and below.

1. Introduction

Quantitative determination of the local low field critical current density, $J_c(H=0)$, in a noncontacting manner is useful in the fabrication of high temperature superconducting (HTS) tapes as a nondestructive evaluation (NDE) probe of processing parameters and spatial uniformity. Transport critical currents are most often measured by contact 4-point probes; however, noncontacting magnetic measurements are desired in many practical situations for their speed and ease of use. The magnetic approach consists of measuring the induced currents in the sample with small source/pickup probe coils that spatially scan over the tape surface [1,2]. Subsequently, the Bean critical state model is relied on to determine J_c from the measured magnetic hysterisis. This approach provides a measure of the critical current density since the intergrain critical current density is usually much smaller than that for intragrain.

Usually, the Bean model is used in a context which ignores sample geometry effects, referred to as demagnetization, due to the intractibility of the calculations required. Plate geometries present the most severe demagnetization effects for external fields directed perpendicularly to the surface. Certain geometries, such as a cylindrically symmetric sample in a uniform field, can be treated by numerical methods [3]. Another approach treats the plate by averging the current distribution over the thickness of the plate and thereby obtains an analytical solution [4]. A new approach is described that combines an analytical description of the problem in the form of an integral equation for the flux front boundary with a numerical method of resolving that equation [5]. This method is applicable in cylindrically symmetric geometries for both the external field and the sample. It can be used to determine the critical state region profile including the effects of demagnetization. The integral equation method has been applied to a sphere in a uniform external field [5] and a plate in a cylindrically normal field [6,7] This paper describes calculational results from the integral equation method for the plate geometry along with experimental measurements on silver clad Pb-BSSCO (2223) tapes.

424

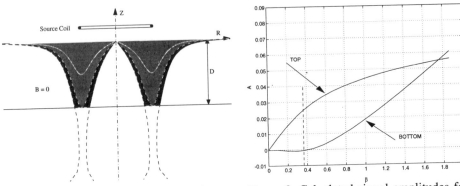

Figure 1. Calculated flux fronts penetrating a tape of thickness D generated above.

Figure 2. Calculated signal amplitudes for top and bottom pickup coils.

Results are presented for measuring coils above and below the tape coaxial to the source coil for a given local critical current density J_c.

2. Calculation of the Flux Front Surface

The source coil above the tape generates a flux front that penetrates into the tape. Eventually, at high excitation currents, this front penetrates completely through the tape and splits into two fronts as shown in figure 1. The approach taken is to calculate the location of the flux fronts based on the shielding ability of the screening currents to produce a region of zero magnetic field below the front. This idea leads to an integral equation for the front location that can be solved by a numerical iteration scheme [8]. The details of the method of solution have been presented elsewhere [4,5]; calculation results and measurements are described here

Exploiting cylindrical symmetry, an integral equation can be written for the flux front

$$\Psi(R,\beta = \frac{1}{J_c r_c^2}), \quad \int_0^\infty dR' \, a_\phi(R, \Psi(R,\beta); R', \Psi(R',\beta)) \frac{-\partial \Psi(R',\beta)}{\partial \beta} = a_\phi(R, \Psi(R,\beta); I, Z_c) \text{w}$$

here $\mu_0 a_\phi(R,Z; R',Z')$ is the vector potential at (R,Z) due to a unit current coaxial loop of radius R' at height Z', all normalized by. the coil radius r_c. J_c is the local critcal current density and I the source coil current. This integral equation can be resolved for the flux front derivative knowing the flux front profile for some initial value of the external field. Since the starting flux front profile for zero external field is just the sample surface, the equation provides an incremental numerical scheme for determining the profile for any external field [5-7]. When the external field becomes large enough for the flux front to reach the bottom plate surface, the front breaks up into two separate profiles. Continuation of the method results in a pair of integral equations that are resolved simultaneously by the same algorithm as with the single profile [6,7].

Once the flux front location is known, the total vector potential and measured signal can be determined for a pickup coil in any position. Figure 2 shows the signals predicted for pickup coils above and below the plate, where the external field is subtracted off above (balanced) and included below (unbalanced). Particularly, the pickup signal below the tape reflects the screening brought about by currents flowing at less than the local critical current density and forms a good indicator of local tape condition. The dashed line of figure 2 indicates the full thickness penetration value of the external field. Complete hysteresis loops, for cycles of the external field, can be obtained for each pickup coil position and are shown in figures 3 & 4. Significantly different hysteresis loops are found for the top and bottom coils due to the

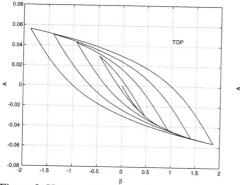

Figure 3 Hysteresis loops at the top balanced pickup coil for several maximum external field values

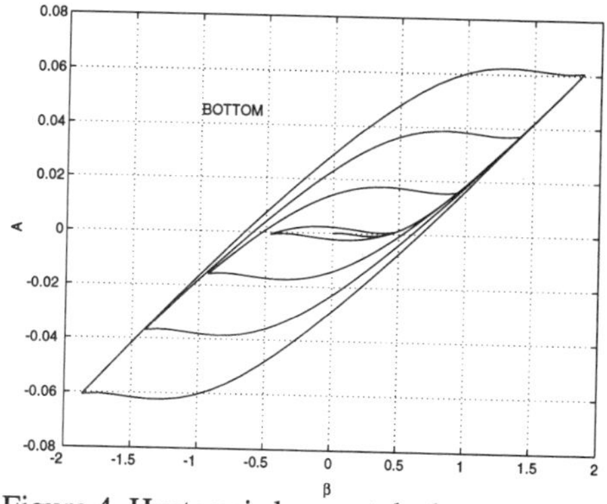

Figure 4 Hysteresis loops at the bottom unbalanced pickup coil for several maximum external field values.

screening currents in the tape. Using the prescription of the Bean critical state, the complete hysteresis response is calculated in the form of an irreversible loop equation from

$$A_{\pm}(R,Z;\beta) = \pm A_0(R,Z;\beta_{max}) \mp 2A_0(R,Z;\frac{\beta_{max} \mp \beta}{2}),$$ where $A_0(R,Z;\beta)$ is the initial curve

shown in figure 2. In general β is a function of time; however, only the quasi-stationary states of the critical state are dealt with in this paper. The time scale for changes in the external field is typically very much longer than that exhibited by flux line motion, so the model always assumes a sequence of stationary states uniquely defined by the history and present value of the external field. Therefore, if the time dependence of β is given, the resultant time dependence of both pickup coil signals can be calculated. Most of the significant changes in the signal waveforms occur for the bottom coil as the flux front penetrates the plate thickness. Harmonic analysis of these signals could lead to a simple method for monitoring the local critical current density through waveform processing.

3. Experimental Measurements

The measurement geometry consisted of a single layer superconductor plate deposited between two outside silver layers. Tape samples of Pb-BiSrCaCuO (2223 phase) produced by the powder-in-tube method [2,9] were measured with a small probe source coil (13 turns of #36 copper wire, 1 mm inside diameter). Balanced opposing pickup coils (5 turns each of #36) were wound over the source coil, producing a small source/pickup probe that could be scanned over the sample surface. An additional pickup coil (5 turns #36 copper wire on 1 mm diameter) was positioned below the tape sample. Lift-off distances for both top and bottom coils were about 0.7 mm. Results are presented for one measurement position on a tape with an average critical current density of 220 A/cm^2.

Figure 5 shows the measured results from the top balanced coil for three different temperatures. These results, which are qualitatively similar to the theoretical results of figure 2, represent the magnetization due only to the induced screening currents within the flux front. No single obvious point signifies the full flux penetration value with which the critical current density can be determined. The results from the bottom coil aresignificantly different. The shielding effect is nearly complete when the external field is insufficient to cause full plate penetration as shown in figure 6. The shielding effect is clearly visible and, by extrapolation, a

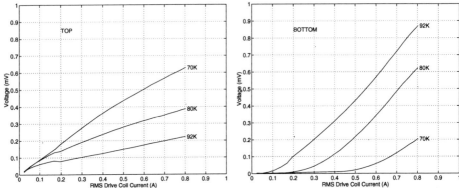

Figure 5 Top coil measured signal amplitudes for different temperatures.

Figure 6 Bottom coil measured signal amplitudes for different temperatures.

unique point corresponding to full plate penetration can be estimated. The local critical current density can then be determined from the corresponding value of β

4. Conclusions

A method has been outlined for calculating the flux front profile for a superconducting sample in either a uniform or nonuniform applied magnetic field possessing azimuthal symmetry. This technique relies upon finding a surface with zero vector potential. This surface is determined by simple integration of its derivative with respect to the external field, found by resolving a linear integral equation of the first kind. Measurement induced voltages and the entire hysteresis loop response can be found by extension of the ZFC magnetization response with changing external field. Other experimentally measured quantities relating to the critical state can be calculated directly from the hysteresis loop if the time dependence of the external field is known.

5. Acknowledgment

Work sponsored by the U.S. Department of Energy, Office of Energy Research, Office of Basic Energy Sciences, under DOE Idaho Operations Office Contract DE-AC07-94ID13223.

6. References

[1] Telschow K L and O'Brien T K 1991 Appl. Phys. Lett. **59** 730-732.
[2] Telschow K L, O'Brien T K, Lanagan M T, and Kaufman D Y 1993 IEEE Trans. Magn. **3** part III 1643-1646.
[3] Navarro R and Campbell L J 1991 Phys. Rev. **B44** 10146
[4] Zeldov E, Clem J R, McElfresh M, and Darwin M 1994 Phys. Rev. **B49** 9802
[5] Telschow K L and Koo L S 1994 Phys. Rev. **B50** 6923-6928.
[6] Telschow K L and Koo L S 1995 accepted for publication in IEEE Trans. Magn
[7] Koo L S and Telschow K L 1995 in preparation.
[8] Gold R 1964 Argonne National Laboratory Report ANL-6984
[9] Samples were provided by M. T. Lanagan at Argonne National Laboratory.

Inst. Phys. Conf. Ser. No 148
Paper presented at Applied Superconductivity, Edinburgh, 3–6 July 1995

Current-voltage characteristics of $Bi_2Sr_2CaCu_2O_x$ tapes with and without Ag sheath

M. Polak, W. Zhang, E. E. Hellstrom and D. C. Larbalestier

Applied Superconductivity Center, University of Wisconsin, Madison, 1500 Engineering Drive, Madison, WI 53706, USA

Abstract. Intrinsic current-voltage characteristics of various $Bi_2Sr_2Ca_1Cu_2O_x$ (2212) tapes were measured with and without the silver sheath. The E-I curves (H=0, T=4K) of bare samples were exponential from 2×10^{-2} to 2 $\mu V/cm$, E then increasing less than exponentially. I_c of the bare samples was slightly reduced compared to the Ag-sheathed ones. Small notches made by laser in bare tapes not only reduced I_c, but also gave rise to a small linear voltage increase. In the current sharing regime the silver sheath carries a current considerably higher than $I_{Ag} = V/R_{Ag}$, where R_{Ag} is the silver resistance estimated experimentally. I_c of bare tapes measured with the magnetic field perpendicular to the tape plane exhibited the well known hysteresis and the tapes had E-I curves dependent on their magnetic history.

1. Introduction

The exact form of the inherent E-J curves of the polycrystalline BSCCO core of Ag-sheathed tapes is of great interest but it is not simple to extract the intrinsic BSCCO core E-J curves from the I-V curves measured on Ag-sheathed tapes. The first issue is to account for current sharing and to identify that flowing in the superconducting core, I_c, and that flowing in the silver sheath, I_{Ag} [1]. Sausaging of the core [2], cracks [3], bubbling [4], inhomogeneities of the local J_c, and second phase all influence the current sharing,. To estimate I_{Ag} requires the silver sheath resistance; fitting procedures [1, 5] or measurement after a strong bending of the tape [6] can be used. However, the electric field E_x is usually inhomogeneous [7] and the measured value $E = V/L_p$ (L_p is the distance between the potential taps) is in fact a mean value of E_x over L_p. Localized inhomogeneities of the BSCCO core can produce a linear component to the I-V curve starting at $I < I_c$ [8]. It is an open question how localized defects affect the I-V relation in a bare tape. Moreover, the current density in 2212 tapes varies within the cross-section of the superconducting core being highest at the silver-superconductor boundary. Thus, the current density calculated in the usual way, $J_c = I/A_s$ (A_s being the BSCCO cross-section) is an average value only. Most data on the E-J curves of 2212/Ag tapes start at electric fields between 10^{-2} and 10^{-1} $\mu V/cm$ [9]. Current sharing between BSCCO and Ag exists at any measurable voltage, but its influence on the form of the E-J curves becomes usually visible at electric fields above 0.1 μV/cm, becoming clearly important at electric fields above 1μV/cm. We also note, that the BSCCO core in Ag-sheathed tapes is under axial compressive stress [10] and it is not clear how the E-J relation changes when this stress is released.

The present study was framed so as to clarify the effect of current sharing and the stress experienced by the core on I_c and the E-I relation. I-V curves were measured with and then without the Ag-sheath under self field conditions at 4.2 K. To study the effect of inhomogeneities, we made small notches in the bare tape and studied their effect on the E-I curve.

2. Experimental Details

The samples were made from three different monocore 2212/Ag tapes (sample 1 from the tape W 28, sample 7 from W 34, samples 2 through 6 and 8 from OX).The tape dimensions and BSCCO core cross-sections are: 0.17 x 2.8 mm and 0.966 10^{-1} mm^2 for tape W 28, 0.16 x 2.91 mm and 0.98 10^{-1} mm^2 (W 34) and 0.17 x 2.73 mm and 0.966 10^{-1} mm^2 (OX). W34 and OX were partial melt processed in 100% O_2,, while 50% O_2 was used for W28. These treatments gave different I_c values. The I_c samples were about 30 mm long and the potential taps were 5-10 mm apart. After measurement of the silver sheathed samples, the sample ends were covered by paint and the silver between the potential taps was completely removed by etching in solution of 5 parts of NH_4OH and 2 parts of H_2O_2(30%). A sketch of the etched sample is shown in the Fig. 1.

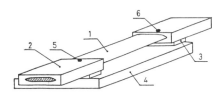

Fig.1 : Sketch of a short sample fixed to its sample holder after Ag was etched off : 1 - bare BSCCO, 2 - Ag-sheathed ends, 3 - insulating plate, 4 - supporting BSCCO plate, 5 and 6 - potential taps

In order to prevent samples breaking under Lorentz forces, we glued some parts of the edges of the bare tape to the support using a small amount of epoxy resin. A low hysteresis magnet made of an AC Nb-Ti wire with a CuNi matrix was used for the measurements of I-V curves in external magnetic fields.

3. Results and discussion

E-I characteristics under self-field conditions at 4.2 K on the sheathed and bare samples are shown in Fig. 2.

The low electric field sections of the E-I curves of both Ag-sheathed and bare tapes are exponential. Current sharing in the sheathed tapes causes

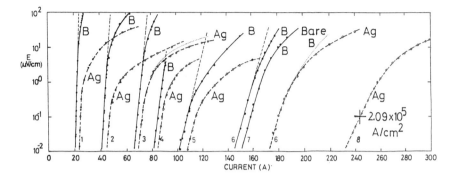

Fig.2: Electric field versus current for the samples 1 through 8 at 4.2 K, zero magnetic field. The dashed lines show the curves of Ag sheathed samples, the full lines those of bare samples. The dotted lines show the electric field versus the current I_n, obtained in the manner explained in the text.

bending of the curves starting at between 0.2 and 0.5 μV/cm. After the silver was removed, the characteristics were exponential up to ~3 μV/cm. (Sample 5 bent over at smaller E) These data indicate that current sharing at ~1 μV/cm is not negligible and that J_c of the BSCCO core is overestimated by 5 to 10% due to current flow in the silver. It is also seen that the E-I curves of bare samples and are shifted towards lower currents. The reduction of I_c values is 14%, 11%, 5%, 3%, 6% and 15% for samples 1, 2, 3, 4, 5 and 7 respectively. We suppose that this reduction is caused by local inhomogeneities, particularly by cracks which open up when the compressive stress is removed, which cannot be shunted by Ag.

To derive intrinsic, core E-I curves from the silver sheathed samples we have to separate I_s from I_{Ag}, using the relation $I_s = I - V/R_{Ag}$. Curves obtained in this way for samples 3 and 7 are shown in Fig 2. We used $\rho_{Ag} = 5\ 10^{-11}\Omega m$, which is the smallest measured resistivity of Ag samples extracted from the tapes. We see that the real current carried by the silver is considerably higher than that calculated. We believe that even small variations of J_c lead to strong variations of the local electric field. As can be deduced from Fig. 2, a change of the local J_c by 5 % produces a change of the local E value by one order of magnitude. As the weaker regions of higher E are shunted by Ag, a corresponding fraction of the current locally flows through the silver, causing only a small increase of the total voltage because the total length of the current sharing path is smaller than the whole distance between potential taps. This means that extraction of the E-I_s curves from measurements on Ag-sheathed tapes using this common method is charged with a considerable error.

A small notch made by a laser cutter produced a linear component to the V-I curve of a bare sample. We believe that this linear component is the signature of c-axis conduction forced by current redistribution around the notch. The influence of magnetic field on the V-I curves of a bare tape are shown in

430

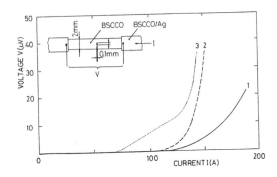

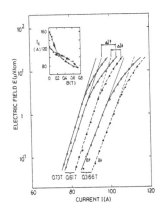

Fig. 3 : V-I curves measured on sample 5: 1 - with Ag, 2 - without Ag and 3 - with a small notch (See insert).

Fig. 4 : E-I curves on bare sample 7 measured in increasing (full lines) and decreasing (dashed) fields.

Fig.4. We see that the magnetic history influences the position of the E-I curve and its bending at higher E.

4. Conclusions

Intrinsic E-I characteristics of a series of BSCCO 2212/Ag tapes having J_c ranging from 3×10^4 to 10^5 A/cm² (4K, H=0)) have been obtained by direct measurement on samples with and without the Ag sheath. Removal of Ag leads to 5-15 % reduction in I_c. The Ag carries considerably higher current then calculated from $I_{Ag} = V/R_{Ag}$. A small notch in the bare tape produced a linear component to the I-V curve of the bare sample. We found that the curvature of E-I plots of bare samples in external magnetic fields depends on magnetic history.

References

[1] Gurevich A, Pashitski A E, Edelman H S, Larbalestier D C, 1993 Appl. Phys. Lett. **62** 1688-1690

[2] Osamura K, Kamo M, Oh S S, Ochiai S, 1994 Cryogenics **34** 303 - 308

[3] Cha Y S, Lanagan M T, Gray K E, Jankus V Z, Fang Y, 1994 Appl. Superc. **2** 47 - 59

[4] Patel S, Haugan T, Chen S, Wong F, Narumi S, Shaw D T, 1994 Cryogenics **34** 537 - 542

[5] Soulen Jr R J, Francavilla T R, Fuller-Mora W W, Miller M M, 1994 Phys. Rev. **B50** 478 -

[6] Mawatari Y, Yamasaki H, Kosaka S, to appear IEEE Trans. on Superconductivity 1995.

[7] Cave J R, Willen D W A, Nadi R, Cameron D, Zhu W, ibid ref 6

[8] Polak M, Larbalestier D C, Parrell J, Zhang W, Hellstrom E, 1995 Spring Meeting of Mat. Res. Soc., San Francisco, April 17-21, paper K 9.9

[9] Shibutani K, Hase T, Egi T, Hayashi S, Ogawa R, Kawate Y, 1994 Appl. Superc. **2** 237 -

[10] Larbalestier D C, Cai X Y, Feng Y, Edelmen H, Umezawa A, Riley G N, Carter, 1993 PysicaC **221** 299

[11] ten Haken B, ten Kate H J, Tenbrink J, ibid ref. 6.

[12] Ries G, Neumuller H W, Schmidt W, Struller C, 1994 7th Int. Workshop on Crit. Currents in Superconductors. Ed. H. Weber

[13] Manufacturer of the tape : Oxford Superconducting Technology.

Inst. Phys. Conf. Ser. No 148
Paper presented at Applied Superconductivity, Edinburgh, 3–6 July 1995
© *1995 IOP Publishing Ltd*

A study of current transfer at 77 K in a high current $(Bi_{2-x},Pb_x)Sr_2Ca_2Cu_3O_{10-\delta}$/silver composite conductor[†]

C M Friend, I Ferguson, I W Kay, L Le Lay, M Mölgg, C Groombridge and T Beales

BICC Cables Superconductivity Group, Hedgeley Road, Hebburn, Tyne and Wear NE31 1XR, UK

Abstract. A $(Bi_{2-x},Pb_x)Sr_2Ca_2Cu_3O_{10-\delta}$ / silver composite conductor with a critical current of 430 Amps at 77 K has been constructed from 100 seven-filament tapes. Tests of the conductor at fault currents show that even at levels of 10 times the critical current, some current still flows in the superconductor.

1. Introduction

The rapid progress over the last two years in the production of $(Bi_{2-x},Pb_x)Sr_2Ca_2Cu_3O_{10-\delta}$/silver (Bi-2223) tapes in long lengths has led to several high-T_c power cable models. Most of these have been ac conductors, e.g. Sumitomo with Tokyo Electric Power Company (TEPCO) [1] and American Superconductor Corporation [2]. It is predicted that within 10-15 years the retrofitting of superconducting cables (replacing underground copper ones) will be commercially viable. There is also a potential market for dc superconducting cables in long distance links between different national grids. BICC Cables, its Italian subsidiary CEAT CAVI srl and Ansaldo Ricerche srl recently designed and commissioned a prototype Bi-2223 dc cable that carried over 11 kA at 31 K [3].

This paper reports on the design and construction of a dc conductor made from 100 multifilamentary Bi-2223 tapes with a total critical current (I_c) of 430 Amps. Analysis of the conductor E-I characteristics provided information on current transfer between the individual tapes. All values of I_c quoted are for 77 K in self field at 1 μVcm^{-1}.

2. Design, construction and testing of the conductor

2.1 Design and construction

The high current-carrying conductor was constructed from one hundred multifilament $(Bi_{2-x},Pb_x)Sr_2Ca_2Cu_3O_{10\pm\delta}$ silver-sheathed tapes, each 7-core and of length 0.52 m. The critical currents (I_c) of the individual tapes ranged from 6.0 - 11.2 A over the whole length. The average I_c was 6.7 A. The tapes were made into ten stacked bundles, each of ten tapes.

---

† This work was part funded by BRITE-EURAM contract no. BRE2-CT92-0207.

432

The tapes in each stack were soldered to each other at both ends and in the middle. All joints, made with InAg solder, covered a length of 2-3 cm. The temperature of the tapes, monitored during soldering, did not rise above 160 °C. The conductor former consisted of a flexible insulating rod of 26 mm diameter with a short section of copper rod, incorporating a current lead connection, attached at each end. Ten equidistant straight channels were cut lengthways along the former into which the tape bundles were placed. The bundles were then soldered to the copper sections, again with InAg solder, at either end. Along the insulating part of the former the bundles were tied down with low temperature tape.

2.2 Testing

For the I_C measurement two configurations for the voltage contacts were used: (a) two wires, 43 cm apart, wound round the conductor in a loop and point soldered to each bundle and (b) two voltage taps attached 41 cm apart to the top of just one of the bundles. The difference in the measured I_C for the two cases was as small as 1%. The conductor was placed directly into liquid nitrogen and the transport current ramped from zero to just above I_C in approximately 40 seconds under computer control. The initial value of I_C was 430 Amps, a substantial current for a conductor of this size. In additional tests the applied current was ramped up to $2I_C$ and held continuously at 200 A for times up to 30 mins. with no noticeable degradation in performance.

The conductor was then put through 20 thermal cycles while I_C was monitored. Each cycle consisted of direct immersion in liquid nitrogen, measurement then removal from the cryogen for at least one hour. The results are plotted in figure 1 showing a gradual decrease in I_C. During cycling a pair of voltage contacts were attached to the middle section of one bundle, 3 cm apart. An I_C of 500 A was measured across this short section. Because of a difference in the size of thermal contraction between the tapes and the former material, buckling of the tape bundles was observed. The strain produced by this buckling was estimated to be high as 0.4% and is the most likely cause for the reduction in I_C with thermal cycling.

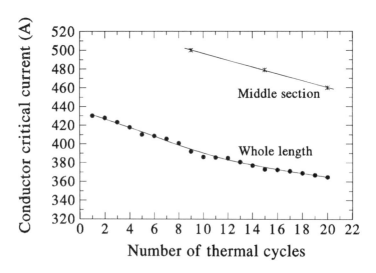

Figure 1. A graph showing the decrease of the conductor critical current with thermal cycling. I_C is measured across the whole length (●) and a 3 cm section in the middle of the conductor (✳).

3. Current transfer in the $(Bi_{2-x},Pb_x)Sr_2Ca_2Cu_3O_{10-\delta}/$ silver tapes

3.1 E-I characteristics

Before their placement in the conductor the I_C of each tape bundle was measured. It was found that I_C was always 70-75% of the sum of the I_Cs for the individual tapes. Figure 2 shows the E-I characteristic of one bundle (I_C=60 A) and that of the conductor and an individual tape (I_C=10.6 A). Also plotted in figure 2 is the predicted characteristic purely for the superconducting core. This was calculated by assuming that the silver sheath acts as a parallel resistive shunt. It is then easily shown [4] that the current in the superconductor $I_{sc} = I_m - \{E_m.A_{sc}/\rho\}$ where I_m and E_m are the measured current and electric field, A_{sc} is the transverse cross-sectional area of the Bi-2223 and ρ is the sheath resistivity. The value 2.9×10^{-9} Ωm for ρ at 77 K was obtained from a silver-only tape made using the same thermomechanical treatments as for Bi-2223 tapes. For A_{sc} the average value calculated from cross-sectional micrographs was used. An error of $\pm10\%$ in the measurement of A_{sc} is typical and the effect of this on the calculated E_m-I_{sc} characteristic is shown in the figure.

3.2 Critical current distributions

An alternative analysis of the V-I characteristics is possible if we assume a continuous distribution of critical currents inside the tape. Following Baixeras and Fournet[5] we find $d^2V/dI^2 \propto P(i)$, the distribution function such that P(i).di is the fraction of material with critical currents in the interval di. For low temperature superconductors P(i) approximates to a Gaussian distribution [6,7]. Figures 3 and 4 plot the calculated critical current distribution (CCD) for a single tape (I_C=10.6 A) and bundle (I_C=60 A) respectively whose E-I characteristics are shown in figure 2. The single tape has an almost Gaussian distribution with what may be a small secondary peak at 25-30 A. In contrast the bundle has a much broader and flatter CCD and a long high-current tail.

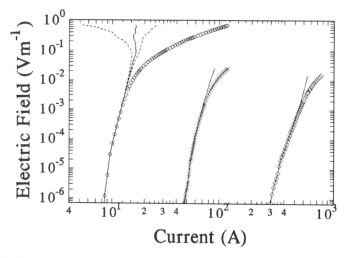

Figure 2. The E-I characteristic for a single tape ($\circ$), a bundle of ten tapes ($\times$) and the conductor ($\diamond$). The solid lines are the calculated characteristics for the superconducting core only. The dashed lines indicate the effect of an error of $\pm10\%$ in the calculation of A_{sc}, the superconducting cross-sectional area.

434

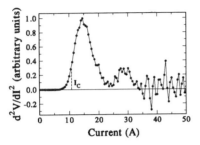

Figure 3. The critical current distribution of a single tape with I_c = 10.6 A.

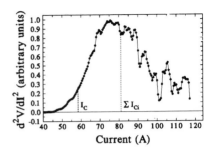

Figure 4. The critical current distribution of a bundle of ten tapes with I_c = 60 A.

4. Discussion and conclusions

When applied to the data in figure 2 the resistive shunt model of 3.1 suggests that even at currents up to $10I_c$ there is still a current I_c flowing in the superconducting filaments. These filaments would still be in the flux creep regime [8] and it is not clear at which point all the current will be transferred to the sheath. A deeper understanding of current transfer will be needed in designing dc cables for fault current protection and minimisation of losses.

In an economic cable the maximum current carrying capacity is required from the volume of tape contained within it. In our conductor the apparent 25% drop in the maximum I_c when the tapes were made into bundles could be due to three factors: (i) damage, this is unlikely as multifilament tapes are relatively robust and the temperature during soldering was kept low. Thermal cycling of the single tapes before stacking was shown to decrease I_c by less than 5%; (ii) self field, this will most affect the tapes near the outside of each bundle. We estimate a self field ($\mu_0 I/2$.width) of the order of 10 mT for each bundle. Measurements at 77 K on similar tapes in such a field have shown a 5-15% reduction in I_c; and (iii) current equalisation [9] between the filaments and individual tapes. There is a large variation in the tape I_cs and as figure 3 indicates, also in the I_cs of the separate filaments. Close to I_c a longitudinal resistivity along the bundle may develop near the current joints as the current tries to equalise itself throughout the tapes. This would be exacerbated by the fact that the tapes in each bundle are largely insulated from each other except at the three solder joints. A lower I_c may then be measured. It is hoped that a narrower distribution in critical currents and placing the tape bundles in metallic conduits would improve the performance of such a conductor though the total I_c will still be limited by self field effects.

References

[1]   1994 *Superconductor Week* **8**(40) 1
[2]   Riley Jr G N, Gannon Jr J J, Miles P K and Parker D R 1994 *Appl. Supercond.* **2** 155
[3]   1995 *Superconductivity News*; also Beales T P 1995 *proc. ISTEC, Maui, June 1995*
[4]   Matthews D N, Müller K -H, Andrikidis C, Liu H K and Dou S X 1994 *Physica C* **229** 403
[5]   Baixeras J and Fournet G 1967 *J. Phys. Chem. Solids* **28** 1541
[6]   Hampshire D P and Jones H 1987 *J. Phys. C* **20** 3533
[7]   Edelman H S and Larbalestier D C 1993 *J. Appl. Phys.* **74**(5) 3312
[8]   Gurevich A, Pashitski A E, Edelman H S and Larbalestier D C 1993 *Appl. Phys. Lett.* **62** 1688
[9]   Wilson M N 1983 *Superconducting Magnets* (Oxford University Press) chapt. 10

Inst. Phys. Conf. Ser. No 148
Paper presented at Applied Superconductivity, Edinburgh, 3–6 July 1995
© *1995 IOP Publishing Ltd*

Jc vs strain performance of Ag/Bi-2223 tapes with different geometrical structures

L. Martini, L. Bigoni, E. Varesi, S. Zannella

CISE Spa, P.O.Box 12081, I-20134 Milan, Italy

L. Gherardi, P. Caracino, S. Spreafico

Pirelli Cavi S.p.a., V.le Sarca 222, 20126 Milano, Italy

Abstract. In this paper we report on the electrical characterisation under mechanical stress of monofilamentary and concentric Bi-2223 silver sheathed tapes having different fill factors and J_c, made with the same powder. The influence of mechanical stress and strain on the current carrying capacity of fully processed samples, as well as of tapes after each thermomechanical cycle, has been systematically investigated. The experimental results on Bi-2223/Ag specimens with different geometries are compared and explained on the basis of a semi-empirical model specifically developed.

1. Introduction

World-wide effort to optimise the manufacturing process of Ag/Bi-2223 tapes has led to a continuous improvement of their superconducting performances, making them more and more promising for a variety of commercial applications [1, 2]. In this perspective, the mechanical properties of these tapes, which are known to be typically rather poor, can turn out to be a critical factor.

The influence of mechanical stress and strain on the current carrying capacity of structurally different tapes has been systematically investigated by means of a specifically developed tensile testing apparatus, which has been designed and constructed so as to allow the simultaneous measurement of critical current, stress and strain.

2. Experimental

2.1. Tapes fabrication

All tapes were fabricated by the Powder in Tube technique, by filling a superconducting powder into silver tubes with 14 mm outer diameter. The powder was a relatively standard type, with nominal cation ratios of Bi:Pb:Sr:Ca:Cu=1.8:0.4:2:2:3; for comparison, a few monofilamentary tapes were also fabricated with a gold doped precursor. This have been

436

obtained by adding gold colloidal particles, precipitated from an $AuCl_3$ solution, to the ceramic powder in a weigth percentage of about 15%, and carefully homogeneising the mixture by repeated grinding in agata mortar.

Concentric wires and tapes, having through their cross section up to three HTS rings separated by silver layers, were fabricated according to [3] by packing the ceramic powder inside the hollow spaces left between silver tubes coaxially positioned. The composite silver tubes were generally cold swaged (with the exception of one tape, which was initially hot extruded) and then drawn, to obtain wires having 2.3-2.9 mm final diameter that were then rolled into tape shape conductors of thickness ranging from 0.5 mm to 1.0 mm. Short specimens (70<l<90 mm) cut from the obtained tapes, were then processed by optimised rolling or pressing and sintering cycles, repeated up to three times, to improve their electrical and mechanical properties.

Pressed and rolled tapes as well as single core, concentric and gold-doped specimens were heat treated at 835°C for 80 hours in Argon/Oxygen athmosphere with an O_2 content of 10% mol. The overall cross section of tapes, as well as the HTS fraction and the fill factor, were evaluated by analising the optical micrographs taken on the polished transversal cross sectional area of specimens by means of a digitizer tablet.

Fig.1 shows the optical cross-sectional micrographs of a monofilamentary and a concentric Bi-2223/Ag tapes. White areas inside the ceramic core of monofilamentary tape identify the gold particles. The fill factor (defined as the ratio of the superconductor to total cross section) was found to range from 10% to 40% in tapes with different structures.

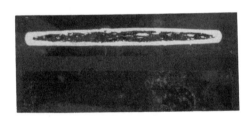 

Fig.1   Optical micrographs of Bi-2223/Ag tapes with different geometrical structures:
a) monofilamentary (gold doped);   b) concentric with three HTS regions.

2.2. Measuring set-up

The self-field transport critical current I_c was measured at 77 K by the usual voltamperometric 4-leads method using the electric field criterion of $E_c = 1\mu V / cm$.

The tensile tests were carried out using a specially designed apparatus [5] allowing for stress-strain and transport critical current versus stress or strain analysys, both at room temperature and in liquid nitrogen bath. A load cell was used to measure the applied force, whereas high sensitivity cryogenic strain gauges were pasted to the tapes to monitor the deformation.

Care was taken in selecting gauges with very low thickness and tensile strength, in order to avoid any significant reinforcement to the tapes. In any case, the voltage taps were always placed so as to be inside the region where axial strain was measured.

3. Results and Discussion

Following the model reported in [5-6], the static equilibrium of a composite (silver sheathed) tape is the result of the mutual interaction between superconducting core and metal sheath occurring during cooling, at the end of the thermal treatment. In this frame, the response of our samples to stress and strain was expected to strongly depend upon the degree of pre-compression imparted by silver to the superconductor. This, in turn, should depend upon the fill factor as well as on the temperature at which the tests were performed.

3.1 Tensile tests in liquid nitrogen

In Fig. 2 the normalized J_c versus axial strain curves for different tapes: 2D (concentric, II° t.c.), 2F (mono, II° t.c.) and 3R (concentric, III° t.c.), all having the same fill factors and similar J_c, are reported. As appears, the "critical strain" ε_c at which J_c starts to drop is practically the same, about .13% for all these tapes.

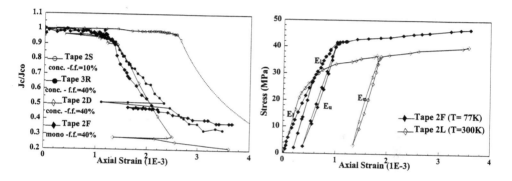

Fig. 2 Relative Jc versus strain   Fig. 3 Stress-strain at 77 and 300 K

  Shown in the same figure is also the performance of tape 2S, which exhibits a significantly higher ε_c value (about .27%). The main reason for such better behaviour, compared to the other samples, is most likely related to the much smaller fill factor of this tape. However, based on previous experience, which had shown [6] a significant improvement of the strain response to be associated to the increase of J_c produced by repeated deformation-annealing cycles in typical monofilamentary samples, this is presumably due also to the higher overall "quality" of the superconductor in such tape, whose high J_c (22 kA/cm^2, to be compared to 3-5 kA/cm^2 for the others) probably reflects also a better densification and geometrical regularity of the ceramic core.

3.2 Tensile tests at room temperature

Significant stresses and strains are expected to be imposed to tapes also at room temperature, during the various stages of cable manufacturing and laying. Tests were then carried out also at about 300 K, by applying strain in steps (increasing strain at each step), and measuring J_c after each.

  Fig. 3 shows the stress-strain curves at room temperature and at 77 K for two monofilamentary tapes. Two important evidences must be especially noted. First of all, the strain tolerance of tapes at room temperature is much lower than at 77 K. Second, the elastic moduli determined for increasing stress (El), and during the tape unloading (Eu), are practically equal at room temperature, whereas in liquid nitrogen roughly El=0.5 Eu. Both features qualitatively confirm that the initial state of pre-compression of the superconductor plays a key role in the mechanical behaviour of the composite. The first effect shows, in particular, that a substantial fraction of the pre-compression at LN$_2$ temperature is due to cooling from 300 to 77 K, in spite of the smaller temperature difference compared to the one after the annealing (about 1100 to 300 K). The second gives indications about the static condition of silver at the beginning of the tests. In fact, the lower elastic response found at 77 K is interpreted as the result of (tensile) prestraining of silver beyond its yield, whereas at room temperature this has apparently not been reached, and the first points in the stress-strain curve of the composite seem to reflect the elastic contribution of both BSCCO and silver.

4. Conclusions

Bi-2223/Ag mono- and multi-core (concentric) tapes, exhibiting self field critical current density J$_c$ at 77 K up to 22 kA/cm^2 have been characterized under tensile stress and strain. Samples having lower fill factor and higher J$_c$ showed a better strain tolerance, presumably due to both better densification and geometrical regularity and a higher pre-compression of the ceramic core. Applying the stress and strain at room temperature was found to change the elastic behaviour of tapes, compared to liquid nitrogen conditions, and to drastically decrease their J$_c$ tolerance to strain. All the results, qualitatively well explained in terms of a "pre-strain" model [5], confirmed that the capability of Ag/BSCCO composites to withstand strains is strongly dependent upon the geometry (especially the fill factor), and the thermal history of the tapes.

Acknowledgements

Thanks are due to F. Curcio and F. Miraglia for technical assistance and helpful discussions. This work was partially supported by ENEL S.p.a. (Italy).

References

[1]   Yamada Y, Sato M, Murase S, Kitamura T and Kamisada J 1992 *Proc. of 5th Int. Symp. on Superconductivity ISS '92*, 718, Kobe, Japan

[2]   Sato K, Shibuta N, Mukai H, Hikata T, Masuda T, Ueiama M, Kato T, Fujikami J 1991 *in Proc. of ISS'91*, Springer, Tokio, 559

[3]   Martini L, Bonazzi S, Majoros M, Ottoboni V and Zannella S 1993 *IEEE Trans. Appl. Superc.*, **3**, 961

[4]   Martini L, Ottoboni V, Zannella S, Caracino P, Gherardi L and Gandini A 1994 *Proc. of 7th IWCC*, 565, Alpbach, Austria

[5]   Gherardi L, Caracino P, Metra P and Vellego G 1994 *in "Processing of Long Length of Superconductors"*, ed. *Balachandran U et al.*, published by The Minerals, Metals and Materials Society 147

[6]   Gherardi L, Caracino P and Coletta G 1994 *Cryogenics* **34** 781-784

Inst. Phys. Conf. Ser. No 148
Paper presented at Applied Superconductivity, Edinburgh, 3–6 July 1995
© 1995 IOP Publishing Ltd

Critical current and hysteretic AC loss in monofilamentary Bi-2223/Ag tapes

K-H Müller, C Andrikidis, H K Liu+ and S X Dou+

CSIRO, Division of Applied Physics, Sydney 2070, Australia

+Centre for Superconductivity and Electronic Materials,
University of Wollongong, Wollongong 2522, Australia

Abstract. We report on magnetic moment measurements of a monofilamentary silver-sheathed BiPbSrCaCuO tape at different temperatures as a
function of the magnetic field applied perpendicular to the tape surface. The
extracted intergranular and intragranular critical current densities both decrease exponentially with increasing magnetic field suggesting that the cause
for the weakness of the transport critical current in strong fields is the posi-
tional disorder of pancake vortices in grains. Measuring the hysteresis losses
reveals that the intergranular loss is greater than the intragranular one.

1. Introduction

A very high intergranular (transport) critical current density J_c^J is achieved for silver-
sheathed BiPbSrCaCuO tapes which is promising for large-scale applications. The high
J_c^J in these tapes results from strong alignment of grains accomplished by pressing and
rolling of the BiPbSrCaCuO powder encapsulated in a silver sheath. Besides the critical
current density the hysteresis loss of tapes is of importance to assess their suitability in
AC applications.

2. Experiment, results and discussion

2.1. Intergranular and intragranular currents

The monofilamentary BiPbSrCaCuO/Ag (2223) tape investigated had an electrically
measured transport critical current density of 19 000 A/cm² at 77 K in self-field. The
magnetic moment m of a short piece of tape was measured at different temperatures
in magnetic fields up to 5.5 T, perpendicular to the tape surface, using a commercial
SQUID magnetometer. The magnetic moment is the sum of the intergranular magnetic
moment m_J, originating from the intergranular current flowing over the entire tape and

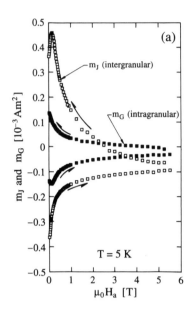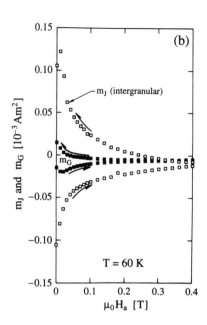

Figure 1 : Upper and lower branches of the intergranular and intragranular magn etic moments, m_J and m_G, as a function of the applied magnetic field H_a. (a) At a temperature of 5 K and (b) at 60 K.

the intragranular magnetic moment m_G, originating from the intragranular currents flowing in the grains. The tape was then severely bent (rolled) to a small diameter of about 1 mm causing microcracks to develop which has been shown to completely interrupt the intergranular current [1]. After straightening the tape the remaining intragranular magnetic moment m_G was measured. The intergranular magnetic moment was obtained from $m_J = m - m_G$.

Figures 1(a) and (b) show the magnetic moment loops $m_J(H_a)$ and $m_G(H_a)$ where H_a is the applied field at temperatures 5 K and 60 K. The upper branch of the m_J-loop shows a peak near zero applied field due to the irreversible behaviour of the intergranular critical current caused by trapped flux in grains. The m_G-loop also shows a peak but in the lower branch which is expected from a critical state model description with a field dependent intragranular critical current density.

The intergranular critical current density J_c^J was calculated from $J_c^J = (m_J^\uparrow - m_J^\downarrow)/(2a^2dL)$ where $m_J^\uparrow$ and $m_J^\downarrow$ are the upper and lower branches of the m_J-loop, a the half-width, d the thickness and L the length of the tape. This formula was derived using the critical state model for a type II superconductor strip in a perpendicular field [2]. The intragranular critical current density J_c^G was calculated from the Bean formula $J_c^G = 3(m_G^\uparrow - m_G^\downarrow)/(4R_G adL)$ where $m_G^\uparrow$ and $m_G^\downarrow$ are the upper and lower branches of the m_G-loop and R_G the average radius of the BiPbSrCaCuO grains.

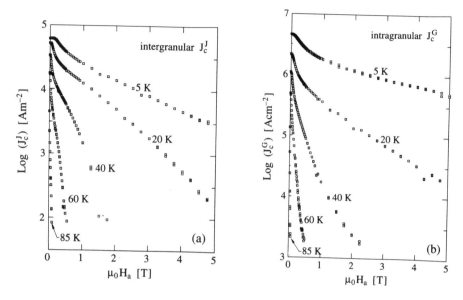

Figure 2 : (a) Intergranular critical current density J_c^J versus applied field H_a for different temperatures. (b) The same for the intragranular critical current density J_c^G.

Figure 2(a) shows the intergranular critical current density J_c^J of the tape ($a = 1.3$ mm, $d = 60$ μm, $L = 5.8$ mm). In fields greater than about 0.3 T, J_c^J decreases exponentially. Figure 2(b) shows the intragranular critical current density J_c^G which also decreases exponentially, similar to J_c^J. To calculate J_c^G an average grain radius of $R_g = 10$ μm was assumed which results in an intragranular critical current density J_c^G which is about a factor 100 larger than J_c^J at 5 K.

One possible interpretation of our experimental data is that the strong field dependence of the intergranular critical current density J_c^J originates from the field dependence of J_c^G which is caused by flux creep. The creep induced disorder of pancake vortices near the grain boundary Josephson junctions randomizes the phase difference of the order parameter across the grain boundary which causes a reduction of the Josephson current.

2.2. Intergranular and intragranular hysteresis losses

The hysteresis loss W of a type II superconductor is the energy per unit volume per AC-field cycle dissipated inside the superconductor. The hysteresis loss W is proportional to the area enclosed by the magnetic moment loop. In the case of a thin superconducting strip like the BiPbSrCaCuO tape, where the field is applied perpendicular to the surface of the strip, Brandt and Indenbom [2] have derived analytical expressions for the magnetic moment under the assumption that the current density, which flows over the entire strip, is field independent. Using their expressions for the magnetic moment, the intergranular hysteretic AC loss is

442

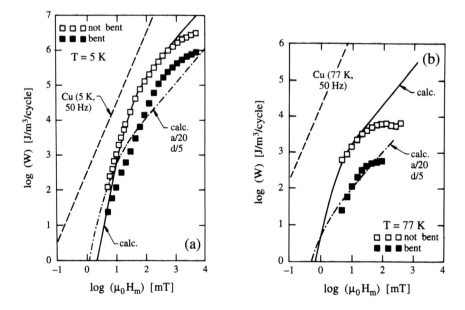

Figure 3 : (a) Total (not bent) and intragranular (bent) hysteresis loss W versus amplitude H_m at 5 K. The dashed curve is the resistive loss in a copper strip at 50 Hz. The dash-dotted curve is the calculation for a multifilamentary tape. (b) The same at a temperature of 77 K.

$$W_J = 2\pi\mu_o \frac{a}{d} H_d H_m \left[-\tanh(\frac{H_m}{H_d}) + \frac{2H_d}{H_m}\ln(\cosh(\frac{H_m}{H_d})) \right] , \qquad (1)$$

where H_m is the AC field amplitude and H_d the characteristic field, $H_d = J_c^J d/\pi$.

Figures 3(a) and (b) show the hysteretic AC loss W as a function of the AC field amplitude H_m at 5 K and 77 K. The intragranular loss is much smaller than the total loss W of the not bent tape [3]. The full curves show the calculation using eq.(1). The dashed curve is the resistive AC loss at 50 Hz for a thin copper strip of cross section equal to the tape core. Here H_m is the self-field amplitude. The dash-dotted curve represents the prediction for the intergranular loss if a filamentary structure with filaments of half-width $a/20$ and thickness $d/5$ is assumed. At high field amplitudes the intergranular loss of a multifilamentary tape is smaller than the intragranular loss, and the total loss is smaller than for a monofilamentary tape.

References

[1] Müller K-H, Andrikidis C, Liu H K and Dou S X 1994 *Phys. Rev.* B **50** 10218–20224

[2] Brandt E H and Indenbom M 1993 *Phys. Rev.* B **48** 12893–12906

[3] Müller K-H, Andrikidis C, Liu H K and Dou S X 1995 *Physica C* (in press)

Inst. Phys. Conf. Ser. No 148
Paper presented at Applied Superconductivity, Edinburgh, 3–6 July 1995
© 1995 IOP Publishing Ltd

Flux Creep and Weak Link Behaviour in Silver Sheathed Bi-2223 Tapes

G Fuchs, T Staiger, P Verges and K Fischer

Institut für Festkörper- und Werkstofforschung Dresden, Postfach, D-01171 Dresden, Germany

A Gladun

Technische Universität Dresden, D-01062 Dresden, Germany

Abstract. Bi-2223/Ag tapes have been investigated by transport measurements in the temperature range between 4.2 K and 77 K for magnetic fields up to 12 T. The results are interpreted in the framework of the 'brick wall' model which includes flux creep and weak link behaviour as current limiting mechanisms. The $j_c(B)$ dependence and E-j characteristics can be described quantitatively by the assumption of a small fraction of strongly coupled grains near the Ag sheath and a large fraction of weakly coupled grains inside the tape. The degradation of the critical current density under the influence of thermal cycles was found to be confined to the field and temperature range governed by the weak link behaviour.

1. Introduction

High critical current densities have been achieved in silver sheathed Bi-2223 tapes, but the factors limiting the critical current in this material are not completely understood up to now. It is well established that the strong influence of magnetic fields on the critical current at 77 K observed for the field direction perpendicular to the tape surface has its origin in thermally activated flux creep of two dimensional pancake vortices in the CuO_2 layers. At lower temperatures, the effect of flux creep is reduced and the grain structure of the samples becomes important. Although the limitation of the critical current at low temperatures is controversially discussed, it is commonly accepted that along the length of the tapes exists only a small fraction of strongly coupled grains. Several models [1,2] have been proposed in which the intragrain current flowing in the strongly coupled grains is the bottleneck for current transport. In the 'brick wall' model [3,4] the critical current at low temperatures is determined by the current across the c axis Josephson junctions, whereas the intragrain current is assumed to be so strong that it does not limit the critical current.

In this paper the critical current density $j_c(B,T)$ and E-j characteristics of Bi-2223/Ag tapes were analysed in the framework of the 'brick wall' model. It is shown that all characteristic features of the results can be explained by this model.

2. Experimental details

Ag sheathed 19 filament Bi-2223 tapes were prepared by the 'powder in tube' technique. The critical transport current and E-j characteristics were measured in the temperature range between 4.2 and 77 K for magnetic fields up to 12 T directed perpendicular to the sample surface. The critical transport current was determined at a criterion of $E_c=10^{-7}$ V/cm using the standard four-point resistive measurement technique.

In order to eliminate the shunting effect of silver on the E-j-characteristics of the superconductor, the current was corrected taking into account the current in the Ag sheath.

3. Results and discussion

In Fig. 1 the critical transport current density of a 19 filament tape is shown as a function of magnetic field for different temperatures.

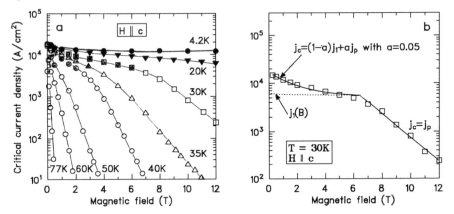

Fig. 1 Critical current density versus magnetic field applied perpendicular to the tape surface.
a) different temperatures, b) fit of the brick wall model to the $j_c(B)$ curve at 30 K:
--- calculated curve (Eq. 3), $\cdots$ $j_1(B)$ used for the calculation

In a wide range of temperatures and magnetic fields (open symbols), the field dependence of the critical current density follows an exponential law

$$j_p(B,T) = j_0 \, exp\left[-\left(\frac{B}{B^*(T)}\right) ln\left(\frac{E_0}{E_c}\right)\right] \tag{1}$$

This behaviour has its origin in thermally activated flux creep of pancake vortices and can be explained by the special form of the activation energy $U/kT = B^*/B \, ln \, (j_0/j)$ resulting in a potential law for the E-j characteristics [5]. The irreversibility field B^* and the fitting parameters E_0 and j_0 can be derived from the E-j characteristics [6].

The weak $j_c(B)$ dependence at critical current densities of about $10^4 A/cm^2$ can be explained by the assumption that the critical current in this region is limited by the critical Josephson currents j_1 of c axis weak links. The field dependence of the critical current across c-axis weak links is given by [4]

$$j_1(B) = j_1(0)\left(\frac{B_0}{B}\right)^v \tag{2}$$

where $B_0 = \Phi_0/L_1^2$ with Φ_0 as the flux quantum and L_1 as the length of the grains in the a,b-directions. The potential law of Eq.(2) describes the $j_c(B)$ curve at 4.2 K. At higher temperatures there is a crossover from the weak link behaviour at low magnetic fields to flux creep behaviour at higher fields which can be understood, when the distribution of grain sizes is taken into account. In the original 'brick wall' model [3] the tapes were assumed to consist of grains of the same size. By Bulaevskii et al.[4] the case of two different grain lengths was considered: L_1 for the short grains deep inside the tape and $L_2 > L_1$ for the long grains near the Ag sheath. The critical currents accross short and long bricks are j_1 and j_2, respectively.

The model predicts in the tape a critical current

$$j_c = \begin{cases} j_p & \text{for } j_p < j_1 < j_2 \\ \\ (1-a)\,j_1 + a\,j_p & \text{for } j_1 < j_p < j_2 \end{cases} \tag{3}$$

where a is the fraction of the long bricks of strongly coupled grains. If $j_p < j_1 < j_2$, then the critical current of the tape is limited by the intragrain current j_p according to Eq.(1). For $j_1 < j_p < j_2$ the current density across the short bricks is j_1, whereas across the long bricks flows a current j_p. Eq.(3) describes the $j_c(B)$ curves in the crossover region quite well, as it is shown in Fig.1b for one example. From the fit in Fig.1b a value of $a = 0.05$ is obtained at 30 K for the fraction of long bricks in the sample.

It is very instructive to study the influence of the current limiting factors on the E-j dependence. In Fig.2 E-j characteristics are shown for three different temperatures. The superconducting current in Fig.2 was calculated by subtracting the sharing current flowing in the Ag sheath from the measured current. This correction is necessary especially at low temperatures, where the resistance of silver becomes small.

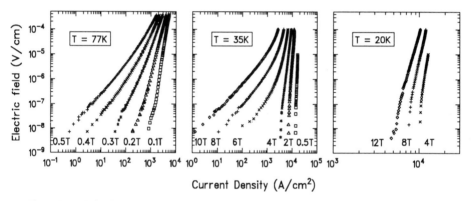

Fig. 2 Electric field versus superconducting current density at T=77 K, T=35 K and T=20 K for different magnetic fields.

At high temperatures the E-j curves follow a potential law which is characteristic for flux creep as mentioned above. A crossover from the potential law to very steep lnE-lnj characteristics is observed at 35 K, when the magnetic field decreases. Moreover, also the lnE-lnj characteristics for 35 K at high fields show this crossover, when the current reaches the value of the critical current $j_1 \approx 10^4 A/cm^2$ of the short bricks. The explanation for the increase of the slope of the lnE-lnj characteristics at j_1 is that currents above j_1 can flow only across the small fraction of the long bricks, whereas the maximum current across the short bricks is j_1.

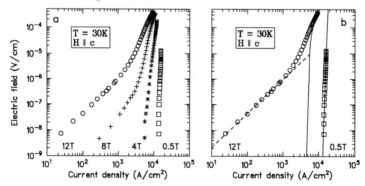

Fig. 3 E-j characteristics at T=30 K. a) different magnetic fields, b) fit of the brick wall model (Eq. 4) to the curves at B = 0.5 T and B = 12 T: ⋯ E(j) for $j < j_1$, --- E(j) for $j > j_1$

With increasing intragrain currents j_p at lower temperatures the crossover field is shifted to higher values and becomes at 20 K higher than the maximum available magnetic field of 12 T. The lnE-lnj characteristics at 20 K show a negative curvature which is, besides of the large slope of the lnE-lnj dependence, the second typical feature of weak links.

The E-j dependence is given by the expression [4]

$$
E / E_0 = \begin{cases} (j / j_0)^p & \text{for } j < j_1 \\ (a\, j_0)^{-p}\, [j - (1-a)\, j_1]^p & \text{for } j > j_1 \end{cases}
\tag{4}
$$

Fig. 3b shows for two magnetic fields at 30 K that experimental E-j curves can be described by Eq.(4) quite well. Here the same parameters were used as derived from the $j_c(B)$ fit.

The degradation of a Bi-2223/Ag tape was investigated under the influence of thermal cycles between room and low temperature. The $j_c(B)$ dependence was measured after each degradation step for different temperatures as shown in Fig. 4.

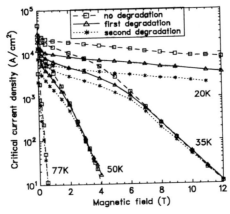

Fig. 4 $j_c(B)$ curves after thermal cycles between room and low temperature

The main result is that the degradation of the critical current is strongly confined to the field and temperature range governed by the weak link behaviour. This can be explained in the framework of the 'brick wall' model taking into account the mechanical strain of the grain connections by the different thermal expansion coefficients of silver and superconductor.

This work was supported by Bundesministerium für Bildung und Forschung, contract no. 13N6102

References

[1] Mawatari Y, Yamasaki H, Kosaka S and Umeda M 1995 *Cryogenics* **35** 161

[2] Hensel B, Grivel J C, Jeremie A, Perin A, Pollini A and Flükiger R 1993 *Physica C* **205** 329

[3] Bulaevskii L N, Clem J R, Glazman L I and Malozemoff A P 1992 *Phys. Rev. B* **45** 2545

[4] Bulaevskii L N, Daemen L L, Maley M P and Coulter J Y 1993 *Phys. Rev. B* **48** 13798

[5] Gladun A, Fuchs G, Fischer K and Verges P 1994 *Physica B* **194-196** 1839

[6] Fuchs G, Vlakhov E S, Nenkov K A, Staiger T, Gladun A 1995 *Physica C* **247** 340

Inst. Phys. Conf. Ser. No 148

Paper presented at Applied Superconductivity, Edinburgh, 3–6 July 1995

Spatial and temporal temperature and voltage signals of a silver-sheathed BSCCO tape undergoing a quench: Experimental and analytical results

Yukikazu Iwasa and Mamoon I. Yunus

Francis Bitter National Magnet Laboratory
Massachusetts Institute of Technology, Cambridge MA 02139 USA

Abstract. We present experimental and analytical temperature and voltage signals, both spatial and temporal, of a silver-sheathed BSCCO-2223 tape quenching along the tape's length. Experimental results, obtained at temperatures of 50 K and 60 K in zero background magnetic field for transport currents up to 70 A, agree well with those based on theory.

1. Introduction

Since the start of the high-temperature superconductivity (HTS) era, we have been investigating design and operational issues for superconducting magnets, particularly stability and protection. At first our effort was confined chiefly to analysis and numerical simulation [1,2]; since the availability of silver-sheathed BSCCO-2223 tapes of lengths at least ~10 cm long in the early 1990s, we have added experimental study: early results of normal-zone propagation along the length of a tape in 1993 [3] and more recent ones in both one- and two-dimensional configurations [4].

We present here results of heating-induced quenching in a current-carrying, silver-sheathed BSCCO-2223 tape at two operating conditions, one at 50 K with a transport current of 70 A and the other at 60 K with 55 A. Spatial and temporal temperature and voltage signals measured, both in zero background magnetic field, agree well with those based on theory.

2. Experimental Technique & Data

Figure 1 shows a drawing of a silver-sheathed BSCCO-2223 tape, 300 mm long, placed in a narrow groove machined in a copper block, whose bottom extension is attached to the cold-head of a cryocooler. As indicated in the figure, thermocouples and voltage taps are attached to the tape along its 285°-arc to create seven measurement zones. Quenching is induced with a current pulse applied over a 25-mm long heater located roughly at the midpoint of the arc.

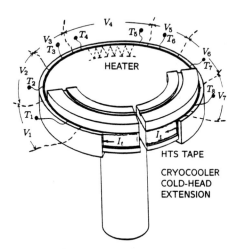

Fig. 1 Drawing of a silver-sheathed BSCCO tape in a copper groove and maintained at T_{op} with a cryocooler.

448

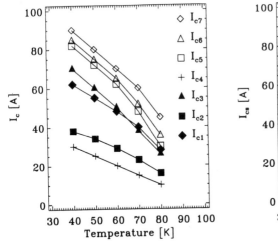

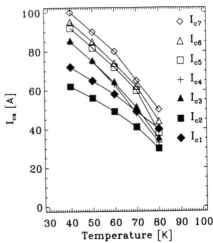

Fig. 2 $I_c(T)$ plots, $E_c = 1\,\mu\mathrm{V/cm}$.

Fig. 3 $I_{cs}(T)$ plots, $E_{cs} = 25\,\mu\mathrm{V/cm}$.

2.1. Critical Current, $I_c(T)$, and Critical Sharing Current, $I_{cs}(T)$

Because quench-induced temperature profiles observed at a given operating temperature, T_{op}, were found to be asymmetric with respect to the heater location, critical current data, $I_c(T_{op})$, for seven zones were taken before each quench event for that operating temperature. At a given T_{op}, E vs I traces indicated a significant variation in I_c (defined at $E_c = 1\,\mu\mathrm{V/cm}$) among seven zones. Apparently this variation is somewhat inherent with these silver-sheathed BSCCO conductors. Figure 2 shows $I_c(T)$ plots for seven zones; the manufacturer's nominal I_c ($E_c = 0.1\,\mu\mathrm{V/cm}$) for this tape is 40 A at 77 K.

Also a large current transition region was observed beyond I_c in each trace; this is often attributed to low index, n, values. Temperature measurement, however, indicated that each zone to remain at T_{op} in this region, at least up to an E-field of $\sim$20 $\mu\mathrm{V/cm}$. In order to neglect this index-induced power generation on quenching, we define a new critical current, "critical sharing current," $I_{cs}(T)$, which in this experiment is determined at $E_{cs} = 25\,\mu\mathrm{V/cm}$. For analytical purpose, the current through the silver, I_m, may be given by: $I_m \simeq 0$ for $I_t \leq I_{cs}(T_{op})$ and $I_m \simeq I_t - I_{cs}(T_{op})$ for $I_t \geq I_{cs}(T_{op})$. Figure 3 shows $I_{cs}(T)$ plots for seven zones for temperatures 40$\sim$80 K. Note that based on this approximation, Joule heating at each zone appears only when its temperature reaches "current sharing" temperature T_{cs}, at which $I_t = I_{cs}(T_{cs})$.

2.2. Quench Initiation and Normal-Zone Propagation

After $I_c(T)$ measurement, a transport current, I_t, was introduced into the tape, and then quenching was triggered with a heating pulse typically of 12 W lasting 0.5 s. Figure 4 shows temperature traces (a) and voltage traces (b, solid) recorded for $I_t = 55$ A and $T_{op} = 60$ K. Transport current, which remained constant, was shut off when a hot-spot temperature, T_{pk}, in the tape reached $\sim$300 K. [In earlier runs, T_{pk} was allowed to reach as high as the tape's melting temperature; we have discovered that $I_c(T)$ data were irreproducible after T_{pk} had exceeded $\sim$400 K.]

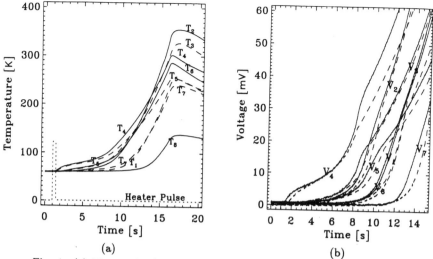

Fig. 4   (a) Measured $T(t)$ traces during quenching.   (b) Measured (solid) and computed (dotted, see Sec. 3) $V(t)$ traces. $I_t = 55\,\text{A}$ and $T_{op} = 60\,\text{K}$.

2.3. Temperature Profiles

Figure 5 shows temperature profiles at selected instances, $T(z,t)$, extracted from the $T(t)$ traces of Fig. 4a. Note that the profile starts symmetric about the heater location—as is expected—and gradually becomes asymmetric. As discussed below, this asymmetry is related to this conductor's nonuniformities of $I_c(T)$ and $I_{cs}(T)$.

3. Analysis and Discussion

3.1. From $T(t)$ Traces to $V(t)$ Traces

Voltage for an jth zone, $V_j(t)$, is given by:

$$V_j(t) = 0 \quad (T_{op} \le T_j \le T_{cs_j}) \quad (1a)$$

$$= \frac{\rho_m(T_j)}{A_m}[I_t - I_{cs_j}(T_j)]$$
$$(T_{cs_j} \le T_j \le T_c) \quad (1b)$$

$$= \frac{\rho_m(T_j)}{A_m}I_t \quad (T_j \ge T_c) \quad (1c)$$

$\rho_m(T_j)$ and A_m are, respectively, the matrix resistivity and cross sectional area.

   The dotted curves in Fig. 4b are computed from Eq. 1 with measured $T(t)$ of Fig. 4a.   Agreement with the solid curves (measured) confirms a successful conversion of $T(t)$ to $V(t)$.

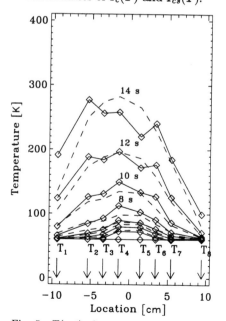

Fig. 5 $T(z,t)$, from traces of Fig. 4a. Dotted curves are computed, see Sec. 3.

3.2. Temperature Profiles

The spatial and temporal temperature of a composite superconducting tape undergoing a quench may be obtained as a solution to a power density equation. In this experiment, in which the tape is under *nearly* adiabatic conditions, we have [5]:

$$C_{cd}(T)\frac{\partial T}{\partial t} \simeq \frac{\partial}{\partial z}\left[k_{cd}(T)\frac{\partial T}{\partial z}\right] + \frac{\rho_m(T)}{A_m A_{cd}}\left[I_t - I_{cs}(T)\right]I_t \qquad (2)$$

$C_{cd}(T)$ and $k_{cd}(T)$ are, respectively, composite tape's heat capacity and thermal conductivity; and A_{cd} is the composite tape's cross sectional area. Note that $k_{cd}(T)$ is essentially equal to silver's thermal conductivity. As was done previously [3], Eq. 2 was solved numerically for $T(t)$ for a given set of operating parameters.

The dotted curves in Fig. 5 are $T(z,t)$ profiles for $I_t = 55\,\mathrm{A}$ and $T_{op} = 60\,\mathrm{K}$ that are direct results of solution to Eq. 2, with measured values of z-dependent $I_{cs}(T)$, shown in Fig. 3, incorporated in computation. Note that when heating is confined only to the heater area and its immediate neighbors, heating is more or less symmetric with respect to the heater location. As the normal zone spreads over the entire tape, "V_2"-zone (Fig. 1), with the poorest $I_{cs}(T)$ (I_{c2} in Fig. 3), experiences the largest dissipation and consequently its temperature rises fastest, causing asymmetric temperature profiles at later times, observed in measurement and collaborated by computation.

4. Conclusion

Key results, *i.e.*, $T(z,t)$ and $V(z,t)$, of our analysis have been confirmed with an experiment, obtained with a silver-sheathed BSCCO-2223 tape undergoing a quench at operating temperatures of 50 K and 60 K in a zero background magnetic field. The particular tape used in the present experiment, for example, has demonstrated that a spatially nonuniform critical current density can lead to overheating at locations other than the quench initiation point. A more detailed account of this work is under preparation and will be submitted soon for publication [6].

The authors express their thanks to the Department of Energy, Superconductivity Technology Program for the support of this work; Pradeep Haldar and James Hoehn Jr. of IGC and Ken-ichi Sato of Sumitomo Electric for silver-sheathed BSCCO tapes.

References

[1] Y. Iwasa, "Design and operational issues for 77-K superconducting magnets," *IEEE Trans. Mag.* **MG-24**, 1211 (1988).

[2] Y. Iwasa and Y.M. Butt, "Normal-zone propagation in adiabatic superconducting magnets over the temperature range 4.2 K to 80 K," *Cryogenics* **30**, 37 (1990).

[3] R.H. Bellis and Y. Iwasa, "Quench propagation in high T_c superconductors," *Cryogenics* **34**, 129 (1994).

[4] Y. Iwasa, H. Lim, and M.I. Yunus, "Stability and quenching in high-temperature superconductors," *IEEE Trans. Appl. Supercond.* (1995).

[5] See, for example, Yukikazu Iwasa, *Case Studies in Superconducting Magnets*, (Plenum Press, New York, 1994).

[6] M.I. Yunus and Y. Iwasa (under preparation).

Inst. Phys. Conf. Ser. No 148
Paper presented at Applied Superconductivity, Edinburgh, 3–6 July 1995
© 1995 IOP Publishing Ltd

Effects of deformation radius during rolling on the properties of Bi-2223/Ag superconductor tapes

D W A Willén, C Breau, W Zhu, D Asselin, R Nadi, and J R Cave

Service Technologie des matériaux, Vice-présidence Technologie et IREQ, Hydro-Québec, 1800 Montée Ste-Julie, Varennes (Québec) J3X-1S1, Canada

Abstract Superconducting Bi-2223/Ag tapes were manufactured using rolling processes with deformation radii in the range of 0.8-42cm for constant reduction rates of 15% strain per pass. The effect of the variation of the radius on the development of work instability in these tapes was studied.   Longitudinal and transverse "sausaging" were observed for large and small radii, respectively. At intermediate radii, a square "biscuit" morphology was observed. The severity of these non-uniformities was significantly reduced for small radii. Cracks in the unsintered powder core at approximately 90° to the tape plane were observed, and are indicative of a tensile fracture mode in the core. This fracture mode is suggested to be a contributor to the development of non-uniformities such as sausaging. The formation of work instabilities is attributed to hardening of the oxide core due to high levels of hydrostatic stress in some rolling configurations.

1. Introduction

Non-uniform morphology of the oxide core is detrimental for obtaining high J_C values in Oxide-Powder-In-Tube (OPIT) high-temperature superconductor wires.   These non-uniformities can be due to work instability during the fabrication process and differences in the mechanical properties of the sheath and superconductor materials.

To eliminate these problems, the use of smaller sized rollers [1,2,3], low reduction per pass [1, 3], dispersion hardening of the Ag-sheath [4], and sequential pressings [5] have been suggested. In the present work, the influence of rolling deformation radii in the range of 0.8-42cm on the creation of non-uniformity, densification, and grain alignment in flat composite tapes has been investigated.

2. Experiment

The superconductor powder was ball-milled and calcined several times to reduce carbon content and particle size. After filling the powder into a pure silver tube, wires were drawn to diameters of 0.82-1.02mm. These wires were then rolled using equivalent true strains of 15% per pass in different rolling configurations in a modified Schlaumann-type mill [6] with exchangeable rollers.

To compare the different experiments, the rollers were stopped so that an imprint of the geometry shown in Fig. 1 could be retrieved from the roll gap. The length, L, width, W, and thickness, H, of the imprints in the deformation zone where measured when the tape thickness was 375μm. Then the characteristic radii of 0.8-42cm were calculated according to $R=L^2/\Delta H$, where $\Delta H=H_1-H_2$. The true plastic strains were measured in three orthogonal directions after each pass through the rollers, down to a final thickness of the tape of (100 ± 3)μm.

Optical microscopy and SEM were used to analyze the powder cores as rolled, as pressed after rolling, and as sintered. The correlation of structural non-uniformities with electric surface potential structures in the flux flow state were obtained with a sliding-contact measurement device [7]. These surface potential measurements display features with length scales from 100μm to 2mm in size and will be presented in detail in a later paper.

452

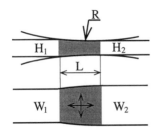

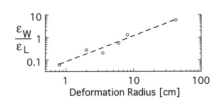

Fig. 1. Schematic of the deformation zone geometry   Fig. 2. Strain ratio for various deformation radii.

3. Results

The ratio of the average strain across the width (ε_W) to the average strain along the length (ε_L) for reductions between 500μm and 100μm thickness is shown in Fig. 2. A large deformation radius caused transverse expansion while a small radius caused longitudinal expansion.

The insides of these tapes, after peeling open the silver sheath, are shown in Fig. 3a-e. It is clear that work instability occurs in the direction(s) of maximum deformation. The largest deformation radius resulted in pronounced longitudinal ridges and a wide tape. Notably, a strain ratio where $\varepsilon_W/\varepsilon_L \approx 1$ gave rise to square "biscuits" in the oxide core. A smaller radius resulted in considerably less severe instabilities and also a less dense core.

Some tapes were heat treated using an accelerated phase formation schedule consisting of short heat treatments and multiple pressings early on in the sintering process whilst the core is less brittle [8]. The cross sections and length sections of three sintered samples are shown in Fig. 4. The severity of the longitudinal ridges of the R=42cm sample are illustrated in this figure. For R=6cm, the cross section is relatively smooth while the length section shows severe sausaging. In contrast, for R=0.8cm, more even length and cross sections are maintained in spite of the large reductions per pass of 15%.

Figure 5a shows the top surface of the powder core after rolling (but before heat treating) with R=42cm, and it appears to be dense and with some degree of alignment of the "platy" powder particles. On the other hand, Fig. 5b shows an identical view of a sample rolled with R=0.8cm, displaying a porous core with randomly oriented particles. After a uniaxial pressing at 5GPa, however, the latter sample looks very similar to that rolled with R=42cm (Fig. 5c), with an increase in density, alignment, and with some small-scale longitudinal cracks.

The fracture of the oxide core has been reported to occur mainly along the planes of maximum shear at 45° to the tape plane [3]. Figure 6, however, show examples of both ~45° cracks and cracks at approximately 90° to the tape plane in unsintered samples. The latter configuration is frequently observed, for example in the necked regions of sausaged samples [8], and is indicative of tensile stresses in the core. In contrast, the core of the R=0.8cm sample shows no cracks after rolling.

These observations show that at a reduction rate of 15% strain per pass, the larger rollers cause a dense core and work instability while the small rollers induce less densification and less work instability. These effects will be discussed in the following section in terms of the geometry of the deformation zone, hydrostatic pressure, and hardening of the powder core.

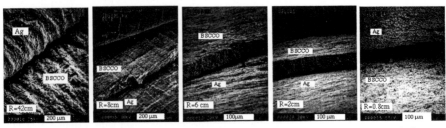

Fig. 3. Tape interiors after rolling with deformation radii of 42, 8, 6, 2, and 0.8cm, respectively.

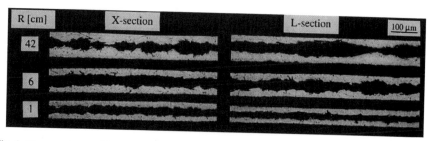

Fig. 4. Cross sections and length sections of tapes after rolling with deformation radii of 42, 6, and 0.8cm, respectively. The tapes were rolled to 100μm thickness, and then pressed and heat treated four times. The R=0.8cm tape was narrower and therefore deformed more during the pressings resulting in a thinner tape.

4. Discussion

The main goal of the shaping process prior to heat treatment is to create a tape with a core that is homogeneous and dense. It is clear that the *strain path* of the external dimensions of the tape will depend largely on the geometry of the deformation zone, and that the material will flow in the direction of least frictional constraint from the shaping tool [1,3,9].

The geometry of the deformation zone can be described by the in-plane aspect ratio, L/W, the roll-gap aspect ratio L/H, and the cross-section aspect ratio W/H. With respect to the strain path, the use of a smaller deformation radius or a smaller reduction ratio will have nearly equivalent effects, serving mainly to reduce the length of the deformation zone, thereby changing the in-plane aspect ratio L/W and consequently the direction of maximum strain. However, the observation that smaller rollers for constant reduction rates and lower reduction rates for constant roller size consistently give a more uniform core [1, 3] is a strong indication that the stress components applied to the powder core are critical parameters. Interestingly, increase in the two remaining aspect ratios, L/H and W/H, for a given coefficient of friction in the roll gap and for a given shear yield stress in the silver sheath will result in an increase in the confinement forces and thereby in the hydrostatic stress.

To determine the causes and mechanisms for the creation of non-uniformity in OPIT tapes during shaping, it is essential to look at the interaction between the two components of this composite material. The mechanical properties of the powder are complex and change during the tape forming process. Its strength depends on the confinement pressures as well as on the applied compaction load [10, 11]. The core starts as a loose powder and is then brought by the hydrostatic stresses into cohesive form and becomes a brittle compact. By comparison, the properties of the silver sheath remain relatively constant during the shaping process.

The above considerations lead us to suggest three possible mechanisms for the formation of non-uniformities, each related to an increasingly hard core due to hydrostatic stress (Fig.7): (1) *powder flow mode* where the core has just reached a hardness that forces the silver sheath to flow into a softer-core region [1], (2) *tensile fracture mode* due to hardening of the core and flow of the silver sheath, (3) *shear fracture mode* where the core has developed sufficient tensile strength to transfer the fracture to the planes of maximum shear [3].

Fig. 5. Top surface of powder cores a) R=42cm (as-rolled), b) R=0.8cm (as-rolled), c) R=0.8cm (pressed). Note the large variations in density and alignment.

Fig. 6. Side view (left) of ~45° cracks across the tape and top view (right) of ~90° cracks along the tape length in as-rolled powder cores.

454

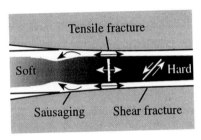

Fig. 7. Schematic side view of the roller gap illustrating mechanisms for the formation of non-uniformities in composite tapes as powder properties evolve: (1) sausaging through powder flow, (2) tensile fracture, (3) shear fracture.

5. Conclusion

The use of a smaller deformation radius during rolling of Bi-2223/Ag OPIT tapes significantly increases the uniformity of the oxide core for a given reduction per pass.

The observation of cracks in the core near to 90° to the tape plane in some samples offers a possible mechanism for the formation of what is commonly described as sausaging in these materials. Large rollers (or large reduction rates) create large hydrostatic stresses that result in hardening of the powder core. The flow of the silver sheath then applies tensile stresses to the brittle core which suffers tensile fracture in the plane of the maximum tensile stress component. Sheath material may then flow into the crack, contributing to the formation of sausaging. As the core hardens further, a sufficient tensile strength develops to transfer the fracture mode to the planes of maximum shear stress at 45° to the tape plane.

As a consequence, the goals of uniform morphology and high density in the powder core of Bi-2223/Ag tapes may be in conflict. It may be necessary to choose a deformation path that introduces the desired strain with low stresses until near final shape is attained in order to maintain uniformity, followed by a limited amount of strain at high levels of the hydrostatic stress to reach a high density. Coincidentally, our best value to date of the critical current density of 24 kA/cm^2 (77K, 0T) [12] was achieved by rolling with R=3cm followed by R=8cm for the last three passes.

References

[1]   Z Hahn and T Freltoft, subm. to *Appl. Supercond: "Proc. and Appl. of BSCCO H.T.S."*

[2]   J O Willis, R D Ray II, J F Bingert, R J Beckman, R J Sebring, D S Phillips, K V Salazar, M G Smith, P A Smith, J Y Coulter and D E Peterson: 1994 *MRS Spring meeting San Francisco* H17.5

[3]   D Á Korzekwa, J F Bingert, E J Podtburg and P Miles 1994 *Appl. Supercond.* 2 No 3/4 p261

[4]   Y Kamisada, T Koisumi, M Satou and Y Yamada 1994 *IEEE Trans. Mag.* 30 No. 4 p1675

[5]   S P Ashworth, B A Glowacki, M P James, R Garre and S Conti 1994 *Appl. Supercond. Conf. Boston* MEB-9

[6]   W L Roberts 1978 *Cold Rolling of Steel.* p31 (New York; Marcel Dekker, Inc)

[7]   J R Cave, D W A Willén, R Nadi, D Cameron and W Zhu 1994 *Appl. Supercond. Conf. Boston* MEC-7

[8]   D W A Willén, C Richer, P R Critchlow, M Goyette, R Nadi, J R Cave, G Quirion and M Aubin 1995 *Supercond. Sci. Technol.* 8 p347

[9]   S E Schoenfeld, S Ahzi, R J Asaro, J F Bingert and J O Willis 1995 submitted to *Phil. Mag*

[10] S E Schoenfeld, S Ahzi, R J Asaro and W R Blumenthal 1995 submitted to *Phil. Mag.*

[11] R Shah, S Tangrila, S Rachakonda, M Thirukkonda, A Gurson and J Kajuch 1995 *Materials Processing in the computer age* ed. V R Voller p279 (Min. Met. & Mat. Soc.)

[12] D W A Willén, R Nadi, D Cameron, B Turcotte, W Zhu, M Trudeau and J R Cave. 1994 *1st Canadian Applied Superconductivity Workshop, Ottawa*

Inst. Phys. Conf. Ser. No 148
Paper presented at Applied Superconductivity, Edinburgh, 3–6 July 1995
© *1995 IOP Publishing Ltd*

IV MEASUREMENTS AND SCALING ANALYSES OF Ag/Bi2223 TAPES

M.G. Karkut, J.O. Fossum, P. Tuset, Wu Ting[#], and K. Fossheim[*]

SINTEF Applied Physics, N-7034 Trondheim, Norway;
[*]Division of Physics, Norwegian Institute of Technology, N-7034 Trondheim, Norway;
[#]Present address: ISTEC, 10-13, Shinonome, Kotu-Ku Tokyo, 135 Japan

Abstract Using a standard four-point technique, we have measured IV characteristics and determined the transport critical current density $J_c(H,T)$ of silver sheathed $Bi_2Sr_2Ca_2Cu_3O_{10}$ tapes (Ag/Bi2223). Magnetic fields from 0 up to 2 Tesla were applied perpendicular to the tape surface (parallel to the c-axis). The IV data have been analysed within the scaling predictions of the vortex glass model and we extract physically reasonable exponents for a 2-dimensional system. In addition, one of us (JOF) has also produced a novel scaling treatment based on the standard flux-creep picture. We show that a flux-creep-model-based single function data collapse can be achieved for the same IV data. We find that a requirement for this collapse is that the pinning potential vary inversely as √H. From these two different scaling analyses we conclude that this type of IV data may not be a primary aid in settling the question of the existence of a vortex glass transition in Bi2223 superconductors.

1. Introduction

There has been steady progress in the manufacture of increasingly longer lengths of Ag/Bi2223 tapes which have critical current densities J_c on the order of 20,000 A/cm^2 at 77K in self-field. However, the main obstacle to using these materials at liquid nitrogen temperatures for bulk applications is the strong magnetic field dependence of J_c. It is therefore of interest to study the transport behavior of Ag/B2223 tapes over a wide range of fields and temperatures for various processing procedures. We have determined the transport $J_c(H,T)$ in applied magnetic fields perpendicular to the tape surface since it is now established [1] that it is this orientation which controls J_c. We have measured IV characteristics of both short and long length, single and multifilament, tapes which have been provided by NKT Research Center A/S as part of the Nordic program on applied superconductivity NORPAS. The main thrust of this paper is our analysis of the IV characteristics on short(pressed) and long(rolled) tapes, both single and multifilament, in two different ways: a) using the scaling predictions of the vortex glass theory [2] and b) using a method [3] based on the standard flux creep/TAFF picture.

2. Samples and experimental details

The Ag/BSCCO tape samples were prepared by the oxide powder-in-tube method at NKT [4]. The tapes were either rolled, semi-continuously pressed, or pressed and their J_c's at 77K and self-field were on the order of 10, 20, and 30kA/cm^2, respectively. We used a standard 4-point

456

technique using pulsed currents to reduce local heating. The voltage contacts were placed 1 cm apart and the J_c criterium was 1 µV/cm. No corrections were made for the Ag sheath since its contribution was less than 1% in the voltage-current-temperature range we considered. In general we terminated the current-pulse sweep when 50 µV was reached. The sensitivity was limited by our current source and was not better than about 0.07µV. Each isothermal current sweep was regulated to better than 75mK and the isotherms were separated, in general, by 2K. The magnetic fields parallel to the c-axis were supplied by a 6.5 Tesla superconducting magnet.

3. Results and discussion

The insert in Figure 1 shows typical electric field -current density, E-J, data taken on Ag/Bi2223 sample S36 in an applied field H = 0.5T. The isothermal sweeps increase in 2K steps from right to left. Similar looking data have been obtained for this sample for applied fields ranging from 0.05T to 2T. We also obtain similar data for all other pressed and rolled samples that we have measured. All of these data can be can be analyzed in the context of the vortex glass (VG) picture which predicts the scaling relations $E\xi_g^{z+1}= \varepsilon_\pm[(J/T)\xi_g^{D-1}]$ for temperatures below (-) and above (+) the VG transition temperature T_g. D is the dimension of the system. TheVG correlation length diverges at T_g as $\xi_g \sim (1-T/Tg)^{-\nu}$, and the characteristic relaxation time $t_g \sim \xi_g^{z}$. At $T=T_g$, the theory predicts a power law relation between E and J of the form $E = J^{(z+1)/(D-1)}$. Thus T_g separates positive type EJ curvature above T_g from negative type curvature below T_g.

Figure 1 presents a scaling plot of the data shown in the inset. The values of the

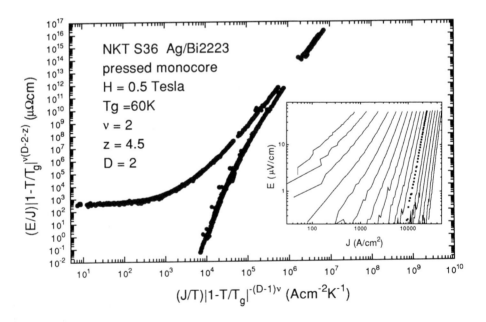

Figure 1. Vortex glass scaling for the EJ curves (shown in the inset) for a Ag/Bi2223 tape. The applied field B parallel to the c-axis is 0.5 Tesla. inset: the isotherms range from 43K to 82K in steps of 2K. The data points are shown explicitly at T = 60K.

parameters v, z and Tg are consistent with recently reported results[5]. We have also been able to scale[3] EJ data from other Ag/Bi2223 tapes we have measured and $T_g(B)$ for applied fields from 0.05 to 2 T are also in agreement with results presented on this sample and with Ref.5. The present experimental data thus provide additional evidence that the VG model is consistent with the EJ characteristics of Bi2223 superconductors. However, being consistenct with the VG predictions does not necessarily prove the existence of a vortex glass state. One reason is that the number of fitting parameters is large. Scaling plots such as Figure 1 can be achieved for a relatively wide range of values of v(B), ζ(B) and $T_g(B)$. Secondly, as discussed by Coppersmith et al. [6], it is possible to produce the features of Figure 1 using a conventional single particle picture [7].

With the above observations in mind, one of us (JOF) has explored, sucessfully, the possibility of finding a simple representation of our Ag/Bi2223 data in terms of the non-interacting single particle that is characteristic of conventional flux creep models. These models predict EJ characteristics of the form

$$E = E_o e^{[-U(B,J,T)/T]} \qquad (1)$$

where the pinning potential can be written as

$$U(B,J,T) = U_o f(J) = C_o h(B) g(T) f(J) \qquad (2)$$

where C_o is a constant and h(B), g(T) and f(J) are model and/or system dependent. Now, to first order, the EJ data (inset Figure 1) can be described by the power law

$$E(J) = A(B,T) J^{n(B,T)} \qquad (3)$$

where the n increases with decreasing temperature and field. Figure 2 shows a plot of logA vs n(B,T) and it is found that there exists a linear relation between logA and n. This linear relation represents a novel finding, and exists in every Ag/Bi2223 sample we investigated. The data in Figure 2 represent one sample and includes fits to about 100 EJ curves taken at temperatures ranging from 35 to 100K and in magnetic fields ranging from 0.05Tesla to 1.5Tesla. By making the identification

$$f(J) = \ln(J_o/J) \qquad (4)$$

and

$$n(B,T) = U_o/T = C_o h(B) g(T)/T \qquad (5)$$

Equation 3 can be tranformed into Equation 1. Thus, within the flux creep picture, the present experimental data are well represented by an empirical pinning potential which depends logarithmically on J. This logarithmic dependence in Bi2223 has been reported by Hergt on torque relaxation data [8].

By extending the above analysis, it was found [3] that by plotting $nB^{0.5}$ vs T, a complete single function data collapse can be achieved for the data above as well as for data of our other investigated samples. By using Equation 5 the field dependence of the pinning potential h(B) ~ $B^{-0.5}$ can be identified. Resistivity data [9] and the torque relaxation data [8] indicate also agree with this finding for h(B).

458

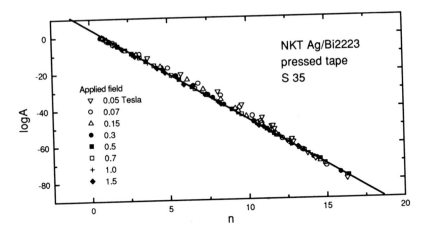

Figure 2. log A(B,T) vs n (B,T) . The linear relationship was found in every Ag/Bi2223 sample we investigated. The significance of this parameter fit is discussed in the text.

4.Conclusions

We have made EJ transport measurements on differently processed Ag/Bi2223 tapes and have sucessfully analyzed the EJ curves using both the vortex glass model and and alternative scaling approach which results in a single function data collapse. The results of the VG analysis agree well with other results in the literature. The alternative scaling, based on flux creep, is consistent with a pinning potential which varies logarithmically with J and inversely with $B^{0.5}$. The two scaling approaches are not necessarily contradictory since the VG approach deals with diverging fluctuation volumes as $T \sim T_g$ whereas the empirical approach deals with the evolution of the pinning potential with changing temperature.

References

[1] M.P. Maley et al. 1992 Phys. Rev. B **45** 7756
[2] D.S. Fisher, M.P.A. Fisher and D.A. Huse 1992 Phys. Rev. B **43** 130
[3] J.O. Fossum et al. submitted to Phys. Rev. B
[4] Q. Li et al. 1993 Physica C **217** 360
[5] Y. Mawatari et al. 1995 Cryogenics **35** 161
[6] S. N. Coppersmith, M. Inui and P.B. Littlewood 1990 Phys. Rev. Lett. **64** 2585
[7] For a review, see P.Kes et al. 1988 Supercond. Sci. Tech. **61** 1658
[8] R. Hergt et al. 1993 Phys. Rev. B **47** 5407
[9] H. Yamasaki et al. 1994 Phys. Rev. B **50** 12959

Inst. Phys. Conf. Ser, No 148
Paper presented at Applied Superconductivity, Edinburgh, 3–6 July 1995
© *1995 IOP Publishing Ltd*

The "railway-switch" model for $(Bi,Pb)_2Sr_2Ca_2Cu_3O_{10}$ silver-sheathed tapes - limitations for the critical transport current by the low intragrain critical current density j_c^c along the c-axis.

B. Hensel[1], G. Grasso[1*], D.P. Grindatto[2], H.-U. Nissen[2], and R. Flükiger[1]

[1] Université de Genève, DPMC, 24 quai Ernest-Ansermet, CH-1211 Genève 4, Switzerland
[2] Laboratorium für Festkörperphysik, ETH Zürich, CH-8093 Zürich, Switzerland

Abstract. The current transport in textured polycrystalline filaments of $(Bi,Pb)_2Sr_2Ca_2Cu_3O_{10}$ silver-sheathed tapes is described by the "railway-switch" model. There is evidence that the critical current density is not limited by the intergrain critical current density across the dominating small-angle c-axis tilt grain boundaries ("railway switches"), but by the low intragrain critical current density j_c^c along the c-axes of the individual grains. A high density of crystalline defects is induced by the mechanical deformation and persists after the heat treatment. This leads to an enhanced performance of the tapes in magnetic fields, as compared to thin films.

1. Introduction

The transport critical current density j_c that can be reproducibly achieved in long lengths of mono- or multifilamentary $(Bi,Pb)_2Sr_2Ca_2Cu_3O_{10}$ [Bi/Pb(2223)] silver-sheathed tapes continously increased over the last years. A strong limitation for j_c persists, however, despite of the efforts that have been put into the refinement of the powder-in-tube (PIT) process. We have proposed the "railway-switch" model for the current transport in the Bi/Pb(2223) filaments [1] and clearly identify the low intragrain critical current density j_c^c along the c-axis as origin for the limitation of the overall transport j_c [2].

The experimental results of this work have been obtained on monofilamentary, silver-sheathed PIT-samples that were prepared in long lengths by rolling as the only tape-forming process. The critical current densities of the tapes were $j_c=20$-$30kA/cm^2$ at T=77K and B=0T. The microstructure has been analysed by Scanning and Transmission Electron Microscopy (SEM/TEM). Measurements of the transport critical current have been performed from T=4.2K to $T_c\approx110K$ in magnetic fields up to B=15T. At T=77K the critical current density j_c^n normal to the tape plane and the current transfer length L_c have been measured.

2. Results

The typical microstructure of the filament of a tape with $j_c(T=77K,B=0T)=25$ kA/cm^2 is shown in Fig.1a (TEM; for details see Ref. [3]). Each platelet is a stack ("colony") of several grains

---

*on leave from Consorzio INFM, Università di Genova, Italy.
Work supported by the Swiss National Science Foundation (PNR 30), the Swiss Priority Program "Materials Research and Engineering" (PPM) and the European Community (Brite Euram II).*

460

with common c-axes. The grains are separated by twist boundaries. Depending on the twist angle the critical current across such twist boundaries is reduced as compared to the intrinsic critical current j_c^c along the c-axis [4]. The contact between neighbouring platelets is most frequently a small-angle c-axis tilt grain boundary ("railway switch") as shown in Fig. 1b. A "brick-wall" type stacking of the platelets is not observed for high quality samples (Fig. 1a).

Fig. 1 Transmission Electron Microscope image of the filament of a $(Bi,Pb)_2Sr_2Ca_2Cu_3O_{10}$ silver-sheathed tape with $j_c(T=77K, B=0T)=25$ kA/cm^2.

Fig. 2 shows the critical current density j_c in external magnetic fields B applied normal (a) and parallel (b) to the tape plane at various temperatures between T=4.2K and T=100K.

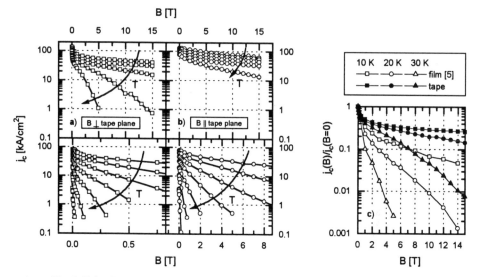

Fig. 2 *Critical current density j_c in magnetic fields normal (a) and parallel (b) to the tape plane at $T=4.2,10,20,30,40K$ (top) and $T=40,50,60,70,80,90,100K$ (bottom). The arrows indicate increasing temperature. Reduced critical current density $j_c(B)/j_c(B=0T)$ of a tape and a high quality thin film [4] for $T=10, 20$ and $30K$ and the field normal to the tape (c).*

Above $T=40K$ an exponential decay of $j_c(B)$ is observed over a wide range of fields while at lower temperatures ($T<40K$) a more complicated behaviour is found and a pronounced hysteresis occurs in $j_C(B)$ (the data in Figs. 2a and 2b have been measured in decreasing fields). The normalized field dependence $j_c(B)/j_c(B=0)$ obtained for Bi/Pb(2223) tapes is significantly enhanced as compared to the one obtained for high quality Bi/Pb(2223) thin films [5] (Fig. 2c). This might find an explanation in the high density of crystalline defects that are induced in the grains by the cold deformation and persist even after the heat treatment [3]. One should, however, keep in mind that the critical current density j_c is almost two orders of magnitude lower in the tapes ($j_c \approx 3 \cdot 10^4 A/cm^2$) than in the thin films ($j_c \approx 10^6 A/cm^2$, both values at $T=77K$ and $B=0T$), mainly due to a strongly reduced effective cross section for the current in the tapes (visible in the "low density" of the grains in Fig. 1a).

At $T=77K$ the critical current I_c along the tape falls to zero in fields below $B=0.5T$ normal to the tape plane. The critical current I_c^n normal to the tape plane, however, persists in this field range, as can be seen in Fig. 3. The rapid decay of I_c in magnetic fields normal to the tape plane is characteristic for the Bi-based High Temperature Superconductors. In the "force-free" configuration (current and field normal to the tape plane) a less strong reduction of I_c^n in magnetic fields is observed. This finding indicates that the intragrain current transport along the c-axes of the individual platelets plays a major role for the overall current transport along the ab-planes of the grains and thus along the tape.

The critical current density j_c^n normal to the tape plane is of the order of $j_c^n=2.5kA/cm^2$ at $T=77K$ and $B=0T$. The anisotropy of the critical current density of the tapes then becomes $j_c/j_c^n \approx 10$. This small value should allow a short-scale current distribution over the filament cross-section. In good agreement with this assumption we find experimentally a very short current-transfer length L_c of only several millimeters.

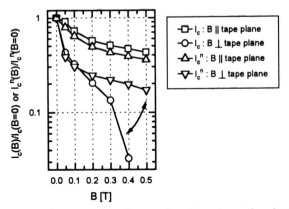

Fig. 3 *Normalized critical currents along the tape plane (I_c) and normal to the tape plane (I_c^n) for magnetic fields applied normal and parallel to the tape plane (T=77K).*

3. Conclusions

From the analysis of the above presented data we must conclude that the dominant intergrain connections via small-angle c-axis tilt grain boundaries ("railway switches") do not limit the critical current density j_c and do not constitute weak links for the supercurrent in Bi/Pb(2223) tapes. From our microstructural observations and values for the critical current density j_c^c along the c-axis of $(Bi,Pb)_2Sr_2CaCu_2O_8$ single crystals [6] it can be estimated that the transport j_c is strongly limited by the low j_c^c. Average aspect ratios of the grains of the order of 20 together with the low $j_c^c=7kA/cm^2$ [6] lead to an inhomogeneous supercurrent distribution in each individual grain and thus strongly reduces the effective cross-section for the current.

For the further development of Bi/Pb(2223) tapes more strong intergrain connections "railway switches" are necessary together with a better alignment of the platelets that leads to a higher j_c(B||tape plane) as the magnetic field component normal to the ab-planes of the grains is then smaller. It should, however, be noted that perfectly aligned grains are not ideal for high critical current densities, as then no more small-angle c-axis tilt grain boundaries can be formed. An improvement of j_c (B⊥tape plane) can only be achieved by overcoming the prevailing lack of pinning in Bi-based High Temperature Superconductors by the introduction of more effective pinning centers by metallurgical treatments or mechanical deformation. We conclude, however, that the just mentioned limiting effects are only of second order and that the main limitation is given by the reduced effective cross-section for the supercurrent. A higher transport critical current density can be expected (and has resulted in the positive development over the last years) when the aspect ratio of the grains will be bigger or when the intragrain j_c^c will be increased, e.g. by the elimination of twist boundaries in Bi/Pb(2223).

References

[1] Hensel B, Grivel J-C, Jeremie A, Perin A, Pollini A, and Flükiger R 1993 *Physica C* **205** 329

[2] Hensel B, Grasso G, and Flükiger R 1995 *Phys. Rev. B* **51** 15456

[3] Grindatto D P, Hensel B, Grasso G, Nissen H-U, and Flükiger R unpublished

[4] Wang Jyh-Lih, Cai X Y, Kelley R J, Vaudin M D, Babcock S E, Larbalestier D C 1994 *Physica C* **230** 189

[5] Yamasaki H, Endo K, Kosaka S, Umeda M, Misawa S, Yoshida S, and Kajimura K 1993 *IEEE Trans. Appl. Supercond.* **3** 1536

[6] Kleiner R and Müller P 1994 *Phys. Rev. B* **49** 1327

Inst. Phys. Conf. Ser. No 148
Paper presented at Applied Superconductivity, Edinburgh, 3–6 July 1995
© *1995 IOP Publishing Ltd*

Lateral distribution of the transport critical current density in Bi(2223) Ag-sheathed tapes

G Grasso*, B Hensel, A Jeremie, and R Flükiger

Université de Genève, DPMC, 24, quai Ernest Ansermet, CH-1211 Genève 4, Switzerland

Abstract. By using a strip-cutting technique, we directly measured the lateral distribution of the transport critical current density of Ag sheathed Bi(2223) tapes prepared by rolling with overall critical current densities in the range of 23-28 kA/cm^2 at 77K and 0T. It has been found that the critical current density of tape strips cut from the sides is reproducibly higher than that of strips from the central part of the sample. For various long tapes prepared by rolling only with a critical current density of 28000 A/cm^2 at 77K and 0T, the local critical current density varies from 23000 A/cm^2 at the center to 53000 A/cm^2 at the sides of the filament. In all cases, a symmetrical behavior of j_c at both sides of the central axis was observed.

1. Introduction

Recently, several groups have shown that the upper limit to the transport critical current density of the Bi(2223) phase is far from being reached. In fact, in certain parts of the filament of Ag sheathed Bi(2223) tapes, local critical current densities at 77K of 110 kA/cm^2 [1], 75 kA/cm^2 [2] and 46 kA/cm^2 [3] have been already reported. These local values are remarkably higher than those measured over the whole filament section (up to 42 kA/cm^2 in rolled tapes [4]).

The aim of the present work is to investigate the lateral critical current distribution inside the Bi(2223) filament of Ag sheathed tapes with various j_c values by direct transport measurements. The measurements have been performed on rolled tapes with average critical current densities at 77K and in self field between 23 and 28 kA/cm^2.

2. Experimental

Bi(2223) monofilamentary tapes have been prepared by the powder in tube (PIT) process. Calcined powders with cation ratio Bi:Pb:Sr:Ca:Cu=1.72:0.34:1.83:1.97:3.13 have been filled inside pure Ag tubes. These tubes were deformed by subsequent swaging, drawing and cold rolling and the resulting tapes were heat treated at temperatures of about 830-840°C for a total time of 200h with intermediate cold rolling steps. More details on the technique are

on leave from Consorzio INFM, Università di Genova, Italy.
Work supported by the Swiss National Science Foundation (PNR 30), the Swiss Priority Program "Material Research and Engineering" (PPM) and the Brite Euram II Project No. BRE2 CT92 0229 (OFES BR060).

464

given elsewhere [5]. The final tapes are about 80-100 μm thick and 2.5-3.5 mm wide. Tapes with lengths up to 10 m have been prepared by this process, but the reaction heat treatments were performed on straight pieces of 0.5 m in length, due to the limited size of the furnace.

In order to measure the j_c distribution inside the oxide core, the tapes have been mechanically cut by means of a razor blade, without taking off the silver sheath. Typically, narrow strips of about 0.2 mm in width have been successively cut away without damaging the superconducting properties of the filaments. The j_cs of the samples have been measured as usual by the standard four probe technique (criterion of 1 μV/cm), with 6 different voltage contacts placed at about 2 mm from each other. In a specially designed sample holder with press contacts instead of soldering, it was possible to perform the j_c measurements on the same tape after each cut. In order to avoid sample heating, all the measurements have been carried out by directly immersing the sample in liquid nitrogen.

3. Results

The lateral j_c distribution has been evaluated by the so-called strip-cutting technique [3]. In order measure the I_c distribution inside the oxide core of a tape, narrow strips of about 0.2 mm in width have been cut starting from one lateral side, as shown in fig. 1. After each cut, I_c of the residual tape has been measured.

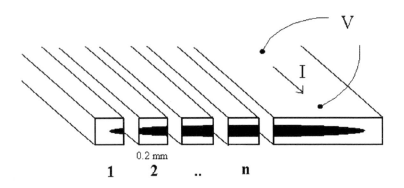

Fig. 1: Schematic description of the technique used to determine the lateral j_c distribution

In fig. 2 (open symbols) the typical decrease of I_c has been plotted versus the number of strips which have already been cut for a tape with average $j_c(77K,0T)=23$ kA/cm^2. The tape cross section S of the residual tape has been evaluated after each cut (full symbols of fig. 2) by means of a standard optical technique. The local j_c has been simply determined as follows: after the cut n, the I_c^n and S^n of the residual tape have been measured, and so we can easily calculate $j_c^n = \dfrac{I_c^n - I_c^{n-1}}{S^n - S^{n-1}}$, which is the local j_c of the strip n. The same operation has been repeated on several tapes prepared from the same batch in order to be sure of the reproducibility of the results [3].

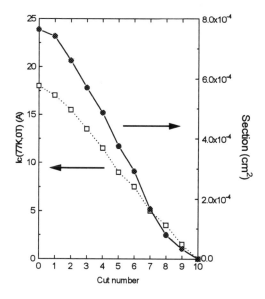

Fig. 2: I_c and section dependences from the number of cuts

This technique has been used on tapes with different critical current densities. The j_c distribution is shown in fig. 3 for two different tapes with $j_c(77K,0T)$=23 kA/cm^2 and $j_c(77K,0T)$=28 kA/cm^2. The main difference between the tapes resides in the starting powder preparation, while the same deformation and treatment parameters have been used for both.

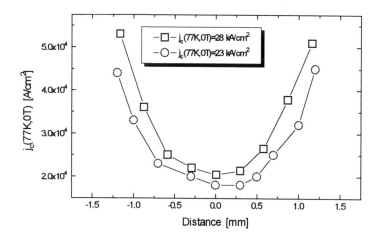

Fig. 3: Lateral j_c distribution for two tapes with different average j_c values

The shape of the j_c distributions is very similar, the local j_c value in the central part being approximately 2.5 times lower than at the sides. Both j_c distributions can be fitted quite well by parabolic-like curves.

The highest local j_c value of 53 kA/cm^2 has been measured at a side of the tape with average j_c value of 28 kA/cm^2.

4. Conclusions

In this work it has been found that, even in high j_c Bi(2223) tapes (average j_c(77K,0T)=28 kA/cm^2), the local critical current density shows a strong position dependence, the value measured at the tape edges reaching 53 kA/cm^2, i.e. about 2.5 times higher than at the tape centre (23 kA/cm^2). Moreover, these results seem to show that the shape of the distribution is not dramatically sensitive to the sample quality.

Two main hypothesis have been already proposed for the explaination of the parabolic-like distribution of j_c: 1) the non-uniform density of secondary phases, and 2) the non-uniform local compression of the filament during deformation, as already reported [3].

The first hypothesis means that the lateral j_c distribution would be influenced only by the Bi(2223) phase purity. The second hypothesis means that the local j_c distribution would be determined only by the deformation technique employed for the tape preparation.

The results presented in this work seem to confirm the latter hypothesis, but more data are needed in order to clarify which is the real mechanism that leads to the non-uniform lateral j_c distribution.

References

[1] Lelovic M, Krishnaraj P, Eror N G, and Balachandran U 1995 *Physica C* **242** 246
[2] Larbalestier D C, Cai X Y, Feng Y, Edelman H, Umezawa A, Riley Jr. G N, and Carter W L 1994 *Physica C* **221** 299.
[3] Grasso G, Hensel B, Jeremie A, and Flükiger R 1995 *Physica C* **241** 45
[4] Sato K, Ohkura K, Hayashi K, Ueyama M, Fujikami J, and Kato T presented at the Intern. Workshop on Advances in High Magnetic Fields, Feb. 20-22, 1995, Tsukuba (Japan), to be published in *Physica B*.
[5] Grasso G, Perin A, and Flükiger R presented at the Cimtec Conf., Florence (Italy), 1-4 July 1994, published in *Adv. in Sci. and Technol.* **8** 437, ed. P. Vincenzini.

STRUCTURAL DEVELOPMENT IN Bi(2223)/Ag TAPES MADE BY PRESS-SINTER PROCEDURE

W Pachla[†], P Kovác[*], H Marciniak[†] and I Hušek[*]

[†] High Pressure Research Center, Polish Academy of Sciences
  ul. Sokolowska 29, 01-142 Warszawa, Poland
[*] Institute for Electrical Engineering, Slovak Academy of Sciences
  Dúbravska cesta 9, 842 39 Bratislava, Slovakia

Abstract. The structural changes that occur during manufacuring of the Bi-2223 superconducting tape by the powder-in-tube method including the cold hydrostatic extrusion process are presented. Hydrostatic extrusion itself does not introduce alingnment of the Bi-2212 phase. Subsequent, drawing and rolling processes generate many internal macro (cracks) and micro (dislocation tangles) defects resulting in the "pseudo-amorphous" microstructure. Strong texturing of Bi-2223 phase can be obtained by the prolonged annealing (a few tens of hours) preceded by the uniaxial cold pressing process (minimum 1GPa).

1. Introduction

One of the main difficulty, which create HT_c ceramics for commercial applications is its brittleness. The main factor enabling to avoid the brittleness is to introduce within its polycrystalline ceramic core the preferred orientation of platelets parallel to the tape surface. The plate-like configuration is needed not only because of the high anisotropy in critical current density but mainly, due to its much higher strain to fracture.

2. Experimental

The composite tapes were fabricated by powder-in-tube method from the powder prepared by spray pyrolisis with a composition of $Bi_{1.7}Pb_{0.3}Sr_{2.0}Ca_{2.0}Cu_{3.0}O_x$ [1]. The powder was calcined, milled and formed into rod by cold isostatic pressing (CIP). Next, the rod was inserted into Ag tube and hydrostatically cold extruded into round wire. The wire was drawn and flat rolled to thickness of ~0.1mm (~2mm in width). The short pieces of the tape (20mm in length) were annealed and uniaxially cold pressed (CP) with various pressure sequences between 1GPa and 5.7GPa with intermediate heat treatments between pressing stages. Details of press-sinter procedure has been described elsewhere [2]. The microstructure of the tapes was examined using XRD and SEM in two main sections : longitudinal - after one, wide

468

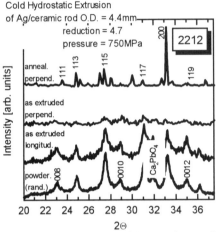

Cold Hydrostatic Extrusion
of Ag/ceramic rod O.D. = 4.4mm
reduction = 4.7
pressure = 750MPa

2212

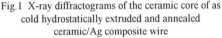

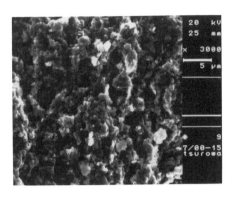

Fig.1 X-ray diffractograms of the ceramic core of as
cold hydrostatically extruded and annealed
ceramic/Ag composite wire

Fig.2 SEM image of the Bi(2223)/Ag tape ceramic
core (longitudinal section) after flat rolling; overall
thickness 0.1mm, width 2.0mm

side of the Ag wall was removed, and transverse - after the tape from longitudinal observation
was bend, and the ceramic core has been fractured.   The critical current I_c was measured by
I-V characteristic at 77K and 0T with criterion $1\mu Vcm^{-1}$ [2].

3. Results

After hydrostatic extrusion the non-superconducting behaviour prevails, the structure is very
fine (finer for higher degree of deformation) and 2212 phase dominates (above 92%), Fig.1.
        The following drawing and rolling operations introduce much larger total reduction
(~94%), although, they are done much more gentle (flat rolling with maximal 5% reduction
per pass).   As-rolled tape characterizes by disordered (random) structure, Fig.2 and
Fig.3(*bottom*).   Such "pseudo-amorphous" microstructure transforms to pure, 2-phase
structure by subsequent thermomechanical treatment (press-sinter procedure).   After short
time annealing 837°C/10h in air ~96% of Bi-2212 content is observed in the ceramic core
,Fig.3(*middle*), with the texture factor of ~3.6 (the ratio intensity of [008] to [115]
reflection).   However, the texturing is evident it is not well developed yet.
        After subsequent cold pressing and prolonged annealing at 837°C/100h in air the Bi-
2223 content and its texture factor increases dramatically (above 97.5% and to about 20,
respectively), Fig.3(top).   Very strong texture in nearly 2223 phase pure structure is obtained.
        The highest measured J_c for one-stage CP tape has reached $4\times10^3 Acm^{-2}$ at 77K and
0T for 2GPa pressure, and shows to be strongly dependend on the post-annealing
temperature, Fig.4. Application of the three-stage press-sinter schedule has shifted J_c for this
tape up to $6.7\times10^3 Acm^{-2}$ for 2GPa-1GPa-2GPa pressing sequence. In comparison, the flat
rolling only, has led to maximal J_c of $4\times10^3 Acm^{-2}$, i.e. was comparable with the one-stage
press-sinter schedule of present results [2].

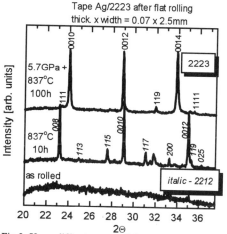

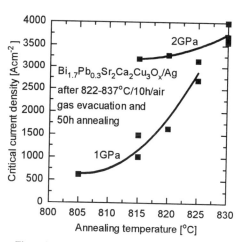

Fig.3 X-ray diffractograms of the Bi(2223)/Ag tape ceramic core; overall tape thickness 0.1mm, width 2.0mm (as in Fig.2); (*bottom*) as-rolled, (*middle*) after annealing 837°C/10h/air, and (*top*) after uniaxial cold pressing at 5.7GPa and annealing 837°C/100h/air

Fig.4 Critical current density dependence on the intermediate annealing temperature for Bi(2223)/Ag press-sintered tape

4. Discussion

During hydrostatic extrusion the powder undergoes enormous deformation. The degree of grains reworking is so high (70-80% in one, single pass), they must break and fracture into finer ones to be able to dislocate within the core, which elongates over 3 times in several tenth of a second. Grains move by constant breakage and relative rotation, leading to very fine, very dense (170HV) and multi-phase (dominates Bi-2212 - about 92%) microstructure. Further annealing of the as-extruded wires reveals the "radial" texture (no [00n] peaks) in Bi-2212 phase, which content drops to 60-70% (with corresponding increase in Bi-2223 phase), Fig.1(*top*). The macrocracks that develop during drawing as a result of hydrostatic tension are commonly observed [2,3]. This is attributed to unfavourable state of stresses within the working zone during drawing and rolling, and to specific mode of deformation which enforces the deformation to be done in hundreds of passes with very small single reduction per pass (usually below 5%). For X-ray analysis material looks like "pseudo-amorphous" one, since a very fine grains (or subgrains created by the tangled dislocation network) create the perfect, isotropic picture with no reflexes (peaks), Fig.3(*bottom*).

After 10 hours treatment at 837°C, Fig.5(*top*) the reaction induced texturing (RIT) process occurs. Texturing of Bi-2212 is not very impressive (texture ratio ~3.6) Fig.3(*middle*), but confirms the others data [4]. The high texture factor of ~20 for Bi-2223 phase after annealing at 837°C for 100h preceded by the 1GPa pressing confirms stimulating role of the latter for more rapid phase transformation from 2212 to 2223 phase and better texturing, Fig.5(*bottom*). It is attributed to an increase in the area contact between grains, reduction in diffusion path and effective reducing of the particle size by crushing the secondary phases and superconducting platelets [5]. The relatively high J_c obtained after one press-sinter stage ($J_c=4\times10^3 Acm^{-2}$) in present research support the importance of pressing for tape manufacturing procedure. In [3] it was demonstrated, that the critical currents of samples receiving no-pressing are poor and are the most severely affected by the magnetic field.

470

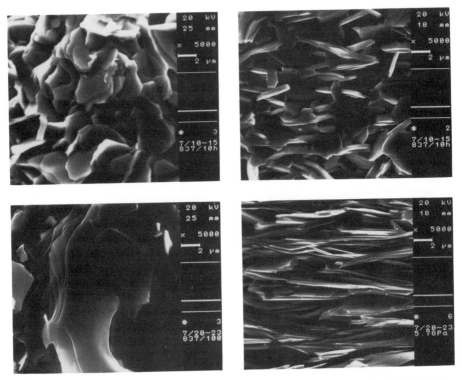

Fig.5 SEM images of the Bi(2223)/Ag tape ceramic core after (*top left*) gas evacuation 837°C/10h/air, longitudinal section, (*top right*) as above, transverse section, (*bottom left*) uniaxial cold pressing 5.7GPa and annealing 837°C/100h/air, longitudinal section, and (*bottom right*) as above, transverse section

5. Conclusions

Hydrostatic extrusion itself does not produce alingnment of the Bi-2212 phase. The weak radial texture (c-axis perpendicular to wire axis) can be revealed in extruded wire after following annealing. Subsequent drawing and rolling introduce heavy degree of structure reworking into the ceramic core, generate many internal macro (cracks) and micro (dislocation tangles) defects and make the ceramic grains much finer, all of these resulting in the "pseudo-amorphous" microstructure. It can be easily recovered to weak textured Bi-2212 phase by the short time heat treatment (below 10h) and to very strong textured Bi-2223 phase by the prolonged annealing (a few tens of hours) preceded by the uniaxial cold pressing.

References

[1] SSC, Inc. *Ceramic Powder Synthesis and Processing* (Information Pamphlet)
[2] Kovác P et al. 1995 *Supercon. Sci. Technol.* **8** 341-346
[3] Briant L et al. 1994 *J. Mater. Res.* **9** 2789-2808
[4] Merchant N et al. 1994 *Appl. Phys. Lett.* **65** 1039-1041
[5] Willén W 1995 et al. *Supercon. Sci. Technol.* **8** 347-353

Inst. Phys. Conf. Ser. No 148
Paper presented at Applied Superconductivity, Edinburgh, 3–6 July 1995
© 1995 IOP Publishing Ltd

Critical currents and microstructure of Tl-1223 high-temperature-superconductor bulk samples and tapes

W. Mexner, S. Heede, J. Hoffmann, K. Heinemann and H.C. Freyhardt

Institut für Metallphysik, Universität Göttingen, Windausweg 2, D-37073 Göttingen

As a first step in the investigation of Tl-1223 high-temperature-superconductors (HTSC) and the further development of tape conductors, the critical currents of polycrystalline $Tl_{0.5}Pb_{0.5}Sr_2Ca_2Cu_3O_9$ (Tl-Sr-1223) bulk samples were investigated with a Faraday magnetometer up to B = 12 T and compared to $Tl_2Ba_2Ca_2Cu_3O_{10}$ (Tl-2223) single crystals. The magnetic hysteresis of Tl-Sr-1223 yields high intragrain critical current densities of up to 10^6 A/cm^2 at low temperatures (4.2 K, 1 T) and 10^4 A/cm2 at 77 K, 1 T. The effective activation energy, $Ueff((j) = Ui\ (j_c/j)^\mu$, as determined by the method of Maley, is higher for the one-intermediate-layer compound Tl-1223 than for the highly anisotropic two-intermediate-layer Tl 2223 compound. A comparison with data for Y-123 and and Bi-2212 from literature indicates, that flux creep in Tl-1223 is 3-dimensional like in Y-123 and might be explained by the 3D collective creep model for not too small currents. For Tl-2223, a 2-dimensional behaviour is expected. The microstructure and critical currents of short tapes of $(Tl/Pb/Bi)_1(Sr/Ba)_2(Ca/Tl)_2Cu_3O_9$, prepared by the powder-in-tube technique, were investigated by TEM, SEM and resistive measurements. The tapes show no pronounced texture and therefore, only relatively small intergrain critical currents of 9000 A/cm^2 (77 K, 0 T), whereas the intragrain current is about 10^4 A/cm^2. The intergrain current density of the Tl-1223 tape is inhomogeneously distributed in the flat-rolled HTSC core. It is higher at the edge than in the middle of the core. The TEM observations reveal no impurities in the grain boundaries and a dislocation density of the order of 10^{10} cm^{-2}.

1. Introduction

In the Tl/Pb-Ba/Sr-Ca-Cu-O system, several superconducting compounds are existing. Tl-2223, which possesses the same crystal structure than Bi-2223, is the compound with the highest T_c of the Tl-HTSC with up to 125 K. In this system, the three superconducting copper oxide planes are separated by two insulating TlO layers. Due to the large distance of 1.14 nm between these planes, a low irreversibility field of about 100 mT at 77 K (anisotropy-constant Γ = 3000) is observed and Tl-2223 is expected to behave like a two-dimensional superconductor similar to Bi-2212. The Tl-1223 system [1,2] is of current interest, because it has nearly the same T_c as Tl-2223, but is less anisotropic ($\Gamma \approx 64$, [3]) due to one missing insulating TlO-plane (d = 0.84 nm). This results in an irreversibility field (fig. 1), which is comparable to Y-123. Another interesting aspect of Tl-1223 is the fact, that the distance between the copper-oxide-planes can be changed from 1.53 nm up to 1.59 nm by a substitution of the small Sr-Ion by the larger Ba-Ion, which causes a systematic change of 10% of the irreversibility field.

472

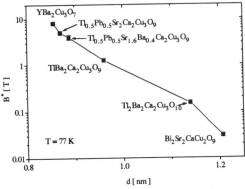

Figure 1: Irreversibility field versus the distance between the superconducting CuO_2-layers. Data for Tl-1223 and Tl-2223 are measured by [3], the other data is taken from Ref. [4].

Due to the large difference in the anisotropy of Tl-1223 and Tl-2223, significant differences in the thermally activated flux creep behaviour can be expected. To get more information about the flux creep in these two systems, the irreversible magnetization and its relaxation with time was measured for a Tl-2223 single crystal and a polycrystalline Tl-1223 bulk sample by a Faraday magnetometer [5]. The critical current density, the apparent activation energy, U_{app}, and the current dependent barriere, $U_{eff}(j)=U_i(j_c/j)^\mu$, i.e. the effective barrier height for the hopping flux

bundles or flux lines, are deduced from this measurements and discussed in terms of the 2D- and 3D-collective-creep and vortex-glass model [6,7,8], in comparison to the behaviour found for Y-123 and Bi-2212. Based on this results, the influence of high mechanical deformation of Tl-1223 in tapes is investigated.

2. Critical currents and flux creep in bulk material

A Faraday magnetometer [5] with a field gradient of up to 2 T/m was used to measure the magnetization, M, of the specimens in a magnetic field up to B = 12 T with a sweep rate of 13.7 mT/s and in a temperature range from 4.2 K up to 77 K. From the width of the magnetic hysteresis loop, ΔM, the critical current density, $j_c(B) = 3\ \Delta M/d$ (d: sample diameter, Tl-2223: d = 0.5 mm, Tl-1223: d = 8 μm), was calculated according to the isotropic Bean model. For the Tl-2223 single crystal all measurements were performed with B parallel to the c-axis.

To measure the decay of the irreversible magnetization, M(t,B), the magnetic field was swept with a constant sweep rate dB/dt of 0.68 mT/s from a low initial magnetic field value to the starting field B, which exceeds the full penetration field and then is held constant. All relaxation curves were investigated for t > 50 s. It was first pointed out by Maley et al. [9], that the current-dependent effective activation energy can be deduced from relaxation measurements at different temperatures, assuming only that the temperature dependence of the pinning potential is weak in comparison to the exponential decay of the magnetization M with temperature. The measurements for different temperatures at a constant magnetic field can be interconnected with only one simple fitting parameter A in order to enlarge the current density window. The activation energy is then given by $U_{eff}(j(M(t))) = - k_B T\ (\ln|dM(t)/dt| +\ln(A))$. With ln(A)=25.8 (T< 40K) for Tl-1223 and ln(A)=26.7 (T<30K) for Tl-2223, it is possible to fit our data onto common curves (Fig. 2).

At low temperatures (around 4.2 K), the intragrain critical current density for polycrystalline Tl-1223 is comparable to Tl-2223 and shows the (for HTSC) well known inverse square root dependence on the magnet field, $j_c \propto 1/\sqrt{B}$. An increase of the temperature up to 40 K results in a drop of j_c by two orders of magnitude for the Tl-2223 single crystal, while for Tl-1223, j_c is only reduced by one order of magnitude. At 77 K, j_c exceeds 10^4 A/cm^2 at fields up to 1 T, which is comparable to the observed current densities in Y-123. But due to the weak links between the grains in this polycrystalline material, the resistively measured intergrain critical current density (2 μV/mm criterion) of 200 A/cm² at 77 K in zero field is more than two orders of magnitude lower.

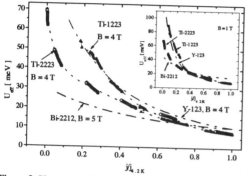

Figure 2: U_{eff} versus j/j(4.2 K) for Tl-1223, Tl-2223, Y-123 (Ref. [11]) and Bi-2212 (Ref. [10]). The inset shows the data for B = 1 T.

The differences in the temperature dependence of the critical currents of Tl-1223 and Tl-2223 and the decrease of the irreversibility field with increasing anisotropy are connected with a change in the effective activation energy. In fig. 2, U_{eff} is plotted versus j/j(4.2K) for Y-123, Tl1223, Tl-2223 and Bi-2212. Data for Y-123 and Bi-2212 are taken from the literature [10,11]. On this reduced current scale, U_{eff} is quite similar for all four systems at high currents (j/j(4.2 K) > 0.5), but at low currents the activation energies diverge. The lowest values are observed for Bi-2212, while the data for Tl-1223 and Y-123 are quite similar. The values for Tl-2223 are between this two extrema. Thus, U_{eff} also reflects the observed variation of the irreversibility line. The similiarity between Y-123 and Tl-1223 indicates, that inspite of the higher T_c of Tl-1223, both systems exhibit a comparable 3-dimensional-flux-creep behaviour. The previous results shows, that Tl-1223 exhibit a high intragrain critical current in a magnetic field and, therefore, is an interesting material for tapes in high magnetic field applications. In the next chapter, the changes in microstructure and critical current due to the the high degree of deformation of the Tl-1223 such tapes is investigated.

3. Microstructure and critical currents of Tl-1223 tapes

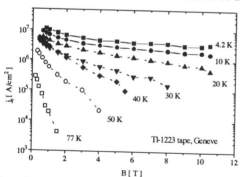

Figure 3: Intragranular current density of a Tl-1223 tape. No intergranular coupling is observed.

To get informations about the intergranular coupling of the grains in the tape, the magnetic hysteresis measurements were performed with a Faraday magnetometer. To determine, whether the shielding currents are flowing through the whole tape size, in a second measurement the tape was cut into half perpendicular to the tape core. The two hysteresis curves for the complete tape and the tape cut into half show no differences. This indicates that the effective supercurrent-carrying radius is much smaller than the tape dimension and seems to be approximately equal to the average particle size inside the tape. Fig. 3 shows the critical current densities, j_c, versus the magnetic field of a tape at different temperatures which were calculated by the Bean-model assuming an average grain diameter of 4 μm. The very high intragrain critical current densities are comparable to those of melt textured YBCO with 211 inclusions. This indicates, that even a small improvement in texturing of the tape would result in a large increase of the intergranular currents. To explain these large intragrain critical current densities, the microstructure of the tapes was investigated by TEM. The TEM investigations reveal no impurities in the grain boundaries. Defects found are dislocation networks with a density of the order of 10^{10} cm⁻² and stacking faults with a density of $4 \cdot 10^5$ cm⁻¹.

474

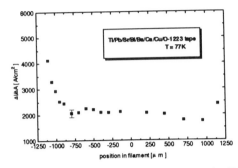

Figure 4: Variation of the critical current density j(x) over the tape-core of a Tl-1223 tape

Transport critical current densities were measured by a standard four probe method with soldered silver contacts. The tape was abraded axially along its long edge and the critical current was measured after each grinding step. From this data, the variation of the current density, j(x), can be calculated within the superconducting core (see fig 4). First the j_c enhancement in the outer region of the tape was thought to be due to a slightly better texturing. However x-ray diffraction measurements reveal that no difference in the texture could be resolved between the inner and the outer region of the tape. Furthermore polefigures as well as rocking curves of the different zones did not indicate any texturing, so that the j_c enhancement might be caused by a higher density of the superconducting core in the edges.

4. Summary

Although Tl-1223 is more anisotropic than Y-123, the rather high values for the critical current density, the irreversibility field at high temperatures and first measurements of the anisotropy-factor indicate, that this compound should be considered as a 3D-superconductor, which is comparable to Y-123. First measurements of Tl-1223 tapes show, that we have still high intragrain critical current densities but due to a small degree of texture only a small intergrain current density, which is higher in the edges of the tape than in the middle.

5. Acknowledgement

We gratefully acknowledge the help of L.Oeltjen and K. Winzer, who supplied the Tl-2223 single crystal and who performed the ac-susceptibility measurements. This work was supported by European Community under the grantnumber BRE2-CT94-0531.

6. References

[1]   R.S. Liu, D.N. Zheng, J.W.Loram, K.A. Mirza, A.M.Campbell, *Appl Phys. Lett* **60** (1992) 1019

[2]   T. Nabatame, J. Sato, Y. Saito, K. Aihara, T. Kamo, and S. Matsuda, *Physica C* **193** (1992) 390

[3]   G. Brandstätter, PhD, University of Vienna 1995

[4]   S.P. Matsuda, A. Soeta, T. Doi, K. Aihara und T. Kamo, *Jpn. J. Appl. Phys.* **31**, 9A (1992) L1229

[5]   W. Mexner and K. Heinemann, *Rev. Sci. Instr* **64** (1993), 3336

[6]   M.V. Feigelman, V.B. Geshkenbein, A.I.Larkin and V.M. Vinokur, *Phys. Rev. Lett.* **63** (1989) 2303

[7]   A.P. Malozemoff, *Physica C* **185-189** (1991) 264

[8]   M.P.A. Fisher, *Phys. Rev. Lett.* **62** (1989) 1415

[9]   M.P. Maley, J.O.Willis, H. Lessure and M.E. McHenry, *Phys. Rev. B.* **42** (1990) 2639

[10] P.H. Kes, Phase Transitions and Relaxation in Systems with Competing Energy Scales, Eds. T. Riste and D. Sherrington, Kluwer Academic Publishers (1993) 71

[11] P.J. Kung, M.P. Maley, M.E. McHenry, J.O.Willis, M.Murakami and S. Tanaka, *Phys. Rev. B* **48** (1993) 13922

Inst. Phys. Conf. Ser. No 148
Paper presented at Applied Superconductivity, Edinburgh, 3–6 July 1995
© 1995 IOP Publishing Ltd

Critical current anisotropy of Ag sheathed 2212 Tl-Ba-Ca-Cu-O superconducting tapes.

F. Chovanec, L.Janšák, P.Kottman, M.Majoroš, P.Ušák, D.Suchoň and M.Jergel

Institute of Electrical Engineering, Slovak Academy of Sciences, 842 39 Bratislava, Slovakia

Abstract. The results of a study of anisotropy in the transport critical current under a magnetic field for Ag-sheathed Tl-Ba-Ca–Cu-O (Tl-2212) tapes produced by the powder-in-tube technique are presented. The critical current anisotropy measured at temperatures 70 K to 90 K in magnetic fields up to 0.5 T as well as in fields up to 5 T at 4.2 K served for the estimation of the degree of grain alignment and its dependence on processing conditions. The lack of significant texturing accompanied by the small I_c anisotropy indicate the three-dimensional charge transfer character in Tl-samples produced by a single stage as well as multiple press and heat process.

1. Introduction

The high irreversiblity field at 77 K and low critical current anisotropy with respect to applied magnetic field are the most attractive properties of Tl-based high temperature superconducting materials. These properties are usually attributed to the stronger c-axis coupling between the groups of Cu-O planes separated by a single Tl-O interlayers in contrast to Bi-containing materials where the Bi-O interlayers are doubled [1]. The three dimensional charge transfer character rather than a two dimensional one of Bi-based materials may be very useful in a wide range of applications.

On the other hand a sharp drop of the critical current density J_c with magnetic field in low field region is reported for Tl-containing Ag-sheathed tapes produced by the powder-in-tube technique (PIT) [2-4]. This effect is associated with weak links between the grains. The lack of preferred orientation of Tl crystallites and any significant texturing is reported in most papers devoted to properties of Tl tapes produced by the PIT technique. In this paper we present the results of a study of anisotropy of the transport critical current I_c under magnetic field in Ag-sheathed Tl-Ba-Ca-Cu-O (Tl-2212) tapes produced by the PIT technology. The critical current anisotropy measured at the temperature range 70-90 K in magnetic fields up to 0.5 T and up to 5 T at 4.2 K served for estimation of the degree of the grain alignment. The dependence of the I_c anisotropy on the processing conditions (rolling, pressing, annealing, temperature and time) has been studied.

2. Sample preparation and experimental technique

The Ag-sheathed Tl-2212 tapes were made by the standard PIT technique. The precursor in the form of $2Ba(NO_3)_2:4H_2O:2.3Ca(NO_3)_2:3CuO$ was first fired at 800 C for 24 hours. After this step an appropriate amount of Tl_2O_3 was added and the homogenized powder mixture was pressed into pellets. The pellets were then annealed in O_2; the annealing parameters were 830 C to 870 C and 30 minutes to 7 hours. The superconducting state of the prepared pellets was indicated by means of the standard resistive T_c measurements. The values of T_c from the interval 85 to 110 K were found. The highest T_c value were found in samples annealed at 865 C. The grain orientation, the unit cell parameters and the phase purity were determined from X-ray powder diffraction data. The X-ray analysis confirmed rather low orientation of grains and more than 90 pure 2212 phases for pellets annealed at 850 C. The 2212 phases were prepared in our case from the 2-2-2.3-3 composition ($Tl_2Ba_2Ca_{2.3}Cu_3O_x$). This fact can be explained by the structure disordering when the 2212 phase is formed from the initial 2223 composition with impurity phases such as CuO, Ca_2CuO_3, $BaCuO_2$ and unreacted Tl_2O_3 [5].

The tapes have been prepared with two kinds of the precursor powder: superconducting powder made by pulverizing the pellets obtained in the way described above and the powder mixture used for the preparation of pellets. Silver tubes with o.d. 5 mm, i.d. 3.5 mm were filled with the powder, rotary swaged to a wire of 3 mm o.d. and cold rolled to the final cross-section 5x0.14 mm^2 and length of about 3 m. The lengths were cut into 30 mm pieces, pressed and rolled to thicknesses up to 0.07 mm and annealed at 835 C - 875 C for 30 minutes to 2 hours. The samples exhibited J_c 2-4 Acm^{-2} at 77 K in the absence of external magnetic field. These measurements indicated that the pellet making step was not necessary for the precursor powder preparation.

There was a great difference in I_c (several times) at low annealing temperatures (835 C) between 'as rolled' and pressed samples. The difference decreases with the increasing annealing temperature and at 870 C there was practically no difference between rolled and pressed samples. It can be supposed that the densification of the oxide core can be obtained by pressing as well as partial melting. This fact is important from the point of view of long lengths of the tape.

The I_c anisotropy in the external DC magnetic field was measured using the standard four probe technique. The I_c of the sample was determined at the $1\mu Vcm^{-1}$ criterion. The DC transport current was perpendicular to the magnetic field direction. It was possible to rotate the tape plane from the perpendicular position to the parallel one with respect to the applied magnetic field direction. The applied field varied from 1 mT to 500 mT in the temperature interval 70 - 90 K and from 50 mT to 5 T at the temperature 4.2 K. Temperature dependence measurements were restricted by the maximum transport current of about 4 A allowed for the continuous flow cryostat used in our experiments.

3. Experimental results and discussion

Two samples were chosen for the anisotropy measurements: 1) sample produced by a single-stage procedure (tape was rolled and annealed at 865 C for 45 minutes); 2) sample produced by multiple pressing and heating (rolled tape was pressed at 2 GPa, heated at 835 C for 45 minutes, pressed again and finally heated at 835 C for 30 minutes). The better grain alignment was expected in sample produced by the multiple process. The

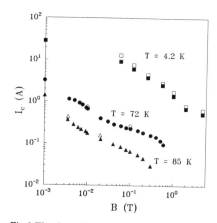

Fig.1:The dependence of I_c on magnetic field
B// (open) and B⊥ (solid) at T 4.2 K, 72 K
and 85 K for pressed sample.

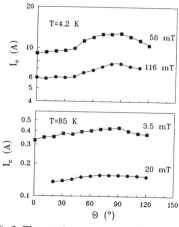

Fig.2: The angular dependence of I_c at 85K
and 4.2K for pressed sample.

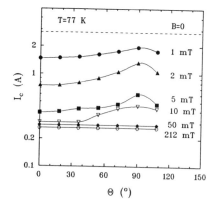

Fig.3: The angular dependence of I_c at 77K for
rolled sample.

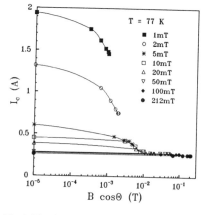

Fig.4: The dependence of I_c on B.cosΘ at 77 K
and various external magnetic fields for rolled
sample.

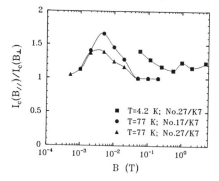

Fig.5: The dependence of the ratio $I_c(B//)/I_c(B⊥)$ on magnetic field for rolled and pressed samples
at 77K and pressed sample at 4.2K.

parameters of the measured samples are given in Table 1. $I_c(B)$ characteristics of the

Table 1.

Sample	Cross-section (mm2)	Procedure	J_c(A/cm2)[77K,0T]	[4.2K,0T]
Rolled	5.3x0.14	single-heat (as rolled)	2600	
Pressed	6x0.125	multiple pressing and heating	2500	21400

pressed sample at temperatures 4.2 K, 72 K and 85 K for two orientations of the external magnetic field (parallel and perpendicular to the tape plane) are given in Fig.1. It is seen from this figure that there is a rather small anisotropy in the whole temperature range 4.2 K - 85 K. Similar results were obtained for rolled sample. The angular dependencies of I_c at various magnetic fields and temperatures are given in Fig.2 for pressed and in Fig.3 for rolled sample. According to Hensel et al. [6] the information about the orientation of grains can be obtained from the effective critical misalignment angle Φ_e defined by $\cos(90 - \Phi_e) = \cos\Theta = B_c^*/B$ where B_c is given by the intersection of the plateau value with the exponential decay of the I_c versus $B\cos\Theta$ dependence. Fig. 4 shows this dependence at the temperature 77 K for the rolled sample. It is seen from this figure that the effective critical misalignment angle Φ_e can be evaluated only for external magnetic fields near the value 5 mT while at higher and lower fields the I_c vs. $B\cos\Theta$ dependence indicates the total non-alignment of grains. The dependence of $I_c(B\|):I_c(B\perp)$ vs. B exhibits local maxima at 77 K as well as 4.2 K, see Fig.5. It can be concluded that the observed anisotropy is caused by the reason other than grain alignment.

4. Conclusions

The small I_c anisotropy in the external magnetic field found in Tl-2212 Ag-sheathed tapes produced by the PIT technique indicates the lack of the preferred grain orientation and the three-dimensional character of charge transfer. The processing conditions and multiple press and heat procedures have no influence on the I_c anisotropy. The small I_c anisotropy and the relatively weak $J_c(B)$ dependence above the low field region at 77 K in comparison with Bi-2223 system indicate the potential application advantage of Tl-2212 tapes.

References

[1] Kim D H, Gray K E, Kampwirth R T, Smith J C, Richeson D S, Marks T J, Kang J H, Talvacchio J and Eddy M 1991 *Physica C* **177** 431

[2] Willis J O, Maley M P, Kung P J, Coulter J Y, Peterson D E, Wahlbeck P G, Bingert J F and Philips D S 1993 *IEEE Trans. on Applied Superconductivity* **3** 1219

[3] Peterson D E, Wahlberg P G, Maley M P, Willis J O, Kung P J, Coulter J Y, Salazar K V, Phillips D S, Bingert J F, Peterson E J and Hults W L 1992 *Physica C* **199** 161

[4] Kung P J, Wahlberg P G, McHenry M E, Maley M P and Peterson D E 1994 *Physica C* 310

[5] Jergel M, Hanic F, Plesch G, Strbik V, Liday J, Falcomy Guajardo C and Contreras Puente G S 1994 *Supercon. Sci. Technol.* **7** 931

[6] Hensel B, Grivel J C, Jeremie A, Perin A, Pollini A and Flükiger R 1993 *Physica C* **205** 329

Inst. Phys. Conf. Ser. No 148
Paper presented at Applied Superconductivity, Edinburgh, 3–6 July 1995
© *1995 IOP Publishing Ltd*

Preparation and physical characterization of Tl(1223) tapes with j_c (77K, 0T) > 10 kA/cm² [†]

R E Gladyshevskii, A Perin, B Hensel and R Flükiger

Département de Physique de la Matière Condensée, Université de Genève, 24 quai E.-Ansermet, CH-1211 Geneva 4, Switzerland

Abstract. Reproducible transport critical current densities of over 10'000 A/cm² (77 K, 0 T) were obtained for Ag-sheathed tapes with pre-reacted Tl(1223) powder, the highest value being 14'000 A/cm². High-purity powders (> 95 wt% Tl(1223), $T_c \sim 120$ K) were prepared by a new method where the Tl(1223) phase is formed by reaction in the tape under high pressure (He). Ag(20 wt% Au) alloy showed to be a good sheath material for an in-tape reaction, in particular when involving partial melting which favors the grain growth. Critical current densities of 9'000 A/cm² (77 K, 0 T) were obtained for such tapes, the field dependence being improved.

1. Introduction

Since the discovery of superconductivity in the Tl-Ba-Ca-Cu-O system [1], several Tl-based phases have been reported and nearly single-phase materials prepared and characterized. Compounds with single TlO layers, like Tl(1223), originally reported for the composition $(Tl_{0.5}Pb_{0.5})Sr_2Ca_2Cu_3O_9$ [2], are considered to allow stronger coupling between the slabs of conducting CuO_2 layers than compounds with multiple TlO layers. Its high T_c, intragrain j_c and irreversibility field make Tl(1223) one of the most promising materials for applications in high magnetic fields at liquid nitrogen temperatures. Large-scale applications require long wires or tapes with high intergrain j_c which persists in a magnetic field, however, up to now, weak links between superconducting grains represent a serious problem, impeding optimal transport properties. Only two groups have reported critical current densities higher than 20'000 A/cm² (77 K, 0 T) in short tapes [3-5], j_c dropping drastically in a magnetic field.

We report here on the preparation and properties of $(Tl,Pb,Bi)(Sr,Ba)_2Ca_2Cu_3O_{9-\delta}$ tapes. It has been shown that substitution of Tl by Bi enhances the formation of the Tl(1223) phase and that substitution of Sr by Ba introduces pinning centers [4].

2. Results and discussion

Superconducting Tl(1223) tapes were produced by the powder-in-tube technique. Small amounts of Pd (6 wt%) or Au (20 wt%) were in some cases added to the silver sheath in order to increase its melting point. The tubes were filled with either a pre-reacted Tl(1223) powder or a mixture of precursors and, subsequently, mechanically deformed by swaging, drawing and rolling to a thickness of 110-200 µm.

A substituted Tl(1223) phase was obtained from a mixture of precursors (Sr, Ba and Ca carbonates calcined with CuO, and Tl, Pb and Bi oxides) of nominal composition

---

[†] Work supported by the Swiss National Foundation and the European Community (Brite Euram Project No. 7055).

$Tl_{0.7}Pb_{0.2}Bi_{0.2}Sr_{1.8}Ba_{0.2}Ca_{1.9}Cu_3O_x$ by heating pellets or tapes up to 990°C under pressures of 25-100 bar He [6]. By working at high pressure Tl losses were avoided during the synthesis. The following impurities were detected from X-ray diffraction (confirmed by EDX analysis): $(Ca,Sr)_2CuO_3$, $(Sr,Ca)CuO_2$, $BaBiO_3$, Tl(1212) and Ca_2PbO_4, the latter two being in particular observed after partial melting in the tapes. By optimizing the preparation conditions powders containing >95 wt% Tl(1223) could be produced. X-ray diffraction diagrams of Tl(1223) powder and tapes are shown in Fig.1.

The typical grain size for samples synthesized at temperatures ~940°C with reaction times shorter than 3 hours was 4-5 μm. Twice larger and better aligned grains were observed in tapes which were maintained at 980°C during 3 hours. Locally textured samples with 25-50 μm-large plate-like grains were formed after partial melting at temperatures above 980°C. The latter samples, however, contained secondary phases, among which a significant amount of Tl(1212), partial decomposition being confirmed by TG/DTA measurements.

Silver in oxygen melts before the Tl(1223) phase and pure Ag cannot be used as sheath material for an in-tape-reaction above 930°C. Ag(6 wt% Pd)-sheathed tapes showed a slight reaction of the sheath with the oxide phase already at 940°C, including formation of Tl(1212), and the surface of even ~180 μm-thick tapes had large holes. Ag(20 wt% Au) alloy sheaths were more resistant to high temperature treatments than Ag(Pd) sheaths. Small holes appeared in thin Ag(Au)-sheathed tapes (< 130 μm) only above 970°C and the tapes contained X-ray pure Tl(1223) even at 980°C.

The majority of the tapes were mechanically pressed and annealed in flowing oxygen at temperatures up to 890°C during 3-5 hours. Higher density and visible texture was observed after this treatment, however, the grain size remained the same and the grain connections were weak.

Fig.2 shows the AC susceptibility versus temperature for Tl(1223) powder and different tapes. The bulk intragrain superconducting transition temperature of the powder was increased from 114.4 to 118.4 K (onset) after annealing in a Ag sheath. Similar values were observed for Ag(Au)-sheathed tapes with in-tape-reacted powder. Further annealing increased T_c to ~121 K, however, the samples were no longer of high purity, due to Tl losses. A broad transition with T_c = 106.4 K was measured for the multiphase Ag(Pd)-sheathed tape, which may also indicate a partial substitution of Cu by Pd.

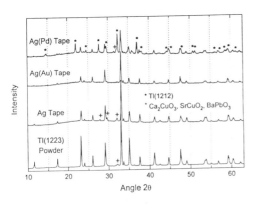

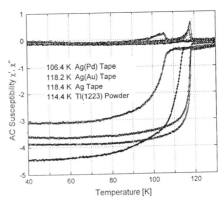

Fig.1. X-ray diffraction diagrams of Tl(1223) powders before annealing, after annealing in a Ag-sheathed tape and formed by reaction in a Ag(Au)- and a Ag(Pd)-sheathed tape (reaction: 940°C, 25 bar He / 0.5 bar O₂; annealing: 870°C, flowing O₂).

Fig.2. AC susceptibility of Tl(1223) powder before annealing, a Ag-sheathed tape with pre-reacted powder, a Ag(Au)- and a Ag(Pd)-sheathed tape with in-tape-reacted powders versus temperature.

Conditions for tape preparation were varied in order to optimize the transport critical current density. j_c increased with increasing reaction temperature, up to a temperature where the powder started to melt (above 980°C for Ag(Au)-sheathed tapes), shorter reaction times being required at higher temperatures. The highest value of j_c for tapes with pre-reacted powder corresponded to a thickness of 100-110 μm, sausaging effect being observed for thinner tapes. For Ag(Au)-sheathed tapes with in-tape-reacted powders, 130 μm thickness yielded the highest values of j_c, holes appearing on the surface of the tapes below this thickness. On the whole, pressed tapes showed higher values of j_c than tapes which had only been rolled; higher density was observed from SEM images for the former. The heat treatment had a strong positive effect on the critical current density, however, for higher temperatures the annealing time had to be shortened in order to avoid loss of Tl. The best results were observed for samples annealed at 870°C for 5 hours.

Reproducible transport critical current densities of over 10'000 A/cm^2 (I_c = 11 A, 77 K, 0 T) were observed for pressed, Ag-sheathed Tl(1223) tapes, the highest value being 14'000 A/cm^2. Critical current densities of 7'000 and 9'000 A/cm^2 (77 K, 0 T) were reached for Ag(Pd)- and Ag(Au)-sheathed tapes with in-tape-reacted powder, respectively. The critical current versus the magnetic field for the latter is shown in Fig.3. The drop of I_c at low field and the strong hysteresis of the critical current indicate weak links between the grains. For the Ag-sheathed tape a factor of 23 was observed between I_c at 0 and 0.5 T. Similar factors characterized all tapes with pre-reacted powder. Improved results were obtained for Ag(Au)-sheathed tapes with in-tape-reacted Tl(1223) powder, for which I_c at 0.5 T was decreased by a factor of 16 with respect to the value at 0 T and lower hysteresis was observed. A small anisotropy of I_c was introduced during the tape preparation and treatment, the ratio of the critical currents parallel and perpendicular to the tape surface exceeding 1.7. Less field dependence is expected for in-tape-reacted powders where the reaction is accompanied by partial melting, however, the low critical current density observed so far, due to the presence of additional phases, made it difficult to study the field dependence.

The transport properties of a Ag(Au)-sheathed tape with in-tape-reacted Tl(1223) powder at different temperatures are shown in Fig.4. It can be seen that I_c increases between 110 and 4.2 K. At 110 and 90 K no significant critical currents could be detected above 10 and 100 mT (increasing field), respectively. The field dependence of the critical current at lower temperatures was similar to that observed at 77 K.

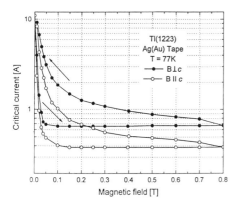

Fig.3. Critical current (77 K) of a Ag(Au)-sheathed tape with in-tape-reacted powder versus magnetic field (I_c criterion 1 μV/cm).

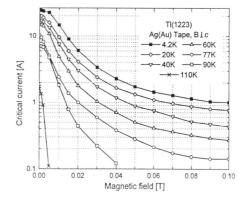

Fig.4. Critical current of a Ag(Au)-sheathed tape with in-tape-reacted Tl(1223) powder versus magnetic field at different temperatures.

482

Magnetic hysteresis for a Ag(Au)-sheathed tape with in-tape-reacted Tl(1223) powder was measured at temperatures between 90 and 5 K with a SQUID magnetometer. The 90 and 77 K loops did not present any significant hysteresis above 1.5 and 3 T, respectively. Fig.5 presents ΔM versus magnetic field for different temperatures with the magnetic field parallel to the surface of the tape. The critical current density, evaluated on the basis of Bean's model using an average grain size of 5 μm, was 8×10^6 A/cm^2 at 5 K and 1T. The value of 5×10^6 A/cm^2 was obtained for Tl(1223) powder with random orientation of the grains under similar conditions.

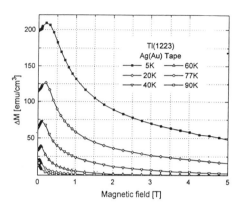

Fig.5. ΔM at different temperatures for a Ag(Au)-sheathed tape with in-tape-reacted Tl(1223) powder versus magnetic field.

3. Conclusion

Our method of in-tape synthesis at high pressure avoids Tl losses, allowing a better control of the composition of the samples. Powders containing >95 wt% Tl(1223) could be produced.

Increasing the reaction time had a positive effect on the grain growth, but a progressive decomposition of the Tl(1223) phase was observed. Locally textured samples with up to 50 μm-large plate-like grains were obtained by in-tape-reaction involving partial melting, the conditions to obtain well textured high-purity samples being, however, not yet optimal.

The interaction of Pd with the Tl(1223) phase at 940°C and above in presence of oxygen, makes Ag(Pd) alloys unsuitable for tape preparation. Ag(Au) alloys seem so far to be the best sheath material for an in-tape-reaction.

A critical current density of 14'000 A/cm^2 (77 K, 0 T) was reached for Ag-sheathed tapes with pre-reacted powder. Values of 9'000 A/cm^2 were obtained for Ag(Au)-sheathed tapes with in-tape-reacted powder, field dependence being improved with respect to the former.

Acknowledgements

We would like to thank Prof. M.Th. Cohen-Adad and Mr. R. Abraham, Laboratoire de Physicochimie Minérale II, Université Claude Bernard-Lyon I, for DTA measurements and Dr. G. Triscone, DPMC, Université de Genève, for SQUID measurements.

References

[1] Sheng Z Z and Hermann A M 1988 Nature 332 138-9.
[2] Subramanian M A, Torardi C C, Gopalakrishnan J, Gai P L, Calabrese J C, Askew T R, Flippen R B and Sleight A W 1988 Science 242 249-52.
[3] Seido M, Sato J, Sasaoka T, Nomoto A, Kamo T and Matsuda S 1992 Proc. Int. Workshop on Superconductivity, Honolulu 236-9.
[4] Ren Z F and Wang J H 1993 Physica C 216 199-204.
[5] Ren Z F, Wang C A, Wang J H, Miller D J and Goretta K C 1995 Physica C 247 163-8.
[6] Gladyshevskii R E and Flükiger R 1995 to be published.

Inst. Phys. Conf. Ser. No 148
Paper presented at Applied Superconductivity, Edinburgh, 3–6 July 1995
© *1995 IOP Publishing Ltd*

Study of Tl-1223 for use in wire and tape fabrication

J. C. Moore, S. Fox, C. J. Salter, A. Q. He[†] and C. R. M. Grovenor

Department of Materials, University of Oxford, Parks Road, Oxford, OX1 3PH, UK.
[†]Laboratory of Atomic Imaging of Solids, Institute of Metal Research, Chinese Academy of Sciences, Wenhua Road, Shenyang 110015, P. R. China.

Abstract. Compositional and microstructural effects have been studied in the $(Tl_{1-x}Bi_x)(Sr_{1-z}Ba_z)_2Ca_2Cu_3O_y$ system. Tapes were fabricated by an insitu process using powders of composition $(Tl_{0.78}Bi_{0.22})(Sr_{0.8}Ba_{0.2})_2Ca_2Cu_3O_y$ and $(Tl_{0.5}Pb_{0.5})(Sr_{0.8}Ba_{0.2})_2Ca_2Cu_3O_y$. The increase in J_c and T_c with time was found to be similar for both systems.

1. Introduction

Tl-1223 compositions of the type $(Tl,M)(Sr,Ba)_2Ca_2Cu_3O_y$ (TlM-1223) where M can be Pb [1], Bi [2] or both [3] are being extensively investigated for superconducting wire and tape fabrication due to their high irreversibility lines at 77K [1, 4, 5]. J_c values of up to $2.1 \times 10^4 Acm^{-2}$ (77K, 0T) have been reported [2] but J_c decreases rapidly in small magnetic fields [5]. The tapes are fabricated using either fully reacted superconducting powder [1, 3] which is given a final heat treatment in order to sinter the grains together, or an insitu process [2] where Tl_2O_3 and Bi_2O_3 and a $SrBaCaCuO_y$ precursor react in a 2 stage heat treatment inside the silver tube. Previous compositional studies of TlBi-1223 [2] have concentrated on improving I_c but we have also studied the effect on microstructure in powder samples and have investigated the insitu reaction rates of TlBi-1223 and TlPb-1223 in tape samples.

2. Effect of composition on microstructure in $(Tl_{1-x}Bi_x)(Sr_{1-z}Ba_z)_2Ca_2Cu_3O_y$

Powders were made by a two step process. A $(Sr_{1-z}Ba_z)_2Ca_2Cu_3O_y$ precursor was fabricated from $SrCO_3$, BaO_2, CaO and CuO powders by heating at 950°C for 48h with 3 intermediate regrinding steps. Tl_2O_3, PbO_2 or Bi_2O_3 were then added in the required stoichiometry and heated at, typically, 940°C for 2h in a sealed alumina crucible.

The Bi content, x, was varied between 0 and 0.4 in $(Tl_{1-x}Bi_x)(Sr_{0.8}Ba_{0.2})_2Ca_2Cu_3O_y$. XRD analysis, Figure 1, showed that 1223 was the majority phase but an increasing amount of $BaBiO_3$ is formed for increased Bi contents. SEM showed cauliflower grains for x=0 and platelike grains for x=0.15 with the grain size increasing from 10 μm at x=0.15 to 20 μm at x=0.4, Figure 2. R/T measurements, Figure 3, revealed a double resistive transition and lower $T_{c(0)}$ for x= 0.3 and higher. HREM studies, Figure 4, showed that the intercalation structure of $(Tl_{0.78}Bi_{0.22})(Sr_{0.8}Ba_{0.2})_2Ca_2Cu_3O_y$ was very regular.

484

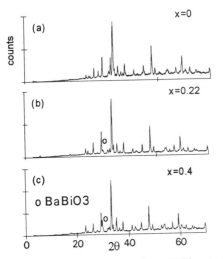
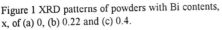

Figure 1 XRD patterns of powders with Bi contents, x, of (a) 0, (b) 0.22 and (c) 0.4.

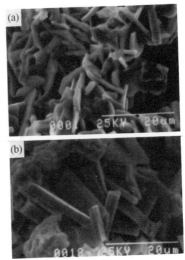

Figure 2 SEM micrographs of powders with Bi contents, x, of (a) 0.15 and (b) 0.4.

The Ba substitution level was varied between 0.1 and 0.3 in $(Tl_{0.78}Bi_{0.22})(Sr_{1-z}Ba_z)_2Ca_2Cu_3O_y$ and $Tl(Sr_{1-z}Ba_z)_2Ca_2Cu_3O_y$. XRD patterns were very similar for all compositions in the $(Tl_{0.78}Bi_{0.22})(Sr_{1-z}Ba_z)_2Ca_2Cu_3O_y$ series but in the $Tl(Sr_{1-z}Ba_z)_2Ca_2Cu_3O_y$ system the formation of 1223 increased from less than 10 vol% for z=0.1 to greater than 90 vol% for z=0.3. No significant change in microstructure was found in the $(Tl_{0.78}Bi_{0.22})(Sr_{1-z}Ba_z)_2Ca_2Cu_3O_y$ system but in $Tl(Sr_{1-z}Ba_z)_2Ca_2Cu_3O_y$ the microstructure became plate like and the grain size increased for z=0.2 and greater, Figure 5. We have found the highest $T_{c(0)}$ of 108K for z=0.3 and 0.4 in $(Tl_{0.78}Bi_{0.22})(Sr_{1-z}Ba_z)_2Ca_2Cu_3O_y$ and for z=0.6 in $Tl(Sr_{1-z}Ba_z)_2Ca_2Cu_3O_y$, presumably due to the increased formation of 1223 in this sample.

These SEM and XRD results suggest that a Bi content in the range 0.15 to 0.4 can be used and that the Ba content can be varied widely to produce phase pure powder containing larger grains. Magnetisation data [6] found very similar J_c values for different Bi substitution contents. However, the R/T data shows that two superconducting phases are produced for Bi contents of 0.3 and larger and that the $T_{c(0)}$ decreased for Ba contents outside the range 0.3-0.4 but these results suggest that there may be some flexibility in the composition for improving the microstructure of the tapes subsequently produced.

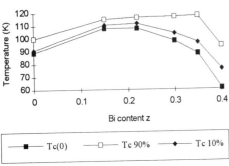

Figure 3 Variation of T_c with Bi content, x.

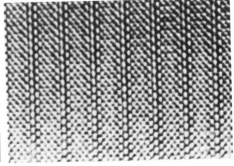

Figure 4 HREM micrograph of a powder with x=0.22.

 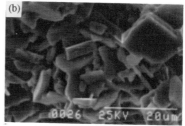

Figure 5 SEM micrographs of powders of composition (a) $Tl(Sr_{0.9}Ba_{0.1})_2Ca_2Cu_3O_y$ and (b) $Tl(Sr_{0.7}Ba_{0.3})_2Ca_2Cu_3O_y$.

3. Effect of powder composition on tape properties

We have fabricated tapes with different thallium site substitutions of $(Tl_{0.78}Bi_{0.22})$, $(Tl_{0.5}Pb_{0.5})$ and Tl with a precursor of composition $(Sr_{0.8}Ba_{0.2})_2Ca_2Cu_3O_y$. We obtain higher J_c values using an insitu reaction process, as also reported by Ren and Wang [2]. In this study an excess of thallium oxide was used but we find a stoichiometric composition gives similar J_c values. We have obtained the highest J_c values using a multiple stage heat and roll process [7]. Typical J_c values are 9000 Acm^{-2} for an insitu process.

EPMA analysis showed that the composition of the superconducting phase in both the insitu reacted and sintered TlBi-1223 tapes was extremely uniform although we observed some variation in the Sr/Ba ratio. However, we did not observe $BaBiO_3$ in the insitu reacted tape. $BaBiO_3$ is present in both the reacted superconducting powder and in the sintered tape. We believe that the formation of this impurity is reduced by a lower oxygen partial pressure in the insitu reaction, suggesting that the insitu reaction may occur at a different rate, and possibly by a different mechanism compared to the powder synthesis. It also suggests that a range of compositions with a Bi content of greater than 0.22 may produce single phase material if processed in this way.

SEM investigations have shown that the insitu reaction results in better connectivity and this probably explains the higher I_c values observed [7]. $I_c(B)$ measurements, Figure 6, showed that the field dependence of the transport I_c for tapes fabricated using the insitu route was less severe than that of the sintered tape although weak links still appear to dominate.

A study of the effect of annealing time on J_c in the insitu synthesis of $(Tl_{0.78}Bi_{0.22})(Sr_{0.8}Ba_{0.2})_2Ca_2Cu_3O_y$ and $(Tl_{0.5}Pb_{0.5})(Sr_{0.8}Ba_{0.2})_2Ca_2Cu_3O_y$ reported much larger J_c values for short annealing times for TlBi-1233 [2]. We have made tapes from both these compositions using a one step reaction process and investigated the rate of reaction by heating at 840°C for different times.

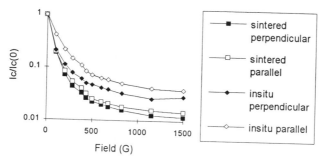

Figure 6 $I_c(B)$ measurements for insitu reacted and sintered tapes.

486

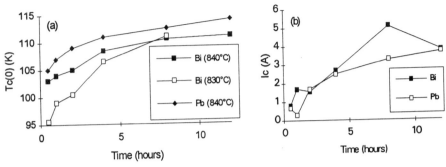

Figure 7 Increase of (a) T_c and (b) I_c with increasing time for the TlBi-1223 and TlPb-1223 systems. (These tapes have lower I_c than the optimum as a one stage process was used.)

Our results show a very similar increase in I_c and T_c for both systems, Figure 7, in contrast to those previously reported [2]. T_c was found to increase more gradually for tapes heated at a slightly lower temperature of 830°C, indicating that the reaction rate is slowed slightly. XRD analysis of these tapes showed that the 1212 phase forms initially reacting to produce pure 1223 after around 2 hours confirming previous TEM observations [8].

4. Conclusions

We have found that the Bi substitution level in $(Tl_{1-x}Bi_x)(Sr_{1-z}Ba_z)_2Ca_2Cu_3O_y$ has the greatest effect on grain size and the formation of the Tl-1223 phase whereas the Ba content has only a minor effect in the presence of Bi. Tapes were fabricated using both the insitu reaction and sintering processes and J_c values were found to be higher for an insitu reaction. The impurity phase $BaBiO_3$ was not found in insitu reacted tapes, which we believe is due to a reduced oxygen partial pressure. This implies that the composition range for insitu tape synthesis, which produces phase pure material, may be slightly different to that identified by powder synthesis. The insitu reaction rate for $(Tl_{0.78}Bi_{0.22})(Sr_{0.8}Ba_{0.2})_2Ca_2Cu_3O_y$ and $(Tl_{0.5}Pb_{0.5})(Sr_{0.8}Ba_{0.2})_2Ca_2Cu_3O_y$ tapes, as determined by I_c and T_c measurements, was found to be similar in contrast to previous results.

5. References

[1] Kamo T, Doi T, Soeta A Yuasa T, Inoue N, Aihara K and Matsuda S 1991 *Appl. Phys. Lett.* **59** 3186-8
[2] Ren Z F and Wang J H 1993 *Physica C* **216** 199-204
[3] Richardson K A, Wu S, Bracanovic D, de Groot P A J, Al-Mosawi, Beduz C, Ogborne D M and Weller M T 1995 *Supercond. Sci Technol.* **8** 239-244
[4] Presland M R, Tallon J L, Flower N E, Buckley R G, Mawdsley A, Staines M P and Fee M G 1993 *Cryogenics* **33** 502-5
[5] Zheng D N, Johnson J D, Jones A R, Campbell A M, Liang W Y, Doi T, Okada M and Higashyma K 1995 *J. Appl. Phys.* **77** 5287-5292
[6] Kilic A, Senoussi S, Traxler H, Moore J C, Collier A J, Goringe M J and Grovenor C R M see this volume
[7] Fox S, Moore J C and Grovenor C R M see this volume
[8] Miller D J, Hu J G, Ren Z and Wang J H, 1994 *J. Electronic Mater.* **23** 1151-4

Acknowledgments This work was funded by EC under the Brite/Euram programme, contract number BRE2-CT93-0455.

Inst. Phys. Conf. Ser. No 148
Paper presented at Applied Superconductivity, Edinburgh, 3–6 July 1995
© *1995 IOP Publishing Ltd*

Influence of sheath material and thermal/mechanical processing on critical current densities in thallium-based superconducting tape

S Fox, JC Moore and CRM Grovenor

Department of Materials, University of Oxford, Oxford OX1 3PH, UK

Abstract An investigation has been made of the influence of processing conditions on the critical current densities observed in (Tl,Bi)(SrBa)CaCuO superconductors. Optimisation of processing parameters (such as sheathing material, mechanical processing and annealing treatments) has led to critical current densities approaching $10^4 A/cm^2$ being achieved at 77K in zero field.

1. Introduction

The successful production of superconducting tapes, particularly where the superconducting phase is produced by in-situ reaction during powder-in-tube processing, requires careful control and optimisation of all the process parameters - from initial tube packing to final annealing treatments. Each of the processing steps has a strong influence on the structure of the superconductor core and hence the maximum attainable current density.

In this study, the variables that have been investigated are: sheath material and cross-section, tape width and thickness, uniaxial pressing, annealing temperature/time and multi-stage annealing treatments.

This study has been conducted with the need to produce significant lengths (>10m) of superconducting tape as a primary consideration and will therefore concentrate on the practical aspects of superconductor process technology.

2. Experimental

Silver-sheathed TlBi-1223 tapes were produced by filling pure (99.99%Ag) and alloy (Ag-0.25Mg-0.25Ni) silver tubes with a mixture of $(Sr_{0.8}Ba_{0.2})_2Ca_2Cu_3O_x$ precursor and Tl and Bi oxides of overall composition $(Tl_{0.78}Bi_{0.22})(Sr_{0.8}Ba_{0.2})_2Ca_2Cu_3O_x$. The tubes were then cold drawn to 1.4 or 1.0mm diameter with intermediate low temperature anneals and finally rolled to 0.12 or 0.15mm with intermediate high temperature treatments.

As the starting dimensions of the pure and alloy silver tubes were different (ø6xø4 and ø4xø3 respectively), some of the pure silver was drawn and drilled to ø4xø3 before powder packing for direct comparison. The influence of wall thickness/core fraction was studied by drilling-out lengths of the ø6xø4mm pure silver tube.

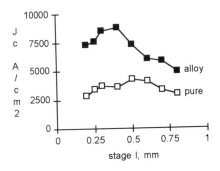

Figure 1 Influence of stage I thickness on Jc in pure
and alloy silver tapes (stage II thickness = 0.15mm).

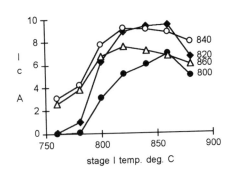

Figure 2 Determination of optimum annealing
temperature for given process times in alloy tape.

3. Results and Discussion

A direct comparison of the pure and alloy silver sheathing materials was made by heat treating identically processed samples at 840°C/0.5h (stage I) and 840°C/7.5h (stage II). This two step process had been found to give optimum results for pure silver [1]. Figure 1 shows that the alloy tapes gave consistently higher critical current densities when compared to pure silver. The dependence of Jc on stage I thickness in the alloy tapes is also apparent from this graph, with a maximum being observed at 0.3 to 0.4mm.

Having established the superiority of the alloy silver as a sheathing material, it was necessary to determine whether the optimum high temperature annealing treatment was the same as that for pure silver. A set of samples was therefore cut from material rolled to the optimum stage I thickness of 0.35mm (see Figure 1) and annealed for 0.5h at temperatures ranging from 760-880°C before rolling to 0.15mm and annealing for 7.5h at temperatures ranging from 800-860°C. Figure 2 shows that for the given process times, a maximum Jc of 8250 A/cm² was achieved with stage I: 820-860°C and stage II: 820-840°C.

As the process times in the experiment to determine optimum annealing temperatures were fixed, it was also necessary to determine the effect of annealing time at a given temperature. On the basis of the results obtained in Figure 2, two sets of samples were heated either for 2-60min followed by 7.5h at 840°C, or 0.5h followed by 0.5-23.5h at 840°C respectively. Figures 3 and 4 show that for optimum Jc, a minimum stage I time of 10min and a minimum stage II time of 6-7h is required. These results indicate that the stage I process time may be extended from 10-60min but the stage II process time should not exceed 12h.

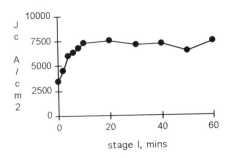
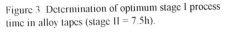

Figure 3 Determination of optimum stage I process
time in alloy tapes (stage II = 7.5h).

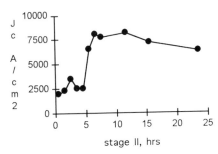

Figure 4 Determination of optimum stage II process
time in alloy tapes (stage I = 0.5h).

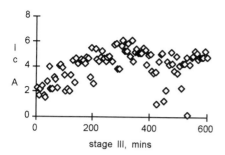

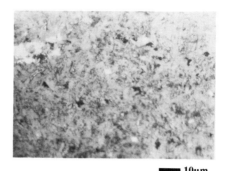

■■ 10μm

Figure 5 Increase in Ic with annealing time for pure silver tape subjected to 3 stage processing treatment.

Figure 6 Light micrograph showing grain structure obtained after 3 stage processing (stage III = 5h).

It has previously been reported that Jc can be significantly improved by a cold pressing treatment before final annealing [1,2]. Samples of pure and alloy silver tape were therefore subjected to a cold pressing treatment of 1GPa. Although improvements of between 25-50% were observed in pure silver tape, the effect was more limited in the alloy silver sheathed material. In order to further investigate the effects of uniaxial pressing, a series of alloy tapes were subjected to higher pressures, ranging from 1 to 15GPa (corresponding to 6-90tonnes) and no improvement in Jc was obtained for pressures greater than 3GPa.

It was reported in ref.[1] that although core compaction, and correspondingly, current density was improved by cold pressing, the tapes became stiff and brittle. The use of a high strength sheath in the present investigation obviates the need for a pressing treatment but unfortunately the alloy tapes also became embrittled after heat treatment. The fact that the tape becomes stiff and brittle after the relatively short stage I anneal precludes a second processing operation in long tapes that have to be coiled and uncoiled during two stage processing. It was for this reason that further investigations, in the present study, reverted to the optimisation of critical current density using a pure silver sheath.

A series of experiments were carried out to determine whether increasing the number of processing steps would result in improved current densities. Figure 5 shows the results from a series of tests on pure silver sheathed tapes that were subjected to a three stage processing treatment. Stages I and II consisted of annealing at 840°C/10min at 0.5 and 0.25mm respectively. The stage III process times ranged from 6 to 600min at 840°C. The maximum Ic of 6.1A corresponds to a Jc of 4900A/cm². The microstructure of the superconductor core is shown in Figure 6. Light microscopy revealed a fine grain size of around 1μm, up to 20% volume fraction of secondary (non-superconducting) phases, and around 10% porosity. It should be noted that the internal diameter of the tube that was used to produce these samples was increased from 4.0 to 4.4mm (giving a wall thickness of 0.8mm).

In order to investigate the combined effects of tube wall thickness, tape width and final thickness, and multi-stage processing in more detail, a series of tubes were drilled to give a range of wall thicknesses. Sections were removed after drawing to ø1.4 or ø1.0mm and at each stage of the rolling heat treatment processing (which was carried out using 3 or 4 steps). These were then mounted and photographed after polishing in order to determine the cross-section and area fraction of the core at each stage.

Figure 7 shows an example of core area measurements for three stage processing of tubes with wall thicknesses ranging from 0.9-0.6mm that were drawn to ø1mm before rolling. The initial core fraction, which ranges from 49 to 64% is in all cases reduced by around 15-20% due to core compaction during the drawing and rolling operations. Figure 8 shows the corresponding current densities that were obtained for these samples.

490

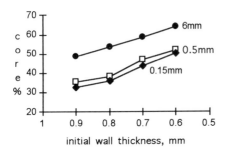

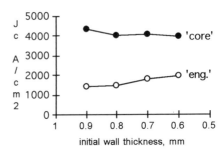

Figure 7 Measurements of superconductor core fraction in pure silver tape (drawn to Ø1mm).

Figure 8 Effect of initial wall thickness on current density for tape shown in Fig. 7.

While the "engineering", or overall, current density increases with core fraction, this appears to be partially offset by a small decrease in "superconductor", or core, current density which may be a result of less efficient compressive stress transfer to the core in thin-walled wire or tape.

Increasing the tape width (by rolling from ø1.4mm instead of ø1.0mm) resulted in useful increases in critical current, corresponding to the increased core cross-section. The maximum critical current in the wider tapes (around 10A) was, however, found to be limited by the self-field generated by the sample under test.

Four stage processing at 840°C (where stages I to III consisted of 3 short, 10min, anneals at thicknesses of 1.0, 0.5 and 0.25mm respectively and stage IV was a final anneal for 7.5h at 0.15/0.12mm) did not result in any significant increases in current density. Although, the slight decrease in superconductor Jc, shown in Figure 8 for ø1.0mm wire processed in three stages, was no longer apparent. Decreasing the tape thickness from 0.15 to 0.12mm resulted in a reduction in Ic commensurate with the reduction in superconductor core area and showed no benefit in superconductor Jc.

4. Conclusions

The use of an Ag-Mg-Ni sheath gives high current density without the need for a cold pressing treatment but is not suitable for multi-stage processing due to brittleness after heat treatment. The current density in silver sheathed tape may be improved by three stage processing. Useful increases in critical current are obtained by increasing the tape width by rolling from larger diameter wire. However, the weak-link dominated behaviour of the material processed so far results in self-field limited critical currents of around a maximum of 10A.

References

[1] Ren ZF and Wang JH 1993 Physica C **216** 199-204
[2] Ren ZF, Wang JH, Miller DJ and Goretta KC 1994 Physica C **229** 137-144

Acknowledgements This work was funded by the EC under the BRITE/EURAM programme, contract number BRE2-CT93-0455.

Inst. Phys. Conf. Ser. No 148
Paper presented at Applied Superconductivity, Edinburgh, 3–6 July 1995
© 1995 IOP Publishing Ltd

491

PREPARATION AND CHARACTERISATION OF Tl₂Ba₂Ca₂Cu₃Oᵧ SUPERCONDUCTING TAPES

M. K. Al-Mosawi, D. M. Spiller, C. Beduz, and Y. Yang

Institute of Cryogenics, University of Southampton, Southampton, SO17 1BJ. UK.

D. M. Ogborne.

Department of Chemistry, University of Southampton, Southampton, SO17 1BJ. UK

Abstract. We present the results of the sintering time dependence up to 20h, and temperature dependence ($820 < T < 865°C$ in O_2) of transport J_c of Ag/Ag alloy sheathed TBCCO tapes produced by the powder in tube (PIT) method. The precursor powder used was 'partially calcined' $Tl_{1.8}Ba_2Ca_{2.2}Cu_3O_y$ comprising of Tl(2212) as the major phase. The influence of intermediate cold rolling and sintering temperature ($820 \leq T_s \leq 850°C$) on transport J_c was studied. The tapes were characterised by transport J_c measurements, XRD, SEM and optical microscopy. Transport J_c's of 4×10^3 Acm^{-2} at 77K (B=0) have been measured using the standard four point technique and a voltage criterion of $1 \mu Vcm^{-1}$. From XRD analysis we have seen an enhancement of alignment of the TBCCO core after the first and second intermediate cold rolling, compared to non rolled tapes. Full conversion to the TBCCO 2223 phase has been achieved at T=820°C, however, at T>850°C complete conversion to the 2223 phase was not possible due to the formation of the secondary phases.

1. Introduction

Up to now most of reported work on Tl based tapes have been focused on Tl-1223 system[1,2] in an attempt to utilise its strong irreversibility. However, a subsequent drawback arises due to the difficulty to align the material. There have been a number of reports on tapes produced with Tl-2223 which appears to have high anisotropy in the grain morphology. Such a property should be advantageous for texturing the material by cold working which is beneficial to obtaining high Jc's.. A partially calcined precursor powder is widely used[3,4] in the (PbBi)2223 system. The PbBi-2223 phase being formed within the tape utilising the constraints of the wall of silver sheath to enhance the alignment of the PbBi-2223 phase. In this paper we report on the use of a 'Partially calcined' precursor TBCCO-2223 powder. We also report the effect of the intermediate cold rolling on transport *Jc* and microstructure of the Tl(2223) tapes.

2. Experimental

The starting powder used in this study was partially calcined $Tl_{1.8}Ba_2Ca_{2.2}Cu_3O_y$ which comprises of Tl-2212 as major phase, plus CuO and CaO. The powder was then packed into a silver tube (OD=3.5, ID=3.0). and then drawn to a wire of 1.2mm diameter. The silver sheathed wire was then inserted into a silver alloy tube (consisted of 99.5% Ag, 0.23% Ni, 0.25% Mg). The tube was then drawn to OD 1.48mm and was rolled to tape 180μm-120μm. The tapes were sintered in 100%O$_2$ at $820 \leq T_s \leq 850°C$ for 10min<ts<20h. Tapes were also treated at 820°C for 1 hour (O2) followed by an intermediate cold rolling and further treated at $820 \leq T_s \leq 850°C$ for time up to 10 hours. Transport current measurements were carried out on 40-50mm long samples by using the standard four probe method at 77K, using a voltage criterion of 1μVcm^{-1}. Microstructural analysis and phase formation analysis were conducted using Optical microscopy, Scanning Electron Microscopy (SEM), and X-ray Diffraction (XRD).

3. Results and Discussion

Figure (1, a-c) shows the XRD traces of the Tl(2223) tapes sintered at $T_s = 820°C$, 845°C and 865°C for sintering times up to 10h. From figure (1a) it can be seen that there is no conversion to the Tl(2223) phase when the tapes were sintered at 820°C for sintering times up to 10h. Figure (1b) shows the XRD traces for tapes sintered at 845°C, it is clear that almost full conversion to the Tl2223 phase was achieved after 3h sintering. For the tapes sintered at 865°C full conversion to the Tl phase was achieved after only 10 min. However closer examination of the XRD traces reveals that as the sintering time is increased there is a significant amount of the secondary phase BaCuO$_2$ formed. From the XRD traces in figure (1, a-c) it is clear that the Tl(2223) phase is randomly aligned.

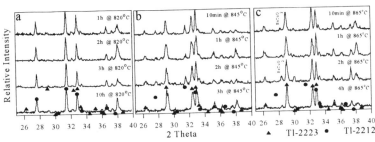

Figure (1) shows the XRD traces of the unrolled Tl(2223) tapes sintered at T_s = 820°C, 845°C and 865°C for sintering times up to 10h.

Figure (2) shows the SEM micrograph of the tapes sintered for 1h at 820°C followed by an intermediate cold rolling. It can be seen that the rolling has broken down the Tl(2212) grain structure obtained after the 1h at 820°C, the grain size of the as rolled tape being ~1-2μm.

Figure (3, a-c) shows the XRD traces of the tapes sintered for 1h at 820°C followed by an intermediate cold rolling and then further sintering at $820°C \leq T_s \leq 865°C$ for sintering times up to 10h. The XRD traces of the rolled tapes sintered at 820°C revealed that the core consisted of the Tl(2223) phase which seems to be inconsistent with the unrolled tapes, figure (1a). The reason for the conversion to the Tl(2223) phase after the intermediate rolling and further sintering at 820°C is not fully understood. The rolling can be seen from figure (2c) to have densified the core and may have improved the homogeneity of the distribution of the liquid phases (CaO and CuO) within the core. The improvement in the homogeneity of the core may increase the surface area of the solid Tl(2212) phase covered by the liquid phase thus increasing the amount of conversion after rolling. XRD traces of the tapes sintered at 845°C and 865°C after the intermediate cold rolling reveal that the conversion rate has increased in comparison to the unrolled tapes. Further, the amount of the secondary phase BaCuO$_2$ is also seen to increase with increased sintering time. However, the XRD traces of the rolled tapes also shows the OO$\underline{14}$ peak of the Tl(2223) core has been enhanced relative to the unrolled tapes. The enhancement of the OO$\underline{14}$ peak is a good indication that the intermediate cold rolling has had the desired effect of aligning the Tl(2212) grains and thus enhancing the Tl(2223) grain alignment after conversion.

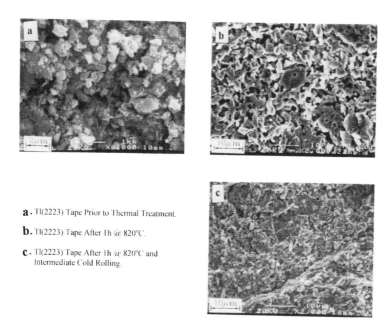

a . Tl(2223) Tape Prior to Thermal Treatment.

b . Tl(2223) Tape After 1h @ 820°C.

c . Tl(2223) Tape After 1h @ 820°C and Intermediate Cold Rolling.

Figure (2) is SEM micrographs showing the effect of the intermediate cold rolling.

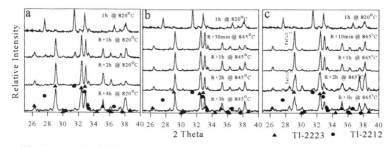

Figure (3) shows the XRD traces of the 'Rolled' tapes sintered for 1h at 820°C followed by an intermediate cold rolling and then further sintering at 820°C ≤ T$_s$ ≤ 865°C for sintering times up to 10h

494

Figure (4a) shows the effect of sintering temperature on the self field transport Jc, 77K for the unrolled tapes. As can be seen the transport Jc of the tapes sintered at 820°C is considerably lower than those sintered at 845°C and 865°C. This is probably due to the lack of conversion at 820°C as seen from the XRD traces in figure (1a). The transport Jc of the tapes sintered at 845°C and 865°C can be seen initially to increase to a peak value at 2h due to the increase in the Tl(2223) phase. After the peak, the decrease in the transport Jc may be due to the increase in the segregation of the secondary phase $BaCuO_2$ to the grain boundaries. The increase in the amount of $BaCuO_2$ can be seen in the XRD traces in figure (1). For the tapes with an intermediate cold rolling, an enhancement in the Jc values can be seen. This improvement can be attributed to the improved grain alignment and core homogeneity.

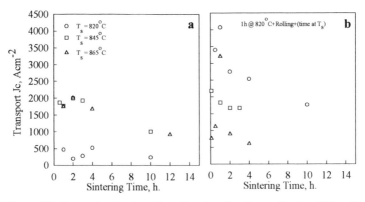

Figure (4) shows the Jc-sintering time for the a) unrolled and b) rolled tapes.

4. Conclusion

It can be concluded that the intermediate cold rolling has increased the density and homogenised the ceramic core. The improvement in the homogeneity of the core may increase the surface area of the solid Tl(2212) phase covered by the liquid phase thus increasing the amount of conversion after rolling. Furthermore, the intermediate cold rolling may be beneficial to enhance the grain alignment, thus improve the Jc. It appears that as the sintering temperature and time is increased there is a significant amount of the secondary phases formed. These secondary phases may effect the grain boundaries, therefore degrade the transport Jc.

5. References

1. P.J. Kung, P.G. Wahlbeck, M.E. McHenry, M.P. Maley and D.E. Peterson, *Physica C 220* (1994) 310-322.
2. B.A. Glowacki and S.P. Ashworth, *Physica C 200* (1992) 140-146.
3. D. M. Spiller, M.K. Al-Mosawi, C. Beduz, Y.Yang and R. Riddle. *EUCAS 1995*, Edinburgh, UK.
4. M. Penny, C.Beduz, Y.Yang, M. Al-Mosawi, R. Scurlock and R. Wroe. *EUCAS 1995*, Edinburgh, UK.

Inst. Phys. Conf. Ser. No 148
Paper presented at Applied Superconductivity, Edinburgh, 3–6 July 1995
© 1995 IOP Publishing Ltd

Effect of thermal cycling on wound HTS conductors

D K Hilton, D K Hamilton, J W Howton and Y S Hascicek

National High Magnetic Field Laboratory, Florida State University, Tallahassee,
FL 32306-4005, USA

Abstract. Bi-2223/Ag and Bi-2212/Ag tape conductors were thermal cycled between room temperature (RT) and liquid helium (LHe) temperature in He ambient then between RT and liquid nitrogen (LN) temperature in air. No degradation in critical current density (J_c) of these conductors was observed. Bi-2223/Ag tape conductors were wound on mandrels of varying diameter at RT and then thermal cycled between RT and 4.2K in He ambient. Apart from degradation in J_c due to bending, no further degradation occurred by thermal cycling in He ambient. Microstructural investigations showed that transverse cracks were present in the oxide core of the wound samples.

1. Introduction

Superconductors are utilized in applications that have to withstand repeated thermal cycling from RT to LHe temperature. Some of the applications for High Temperature Superconductors (HTS) may be at LN temperature of 77K, but they will still have to be cycled to this temperature from RT. There have been some reports that indicated that the J_c of HTS conductors would degrade by thermal cycling [1]-[4]. It was shown by the present authors that the J_c of powder in tube processed (PIT) Bi-2223/Ag and Bi-2212/Ag tape and wire conductors did not degrade if the thermal cycling was carried out in helium ambient [5]-[8]. It is also known in the scientific community that the J_c of some of the HTS coils degrade in time due to aging, thermal cycling environmental effects, or a combination of all or some of the above. A systematic study of these effects was undertaken at The National High Magnetic Field Laboratory. The results of thermal cycling between RT and LHe temperature in helium ambient, in air (to 77K only) of short samples of HTS conductor, and wound HTS conductors are presented.

2. Experimental

The Ag-sheathed BSCCO wire and tape samples used in this investigation were produced by Intermagnetics General Corporation (IGC) by the powder-in-tube (PIT) process [9]. The ratio of silver to BSCCO is about 3:1 by volume. Care was taken not to introduce any tension or bending in these straight samples.

Fully heat treated short samples of about 5 cm in length 0.1 mm in thickness and 5 mm in width, Bi-2223/Ag tape conductors were tested for their $J_c(T)$ profile at zero applied field in a helium storage dewar by a standard four probe technique with a 1 μV/cm criterion. Once their virgin $J_c(T)$ profiles were obtained, the samples were cycled in the same storage dewar up to 55 times between RT and 4.2K. $J_c(T)$ profiles showed no degradation by thermal cycling in helium ambient. Thermal cycling was then carried out by gently immersing the same sample in liquid nitrogen in a thermal flask in air up to 40 times. After liquid nitrogen thermal cycles, a last $J_c(T)$ profile was measured in He ambient down to 4.2K.

Furthermore, after measuring the virgin $J_c(T)$ profile, a short straight Bi-2223/Ag fully heat treated tape conductor was wound at room temperature on mandrels of decreasing diameter and thermal cycled at each diameter in helium ambient between RT and 4.2K. Samples of longitudinal sections of the wound Bi-2223/Ag tape samples were prepared by standard metallographic techniques and examined with Scanning Electron Microscope (SEM).

3. Results and discussions

Effects of thermal cycling on the Jc of silver sheathed BSCCO wire and tape conductors were presented elsewhere[8]. Here J_C versus temperature profiles of a Bi-2223/Ag are presented in Fig. 1. As seen from Fig. 1 there is no degradation in J_C due to thermal cycling. These samples were chosen such that they did not have the notorious pin holes or bubbles. It is shown that degradation in J_C by thermal cycling is not an intrinsic problem of these composite HTS conductors, but if the pin holes or the bubbles are present the samples will degrade by thermal cycling.

Fig. 2 shows a Jc versus temperature profiles for a Bi-2223/Ag monocore tape sample as it was thermal cycled between RT and LHe temperature at varying strain levels in He ambient. Bending surface strain dependence of J_C of these conductors have been presented elsewhere [6], [7] and is well known. But as seen from these curves in Fig. 2, apart from degradation in J_C due to bend strain, no additional degradation in J_C due to thermal cycling was observed.

Microstructural investigations by SEM on the longitudinal sections of wound HTS conductors revealed that there are transverse cracks in the oxide core. In all cases cracks initiated at the tension surface of the oxide core and propagated radially. Occasionally small normal cracks (Normal to the main radial crack direction) were observed. Fig.3 shows a typical SEM micrograph of a longitudinal section of a wound Bi-2223/Ag tape sample. Fig. 4 shows an SEM micrograph of one of these transverse cracks in a longitudinal section of a Bi-2223/Ag tape sample. Note the fine normal crack running towards the top of the page in the middle of the micrograph.

4. Conclusions

No degradation in critical current density (J_C) was observed when Bi-2223/Ag and Bi-2212/Ag tape and wire conductors were thermal cycled between room temperature (RT) and liquid helium temperature in He ambient then between RT and liquid nitrogen temperature in air. Bi-2223/Ag tape conductors were wound on mandrels of varying diameter at RT and then thermal

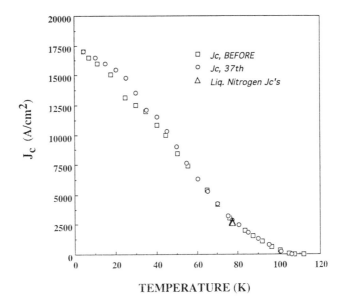

Fig. 1. Jc versus temperature profiles for a Bi-2223/Ag tape HTS conductor before and after thermal cycling from RT to 4.2K and 77K (Note that all of the 77K data appear as one data point at 77K).

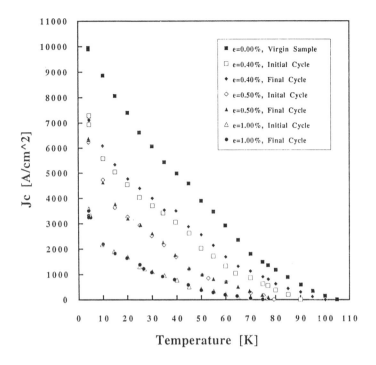

Fig. 2 Jc versus temperature profiles for a Bi-2223/Ag wound tape conductor before and after thermal cycling at varying winding radii. Only thermal cycles at there strain levels are shown here for clarity.

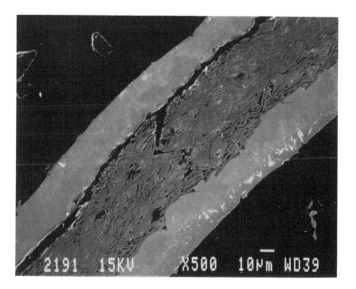

Fig. 3 A typical SEM micrograph of a longitudinal section of a wound conductor. Sections were prepared after the sample was wound on the smallest mandrel (1.0% surface strain at RT).

498

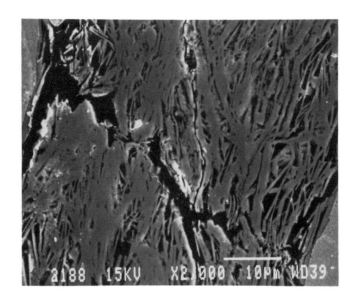

Fig. 4 An SEM micrograph of one of the radial cracks in the longitudinal section of a wound Bi-2223/Ag tape sample (1.0% surface strain at RT). Note the fine normal crack running towards the top of the page in the middle of the micrograph.

cycled between RT and 4.2K in He ambient. Apart from degradation in J_c due to bending, no further degradation occurred by thermal cycling in He ambient. Therefore thermal cycling not only does not degrade Jc of these HTS conductors, but also does not even cause the propagation of existing cracks in the wound samples, provided that no pin holes and bubbles are present. Contrary to the belief, the degradation of Jc is not the intrinsic property of these composite HTS conductors. Microstructural investigations showed that transverse cracks were present in the oxide core of the wound samples.

Acknowledgment

The authors would like acknowledge the practical help with the SEM work by T J Fellers. This project is supported by the State of Florida and the National Science Foundation through NSF Cooperative Grant No.DMR 9016241.

References

[1] Ochiai S, Hayashi K and Osamura K 1991 Cryogenics **31** 954-961
[2] Jenkins R G et al 1993 Cryogenics **33** 81-85.
[3] Willis J O et al HTS Wire Development workshop, US DOE conf-940278, Feb. 16-17, 1994.
[4] Huang S et al 1995 accepted for pub in Materials Lett.
[5] Van Sciver S W et al Advances in Superconductivity-V, Proc the 5th Intern Symp on Superconductivity (ISS'92), Nov. 16-19, 1992. Springer-Verlag, pp. 818-22.
[6] Hascicek Y S et al 1994 IEEE Trans.Magnetics **30** 2229-2232
[7] Hascicek Y S et al 1993 Processing of Long Length of Superconductors TMS pub. Ed Balachandran, Collings and Goyal
[8] Hilton D K Hascicek Y S 1995 accepted for pub. in IEEE Trans. Appl. Superconductivity.
[9] Haldar P et al 1993 IEEE Trans. Appl. Superconductivity **3** 117-1130

Inst. Phys. Conf. Ser. No 148
Paper presented at Applied Superconductivity, Edinburgh, 3–6 July 1995
© 1995 IOP Publishing Ltd

Growth of HTSC films with high critical currents on polycrystalline technical substrates

A I Usoskin[1], H C Freyhardt[1,2], F García-Moreno[1,2], S Sievers[1], O Popova[1] K Heinemann[2], J Hoffmann[2], J Wiesmann[2] and A Isaev[2]

[1] Zentrum für Funktionswerkstoffe gGmbH, Windausweg 2, D-37073 Göttingen;

[2] Institut für Metallphysik, Universität Göttingen, Hospitalstr. 3-7, D-37073 Göttingen.

Abstract. Quasi-equilibrium substrate heating and an improved technique of target ablation were employed to grow HTSC films by pulsed laser ablation. A 0.15 mm thick polycrystalline yttria-stabilized zirconia ribbon and polycrystalline Ni foils were used as substrates. Both of these substrates were covered with YSZ buffer layers, which were produced by ion-beam-assisted deposition. High levels of j_c of $\geq 5 \cdot 10^5$ A/cm^2 were observed for HTSC films with a thickness of ≥ 400 nm. The formation of a "dead" layer at the initial stages of YBCO film growth was considered as a reason of deterioration of j_c at lower thicknesses.

1. Introduction

A lot of efforts are undertaken to grow high quality HTSC films on non-single-crystalline substrates, which allow electrical and power engineering applications. The highest values of critical current densities of $\geq 8 \cdot 10^5$ A/cm^2 at 77 K for YBa$_2$Cu$_3$O$_{7-x}$ (YBCO) films on Ni substrates with textured buffer layers have been demonstrated by Wu et al. [1]. Nevertheless, critical current densities for films grown on such "technical" substrates are still significantly below those for films deposited on single crystalline substrates. Apparently, a further development of the methods of high-j_c films preparation is necessary not only for metallic technical substrates but also for dielectric ones as they can provide new areas of HTSC film applications such as magnetic field transducers, cheap bolometers, HF cavities, field controlled power devices, fault current limiters, etc..

In the present paper we report on new results recently obtained for YBCO film growth on polycrystalline yttria-stabilized zirconia (YSZ) flexible ribbons and on nickel foils.

2. High-j_c film deposition

A 0.15 mm thick polycrystalline yttria-stabilized zirconia (YSZ) ribbon and polycrystalline Ni foils were used as substrates. The mirror-like surface of the substrates were prepared either by chemical mechanical polishing in the case of YSZ ribbons or by electrochemical polishing in the case of Ni foils.

Both of these substrates were covered with YSZ buffer layers, which were produced by ion-beam-assisted deposition (IBAD) [2]. The full width at half-maximum of x-ray Phi-scan of the (111) peaks of such textured layers was typically 25°-30°. For the Ni foil, a non-textured YSZ buffer film prepared by pulsed laser deposition was also used.

A conventional pulsed laser deposition process using an excimer laser XP 2020 (308 nm, 2J/pulse) was employed for high j_c film preparation. However, two of the principle features in which the present method differs from others will be shortly described.

The first one is a usage of a quasi-equilibrium heating technique [3] which is capable of a precise maintenance of the substrate temperature with an accuracy of 0,5°C, independently on variations of the IR emissivity of the film during its deposition. Another problem, which is solved by this method, is that no mechanical bonding of a substrate with the heater is required, and, therefore, long pieces of tapes can be used as substrates.

The second one is the use of devices providing an improvement of surface uniformity during target ablation. As is well known, a deterioration of the flatness of the ablated target surface results in variations of the angular direction of the laser plume. Due to a non-uniform erosion of the target material the angular deviation can reach 5°-10° for 4 cm^2 targets after 30 000 ablation pulses. Obviously, this creates a serious problem for the deposition of thick films as well as for HTSC film coverage of large surfaces, because of instabilities of the deposition rate, and a deterioration of the film quality when not a central part of laser plume is used for the deposition [4].

Two different kinds of scanning of disk-like target were examined for a long-term ablation process. The first of them provided a 3-way target scanning with a laser beam due to (a) a target rotation, (b) a smooth rotational motion of the laser spot over the target surface, and (c) a periodical variation of the position of the last rotation center provided by a movement of the target in the plane of its ablated surface. The second, 2-way target scanning arrangement, included only (a) a target rotation and (d) an "in-plane" target oscillations. To get the maximal smoothness of the target ablation in this case, the following approximation of the radial dependence for an oscillatory motion (d) was used:

$$r(\tau) = R_0 \frac{\sqrt{|\sin \pi f \tau|}}{1 + A |\sin 2 \pi f \tau|} .$$

to lead to Here r denotes the radial position of the central part of laser spot relative to the target rotation axis at the time τ, R_0 is the radius of the target, f is the frequency of the target oscillation, and A is a constant which equals to ~ 0.20. In a comparison of these two schemes of target scanning, the 2-way arrangement was shown to lead to minimal deflections of the laser plume of less than 1° after 50 000 laser shots with the energy density of 2.5 J/cm^2 and with an ablation spot diameter of 3 mm over an YBCO target with $R_0 = 1$ cm, while at the optimal parameters for a 3-way technique only 3° deflections can be reached. Therefore, the 2-way scanning device was finally used for the YBCO film deposition.

3. Critical currents

Measurements of the superconducting parameters were performed by a 4-probe pulsed d.c. technique with a duration of the current the pulse of ~ 30 μs, as well as by a magnetic-shielding a.c. method operating at the frequency of 100 KHz. Thickness dependences of the measured critical current densities are shown in figure 1. It is obvious that a significant difference in the j_c behaviour for SrTiO$_3$ and technical substrates is observed. For the latter ones a characteristic decrease occurs of j_c at thicknesses below 400 nm, while in the case of SrTiO$_3$ no reduction of j_c can be found, at least, in the considered range of thicknesses. One can conclude that at the initial stages of YBCO film growth on Ni or YSZ substrates some kind of inhomogenous structure is formed. It is possible to expect that in between the substrate (with or without buffer layer) and the high-j_c film which is created at the further deposition a "dead" layer appears. Such a layer can confine the main "transporting" part of HTSC film to a smaller thickness as the measured one. Further investigations are required to clarify this problem.

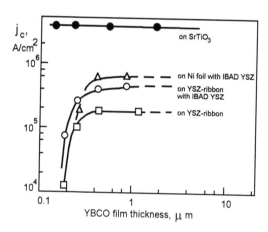

Figure 1. Thickness dependence of j_c for YBCO films various substrate/buffer systems

Maximal values of critical parameters measured for various combinations of substrates and buffer layers are shown in table 1. The typical structure of YBCO films which is shown in Figure 2 seems to be very dense and poreless. The full width at half-maximum of x-ray Phi-scan of the (103) peaks for such a film on YSZ ribbon with IBAD-YSZ buffer layer corresponds to $\sim 20°$.

Table 1. j_c and T_c of YBCO films grown on various substrate/buffer systems

Substrate Buffer	YSZ ribbon –	YSZ ribbon IBAD-YSZ	Ni foil –	Ni foil YSZ	Ni foil IBAD-YSZ
j_c (77 K)	$2 \cdot 10^5$ A/cm^2	$5 \cdot 10^5$ A/cm^2	–	$1 \cdot 10^4$ A/cm^2	$6 \cdot 10^5$ A/cm^2
T_c	87 K	90 K	< 77 K	83 K	89.5 K

502

YBCO film

YBAD-YSZ
buffer layer

YSZ ribbon

2 μm

Figure 2. Cross-section of YBCO film on YSZ polycrystalline substrate with
IBAD-YSZ buffer layer (SEM picture)

The measurements demonstrate a considerable improvement of the quality of HTSC films deposited on technical substrate materials, especially in the case of Ni foil when an IBAD-YSZ buffer is used. The observed j_c values for Ni foils are in agreement with those previously reported ($5 \cdot 10^5$ A/cm^2 at 75 K by Wu et al. [1]. The results indicate that similar critical current densities can be reached for dielectric YSZ ribbons. With $5 \cdot 10^5$ A/cm^2 they are the highest ones reported for these substrates.

In summary, we have demonstrated that HTSC films with critical current densities of $\geq 5 \cdot 10^5$ A/cm^2 at 77 K can be deposited not only on metallic foils but also on flexible dielectric ribbons. The introduced improvements of the pulsed laser deposition technique allow the development of a long-term deposition process for YBCO films on large areas of technical substrates of various shapes.

This work is supported by the BMBF (projects No. 13N6482 and No. 13N6565) and by Siemens AG.

References

[1] Wu X D, Foltyn S R, Arendt P, Townsend J, Adams C, Campbell I H, Tiwari P, Coulter Y and Peterson D E 1994 *Appl. Phys. Lett.* **65** (15) 1961-1963

[2] Iijima Y, Tanabe N, Kohno O and Ikeno Y 1992 *Appl. Phys. Lett.* **60** 769-771

[3] Usoskin A I, Freyhardt H C, Neuhaus W and Damaske M 1994 *Critical Currents in Superconductors* (Singapore: World Scientific) 383-386; Patent DE No. 42 28 573 Cl. C23C14/28 1994

[4] Davis M F, Wosik J, Forster K, Deshmukh S C, Ramoersad H R, Shah S, Siemsen P, Wolfe J C and Economou D J 1991 *J. Appl. Phys.* **69** (10) 7182-7188; Patent US No. 5 015 492 C23C14/00 1991

Inst. Phys. Conf. Ser. No 148
Paper presented at Applied Superconductivity, Edinburgh, 3–6 July 1995
© *1995 IOP Publishing Ltd*

Biaxially textured YSZ and CeO$_2$ buffer layers on technical substrates for large-current HTS-applications

J Wiesmann[1], J Hoffmann[1], A Usoskin[2], F García-Moreno[2], K Heinemann[1,2] and H C Freyhardt[1,2]

[1]Institut für Metallphysik, Universität Göttingen, Hospitalstr. 3-7, D-37073 Göttingen
[2]Zentrum für Funktionswerkstoffe gGmbH, Windausweg 2, D-37073 Göttingen

YSZ (Yttria Stabalized Zirconia) and CeO$_2$ buffer layers were prepared with a dual ion gun system (IBAD) on different metallic and ceramic substrates. The best YSZ films show in-plane alignment with half widths (FWHM) of about 20° in (111)-Φ-scan. The quality depends strongly on the surface roughness of the substrate. A buffer deposited on an area of 3×5 cm^2 without moving the substrate holder showed a FWHM of better than 30°. CeO$_2$ could also be successfully aligned, with an in-plane texture characterized by a FWHM of 40°, in low oxygen partial pressures. Thus the window of deposition parameters for a biaxial alignment becomes much narrower as compared to YSZ. All buffer layers exhibit good diffusion barrier properties. On YSZ buffers, YBCO films with j$_c$ > 10^5 Acm^{-2} could be deposited. The best film on nickel had a j$_c$ of (4..6)·10^5 Acm^{-2}.On YSZ-tapes with YSZ buffer a favourable dependence of j$_c$ on the external B field was measured (j$_c$ (B=2T) > 10^4 Acm^{-2}).

1. Introduction

For the applicability of HTS films in large-current devices, it is necessary to grow them on large-area technical, metallic and non-metallic, substrates. In general a buffer layer between substrate and HTS film is necessary to prevent interdiffusion and to transmit a texture to the HTS film. This is the prerequisit for reaching a high j$_c$ value in the self field of the superconductor. A successfully prepared three-component system, based on a technical untextured substrate (Hastelloy), a thin biaxially-textured YSZ layer and a YBa$_2$Cu$_3$O$_{7-x}$ film, was reported for the first time by Ijima et al. in 1992 [1]. Till now Foltyn et al. reached the best j$_c$ values of 1.3·10^6 A/cm^2 in YBCO on Nickel with an IBAD-YSZ buffer [2]. The IBAD (Ion-Beam-Assisted-Deposition) technique with two Kaufman ion sources is used in the present investigation for the deposition of the buffer layers.

2. Experimental

As technical substrates, Hastelloy plates (alloy based on Ni,Cr,Mo,Fe), nickel foils, YSZ tapes and polycrystalline Al$_2$O$_3$ plates were used. To reach the desired surface roughness of below 60 nm, the Hastelloy was mechanically polished finally with an alumina suspension of a grain size of 40 nm. Commercially polished YSZ tapes and Al$_2$O$_3$ plates with surface roughnesses below 30 nm were only cleaned. The nickel foils were electrochemically polished to roughnesses of below 100 nm. Further experiments were made on single crystalline silicon substrates with a native SiO$_2$ layer. Silicon is suitable for fundamental investigations of the

504

growth mechanisms during IBAD because of its smooth surface. In order to improve the adhesion between substrate and buffer and to remove some ad-atoms from the surface, the substrates were etched in addition in-situ with an Argon-ion beam of 400 eV for a few minutes before the buffer deposition. The CeO_2- and YSZ- buffer layers were deposited in a dual-ion-source system (Figure 1). The two ion sources are of the Kaufman type with diameters of 2.5 cm. The first one produces Xe ions, which are focussed onto the water-cooled target and the second one directs an Ar-ion beam onto the substrates in a distance of 20 cm away from Ar source. The assisting Ar beam has an energy of 300 eV and a density at the substrates of 380 µA/cm^2. The angle between the beam and the substrate normal is fixed at 55°.

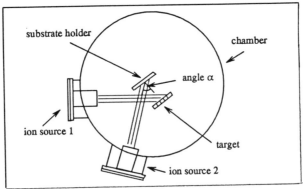

Figure 1: Dual ion gun system

The deposition rate for the buffer film was reduced from 0.12 nm/s by about 50% due to the additional resputtering from the IBAD source. The film thickness for all sample was around 250 nm. For complete oxidation of the layers an oxygen partial pressure of $1\cdot10^{-4}$ mbar was maintained during the process. The total pressure in the chamber reached $3\cdot10^{-4}$ mbar.

3. Results

The YSZ layers grown by this IBAD deposition technique reveal a (100) texture with a FWHM of the rocking-curve of about 6° for deposition temperatures lower than 100°C. The lattice constant increases from 5.22 Å without to 5.27 Å with IBAD.

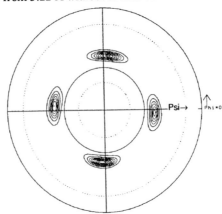

Figure 2: (111)-polefigure of a biaxially textured YSZ film on Hastelloy, 4 peaks at $\psi = 55°$

The unit cell is extended because of the incorporation of Ar atoms. In EDX measurements 2 at-% Ar could be detected in the buffer layers. (111) pole figures were recorded for (200)-textured YSZ buffers on various substrate materials. The pole figure of a layer on Hastelloy (see Figure 2) shows four peaks with a FWHM of about 23°, which demonstrates the in-plane alignment. The assisting ion beam during deposition is always parallel to <111> direction in YSZ. The in-plane texture of YSZ on different kinds of substrates is summarized in Table 1. The degree of texture depends on the surface roughness. The best results are observed for depositions on the very smooth silicon wafer. The well polished dielectric ceramic substrates are a little bit better than the metallic ones. But on all used substrate materials a FWHM below 35° can be obtained, whereby the film thickness is not changed. In an area of 3×5 cm^2 the buffers exhibit a texture better than 30° on ceramic and 35° on metallic substrates, respectively.

Table 1: In-plane texture of YSZ buffers on different kinds of substrates

Substrate	roughness [nm]	FWHM Φ-scan
Silicon	< 5	19
Hastelloy	60	23
Al$_2$O$_3$	30	21
YSZ-tape	30	<25
Ni	100	30

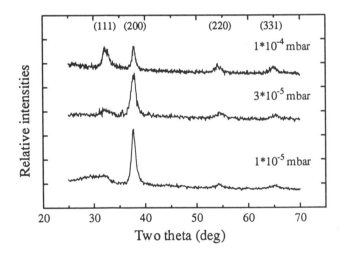

Figure 3: CeO$_2$ buffer layer, deposited at different oxygen partial pressures

CeO$_2$ buffer layers deposited by employing the same IBAD parameters as for YSZ developed a (111) texture. But a reduction of the oxygen partial pressure ,p$_O$, down to 10^{-5} mbar led to the preferred (200) texture (see Figure 3) with an in-plane alignment of around 40°. In comparison to YSZ, the window of deposition parameters for achieving an in-plane texturing is much narrower.

It could be furthermore shown that the in-plane alignment of the grains improves the properties of the buffer layer as a diffusion barrier. AES-depth-profile analyses on the system YBCO / YSZ or CeO$_2$ / Hastelloy show that no interfacial reactions occur [3,4].

506

The superconducting properties of YBCO films on various substrates are summarized in Table 2 [5].

Table 2: Superconducting properties of YBCO on YSZ / technical substrate

Substrate	T_c [K]	j_c [A/cm^2] (77K) at 0 T	at B-field
Hastelloy	90	$1 \cdot 10^5$	$1 \cdot 10^4$ (0.3 T)
YSZ-tape	90	$4 \cdot 10^5$	$1 \cdot 10^4$ (2 T)
Ni	89.5	$5 \cdot 10^5$	-

4. Discussion and conclusion

Highly textured YSZ buffer could be prepared on different ceramic and metallic substrates by an IBAD-deposition. Allthough the critical current densities of YBCO-films on this buffers are rather high (i.e. $j_c = 5 \cdot 10^5$ Acm^{-2} on Ni) a further improvement up to densities of $j_c > 10^6$ Acm^{-2} seems to be possible for two reasons: Firstly a reduction of the surface roughness of the substrates improves the in-plane texture and thus should result in higher critical current densities. Secondly, the growth mechanism for IBAD layers, proposed by Sonnenberg et al. [6], predicts an improvement of the texture with film thickness. The experimental data, published by [1] and [2], seem to be in agreement with this prediction, because the in-plane texture of the YBCO films was always better than the one of the YSZ films. So far mainly films with a thickness of only 250 nm were investigated, but first results on thicker films indicate indeed a reduction of the FWHM for these samples. Up to now, the buffers could be in-plane aligned in an area of 3×5 cm^2. For an extension of the experiments to long tape lengths a movement of the substrates will soon become possible and for larger substrates up to 400 cm^2 the used deposition process will be scaled up. The additionally investigated buffer material CeO$_2$ can represent an alternative to YSZ because of the smaller misfit to YBCO and better diffusion properties due to the missing structural vacancies on the O sites. Allthough the range of parameters for a successful deposition is much smaller an in-plane texture could be found that will be optimized in the near future. This work was supported by the BMBF under grant numbers 13N6009 and 13N6482. Financial support by the Siemens AG is gratefully acknowledged.

5. References

[1] Iijima Y, Tanabe N, Kohno O and Ikeno Y, 1992, *Appl. Phys. Lett.* **60** 769
[2] Foltyn S, 1995, *HIGH T$_c$ Update vol. 9* **10** 1
[3] Wiesmann J, Heinemann K, Damaske M, Usoskin A, Freyhardt H C, Simon T and Neuhaus W, 1993, *Appl. Supercond. vol. 1* (Oberursel: DGM Informationsgesellschaft 627-629
[4] Simon T, 1995, PhD thesis (University of Göttingen)
[5] Usoskin A, this conference
[6] Sonnenberg N, Longo A S, Cima M J, Chang B P, Ressler K G, McIntyre P C and Liu Y P, 1993, *J. Appl. Phys.* **74** 1027

Inst. Phys. Conf. Ser. No 148
Paper presented at Applied Superconductivity, Edinburgh, 3–6 July 1995
© 1995 IOP Publishing Ltd

Improvement on J_c transport of the quaternary $Pb_{0.6}Sn_{0.4}Mo_6S_8$ Chevrel phase wire

N. Cheggour, A. Gupta, M. Decroux, J.A.A.J. Perenboom[*], P. Langlois[**], H. Massat[**], R. Flükiger and Ø. Fischer

DPMC - GAP, Université de Genève, 1211 Genève 4 (SUISSE)
[*]High Field Magnet Laboratory, University of Nijmegen, 6525 ED Nijmegen (The Netherlands)
[**]Lab. de Physicochimie des matériaux, CNRS UPR 211, 92195 Meudon (France)

Abstract. Chevrel phase wire with high quality powder was made under very careful conditions. To compare the effect of the Hot Isostatic Pressing (HIP) to the normal annealing upon J_c performances, small coils of about 1 meter wire length were heat-treated at 900 °C for 0.5 hour, under 1.9 kbar argon pressure or at ambient pressure. They were measured in a magnetic field from 6 up to 15 tesla. The critical current of the HIP treated coil increased by almost a factor 2.5, reaching the record value of 5.0×10^8 A/m^2 at 15 tesla and 4.2 kelvin, and staying higher than 10^9 A/m^2 up to 10 tesla. The effect of the HIP treatment was essentially a superconducting powder compaction.

1. Introduction

The production of high magnetic fields is one of the important applications of the superconductivity. After the NbTi and $(Nb,Ta)_3Sn$ magnets, there is now a need to develop a third superconducting wire to generate magnetic fields above 20 tesla. With a critical field of 55 tesla at 4.2 kelvin, the $(Pb,Sn)Mo_6S_8$ Chevrel phase (CP) material satisfies largely the first requirement for such applications [1]. However, with respect to the best results reported so far [2,3], important work has still to be done that the overall critical current reaches the practical value of 100 A/mm^2 at the operating field. Our effort was concentrated on making the powder with micron size grains without any mechanical grindings, under controlled and reproducible conditions to reduce oxygen and carbon contamination. In this paper, we report the benefit of this approach together with the wire heat treatment under HIP conditions. We will evoke, in the case of the studied wire, the effect of a powder densification during the first stages of the billet deformation. Finally, we will briefly comment very recent critical current density results at high magnetic fields, obtained on similar type of wire.

2. Wire manufacturing and J_c measurements

We synthesise the $Pb_{0.6}Sn_{0.4}Mo_6S_8$ powder starting from Mo_6S_8 stoichiometric precursor. This powder has a very good quality, with a regular grain size of about 2 microns [4]. In the wire studied in this work, 0.2% at of Sn were added to the powder . The wire is made from a CP/Nb/Cu billet, using the powder in tube method. A precompaction of the powder allows to achieve an initial density of 65%. The cold isostatic extrusion is carried out to reduce the billet diameter from 20 to 10 mm, followed by the conventional drawing process at ambient temperature. At a diameter of 2 mm, the wire is reinforced by a stainless steel tube and drawn down to 0.6 mm. For the heat treatment, 1.3 meter of wire is winded on a stainless steel holder of 44 mm diameter. This small coil is then carefully transferred to an insulating holder for the J_c measurement. To compare the effect of the HIP to the normal annealing upon J_c performances, two samples were heat-treated at 900 °C for 0.5 hour, at ambient pressure (coil #1) or under 1.9 kbar argon pressure (coil #2).

3. Results and discussion

3.1. Powder density

Transversal micrographs for both samples are shown in figures 1 and 2. The darkness of the CP core for the non HIPed sample reveals a rather low powder densification. From longitudinal micrographs one can observe a multitude of cracks along the CP component which contribute to this low density. With the HIP treatment, the presence of cracks is strongly reduced, allowing an almost fully densified core. The superconducting cross section after the HIP treatment is about 55% of that of the non HIPed sample, for which the superconducting area did not change with respect to the non annealed wire. Assuming that the powder density is about 100% for the HIPed sample and there is no loss of mass during the annealing, the CP powder density after the wire deformation can be approximately deduced from the ratio of the CP cross section between the HIPed and non HIPed sample. This density is found to be of only 55% or even less, a value which is finally smaller than the initial density after the powder precompaction.

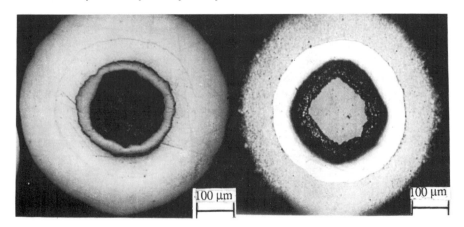

Figures 1 & 2: Transversal micrographs of the non HIPed and HIPed samples

The CP grains may not reach their domain of plasticity at ambient temperature. During the drawing process, the powder moves inside the Nb tube without undergoing substantial deformation. The extrusion as a first operation of the billet deformation allows to achieve a CP powder density between 80 to 90% [5]. With such high density, the "powder mobility" is strongly reduced. Free spaces in a form of cracks tend to develop during the drawing process, leading to a continuous decrease of the powder density.

As we will see later, the CP density is a very important parameter which strongly influence the J_c performances. According to these considerations, the powder densification has to be performed during the heat treatment or eventually at the last stages of the deformation process rather than at the beginning.

3.2. J_c results

The J_c dependence on magnetic fields up to 15 tesla, at the temperatures 4.2 and 1.8 kelvin, are shown in figure 3 for both coils #1 and #2. The coil #1 has a J_c at 15 T and 4.2 K of 2.1 x 10^8 A/m^2. For the HIPed coil (#2), J_c is increased by a factor 2.5, reaching at the same field 5.0 x 10^8 A/m^2 and 8.1 x 10^8 A/m^2 at 4.2 and 1.8 K respectively. These values are the highest ever reported in CP wires and demonstrate the importance of the powder densification, which increase the current flow cross-section.

The Kramer plot leads to $\mu_0 H_{c2}$ values between 31 to 33 T for both samples, which stay far from the expected bulk value. Moreover, the normalised volume pinning force dependence on field stays the same for the two treatments. We can conclude that in the case of the studied wire, the effect of the HIP treatment with the specifications mentioned above is purely a CP powder densification, without significant changes neither on the grain boundary limitations nor on the pinning mechanisms.

The U(I) measurement at 15 tesla and 4.2 K was performed up to relatively high voltage for coil #2 [6]. After a very short and steep round transition, U(I) characteristic is linear up to a current 64% higher than the critical current defined by 1 μV/cm criterion. Its extrapolation to zero voltage does not cross the origin of the axis and consequently, the U(I) linear behaviour does not correspond to an ohmic regime. This means that the transition to the resistive state was not complete. Some regions of the wire are still superconducting, with a J_c as high as 8 x 10^8 A/m^2. This shows that the wire has even better potential than the reported values in the present work.

3.3. Fulfilment of practical requirements for the high field production

On similar wires, we obtained very recently J_c values of 3.1 x 10^8 A/m^2 and 5.4 x 10^8 A/m^2 at 24 and 20 tesla respectively, at the temperature of 1.9 K. These record performances confirm the good potential of the CP wires at very high fields. Moreover, the overall J_c reached 60 and 105 A/mm^2 at the same fields and temperature, demonstrating very clearly that the CP wires are well adapted for the high field applications.

510

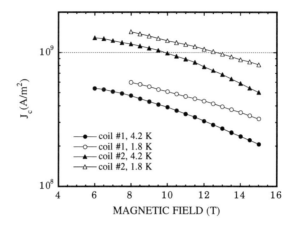

Figure 3 : $J_C (\mu_0 H)$ for coil #1 & coil #2, at 4.2 & 1.8 K

4. Conclusion

In this work, we have reported J_c record values on Chevrel phase wires. A factor 2.5 J_c improvement, shown up to 15 tesla, was achieved with the HIP treatment as compared to the normal annealing. The effect of this treatment was purely a densification of the CP powder. On the other hand, the CP powder density after the deformation process was found to be as low as 55%, smaller than the initial value. Because of the very bad ductility of the CP grains, the powder densification has to be achieved during the heat treatment or eventually at the last stages of the deformation process rather than at the beginning. Finally, we have reported very new results obtained on similar wires up to 24 tesla. With an overall critical current density of 60 and 105 A/mm^2 at 24 and 20 tesla respectively, these performances have to be considered as an important breakthrough for applications. We have demonstrated experimentally and for the first time, that the Chevrel phase wires fulfil the essential requirements for the achievement of the third generation of the superconducting magnets.

Acknowledgement

We are grateful to F. Liniger for the wire micrographs and J. L. Mas for the HIP treatment.

References

[1]   Ø. Fischer and B. Maple, eds., Superconductivity in Ternary Compounds I and II in Current Physics, Vol. 32 and 34, Springer-Verlag, Berlin, (1982).

[2]   W. Goldacker et al Proceedings ICMAS-90, 99-104 (1990).

[3]   G. Rimikis at al, IEEE Trans. Magn., Vol. 27, n° 2, 1116 (1991).

[4]   N. Cheggour et al, Proc. ICMAS-93, 403-408 (1993).

[5]   P. Rabiller, PhD Thesis n°655, University of Rennes (1991).

[6]   N. Cheggour et al, to be published.

Inst. Phys. Conf. Ser. No 148
Paper presented at Applied Superconductivity, Edinburgh, 3–6 July 1995
© *1995 IOP Publishing Ltd*

Inhomogeneities of fine-filamentary ac superconductors

M Majoroš D Suchoň

Institute of Electrical Engineering, Slovak Academy of Sciences, Dúbravská cesta 9, 842 39 Bratislava, Slovak Republik

M Polák

Applied Superconductivity Center, University of Wisconsin, 1500 Johnson Drive, Madison, WI 53706, USA

Abstract. Current-voltage characteristics of a 5 m long commercial fine-filamentary composite with filament diameter d_f=0.6 μm as well as those of 33 sections of this wire were measured. They have an exponential form between 0.01 μV/cm and 1 μV/cm and a downward curvature (in log-lin scale) was observed on some sample sections at E< 0.01 μV/cm. The variation of the critical currents as well as of the slope parameter were determined.

1. Introduction

The critical current density J_c of type-3 superconductors is always specified in terms of a measurement criterion. Its value is not uniquely defined and different criteria lead to different values of the critical current. Fine-filamentary composite wires for ac applications exhibit broad resistive transitions which vary with the applied magnetic field [1-3]. It has been appreciated that a broad resistive transition is caused by sample inhomogeneities and there has been considerable interest in expressing the resistive transition quantitatively in terms of the spatial distribution of the critical current density [1], [4], [5], [6] (and references therein). In this paper we present the current-voltage characteristics measured in the range from 0.001 μV/cm up to 1 μV/cm on 33 different sections (each 14.8 cm long) along a 488.4 cm long ac fine-filamentary wire with diameter 0.12 mm.

2. Experiment

The sample was made of a commercial NbTi fine-filamentary wire with CuNi matrix manufactured by Alsthom Atlantique. The parameters of the wire are shown in Table 1. The sample was uninsulated, 488.4 cm long, bifilarly wound on the sample holder shown in Fig.1. This was made of fibre glass epoxy with cooling channels. The wire was placed into triangular grooves machined on the surface (see the detail in Fig. 1) in order to ensure good cooling conditions as well as to prevent the winding movement. 33 pairs of potential wires (0.1 mm thick copper wire) were soldered to the sample as shown in Fig. 1.

Table 1: Sample parameters

Ref	Matrix filament zone and shield	Matrix central zone	Nominal diameter (mm)	Number of filaments	Filament diameter (μm)	Matrix Cu/CuNi/ NbTi	Twist pitch (mm)	Weight (kg/km)
CCN 14000LL	CuNi	Cu+ CuNi	0.12	14 496	0.6	0.83/1.16 /1	0.8	0.09

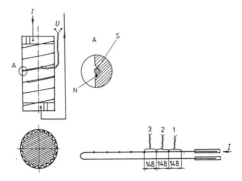

Fig. 1 The sample holder for current-voltage characteristics measurements (S - sample, N - potential tap).

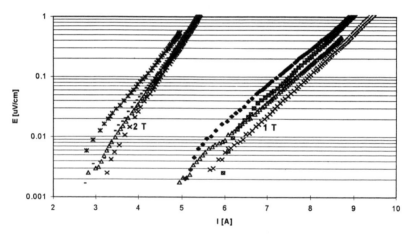

Fig. 2. Scatter of E-I characteristics together with one corresponding to the whole sample length
(open triangels). E-I characteristics measured on a few depict pieces of the sample are also shown.

The length of each section was 14.8 cm. The sample was placed in homogeneous superconducting solenoid. During the measurements the measured section of the sample was placed in the center of the magnet, so that the magnetic field homogeneity along the measured sample length was better than 0.1 %. At measurements of the whole sample the magnetic field homogeneity was better than 1 %. It was experimentaly tested that the field difference of 1 % has a small influence on measured voltage-current characteristics The current-voltage characteristics were measured at two magnetic fields (B=1 T and B=2 T) perpendicular to the wire axis. The standard four-probe method using a Keithley nanovoltmeter was adopted for sensitive current-voltage measurements. All measurements were carried out at 4.2 K in liquid helium bath in zero-field cooling conditions. Maximum power dissipated in the sample during the measurements was lower than 1 mW per cm^2 of the sample surface, which corresponds to the temperature increase of the sample $\ll$ 0.1 K and is about 10^3 lower than that corresponding to the onset of film boiling of the liquid He. Special attention was paid to the suppression of the thermo e.m.f. Under these conditions we believe that obtained curves are static voltage-current characteristics of the composite.

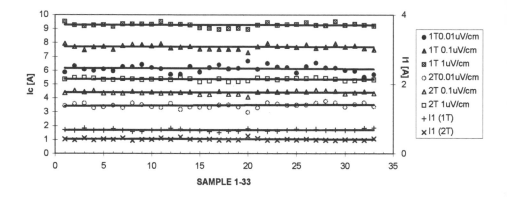

Fig. 3. Variation of I_c and I_1 at different criteria and magnetic fields on different sections of the sample length (filled symbols - I_c(1T), open symbols - I_c(2T), crosses - I_1, lines - least square approximations). The numbers on horizontal axis denote the individual sections.

3. Results and discussion

The current-voltage characteristics measured on the whole sample as well as on some depicted sections of the wire at 1 T and B=2 T are shown in Fig. 2. It is convenient to display the electric field E rather than the voltage as a function of the current I. One can notice quite large scatter of E-I characteristics on the different sections of the sample around the E-I characteristic corresponding to the whole sample. In the range from 0.01 µV/cm up to 1 µV/cm the curves have an exponential form $E=E_o exp[(I-I_c)/I_1]$, where I_c is the critical current determined at $E=E_o$ and I_1 characterises the slop of the curve. The ratio I_1/E is proportional to the differential conductivity of the superconductor. In the range of electric fields below 0.01 µV/cm a downward curvature of E-I characteristics on some sections of the sample was observed (Fig. 2). One can notice that at both fields I_c as well as I_1 vary. On the other hand comparing E-I curves at different magnetic fields one can see that they tend to intersect if extrapolated to very low electric fields (< 0.01 µV/cm). This behaviour was recently explained [1] by strong defects like breaks of the superconducting filaments. In Fig. 3 we show the variation of I_c and I_1 at different criteria: 0.01 µV/cm, 0.1µV/cm and 1µV/cm and magnetic fields B=1 T, B=2 T. The lines represent the least square approximations of the measured data. In Table 2 the mean values of I_c ($<I_c>$) and I_1 ($<I_1>$) as well as mean root square deviations of I_c and I_1 are compared with those on full whole sample. One can notice quite good accordance between $<I_c>$, $<I_1>$ and the corresponding values measured on the whole sample. This supports the idea of normal (Gaussian) I_c and I_1 distribution. Slight deviations may be caused by the fact, that during the measurement of the whole sample the magnetic field seen by it was more inhomogeneous. The variation of local I_c along the sample length is within around 10% of the value measured on the full sample length, if we use 0.01 µV/cm as a critical current criterion, within about 5 % for I_c at 0.1 µV/cm and within about 3 % for I_c at 1 µV/cm. The variation of I_1 is within about 25 % at 1T and 37 % at 2T. The sample is quite homogeneous and the variation of I_c has a decreasing tendency with increasing criterion of E. It seems that a correlation between I_c determined at different criteria exists, but it is more difficult to find a correlation between I_c determined at different criteria and I_1 from the present measurements. Since the pioneering papers of Baixeras and Fournet [4] and Jones, Rhoderick and Rose-Innes [5] about the influence of macroscopic inhomogeneities on pinning centre distribution, many papers have dealt with their effect on the current-voltage characteristics of type-3 superconductors ([6] and references therein). In this model E-I curves inherent to

514

Table 2: Mean values $<I_c>$, $<I_1>$ and the maximum diffrences (ΔI_c, ΔI_1) measured on different parts of the sample in comparison with measurements on full length sample (I_c, I_1).

B (T)	0.01 µV/cm			0.1 µV/cm			1 µV/cm					
	$<I_c>$ (A)	ΔI_c (A)	I_c (A)	$<I_c>$ (A)	ΔI_c (A)	I_c (A)	$<I_c>$ (A)	ΔI_c (A)	I_c (A)	$<I_1>$ (A)	ΔI_1 (A)	I_1 (A)
1	6.16± 0.009	0.5	6.07	7.71± 0.009	0.35	7.48	9.26± 0.002	0.27	9.07	0.67± 0.009	0.175	0.69
2	3.4± 0.048	0.31	3.46	4.4± 0.003	0.21	4.44	5.37± 0.017	0.15	5.39	0.42± 0.066	0.155	0.41

homogeneous fractions of the wire are supposed to have the following form: E=0 for I<I_c, E≈(I-I_c) for I>I_c. In this case it is possible to deduce the critical current distribution function f(I_c) from the measured E-I curves [4,5]. As pointed out in [6], f(I_c) cannot be obtained by the procedure [4,5] if the inherent form of E-I curve is a non linear one. In our case the sections have not only different I_c but also different slope characterized by I_1. We suppose that the distribution function could depend on both I_c and I_1.

4. Conclusions

We have measured E-I curves of a 488,4 cm long piece of fine-filamentary NbTi wire for ac use, as well as similar curves on 33 sections of this wire, each of them having the length of 148 mm. All measured E-I curves were exponential for 0.01 µV/cm <E<1 µV/cm. A downward curvature of E-I curves was observed on some sections of the wire at E<0.01 µV/cm. The I_c values measured on individual sections at 0.01 µV/cm varied by ± 10% around the I_c value measured over the whole length. The higher is the I_c criterion the smaller is the variation of I_c. The parameter characterizing the slope of E-I curves, I_1, varied by ± 25 % about the value measured on the whole sample.

References

[1] Polák M, Hlásnik I, Rakhmanov A L and Ivanov S S 1995 *Supercond. Sci.& Technol.* **8** 112-118
[2] Polák M, Hlásnik I, Fukui S, Ikeda N and Tsukamoto O 1994 *Cryogenics* **34** 315-324
[3] Polák M, Majoroš M and Baev V P 1992 *Supercond. Sci. & Technol.* **5** 419-422
[4] Baixeras J Fournet G 1967 *J. Phys. Chem. Sol.* **28** 1541-1547
[5] Jones R G Rhoderick E H Rose-Innes A C 1967 *Phys. Lett.* **24A** 318-319
[6] Takács S 1988 *Cryogenics* **28** 374-380

Inst. Phys. Conf. Ser. No 148
Paper presented at Applied Superconductivity, Edinburgh, 3–6 July 1995
© 1995 IOP Publishing Ltd

515

Temperature effects on Nb-Ti strands extracted from coextruded Al stabilized cables

R. Garré, S. Conti, A. Baldini, S. Rossi, P. Ricotti

Europa Metalli Superconductors Division, Piazzale Luigi Orlando, Fornaci di Barga (Lucca) Italy

Abstract In the framework of the high energy physics projects the use of large magnets for particle detection realized by coextruded Al or Al alloy stabilized cables is foreseen. For the processing of this type of cable it is necessary to operate at high temperatures in order to guarantee a good bond between the aluminium and the copper surface of the superconductor. Such hot working could affect the electrical transport properties of the superconducting cables. Europa Metalli Superconductors Division has developed coextruded Al cables for the particle detector magnet KLOE (DAPHNE project - INFN - Italy) and for the particle detector magnet ATLAS to be used in LHC project (CERN). The present study investigates the effects of different heat treatments on the critical currents of extracted superconducting strands from Al matrix. The degradation of the critical current as a function of an external applied magnetic field is examined for the different strands.

1. Introduction

The new generation of colliders for High Energy Physics foresees the use of large magnets for particle detection. Various parameters such as quench protection, stabilization, spooling techniques and the weight of the magnet must be considered during the design of the conductor for the realization of a magnet. For the superconducting strands, copper is the material which best satisfies the criteria of quench protection and stabilization. However, for this application the superconducting Nb-Ti cable is stabilized in pure aluminium due to the advantages of weight reduction and, above all, particle transparency. The Nb-Ti superconducting cable used is a Rutherford type which, by means of a coextrusion process is cladded in the aluminium matrix. This is a fundamental operation as a good bonding between the cable and the stabilizing matrix must be guaranteed. At the same time, it is also particularly delicate, as the Niobium Titanium should avoid excessive heating which may drastically reduce the electrical transport properties of the superconductor [1].

The EM Superconductors Division, on the basis of past experiences gained during the ZEUS detector project for DESY-Hamburg, has developed a coextrusion process successfully used for the realization of the KLOE detector (DAPHNE project - INFN , Italy) and a prototype cable for the ATLAS detector to be installed at the Large Hadron Collider at CERN in Ginevra.

Several samples, extracted from the aluminium matrix of the two above mentioned cables, were heat treated at various temperatures to determine the degradation of critical currents.

516

2. Conductors

table 1

KLOE CONDUCTOR	
Cable dimension	: 10 x 5 mm
Number of strands	: 8
Strand diameter	: 0.7 mm
Cu: Nb-Ti ratio	: 1.8
ATLAS CONDUCTOR	
Cable dimension	: 70 x 7 mm
Number of strands	: 32
Strand diameter	: 1.3
Cu Nb-Ti ratio	: 1.7

Table 1 reports the characteristics of the cable and the wire used for the realization of the two conductors. As already mentioned, the finished conductor was produced by means of a coextrusion process in which the superconducting cable, by means of a dye, is cladded with the high purity aluminium matrix, obtaining the finished product at the dye exit. The main characteristic of this coextrusion process is the metallic bonding between the aluminium and the stabilized copper superconducting cable. A metallic diffusion is in fact obtained between the copper and aluminium under suitable temperature and pressure conditions. Nevertheless, the temperature and the extrusion speed must be carefully kept under control to avoid a Nb-Ti alloy degradation. Figure 1 reports a cross sections of the KLOE cable.

Fig. 1: Cross-section of the KLOE cable

3. Measurements

Various wire samples, extracted from the aluminium matrix, were heat treated for 8 minutes at temperatures within a 300 °C and 430 °C range. Critical current measurements (4.2 K, criteria of 10^{-12} Ωcm) were carried out on the samples, also in the presence of an external magnetic field in order to define in greater details the parameters for the coextrusion process.

KLOE Cable - The critical current density of samples heated at different temperatures is reported in figure 2. The external applied magnetic field was 3 T. Critical current densities of various samples, heat treated at 430 °C were then measured at different values of magnetic fields (up to 8 T).The ratio between the critical current densities of heat treated samples and the critical currents of samples as extracted from the Al matrix is reported in figure 3 as a function of magnetic field.

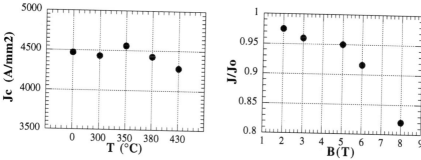

Fig. 2: Critical current densities (external field **B** = 3 T) of various samples heat treated at different temperatures for 8 minutes.

Fig. 3: Ratio between critical current densities of heat treated and not heat treated samples with an external applied magnetic field up to 8 T.

ATLAS cable - The critical current densities of samples extracted from the Al matrix and heat treated at different temperatures were compared with the critical current densities of samples not heat treated. The ratio of these values is reported in figure 4. The measurements were carried out in a magnetic field of 8 T. Critical current measurements were carried out on these same samples at different magnetic fields between 3 T and 8 T. Figure 5 reports the behaviour of the ratio of J/Jo as a function of the magnetic field, where J is the average current density of the heat treated wires and Jo is the average density of the wires extracted from the aluminium matrix.

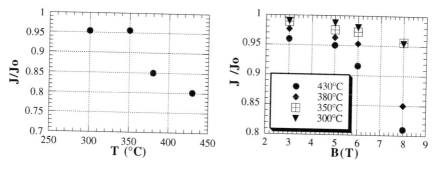

Fig. 4: Ratio between critical current densities of heat treated and not heat treated samples with an external applied magnetic field of 8 T for various temperatures.

Fig. 5: Ratio between critical current densities of heat treated and not heat treated samples with an external applied magnetic field up to 8 T.

4. Results

Figure 2 shows that heat treatment at temperatures below 400 °C have little effect on the electrical transport properties. Figure 4 and 5 shows that the critical current degradation mainly depends on the external applied magnetic field. At a 3 T magnetic field the maximum critical current reduction of the strand is below 5 % regardless of heat treatments. As shown in fig. 3, for magnetic fields less than 3 T the critical current degradation is considered unreduced. At higher magnetic fields (5T - 8T) a considerable difference is noted among samples heat treated at different temperatures. At 8 T, see figure 5, the critical current reduction is within the region of 5 % for samples heat treatd at 300 °C and 350 °C and is higher than 15% for samples heat treated at 380 °C and 430 °C.

5. Conclusions

A study was carried out on the degradation of critical currents of Al and Al stabilized Nb-Ti superconductors as a function of temperatures in the range 300 °C - 430 °C.
Several superconducting wires taken from cables for KLOE and ATLAS detector magnets were heat treated at various temperatures and characterized by means of critical current measurements at different fields. Results showed that critical current degradation is strongly dependent on heat treatment temperatures at magnetic fields greater than 3 T. For magnetic fields lower than 3 T critical current degradation is in any case lower than 5% .

References
[1] E. Baynham, V. Edwards, R.K. Maix et al. "The aluminium stabilized superconductors for the DELPHI magnet" - Proceedings of the 9th Int. Conference on Magnet Technology, Zürich (CH) 9-13 sept. 1985

Inst. Phys. Conf. Ser. No 148
Paper presented at Applied Superconductivity, Edinburgh, 3–6 July 1995
© 1995 IOP Publishing Ltd

Quench propagation in cabled superconductor: the effect of the current redistribution among strands

N A Buznikov[*], A A Pukhov[*], A L Rakhmanov[*] and V S Vysotsky[**]

[*] Scientific Center for Applied Problems in Electrodynamics, Russian Acad. Sci., Izhorskaya 13/19, 127412 Moscow, Russia
[**] Plasma Fusion Center MIT, 02139, Cambridge, MA, USA

Abstract. The influence of current redistribution among strands on the quench in a superconducting cable is studied. A theoretical model is developed taking into account non-uniformities in strands. Three different current redistribution modes are shown to exist, depending on the initial current I_0. It is found that at I_0 lower than some threshold current I_0^* the inductively initiated quench of the whole cable is impossible. The obtained analytical dependence of I_0^* on the cable length is in a good agreement with the experimental data.

1. Introduction

The quench in a superconducting cable (SC) is accompanied by current redistribution among strands [1,2]. The first strand, being quenched, induces current rising in adjacent strands. Under certain conditions, current increase results in the quench origination in these strands. Three modes of quench have been observed in SC, depending on the initial current per strand I_0 [1-4]. At low I_0 the current redistribution due to normal zone propagation in one strand does not lead to the quench of the whole SC. In this mode the total current remains constant. If current I_0 exceeds some threshold I_0^*, then, the current redistribution results in the normal zone nucleation in strands. The SC quenches completely due to the normal zone propagation and the total current decreases with the extinction time $t_q \approx$ 1-10 ms. At larger currents $I_0 > I_0^{**}$ the quench of one strand gives rise to fast quench of SC with $t_q \approx 0.1$ ms [1].

In the present paper the current redistribution is studied using two-strand model accounting for the interaction between normal zone and thermomagnetic disturbances initiated by varying current. It is shown that strand non-uniformities affect quench dynamics in multistrand cables. The proposed model explains the phenomena observed in [1-4].

2. Model description

A SC is simulated by magnetically coupled circuit consisting of two strands. Kirchhoff laws for this circuit have the form:

$$\begin{cases} L_1\dot{I}_1 + M\dot{I}_2 + R_1I_1 = \varepsilon - r(I_1 + I_2) \\ M\dot{I}_1 + L_2\dot{I}_2 + R_2I_2 = \varepsilon - r(I_1 + I_2) \end{cases} \tag{1}$$

where I_1, I_2 are strand currents; $\dot{I}_1$, $\dot{I}_2$ are the current rates; L_1, L_2 are the strand inductances; $R_1(t)$, $R_2(t)$ are the strand resistances, M is the mutual inductance; ε is EMF of power supply; r is the current leads resistance. The time derivative of $R_1(t)$ is defined by

$$\dot{R}_1(t) = 2\rho \cdot n_1 \cdot v\left[I_1(t), \dot{I}_1(t)\right] / A, \tag{2}$$

where ρ is the strand resistivity, A is the cross-section area, n_1 is the number of "weak" spots at which normal zone nucleates and v is the normal zone velocity. The similar expression can be written for strand 2. The analytical formula for $v(I,\dot{I})$ was found in [5]:

$$v = v_{ad} \cdot i / \sqrt{1 - i - q_s}, \tag{3}$$

where $v_{ad} = j_s C^{-1}\left[\rho k / (T_c - T_0)\right]^{1/2}$, I_s is the critical current, $i = I/I_s$, h is the heat transfer coefficient, T_0 is the coolant temperature, T_c is the critical temperature, P is the strand perimeter, $j_s = I_s/A$, C is the specific heat, k is the thermal conductivity and $q_s(I,\dot{I})$ is the dimensionless heat release in the superconducting state which can be expressed as [5]:

$$q_s = -(|\dot{I}|/\dot{I}_0) \cdot (1 - i_q) \cdot \{F(i) + \log[1 - F(i)]\}, \tag{4}$$

where $i_q = I_q/I_s$, $\dot{I}_0 = 16\pi h j_1 (T_c - T_0) / \mu_0 D j_s^2$, D is the strand diameter and j_1 is the parameter of the exponential I-V characteristic of the superconductor $E = E_0 \cdot \exp[(j - j_s)/j_1]$ [6]. Function $F(i)$ depends on the sign of $\dot{I}$ and on the values of the maximum and the minimum currents in the preceding variation cycle. At the initial stage of the current redistribution process $F(i) = i$ at $\dot{I} > 0$ and $F(i) = (i_0 - i)/2$ at $\dot{I} < 0$ [5], where $i_0 = I_0/I_s$.

The strand quench current I_q is determined by the equation [6]:

$$F(i_q) + \log[1 - F(i_q)] + \dot{I}_0 / |\dot{I}| = 0. \tag{5}$$

We assume that phase nucleation site is related to local decrease of heat transfer coefficient h^* $< h$ and is characterized by the inhomogeneity coefficient $f = h^*/h$. The local quench current at "weak" spot $i_q^* = I_q^*/I_s < i_q$ is determined by (5) with substitution $i_q \to i_q^*$ and $\dot{I}_0 \to f\dot{I}_0$ [7].

3. Current redistribution in the SC

The quench in a two-strand SC is simulated numerically using Eqs. (1)-(5). To initiate quench the normal zone nucleus is set in strand 1 at $t = 0$. Strand 1 is assumed to be homogeneous and strand 2 to have a single "weak" spot with the inhomogeneity coefficient $f = f_2$. The values $I_1(t)$ and $I_2(t)$ calculated numerically are shown in Fig. 1 at different $I_0 = \varepsilon/2r$. Three current redistribution modes are found similar to that observed in experiment [1].

I. *Current sharing mode.* At low I_0 the current transfer to strand 2 does not result in its quench (see Fig. 1a). Strand 2 remains superconducting because current I_2 is less than the local quench current I_{q2}^* at the "weak" spot. In this mode the normal zone disappears and superconductivity recovers.

II. *Slow quench mode.* At initial current in the strands I_0 exceeding some threshold I_0^*, the normal zone nucleates at the "weak" spot in strand 2 at $I_2 = I_{q2}^*(f_2)$ (see Fig. 1b). An

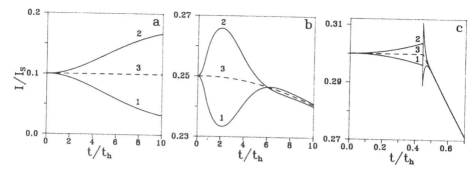

Fig. 1 The current redistribution modes in the SC ($L_1=L_2=880$ μH, $M=877$ μH, $A=1.96\cdot10^{-7}$ m^2, $\rho=1.35\times10^{-7}$ Ohm·m): $I_1(t)$ (1); $I_2(t)$ (2); averaged current in SC $[I_1(t)+I_2(t)]/2$ (3). a - current sharing mode, $I_0/I_s=0.1$, $f_2=0.7$. b - slow quench mode, $I_0/I_s=0.25$, $f_2=0.7$. c - fast quench mode, $I_0/I_s=0.3$, $f_2=1$.

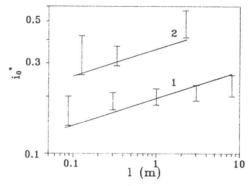

Fig. 2 The dependence of the threshold current I_0^* on a cable length l: curve 1 - formula (8) at $f_2=1$, vertical bars - experimental data [3] ($I_s=140$ A, $A=2.54\cdot10^{-8}$ m^2, $\rho=2.5\cdot10^{-7}$ Ohm·m, $\dot{I}_0=1.4\cdot10^4$ A/s, $v_{ad}=70$ m/s, $a=2.4\cdot10^{-4}$ m); curve 2 - formula (8) at $f_2=1$, vertical bars - experimental data [4] ($I_s=100$ A, $A=6.16\cdot10^{-8}$ m^2, $\rho=2\cdot10^{-7}$ Ohm·m, $\dot{I}_0=1.2\cdot10^4$ A/s, $v_{ad}=5$ m/s, $a=2.8\cdot10^{-4}$ m).

increase of R_2 results in the current transfer to strand 1. In this mode the normal zone propagates in the both strands with the characteristic velocity of the order of v_{ad}, so the extinction time is $t_q \approx l/v_{ad}$, where l is the cable length.

III. *Fast quench mode.* At sufficiently large initial currents $I_0 > I_0^{**}$ the fast quench mode occurs (see Fig. 1c). In this mode strand 2 quenches uniformly over its length due to the development of the thermomagnetic instability at the strand quench current $I_2 = I_{q2}(\dot{I}_2)$. Then, the current transfers to strand 1, which also quenches uniformly at $I_1 = I_{q1}(\dot{I}_1)$. Strand currents decrease fast with the extinction time $t_q \approx 2M/(R+2r)$ being much less than that in the slow quench mode (where $R = \rho l / A$ is the strand resistance in the normal state).

For current sharing mode Eqs. (1) and (2) can be solved analytically [8]. Assuming that $L_1 = L_2 = L$, $R_1(t) \ll r$, $L_{eff} = (L^2-M^2)/L \ll L$, $q_s \ll 1$ and $n_1 = 1$ we find:

$$i_1(\tau) = i_0 / \cosh^2\left(\sqrt{\gamma i_0}\,\tau\right) , \qquad i_2(\tau) = 2i_0 - i_1(\tau), \qquad (6)$$

where $\tau = t/t_h$, $t_h = CA/hP$, and $\gamma = \rho v_{ad} t_h^2 / AL_{eff}$. The local quench current is determined by (5) with $F(i_{q2}^*) = i_{q2}^*$ and $h^* = f_2 h$. Using (6) we get at $i_{q2}^* \ll 1$:

$$i_{q2}^* = \left(i_0^2 AL_{eff} f_2^2 / \rho v_{ad} I_s^2 i_0^3 \right)^{1/4} \times \cosh^{3/2}\left(\sqrt{\gamma i_0}\,\tau\right) / \sinh^{1/2}\left(\sqrt{\gamma i_0}\,\tau\right). \tag{7}$$

The threshold current I_0^* is determined by the conditions $i_2 = i_{q2}^*$, $di_2/d\tau = di_{q2}^*/d\tau$ at some moment of time. Using (6) and (7) we find the expression for $i_0^* = I_0^*/I_s$:

$$i_0^* = 1.07 \cdot \left(i_0^2 AL_{eff} f_2^2 / \rho v_{ad} I_s^2 \right)^{1/7}. \tag{8}$$

Current I_0^* is the characteristic threshold above which thermal disturbances gives rise to the quench of SC. Eq. (8) allows to describe an increase of i_0^* with a cable length observed in [2-4]. Indeed, $i_0^* \propto L_{eff}^{1/7}$, whereas L_{eff} is proportional to l: $L_{eff} = (\mu_0 l / \pi) \cdot [\log(a/r) + 0.25]$ [2], where a is the distance between strands. In Fig. 2 the experimental data [3,4] are compared with the dependence of $i_0^*(l)$ given by (8).

4. Discussion

There are three modes of current redistribution in SC. At low current the normal zone nucleation in one strand does not result in the quench of the whole SC and normal zone in this strand disappears (current sharing mode). If the initial strand current exceeds the threshold I_0^*, then, the current redistribution due to the nucleation of normal zone in one strand results in the quench propagation in the nearest neighbor strands (slow quench). At a larger current $I_0 > I_0^{**}$ SC quenches due to the development of the thermomagnetic instability (fast quench).

Our model allows to describe the quench in SC having several nonuniformities. The current redistribution due to normal zone propagation in strand 1 results in the normal zone nucleation in strand 2 at "strongest" nonuniformity. This process leads to additional current redistribution and to nucleation of new normal regions in strand 1. So, each step of current rising in strands is accompanied by normal zone nucleation at the increasing number of "weak" spots. Such multiple normal zone nucleation results in an abrupt increase of strand resistance and may be interpreted as an increase of the observed normal zone velocity [1,2].

The work is supported in part by Russian State Program on High-T_c superconductivity (Grant No 93027).

References

[1]   Vysotsky V S, Tsikhon V N and Mulder G B J 1992 *IEEE Trans. on Magn.* **28** 735
[2]   Iwakuma M, Kanetaka H, Tasaki K, Funaki K, Takeo M and Yamafuji K 1990 *Cryogenics* **30** 686
[3]   Mulder G B J, Krooshoop H J G, van de Klundert L J M and Vysotsky V S 1992 *IEEE Trans. Magn.* **28** 743
[4]   Vysotsky V S and Tsikhon V N 1992 *Cryogenics* **32** (ICEC suppl.) 419
[5]   Buznikov N A, Pukhov A A and Rakhmanov A L 1994 *Cryogenics* **34** 761
[6]   Mints R G and Rakhmanov A L 1988 *J. Phys. D: Appl. Phys.* **20** 826
[7]   Buznikov N A, Pukhov A A and Rakhmanov A L 1994 *IEEE Trans. on Magn.* **30** 1994
[8]   Mulder G B J, van de Klundert L J M and Vysotsky V S 1992 *IEEE Trans. on Magn.* **28** 739

Inst. Phys. Conf. Ser. No 148
Paper presented at Applied Superconductivity, Edinburgh, 3–6 July 1995
© 1995 IOP Publishing Ltd

Minimum propagation current in composites with small contact electrical resistance, the model with current sharing

Alexander A. Akhmetov[1], Kunishige Kuroda and Masakatsu Takeo

Department of Electronics, Kyushu University, 6-10-1 Hakozaki, Higashi-ku, Fukuoka 812, Japan

Abstract. General method is applied to calculate the modification of the minimum normal zone propagation current in the superconducting composites due to the presence of a small contact electrical resistance. Current sharing between the superconductor and the matrix is taken into account.

1. Introduction

If the composite superconductor carrying transport current I can be characterized by three stationary states, *i.e.*, the stable normal and superconducting ones and the unstable resistive state, the normal zone (NZ) propagation in it is of autowave nature. A normal front, transforming the composite from the superconducting to the normal state, moves along it with the constant velocity $V(I)$. At one particular current, I_p, the normal front is motionless [1]; $V(I_p) = 0$. Many factors affect the minimum propagation current (MPC) I_p [1,2]. Among them, an electrical contact resistance (ECR) between the superconductor proper and the matrix of normal metal. If the contact resistance is high, it drastically changes the conditions of the normal zone existence and propagation [3,4]. Here, the influence of the small ECR on I_p is considered in the model with the current sharing.

2. Basic equations

Equations, describing an evolution of the NZ in time τ and along the coordinate z in the composites with ECR

$$\epsilon \frac{d^2 j_n}{d z^2} - r^2 j_n + \mathcal{E}_s = 0, \tag{1}$$

$$\frac{\partial \Theta}{\partial \tau} = \frac{\partial^2 \Theta}{\partial z^2} - \Theta + \alpha i^2 j_n^2 + \epsilon \alpha \frac{i^2}{r^2} \left(\frac{d j_n}{dz}\right)^2 + \alpha \frac{i^2}{r^2}(1 - j_n)\mathcal{E}_s, \tag{2}$$

[1] E-mail: akhmetov@ele.kyushu-u.ac.jp

are written here as in [3] for the dimensionless temperature Θ and current density in the normal matrix j_n. Dimensionless current $i = I/I_c$, where I_c is the critical current, stability parameter α [1] and the ratio of the resistivities per unit length of the matrix and the superconductor being in the normal state, $r^2 \ll 1$, are the parameters of the problem. Only one of them, *i.e.*, the current i can be effectively changed during the experiment. Parameter ϵ is proportional to the thickness and resistivity of the barrier separating the superconductor and the matrix. For $\epsilon = 0$, the MPC i_p depends mostly on α [1]. The electric field in the superconductor $\mathcal{E}_s$ is given by

$$\mathcal{E}_s = \begin{cases} 1 - j_n, & \text{if } 1 < \Theta; \\ b\big(1 - j_n + (\Theta - 1)/i\big)/(1 - (1 - b)\Theta), & \text{if } 1 - i + ij_n \le \Theta \le 1; \\ 0, & \text{if } 1 - i + ij_n > \Theta; \end{cases} \tag{3}$$

where b is the parameter proportional to the external magnetic field [5].

3. Application of the general method

The general method [6] would be directly applicable to the problem if the right parts of Eqs. (1) and (2) could be written as the smooth functions $F_1(\Theta, j_n, i, \alpha, r)$ and $F_2(\Theta, j_n, i, \alpha, r)$ correspondingly. Then, for the immobile NZ and $\epsilon = 0$, Eqs. (1) and (2) can be transformed to one equation of the type

$$\frac{\partial^2 \Theta}{\partial z^2} + \Phi(\Theta, i, \alpha, r) = 0, \qquad \Phi \approx \begin{cases} \alpha i_p^2 - \Theta, & \text{if } 1 < \Theta; \\ \alpha i_p(\Theta + i_p - 1) - \Theta, & \text{if } 1 - i_p \le \Theta \le 1; \\ -\Theta, & \text{if } 1 - i_p > \Theta; \end{cases} \tag{4}$$

Condition of the NZ immobility [1,2,5,6] is

$$S(\Theta_m, i_p, \alpha, r) = 0, \qquad S(\Theta, i, \alpha, r) = \int_0^\Theta \Phi(\Theta, i, \alpha, r) d\Theta, \tag{5}$$

where $\Theta_m = \alpha i^2/(1 + r^2)$ is the temperature of the NZ far from the normal front.

The presence of the small ECR changes the value of $S(\Theta_m, i_p)$ from zero by $\Delta S^{(\epsilon)} \approx \epsilon \Psi_n(\alpha, r, i_p)$. To compensate the change, MPC has to be altered to produce $\Delta S^{(i_p)} \approx \Delta i_p \Psi_d(\alpha, r, i_p)$, which has to be equal to $-\Delta S^{(\epsilon)}$ to have the NZ immobile [6]. It follows that

$$\Delta i_p \approx -\epsilon \frac{\Psi_n(\alpha, r, i_p)}{\Psi_d(\alpha, r, i_p)}. \tag{6}$$

Reference [6] provides the way to express the functions $\Psi_n(\alpha, r, i_p)$ and $\Psi_d(\alpha, r, i_p)$ through the functions $F_1(\Theta, j_n, i, \alpha, r)$ and $F_2(\Theta, j_n, i, \alpha, r)$, which would be enough to solve the problem if it were not for the particularities of Eqs. (1) through (2).

4. Correction of the general method and results

Necessary corrections of the method [6] have to include both the term $\epsilon \alpha i^2 (dj_n/dz)^2/r^2$ which in Eq. (2) describes the heat generation due to the current flowing through the contact resistance, and the existence of the current-dependent boundary between the resistive and superconducting states in Eq. (3). Let us begin with the second alteration.

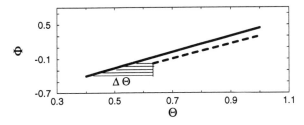

Figure 1. Dependences $\Phi = \Phi(\Theta)$ for $\epsilon = 0$ (——) and $0 < \epsilon << 1$ (----).

If there were not for ECR, at $\Theta < 1 - i$ all current would flow through the super-conductor. At $\Theta = 1 - i$ the current sharing would begin. When $0 < \epsilon << 1$ appears, the situation becomes different. The change in the current j_n in the most of the resistive region is proportional to ϵ [6]. However, in the small region close to the point $\Theta = 1 - i$ the change in the current j_n is much more drastic and greatly affects the boundary between the resistive and superconducting regions. To estimate it, note, that the solution of Eq. (1) in the superconductive region as a function of coordinate is simply

$$j_n = j_0 \exp\left(-\frac{r}{\sqrt{\epsilon}}z\right), \qquad \frac{dj_n}{dz} = -\frac{r}{\sqrt{\epsilon}}j_0 \qquad (7)$$

where j_0 is the constant to be defined later.

On the other hand, the same derivative can be expressed (see [5,6]) in two terms

$$\frac{dj_n}{dz} = \frac{dj_n}{d\Theta}\frac{d\Theta}{dz} \approx \frac{1}{i_p}\sqrt{-2S(\Theta, i_p, \alpha, r)}. \qquad (8)$$

Using Eqs. (4) and (5) one finds that in the superconducting region

$$S(\Theta) = -\frac{\Theta^2}{2}. \qquad (9)$$

At the beginning of the resistive region $\Theta = 1 - i + ij_n$, and combining Eqs. (7) through (9) it follows that if $\epsilon << r^2$ then

$$j_0 \approx \frac{(1 - i_p)\sqrt{\epsilon}}{i_p r}. \qquad (10)$$

In turn, presence of the nonzero current j_0 shifts the boundary between the super-conducting and resistive region from the initial position $\Theta = 1 - i$ on

$$\Delta\Theta \approx \frac{(1 - i_p)\sqrt{\epsilon}}{r}. \qquad (11)$$

This effect is illustrated in Fig. 1 where the function $\Phi(\Theta)$ in the resistive region is shown schematically before and after appearance of the small ECR. The shaded tri-angular area represents the change in the function $S(\Theta_m, i_p, \alpha, r)$ due to the shift of the boundary. From the simple geometrical reasons this change can be estimated as

$$\Delta S_b^{(\epsilon)}(\Theta_m, i_p, \alpha, r) \approx -\frac{1}{2}\frac{\alpha i_p}{r^2}(1 - i_p)^2\epsilon. \qquad (12)$$

526

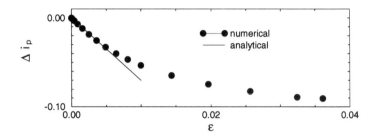

Figure 2. The change in the MPC due to the presence of small ECR.

The latter term has to be incorporated in the numerator of Eq. (6) to describe the changes due to the modification of the boundary conditions.

Let us now consider the heat generation due to the current's flow through the ECR. Using Eq. (8) it follows that this results in the change in the function S in the point Θ_m which can be expressed as

$$\Delta S_h^{(\epsilon)}(\Theta_m, i_p, \alpha, r) \approx -\epsilon \frac{2\alpha}{r^2} \int_{1-i_p}^{1} S(\Theta, i_p, \alpha, r) d\Theta. \tag{13}$$

Using the second of Eqs. (4) to obtain dependence $S = S(\Theta, i_p, \alpha, r)$ and then fulfilling integration in Eq. (13) it follows that

$$\Delta S_h^{(\epsilon)}(\Theta_m, i_p, \alpha, r) \approx \epsilon \frac{\alpha i_p}{r^2} \left(1 - i_p + \frac{1}{3} i_p^2 - \frac{1}{3} \alpha i_p^3 \right). \tag{14}$$

Last, combining all appropriate terms in the numerator of Eq. (6), one gets the alteration in the MPC due to the small ECR

$$\Delta i_p \approx -\epsilon \frac{\alpha i_p}{3r^2} \frac{3 - 3i_p + i_p^2 - \alpha i_p^3}{4\alpha^2 i_p^3 + 3\alpha i_p^2 - 4\alpha i_p + 2i_p - 2}, \tag{15}$$

At small ECR, the results of numerical calculations made by FDM for $\alpha = 4$ and $r = 0.1$ and shown in Fig. 2, are in a good agreement with those calculated by Eq. (15).

References

[1] Wlison M N 1983 *Superconducting magnets* (Oxford: Oxford University Press)

[2] Altov V A, Zenkevich V B, Kremlev M G and Sytchev V V 1977 *Stabilization of superconducting magnetic systems* (N.Y.: Plenum Press)

[3] Akhmetov A A and Mints R G 1985 *J. Phys. D: Appl. Phys.* **18** 925–38

[4] Akhmetov A A and Baev V P 1984 *Cryogenics* **24** 67–72

[5] Gurevich A V and Mints R G 1987 *Thermal waves in normal metals and superconductors* (Moscow: IVTAN Press)

[6] Akhmetov A A 1994 *Phys. Rev. (E)* **50** 3271–3

Inst. Phys. Conf. Ser. No 148
Paper presented at Applied Superconductivity, Edinburgh, 3–6 July 1995
© *1995 IOP Publishing Ltd*

Decay of long current loops in the superconducting cables

A.A. Akhmetov [1]**, K. Kuroda, T. Koga, K. Ono and M. Takeo**

Department of Electronics, Kyushu University, 6-10-1 Hakozaki, Hihashi-ku, Fukuoka 812, Japan

Abstract. The currents organized in the long loops were initiated in the samples of the flat two layer superconducting cables. Characteristic time constants of the current's decay were measured and fitted by numerical calculations.

1. Introduction

It is fairly possible that the delayed redistribution of the current between different strands of the superconducting cable is responsible for the effect observed in the experiments with the dipole magnets. It was found [1] that sometimes, after completing of the magnet's discharge, the magnetic field does not disappear but forms a periodic pattern along the length of the magnet with the period equal to the cable twist pitch. This pattern which can persist without significant attenuation for more than twelve hours [1] is, probably, due to the very slow decay of the current circulating in the loop composed of many inductances connected in series and resistances connected in parallel [2].

In general, multi-element circuits are characterized by the set of different time constants which together govern the rates of the current change in the elements. However, with time passing the influence of the short-living processes diminishes and only the terms described by the largest time constants remain. The present work is performed with the aim to try to reveal the maximum time constants of the current redistribution between the strands of the samples of flat two-layer superconducting cables. Also, the relation between the parameters of the cable and the maximum time constant is studied both theoretically and in the experiment.

2. Experiment

2.1. The samples

Two kind of cables were investigated. The first model cable ($C1$) was made of just three composite strands with the matrix of pure copper and the outher diameter $0.049\,cm$

---

[1] E-mail: akhmetov@ele.kyushu-u.ac.jp

528

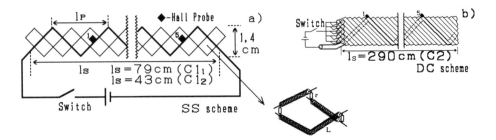

Figure 1. The samples and experimental setups.

soldered together in corresponding points to create a sparse flat cable with the width $1.4\,cm$ and the pitch length $2.6\,cm$ (see Fig. 1(a)). With no external magnetic field, the Pb-Sn solder was in the superconducting state, so the crossover resistances between the strands r were determined only by the resistance of the matrix. There were two samples of this kind; one with the length $l_s = 79\,cm$ ($C1_1$) and another one with the length $l_s = 43\,cm$ ($C1_2$). A piece of an eleven-strand flat superconducting cable with the length $l_s = 290\,cm$ was used as the sample of different type ($C2$) (see Fig. 1(b)). In this cable the strands with the Cu-Ni barriers, Cu sheaths and the outer diameters $0.12\,cm$ were shaped into the flat form with the rectangular cross-section $0.67\,cm \times 0.21\,cm$ and the pitch length $5\,cm$.

2.2. Experimental setup

The Hall probes (HP's) were placed around the samples so that they were able to pick up the values of the current in five points of one particular strand (**the Strand**). Two different experimental setups were used for all samples. In accordance with the first scheme, which is depicted in Fig. 1(a), the current went along **the Strand**, while the rest of the cable served as a shunt (same strand or **SS** scheme). According to the other method, the current went in the sample through **the Strand** (see Fig 1(b)) and got out of the sample through the rest of the cable (distributed current or **DC** scheme. In both cases, after the constant current distribution had been achieved, the current was switched off for the time shorter than $10^{-4}s$ and the signals from the HP's were recorded during the discharge. Also, the HP's signals have been recorded during the energization of the samples $C1_1$ and $C1_2$.

2.3. Results

2.3.1. Energization of the C1 samples. After the current source had been turned on, the transport current I_t flowing through the samples of the $C1$ type have reached the chosen value $I_m = 40\,A$ for the time $t_1 \approx 60\,s$. However, it took a much longer time for the signals from the HP's to achieve an equilibrium. Figs. 2(a) and (b) show the time dependences of the currents in **the Strand** for the sample $C1_2$ and **DC** and **SS** schemes correspondingly. It is seen that, regardless of the scheme used, the process of the redistribution of the current I_m between the strands is completed only after about $10^3\,s$. Exactly the same qualitative results were obtained for the sample $C1_1$. The only difference was that the time to achieve equilibrium was approximately four times larger.

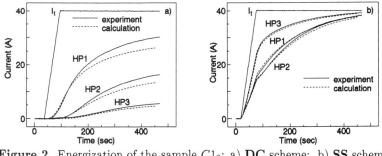

Figure 2. Energization of the sample $C1_2$: a) **DC** scheme; b) **SS** scheme.

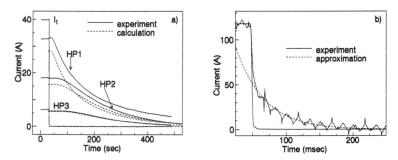

Figure 3. The currents in **the Strand** during the discharge: a) $C1_2$; b) $C2$.

Also, it is seen in Figs. 2 (a) and (b) that after the equilibrium has been achieved, in the case of the **SS** scheme the current in **the Strand** was uniform, while in the case of the **DC** scheme it decreased with the increase of the distance from the current leads.

2.3.2. Discharge of the samples. The currents in **the Strand** recorded after the constant external current I_m has been cut off from the samples $C1_2$ (**DC** scheme) and $C2$ (**SS** scheme, the first HP) are shown in Figs. 3 (a) and (b) respectively. Qualitative similarity of the processes of the current decay in the three-strand model cable and the industrial 11-strand cable is seen in Figs. 3 (a) and (b) despite of the quantitative difference in the corresponding rates of more than four orders of magnitude. Analogous results were obtained for the sample $C1_1$, except for the rate of the current's decrease was approximately four times smaller than that of the current in the sample $C1_2$.

Analysis of the curves shown in Figs. 3 (a) and (b) indicates that after some time has gone, the dependences of the currents on time can be approximated as $I = I_0 \exp(-t/t_0)$, where I_0 is the constant which is different for every combination of the sample and the experimental setup, and t_0 is the time constant which does not depend on the experimental scheme. For the samples $C1_1$, $C1_2$ and $C2$ the time constants are found to be $\approx 650\,s$, $\approx 170\,s$ and $\approx 3 \cdot 10^{-2}s$ correspondingly.

3. Analysis and discussion

A set of Faraday's equations have to be applied to M elementary loops created by the intersecting or bending strands of the cable to calculate the rate of the change in the current in any point of the sample. These equations imply the absence of the time dependent magnetic flux penetrating the loops and in the symbolic form can be written as

$$\vec{R}\vec{I} + \vec{L}\frac{d}{dt}\vec{I} = \left(\vec{a} + \vec{b}\frac{d}{dt}\right)I_t(t),\tag{1}$$

where $\vec{I}$ is the one-column matrix composed of M currents circulating around the loops, $\vec{R}$ and $\vec{L}$ are the the square matrices of the rank M which contain the interstrand resistances r and inductances L per the side of the loop with the proper coefficients correspondingly, and $I_t(t)$ is the time dependent current imitating the current source. The one-column matrices $\vec{a}$ and $\vec{b}$ provide the resistive and inductive parts correspondingly of the voltage drop associated with the external current source. They are different for the **DC** and **SS** experimental schemes. In the latter case $\vec{a} \equiv \vec{0}$.

First, Eq. (1) was solved for the sample $C1_2$ with $M = 97$. The current I_t had been increased from zero to I_m for the time $t_1 = 60\,s$ and then was kept constant till $t = t_2 \gg t_1$. It was found that calculated currents in **the Strand** agreed with the experimental ones if the ratio $t_e = L/r$ was equal to $0.175\,H/\Omega$ (see Figs. 2(a) and (b)). The solution for the **DC** scheme shows that at $t \approx t_1$, the current flows from one current lead to the other one by the shortest possible path. At $t \gg t_1$ it gradually diffuses into the sample. Last, at $t \gg 10^2 s$ the current I_m overflows from **the Strand** to the rest of the cable on the whole length of the sample.

On the other hand, in the case of the **SS** scheme the current I_m at $t \approx t_1$ goes along the sample through the whole cross-section of the cable. At $t \gg t_1$ it diffuses in **the Strand**.

The discharge of the sample $C1_2$ calculated for $t_e = 0.175\,H/\Omega$ agrees with the experiment as well (see Fig. 3(a)). The calculations were not done for the other two samples because the value M was too large. However, calculations made for the samples shorter than $C1_2$ show that the time constant t_0 is proportional to the square power of the sample length. The ratio of the experimental results for the samples $C1_2$ and $C1_1$ agrees with this statement.

In author's opinion, the huge difference in the values of t_0 for the samples of types $C1$ and $C2$ is mainly due to two reasons. The single contact resistance r for the sample $C2$ is much larger than that for the samples of the $C1$ type because of the $Cu - Ni$ barriers inside the strands and the fact that the contacts between the strands are not soldered. Also, associated inductances [2] for the tightly packed strands of the sample $C2$ are much smaller than those for the strands in the sparse samples of the $C1$ type.

4. Conclusion

With the time passing, the largest time constant begins to define exclusively the rate of the current redistribution between the elements of the sample of the multi-strand superconducting cable. Given other parameters fixed, the largest time constant is proportional to the square power of the length of the sample.

References

[1] Ghosh A K, Robins K E and Sampson W B 1992 *Supercollider* vol 4 ed J Monte (N.Y.: Plenum Press) 765-768

[2] Akhmetov A A *et al* 1993 *Supercollider* vol 5 ed P Hale (N.Y.: Plenum Press) 443-446

Inst. Phys. Conf. Ser. No 148
Paper presented at Applied Superconductivity, Edinburgh, 3–6 July 1995

In-situ TEM observation of the decomposition of Y-124 into Y-123 and CuO

M.Reder, D.Müller, K.Heinemann and H.C.Freyhardt

‡ Institut für Metallphysik, Universität Göttingen, Hospitalstraße 3/7, 37073 Göttingen (Germany)

Abstract. $YBa_2Cu_4O_8$ (Y-124) decomposes into $YBa_2Cu_3O_{7-\delta}$ (Y-123) and CuO at high temperatures, e.g. bulk Y-124 is stable up to 850-900°C at $p(O_2)$=0.2 bar. In earlier investigations Y-123 was obtained by annealing of Y-124 in air at temperatures above 900°C. It was expected that this process might create Y-123 with finely dispersed CuO precipitates suitable for flux pinning. In fact, samples of thermally decomposed Y-124 exhibit a critical current density, j_c, which is enhanced with respect to the starting material as well as to pure Y-123.

The decomposition process was carried out directly within the transmission electron microscope (TEM) in order to clarify whether this enhancement of j_c is caused by CuO precipitates or by additional defects like stacking faults and associated partial dislocations. For this purpose, the intensity of the electron beam was increased by opening the condensor aperture of the microscope. The low oxygen partial pressure within the TEM lowers the decomposition temperature of Y-124. Thus, the enhanced heating of the sample with the electron beam within the TEM is sufficient to cause the decomposition into Y-123 and CuO. The process can be controlled up to a certain extent by focussing the electron beam and the decomposition can be observed in different stages. The most important advantage of this method of decomposing Y-124 is the possibility to investigate the same grains in different stages of the decomposition. Despite of very different conditions compared to the experiments mentioned above it was possible to obtain areas of twinned Y-123 by in-situ decomposition. Fine CuO precipitates were identified by EDS and electron diffraction. Therefore, it seems very probable that CuO precipitates play an important role for the observed enhancement of j_c of decomposed samples.

1. Introduction

The controlled enhancemant of the critical current density, j_c, is one of the most important requirements in the development of high T_c superconductors for technical applications. j_c is determined by pinning of the magnetic flux lines at microstructural defects like dislocations or small inclusions. The decomposition of $YBa_2Cu_4O_8$ (Y-124) into $YBa_2Cu_3O_{7-\delta}$ (Y-123) and CuO at high temperature (850-900°C at $P(O_2)=0.2$ bar) leads to Y-123 matrix with an enhanced intragranular j_c compared to sintered Y-123 [1,2].Therefore, it was expected that this process might create Y-123 with finley dispersed CuO precipitates suitable for flux pinning. In order to clarify whether this enhancement of j_c is caused by CuO precipitates or by additional defects like stacking faults and associated partial dislocations which were observed in the decomposed samples the decomposition process was carried out directly within the transmission electron microscope (TEM).

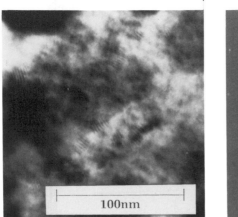
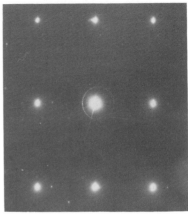

Figure 1) *a) Small CuO particles in the Y-123 matrix after decomposition identified by Moiré pattern and b) electron diffraction pattern in [031] orientation*

2. Experimental

Y-124 was prepared by sintering of Y-123 and Cu at 825°C and $p(O_2) = 1$ bar. Details have been presented in [1].
The samples for the TEM investigations have been prepared by mechanically thinning and ion milling at 4 kV under an angle of 13°. In order to minimize artefacts due to the ion bombardement the samples were cooled by liquid nitrogen during the ion milling. The TEM investigations were performed with a Philips 400T working at 120 kV. The analysis of the chemical composition was carried out with an energy dispersive x-ray spectrometer (EDS). The thermal heating of the samples in the TEM was achieved by increasing the intensity of the electron beam by opening the condensor aperture of the microscope.

3. Results

Pure Y-124 samples were studied in the TEM. The structure of Y-124 is orthorhombic like that of Y-123 but additional layers of CuO enlarge the unit cell of Y-124. As the orthorhombicity is less marked no twinning parallel to (110) and ($1\bar{1}0$) planes like in Y-123 occurs in the Y-124 phase [3]. The low oxygen partial pressure within the TEM lowers the decomposition temperature of Y-124. Thus, the enhanced heating of the sample with the electron beam within the TEM is sufficient to cause the decomposition into Y-123 and CuO. Electron diffraction confirms that the decomposition from the Y-124 phase into the Y-123 phase has been successfully achieved in the TEM. Similar experiments were reported in the literature to transform orthorhombic Y-123 into tetragonal Y-123 in which a detwinning was achieved by heating the orthorhombic Y-123 phase under the electron beam [4].

 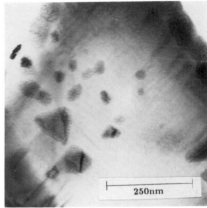

Figure 2) a) *Twin lamellas in orthorhombic Y-123 after the decomposition of Y-124 near [001] zone axis, b) Y-123 matrix with particles of about 100 nm diameter (same area as in a)*

The CuO-particles have been identified by electron diffration and Moiré pattern (Fig.1). The particles identified by Moiré pattern have an average size of 10 nm in an early stage. The periodicity, D, of the Moiré fringes amounts to 21 Å. These fringes are caused by double diffraction of the Y-123 (113) diffraction spot and the CuO (012) spot.

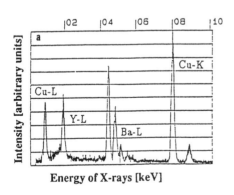 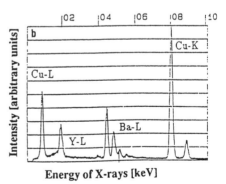

Figure 3) *Energy dispersive x-ray spectrum a) of the Y-123 matrix after decomposition and b) of a CuO particle with 100 nm diameter*

534

According to the diffraction pattern the angle, α, between these spots is about 1.5 - 2°. D is calculated to 20.9 Å - 21.5 Å which is in good agreement to the measured value of 21 Å using $D=d_1d_2/\sqrt{d_1^2 + d_2^2 - 2d_1d_2cos\alpha}$ [5].

The process of the in-situ decomposition can be controlled up to a certain extent in the TEM by focussing the electron beam. Thus, the decomposition can be seen in different stages. After the decomposition the Y-123 phase is orthorhombic, which can be clearly seen by the appearence of small twin lamellas (Fig.2a). The streaking in [110] and [1$\bar{1}$0] direction in the electron diffraction spots is due to a marked thickness of the twin boundary [6]. Particles in an later stage with a size of about 100 nm (Fig.2b) have been analysed by EDS. These particles are Cu-rich (Fig.3).

4. Conclusion

The decomposition of Y-124 leads to a orthorhombic Y-123 matrix and finely dispersed CuO. Therefore, it is reasonable to assume that the small CuO particles play an important role for the enhancement of j_c in comparison to sintered Y-123. Thus, it is possible to control j_c by controlling the microstructure due to the decomposition of Y-124 into Y-123 and CuO. The biggest problem seems to be the high velocity of the phase transformation.

Acknowledgement

This work was supported by the BMBF under grant number 13N5493A and 13N6566.

References

[1] J.Krelaus, B.Ullmann, K.Heinemann and H.C.Freyhardt , *Applied Superconductivity 1993*, ed. H.C. Freyhardt, Proc. EUCAS (1993) 783

[2] S.Jin et al, Appl.Phys.Lett. **56** (1990) 1287

[3] T.Krekels et al, Physica C **169** (1990) 457

[4] G. Van Tendeloo and S.Amelinckx, Phys.Stat.Sol. **103** (1989) K1

[5] J.W.Eddington, **Practical Electron Microscopy in Material Science**, MacMillan, London and Basingstoke (1976)

[6] Y.Zhu, M.Suenaga and A.R. Moodenbaugh, Phil.Mag.Lett. **62** (1990) 51

Inst. Phys. Conf. Ser. No 148
Paper presented at Applied Superconductivity, Edinburgh, 3–6 July 1995
© *1995 IOP Publishing Ltd*

Relaxation measurements of VGF melt-textured YBa$_2$Cu$_3$O$_{7-\delta}$ with varying stoichiometry

J Krelaus†, M Ullrich‡ K Heinemann† and H C Freyhardt††

† Institut für Metallphysik der Universität Göttingen, Windausweg 2, D-37073 Göttingen, Germany

‡ Zentrum für Funktionswerkstoffe gGmbH, Windausweg 2, D-37073 Göttingen, Germany

Abstract. Relaxation measurements are presented of melt-textured YBa$_2$Cu$_3$O$_{7-\delta}$ (Y-123) samples grown by the vertical gradient freeze technique and of Y-123 powder material. The melt-textured samples are "quasi single crystals" consisting of stoichiometric Y-123, Y-123 with 25 mol-% Y-211 additions and Y-123 with 25 mol-% Y-211 as well as 1.8 mol-% PtO$_2$. The measurements are performed by means of a Faraday magnetometer with an applied magnetic field B_a($\parallel$ c) of 4T in the temperature range from 4.2 to $\approx$ 60 K. A logarithmical decay of the magnetization is observed up to 10^4 s after an initial 10 s long non-logarithmical decay region. Melt-texturing leads to a drastic enhancement of the $U(j/j_0)$ values compared to the Y-123 powder. Y-211 addition leads to a further enhancement of these values, whereas the addition of PtO$_2$ has no observable effect on the normalized $U(j/j_0)$ curve.

1. Introduction

It has been shown that the vertical gradient freeze (VGF) method is suitable for melt-texturing YBa$_2$Cu$_3$O$_{7-\delta}$ (Y-123) [1]. Y$_2$BaCuO$_5$ (Y-211) in excess leads to several μm large Y-211 inclusions within the Y-123 material associated with many structural defects. Further addition of PtO$_2$ decreases the average Y-211 diameter by about one order of magnitude and significantly changes the microstructure of the material [2]. At temperatures close to T_c, the Bean critical current density of the non-stoichiometric samples is significantly enhanced over that of the stoichiometric samples [3]. The present work investigates the effect of second phase additions on the relaxation behaviour of VGF melt-textured Y-123. The relaxation data obtained by using a Faraday magnetometer are compared to published results of SQUID measurements on melt-powder-melt-growth Y-123 [4, 5]. Furthermore, the influence of PtO$_2$ on the relaxation behaviour is examined.

2. Samples

The polycrystalline material is commercial Y-123 powder (Solvay Barium Strontium GmbH). The data of the samples are given in table 1. The demagnetization factors were obtained from virgin hysteresis loops at $T = 4.2$ K.

Table 1. Composition and geometry of the samples

Sample name	POL	MT	MT+	MT++
Compound	polycryst. (powder)	melt-textured stoich. Y-123	"MT" + 25 mol% Y-211	"MT+" +1.8 mol-% PtO_2
Mass	92.75mg	$19\mu g$	$29~\mu g$	$29~\mu g$
Screening length	$3.2\mu m$	$230\mu m$	$270\mu m$	$250\mu m$
Demagn. factor	(0.3)	0.874	0.800	0.847

3. Relaxation measurements

Relaxation is measured at applied magnetic fields, B_a, of 4T which is adjusted with $\dot{B}_a = \pm 36.5$ mT/s respectively. B_a is swept to the maximal value of 10 T to ensure a complete flux profile reversal within the samples. The logarithmical regime of the relaxation starts at about 10 s. Measurements are performed up to 10^4 s. The hysteresis at $B_a = 4$T is calculated from the relaxation data as the difference of the average of the two relaxation curves in the interval 10 s – 30 s. Since the magnetization signal of all samples is smaller than $1T/\mu_0$ and since the hysteresis loops are quite flat at 4T (i.e. $M(4T \pm 1T) \approx M(4T)$), no demagnetization correction is performed. $U(j)$ is determined as follows [7, 8, 4]: First, from each relaxation curve $M(T,t)$ a data set $U_T = k_B T \cdot (C + \ln(dM/dt))$ vs. the Bean current density $j = 2 \cdot M/(\mu_0 \cdot 0.5D)$ is created ([6], D is the screening length). Then, the parameter C is determined by supposing continuity of $U_T(j)$ for $T < 20$ K (i.e. $G(T) \approx 1$, see below). Table 2 collects the C-values determined in this manner. For the non-stoichiometric samples MT+ and MT++ these values agree with $C = 16$ resp. 15 given in [4, 5], whereas for the stoichiometric samples the much smaller value 2 is found. $G(T)$ describes the explicit temperature dependence of $U_T(j)$: $U_T(j) = G(T) \cdot U(j)$. This function is determined by assuming continuity of $U(j) = U_T(j)/G(T)$ for higher T. The common assumption of $G(T) = (1 - (T/T_x)^2)^m$ [9] uses the two free parameters $T_x \approx T_{irr}$ and $m = 1 .. 2$. In our calculations T_x is chosen as $T_c \approx 93$ K. For temperatures not too close to this value, the scaling is quite insensitive to m. A measurement of the relaxation on a large sample of the same composition as MT++ at high enough temperatures yields $m \approx 2$ in qualitative agreement with the value 1.7 given by [4] for a corresponding sample. For simplicity, this value is chosen for all our samples.

Fig. 1 depicts the $U(j)$ curves at $B_a = 4$ T ($\dot{B}_a > 0$). The measurements with $\dot{B}_a < 0$ do not show significant differences. MT++ possesses the highest absolut j

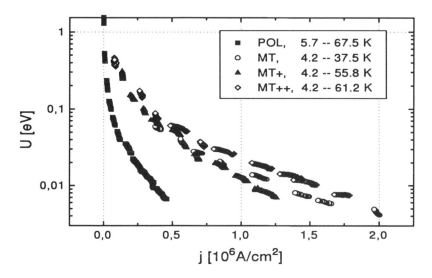

Figure 1. $U(j)$ curves at $B_a = 4$ T ($\dot{B}_a > 0$).

leading to the highest $U(j)$ values, POL exhibits the lowest j and $U(j)$. Between MT and MT+ a crossover can be observed: for low temperatures (high j), MT shows the higher U values whereas for high temperatures it is reversed. This is an effect of the j_c enhancement as $T/T_c \to 1$ due to the doping with Y-211.

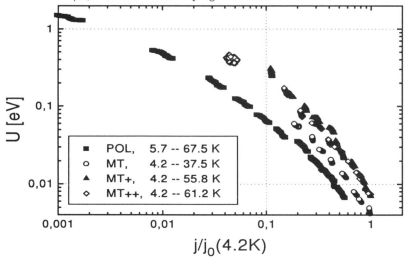

Figure 2. Logarithmical plot of $U(j/j_0(4.2\text{K}))$ at $B_a = 4$ T ($\dot{B}_a > 0$).

In Fig. 2 $\log(U)$ is plotted vs. $\log(j/j_0(4.2\text{K}))$. $j_0(4.2\text{K})$ is calculated from the measured starting value of the magnetization at 4.2 K taken as an approximation for the Bean critical current density j_c without thermal activation. Again, POL shows the lowest U curve. U of MT is a factor ≈ 1.5 higher. The non-stoichiometric samples MT+

and MT++ almost coincide and show a further U enhancement by a factor ≈ 2 with respect to MT.

In table 2 the values $-d\ln(U)/d\ln(j) = \mu$ are given in addition. They tend to be higher than the theoretical predictions $\mu = 7/9$ for small j and $\mu = 3/2$ for high j from the collective creep model [10, 11]. This behaviour was also found by [4].

Table 2. Parameters describing the scaled $U(j)$ curves.

	POL	MT	MT+	MT++
C	2	2	12	14
$\mu(j/j_0(4.2\text{K}) = 0.2)$	0.9	1.2	1.3	1.4
$\mu(j/j_0(4.2\text{K}) = 0.7)$	1.9	1.9	1.7	1.7

4. Summary/Acknowledgement

Relaxation measurements on VGF melt-textured Y-123 with or without excess Y-211 lead to results that are comparable to similar experiments in the literature. Normalized $U(j/j_0(4.2K))$ curves of melt-textured stoichiometric Y-123 are by a factor of ≈ 1.5 higher than values for polycrystalline material. Addition of Y-211 leads to a further enhancement by a factor ≈ 2. PtO_2 doping of these samples, however, does not affect the normalized U curves. Our VGF melt-textured samples exhibit current densities that are about 5 times higher than the values given in [4]. Corresponding to that, also the absolute U values are slightly enhanced.

The financial support by the German BMBF (grant nos. 13N5493A and 13N6566) and by the European Union (project no BRE2-CT04-1011) is gratefully acknowledged.

References

[1] Ullrich M and Freyhardt H C 1993, in *Superconducting Materials* (Eds. J Etourneau, J B Torrance and H Yamauchi) IITT International, Gournay sur Marne (France), p 205

[2] Müller D, Ullrich M, Heinemann K and Freyhardt H C 1994, in *Critical Currents in Superconductors* (Ed. H W Weber, World Scientific, Singapur) p 443

[3] Ullrich M, Müller D, Heinemann K and Freyhardt H C, in *Critical Currents in Superconductors* (Ed. H W Weber, World Scientific, Singapur) p 46

[4] Kung P J, Maley M P, McHenry M E, Wills J O, Murakami M and Tanaka S 1993 *Phys. Rev. B* **48**(18) 13922

[5] Wen H H, Rong X S, Yin B, Che G C and Zhao Z X 1995 *Physica C* **242** 365

[6] Bean C P 1962 *Phys. Rev. Lett.* **8**(6) 250, *Rev. Mod. Phys.* **36** 31

[7] Beasley M R, Labusch R, Webb W W 1969 *Phys. Rev.* **181** 682

[8] Maley M P, Willis J O, Lessure H and McHenry M E 1990, *Phys. Rev. B* **42**(4) 2639

[9] McHenry M E, Simizu S, Lessure H, Maley M P, Coulter J Y, Tanaka I and Kojima H 1991, *Phys. Rev. B* **44** 7614

[10] Feigel'man M V, Geshkenbein V B, Larkin A I, Vinokur V M 1989 *Phys. Rev. Lett.* **63** 2302

[11] Fisher M P A 1989 *Phys. Rev. Lett.* **62** 1415

Inst. Phys. Conf. Ser. No 148
Paper presented at Applied Superconductivity, Edinburgh, 3–6 July 1995
© *1995 IOP Publishing Ltd*

Screening properties of Bi-2212 superconducting tubes

J R Cave
Vice-présidence Technologie et IREQ, Hydro-Québec, 1800 montée Ste-Julie, Varennes (Québec)
J3X 1S1, Canada

D R Watson and J E Evetts
Dept. Materials Science and Metallurgy, University of Cambridge, Pembroke Street, Cambridge
CB2 3QZ, UK and IRC in Superconductivity, West Cambridge Site, Madingley Road,
Cambridge, CB3 0HE, UK

Abstract The AC screening properties of tubes formed from composite reaction textured (CRT) Bi-2212 have been investigated at low frequencies at 77K. Due to the shallow E-J characteristics typical of these materials in this regime the usual critical state analysis of the flux profiles does not strictly apply. In this case solutions of the magnetic diffusion equation with a field dependent resistivity are shown to be more appropriate. These results are relevant to the behaviour of these materials in practical devices such as fault current limiters.

1. Introduction

Bulk Bi-2212 superconducting material shows promise for applications such as current leads, coils, screening enclosures and fault current limiters. In order to progress to these applications we need to fully characterize such properties as J_c (the critical current density) and ρ_f (the flux-flow resistivity) under a variety of situations (e.g. various temperatures and fields, AC and DC currents). In this article we present results using an AC screening technique in which the field penetrating to the centre of a hollow superconducting cylinder is measured [1].

2. Sample preparation

The samples used in this study were in the form of tubes or stacked rings produced using the CRT process [2-4]. These references describe how precursor materials consisting of Bi-2212 powder (particle size 2-4μm, Hoechst, AG) and MgO-fibres (produced in house) were used to prepare perform material with a technique referred to as elastomer processing (EP) [2]. This technique involves roll milling DuPont® ethylene/methlacrylate copolymer, plasticizer, Bi-2212 powder and 10wt% MgO-fibres. The resulting material was then warm pressed yielding uniform sheets 1-5mm thick with MgO-fibres randomly aligned in the sheet plane.

Two types of cylinders were prepared from the green sheets; to make a modular cylinder (consisting of a stack of rings) an annulus shaped die was used to stamp rings from the elastomer processed sheet whereas for a monolithic cylinder a strip of EP material was wound around a cylindrical mandrel several times thereby building up the desired wall thickness.

Preform cylinders were subsequently subjected to binder-burn off and CRT heat sequences [3,4]. Finally, thin walled brass or stainless steel cylinders were slid over the CRT cylinders for reinforcement. Fig 1 shows reinforced modular and monolithic cylinders.

3. Screening measurements

We have investigated the screening characteristics over a range of low frequencies (1-200Hz) at liquid nitrogen temperatures (77K). Typical results for the central field amplitude, B_i, versus the applied AC field amplitude, B_0, are shown in Fig. 2 for a stack of CRT rings inside a brass cylinder.

Fig. 1. Reinforced tubes fabricated from CRT Bi-2212 material (left; brass, h=48mm and right; stainless steel).

The screening threshold field at which significant flux begins to penetrate into the central region of the hollow tube increases with frequency. This is because of the shallow E-J curve in these materials and the low flux-flow resistivity giving rise to skin-depth screening effects. As there are also associated phase shifts between the drive field and the field in the centre of the tube we have employed a whole waveform technique (with averaging) using a digital oscilloscope to obtain the central field amplitude and its phase rather than a phase sensitive detector.

4. Screening theory

For low temperature superconductors and for HTS materials under certain conditions [5] the critical state analysis of this type of experiments yields good agreement. However, in materials where the E-J curve is shallow this may not be the case. Fig. 3 shows schematic E-J curves described by the often used empirical relationship $E\sim J^n$. Rhyner [6] has shown that for n below about 8 the superconductor behaves more like a low resistivity linear material.

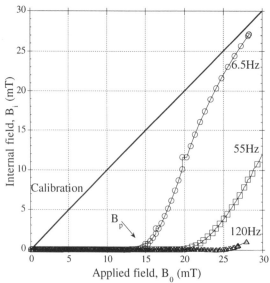

Fig. 2. Screening properties of a stack of CRT rings (16.6mm OD, 3.7mm wall and 1.5mm thick for a total height of 48mm) inside a reinforcing brass tube at 77K.

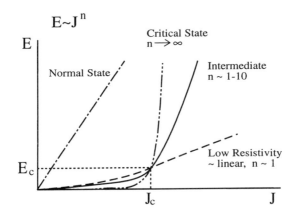

Fig. 3. Schematic E-J curves for a type II superconductor.

For a low resistivity linear material Chen [7] has developed the appropriate solutions of the magnetic diffusion equation. In particular, the solution for the internal field, B_i, and its phase in the central region of a hollow tube, see Fig 4, is given by (inner radius, r_b, outer radius, r_a, AC drive field B_0, skin depth, δ)

$$b(r=r_b) = \frac{(K_1(k\,r_b)\,I_0(k\,r_b)+I_1(k\,r_b)\,K_0(k\,r_b))}{(K_1(k\,r_b)\,I_0(k\,r_a)+I_1(k\,r_b)\,K_0(k\,r_a))} \qquad B_i = B_0\,\mathrm{Mod}[b(r=r_b)] \quad \text{Eqns (1)}$$
$$\text{Phase} = \mathrm{Arg}[b(r=r_b)]$$

where $k = (1+i)/\delta$ with $i=\sqrt{-1}$ and K and I are modified Bessel functions.

To take into account the decrease in J_c and increase in resistivity as the applied field, B_0, increases we introduce an approximate analysis using an effective field dependent resistivity and skin depth

$$\rho=\rho_0(B/B_p)^\gamma \text{ and } \delta=\sqrt{(2\rho/\mu_0\omega)} \qquad\qquad \text{Eqns (2)}$$

Fig. 5. shows good agreement between theory and experiment for brass reinforced CRT rings (at low applied fields the phase angle tending to 0° is due to a small leakage flux).

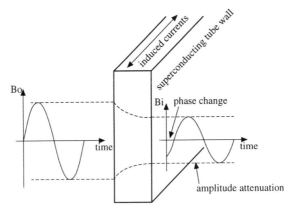

Fig. 4. Schematic of amplitude attenuation and phase change of the central field, B_i, with applied field, B_0 for a hollow low resistivity cylinder.

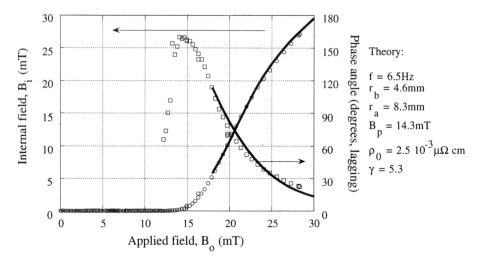

Fig. 5. Comparison of the experimental data (points) in Fig. 2. with theory (solid lines), Eqns (1), using the effective field dependent resistivity from Eqns (2).

5. Discussion and conclusions

At 77K the intrinsic $J_c(B)$ curve is very sensitive to magnetic fields in the mT range. This accounts in part for the low overall values of J_c calculated from $B_p/(\mu_0 w)$ for the data in Fig. 2, B_p is the penetration field and w is the wall thickness. Here, $J_c(14mT, 6.5Hz)=300A/cm^2$, $J_c(20mT, 55Hz)=430A/cm^2$ and $J_c(25mT, 120Hz)=540A/cm^2$. Another contributing factor is that CRT yields highly textured microstructures in bulk Bi-2212/MgO-f composites, with the best c-axis texture observed with a random 2 dimensional planar distribution of MgO-fibres. This results in strongly anisotropic J_c values of ~2500A/cm² in the fibre plane versus <500A/cm² perpendicular to the fibre plane [1]: a cylinder with the optimum MgO-f orientation with w=1.0mm had $J_c(23mT, 17Hz)=1825A/cm^2$.

These results show that at power frequencies some HTS materials can be more appropriately described using the magnetic diffusion equation rather than the critical state model. Gurevich [8] reviews recent developments in non-linear flux diffusion in HTS material. The implications for applications is that the regime in which the superconductor is acting in must be identified and understood in order to develop fully optimised devices.

References

[1]   J R Cave, M Mautref, C Agnoux, A Leriche and A Février 1989 Cryogenics **29** p341
[2]   D R Watson, M Chen, D M Glowacka, N Adamopoulos, B Soylu, B A Glowacki and J E Evetts, 1995 IEEE Trans. on Appl. Supercon. [ASC'94, in press]
[3]   D R Watson, M Chen and J E Evetts 1995 Supercond. Sci. and Tech. **8**, p311
[4]   M. Chen, D M Glowacka, B Soylu, D R Watson, J K S Christiansen, R P Baranowski, B A Glowacki and J E Evetts 1995 IEEE Trans. on Appl Supercond. [ASC'94, in press]
[5]   J B W Ricketts, K-H Müller and R Driver 1991 Physica C**183**, p17
[6]   J Rhyner 1993 Physica C**212**, p292
[7]   Q Y Chen 1991 in Magnetic Susceptibility of Superconductors and Other Spin Systems ed. R A Hein et al., Plenum Press p81
[8]   A Gurevich, 1995 Int. J. Mod. Phys. B, Vol 9, p1045

Inst. Phys. Conf. Ser. No 148
Paper presented at Applied Superconductivity, Edinburgh, 3–6 July 1995
© *1995 IOP Publishing Ltd*

543

Physical and mechanical characterisation of YBCO samples obtained by new TLDMG process

F. Abbattista, R. Albanese, M. Vallino

Dept. of Science of Materials, Politecnico Torino, c.so Duca Abruzzi 24, 10129 Torino, Italy

R. Gerbaldo, G. Ghigo, L. Gozzelino, E. Mezzetti, B. Minetti

Dept. of Physics, Politecnico di Torino, c.so Duca degli Abruzzi 24, 10129 Torino, Italy

A. B. Mossolov

Dept. of Mechanics, Politecnico di Torino, c.so Duca degli Abruzzi 24, 10129 Torino, Italy

Abstract. This paper deals with a comparison between physical and mechanical properties of YBCO melt grown samples by means of a theoretical model. A new method to obtain the precursors for the MG process is presented and denominated Temporary-Liquid-Densification (TLD). Magnetic hysteresis cycles and a.c. susceptibility measurements were performed to evaluate the critical current density values. Mechanical characterisation in zero field cooling and field cooling conditions, showing hysteretic behaviour and creep of the levitation force was also performed. The correspondence between physical and mechanical properties was obtained by means of an analytical model for levitation force. This model is based on the Bean critical state concept and takes into account the inhomogeneity of the material. A quantitative correspondence with experimental results was found.

1. Introduction

Probably the most promising mechanical applications of bulk HTS are Superconducting Magnetic Bearings (SMB). These applications require large critical current densities for fields up to 1 T in bulk pellets. As it is well known, in sintered materials, because of weak links and high angle grain boundaries, critical current densities are too low. In order to obtain larger current carrying capability, in the last years several texturing techniques were developed [1,2]. Good grain alignment is usually reached with a melt growth (MG) technique. For the optimisation of this process, high density pre-sintered pellets are needed. In this paper a new method to obtain such a precursor for the MG process is presented and denominated Temporary Liquid Densification (TLD).

Moreover, to design practical applications a model which describes the levitation force, taking into account chemical, physical and geometric properties of the pellets, is needed. An analytical model (details are given in ref. [3]), based on the Bean critical state concept, is used here to compare physical and mechanical characterisations on samples prepared with the TLD-MG technique. The good quantitative correspondence shown below seems to be very promising.

544

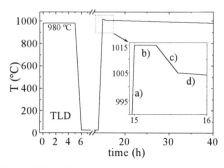

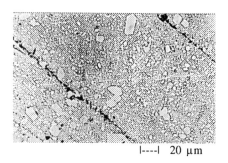

Fig.1 - TLD-MG heat treatment. The procedure is also described in the text.

Fig.2 - Optical micrography of a pellet prepared with the TLD-MG technique.

2. Preparation

Starting from high purity $Ba(NO_3)_2$, CuO and Y_2O_3, the phases $BaCuO_2$ and $Y_2Cu_2O_5$ were preliminary prepared according to the reactions:

$$Ba(NO_3)_2+CuO \rightarrow BaCuO_2+2NOx+nO_2$$
$$Y_2O_3+2CuO \rightarrow Y_2Cu_2O_5$$

After X-ray analysis, these phases were used to prepare mixtures ($BaCuO_2$ + $Y_2Cu_2O_5$ + Y_2O_3) for obtaining biphase solid ($Ba_2YCu_3O_{7-x}+BaY_2CuO_5$) according to the reaction:

$$Y_2Cu_2O_5+nY_2O_3+(4+n)BaCuO_2 \rightarrow 2Ba_2YCu_3O_{7-x}+nBaY_2CuO_5$$

After grinding and homogenising, the powders were pelleted (30 mm diameter, 6 mm thickness) under axial pressure at P = 500 MPa and then kept in a furnace preheated at 970-980 °C for five hours (fig.1). The presence of ternary liquids at T < 970 °C of the non-binary $BaCuO_2$-$Y_2Cu_2O_5$ and binary $BaCuO_2$-Y_2O_3 systems facilitates the formation of solid phases $Ba_2YCu_3O_{7-x}$ and BaY_2CuO_5 with the gradual disappearance of the liquid phases (Temporary Liquid Densification, TLD). A microcrystalline sintered pellet of high density and perfect phase distribution was therefore produced. Because of the relatively low temperature of the process and the absence of free CuO in the original mixture, the loss of oxygen in this process is very low. The sintered pellet obtained in this way was subjected to a MG process according to the following procedure, also sketched in fig.1: a) rapid heating to 1015°C (60 min); b) maintenance at 1015°C (20 min); c) cooling from 1015°C to 1005°C (20 min); d) slow cooling from 1005°C to 980°C (1 °C/h). Spontaneous cooling in the furnace to 400 °C followed. The oxygen treatment was performed at 430 °C for 5 days. Fig. 2 shows a optical micrography of a pellet prepared according to the technique presented here. A homogeneous distribution of 211-phase inclusions within large 123 oriented domains is visible.

3. Characterisation

3.1 Critical current density measurements
In order to check the quality of the samples prepared by the described TLD-MG technique, four smaller samples were cut from different regions of the pellet.

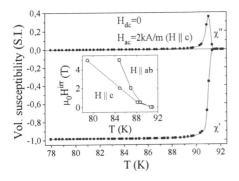 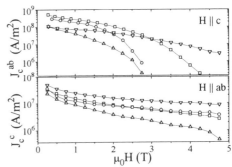

Fig.3 - Magnetic a.c. susceptibility measurement. In the inset the irreversibility lines for two different field orientations are reported.

Fig.4 - J_c values as a function of magnetic field at 77K for all the samples and two field orientations (H‖c upper figure; H‖ab lower figure).

The samples were shaped as slabs of about $2.5 \times 2.5 \times 0.3$ mm^3 with the largest surface parallel to ab-planes. Magnetic measurements were performed by means of a Lake Shore 7225 susceptometer-magnetometer.

Typical a.c. susceptibility measurements concerning these slabs are presented in fig.3. The imaginary part of the susceptibility shows only the intragrain peak, which means that superconducting currents are not limited by weak links and flow through the whole sample. Moreover, the transition is very sharp and almost perfect diamagnetism is reached at temperatures of about 86 K. The maximum of the out-of-phase peak was also used to define the irreversibility temperatures for applied d.c. fields up to 5 T. The irreversibility lines for two field orientation are presented in the inset of fig.3. These lines show the anisotropy of the material, in accordance with the hypothesis of good alignment of 123 regions in the sample.

The critical current densities inferred from d.c. magnetisation hysteresis cycles up to 5 T at 77 K, for all the samples and for two different field orientations, are presented in fig.4. J_c values were calculated by means of the extended Bean model [4], using the macroscopic dimensions of the sample. As it is well known, because of the anisotropy of the material, two components of the critical current density should be distinguished: the critical currents flowing in ab-planes J_c^{ab} and the critical currents flowing through planes J_c^c. From the point of view of applications as SMB, we are interested in crictical current density values in relatively low fields (up to 1 T). Although some spread was observed, J_c values are comparable, i.e. about $2 \cdot 10^8$ A/m^2 for J_c^{ab} and $3 \cdot 10^7$ A/m^2 for J_c^c.

3.2 Levitation force measurements

The levitation force, due to interaction between a permanent magnet and the superconductor pellet, was measured in Zero Field Cooling (ZFC) and Field Cooling (FC) conditions. The permanent magnet was a NdFeB ring (ϕ_i=10 mm, ϕ_e= 20 mm, h= 5 mm) with a maximum field of 0.4 T. In order to prevent the influence of creep, the levitation force was always automatically measured 5 s after every new displacement between the pellet and the magnet. In ZFC conditions, during the first motion of the magnet toward the pellet, we found values of repulsive force higher than during next steps. Moreover an attractive force was found during the removal of the magnet, leading to a hysteretic behaviour in the levitation force (fig. 5). In FC conditions no differences were observed between results obtained from the first and further motions toward the pellet.

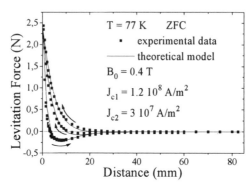

Fig.5 - Hysteresis in the levitation force vs. distance between the permanent magnet and the superconductor pellet. Data were fitted by means of the theoretical model described in the text (main parameters are reported).

3.3. Comparison between physical and mechanical properties by means of a theoretical model

Experimental results were compared with data calculated by a model based on the Bean theory. Basically, the levitation force is related to critical currents flowing in the superconductor by the formula

$$F = \mu_0 \int_V J_c \times H \, \mathrm{d}v$$

Following ref. [5] it is possible to rewrite this equation as

$$F = C \, M \, V \, dH/dz$$

where M is the average magnetization of the HTSC, V is the volume of the superconductor and C accounts for the geometry of the system and the distribution of magnetic field. Moreover, in order to take into account the anisotropy of the material, the pellet is modelled as consisting of two superconductive phases with two different critical current densities J_{c1} and J_{c2} [3]. J_{c1} and J_{c2} can be associated with critical current densities flowing in ab-planes (J_c^{ab}) and along c-axis (J_c^c), respectively. Critical current density values deduced by means of other techniques can thus be used to predict the levitation force. On the other hand, from levitation force experimental data it is possible to evaluate the parameters J_{c1} and J_{c2} (see fig.5) and to compare these average values with J_c^{ab} and J_c^c obtained by magnetic measurements (see fig.4). In this case, a good qualitative correspondence was found. This result is a further proof of the fact that the whole pellet consists of good quality grains, with average physical characteristics described in § 3.1.

References

[1] Murakami M., Oyama T., Fujimoto H. et al. 1991, *IEEE Trans. on Magnetics* **27** (2)
[2] Bornemann H.J. et al. 1994 *Advances in Superconductivity* (Springer-Verlag) p. 1311
[3] Mossolov A., Genta G., Albanese R. 1995 *Superconductivity and Superconducting Materials Technology*, edit by P.Vincenzini, (Faenza (I): Techna) p.671
[4] Gyorgy E.M., van Dover R.B., Jackson K.A., Schneemeyer L.F. 1989 *Appl. Phys.Lett.* **55**, 283
[5] Brandt E.H., 1990, *Am. J. Phys.* **58**, 43

Inst. Phys. Conf. Ser. No 148
Paper presented at Applied Superconductivity, Edinburgh, 3–6 July 1995

Investigating the characteristics of high-temperature superconductors by means of MMMA

B Andrzejewski[†], B Czyżak[†], J Stankowski[†], L Kevan[‡]

† Institute of Molecular Physics, Polish Academy of Sciences, Smoluchowskiego 17, PL-60179 Poznań, Poland
‡ Department of Chemistry and Texas Center for Superconductivity, University of Houston, Houston, Texas 77204-5641

Abstract. Attempts to apply the technique of microwave absorption to characterising superconducting materials have been made ever since Bednorz and Müller first discovered the phenomenon of high-temperature superconductivity (HTS). In this report we would like to show a few examples of how some basic HTS properties can be investigated by means of the magnetically modulated microwave absorption technique (MMMA).

1. Introduction

The discovery of extremely strong microwave absorption in HTS in the late 1980s [1,2] generated a broad interest in the subject and brought along a host of studies and communications in the following years. Unfortunately, the enthusiasm waned when it turned out that interpretation of the abundant data provided is very complicated and equivocal. Moreover, an adequate description of the process of microwave absorption in superconductors was not yet, and still is not, successfully completed. These obstacles hampered early attempts to use the microwave absorption technique to characterise the basic properties of newly discovered superconductors. Still, microwave absorption caught on very quickly and started to be implemented also as a standard test for superconductivity [3], even though Oseroff et al. pointed out that in rare-earth copper oxides in the range of low magnetic fields the presence of microwave absorption does not necessarily prove superconductivity [4]. What made the new technique so popular were its undeniable attractions: very high sensitivity, easy measuring procedure, and cheapness and availability of the electron paramagnetic resonance (EPR) spectrometers, typically used in microwave absorption experiments.

Below, we shall discuss methods of measuring the critical temperature T_c, testing the quality of thin films and single crystals, and studying the dynamics of the processes accompanying the synthesis of Rb_3C_{60} fullerides. Then we shall go on to talk about the possibilities of applying magnetically modulated microwave absorption to measuring the critical fields and intergrain critical supercurrent J_c.

2. The MMMA technique

In most MMMA studies, a standard EPR spectrometer is applied which uses the microwave band X (or, less frequently, higher frequency bands K or Q). The DC magnetic field in which the superconductor is placed is modulated by an AC magnetic field of a frequency of 100 kHz and amplitude of several Oe. The sample is positioned in a microwave cavity in the place of the maximum magnetic microwave field. The configuration of all the magnetic fields involved is the same as in the EPR experiments.

548

3. Examples of MMMA applications

3.1. Measurements of the critical temperature

As one of its first applications MMMA was used to measure the critical temperature T_c of a superconductor. This was possible thanks to the fact that a transition into the superconducting phase leads to a strong microwave dissipation in HTS. The advantage of such microwave measurements lies in their high precision allowing one to gauge the transition temperature T_c with 0.1 K accuracy; another positive fact about the MMMA technique is that it is a contactless method. All the known pluses of MMMA encouraged us to design a new technique of measuring the temperature of superconducting transition T_c [5]. The innovatory feature of this technique is the replacement of the DC magnetic field in which the tested sample is placed with a field of an unchanging strength but an alternating sense. Periodical changes of the magnetic field cause the registered MMMA signal to assume a specific shape that makes it possible to measure precisely the critical temperature T_c (Fig. 1).

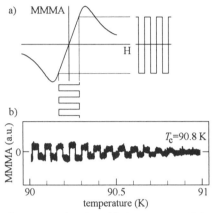

Fig. 1 Measuring the critical temperature T_c by means of MMMA.
a) The superconductor is placed in a magnetic field H which changes its sense periodically.
b) If the temperature of the superconductor changes with time, a characteristic "step-like" MMMA line can be noticed. Its disappearance indicates a transition into the normal state.

3.2. Testing the quality of superconducting materials

Later, MMMA was also applied to studying the quality of superconducting materials - especially thin films and single crystals. Generally speaking, it was found out that the best superconducting materials - those with the least defects in the crystal lattice - are characterised by low microwave absorption. An interesting example is that of the use of MMMA to investigate the properties of such fine superconducting samples as BiSrCaCuO whiskers [6]. While in this case the sensitivity of SQUID magnetometers was found insufficient, the microwave technique allowed us to confirm the superconductivity of the whiskers and measure their critical temperatures T_c. It also provided us with much valuable information on their structural defects. The best, regular and periodic, MMMA lines, were observed when (a,b) plane of a whisker was perpendicular to the direction of the magnetic field (with the c axis aligned parallel to the field) (Fig. 2). In the lower magnetic field range, magnetic flux would not then penetrate into the sample and in effect no microwave absorption was registered. In contrast, a periodic MMMA signal occurred for fields exceeding 1.5 Oe, which value was

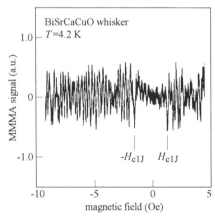

Fig. 2. The regular MMMA signal observed for superconducting whiskers. In the low field range microwave absorption does not occur, which proves the fact that for $H<H_{c1J}$ magnetic flux does not penetrate into the Josephson junctions.

identified with the lower Josephson critical field H_{c1J}. The periodicity of the MMMA lines in itself results from quantisation of the magnetic flux on the twin-boundaries of a monocrystal. The surfaces of the defects may be estimated on the basis of the flux quantisation condition: $\Phi_0 = SB$, where Φ_0 is the flux quantum, S stands for the surface of the superconducting loop delineated by the crystal lattice defects, and B represents the MMMA signal period. Accordingly, the periodicity values 0.19 Oe and 3.75 Oe measured in the whiskers correspond to defect surfaces equal 100 μm^2 and 5 μm^2.

3.3. Studying of the chemical processes

Thirdly, owing to its high measuring sensitivity, MMMA was used to study the dynamics of the chemical processes of superconductor synthesis. More precisely, we undertook to characterise the diffusion processes responsible for the formation of the superconducting phase in Rb_xC_{60} fullerides [7]. In the synthesis of the fulleride the alkali metal azide RbN_3 was used. Selecting this procedure, in which the rubidium covers evenly all the surface of the fullerene grains immediately after the thermal decomposition of the azide, determines the kinetics of the reaction. The fullerene grains thus enclosed in the alkali metal constitute the nuclei of the product phase. Because of the diffusion of the rubidium, the thickness of the product phase on the grain surface increases with time. As the number of the fullerene grains N remains constant throughout the process, the reaction can be described with the Avrami model [8]:

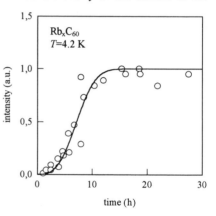

$$\alpha = 1 - \exp\left(-\frac{\pi}{3}NU^3t^3\right) \text{ and } \alpha = 1 - \frac{V_s(t)}{V_s(0)} \quad (1)$$

where α is the reaction progress coefficient, while $V_s(0)$ and $V_s(t)$ represent, respectively, the initial amount of the substrate-phase and the amount of it after the elapse of t. U defines the Rb diffusion speed. Formula 1 rightly describes the relation of

Fig. 3 An increase in MMMA intensity as a function of the time elapsed from the onset of the synthesis of the Rb_xC_{60}. The solid line represent the fit obtained on the basis of Eq. 1.

the intensity of the MMMA signal and time (Fig. 3), because the intensity of microwave absorption remains in straight proportion to the amount of the superconducting phase [9].

3.4. Measuring the critical fields and currents in HTS

Recently, MMMA has been used by us to examine the critical states of HTS [10]. First, zero field cooling of the ceramic superconductor YBaCuO to the temperature of $T=4.2$ K had to be effected. Then measurements were taken of the hysteresis width ΔH of the microwave absorption signal in relation to the magnetic field sweep amplitude H_{max}. The results given in Fig. 4 show that flux penetration proceeds in two stages. Firstly, in the range above the lower Josephson critical field H_{c1J}, the magnetic field starts to penetrate into the intergrain regions, which is reflected in the widening MMMA hysteresis. Then, in the field range for which the hysteresis width is in a plateau, in the entire volume of the sample the intergrain critical state is established. In the second stage, for fields exceeding the lower ceramic grains critical field H_{c1G}, widening of the hysteresis is again observed, which is caused by the penetrating action of the flux and by the spreading of the critical state into each individual ceramic grain.

550

Assuming that the hysteresis width ΔH is proportional to the mean field trapped in the superconductor during a single magnetic field sweep $<H>_{trap}$, we get:

$$\Delta H = p\langle H\rangle_{trap} \quad \text{and} \quad \langle H\rangle_{trap} = \frac{1}{V}\int_V d^3 r H(r) \quad (2)$$

where p is a coefficient dependent on the structure and geometry of the sample and V is it's volume. The local intensity of the magnetic field $H(r)$ can be calculated on the basis of generalised critical state model [11]. Fig. 4 presents the best fit between model (2) and the empirical data received in our experiment. It was found for the density of the intergrain critical current $J_c=400$ A/cm^2, the critical exponent $n=1$ and the characteristic field $H_0=3$ Oe.

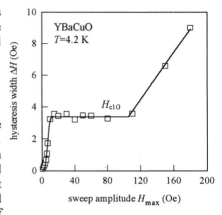

Fig. 4 The relation between the MMMA hysteresis width ΔH and the amplitude of the magnetic field sweep H_{max}. The solid line shows the best fit between the empirical data and Eq. 6.

4. Summary

Observations made on the basis of the experiments make us believe that the technique of microwave absorption, or of magnetically modulated magnetic absorption, is complementary to other magnetic measuring techniques. The advantage of MMMA over the magnetic methods stems from the wide range of information it can provide, and from its fine measuring sensitivity (surpassing that of SQUID magnetometers).

Acknowledgements

Work partly supported by National Committee for Scientific Research under grant No. 2P03B 009 08.
Mr. Bartlomiej Andrzejewski is a holder of a scholarship awarded by the Foundation for Polish Science.

References

[1] Durny R, Hautala J, Ducharme S, Lee B, Symko O G, Taylor P C and Zheng D J and Xu J A 1987 *Phys. Rev. B.* **36** 2361-2363
[2] Stankowski J, Kahol P K, Dalal N S and Moodera J S 1987 *Phys. Rev. B.* **36** 7126-7128
[3] Hebard A F, Rosseinsky M J, Haddon R C, Murphy D W, Glarum S H, Palstra T T M, Ramirez A P and Kortan A R 1991 *Nature* **350** 600-601
[4] Oseroff S B, Rao D, Wright F, Vier D C, Schultz S, Thompson J D, Fisk Z, Cheong S-W, Hundley M F and Tovar M 1990 *Phys. Rev. B.* **41** 1934-1948
[5] Stankowski J, Krupski M, Czyżak B and Baszyński J 1988 *proc. 24th AMPERE Congress* Poznań 1988 Ed. N. Piślewski p. 819-824
[6] Czyżak B, Andrzejewski B, Szcześniak L, Danilova N and Stankowski J 1995 *Appl. Magn. Res.* **8** 25-33
[7] Stankowski J, Kevan L, Czyżak B and Andrzejewski B 1993 *J. Phys. Chem.* **97** 10430-10432
[8] Avrami M 1939 *J. Chem. Phys.* **7** 1103
[9] de Biasi R S and Araujo S M V 1995 *Phys. Rev. B.* **51** 8645-8646
[10] Andrzejewski B, Czyżak B and Stankowski J 1994 *Physica C* **235-240** 2044-2045
[11] Xu Ming, Shi Donglu and Fox R F 1990 *Phys. Rev. B.* **42** 10773-10776

Inst. Phys. Conf. Ser. No 148
Paper presented at Applied Superconductivity, Edinburgh, 3–6 July 1995
© *1995 IOP Publishing Ltd*

Surface Degradation of $YBa_2Cu_3O_{7-x}$ Powders Studied by XPS and A.C. Susceptibility

I.Sargánková, P.Diko

Institute of Experimental Physics, Slovak Academy of Science, Watsonova 47, 04353 Košice, Slovakia.

J.D.Tweed, C.A.Anderson, N.M.D.Brown

Joint Ceramics Research Centre, Surface Science Laboratory, University of Ulster, Cromore Road, Coleraine, Co. Londonderry, Northern Ireland, BT52 1SA

Abstract. The preparation of powder fractions of $YBa_2Cu_3O_{7-x}$ by sedimentation in 2-butanol raises questions of particle surface degradation during sedimentation and the possibility of regeneration by re-annealing. In this work we studied this behaviour by XPS and a.c. susceptibility measurements. Characteristic peaks for monitoring changes due to degradation and regeneration were found in the Y3d (c. 160 eV) and C1s (c. 285 eV) regions. Our results suggest that annealing close to the sintering temperature can regenerate the particle surfaces.

1. Introduction

The sensitivity of $YBa_2Cu_3O_{7-x}$ (123) to atmospheric CO_2 and H_2O poses a serious problem in the general use of this material. The especially high reactivity exhibited by the superconducting cuprates has been attributed to the high formal oxidation state of the copper ($+2.33$) and to the presence of alkaline earth elements in the structure which react to form carbonates with high thermodynamic stability [1-3]. Yan et al [4] suggest that the superconducting phase decomposes in water according to a two step reaction:

$2YBa_2Cu_3O_{7-x} + 3H_2O \rightarrow 5CuO + 3Ba(OH)_2 + Y_2BaCuO_5 + 1/2O_2;$

$Ba(OH)_2 + CO_2 \rightarrow BaCO_3 + H_2O.$ Flavell et al [4] established the following as the dominant reactions with CO_2:

$2YBa_2Cu_3O_{7-x} + 3CO_2 \rightarrow Y_2BaCuO_5 + 3BaCO_3 + 5CuO + 1/2O_2$ (at 950°C);

$2YBa_2Cu_3O_{7-x} + 4CO_2 \rightarrow Y_2Cu_2O_5 + 4BaCO_3 + 4CuO + 1/2O_2$ (at lower temperature), strongly catalyzed by water vapour. The XPS and HREM studies of Zhao et al [5] indicate a structural model of 123 decomposition: in the initial stage of degradation in humid environments an amorphous surface layer and planar defects form via anion and cation exchange reactions.

The degradation of cuprates is a major problem and the stability of 123 in different organic solutions has been studied [6]. In this work we have focused on the degradation of 123 powder in 2-butanol during sedimentation, regeneration of separated powder fractions by heat treatment in an oxygen flow and controlled degradation of regenerated powders in moist CO_2. We used X-ray photoelectron spectroscopy (XPS) and a.c. susceptibility measurements as our principal methods of characterisation.

2. Experimental

Fractions with narrow particle size distributions were prepared by sedimentation of milled, coarse $YBa_2Cu_3O_{7-x}$ ceramic in 2-butanol. The 123 ceramic was prepared by stepwise synthesis using a sol-gel method described previously [7]. XPS experiments were carried out in a Kratos XSAM 800 spectrometer on separated powder fractions after sedimentation (s), regeneration (r) in oxygen flow and degradation in moist CO_2 (d). After sedimentation, part of each fraction was regenerated in an oxygen flow by heat treatment at 870°C for 10 hrs and at 450°C for 6 hrs. Regenerated powders were then degraded in moist CO_2 at 150°C for 1 hour. After both processes the samples were immediately analysed by XPS and characterised by a.c. susceptibility. The a.c. susceptibilities were measured in the standard way using an APD He cryostat coupled with the associated signal electronics and phase sensitive detector system interfaced to and controlled by an appropriate suite of PC-based software [8]. Powder morphology was checked by scanning electron microscopy (SEM, Hitachi S-2300).

3. Results and discussion

In the experiments described fractions with mean particle sizes of 1.7, 4.0, 12.3 19.4 and 30.7 μm were processed, analysed and characterised.

3.1. XPS measurements

Powder fractions were measured after sedimentation, regeneration and degradation. Great care was taken to ensure that samples received minimal exposure to the outside atmosphere prior to testing. Samples were kept in a polythene bag containing oxygen after annealing and were transferred to the spectrometer via a hood flushed with nitrogen. The spectrometer was operated at a base pressure of 10^{-9} torr with an Mg-anode providing non-monochromatized Kα X-radiation. Spectra were calibrated by assigning a binding energy of 285 eV to the adventitious C1s peak. In our measurments the most significant changes in the spectra after the various processes were observed in the Y3d, O1s and C1s regions (Figure 1). Normally the Y3d peak is a doublet with 156 eV and 158 eV

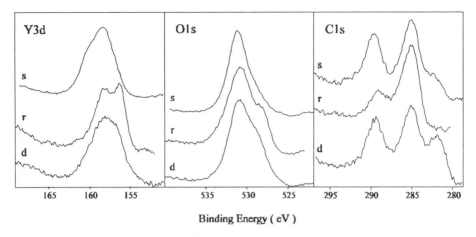

Figure 1 XPS spectra of $YBa_2Cu_3O_{7-x}$ powder after sedimentation (s), regeneration (r) and degradation (d).

components [9]. This doublet was observed in the spectra of the regenerated powders (Figure 1) while this region did not show two intensity maxima in the spectra of the decomposed surfaces. From deconvolution of the observed peak three contributions to the binding energy centred at 156 eV, 158 eV, and 160 eV were established. The third contribution diminished after regeneration and is attributed to a degradation product (Y_2BaCuO_5, $Y_2Cu_2O_5$). In the O1s region the intrinsic material spectrum is dominated by a peak centered at 528 eV. All three geometrically distinguishable oxygen site types present in intrinsic 123 (i.e. at chain, plane and apex locations) are included within this peak . The intensity at around 531 eV is attributed to extrinsic surface components [10,11]. Figure 1 shows that after sedimentation the main contribution of the O1s region is at 531 eV. After regeneration the shoulder at lower energy appeared (around 528 eV). The effective sampling depth of XPS is around 50Å, it is argued that the decomposed layer after sedimentation is thicker than the sampled depth thus no contribution from intrinsic material is seen after degradation. In the C1s region two typical contributions at 285 eV for adventitious hydrocarbon and at 289 eV for carbonate species [10,11] are found. Moreover it was observed that the third peak at lower energy (282 eV) vanished after regeneration and then reappeared after degradation in moist CO_2 (Figure 1). This peak may be due to adsorbed CO_2.

3.2. A.C. Susceptibility

Figure 2 shows that the volume susceptibility of the powders after regeneration changed significantly for powders with a lower mean particle size. Changes in the volume of the degraded surface layer are more significant for small particles than for large ones. The anomalous behaviour of powders B and A can be associated with the relationship between the diameter of the superconducting volume in the particle before/after regeneration and

554

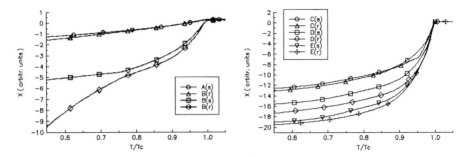

Figure 2 Real susceptibility for $YBa_2Cu_3O_{7-x}$ powders before (s) and after (r) regeneration. The curves are labled according to the mean particle size: $a=1.7\mu m$, $B=4.0\mu m$, $C=12.3\mu m$, $D=19.4\mu m$, $E=30.7\mu m$.

the London penetration depth. Before regeneration, small particles are undetectable due to the penetration depth (about $1\mu m$). In the case of powder B, after regeneration a large part of each particle contributes to the measured signal, while in the fraction with the smallest particles (powder A) a significant fraction of the particle volume is penetrated by the field.

4. Conclusions

The XPS measurements have shown that the products of degradation of $YBa_2Cu_3O_{7-x}$ in 2-butanol and moist CO_2 are identical. The Y_2BaCuO_5 was detected as the main product of degradation. The Y3d and C1s regions of XPS spectra were identified as the most suitable for monitoring surface changes. Surface regeneration was also detected by a.c. susceptibility measurements.

References

[1] Zhou J P et al 1993 *Chem. Mater.* **5** 361-365
[2] Brow R K 1988 *Research Update, Ceramic Superconductors II, The Amer. Ceram. Soc.* 598-606
[3] Flavell W R et al 1993 *J. Alloys and Compounds* **195** 535-542
[4] Yan M F et al 1987 *Appl. Phys. Lett.* **51** 532-534
[5] Zhao R and Myhra S 1994 *Physica C* **230** 75-81
[6] Spyker R L and Kozlowski G 1991 *J.Appl.Phys.* **70** 6492-6494
[7] Sargankova I et al. 1994 *IEEE Transactions on Magnetics* **30** 1181
[8] McDowell J C 1994 *DPhil. thesis: Preparation, Characterisation and Stability of Selected High Critical Temperature Ceramic Superconductors* University of Ulster
[9] Briggs D and Seah M P *Practical Surface Analysis* **1** 2nd edition (Salle + Sauerländer)
[10] Brudle C R and Fowler D E 1993 *Surface Science Reports* **19** 143-168
[11] Egdell R G et al 1989 *J Solid State Chemistry* **79** 238-249

Inst. Phys. Conf. Ser. No 148
Paper presented at Applied Superconductivity, Edinburgh, 3–6 July 1995
© *1995 IOP Publishing Ltd*

The field and temperature dependencies of the intra-grain and inter-grain critical current densities in $(Tl_{0.5}Pb_{0.5})(Sr_{0.8}Ba_{0.2})_2Ca_2Cu_3O_x$ ceramics

S. Senoussi, A. Kilic.

Laboratoire de Physique des Solides (URA 2 associée au CNRS), Université Paris-Sud, 91405 Orsay Cedex, France.

S. K. Wivel and C.R.M. Grovenor

Department of Materials, Oxford University, Parks Road Oxford, OX1 3PH, UK

We report investigations of the magnetization and the associated critical current density of a $Tl_{0.5}Pb_{0.5})(Sr_{0.8}Ba_{0.2})_2Ca_2Cu_3O_x$ ceramic sample in T and H varying from 10^{-2} to 7×10^4 Oe and 4.2 K to $T_C \approx 110$ K. At high enough temperature, the M vs H relationship exhibits 5 different regimes reflecting various modes of penetration of flux through the material: (1) At H≤ 2 Oe, the M(H) function is linear, reversible and controlled by the macroscopic demagnetizing factor N of the pellet. (2) As H is further increased one observes a tiny hysteresis cycle in a narrow field domain of about 40 Oe. (3) At still higher H and up to $H_{c1,g}$ of the grains, the M(H) relationship becomes again linear and reversible but with a slope reduced by a factor of four as compared to the initial one. (4) For $H > H_{c1,g}$ and up to the irreversibility line we observe the usual regime of strong irreversibility and deduce the inter-grain current density using a modified Bean model that accounts for the anisotropy.

Introduction

It has been shown sometime ago that the magnetic properties of YBCO polycrystalline materials were determined by the interplay between four critical current densities of different physical origins [1-3]. To help analyzing the present data we illustrate in figure 1 and 2 a typical example of such a behavior.

(1) It is seen (Fig.1a) that for H smaller than the Josephson first critical field ($H_{c1}{}^{W}$) the distribution of currents and fields within the sample are restricted to the Josephson penetration depth λ_j in the junction and to the London penetration depth λ_L around the grains. Here (Fig.2a) M is imposed by the macroscopic demagnetizing factor N_p, of the pellet. In the limits R>> λ_j and H<< $H_{c1}{}^{W}$, one has :

$$M = -\left(1 - \frac{2\lambda_J}{R}\right)\frac{H}{4\pi(1 - N_P)} \; , \; (\lambda_J << R, \; H << H_{cl}^W) \qquad (1)$$

Typical values for YBCO ceramics are 5 to 50 μm for λ_j and 0.1 to 2 Gauss for $H_{c1}{}^{W}$.

(2) Figure 2b shows schematically that as H is increased above $H_{c1}{}^{W}$, magnetic flux and associated currents penetrate the sample through the weak links. This gives rise to current loops extending over the whole

556

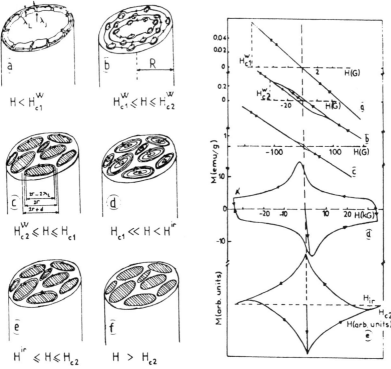

Figure 1: The field and current penetrate the sample in several steps. a) H is smaller than the first critical field of the junctions. b) Intergranular and London currents coexist in the pellet. c) The intergranular current vanishes and M is imposed by the London currents. d) The field and currents enter into the grains.

Figure 2: Evolution of the hysteresis cycle of a typical YBaCuO pellet as a function of the maximum cycling field. The various cycles displayed here correspond to the four situations of Fig.1 for the distribution of J_c within the sample.

sample as well as to London currents circulating around the individual grains.

In this field region, the magnetic behavior is extremely complicated and the magnetization is influenced by the demagnetizing factors of both the macroscopic pellet and the individual grains [3,4]. Nevertheless, it is possible to extract the intergrain critical current from the small magnetic loop of Fig.2b by the following formulae (which represents a severe modification of the Bean model [1-3]).

$$J_{tr}(H) \approx 15 \frac{M_{tot}^+ - M_{tot}^-}{R} \frac{\chi(H_{c1}^w)}{\chi(H_{c1}^w) - \chi(H_{c2}^w)} \qquad (2)$$

Here, $\chi(H_{c1}{}^w)$ and $\chi(H_{c2}{}^w)$ represent the magnetic susceptibilities for $H < H_{c1}{}^w$ and $H_{c2}{}^w < H < H_{c1}{}^g$ respectively. At this point, it is important to emphasize that the factor $\chi(H_{c1}{}^w)/[\chi(H_{c1}{}^w) - \chi(H_{c2}{}^w)]$ comes out from numerical calculations. It accounts for the fact that the irreversible magnetization of Fig.2b depends not only on J_{tr} but also on the total number n_w of the weak links within the material.

(3) At fields larger than the Josephson decoupling field but lower than the first critical field $H_{c1,g}$ of the grains, M is governed exclusively by the London currents (Fig.1c). In this field domain M is again linear and reversible but with a reduced slope. Now, it depends on the demagnetizing factors of both the pellet and the individual grains[4,5], the density of the actual superconducting material, the anisotropy as well as the ratio between the grain sizes and the London penetration depth. For large enough grains we can write

$$ M = -a\left(1 - \frac{2\lambda_J}{R}\right)\frac{H}{4\pi(1-\langle N \rangle)} \quad , \quad \left(\lambda_J << r_g, \; H_{c2}^w << H << H_{c1}^g\right) \quad (3) $$

with $\langle N \rangle$ representing some average [4] of the demagnetizing factors N_p and N_g. For highly compact and not very anisotropic specimens one would have $\langle N \rangle \approx N_p$ while for very diluted granular samples $\langle N \rangle \approx N_g$. The factor a is intended to account for the anisotropy of the material. It is of the order of one for moderate anisotropy so that the conditions $\lambda_L << r_g$ and $H < H_{c1,g}$ are satisfied. For highly anisotropic superconductors, however these two conditions are extremely difficult to fulfill for grains having their c-axis directed far away from the H direction. In this limit the factor a is of the order of 1/3

(4) Finally, at still higher fields, $H > H_{c1,g}$, vortex lines and associated intragrain currents enter the interior of the grains (Figs.1d and 2d). Fig.3 illustrates how the low field hysteresis cycles of a ceramic sample $(Tl_{0.5}Pb_{0.5})(Sr_{0.8}Ba_{0.2})_2Ca_2Cu_3O_x$ evolve with temperature for T=4.2 K, 40 K and 77 K. The sample was prepared by conventional solid state synthesis from pressed mixed oxides. The intergrain critical current density deduced from this figure at H=O and using Eq. 2 is 120 A/cm^2 at 4.2K. Fig.4 illustrates the high field hysteresis cycle of the same ceramic sample. The insert represents the critical current density deduced by application of the usual Bean model. Because of the very high anisotropy of the material this current would represent only 1/3 of the in plane value. We note that the behavior of figs 3 and 4 resembles that of YBCO (Fig.2) but with the four magnetic regimes less clearly defined here. Note that the difference between the initial and final slopes dM/dH of fig 3 which differ by a factor of nearly 4.

558

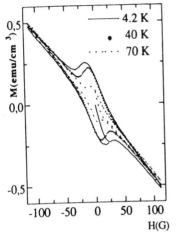

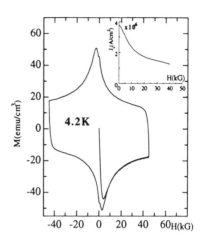

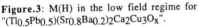

Figure.3: M(H) in the low field regime for "(Tl$_{0.5}$Pb$_{0.5}$)(Sr$_{0.8}$Ba$_{0.2}$)$_2$Ca$_2$Cu$_3$O$_x$".

Figure 4: The same as fig.3 but at high fields. Insert : J$_c$(H) deduced from the Bean model

In conclusion, despite the fact that our pellet contains severe weak-links, the inter-grain J$_c$ is quite high (120 A/cm^2) and comparable to what is seen in YBCO(123) of similiar dimensions. Moreover, we also find that this sample exhibits, almost the same intergrain J$_c$ as Y$_1$Ba$_2$Cu$_3$O$_7$ for the T and H explored (T≤77 K, H≤50 kG). Therefore, the pinning properties of this Tl-(1223) sample are much more similiar to those of YBCO than Bi-(2212); the critical current of which drops dramatically with both T and H.

Acknowledgements: This work is in part supported under a Brite Euram project, BRE2-CT93-0455, one of us (A.K.) would like to thank TUBITAK/TURKEY for financial support. We also thank K. Frikach for his help

References:

[1] S. Senoussi, J. de Phys. III, 2 (1992) 1041

[2] S. Senoussi, M. Ousséna and S. Hadjoudj ; J. Appl. Phys. 63 (1988), 4176.

[3] S. Senoussi, S. Hadjoudj, R. Maury, An. Fert, Physica C, 165, (1990),

[4] U. Yaron, Y. Kornyushin and I. Felner Phys. Rev. B,46,(1992)14 823

[5] M.L. Hodgdon, R. Navarro and L.J. Campbell, Europhys. Lett. 16, (1991) 677

Inst. Phys. Conf. Ser. No 148
Paper presented at Applied Superconductivity, Edinburgh, 3–6 July 1995
© 1995 IOP Publishing Ltd

The effect of twin planes in the pinning mechanism of YBa$_2$Cu$_3$O$_{7-\delta}$ single crystals

K Deligiannis , M Oussena , P A J de Groot , A V Volkozub

Physics Department, University of Southampton, Southampton SO17 1BJ, UK

R Gagnon , L Taillefer

Physics Department, McGill University, Montréal (Québec), Canada H3A 2T8

Abstract. We demonstrate, by studying the magnetic hysteresis of twinned and detwinned YBa$_2$Cu$_3$O$_{7-\delta}$ single crystals, that twin planes, contrary to the usual belief, can greatly decrease the effective pinning in the field regime near the hysteresis maximum. Twin planes act as flux channels: an easy way for vortices to enter into the sample. By varying the angle of the twin planes with the magnetic field, we have found that the hysteresis and, thus, the critical current is greatly affected in a wide field and temperature range.

The pinning mechanism of the high-T$_c$ superconducting oxides plays a crucial role for the practical use of these materials. Therefore, defining the extent and the limits of pinning as determined by the interplay of random (point defects) and coherent (columnar defects, layered structure) disorder is a major priority. Of special interest is the case of YBa$_2$Cu$_3$O$_{7-\delta}$ (YBCO) where coherent disorder appears naturally in the additional, planar form of twin planes.

In spite of the numerous theoretical and experimental investigations the effect of twin planes on pinning is anything but clear. Techniques vary: Bitter decoration, resistivity, magnetisation, torque measurements, magneto-optical flux visualisation; so do interpretations: suggestions that twin planes decrease the order parameter [3,4] are contradicted by observations of enhanced pinning in their presence [5-9]. In this paper we present the first clear experimental evidence, from magnetic measurements in an applied field of up to 12 T, that twin planes *do decrease* pinning, provided that certain conditions are met [1].

The crystals investigated were grown as described in [10] by a conventional self-flux method. Crystals have T$_c$ = 93.8 K , ΔT$_c$ < 0.3 K and oxygen content of 6.92; crystal A (363μg) has dimensions 1.15x0.8 mm^2 and is microtwinned, while crystal B (81μg) 0.78x0.79 mm^2 and contains only one direction of twin planes. Measurements were carried out on a 12T VSM and a 6 T SQUID magnetometer.

It is known that the magnetic hysteresis of a twinned crystal, for an applied field H$_a$ along the c axis, can present essential differences compared with this of a detwinned and same oxygen content sample [1]. Once scaled by the characteristic size R in order to make direct comparisons between different crystals [11], the magnetic hysteresis of the twinned crystal

560

fails to exhibit the characteristic peak for YBCO. Twin planes provide for vortices an easy way to enter into the sample and it is this process of vortex channelling *along* the twin planes that causes the decrease of pinning. The observed flattening and thus pinning depression in the field regime of the peak narrows with increasing temperature, disappearing at about 75 K, and depends strongly on the pattern of the twin planes.

The latter is verified by Fig. 1. Crystal A contains a pattern of twin planes. Its magnetic hysteresis presents the pronounced flattening at the position of the peak. Crystal A1 (0.71x0.47 mm^2, 120μg) is a part of A which contains twin planes of only one direction (simply twinned). Surprisingly, crystal A1 exhibits a considerable decrease of the hysteresis width in the intermediate field regime: for the first time a part of a single crystal has a *lower* critical current density than the whole. The boundaries between different domains of twin planes provide strong pinning sites, interrupting the easy motion of vortices and, thus, quench channelling. It is the absence of these columnar-like defects at the boundaries between different domains that enhances channelling in simply twinned crystals, as A1.

Since channelling takes place in directions along the twin planes it should rely crucially on vortices being locked-in to the twin planes. It follows that removing the effect of twin planes, by e.g. tilting the applied field, should lead to a decrease in channelling and an increase in pinning. Indeed, this is the case [1]: rotating the field in respect to the c-axis (plane of rotation vertical to the twin planes) the flattening in the hysteresis gradually retreats giving back the expected, well defined peak. A clear demonstration of this is shown in Fig. 2. Measurements were carried on crystal B in a 6 T SQUID magnetometer and at constant temperature of 5 K. The sample was mounted on a step formed by bending oxygen free wire, with the normal to the twin planes along it, and introduced in a thin quartz tube. The angle θ between the step and the tube axis (equivalently between H_a and twin planes) was measured using a goniometer and a microscope with an uncertainty of less than 0.5°. Because of the anisotropy Fig. 2 has been plotted as the projection of the component of the magnetization M

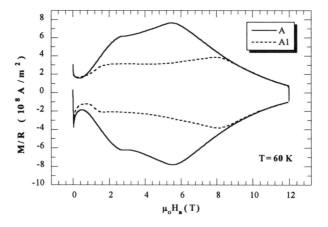

Figure 1: Measurements were carried out at 60 K. Crystal A contains a pattern of twin planes; crystal A1 is a part of A with only one domain of twins. It is obvious that in the case of A1 pinning is strongly reduced.

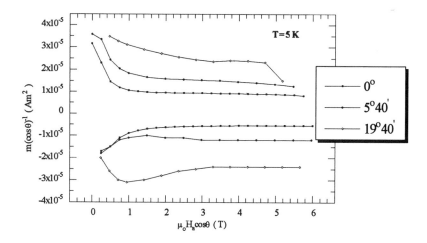

Figure 2: The magnetic hysteresis of B (simply twinned) at 5 K for various angles of H_a in respect to the twin planes (plane of rotation vertical to twin planes).

along the c-axis, $M_\perp = M/\cos\theta$, as a function of the component of H_a along the c axis, $H_\perp = H_a\cos\theta$, for angles up to 20°. As it is seen, increasing θ increases the width of the hysteresis in a broad field regime: it becomes harder for vortices to use the twin planes channels in order to enter the sample and channelling weakens.

At the point where, according to what is stated above, the vortices are no longer locked-in to the twin planes, we should have recovered the static response of the untwinned regions. The additional proof is provided by Fig. 3.

Fig. 3 displays the temperature dependence of the critical current (hysteresis width) in crystal B for different angles; on the right a scaled for size comparison is made, between a detwinned crystal of the same oxygen content at $\theta = 0°$ and crystal B at $\theta = 19.6°$ where no channelling is observed and therefore no lock-in situation occurs. It is clear that although there is some extra amount of disorder in the case of the twinned crystal, as expected, once we surpass the lock-in, its behaviour is qualitatively a faithful reproduction of the detwinned one. Also in Fig. 3 (left) it can be seen that: a) the critical current gradually builds up *removing* the effect of twin planes and b) the temperature dependence of the critical current increases with θ. From the very weak dependence at small angles, where channelling dominates and, thus, the pinning potential due to the restricted dimensionality of locked-in vortices decays only algebraically with temperature, we pass gradually to an intense dependence at the situation where point defects dominate and the pinning potential, much more vulnerable to thermal fluctuations, decays exponentially with temperature [2].

In conclusion, we demonstrated that twin planes decrease the order parameter and the overall pinning, for motion along their direction. Since channelling occurs due to the easier motion of vortices in the twin planes, it follows that it is observed at the temperature regime

562

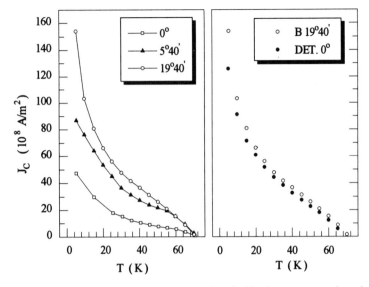

Figure 3: Measurements were taken at $H_\perp = 4$ T. Left: The J_C temperature dependence for different angles of applied field and twins. Right: Comparison between the J_C behaviour of a detwinned crystal at 0° and B at 19° 40' (same oxygen content).

where the effective pinning in the untwinned regions is stronger than the effective pinning in the twin planes [1]. By the temperature and angle dependence of the critical current the importance of vortices being locked-in to the twin planes in channelling is proved. Though further work is needed, our studies clarify for the first time the role of planar defects in a wide temperature and field regime.

References

[1] M. Oussena et al, *Phys. Rev. B (Rapid Comm.)* **51**, 1381 (1995)

[2] G. Blatter et al, *Rev. Mod. Phys.* **66**, 1125 (1994)

[3] G. Deutcher and K. A. Müeller, *Phys. Rev. Lett.* **59**, 1745 (1987)

[4] C. A. Duran et al, *Nature* **357**, 474 (1992); C. A. Duran et al, *Phys. Rev. Lett.* **74**, 3712 (1995) (comm.). Notice that these are very low field results, not necessarily valid for intermediate fields.

[5] A. A. Abrikosov et al, *Supercond. Sci. Technol.* **1**, 260 (1989)

[6] I. N. Khlustikov and A. I. Buzdin, *Adv. Phys.* **36**, 271 (1987)

[7] V. K. Vlasko-Vlasov et al, *Phys. Rev. Lett.* **72**, 3246 (1994)

[8] L. A. Dorosinskii et al, *Physica C* **219**, 81 (1994)

[9] W. K. Kwok et al, *Phys. Rev. Lett.* **64**, 966 (1990)

[10] R. Gagnon et al, *Phys. Rev. B* **50**, 3458 (1994); M. Oussena et al, *Phys. Rev. Lett.* **72**, 3606 (1994)

[11] R. L. Peterson, *J. Appl. Phys.* **67**, 6930 (1990)

Inst. Phys. Conf. Ser. No 148
Paper presented at Applied Superconductivity, Edinburgh, 3–6 July 1995
© 1995 IOP Publishing Ltd

Synthesis of Bi-Pb-Sr-Ca-Cu-O superconducting fiber by using sodium alginate

A.Takada*, T.Shimizu**

Dept. of Electrical Engineering*, Dept. of Industrial Chemistry**, Hakodate National College of Technology, Hakodate 042 Japan;

I.Shimono

Hokkaido Industrial Technology Center, 379 Kikyo-cho, Hakodate 041 Japan;

H.Konishi

Dept. of Functional Machinery, Shinshu University, Ueda, Nagano 386 Japan.

Abstract. The bismuth based high-Tc superconducting fiber was synthesized using sodium alginate. The thermal pyrolysis of a precursor which was obtained from an ion-exchange of the sodium alginate is studied. Sintering precursor at 860℃ for 30h, the Bi-Pb-Sr-Ca-Cu-O superconducting fiber in which the mixed phase (2212 and 2223) formed principally was obtained. The solid state reactions associated with a formation of the superconducting phase via intermediate compounds is also discussed briefly. In addition, Ag-coated fiber was synthesized successfully.

1. Introduction

Up to date, much efforts have been devoted to prepare a good high-Tc superconducting material. Many studies have also been made on preparation of the specimen in the shape of wire, tape and fiber for high field magnets. Among the synthetic techniques of these specimen the "powder in tube" method is usually used, though many kinds of chemical techniques for the preparation of the fiber specimen without powder material have also been developed. A principal advantage of the chemical technique is summarized its brief manner. In these techniques, it is, however, difficult to prepare the fiber which has a uniform diameter. Moreover, there is a limit of the length of the specimen which one can synthesize.

We report on a framework for the synthesis of the bismuth based high-Tc superconducting fiber by using sodium alginate[1] on the basis of the ion-exchange property of this alginate. In our method, the length of the fiber is not limited in principle and a uniform diameter is available. Furthermore Ag coated fibers were also prepared successfully.

To our knowledge, Konishi *et al.* firstly synthesized successfully the yttrium-based superconducting fiber by using sodium alginate.[2] Their study is interesting in the viewpoint of not only preparation technique but also the use of a polyelectrolyte.

2. Synthesis and thermal analysis of precursor

Following the preparation of a nitrate solution involving Bi-, Pb-, Sr-, Ca- and Cu- ions which prepared by pure nitrates; $Bi(NO_3)_3 \cdot 5H_2O$, $Pb(NO_3)_2$, $Sr(NO_3)_2$, $Ca(NO_3)_2 \cdot 4H_2O$ and $Cu(NO_3)_2 \cdot 3H_2O$, 5wt% viscous aqueous solution of the sodium alginate was fiberized into the nitrate solution from a nozzle (2mm in diameter) and soaked in the solution for 1 hour. Then these five kinds of cations were incorporated into the sodium alginate, instead of sodium ions in the alginate were excreted into the nitrate solution. This process induces the gelatin of the alginic acid. For a precise control of the cation ratio in this gel fiber, i.e., to avoid additional accumulation of cations into the

564

gel, water rinsing of the gel was done. Consequently, the precursor fiber was obtained by stretching and drying the gel fiber. Even if using our small equipment, a long precursor with more than 1.5m in length and ~0.6mm in diameter was easily synthesized. Sintering the precursor in air at 810~860℃, the Bi-Pb-Sr-Ca-Cu-O (BPSCCO) superconducting fiber with 200μm in diameter was obtained. In this method there is no limitation of length of the fiber that one can synthesize in principle.

It has been known that the alginate consists of a block-copolymer[1] which has an accumulation property for cations. Although this ion-exchange property for the sodium alginate has been studied for individual Pb-, Sr-, Ca-, Cu-ions etc., there is no report on its accumulation property for neither bismuth or a mixture of various cations.

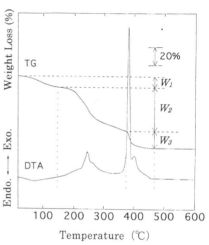

Fig.1. DTA-TG curves of precursor for temperatures below 600℃.

In these circumstances we determined a composition for these multivalent cations in the nitrate solution based on the atomic absorbent spectrum analysis of the resultant precursor, so as to obtain the precursor with the nominal composition of high-Tc phase, i.e., (Bi,Pb):Sr:Ca:Cu=2:2:2:3 which was exactly modified to be Bi:Pb:Sr:Ca:Cu =1.84:0.34:1.91:2.03:3.06 for a starting composition.[3] Consequently we obtained Bi:Pb:Sr:Ca:Cu =0.43:0.22:2.05:3.14:2.91 as the composition of the nitrate solution.

Fig.1 shows a differential thermal analysis (DTA) and thermal gravimetric (TG) data for the precursor at temperatures below 600℃. In this figure, weight losses were monitored for the three temperature regions in which exothermic or endothermic reaction occurred. Thus we defined these losses to be W_1, W_2 and W_3, respectively, as shown in this figure.

On the basis of infrared ray spectra for the precursor, the following considerations are available. Firstly, up to about 150℃, an endothermic reaction with small weight loss (W_1) occurs, which resulted by evaporation of moisture from the precursor. In the higher temperature region (150~380℃) at which the loss (W_2) occurred, the structure of the alginate was decomposed and organic matter was evaporated as a gas. Lastly the resultant organic compound formed in the previous region was burned at the temperature above 380℃, resulting in the loss W_3 occurred.

As shown in Fig.2, W_1, W_2 and W_3 are replotted against with concentration of the total multivalent cations (metal ions) in the nitrate solution. It is evident that there is a relative strong dependence of W_2 on concentration rather than W_1 or W_3. Hence we considered that W_2 is closely related to the content of the cations in the precursor. As increase in con-

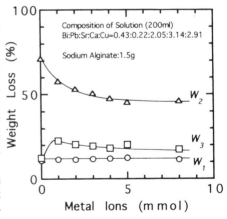

Fig.2. Dependence of the concentration of solution on the weight loss of the precursor.

tent of the cations in the solution, W_2 decreases gradually and approaches to a constant value. This situation suggests a saturation of the accumulation of the cations for the alginate at higher concentration region above ~ 0.165 mol/l. We adopted this value as the optimized concentration of the nitrate solution.

3. Solid state reaction and formation of superconducting phase

Fig. 3 shows DTA-TG trace for the precursor above 500℃. From this figure, it is suggested that several endothermic reactions appear below the melting temperature which are marked by arrows. As an attempt to clarify a formation process of superconducting phase via intermediate compounds, X-ray diffraction patterns for the specimen, which obtained at various temperatures marked by arrows in Fig.3, were studied. From this analysis, it is revealed that $SrCO_3$ and $CaCO_3$ remained up to the temperature near 800℃ and a portion of CuO seemed to be suitable up to 825℃. $Bi_2Sr_2CuO_x$, which is called "2201 phase", seems to be formed at near 770℃. It can therefore be presumed that the low Tc phase (2212 phase) was formed via a reaction among 2201 phase, CuO and $CaCO_3$, as indicated by the follow equation

$$Bi_2Sr_2CuO_6 + CaCO_3 + CuO \rightarrow Bi_2Sr_2CaCu_2O_8 .$$

In the above equation, $CaCO_3$ should be replaced with Ca_2CuO for the reaction in conventional method using powder material. This difference may be due to the relatively high contents of the organic compound in our precursor. The main products obtained at 825℃ is 2212 phase and the remainders as impurities were Ca_2PbO_4, CuO and 2201 phase.

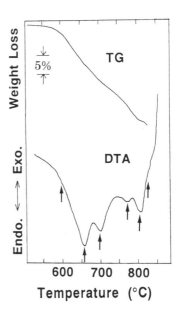

Fig.3. DTA-TG curves of precursor for temperatures above 500℃.

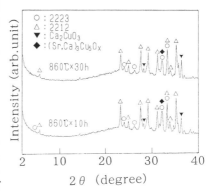

Fig.4. XRD patterns for precursors sintered at 860℃ for 10h and 30h, respectively.

Fig.4 shows the XRD pattern for the specimen obtained at 860℃ for 10 and 30 hours. The principal phases formed in the specimen were 2212 and 2223 except for Ca_2CuO_3 or $(Sr,Ca)_3Cu_5O_x$. In this sintering condition a partial melting was observed in the specimen.

There have been many reports on the enhancement of the volume fraction of 2223 phase. In addition, the process of the solid state reaction in BPSCCO system has been studied. However, these condition depends upon the individual synthesis process. In spite of these situation, it has been revealed either a presence of a liquid phase associated with a formation of 2223 phase or conversion of 2212 phase into 2223 phase.

Chen *et.al.* proposed that the liquid phase results in a higher internal diffusion of Ca_2CuO_3 or CuO exceeds solid-state diffusion.[4] Also the catalytic reaction of Ca_2PbO_4 helps to form and stabilize a structure of the 2223 phase. Although our result is in consistent with their consideration, the catalytic reaction was not observed.

The dc four terminal measurement was made to trace the temperature dependence of the electrical resistance for the sample obtained for the sintering at 860℃ for 30h. Decreasing temperature, resistance decreased rapidly at both temperatures of 108K and 81K (zero resistance), corresponding to the transition of 2223 and 2212 phases, respectively.

Fig.5 shows the cross-sectional view of the sintered fiber. In this study the sintering time of 30h seems to be preferable rather than a sintering for long period. For example, sintering for more than

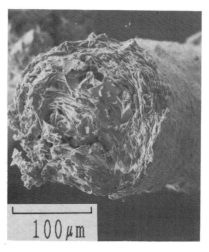

Fig.5. SEM cross-sectional view of the broken fiber

50h resulted in a poor growth of the superconducting phase. A possible consideration is due to the decomposition of the superconducting phase which may be resulted by the evaporation of the small amount Pb for the long period sintering. One should note that this situation is against with the remarkable improvement of the superconducting property of bulk sample for long time sintering, obtained by the conventional solid state reaction method.[5] We consider that this decomposition of the superconducting phase is easily proceeded due to the porous structure of our fiber.

For the preparation of a high-Tc suprconducting wire, the Ag coated fiber is preferred as a practical material. In our method, it can be carried out easily. The Ag coated BPSCCO fiber was synthesized successfully by sintering precursor coated with Ag paste on its surface. A preliminary experiment showed that the fiber obtained from the sintering at 830℃ mostly consisted of 2212 phase.

4. Conclusions

We synthesized the Bi based superconducting fiber by using sodium alginate on the basis of the ion-exchange property of the alginate. The solid state reaction for sintering of the precursor was studied by X-ray diffraction analysis and DTA. The temperature 860℃, the period 30h as sintering conditions seems to be preferable rather than the sintering for long period. The detailed study such as measurement of the critical current density of the specimen, especially for the Ag-coated fiber, are in progress.

References

[1] Haug A and Smidsrod O 1970 *Acta Chem.Scand.* **24** 843-854
[2] Konishi H, Takamura T, Kaga H and Katsuse K 1989 *Jpn.J.Appl.Phys.* **28** L241-L243
[3] Tanaka A, Kamehara N, Niwa K 1989 *Appl.Phys.Lett.* **55** 1252-1254
[4] Chen F H, Koo H S and Tseng T Y 1991 *Appl.Phys.Lett.* **58** 637-639
[5] Nobumasa H, Shimizu K, Kitano Y and Kawai T 1988 *Jpn.J.Appl.Phys.* **27** L846-L848

Inst. Phys. Conf. Ser. No 148
Paper presented at Applied Superconductivity, Edinburgh, 3–6 July 1995

Microwave Properties of Bulk Polycrystalline YBCO

B. A. Tonkin[1] and Y. G. Proykova

Physics Division, University of Portsmouth, Park Building, King Henry I Street, Portsmouth, Hants. PO1 2DZ (UK)

Abstract. The surface resistance of viscous processed and powder processed bulk polycrystalline $YBa_2Cu_3O_{7-x}$ specimens have been measured as a function of frequency (27-38GHz) and temperature (20-150K) using an endwall replacement technique. The surface resistance of the viscous processed material within the 9-38GHz frequency region at 77K gives a crossover frequency of 8.2GHz. The powder processed material has the higher crossover frequency of 11GHz.

1. Introduction

Microwave measurements of the surface resistance of high-Tc superconductors coupled with studies of their microstructure are important and necessary steps in determining the major causes of microwave power losses.

In this paper we present and discuss the frequency and temperature dependence of the surface resistance of viscous processed and powder processed $YBa_2Cu_3O_{7-x}$ (YBCO) bulk polycrystalline specimens in the 27-38GHz frequency range (Ka-band) and 20-150K temperature range together with additional X-ray diffraction and SEM characterisation of the specimens.

2. Experimental

The microwave measurements were carried out in a right cylindrical gold plated transmission cavity using an endwall replacement technique [1]. The surface resistance of the specimen is found from the change in the loaded Q-factor of the cavity using

$$R_{exp}(T) = R_{Cu}(T) + \frac{1}{G}\Delta\left(\frac{1}{Q(T)}\right) \tag{1}$$

where $R_{Cu}(T)$ is the surface resistance of an oxygen free high conductivity copper endwall and G is a geometry factor. The automated microwave measurement system enables the surface resistance of disks with diameters between 26 and 44 mm to be measured in the frequency range 27-38GHz, using the TE_{012}, TE_{021} and TE_{013} modes, and over the 20-150K temperature range.

---

[1]Also at the Department of Electrical and Electronic Engineering

568

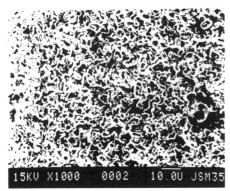

Figure 1(a). SEM micrograph of specimen A - viscous processed material.

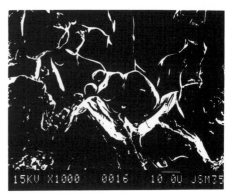

Figure 1(b). SEM micrograph of specimen B - powder processed material.

The samples were synthesised using two methods : (i) viscous processing and extrusion [2] and (ii) powder processing, a solid state reaction with repeated grinding and sintering, a procedure consisting briefly of the following: Stoichiometric amounts of oxide powders were intimately mixed after sieving, calcined at 800°C then repeatedly ground, sieved and sintered at 980°C in flowing oxygen until x-ray diffraction characterisation showed an impurity level less than 1%. For microwave characterisation the powder was uniaxially pressed into a disk 32mm in diameter and sintered. Finally, in order to reduce the surface roughness the specimen surface was lightly polished with fine abrasive paper and then ultrasonically cleaned.

X-ray diffraction characterisation of the disks was carried out to identify any impurities, give an evaluation of the oxygen content (from the c-axis lattice parameter) and the level of any grain alignment, or texturing, of the specimens, a measure of which may be obtained from the ratio of the 005 and 013 diffraction peak intensities. Oxygen content was also determined by iodometric titration of a small part of each specimen. The density of the specimens were determined from mass and dimensions. Characterisation by SEM was used to determine the average grain size and allow an examination of the grain boundary quality.

3. Results and Discussion

The disk specimen manufactured by viscous processing (referred to as specimen A) was determined to have an oxygen deficiency $x = 0.24\pm0.1$, a density 75% of the theoretical value (6.3g/cm) and a few percent $BaCuO_2$ impurities. The disk specimen produced by repeated grinding and sintering (referred to as specimen B) was found to have approximately the same density and a similar level of the same impurities but with a slightly higher oxygen content, $x=0.13\pm0.1$. Specimen A exhibited the diffraction pattern expected from a completely random grain orientation, whereas the diffraction pattern of specimen B indicated that the grains within the specimen were partially c-axis aligned.

An SEM image of the surface of specimen A, figure 1(a), shows a large number of small but well connected grains. An analysis is very difficult with this type of morphology since it is difficult to ascertain the true size of grains. However, from the viewpoint of the microwaves the area of the grain is precisely that seen in the flat image and ignores any hidden orientation below the surface. This leads to an average grain size ~3μm with a small standard deviation. Although this may not give a true description from the microstructural viewpoint it does

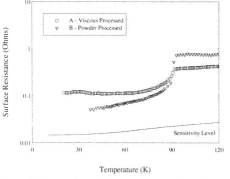

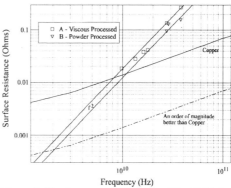

Figure 2. Comparison of the temperature dependence of the surface resistance of specimen A and B at 27.4GHz. The solid line represents the sensitivity limit.

Figure 3. The frequency dependence of R_{exp}(77K) for specimen A and B illustrating the crossover and upper operating frequencies.

represent a much more realistic description from the incident microwave viewpoint. This average represents a measure of the diameter of a circular grain whose area equals the mean area of all the grains in the picture. As seen in the image, a large porosity is observed. The connected pore volume is important for complete oxygenation of the specimens since the critical current for dense material has been observed to decrease due to incomplete oxygenation [3].

Figure 1(b) illustrates the granular micro structure of specimen B which intersperse the large polished areas at the same magnification as for specimen A. The surface damage caused by the polishing is clearly visible as is some of the resultant fine particles. It is clear from the image that the grain size is significantly larger than the viscous processed material with an average size ~30μm. The size of the polished areas are even larger although this may not represent the true grain size. The large number of small particles formed by the polishing (and not removed by ultrasonic cleaning) are likely to influence the microwave losses.

Figure 2 compares the measured temperature dependence of the microwave surface resistance of specimen A and specimen B at the same measurement frequency, 27.4GHz, together with the present measurement sensitivity limit. The onset transition temperature T_C was found to be 92K for specimen A and 92.5K for specimen B, which agrees well with the results from d.c. methods [4] and the slight difference in the deduced oxygen contents. In the superconducting state, the measured surface resistance, R_{exp}, of both specimens is seen to monotonically decrease toward some finite value at T=0K.

Above T_C there is a linear variation of R_{exp} which continues to room temperature for both types of specimen.

The larger grained powder processed YBCO is clearly superior to the viscous processed material in the superconducting state despite the surface damage from the polishing. However, in the normal state the reverse is observed which suggests that the small particles become superconducting. One possible explanation of the large normal state surface resistance of the powder processed specimens is related to their comparatively large grain boundary resistance. The grain boundary resistance, R_{GB}, can be defined as the normal state surface resistance, R_N, extrapolated to T = 0K, assuming that

$$R_N = R_{GB} + AT \qquad (2)$$

where A is a constant. It has been observed that R_{GB} of the powder processed material is twice as large as that found for viscous processed material. This may be directly related to the polishing damage.

R_{exp} at 77K for specimen A together with the values of the measured surface resistance at 77K found previously over the frequency range 9-18GHz [5] are plotted in Fig. 3 as a function of frequency. Also plotted in figure 3 is the surface resistance value for copper at 77K and an order of magnitude improvement, often proposed as the operating level of devices. The surface resistance value allows the cross-over frequency to be determined while the order of magnitude improvement value gives an upper operating frequency. These were found to be 8.2GHz and 1.6GHz respectively for viscous processed YBCO (specimen A) and 11GHz and 2GHz for powder processed YBCO (specimen B). This illustrates that a large improvement in the cross-over frequency is required to increase the upper operating frequency significantly.

4. Conclusions

The microwave surface resistance of viscous processed and powder processed bulk polycrystalline $YBa_2Cu_3O_{7-x}$ (YBCO) specimens has been measured as a function of temperature and frequency at Ka-band.

The partially aligned larger grained powder processed YBCO specimen was found to have a lower microwave surface resistance than the randomly oriented viscous processed YBCO specimen.

The measured surface resistance values for the viscous processed specimen are consistent with previous microwave measurements and show an ω^2 dependence from 9-38GHz with a crossover frequency of 8.2GHz and an upper operating frequency of 1.6GHz. The powder processed specimen shows a similar frequency dependence of the measured surface resistance with a slightly superior performance; a crossover frequency of 11GHz and upper operating frequency of 2GHz.

Acknowledgement

The authors would like to thank Dr. N. McN. Alford and Dr. T. W. Button of ICI Superconductors, Billingham, UK, who kindly provided the viscous processed YBCO specimens.

References

[1]   Tonkin B A and Proykova Y G 1995 *IEEE Trans. Appl. Supercond.* **5** (in press)
[2]   Alford N McN, Birchall J D, Clegg W J, Harmer M A, Kendall K and Jones D H 1988 *J. Mater. Sci.* **23**, 761-768; Alford N McN, Button T W, Peterson G E, Smith P A, Davis L E, Penn S J, Lancaster M J, Wu Z and Gallop J C 1991 *IEEE Trans. Magn.* **27** 1510-1513
[3]   Alford N McN, Birchall J D, Clegg W J, Harmer M A and Kendall K 1988 Man Yan (ed.) *Ceramic Superconductors II* Amer. Ceram. Soc. Publ. 232-240
[4]   Cava R J, Batlogg B, van Dover R B, Murphy D W, Sunshine S, Siegrist T, Remeika J P, Rietman E A and Espinosa G P 1987 *Phys. Rev. Letts.* **58** 1676-1679
[5]   Tonkin B A and Proykova Y G 1993 *Supercond. Sci. Tech.* **6** 353-359

Inst. Phys. Conf. Ser. No 148
Paper presented at Applied Superconductivity, Edinburgh, 3–6 July 1995
© *1995 IOP Publishing Ltd*

Frequency dependence of AC susceptibility in YBCO sintered samples

M. Polichetti, I. D'Acunto, S. Pace, L. Reggiani,
A.M. Testa, D. Di Gioacchino*

Dept. of Physics, University of Salerno , Salerno (ITALY)
* I.N.F.N. - LNF, Rome, (ITALY)

Abstract. The intragranular contribution to the imaginary part of AC susceptibility $(\chi")$ of YBCO sintered pellets has been investigated as a function of the field frequency (f = 17 Hz + 1070 Hz), for a narrow temperature range around the transition temperature . Different frequency dependences of $\chi"$ have been found as the temperature varies. For temperatures T<91.9 K $\chi"$ is found to increase linearly with *ln* f, whereas for T≥ 91.9 K a linear frequency dependence is found. Moreover, for T<92.3K the zero frequency limit approaches to a finite value, whereas for T>92.3K such limit is zero.

1. Introduction

Different experimental methods [1-3] have been extensively employed in the study of dissipative effects in the mixed state of high T_C superconductors. In particular, AC techniques such as AC susceptibility [4], AC transport measurements, mechanical oscillator [5,6], provide informations useful for the description of the fluxon dynamics in these materials. However, the interpretation of experimental data is quite difficult due to the interplay amongst the intrinsic structural anisotropy, thermal fluctuations and disorder, leading to the formulation of novel concepts (i.e. irreversibility line [2], thermally activated flux flow [7], entangled vortex liquid [8], vortex glass [9,10]). As far as the AC susceptibility is concerned, it is well known that its real part (χ') is a measurement of the magnetic shielding generated by supercurrents, while the imaginary part $(\chi")$ is related to losses. Such losses are determined by two contributions : a) frequency independent hysteretic losses due to pinning phenomena; b) frequency dependent AC losses due to resistive effects (normal , flux flow, creep, ..). In granular materials both intergranular and intragranular currents contribute to the measured susceptibility : however, in these systems, the low field magnetic behaviour at temperatures well below the critical temperature (T_C) is ruled by the intergranular Josephson coupling between the grains, whereas in proximity of T_C the behaviour is determined by the isolated grains. Therefore, by selecting the proper temperature range and samples with weak intergranular coupling, it is possible to separate the intragranular contribution from the intergranular one. In a previous paper [11] on

AC susceptibility near the transition temperature of YBCO sintered samples, we have reported the existence of a difference between the onset temperatures of χ' and χ''. Such a difference has been found to decrease for increasing frequencies, and to revert its sign for frequencies higher than 300 Hz. In this paper the results of AC susceptibility measurements near T_C of YBCO sintered pellets are reported. In particular, the intragranular contribution to χ'' was studied as a function of the field frequency (f = 17 Hz + 1070 Hz) in a narrow temperature range around T_C ($\cong$ 92.0 K). In sec.2 the sample preparation technique and the experimental apparatus for AC susceptibility are briefly summarized. In sec.3 susceptibility data for the samples investigated are analyzed and discussed.

2. Experimental set-up

YBCO pellets have been prepared following a modified citrate pyrolysis procedure, with only two thermal threatments (calcination and sinterization) carried out in oxygen atmosphere . The sample dimensions are (5 x 3 x 15) mm^3. The complex magnetic susceptibility was measured as a function of temperature by means of a non-commercial 3-coils susceptometer in the bridge configuration. The voltage induced in two pick-up coils, connected in series and wrapped in opposite direction, was measured by a lock-in amplifier (EG&G 5302), being χ' and χ'' proportional to the in-phase and out-of-phase component of the measured voltage respectively. The sample temperature was measured by a Pt-thermometer and it was controlled by a thermoregulator (Lake Shore DRC 93 CA); temperature inhomogeneities along the samples have been found to be less than 0.02 K . Measurements were carried out by applying the AC magnetic field at room temperature, then cooling the samples down to T = 78 K. After samples thermalization at 78K , data were collected up to 110 K with a warming rate of 0.04 K/min. All the measurement procedures are completely automatized, therefore their repeatability is assured.

3. Results and discussion

The characterization of the superconducting behavior of the samples was performed by measuring χ' and χ'' as a function of temperature at f = 107 Hz for different field amplitudes. The samples investigated were selected just because of their weak coupling between the grains, allowing us to focus the attention on the intragranular contribution to the susceptibility. Indeed, in these samples, χ'' develops two well separated peaks (inter- and intragranular), with the intergranular one shifting at lower temperatures as the AC field amplitude increases in the range (1,5,10,15, 20 Oe). Then, in order to investigate the flux dynamics, $\chi''(T)$ data recorded in a narrow range around T_C ($\cong$ 92.0 K, defined as the temperature corrisponding to 10% of the full diamagnetic signal) have been analyzed in detail as a function of frequency (17, 68, 107, 340, 1070 Hz), for different AC fields (5,10,20 Oe). Curves of χ'' vs frequency at constant temperature and for an AC field of 20 Oe are reported in fig 1 for the sample denoted as 2PS, where plotted data have been corrected for the contribution due to the background signal. In particular, the background subtraction was performed by recording the signal without the sample and fitting the data in the range 93-110 K with a linear relationship. Then the susceptibility was measured in presence of the sample, performing again a linear fit to measured data in the same temperature range. Therefore, the sample contribution to the imaginary part of the susceptibility can be written as :

$$\chi''_s (T) = \chi''_{meas} (T) - \chi''_{back} (T)$$

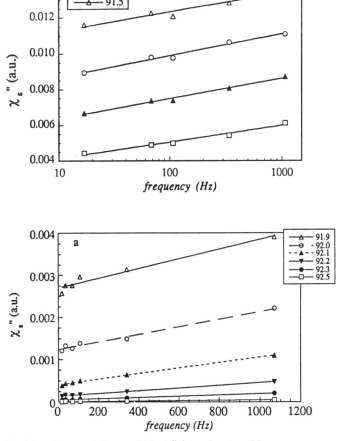

Fig.1 Imaginary part of susceptibility (χ''_s) as a function of frequency at 20 Oe, for different temperatures. (a): T≥ 91.9K; (b) T ≤ 91. 8 K .

Three different behavior have been found in different temperature ranges :

a) for T > 92.3 K χ''_s shows a linear frequency dependence, vanishing in the limit of zero frequency ; b) for 91.9 K $\leq$ T $\leq$ 92.3 K a linear frequency dependence is also found for χ''_s (fig.1a), but the values extrapolated in the limit of zero frequency are finite; c) for T< 91.9 K experimental data can be satisfactorily fitted with a logarithmic law, as shown in fig. 1b.

In the case (a), it is known that for a material with a constant differential resistivity dE/dJ the low frequency limit of χ'' is [12] : $\chi'' = k \, \sigma(T) \, f$, where $\sigma(T)$ is the electrical conductivity and k a constant dependent on the sample geometry. Such a linear relationship is valid in the limit $\delta/d >> 1$, where δ is the normal skin depth and d the sample size (the grain size in our case).This condition is satisfied in our case. Therefore, for temperatures above the mean field transition temperature ($T_{CO} = 92.28K$), the linear frequency dependence of χ'' is determined by the excess fluctuation conductivity [13]. For the cases (b)-(c), in spite of the different frequency dependence, the χ'' values extrapolated in the zero frequency limit are finite, indicating that a *residual* hysteresis still persists.

References

[1] Palstra T.T.M., Batlogg B., Schneemeyer L.F., Waszczak J.V., 1988 Phys.Rev.Lett. **61**, 1662 .

[2] Goldfarb R.B., Lelental M., Thompson A.C., 1989 *Magnetic Susceptibility of Superconductors and Other Spin Systems* (Plenum Press, New York) p.49 .

[3] Wu H., Ong N.P., 1993 Phys.Rev. Lett. **71**, 2642 .

[4] Malozemoff A.P.,Worthington T.K., Yeshurun Y., Holtzberg F., 1988 Phys.Rev.B 38 7203 .

[5] Gammel P.L., Schneemeyer L.F., Waszczak J.V., Bishop D.J., 1988 Phys.Rev.Lett. **61**, 1666 .

[6] Farrell D.E., Rice J.P.,Ginsberg D.M., 1991 Phys.Rev.Lett. **67**, 1165 .

[7] Kes P.H., Aarts J., van den Berg J.,van der Beek C.J., Mydosh J.A., 1989 Supercond.Sci.Technol. **1, 242** .

[8] Nelson D.R., 1988 Phys.Rev.Lett. **60**, 1973 .

[9] Fisher D.S., Fisher M.P.A.,Huse D.A., 1991 Phys.Rev.B **43**, 130 .

[10] Nelson D.R.,Vinokur V.M. , 1992 Phys.Rev.Lett. **68**, 2398 .

[11] Polichetti M., Pace S., Saggese A., Testa A.M., Federico C., Celani F., Boutet M., Di Gioacchino D., 1994 Physica C **235-240**, 3217 .

[12] A.F.Khoder, M.Couach , 1989 *Magnetic Susceptibility of Superconductors and Other Spin Systems,* (Plenum Press, New York) p.213 .

[13] Polichetti M., Pace S., **Reggiani L. Testa** A.M., presented at IV ECERS 95 Conference, Riccione (Italy), Oct . 1995

Inst. Phys. Conf. Ser. No 148
Paper presented at Applied Superconductivity, Edinburgh, 3–6 July 1995
© 1995 IOP Publishing Ltd

Neutron Irradiation Effects on Critical Current Densities in Tl-1223 and Tl-2223 Superconductors*

G. Brandstätter, X.Yang, F.M. Sauerzopf, and H.W. Weber

Atominstitut der Österreichischen Universitäten, A-1020 Wien, Austria

Abstract. Tl-2223 and Tl-1223 single crystals as well as polycrystalline Tl-1223 ceramics were investigated by SQUID magnetometry. The samples were subjected to sequential neutron irradiation up to a fluence of $16·10^{21}$ m^{-2}. At low temperatures J_c is very high in both materials ($\sim 10^{10}$ Am^{-2}), but disappears at ~ 40 K in Tl-2223, whereas Tl-1223 shows critical currents up to 93 K. After irradiation J_c is tremendously enhanced in both materials. The irreversibility line of the Tl-2223 compound has a very flat slope down to low temperatures, whereas Tl-1223 shows a rapid increase already at high temperatures. After irradiation the irreversibility lines of both materials are shifted to higher fields and temperatures.

1. Introduction

The two most interesting compounds of the large family of Tl-based high temperature superconductors are those with the 1223 and the 2223 structure, i.e. with one or two Tl-O-layers between the CuO$_2$-planes. The first has a very high irreversibility line similar to YBCO-123 making it a candidate for future practical applications, the other has a very low lying irreversibility line, which offers the possibility of investigating the reversible mixed state properties in a large temperature range.

At low temperatures both compounds show very high critical current densities J_c, which are comparable to or even higher than those found in YBCO-123. In this paper we report on neutron irradiation effects on the irreversible properties of both materials.

2. Experimental

Single crystals with the nominal composition Tl$_2$Ca$_2$Ba$_2$Cu$_3$O$_{10}$ (Tl-2223) and (TlPb)(SrBa)Ca$_2$Cu$_3$O$_9$ (Tl-1223) as well as (TlPb)Sr$_2$Ca$_2$Cu$_3$O$_9$ polycrystalline ceramics were examined by SQUID-magnetometry. A detailed description of the preparation procedures employed for these compounds is given in references [1, 2, 3]. The crystals are

---
* Partial support by the Austrian Science Foundation (grant #9194) and by the EU Brite Euram II Program (grant #BRE2-CT94-0531) is gratefully acknowledged.

small platelets with dimensions 1030·430·65 μm³ (Tl-2223) and 380·350·108 μm³ (Tl-1223), respectively. From the polycrystalline pellet a cubic piece with dimensions 2·2.5·1.84 mm³ was cut. The average grain size determined from SEM photographs is 4.2 μm. The samples were mounted onto small U-shaped aluminum holders, which fit into an aluminum rod and allow a reproducible and accurate orientation with respect to the field. Hysteresis loops were measured in an 8 T-SQUID magnetometer up to an applied field of 8 T; zero-field-cooled (ZFC) and field-cooled (FC) measurements were carried out both in the 8 T- and in a 1 T-SQUID magnetometer with H∥c. The superconducting transition temperature was measured in an applied field of 1.1 mT with H∥ab. The samples were sequentially subjected to fast neutron irradiation to fluences of $2 \cdot 10^{21}$ m^{-2}, $4 \cdot 10^{21}$ m^{-2}, $8 \cdot 10^{21}$ m^{-2}, and $16 \cdot 10^{21}$ m^{-2} (E>0.1 MeV) for Tl-2223, and to $2 \cdot 10^{21}$ m^{-2} (E>0.1 MeV) for Tl-1223.

3. Results

The superconducting transition temperature T_c was measured in the unirradiated state as well as after each irradiation step. Similar to one of our previous experiments on Tl-2223 single crystals [4] a small plateau can be observed at low fluences (121.5 K, 120.5 K, 118 K, 117.3 K, and 113.5 K), but generally T_c decreases linearly with a slope of -4.8 K per 10^{22} neutrons/m², which is more pronounced than in YBCO-123 [5]. The Tl-1223 single crystal shows a T_c of 107 K, which is lower than that of the polycrystalline sample (114 K) and values published by other groups [6]. This difference is attributed to different oxygen contents in the samples. After the first irradiation step a decrease of T_c to 105.2 K was observed, which may result in a similar plateau as found in Tl-2223.

J_c of the single crystals was calculated as a function of the local induction B using an anisotropic Bean model, which takes demagnetization effects into account [7]. In case of the ceramic sample, the intragrain critical current densities were determined as a function of the *applied* field with a simple Bean model. In the unirradiated state J_c of Tl-2223 is strongly temperature and field dependent and disappears already above 40 K. The Tl-1223 single crystal shows fishtails, which are pronounced even at 10 K. Thus, J_c is kept nearly constant over a wide local induction range, e.g. at 10 K $J_c \approx 2.3 \cdot 10^{10}$ Am^{-2} at least up to 8 T. In comparison to Tl-2223, the temperature dependence of J_c is weak, since it drops below 10^9 Am^{-2} only above 60 K. At 77 K, significant J_c's of about $2 \cdot 10^8$ Am^{-2} at 0.5 T can be found in the single crystal. J_c of the polycrystalline sample extends up to 93 K ($2.7 \cdot 10^8$ Am^{-2} at 0.5 T). After each irradiation step, J_c increased systematically. At 40 K and 1 T, the increase in J_c of Tl-2223 after the last irradiation step corresponds to an enhancement factor of 61 (see Fig. 1). Tl-1223 shows a similar behavior. However, the increase in J_c at low inductions is much more pronounced than in Tl-2223, since the fishtail disappeared. After the first irradiation step the enhancement factor varies between 5 (high inductions) to 20 (low inductions) (see Fig. 2).

For the single crystals the irreversibility lines (Fig. 3) were determined using a method based upon the distortion of the SQUID response curves. It is described in detail in references [8, 9]. Tl-2223 has a very large reversible regime, almost down to $T/T_c \approx 0.5$, which leads to a very small initial slope of the irreversibility line. A rapid increase can be observed below t≈0.3. Tl-1223 shows a similar temperature dependence of the irreversibility line, but the rapid increase already occurs at $T/T_c \approx 0.8$. After each irradiation step the

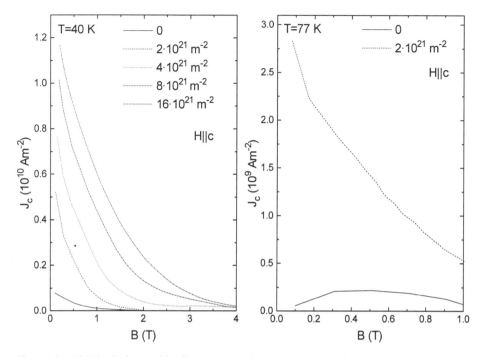

Fig. 1: J_c in a Tl-2223 single crystal irradiated to different fluences.

Fig. 2: J_c in a Tl-1223 single crystal before and after irradiation.

irreversibility line is shifted to higher fields and temperatures. In the case of Tl-2223, there is almost no change between the last two irradiation steps, which indicates that $16 \cdot 10^{21}$ m^{-2} is probably the highest fluence the material can sustain without a major degradation of superconductivity. Compared to the irreversibility line of Tl-1223 in the unirradiated state, those of Tl-2223 are still by far lower (see Fig. 3). In the case of Tl-1223 a very large shift to higher temperatures is observed after the first irradiation step. This shift is much larger than in Tl-2223 or in YBCO-123 [9]. In each case the irreversibility line can be described by an exponential law.

4. Summary

Tl-2223 and Tl-1223 single crystals and cermics were investigated by SQUID magnetometry. Critical current densities and the irreversibility lines were determined before and after each irradiation step. We find a tremendous increase of J_c in Tl-2223 after the last irradiation step (enhancement factors up to 61), but the J_c values of the related Tl-1223 compound cannot be reached. The Tl-1223 single crystal shows fishtails in the unirradiated state, which disappear after the first irradiation step. J_c is enhanced by factors up to 20 at low inductions. The irreversibility line of both compounds is shifted to higher fields and temperatures after fast neutron irradiation.

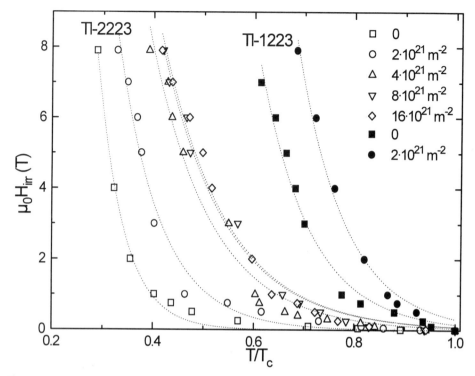

Fig. 3: Irreversibility lines of Tl-2223 and Tl-1223 single crystals after irradiation to different fluences, H∥c.

5. References

[1] Brandstätter G, Sauerzopf F M, Weber H W, Mexner W, Freyhardt H C, Aghaei A, Ladenberger F, and Schwarzmann E 1993 *Applied Superconductivity* edt. by Freyhardt H C (DGM Informationsges. Verlag) 109-12

[2] Winzer K 1992 *Annalen d. Physik* 1 479

[3] Mexner W, Heede S, Heinemann K, Freyhardt H C, Ullmann B, Ladenberger F, and Schwarzmann E 1994 *Critical Currents in Superconductors* edt. by Weber H W (Singapore: World Scientific) 513-16

[4] Brandstätter G, Sauerzopf F M, Weber H W, Aghaei A, Schwarzmann E 1994 *Physica C* **235-240** 2797-8

[5] Sauerzopf F M, Wiesinger H P, Weber H W, Crabtree G W 1995 *Phys. Rev. B* **51** 6002-12

[6] Aihara K, Doi T, Soeta A, Takeuchi S, Yuasa T, Eido M S, Kamo T, Matsuda S 1992 *Cryogenics* **32** 936-39

[7] Wiesinger H P, Sauerzopf F M, Weber H W 1992 *Physica C* **203** 121-28

[8] Suenaga M, Welch D O, Budhani R 1992 *Supercond. Sci. Technol.* **5** 1-8

[9] Sauerzopf F M, Wiesinger H P, Weber H W, Crabtree G W 1992 *Advances in Cryog. Eng.* **38** 901-5

Inst. Phys. Conf. Ser. No 148
Paper presented at Applied Superconductivity, Edinburgh, 3–6 July 1995
© 1995 IOP Publishing Ltd

Flux Visualization as Sensitive Tool to Investigate Defect Structures in Superconductors

Koblischka M R†[1]

†Free University, Faculty of Physics and Astronomy, De Boelelaan 1081, 1081 HV Amsterdam, The Netherlands.

Abstract. The visualization of flux patterns in superconductors is shown to be a very sensitive tool to investigate the influence of defects within the samples on the magnetic moment and the related critical current density. Especially the High-Resolution Faraday (HRF) technique which combines a relatively high spatial resolution of 0.5 μm with the possibility of dynamic observations, is optimally suited for such investigations [1]. Defects within the sample are demonstrated to enhance the flux penetration, so that the normal geometry-dependent flux distribution is severely disturbed. Consequently, the distributions of currents flowing in the sample is changed. Magnetic patterns reveal defects which are *not* visible in ordinary microscopy. Such an analysis is shown on $YBa_2Cu_3O_7$ (YBCO) thin films, on single crystals of $DyBa_2Cu_3O_{7-\delta}$ (DyBCO) and $Bi_2Sr_2CaCu_2O_8$ (Bi-2212). Especially on thin films defects induced by chemical etching or by the substrate are shown to play an important role. Even defects like intergrowths of foreign phases can be detected.

1. Introduction

Most of the magnetic measurements on high-T_c superconductors are of the integral type, i. e. they are sensitive only to the *total* magnetic moment of a sample. This is the case for various kinds of magnetometers and AC-susceptibility techniques. A major disadvantage of these techniques is that they are completely insensitive to the current- and flux-*distributions* in the superconductor. On the other hand, the magneto-optical visualization of flux in a superconductor is a powerful tool to study directly the interaction of defects in the sample with the flux lines [1]. From such observations, one can determine important characteristics locally such as pinning forces and current densities [1, 2].

[1] present address: Groupe de Physique appliquée, Université de Genève, Rue d'Ecole de Medecine 20, CH-1211 Genève 4, Switzerland.

2. Faraday effect and magneto-optical setup

The flux visualization is based on the Faraday effect in a thin magneto-optical layer (MO). The local flux is determined from the rotation of the polarization plane of linearly polarized light within the magneto-optical layer. In regions without flux, the light is reflected without rotation of the polarization plane. This light is thus not able to pass the analyzer which is set in crossed position with respect to the polarization plane of the incident light. In this way, the Meissner phase remains dark, whereas the Shubnikov phase is imaged bright. The MO can be evaportated directly onto the sample surface together with an extra reflection layer, however, for a destruction-free investigation, the MO is evaporated onto a glass substrate which can be laid onto the sample. The maximum spatial resolution is obtained using EuSe as MO. In order to perform invetigations also at higher temperatures ($T > 20$ K), a garnet indicator can be used. Additionally, a freezing-in technique (applying an external field to a zero-field cooled state, heating up to the target temperature and subsequent fast cooling) with EuSe films could be applied in order to observe flux distributions up to the irreversibility temperature [1, 3]. The great advantage of the MO technique is the large range of applicability, e. g. from large samples ($\approx$ cm) down to the study of small defects in the micrometer range. Even smaller defects can be visualized by their influence on stray fields. The magneto-optical observation techniques are very sensitive to the enhanced stray fields along typical defects, but also to variations in the thickness of the superconductor [4].

3. Results and discussion

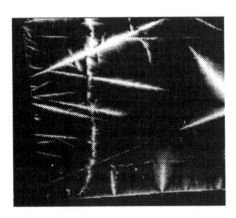 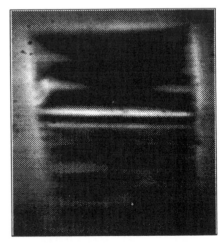

Figure 1. (left): YBCO thin film on substrate with defects; $T = 4.2$ K, $\mu_0 H_a = 100$ mT. (right): Bi-2212 single crystal containing many defects; $T = 1.5$ K, $\mu_0 H_a = 1$ T. The shielding currents are flowing only in small loops.

In figure 1, two examples of defect-containing samples are presented. The left figure shows an YBCO thin film ($T = 4.2$ K, $\mu_0 H_a = 100$ mT) on a substrate which contained a lot of defects.

Due cutting of the substrate, many additional defects have been produced. Such defects enhance the flux penetration enormously; when vortices penetrating along different defect lines merge together, the currents change their length scale. In this way, small current loops are formed. The usually observed geometry-dependent flux pattern [5] is completely suppressed here. A well polished substrate is therefore very important to obtain a qualitatively good sample. Note that many thin film samples look at least partly like this.

Figure 1 shows also an example of a Bi-2212 single crystal (right). The flux pattern is completely disturbed by defects leading to the formation of small current loops. Additional X-ray measurements have shown that there is a tilt of the c-axis splitting the crystal into various domains. This c-axis tilt may be caused by intergrowth of some unit cells of Bi-2223 [6]. The effect of all other optically visible defects like growth steps, twin boundaries etc. can be studied directly. Due to the large demagnetization factors of the high-T_c samples, the influence of even very small defects on the flux distributions is considerably enhanced as compared to large cylindrical samples. In figure 2, schemat-

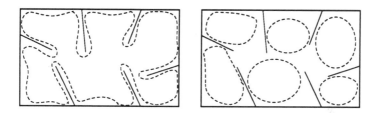

Figure 2. Schematical drawing of current flow around defects in a superconductor; (left): critical currents at small fields, (right): loops formed by the shielding currents due to the presence of defects.

ical sketches of the current flow around defects is presented. At low fields, flux lines penetrates only a small belt around the sample. Along some defect lines, flux rushes deeper into the sample. Such defect lines can be e. g. twins [8] or a network of grain boundaries [7]. With raising the external magnetic field further, the flux penetration along the defects can merge together long before a fully penetrated state is reached. Within the flux carrying belt, critical currents are flowing. Now, the penetrating flux leads to formation of current loops of shielding currents, and the critical currents are changing their length scale. Integral measurements of magnetization and critical current will lead to a severe underestimation because of using the wrong length scale.

In figure 3, two flux distributions obtained on a DyBCO single crystal are presented. The left image is obtained at $T = 20$ K and $\mu_0 H_a = 55$ mT, the right one at $T = 20$ K and $\mu_0 H_a = 218$ mT. Two defects lead to an enhanced flux penetration which changes the flux pattern also close to the full penetration field. Such defect-induced changes in the flux patterns persist to high magnetic fields. This is a demonstration of the so-called magnetically induced granularity.

582

Figure 3. Flux patterns of a DyBCO single crystal at $T = 20$ K and (left):
$\mu_0 H_a = 73$ mT; (right): $\mu_0 H_a = 246$ mT. These images correspond to the
schematic drawings of figure 2.

In this paper examples of typical defects found in high-T_c superconductors are presented. Due to the large demagnetization factors, the influence of defects on the flux distributions is enormous; thus enabling one to visualize also the effect of very small defects which are not visible in SEM or ordinary microscopy. The magneto-optical investigation technique may become a standard method for quality checks of superconducting samples.

I would like to thank my colleagues Th. Schuster, H. Kuhn (MPI Stuttgart), and R. J. Wijngaarden (Free University Amsterdam) for helpful discussions.

References

[1] Koblischka M R and Wijngaarden R J (1995) *Supercond. Sci. Technol.* **8** 199, and references therein

[2] Huebener R P, 'Magnetic Flux Structures in Superconductors',in Solid State Sciences 6, Springer, New York (1979), and references therein.

[3] Indenbom M V *et al* (1993) *Physica C* **209**

[4] Klein W and Kronmüller H (1979) *Phys. Status Solidi* (a) **167** 109

[5] Schuster Th *et al* (1994) *Phys. Rev. B* **49** 3443

[6] Koblischka M R *et al* (1995) *Physica C* (in press)

[7] Koblischka M R, Schuster Th and Kronmüller H (1994) *Physica C* **211** 263; (1994) *Physica C* **219** 205

[8] Schuster Th *et al* (1993) *J. Appl. Phys.* **74** 3307; Welp U *et al Physica C* **235-240** 49

Inst. Phys. Conf. Ser. No 148
Paper presented at Applied Superconductivity, Edinburgh, 3–6 July 1995

583

Specific Heat Of $PbMo_6S_8$ and $(Pb_{0.7}Gd_{0.3})Mo_6S_8$ Superconductors in high magnetic fields

S Ali, H D Ramsbottom, D N Zheng and D P Hampshire

Superconductivity Group, Department of Physics, University of Durham, Durham, UK.

Abstract. Specific heat measurements have been completed on $PbMo_6S_8$ (PMS) and $(Pb_{0.7}Gd_{0.3})Mo_6S_8$ (Gd-PMS) from 4.2 K up to T_C in magnetic fields up to 10 T using the quasi-adiabatic heat pulse method and A.C. temperature technique. The properties of bulk PMS and Gd-PMS fabricated at ambient pressure are compared with material fabricated at a pressure of 2×10^8 N.m^{-2} (2000 bar) using a Hot Isostatic Press (HIP). The volumetrically averaged upper critical field ($B_{C2}(T)$) has been determined as a function of temperature for these materials.

The Gd-PMS shows evidence for a two-phase separation into a low temperature phase which is probably magnetic in origin and a PMS-like superconducting phase where the Gd may act primarily as an oxygen getter. Both PMS samples show high upper critical field values ($T_C \approx 13K$, $B_{C2}(0) \approx 55$ T) which were significantly higher than the irreversibility lines determined from complementary transport and magnetic measurements. The results show strong similarities with the high temperature oxide superconductors.

1: Introduction:

The lead Chevrel-phase $PbMo_6S_8$ is well known for its high upper critical field (B_{C2}) and hence is a potential candidate for high field applications [1]. It has been reported that adding rare-earth elements to PMS increases the B_{C2} and T_C [2].

In contrast to most transport measurements [1,2], the specific heat provides a true volumetric measurement of superconducting critical parameters. This paper provides calorimetric measurements on $PbMo_6S_8$ (PMS) and $(Pb_{0.7}Gd_{0.3})Mo_6S_8$ (Gd-PMS) fabricated at ambient pressure (unHIP'ed) and at 2000 bar (HIP'ed) in order to evaluate whether substituting Gd improves the high field performance of PMS.

2: Fabrication procedure

To obtain the Chevrel phase ceramic samples, a solid state reaction technique was used. Elemental Pb, Gd, Mo and S were mixed in the appropriate atomic ratios and reacted in a sealed quartz tube under vacuum at 450 °C for 4 hours followed by 8 hours at 650 °C. The material was then ground and pressed into discs. The discs were reacted again in an evacuated sealed quartz tube, at 1000 °C for 44 h. The HIP'ed samples were then reacted at 2000 bar and 800 °C for 8 h in the hot isostatic press.

584

3: Experimental

The specific heat measurements were made using a quasi adiabatic heat pulse method [3] and an A.C. technique [4]. A comparison of results obtained on Cu and NbTi with literature values suggests that the accuracy of the calorimeter is approximately 3%. Measurements on HIP'ed and unHIP'ed PMS and Gd-PMS were completed at 0, 2.5, 5, 7.5 and 10T from 4.2 K up to 16 K.

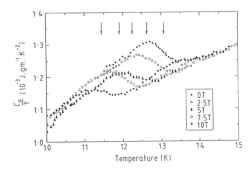

Figure 1. Specific heat capacity/ temperature as a function of temperature for HIP'ped PMS in high magnetic fields. The arrows denote the transition at each magnetic field

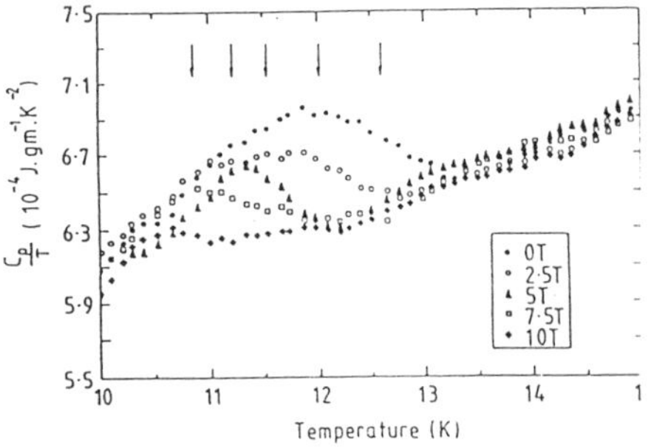

Figure 2 Specific heat capacity/ temperature as a function of temperature for unHIP'ped PMS in high magnetic fields. The arrows denote the transition at each magnetic field

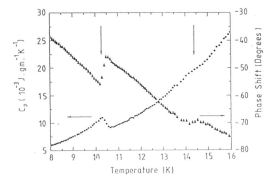

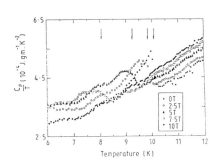

Figure 3 Heat capacity and phase shift as a function of temperature for unHIP'ed Gd-PMS in zero-field. There is a clear indication of two phases.

Figure 4 Specific heat capacity/ temperature as a function of temperature for HIP'ed PMS in high magnetic fields. The arrows denote the transition at each magnetic field

Material	T_C	dB_{C2}/dT	$B_{C2}(0)$
PMS HIP'ed	12.9	-6.8	61
PMS unHIP'ed	12.6	-5.7	50

Table 1 T_C, dB_{C2}/dT (at T_C) and $B_{C2}(0)$ for the Chevrel phase samples. $B_{C2}(0)$ is calculated from W-H-H theory assuming no paramagnetic limiting

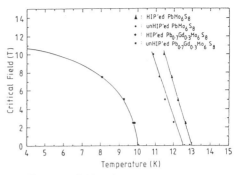

Figure 5 Critical field versus temperature for PMS and Gd-PMS

4: Results

Figures 1 and 2 display the data for the HIP'ed and unHIP'ed PMS samples. They are presented as (specific heat capacity)/(temperature) to emphasize the superconducting anomaly. In fig. 3, zero-field data are presented for the unHIP'ed Gd-PMS obtained using the A.C. technique. The left-hand axis gives the heat capacity of the sample, the right-hand axis gives the phase difference between the ac. power and the ac. temperature variations. In addition to the low temperature phase transition at approximately 10.4K, there is evidence for second phase material with a T_C of about 14.2 K. Complementary resistivity data on a $GdMo_6S_8$ sample shows resistive behaviour at all temperature above 4.2K suggesting the low temperature phase transition is magnetic in origin. In figure 4, data for the HIP'ed Gd-PMS are presented. Only the low temperature anomaly can be clearly observed in high magnetic fields. Similar data were obtained for the unHIP'ed Gd-PMS sample.

In table 1 and fig. 5, analysis of the data is provided. The critical field (magnetic and superconducting) values for the HIP'ed and unHIP'ed Gd-PMS are very similar. The curves we have drawn for these samples takes account of measurements at 10 T in which no specific heat anomaly was found above 5.5 K. Using W-H-H theory [5], a value of $B_{C2}(0)$ has been calculated for both HIP'ed and unHIP'ed PMS samples.

5: Discussion

Although the HIP'ed PMS sample has a similar value of dB_{C2}/dT to the unHIP'ed sample, T_C is slightly higher. The higher T_C probably results from better homogenization of the sample during the HIP process. The critical field values for the HIP'ed and unHIP'ed Gd-PMS for the low temperature phase are almost identical. It is of note that the critical temperature for the higher T_C phase (~14.2K) in Gd-PMS is larger than T_C found in the PMS samples.

It has been reported that rare-earth additions to PMS were found to increase T_C from 12.5 K. to 14 K. with a commensurate increase in $B_{C2}(T)$ [2]. Increases in T_C up to 14.5 K, have also been achieved by reducing the oxygen incorporated in the PMS. 1.5 at% oxygen

586

in PMS can reduce T_C by 2 K [6]. The two phases we have observed in the Gd-PMS specific heat measurements, suggest that in Gd-PMS two processes can occur; the Gd can replace the Pb. A low temperature phase transition which is probably magnetic in origin can occur in this Gd-rich material. Alternatively the Gd can serve as an oxygen getter which leads to regions of superconducting PMS with very low oxygen content and T_C above 14 K.

Complementary magnetic and transport measurements have been completed on the PMS samples [7]. An irreversibility line B_{IRR} was found significantly below $B_{C2}(T)$. A marked difference between B_{IRR} and $B_{C2}(T)$ is a well established result in high T_C oxide superconductors. The transport measurement have shown typically a seven-fold increase in the critical current density of the HIP'ed PMS sample over the unHIP'ed sample. The data on PMS in this paper demonstrate that this improvement cannot be attributed to changes in the bulk superconducting critical parameters. Better connectivity between the grains and improved grain boundary structure are more probable explanations.

6: Conclusion

Very high upper critical fields ($B_{C2}(0) \approx 55$ T) have been found in PMS samples. The HIP process increases T_C probably because of improved homogeneity.

Calorimetric evidence has been found for two superconducting phases in Gd-PMS. The temperature dependence of the low temparature phase transition has been measured - it is probably magnetic in origin. The high T_C phase may be regions of PMS with low oxygen content where Gd acts as an oxygen getter.

Acknowledgements

The authors acknowledge: P. Russell for help with the production of the drawings; A. Crum of Engineered Pressure Systems for the use of the H.I.P.; the support of the Govt. of Pakistan, EPSRC (GR/J39588) and The Royal Society, UK.

References

1) Odermatt R, Fischer Ø, Jones H, and Bongi G. *J. Phys. C.* 7 (1974) L13
2) Fischer Ø, Jones H, Bongi G, Sergent M and Chevrel R. *J. Phys. C.* 7 (1974) L450
3) Fagaly R L and Bohn R G , *Rev. Sci. Instrum.*, 48 (1977) 1502
4) Sullivan P F and Seidal G, *Phys. Rev.* 173 (1968) 679
5) Werthamer N R, Helfand E and Hohenberg P C, *Phys. Rev.* 147 (1966) 295
6) Rikel M O *Ph.D Thesis*. Lebedev Physical Institute, Moscow (1988)
7) Hamidi H, Zheng D N, Hampshire D P (1995) Presented at this conference

Inst. Phys. Conf. Ser. No 148
Paper presented at Applied Superconductivity, Edinburgh, 3–6 July 1995
© 1995 IOP Publishing Ltd

Thermal conductivity (and thermoelectric power) of neutron irradiated Bi-based High Critical Temperature Superconductors

M Pekala (*)

Department of Chemistry, University of Warsaw, PL-02-089 Warsaw

H   Bougrine (*) and **M   Ausloos**

SUPRAS, Institut de Physique, B5, Université de Liège, B-4000 Liège
(*) also Institut d'Electricité Montefiore, B28, Université de Liège, B-4000 Liege

Abstract. Ceramic superconductors with a nominal composition $Bi_{1.5} Pb_{0.5} Sr_2 Ca_2 Cu_3 O_{10}$ were irradiated by fast neutrons (E > 0.1 MeV) with fluences up to $2 * 10^{18}$ cm^{-2}. The thermal conductivity κ and the thermoelectric power were simultaneously measured by means of a steady flow method. The thermal conductivity magnitude and temperature variation are   comparable to data previously reported by other authors. We evaluate that the electron thermal conductivity contributes to no more than 5 % of the measured κ which is thus dominated by the phonon component. Neutron irradiation increases the density of structural defects. This gradually reduces κ absolute values in the normal   state. The highest fluences reduce the magnitude of κ and the height of the maximum in the superconducting state. This is accompanied by an electrical resistivity increase. Since the electrical resistivity changes much faster with irradiation fluence than the thermal conductivity does we conclude that the irradiation affects more the electron than the phonon scattering. From the thermoelectrical power data we calculate the Fermi level value which is insensitive to the irradiation.

1.   Introduction

Devices based on high critical temperature superconductors (HTS) will be made of polycrystalline ceramics, not withstanding the use of thin films. One pertinent property is the critical current density. In order to increase it and to reduce its decrease in a magnetic field it has been thought that one could introduce artificial pinning centers. One way is to irradiate the materials and introduce columnar or so defects. Other types of important properties are those pertaining to the class of "dynamic properties". Thus physical effects in an electrical and/or thermal gradient are of fundamental interest. Here we characterize the "electrical and thermal responses" of Bi-based 2223 irradiated HTS.

2. Experimental

Five ceramic superconductors with a nominal composition $Bi_{1.5}Pb_{0.5}Sr_2 Ca_2Cu_3O_{10}$ containing the Bi-2223 as a major phase were irradiated by fast neutrons (E > 0.1 MeV) with fluences E1 = 6 , E2 = 15 , E3 = 30 , E4 = 60 and E5 = 200 (* 10^{16} cm^{-2}). Influence of thermal neutrons was eliminated by a cadmium shield. The thermal conductivity κ was simultaneously measured with the thermoelctric power [1] by a steady flow method, creating a temperature gradient by means of a small heater (5 to 50 mW) attached to one end of the sample, whereas the second one was connected to the heat sink. The relative accuracy was about 1 %.

3. Results and discussion

The thermal conductivity κ (Fig. 1a) of the unirradiated E0 sample equals 1.9 W/Km and exhibits a local minimum at the superconducting transition ($T_c \simeq 110K$), then rises in the normal state up to 3 W/Km at 270 K. Such a κ magnitude and temperature variation are comparable to data reported in refs.[2-8]. Applying a Wiedemann - Franz relation

$$\kappa = L T / \rho \qquad (1)$$

and assuming that the inelastic electron scattering may be neglected for the electrical resitivity ρ, we evaluate that the electron thermal conductivity contributes to no more than 5 % to κ. This proves that the thermal conductivity is dominated by the phonon component. Below T_c the slight maximum of κ (1.9 W/Km at 88 K) is caused by an increase of the electron thermal conductivity [9]. An increase in phonon mean free path below T_c has also been given as an explanation. However the relatively weak maximum in Bi-based materials seems better described in terms of an electronic anisotropy picture [9].

The neutron irradiation modifies the absolute values in the normal state, roughly by a factor two when the fluence increases up to 2 * 10^{18} cm^{-2}. A mild decrease in the slope of κ vs T can be seen (Fig. 1a). Much more dramatic changes occur in the superconducting state : the κ maximum diminishes with increased fluence and disappears for 2 * 10^{18} cm^{-2}. Additionally the minimum at T_c is flattened and eventually washed out at the highest fluences.

The observed behavior under neutron irradiation matches that reported for other HTS [10-12]. The neutron irradiation causes a large density of structural defects and is accompanied by an increase in the corresponding thermoelectric power (Fig. 1b) and electrical resistivity (Fig. 1c) of the irradiated superconductors. Since the electrical resistivity changes much faster with irradiation fluence than the thermal conductivity does, we conclude that the irradiation introduces defects more sensitive to the electron than the phonon scattering.

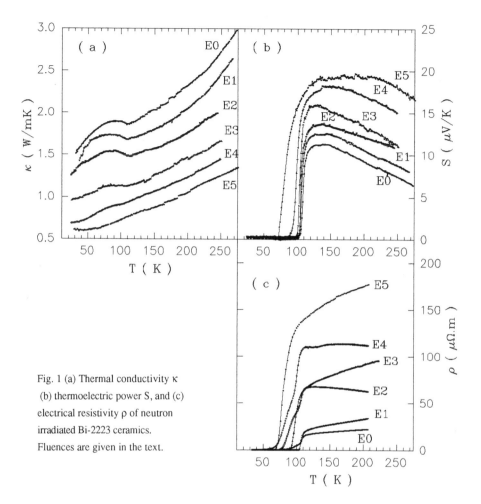

Fig. 1 (a) Thermal conductivity κ (b) thermoelectric power S, and (c) electrical resistivity ρ of neutron irradiated Bi-2223 ceramics. Fluences are given in the text.

The temperature dependence of the thermoelectric power (TEP) of the irradiated and unirradiated samples (Fig. 1b) fits to the TEP general pattern of the Bi2223 HTS[12]. The approximately linear variation of TEP with temperature in the normal state enables us to calculate the Fermi energy from a relation in ref.[7] for anisotropic systems. It gives E_F = 1.9 ± 0.2 eV in reasonable agreement with data reported by Cloots et al. [7]. Values of E_F do not seem to be much altered by the neutron irradiation since the slope in the normal state is quasi unaffected. As for the resistivity (Fig. 1c) the transition interval does not exceed 5 K for the unirradiated sample whereas it broadens to more than 30 K at highest fluence. Moreover the TEP maximum smears out and shifts from 130 to 180 K at the highest fluence. For both latter coefficients the intragrain superconducting transition temperature monotonically shifts with increasing fluence from 106 K for the unirradiated sample to ca. 70 K for the most irradiated one. The mean rate of superconductivity degradation being about 16 K/10^{18} cm^{-2} is comparable to the 9.3 K/10^{18} cm^{-2} value reported by Wisniewski et al. [11]. This rate is remarkably higher than for the Y123 HTS

[12] but similar to that for Y124 [10]. If we apply Obertelli et al. [13] relation between T_c, the room temperature values of TEP and the hole concentration p controlling these parameters we conclude from the low room temperature values of TEP (between 4.5 and 14 μV/K) that the samples are underdoped. The hole concentration varies from 0.110 for the unirradiated sample to 0.086 at highest fluence. This effect should be ascribed almost entirely to the oxygen removal during the irradiation since the relatively light oxygen atoms are most easily removed from HTS. It is interesting to notice that the E0-T_c corresponds well with that for an oxygen deficiency ca. 0.2 in Van Driessche et al. [14].

Acknowledgement

Part of this work has been financially supported through grant BW/95. This work is also part of the SSTC (Brussels) contract SU/02/013 and the ARC (94-99/174) grant. NATO support is acknowledged through HTECH 5-2-05RG/930344. MA received a British Council travel grant for attending EUCAS95. We thank Prof. H.W. Vanderschueren for allowing us to use the Measurement and Instrumentation in Electronics Laboratory (MIEL)

References

[1] Bougrine B and Ausloos M 1995 *Rev. Sci. Instrum.* **66** 199-206

[2] Aliev F G, Moshchalkov V V and Pryadun V V 1989 *Solid State Commun.* **162-164** 572-3

[3] Mori K, Sasakawa M, Igarashi T, Isikawa Y, Sato K, Noto K and Muto Y 1989 *Physica C* **162-164** 512 -3

[4] Efimov V B, Mezhov-Deglin L P and Shevchenko S A 1990 *Physica B* **165-166** 1203-4

[5] Dey T K and Barik H K 1992 *Solid State Commun.* **82** 673-8

[6] Ikebe M, Fujishiro H, Naito T and Noto K 1994 *J. Phys. Soc. Japn* **63** 3107-14.

[7] Cloots R, Bougrine H, Houssa M, Stassen S, D'Urzo L, Rulmont A and Ausloos M 1994 *Physica C* **231** 259-70

[8] Pekala P, Bougrine H, Ausloos M, Lada T and Morawski A 1994 *Molec. Phys. Rep.* **7** 249-53

[9] Houssa M and Ausloos M 1995 *Phys. Rev. B* **51** 9372-74; 1994 *Physica C* **235-240** 1483-4

[10] Pekala M, Morawski A, Lada T and Maka E 1994 *Phys. Stat. Sol.(b)* **186** 487-92

[11] Wisniewski A, Baran M, Koziol Z, Przyslupski P, Piechota J, Puxniak R, Pajaczkowska A, Pekala M, Pytel B and Pytel K 1990 *Physica C* **170** 333-42

[12] Kaiser A B and Uher C 1991 *High Temperature Superconductors*, vol. 7, ed. Narlikar A V (NewYork : Wiley) p. 353

[13] Obertelli S D, Cooper J R and Tallon J L 1992 *Phys. Rev. B* **46** 14 928 - 931

[14] Van Driessche I, Cattoir S and Hoste S 1994 *Appl. Supercond.* **2** 101-110

Inst. Phys. Conf. Ser. No 148
Paper presented at Applied Superconductivity, Edinburgh, 3–6 July 1995
© 1995 IOP Publishing Ltd

Inhomogeneous distribution of the surface electric field on PbBi2223 tapes: implications for the measurement of total AC losses

Y. Yang, T. Hughes* , D.M. Spiller, C. Beduz, M. Penny and R.G. Scurlock

Institute of Cryogenics, Southampton University, Southampton. SO17 1BJ

Abstract This paper examines the significant variation in the loss and inductive components of the transverse surface electric field on two PbBi2223 tapes with cores of thickness 30μm and 100μm. A serious consequence of this variation in electric field is that the self-field losses cannot be simply calculated using $I \times E$. The inhomogeneous distribution of electric field can be attributed to the large aspect ratio (>1:10) of the superconducting core which distorts the surrounding magnetic self field of the sample. We show that an accurate measurement of the losses can be made when the voltage leads are closed at a distance three times the tape half width. The measured losses and the distribution of the electric field for each tape were compared with calculations for two types of tape-like geometry: (i) ellipse and (ii) thin-strip. We also demonstrate the influence of the tape geometry on the loss behaviour. It is shown that the *thin* (30μm) tape can be described in all respects by the "thin-strip" model. The loss behaviour of the *thick* (100μm) tape cannot be adequately described by either geometry but the spatial variation of the inductive and loss electric fields are more consistent with the "thin-strip" geometry.

1. Introduction

Advances in methods of manufacturing Ag sheathed PbBi2223 superconducting tapes with high critical current densities ($J_c > 10^4 Acm^{-2}$) have stimulated new interest in the engineering exploitation of high temperature superconductors. Many factors need to be considered in appraising the economic and technical application of these materials. With high power cables in mind we have studied the self field ac. losses in superconducting tapes. Traditionally, ac losses in superconductors have been measured by calorimetric or magnetisation methods. However calorimetric methods based on nitrogen boil off lack the sensitivity required when using short samples. Magnetisation measurements cannot be interpreted as self-field losses because of the complicated shape and the inhomogeneity and anisotropy of the tape structure. Therefore, in most cases, self field losses have been found from measurement of the axial component of the surface electric field [1,2]. An important feature of the axial electric surface field is that it varies around the periphery of the tape [3,4]. In this paper we explore the nature of the variation of electric field near a current carrying superconducting tape, relate this distribution to the geometry of the tape and choose a a procedure for measurement which accuratly reflects the heat produced in the tape by the flow of current.

---

* Supported by National Grid Company UK

592

2. The inhomogeneous distribution of the electric field

In two recent papers [5,6] it has been shown that the inhomogeneous electric field around a tape is due to the spatial distortion of the magnetic self-field i.e. the magnetic field is no longer distributed in a cylindrically symmetric manner around the tape. The calculations for both magnetic and electric fields can be made for two kinds of "tape-like" geometry (i) the tape is assumed to be of elliptic cross-section or (ii) a thin strip. Both of these kinds of geometry were originally formulated by Norris [7]. The loss-current dependence for the kinds of tape-like geometry is fundamentally different because of the differences in field penetration. The loss per cycle for a thin strip is proportional to $(I/Ic)^4$ in contrast to $(I/Ic)^3$ for an ellipse.

3. Experimental studies

3.1. Methods

Self field ac losses were measured using a set-up described previously[1], which allows simultaneous measurements of both the inductive and loss components of the electric field. The Ag sheathed PbBi2223 tapes were produced within the Institute of Cryogenics, using the standard powder in tube method. The critical current density, Jc, of these tapes were 16,000Acm^{-2} (at 1μVcm^{-1}, 77K 0T). Each sample was connected with five pairs of voltage taps, each pair closed at different lateral locations (x=1.2a, 1.4a, 3.1a, 4.5a and 9.4a where a is the tape half-width). In order to eliminate the contribution of eddy current losses[4], all the measurements were performed at 47Hz.

3.2. Results and discussion.

As the two geometries exhibit different current dependence for the *loss per cycle* [7], this allows us to make a comparison between the measured values and those calculated using the two models. This comparison is shown in Fig 1a-b where the measured values (symbols) for the two tapes are shown with the calculations (solid lines) for the two geometries.

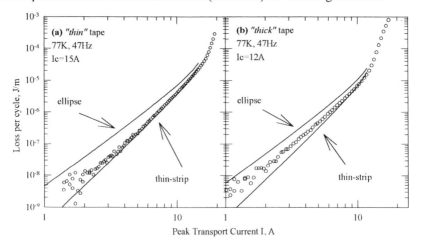

Figure 1a-b. Current dependence of the Loss per cycle, calculated and measured values.

It can be readily seen that the measured values for the *"thin"* ($d\sim30\mu m$) tape corresponds to the calculated values using the *thin-strip* geometry. However the measurements for the *"thick"* ($d\sim100\mu m$) tape lie in between the predictions of the two geometries and exhibit a current dependence with an exponent between 3 and 4 again in between that expected from the two types of geometry. It is evident that neither geometry accurately describes the loss-current behaviour of the *"thick"* tape.

To further examine whether the tapes under measurement could be adequately described by the *thin-strip* geometry the spatial variation of the *loss* and *inductive* components of the electric field was studied. In Figures 2a-b the inductive components $E'(x)$ for different currents are plotted as a function of lateral position x/a, for the two tapes. The values are scaled by the value at $x=1.2a$ and $1.4a$ respectively. The lines represent the calculations for the thin-strip geometry for the corresponding currents. The dependence for an ellipse is not displayed here but there was no correlation between the measured and calculated values for this geometry. It is interesting that the measured values for both the *"thin"* and *"thick"* tapes correspond to the calculations using the thin-strip geometry.

A similar analysis was undertaken for the *loss* component of the electric field with $E''(x)$ being scaled to the value at $x=\infty$ to give $E''(x)/E''(\infty)$ which is then plotted as a function of lateral distance for different currents as in Figure 3a-b. The lines again represent the calculated values for the thin-strip geometry. As in the analysis for the inductive component both tapes only show a good agreement with the predictions for the thin-strip model. It can readily be seen that an accurate measurement of the losses can be made if the voltage leads are closed at a lateral position of approximately three times the half width of the tape. At this position the loss component of the electric field is approximately constant e.g. an error of about 5%.

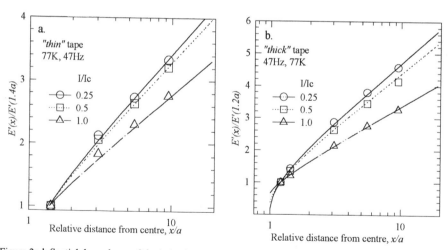

Figure 2a-b Spatial dependence of the inductive electric fields, measured and calculated values (thin-strip).

594

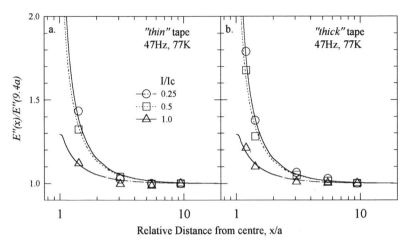

Figure 3a-b Spatial variation for the loss electric field, measured and calculated values (thin-strip).

4. Conclusions

It has been shown that an accurate measurement of the losses can be performed if the voltage loop is closed sufficiently far from the tape axis. Both experimental evidence and calculations indicate that a "sufficient" distance is approximately three times the tape half-width. At this distance "active" compensation of the inductive signal was not required and the accuracy of the measurement is within 5%.

The *"thin"* tape was shown to be adequately described by the thin-strip model in terms of both it's loss-current behaviour and the spatial variation of the inductive and loss electric fields. The *"thick"* tape however showed a loss-current behaviour which could not be described by either model, but lay in between the predictions of both models. However the spatial variation of both the inductive and loss electric fields were accurately described by the thin-strip geometry. This would seem to indicate a gradual transition between the two geometries as the thickness of the tape is increased. More detailed investigation of this *"thickness"* effect is underway.

References

[1] Zannella S, Jansak L, Majoros M, Selvamanickam V and Salama K *Physica C* (1993) 205 pp.14-20

[2] Hughes T, Yang Y, Beduz C, Yi Z, Jansak L, Mahadi A E, Stoll R L, Sykulski J K, Harris M R and Arnold R, *Physica C*, **235/240** 3423 (1994)

[3] Fukunaga T, Maruyama S and Oota *Advances in Superconductivity—VI* Fujita Shiohara (Eds.) **Vol 2** 633 (1994)

[4] Yang Y, Hughes T, Beduz C, Spiller D M and Scurlock R G, *IEEE Trans. Appl. Supercon.* (1995) in press.

[5] Yang Y, Hughes T, Beduz C, Spiller D M and Scurlock R G, submitted to *Physica C,* (1995)

[6] Ciszek M, Campbell A M and Glowacki B A, *Physica C* **223** 203 (1994)

[7] Norris W T, *J. Phys.D* **3** 489 (1970)

Inst. Phys. Conf. Ser. No 148
Paper presented at Applied Superconductivity, Edinburgh, 3–6 July 1995

AC losses of 2223 BPSCCO Ag-sheathed tapes and cables

J. Wiezoreck[1,2], M. Leghissa[1], G. Ries[1], H.-W. Neumüller[1], M. Lindmayer[2]

[1]Siemens AG, Research Laboratories, 91050 Erlangen, Germany
[2]Institut für Elektrische Energieanlagen, Technische Universität Braunschweig,
  38106 Braunschweig, Germany

Abstract
This paper presents loss data of multifilamentary 2223 BPSCCO Ag-sheathed tapes prepared by the powder-in-tube technique and a multistrand test conductor of 1.3 m length.
DC and AC transport measurements have been performed at liquid nitrogen temperature in zero external field. It was found that the measured AC losses strongly depend on the position of the voltage taps. Therefore a method has been developed to receive unambitious results.
Investigations of the losses are presented as a function of the current amplitude and frequency for tapes exhibiting different critical current densities. At power frequencies and amplitudes below the critical current the losses show a $P\sim I^n$ dependence with $n \approx 3$ as predicted by the Bean model. In a frequency range up to 160 Hz no eddy current losses were found. As expected a linear frequency dependence was observed. The experimental results are in good agreement with theoretical calculations using a model for elliptical wires. Losses of the 1-layer multistrand cable conductor could be expressed as the sum of each strands losses.

1. Introduction

Ag-sheathed 2223 BPSCCO superconducting tapes show good promise for practical applications and have been produced in long length up to 1080 m [1]. One of the expected use is the application in High Tc power transmission cables.
The economic efficiency of electric power cables made of High Tc Superconductors strongly depends on their AC losses.

2. Experimental

Fabrication of High Tc wires
The samples have been produced by the conventional powder-in-tube technique as described in [2]. We have investigated tapes of 19 and 49 filaments with I_c up to 27.5 A and j_c up to 12 kA/cm^2.

Fabrication of the cable conductor
The cable conductor was made by winding 2223 BPSCCO Ag-sheathed tapes around a flexible high-density polyethylene former with an outer diameter of 40 mm. The twist pitch is 330 mm. Table 1 shows some more information about the cable conductor. It was found that the degradation of the critical currents due to strain and torsion during bending is negligible.

596

Specifications cable conductor:

total length:	1.35 m
former:	tube of high-density PE
outer diameter former:	40 mm
number of layers:	1
twist pitch:	330 mm
number of strands:	33
critical current of single strands (unwound)	2.9-3.2 A
number of filaments	19

Table 1

AC losses are determined by means of voltage taps. The inductive part of the voltage drop is reduced by a special pair of compensation coils. A lock-in amplifier evaluates the resistive (lossy) part of the signal. All experiments have been performed at 77 K in zero external field.

3. Results and discussion

Figure 1 shows the frequency dependence of AC losses for a single tape. The losses rise linearly with frequency for current amplitudes below the critical value. This emphasizes the hysteretic nature of losses. In the frequency range 25-160 Hz no eddy current losses were found. This coincides with measurements in [3] where no eddy current losses up to frequencies of 1 kHz appeared. Eddy current losses would lead to a $P\propto f^2$ dependence with P being the losses and f the frequency.

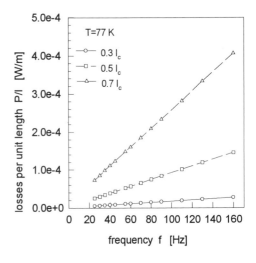

Fig. 1: Amplitude dependence of AC losses for a single tape (I_c=7.5 A, 19 filaments)

In literature it has been pointed out that for single tapes measured losses strongly depend on the configuration of the voltage taps [4, 5]. If the voltage leads are attached on the surface of the conductor measured losses rise from the centre of the wide side of the tape to the edges. The losses vary by more than one order of magnitude [4, 5]. This is due to the hysteretic nature of AC losses and the high aspect ratio of the tapes. Losses are caused by the change of flux with time in the superconductor. Edge contacts pick up more flux than centre contacts so the indicated losses rise [6]. Figure 2 shows loss data for centre contacts with different distances h to the surface of the conductor. Measured losses rise with h and show saturation at h≈2.5 mm. As expected by the Bean model losses rise with the third power of the current amplitude [7]. Figure 3 shows a method of receiving loss measurements independent of the voltage taps positions. If the loop enclosed by the voltage leads is large enough losses for edge and centre contacts agree with each other.

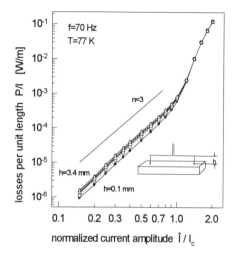

Fig. 2: Amplitude dependence of AC losses for different configurations of voltage taps (I_c=7.5 A, 19 filaments)

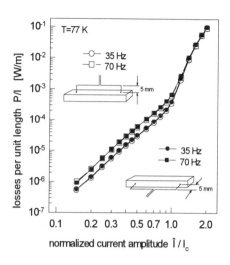

Fig. 3: Amplitude dependence of AC losses for different configurations of voltage taps (I_c=7.5 A, 19 filaments)

Figure 4 compares measured and calculated losses for a single tape. The calculations carried out by Norris [8] apply to monocore conductors with elliptical cross sections and result in:

$$\frac{P_h}{l} = \frac{\mu_0 \cdot I_c^2}{\pi} \cdot \left\{ (1-F)\cdot \ln(1-F) + (2-F)\cdot F/2 \right\} \cdot f \tag{1}$$

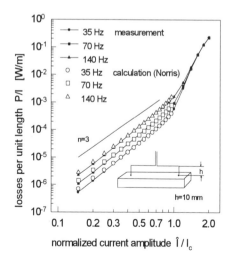

Fig. 4: Measured and calculated losses for a single tape (I_c=9.0 A, 19 filaments)

with

$$F = \hat{I}/I_c \tag{2}$$

P_h: hysteresis losses
$\hat{I}$: current amplitude
f: frequency
l: length of the conductor

Despite the fact that our samples were multifilamentary tapes measured and calculated data are in a very good agreement. For currents above I_c the losses rapidly increase due to appearance of a resistive voltage signal.

598

Figure 5 shows measured and calculated data for the cable conductor. Measured losses agree with calculated ones. The calculation is based on the simple sum of losses of every single strand determined by equation (1).

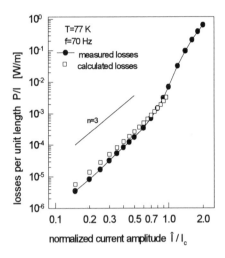

Fig. 5: Measured and calculated losses for the 1.3 m 1-layer multistrand cable conductor (I_c=103 A)

4. Conclusion

Transport AC losses of single 2223 BPSCCO Ag-sheathed tapes and a single layer multistrand cable conductor were measured at liquid nitrogen temperature and zero external field within a frequency range of 25 to 160 Hz. For single tapes a method providing loss data independent of the voltage taps positions on the surface of the conductor was found. For power frequencies the total energy dissipation of the superconducting tapes is mainly determined by hysteresis losses. Measured and calculated losses are in a good agreement when describing the tapes as wires with elliptical cross-sections. The losses of a single layer cable conductor could be expressed as the sum of each strands losses.

Financial support by the german BMBF (contract no. 13N6481) and by BRITE EURAM II (contract no. BRE2-CT92-0207) is gratefully acknowledged.

5. References

[1] J. Fujikami, N. Shibuta, K.Sato, H. Ishii and T. Hara, Applied Superconductivity Vol.2, No 3/4, (1994) 181-190
[2] M. Wilhelm, H.-W. Neumüller and G. Ries, Physica C **185-189** (1991) 2399-2400
[3] T. Fukunaga, S. Maruyama and A. Oota, Advances in Superconductivity VI, Proceedings of the 6th International Symposium on Superconductivity (ISS´93) (1994) 699-702
[4] M. Ciszek, A.M. Campbell and B.A. Glowacki, Physica C **233** (1994) 203-208
[5] Y. Yang, T. Hughes, C. Beduz, D. M. Spiller, Z. Yi and R. G. Scurlock presented at ASC 16.-21.10.1994 in Boston
[6] A.M. Campbell, presented at ASC 16.-21.10.1994 in Boston
[7] C.P. Bean, Rev. Mod. Phys. **36** (1964) 31-39
[8] W.T. Norris, J. Phys. D. **3** (1970) 489-507

Inst. Phys. Conf. Ser. No 148
Paper presented at Applied Superconductivity, Edinburgh, 3–6 July 1995
© *1995 IOP Publishing Ltd*

599

Critical currents and transport AC losses in tube-in-tube Ag/Bi-2223 tapes

M. Ciszek*, M.P. James†, B.A. Glowacki†, S.P. Ashworth, A.M. Campbell, R. Garré‡ and S. Conti‡

IRC in Superconductivity, University of Cambridge, Madingley Road, Cambridge CB3 OHE, UK
‡Centro Ricerche Europa Metalli- LMI, 55052 Fornaci di Barga (LU), Italy.

Abstract. The critical current and transport AC losses of a 'Tube-in-Tube' Ag/Bi-2223 tapes are presented. These conductors include a silver rod down the centre of a powder in tube conductor. This increases the silver to superconductor surface area and thus also the alignment at the centre of the conductor ceramic core. AC transport losses were measured by means of electrical method in the frequency range 30-180 Hz. The AC losses are potentially up to an order of magnitude less than for the more usual oxide powder in tube (OPIT) tape geometries and fit the models for current carrying 'strips' rather than 'elliptical' wire.

1. Introduction

The future application of high temperature superconducting wires depends on the production of long lengths of high quality materials. Of the possible processes that produce conductors for use at 77 K, the oxide powder in tube (OPIT) process using Bi-2223, seems to offer the most promise. There is considerable evidence that texture development is more pronounced at the interface between the ceramic core and the silver sheath. Close to the silver sheath the conversion to the Bi-2223 is more complete and the grains are larger and more highly aligned, consequently the critical current density is highest close to the silver sheath and at the edges of the ceramic core [1]. 'Tube-in-Tube' conductor, with a silver rod in the centre of the ceramic core, aims to maximise the effect seen at the edges of a standard OPIT process by increasing the silver to superconductor surface area and hence alignment at the centre of the conductor.

The practical applications of superconductors are limited by the magnitude of the electric currents which can be transported with a very low level of energy loss. The energy dissipation in superconductors is connected to the intrinsic properties of these materials. Numerous parameters influence the AC losses in superconductors, among them the most important are pinning force, surface energy barriers, magnetic flux flow and flux creep. Extra complications arise due to the granular nature and the presence of weak spots such as microcracks and intergrowths in most real superconductors. Models exist which describe losses in classical low temperature superconductors, the situation is more complex in high temperature due to the very complex morphology and inhomogeneous current distribution.

2. Experimental

The superconducting powder used in the tapes was prepared using a method developed at EM-LMI Research Center, based on the co-precipitation of metallic oxalates. The tapes were produced using the Tube-in-Tube technique, described elsewhere [2].

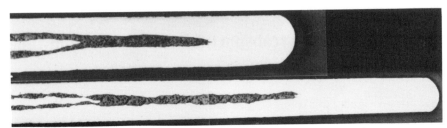

Fig.1. Optical micrographs of rolled (top) and pressed (bottom) samples.

After rolling to 286 μm in thickness the tapes were cut to sections of 4 cm in length, and pressed at 1.5 GPa (SP820, SP840) or rolled with a 5% reduction in thickness per pass (SR820, SR840), then sintered at temperatures of 820 or 840°C in a flowing 10%O$_2$/90% Ar atmosphere. The dimensions of the pressed tapes are 200 μm×4.4 mm and for rolled tapes 265 μm×2.9 mm. The Ag/superconductor ratio was approximately 6:1. The pressed tapes show characteristic cracks parallel to the rolling direction. At the center of the tape, where the rod is located, the external Ag sheath contacts the internal Ag rod at the points where the silver is pushed into the cracks, giving the effect of breaking the tape into lobes. Half of the cross sections of the tapes are shown in Fig.1.

Samples 3×4 mm^2 were cut for measurement of local transverse (J_{ct}) and longitudinal (J_{cl}) DC critical current densities in the plane of the tapes as described elsewhere [3] using a $E=1\mu V/cm$ criterion. The data are summarised in Table 1.

It is common to measure AC transport current losses by means of voltage taps on the conductor but it has recently been pointed out that this voltage is very dependent on the position of the contacts [4-6]; there is far more flux crossing the edge contacts than the centre ones and the voltage will be greater by a factor approximately equal to the aspect ratio of the tape. To obtain a model independent measurement of the loss the loop containing the voltmeter is extended to a point well away from the tape so that the lines of force are circular. The electric field will be the same at any point on these lines so a single measurement will suffice, but there is the considerable disadvantage of a large inductive signal [7-9].

Table 1. DC critical currents density data (A/cm^2)

Samples	Jcl	Jct	Jcl/Jct
S Roll-820	1590	2880	0.55
S Press-820	5000	850	5.90
S Roll-840	3700	8200	0.45
S Press-840	10700	3550	3.00

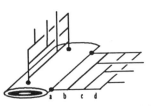

Fig.2. The arrangements of the potential leads for AC transport current loss measurements on Ag sheathed BSCCO-2223 tape. The distances ab, bc and cd are about 5 mm.

We have measured the voltage from a series of contacts with different loop sizes as shown in Fig.2. For practical purposes, a loop width greater than twice the tape width is sufficient to yield unambiguous loss measurements in good agreement with theory [10]. AC loss measurements were carried out for frequencies from 30 to 180 Hz. The lossy component U''_{rms}, of the voltage signal, at the fundamental frequency was measured using a Lock-in amplifier. In the case of large loops the inductive component of the measured voltage is more than two orders of magnitude higher than the loss signal and special precautions should be taken to set the phase and to avoid common mode signals. A specially designed compensating transformer was used as well as grounding only in one common point. All measurements were carried out at liquid nitrogen temperature. The sensitivity of our apparatus was better than 10^{-8} V and the voltage was turned into an apparent transport loss using the usual expressions: $Q_t=(L f)^{-1} I_{rms} U''_{rms}$, where Q_t -self-field loss, f is the frequency, I_{ms} is the transport current and U''_{rms} is the loss voltage measured between potential leads (separation L). The data show that to measure correctly the loss voltage it is enough to increase the distance between sample surface and voltage wires to a few times the tape half width, in our case 5-6 mm separation was sufficient (a-b loop in Fig.2).

3. Results and Discussion

3.1. DC Critical Currents

Table 1 contains the longitudinal and transverse critical currents densities for the tape samples. Transverse microcracks, characteristic of rolled samples, limit the longitudinal J_{cl} of the rolled tapes as these cracks run perpendicular to the current flow direction. On pressing, strain is induced in the transverse direction and cracking occurs in the longitudinal direction and hence the longitudinal J_{cl} of the sample increases. Pressing the tape increases the silver to superconductor interface area, the density and the grain alignment. Note that the longitudinal J_{cl} of the pressed tapes are higher than the transverse J_{ct} of the rolled tape. Increasing the sintering temperature also aids densification, grain size and alignment and the critical current density increases accordingly.

3.2 AC Losses

The measured AC loss of the sample as a function of the AC peak current (normalised to the DC critical current) for a number of frequencies are presented in Fig.4. For all samples the losses per cycle are independent of frequency. According to Norris [10] for a circular or elliptical superconducting wire self-field losses are given by

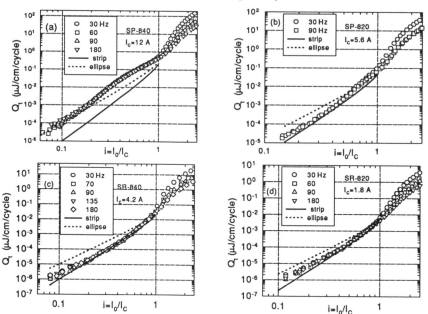

Fig.4. Transport loss Q_t per unit length and per cycle as a function of the current amplitude ($i=I_0/I_c$) for different frequencies, as measured on the edge of the tape for larger potential wire loops. Solid and dashed lines are theoretical curves, for an ellipse and a strip, according to Eqs. 1 and 2, respectively.

$$Q_t = \frac{\mu_0 I_c^2}{\pi}\big((1-i)\ln(1-i)+(2-i)/2\big) \tag{1}$$

and for a thin superconducting strip

$$Q_t = \frac{\mu_0 I_c^2}{\pi}\big[(1-i)\ln(1-i)+(1+i)\ln(1+i)-i^2\big] \tag{2}$$

where $i=I_0/I_c$.

At low currents the ellipse follows I_0^3 whilst the strip losses are proportional to I_0^4. This difference in power law can be attributed to the different topologies of the self field penetration into the ellipse and strip. Below full penetration the ellipse preserves a magnetic field free region within the sample, the strip does not. Data for samples SR-840, SR-820 and SP-820 follows the 'strip' losses closely up to critical current (with no adjustable parameters). Previously published data on mono-core and multi-filamentary tapes [11-13] has always shown 'ellipse' type losses. As a consequence for a given critical current the losses in these tube-in-tube tapes are substantially lower than for tape following the 'ellipse' equation. Sample SP-840 shows an 'ellipse' type behaviour, along with a high loss anomaly. This is attributed to the extensive deformation of the superconducting core produced in this sample.

4. Conclusions

Experimental results show that energy losses are hysteretic in nature in the range of applied frequencies. Electrical measurements of losses using the voltage due to a transport current give results dependent of contact configuration, results with increasing loop sizes tend towards the same limit which is in good agreement with theory. The AC losses of tube-in-tube rolled tapes are lower than for mono-core or multi-filamentary tape of the same critical current and so represent a technologically important material requiring further development. However the correlation between type of AC losses ('strip' or 'ellipse') and conductor microstructure requires further investigation.

References

*On leave from Institute for Low Temperature and Structure Research, Polish Academy of Science, Poland

†Also at Department of Science and Metallurgy, Cambridge University , Cambridge CB2 3QZ, UK

[1] Larbalastier D C, Cai X Y, Feng Y, Edelman H, Umezawa A, Riley Jr G N, Carter W L, 1994 Physica C 221 299-303
[2] R. Garré R, Conti S, Crincoli P, Lunardi G, Salotti G, Tonarelli N, 1993 Appl. Supercond. 1 205-207
[3] Glowacki B A, Jackiewicz J, 1994 J. Appl. Phys. 75 2992-2997
[4] Fukunaga T, Maruyama S, Oota A, 1994, Adv. in Superconductivity, Tokyo, Spring Verlag, 633-636
[5] Ciszek M, Campbell A M, Glowacki B A, 1994 Physica C 233 203-208
[6] Yang Y, Hughes T, Beduz C, Spiller D M, Yi Z, Scurlock R G, 1995 IEEE Appl. Superconductivity 5 (in press)
[7] Campbell A M, 1995 IEEE Trans. Appl. Superconductivity 5 (in press)
[8] Hlasnik I, Jansak L, Majoroš M, Kokavec J, Chovanec F, Martini L, Zanella S, presented at MT-14 Conf. Tampere, Finland, 11-16 June 1995
[9] Clem J R, Workshop on AC losses, April 17-18, 1995, San Francisco, USA (unpublished)
[10] Norris W T, 1970 J. Phys. D: Appl. Phys. 3, 489-507
[11] Fukunaga T, Abe T, Oota A, Yuhya S, Hiraoka M, 1995, Appl. Phys. Lett. 66 2128-2130
[12] Ashworth S P, 1994 Physica C 229 423
[13] Vellego G, Metra P, 1995 Supercond. Sci. Technol. 8 476-483

Inst. Phys. Conf. Ser. No 148
Paper presented at Applied Superconductivity, Edinburgh, 3–6 July 1995
© 1995 IOP Publishing Ltd

Inductive superconducting current limiter: state of art and prospects

V Meerovich, V Sokolovsky, G Jung, S Goren

Physics Department, Ben Gurion University of the Negev, P.O.Box 653, 84105 Beer-Sheva, Israel

Abstract. This paper describes the contemporary status of an inductive current limiter based on the superconducting to normal state transition in a superconductor. This device consists of a primary normal metal coil coupled via a ferromagnetic core to a secondary short-circuited high-temperature superconducting (HTSC) coil that can be performed in the form of a set of rings or cylinders. Investigations carried out recently in several laboratories showed feasibility of employing HTSC materials to build a current limiting device. However, there is a number of problems related both to the device's design and application. The problems of inhomogeneity of superconducting elements, overheating and recovery of superconducting state are considered. The simulation results of operation of a 13.8 kV distribution class device are presented. In conclusion, the prospects of a HTSC current limiter are discussed.

1. Introduction

Superconducting fault current limiters (FCLs) have been discussed for many years. Their main advantages are: negligible influence on the network under normal conditions, high impedance under fault conditions, practically instantaneous limitation, automatic response without external trigger. There are two basic concepts of a superconducting FCL based on the superconducting-normal state transition: resistive and inductive. A simplest resistive device constitutes a superconductor connected in series with the protected circuit*. An inductive device is based on magnetic coupling between a superconducting element and the protected circuit. The inductive FCL consists of a primary normal metal coil coupled via a ferromagnetic core to a secondary short-circuited superconducting coil. The secondary coil can be assembled in the form of a set of HTSC rings or cylinders [1,2].

At present, several firms and research groups (ABB, Switzerland; VPTI Hydro-Qu'bec, Canada; CRIE, Japan; Daimler Benz AG, Germany; Krzhizhanovsky Power Engineering Institute, Russia; Ben-Gurion University, Israel, etc.) have built prototypes and shown a feasibility of the conception of FCL based on HTSC [3]. Most of works is performed in the framework of large projects and related to the inductive type of FCL. The main difference between various inductive FCL designs consists in the arrangement of superconducting rings or configuration of a magnetic core. The position of a ring with respect to the primary coil, inside or outside, is of no principal importance.

There are several companies which produce large rings and cylinders with high critical current density J_c. Sizes of these rings are sufficiently large to build FCLs for distribution networks of 6-15 kV (50-70 cm in diameter). One would think that all is ready for building a real power device. However, the experiments even with middle-scale prototypes reveal a number of problems that call for a technological solution before the design of a full-scale FCL will be possible.

---

* This device was first proposed by Kalashnikov (Russia) in 1936.

604

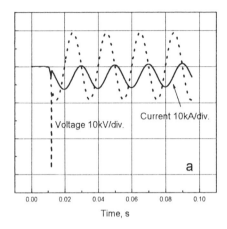
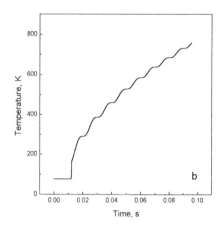

Fig. 1. Computer simulation of 13.8 kV/600 A FCL operation: a) fault current and voltage drop across FCL; b) temperature of weak section. Inhomogeneous ring with J_c=2000 A/cm^2, 5% of the ring circumference is a weak section with reduced J_c=1600 A/cm^2.

2. Quenching and problem of inhomogeneities

In a difference to small scale models, the increase of the device power and J_c of rings result in a pronounced influence of thermal processes in the ring on quenching characteristics. The quenching process in HTSC materials has important peculiarities in comparison with low-Tc materials. The velocity of a normal zone propagation in HTSC materials is very low (of the order of 10^{-2} m/s), as compared to low-Tc superconductors (of the order of 10^3 m/s). Therefore, as shown in [4,5], even small inhomogeneities in the critical current distribution along the superconductor can cause the formation of a resistive zone heated up to a high temperature. Thus, small section possessing lower critical current can be overheated and destroyed.

We have carried out computer simulations of the operation of a 13.8kV/600A FCL employing inhomogeneous HTSC rings with "weak" sections, possessing lower critical current than the rest of the ring. The device contained a set of HTSC rings of 0.6 m in diameter and 1 m in height. Fig. 1 represents the numerical results for the FCL with an inhomogeneous ring having J_c=10^4 A/cm^2 and including a weak section with 30 % reduction of the J_c. We obtain a swift switching, an impedance builds up to the maximum value already during the first half-period of the AC current. Transient fault current is strongly limited but overvoltages appear at the activation. The temperature of the weak section increases to the extent (Fig. 1,b) that can cause the destruction of the rings. Large inhomogeneities lead to overheating of weak sections even at low critical current densities (Fig. 2). The intensity of heating depends on J_c^2 . The higher the J_c, the more pronounced is the influence of inhomogeneities. At a very high J_c, even in a homogeneous ring overvoltages are developed due to too fast resistance increase. An alternative is to use the rings with a low J_c and not to exceed the critical temperature [1,5]. However, the resistance of the rings in the flux flow regime is insufficient for effective current limitation.

In order to study quenching in a real device, we have built and tested several middle-scale prototypes employing HTSC rings with critical current densities above 1000 A/cm^2 [4,6]. We observed experimentally overheating resulting in eventual destruction of HTSC rings. Overvoltages appeared in a circuit with a device employing rings having the quenching current density of 9700 A/cm^2 [7].

3. AC losses

Large AC losses in the superconducting state make the use of bulk materials with a low J_c problematic. For 13.8kV/600A FCL, the thickness of rings with J_c= 2000 A/cm^2 must be 5 mm and AC losses are estimated at 270 W. This value is prohibitively high for real applications. This problem was solved for low-temperature superconductors by using multifilamentary wires in CuNi matrix. The technology of HTSC rings is not ready get for a similar solution. Another solution is to increase J_c by at least one order of magnitude. However, as shown above, the requirements for the homogeneity of a ring increase accordingly.

4. Recovery after fault

The device must return to the initial conditions in about 1s after a circuit-breaker interrupts a short-circuit current. Overheating and a low thermal diffusivity result in a slow recovery of the superconducting state in rings after a fault. Fig. 3 shows the theoretical distribution of temperature in a 5 mm thickness HTSC ring cooled on its one side. During 1s after the interruption of the current, the temperature difference between the ring and coolant decreases about three times on the cooled surface and only 12% in the middle of the cross-section of the ring. This means that thick rings are not fully cooled by the moment of the recovery of normal conditions in the circuit. Therefore, thick bulk rings can be used only at the following conditions: 1) maximum temperature of the ring must exceeds the critical temperature only by several degrees; 2) critical current of rings must be an order of magnitude bigger than the current under normal operating conditions. The use of thin rings with high J_c overcomes the recovery problem [8] but other problems arise due to a fast quenching resulting in overvoltages and overheating.

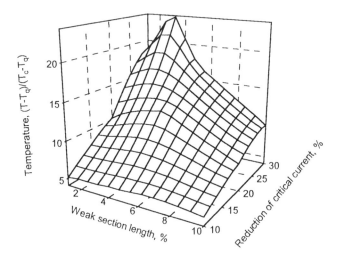

Fig. 2. Overheating of a weak section of a ring. T_q is the temperature of coolant, T_c is the critical temperature. Length of a weak section is given in percents of the ring circumference.

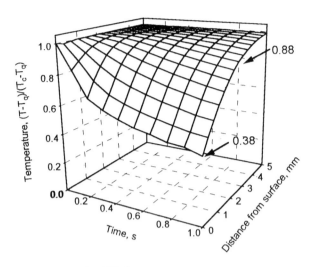

Fig. 3. Cooling of 5 mm thickness bulk HTSC ring

5. Summary

Some problems connected with the use of HTSC rings and cylinders in FCLs were discussed. It is necessary to protect HTSC rings from overheating and destruction, to provide prolonged and stable operation of the device and a quick recovery of the superconducting state in the rings after fault. The proper rings must possess low AC losses in the superconducting state. Note, that most of the problems considered above refer also to resistive current limiters. To build the reliable device, one needs large homogeneous HTSC rings with high values of J_c. To our best knowledge, the technology providing rings with required properties does not exist yet. There are apparently two ways to follow. One relays on further improvement of the properties of produced rings. Another approach to this problem lies in new technological solutions, like the use of parallel rings, shunting resistors or voltage limiters for reduction of overheating and overvoltages [5,8].

The application field and parameters of real devices can be determined on the basis of the investigation of the operation of full-scale prototypes and FCL's influence on the electromagnetic and electromechanical processes in power systems. It is necessary to determine technical and economical gain from the device application. Such a survey was not yet carried out for any of the proposed superconducting current limiters.

References

[1] Bashkirov Yu A, Fleishman L S, Patsayeva T Yu, Sobolev A N and Vdovin A B 1991 *IEEE Trans. on Magn.* **27** 1089-1092
[2] Meerovich V and Sokolovsky V L 1990 *Proceedings of the workshop on magnetic electronics, Krasnoyarsk* **176-177** (in Russian).
[3] Giese R F 1995 *Fault Current Limiters-A Second Look* (Argonne National Laboratory: Report)
[4] Meerovich V, Sokolovsky V and Bock J 1995 *IEEE Trans. on Appl. Supercond.* **5** (in press)
[5] Lindmayer M and Schubert M 1993 *IEEE Trans. on Appl. Supercond.* **3** 884-888
[6] Meerovich V, Sokolovsky V, Jung G and Goren S 1994 *Applied Superconductivity International Conference* (Boston, USA)
[7] Meerovich V, Sokolovsky V, Gawalek W and Görnert P 1995 *Proceeding of this Conference*
[8] Ichikawa M and Okazaki M 1994 *Applied Superconductivity International Conference* (Boston, USA)

Inst. Phys. Conf. Ser. No 148
Paper presented at Applied Superconductivity, Edinburgh, 3–6 July 1995
© 1995 IOP Publishing Ltd

HTSC switch controlled by magnetic field

V Meerovich[†], V Sokolovsky[†], W Gawalek[‡] and P Görnert[‡]

[†] Physics Department, Ben Gurion University of the Negev, P.O.Box 653, 84105 Beer-Sheva, Israel

[‡] Institut für Physikalishe Hochtechnologie e. V. Bereich 1, Helmholzweg 4, D-07743 Jena, Germany

Abstract. The applicability of melt-textured YBCO materials for fabrication of a HTSC switch controlled by a magnetic field is studied experimentally. A magnetic field pulse is applied to the HTSC sample in the direction of a transport current. The pulse initiates the switching but a pronounced resistance appears as a result of thermal processes. The dependence of the switching time up to the normal state on a transport current and a control pulse is investigated. With an increase of the amplitude or duration of the control pulse, the switching time is reduced and becomes less than 1 ms.

1. Introduction

Superconducting power switches are of interest because of their possible applications: (a) as elements of devices for fault current limitation in power systems; (b) in circuits connecting a superconducting magnetic energy storage to a power system; (c) in power converters; (d) as elements of protection systems for fast evacuation of energy from power magnetic systems at fault. Various low-temperature superconducting switches controlled by current, light, electromagnetic pulse, magnetic field, and heating have successfully been tested [1,2].

In contrast to the low-Tc superconductors, the high-Tc superconductors (HTSC) have a higher thermal and magnetic stability due to high thermal capacity ($\sim 10^6$ J/m^{3}K) and high critical magnetic field ($\sim$20 T at 77 K). Therefore the transition caused by the thermal or magnetic field requires high power sources of thermal or magnetic energy.

However, it is possible to use a significantly low magnetic field to control a HTSC switch. This possibility is based on a strong degradation of critical currents of HTSC bulk materials in an external magnetic field. This allows one to use a field of 10-300 mT to initiate a switching of a current-carrying superconductor into the resistive state. The switching occurs as a result of both action of an external magnetic field and a current in the superconductor. A HTSC switch controlled by a weak magnetic field perpendicular to the current flow has been demonstrated in [3]. Because of a low critical current density (160 A/cm^2) of the sample used in [3], the transition to the resistive state was not accompanied by heating. In contrast, the transition in superconductors with a high critical current density (10^3-10^4 A/cm^2) leads to strong heating. In this case, the switching process includes two stages. First, the switching is initiated by a magnetic field pulse reducing the critical current of the superconductor. Second, appearance of the resistance leads to the heating and normal state propagation.

The power switch concept requires a superconducting material with a high normal state resistivity and a high critical current density. For this purpose, melt-textured YBa$_2$Cu$_3$O$_{7-x}$ (MT-YBCO) is a potential material with excellent properties. This paper reports the experimental investigation results of switching in a bulk MT-YBCO superconductor effected by a magnetic field directed parallel to the transport current in a superconducting sample.

608

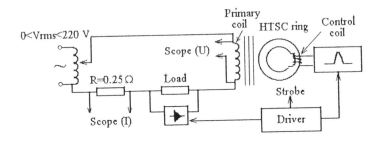

Fig. 1. Experimental circuit.

2. Material processing

Bulk cylindrical $YBa_2Cu_3O_{7-x}$ blocks up to 5 cm in diameter are fabricated by a modified melt texturing process [4]. First, commercial YBCO powder with an addition of Y_2O_3 was pressed into blocks. For the refinement of 211 inclusions, up to 1wt% Pt-oxide is added, thus providing more effective pinning in the material. Blocks were melt textured in a chamber furnace, heated on six sides, in defined low temperature gradients. In the temperature window for the peritectic growth, the blocks are cooled down with 0.5-2.0 K/h. The blocks consist of textured domains in the range of about 1 cm^3. Domain structure of the blocks was characterized on polished cross-sections with polarization microscopy. For characterization, quasi-single crystalline domains were extracted from the blocks and cut into different shape samples with well-defined crystal orientation. The magnetic properties were measured using an Oxford Instruments 3001 vibrating sample magnetometer and an a. c. susceptometer. Critical current densities J_c were obtained by magnetization measurements both for the fields parallel and perpendicular to the crystal c-axis. At 77 K, values of J_c larger than 10^4 A/cm^2 were found up to 2 T magnetic field parallel to the c-axis, while J_c is smaller than 10^3 A/cm^2 when the field is perpendicular to the crystal c-axis. A 1.5-2 times critical current degradation is observed up to a magnetic field of about 0.2 T depending on the crystal orientation; at higher fields the critical current remains constant [4]. The critical temperature of the sample was about 90 K.

3. Experimental sample and procedure

The cylindrical block of 4.6 cm in diameter and 1.8 cm in height obtained by the method described above was used for the preparation of an experimental sample. First, a ring of 4.6 cm in outer diameter and 0.6 cm in wall thickness was cut out from the block. Further, a part of the ring of 0.9 cm in length was thinned down to 0.2 cm, thus forming a section with a substantially lower critical current than the other parts of the ring (Fig.1). A control coil with 60 turns was wound around the switch. A current flowing in the coil produces a longitudinal magnetic field parallel to the induced current in the ring. The ring forms a secondary coil of a transformer consisting of a laminated iron core surrounded by a primary 100 turn copper coil. The primary coil is connected in series to a load resistor and supplied with a 50 Hz alternating current (Fig. 1). The transformer is placed in a cryostat with liquid nitrogen at 77 K.

The weak section in the ring can be considered as a switch inserted in a secondary coil of a transformer and models many cases of switch application. A ring shaped sample with a weak section allows one to investigate a high-current switch without applying contacts.

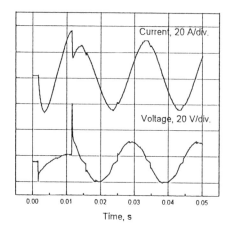

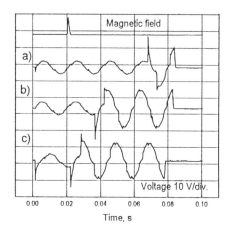

Fig. 2. Current in the primary coil and voltage across the coil without control pulse.

Fig. 3. Voltage across the coil at different amplitudes of control field pulse: (a) 0.6 T; (b) 0.85 T; (c) 1.2 T. Current in the circuit is 25 A_{peak}.

Experimental procedure was the following. A current of low magnitude (20-100 mA) was set in the primary coil inducing a current well below the critical value in the ring. With the electronic switch, the load resistor was short-circuited, resulting in a sharp increase of the current in the circuit. The short-circuit was continued over about 100 ms, following which the electronic switch was opened. A home-made pulse generator was used to give single current pulses of up to 100 A and 1-5 ms duration in the control coil. The scope traces of current in the primary coil, voltage across the coil and control field pulse were recorded on a storage oscilloscope and sent to a computer interfaced acquisition system. Such a procedure allows us to avoid heating of the primary and control coils and heating of the superconductor due to hysteresis losses.

4. Results and discussion

Fig. 2 shows scope traces of the current and voltage across the primary coil at a short-circuit in the test circuit without a control magnetic field. Quenching occurs at a current of about 35 A corresponding to a current of 3500 A and a current density of about 9700 A/cm^2 in the switch. When a magnetic field is applied, quenching is observed at a current in the primary coil beginning with 20 A_{peak}. The pulse of a magnetic field initiates the transition process but the quenching can be delayed for several a. c. periods after the pulse (Fig. 3, a-c). As the control magnetic field amplitude or duration increase, the delay time is reduced. A resistive zone in the switch originates under the influence of the control pulse. A further growth of the zone and its resistance occur as a result of heating. This explains the delay observed [5]. The delay can not be observed in non-textured samples with low critical current densities [3].

For calculation of the current in the ring and voltage drop across the switch, we use the method proposed in our previous paper [6] based on an equivalent circuit of a double-wound transformer. The typical calculated wave forms are given in Fig. 4,a and allow one to obtain a voltage-current characteristic of the switch (Fig. 4,b). Using this characteristic and temporal dependence of the critical current [4], an increase of the temperature of the superconductor

610

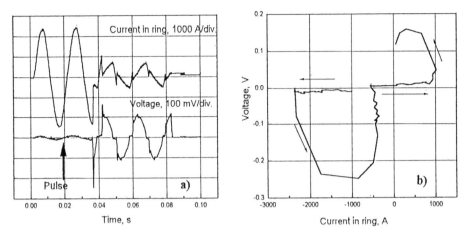

Fig.4. Analysis results for Fig. 3, b: (a) current in the switch and voltage across the switch; (b) voltage-current characteristic during time interval from 0.033 to 0.043 s.

can be estimated. The increase in temperature is thus about 8 K during the time interval between 0.037 s and 0.0395 s for the process shown in Fig. 4.

In conclusion, it was shown that a pulse of a longitudinal magnetic field can be used for controlling the HTSC switch based on MT-YBCO materials. Response time of the switch depends on the amplitude and duration of the control pulse.

References

1] Shevchenko O A, Mulder G B J, Markovsky N V and ten Kate H H J 1992 *Cryogenics* **32** ICEC Supplement 447-450
2] Voronov B M, Gershenson E M, Golzman G N, Dzardanov A L and Malikov S V 1991 *Superconductivity: Physics, Chemistry, Technique* **4** 390-394 (in Russian)
3] Tzeng Y, Cutshaw C, Roppel T, Wu C, Tanger C W, Belser M, Williams R, Czekala L, Fernández M and Askew R 1989 *Appl. Phys. Lett.* **54** 949-950
4] Gawalek W, Habisreuther T, Straßer T, Wu M, Litzkendorf D, Fischer K, Görnert P, Gladun A, Stoye P, Verges P, Ilushin K V and Kovalev L K 1994 *Appl. Superconduct.* **2** 465-478
[5] Meerovich V, Sokolovsky V and Bock J 1995 *IEEE Trans. on Appl. Supercond.* **5** (in press)
[6] Meerovich V, Sokolovsky V, Shter G E and Grader G S 1994 *Appl. Supercond.* **2** 123- 126.

Inst. Phys. Conf. Ser. No 148
Paper presented at Applied Superconductivity, Edinburgh, 3–6 July 1995
© *1995 IOP Publishing Ltd*

Preparation of superconducting YBCO films on poly-crystalline YSZ-substrates by single-source MOCVD

L Klippe and G Wahl

Institut für Oberflächentechnik und Plasmatechnische Werkstoffentwicklung (IOPW), Technische Universität Braunschweig,
Bienroder Weg 53, 38108 Braunschweig, Germany

Abstract. A new single source MOCVD technique for the deposition of YBCO layers has been developed, using a band mechanism to feed the precursors into an evaporation zone. This method has been developed to synthesize YBCO layers in a continuous process with high rates at a constant precursor stoichiometry.

Superconducting layers have been synthesized on polycrystalline YSZ substrates with $T_c>85K$ and $j_c>10^4$ A/cm². The temperature dependence of preferential film growth has been investigated.

1. Introduction

Manufacturing a resistive current limiter requires the deposition of superconducting films on large area substrates (i.e. long tapes or large plates). Its high throwing power and the potential of performing long deposition periods make CVD a suitable technique for large scale industrial applications. Our work is concerned with the development of a continuously driven CVD process for the deposition of YBCO layers from metal-β-diketonates.

Conventional multi-source CVD techniques inherit the so-called "barium-problem" (deviations of the evaporation rate of $Ba(thd)_2$, i.e. due to aging effects), leading to changes of the film stoichiometry after long process times and also to a high sensitivity to variations in the precursor treatment. To overcome this, we went over to using single-source techniques [1], such as ultrasonic nebulization and precursor evaporation from a band as described below.

Sintered Y_2O_3-stabilized ZrO_2 offers some favourable features as substrate material for the deposition of YBCO films. It is stable in oxygen atmosphere at very high temperatures and shows only minor chemical interactions with the layer material. It can in principle be manufactured in large size and it is - compared to single crystals - relatively inexpensive. Even thin flexible foils are available. Its high electrical resistivity at operation temperatures is also an advantage for current limiter applications.

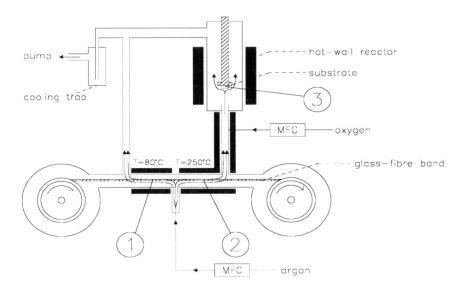

Figure 1 MOCVD system with band evaporator

2. Experimental

In the band evaporator system (Figure 1) the precursors are fed mechanically into the evaporation chamber. First of all the precursor substances $Y(thd)_3$, $Ba(thd)_2$ and $Cu(thd)_2$ are dissolved in diglyme (diethyleneglycoldimethylether) and the solution is put on a glass fibre band. After evacuating the system the band is in step 1 pulled through a drying zone (T=50-80°C). Herein the solvent is evaporated from the band and by a separated argon gas flow conducted into a cooling trap. After passing the drying zone the band is coated with a thin layer of precursors, which evaporate completely in step 2 at T = 230-280°C. By an argon gas flow the precursors are transported into the reactor, where the deposition takes place (step 3). The substrate is positioned close to the reactor inlet in an stagnation flow arrangement. After the deposition, while the furnace is cooling down, the film is annealed for 30-60 min in oxygen atmosphere.

Table 1   Experimental conditions

deposition temperature	800-900 °C
deposition rates	0.1-10 µm/h
line temperatures	250 °C
total pressure	500-1000 Pa
reactor argon gas flow	315 sccm
oxygen gas flow	220 sccm

The gas flow conditions in the evaporation zone give the opportunity to achieve high precursor evaporation rates and thereby high growth rates (>10µm/h). Like other single source methods, band evaporation give a constant ratio between the gas phase concentrations of the different precursors over long process periods. In contrast to the ultrasonic nebulization

[2] and to other liquid single source techniques (i.e. [3]) the presence of solvent vapour in the reactor can be avoided.

Instead of impregnating the whole band before evacuating the system, the solution feeding can be carried out continuously by a mass flow controller for liquids, so that the deposition process can be run steadily.

The substrate material used so far is a polished stabilized Zirconia tape with 8 mol-% Y_2O_3 (KERAFOL, Germany). Recently also flexible tapes containing 3 mol-% Y_2O_3 (Mitsui, Japan) have been employed.

3. Results

The morphology and the structure of the films were analyzed by SEM and XRD. The XRD-patterns show a temperature dependence of the growth direction (Figure 2). Films grown at $T_{dep} = 800\ °C$ give powder-like diffraction spectra, while the films grown at $T_{dep} \geq 850°C$ show a strong preferential c-axis orientation. Pole figures of the c-axis oriented films (Figure 3) indicate randomly distributed a-b-axes.

The scanning electron micrograph of a layer grown at 850°C (Figure 4, right) shows a smooth ground layer (thickness 0.5 μm) of c-axis oriented YBCO containing small holes and with some precipitations on its surface. The precipitations are assumed to be formed by deviations from 1-2-3 stoichiometry. Their density per surface area increases with the films thickness until at about d = 3μm it is completely covered with misoriented grains.

The SEM graph of a layer formed at 825°C (Figure 4, left) shows surface sections, which are a several microns in diameter. Inside these sections the film appears to have a shingle-like surface morphology. We assume, that the YBCO film grows locally aligned on each single grain of the substrate. The typical grain diameter of the YSZ tapes is 3μm. Samples prepared at 800°C do not show this effect. They have a fine granular structure.

We interpret the above described growth behaviour as an effect of the temperature dependent surface mobility of crystallization nuclei. Results obtained from first deposition experiments on flexible YSZ tapes indicate, that a transition to c-axis oriented growth occurs at higher temperatures.

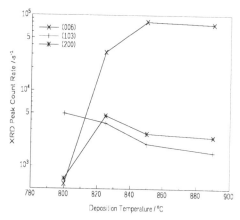

Figure 2: XRD peak intensities of YBCO films grown at different deposition temperatures

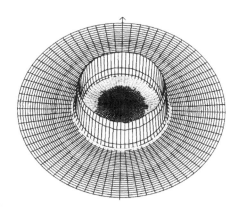

Figure 3: (103)-pole figure of a layer deposited at 850°C

614

Measurements of the superconducting properties of the layers with a preferential c-axis orientation yielded values of critical temperatures $T_c > 85K$ and of critical current densities $j_{c,77K} > 10^4$ A/cm². Non-c-axis oriented films have significantly lower values and also deviations from the 1-2-3 stoichiometry cause T_c and j_c to decrease.

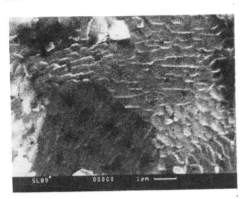 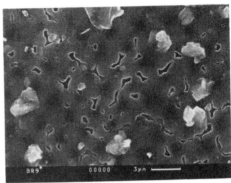

Figure 4: SEM images of layers grown at T_{dep} = 825°C (left) and 850°C (right) respectively.

4. Discussion

It is known, that a precondition for achieving good superconducting properties is generally a strong preferential c-axis orientation of the YBCO film. It was shown, that on polycrystalline substrates this can be achieved by the influence of the deposition temperature. Some technical applications like the current limiter do not require critical current densities larger than 10^5 A/cm², so that films without in-plane orientation may already fulfill the demands. It appears, that in this case more important than a perfect layer structure is the technical feasibility of the deposition process.

Our major future objective will be the reduction of the density of precipitations on the film in order to grow thick layers (d > 3μm) and to increase the deposition rate, while still yielding a preferential c-axis orientation.

References

[1] Klippe L, Stolle R, Decker W, Nürnberg A, Wahl G, Gorbenko O Yu, Erokhin Yu Yu, Graboy I E, Kaul S A 1993 , *Proc. of the first Europ. Conf. on Applied Superconductivity, Göttingen* (EUCAS '93) vol 1 (Oberursel, Germany: DGM Informationsgesellschaft) p 407

[2] Weiss F, Fröhlich K, Haase R, Labeau M, Selbmann D, Senateur J P, Thomas O 1993 J. de Physique IV **C3 3** 321

[3] Yamaguchi T, Ijima Y, Hirano N, Nagaya S, Kohno O, 1994 Jpn. J. Appl. Phys. **33** 6150

We acknowledge financial support from the german BMFT (until 1993, project no. 13N6227) and from the EU (Brite-EuRam II project no. 7887).

Inst. Phys. Conf. Ser. No 148
Paper presented at Applied Superconductivity, Edinburgh, 3–6 July 1995
© 1995 IOP Publishing Ltd

Current limiters using high temperature superconducting wire: applications and realization concepts

M. Kleimaier

RWE Energie Aktiengesellschaft, Dept. E-EG, Kruppstr. 5, 45128 Essen, Germany

C. Russo

American Superconductor Corporation, Westborough, MA 01581, USA

Abstract: Superconducting current limiters are of general interest for utilities, as these devices could be easily integrated in existing grids. In this paper potential areas of application together with their specific requirements are described and possible solutions are discussed.

1. Introduction

Short-circuit or fault currents in an electric grid should be kept as high as needed, but as low as possible. High fault currents are necessary to provide a well defined triggering criterion to the protection system. High short-circuit power reduces perturbations coming from high power fluctuating loads and is especially necessary for the stability of the grid and the safe operation of a power plant.

Unfavorable properties of the fault current are the mechanical and thermal strains to conductors and devices having to carry these high short-circuit currents in case of a fault. Indirect implications like influences on telephone cables and contact voltages caused by potential-rise in the grounding system may cause problems. Evidently, it's today's status to make a reasonable compromise, guided by technical and economical arguments.

The development of superconducting fault-current limiters could enable an increase of short-circuit power under normal operation, without increasing the fault currents under short-circuit conditions.

2. Applications for superconducting fault-current limiters

Conventional fault-current limiters are only applicable with operating voltages below 30 kV. Additionally, they are not inherently fail-safe as they require an external trigger signal. It is expected that superconducting fault-current limiters will operate 100 % reliably, and that they can be built for the highest voltage levels. Under these conditions, a number of interesting fields of operation can be envisaged for the future:

- power quality
- use of transformers with low u_k
- ultra-high-speed bus decoupling for 220 kV/380 kV grids
- coupling of 110 kV subgrids

- eliminiation of bottlenecks in old installations
- connecting high-voltage motors and auto-producers
- lower strain to installations and devices
- dimensioning of new installations for smaller short-circuit currents
- use of fault-current limiters in combination with other superconducting devices

2.1 Power quality

Large power electronics devices, arc-furnaces and welding-machines require a connection to a grid having a high short-circuit power. For this reason it may happen today that such a load has to be connected to a grid having a higher voltage level than dictated by its energy consumption. Coupling two busbars of a medium-voltage or of a 110-kV-substation for getting a higher short-circuit power is normally not allowed, as the permissible short-circuit currents would be exceeded. Coupling the busbars via a current limiter could solve such problems.

Furthermore, a current limiter installed in a busbar coupling could be helpful for the supply of sensitive loads. By a suitable occupation of the busses, sensitive loads can be separated from general supply via a current limiter. In this case the perturbations generated by faults within the general supply grid are reduced at the sensitive loads to small and short sags. Nevertheless, the load is evenly distributed on both transformers and the losses are minimized.

2.2 Connecting high-voltage motors and non-utility generators

Motors and generators having a high power rating (> 500 kW) are built for high voltages (6 kV or 10 kV). If these devices are connected to the medium-voltage grid without an intermediate transformer the motors and generators contribute substantially to the resulting short-circuit currents (unless fed by power electronics). If with this contribution the short-circuit current level rises above the rated value at any point of the grid, the owner of the motor or generator has to take measures to reduce its contribution to the short-circuit current to an acceptable level. Current limiters could be used in these cases.

2.3 Lower strain to installations and devices

Although each device is rated for the expected level of short-circuit currents, it may happen that a subsequent fault occurs especially if the short-circuit occurs nearby a substation. E.g., for transformers, it is known that the ability to withstand short-circuit currents deteriorates with time due to aging effects in the materials. In older transformers, this reduced ability to withstand short circuits can, especially in case of faults close to the transformers, cause subsequent faults in the transformer. The use of current limiters may drastically reduce the risk of these damages.

2.4 Use of fault-current limiters in combination with other superconducting devices

Superconducting devices like cables or transformers should be protected from quenching. Quenching is the transition from superconducting to normal state. If such a device quenches, the subsequent temperature rise and the high thermal mass would result in long duration for the re-establishment of superconductivity. To prevent these installations from quenching, current limiters are necessary.

3. Requirements for a current limiter

A current limiter has to fulfill the following requirements (list not exhaustive):

- The action has to be 100 % fail-safe on opening, as any underfunction would result in enormous damage.
- In case of a fault, the current has to be limited to a tolerable level already during the rise (within 2 ms to 3 ms). The trigger level and the residual current may change from location to location.
- The limiting action should be soft, i.e., without causing overvoltages.
- Installed in a busbar coupling, the current limiter has to be able to withstand the limited fault current for a hold time of at least 100 ms until it is released by a conventional switch. The trigger level for this application can be designed to be very low (less or equal to the rated current I_{rTLV} of the feeding transformer). The residual current could also be very low, theoretically even lower than the trigger level, as the coupling breaker may open unselectively. If the breaker fails at opening, the current limiter has to act as a fuse in a medium voltage application. Presumably, this fusing function is not possible for high-voltage operation. But even in this case, the damage has to remain limited if the breaker fails; that means, no damage to other installations and no hazard to people by "fragmentation" of the current limiter.
- Installed in a feeder or transformer bay, the current limiter must not have any impact on the selectivity of downstream protection systems. Therefore, the current limiter has to be able to carry the residual current for a longer time (for a medium voltage application about 2 s). Both trigger as well as residual current may not get too small in order to ensure that downstream protection systems still have a sufficient triggering criterion ($> 2...3 \times I_{rTLV}$).
- A reset should be possible within a few seconds. For certain applications even shorter recovery times would be needed.
- Installed in a feeder or transformer bay, the automatic reclosing has to be considered. In this case, the current limiter has to operate twice or even three times.
- The current limiter should be able to supply a trigger signal to a conventional switch in series with the respective current limiter. This switch has to interrupt the residual current which may otherwise destroy the current limiter thermally. The switch is also needed for switching under normal operating conditions.

4. Realization Concepts

Fault-current limiters based on high temperature superconducting tape material appear feasible and beneficial. The medium voltage busbar coupling is a particularly promising application, and it offers a short-term proof-of-principle possiblity.

Expansion towards higher voltages (in a transmission fault-current limiter) or higher currents (in a feeder position) need significant but achievable development.

There is a large number of limiter concepts which will fulfill the RWE specifications. First cost considerations, and the requirements for reliability dictated the simplest limiter possible. Therefore, a resistive limiter was chosen for further analysis. The properties of this limiter are described below. The analysis includes a limiter in both the busbar coupling and the feeder position.

The straight resistive limiter in a busbar coupling could be made in a simple, practical manner with wire that is close to the present state of the art. Further advances expected in the next few years could make high-current limiters for series circuit protection a practical possibility as well.

Table 1: Total Length of Composite HTS Conductor as a Function of J_c, I_r, and ρ_m for a 77 K Straight Resistive 2 Ω Limiter

J_c (A/cm²)	I_r (x I_{rTLV})	ρ_m (77 K) ($\mu\Omega \cdot$ cm)	conductor length (m)
10,000	1.2	1.0	188,000
10,000	0.6	1.0	47,000
10,000	0.2	1,0	5,200
12,000	1.2	3.0	43,500
12,000	0.6	3.0	10,800
12,000	0.2	3.0	1,200
15,000	1.2	5.0	16,900
15,000	0.6	5.0	4,200
15,000	0.2	5.0	470

The proposed material for the limiter is American Superconductor's multifilamentary composite conductor. This material has mechanical properties that are far superior to those of bulk HTS materials. The tape is rugged enough to be used in a utility device, an application where lifetimes of 40 years or more are required.

A design for a bus-tie limiter having a rated current I_r of 20 % of transformer rated current I_{rTLV} has been evaluated and found to: trigger at $< 1,5 \cdot \sqrt{2} \cdot I_r$, quench quickly ($< 3$ ms), limit transient current to twice the trigger level, limit current to 1.5 I_r initially and 0.75 I_r at the end of the hold time, achieve hold times of 200 msec, recover from a fault in under 15 seconds, and operate for days with a liquid nitrogen reservoir after a cooling system failure.

The maximum temperature allowed is 300 K during the fault. During this temperature rise the resistivity at least doubles, and limits the fault current to values roughly half of those calculated at 77 K. Results are shown in Table 1.

5. Conclusions

Superconducting current limiters have a number of conceivable areas of applications in each voltage level. However, the technical and economical benefits have to be proven in each case. On the way to commercially viable curent limiters, the first resonable step seems to be the development of prototypes for applications in the medium voltage grid especially for coupling busbars. This voltage level offers also the advantage that voltage problems can be controlled rather easily.

The costs will finally decide whether the market for superconducting current limiters will be a niche market or grow into a large field application.

Inst. Phys. Conf. Ser. No 148
Paper presented at Applied Superconductivity, Edinburgh, 3–6 July 1995
© 1995 IOP Publishing Ltd

619

Current limiting properties of superconducting YBCO films

R. Wördenweber, U. Krüger*, J. Schneider, R. Kutzner, and G. Ockenfuß
Institute of Thin Film and Ion Technology (ISI), Research Centre Jülich (KFA),
52425 Jülich, Germany
* present address: Siemens AG, ZPL1 TW32, Siemensdamm 50, 13629 Berlin

Abstract. The current limiting properties of large area epitaxial c-axis oriented $YBa_2Cu_3O_7$ films on sapphire with critical current densities $J_c > 2MA/cm^2$ are examined. Measurements of the evolution of current, electrical field, power and resistance in time demonstrate the capability of bare and Au-shunted YBCO strip lines to restrict the current within less than 0.5 msec to values below J_c if exposed to dc and ac excess voltages. However, only up to 5% of the normal state resistance is acchieved in the YBCO layer. The thin film demonstrators are stable even for pulses up to several kW and voltages above 100V. An upscaling up to ~90kW seems to be possible with the described technology.

1. Introduction

Due to the development of high-T_c superconductors (HTS) there exists a renewed interest in superconducting ac fault current-limiting devices (SCLD) during the past few years. Utilising the large difference in conductivity and the ultra fast switching between the superconducting and normal state, resistive current limiters can be produced from superconducting bulk wires or thick films. This principle has been used for SCLD bulk demonstrators [1-3] and thin film low-powers hf-limiters [4]. The superconductor naturally and nearly instantaneously limits a fault current if it exceeds the critical current of the superconductor. Whereas classical normal devices have limitations in voltages, SCLD's are expected to be suitable for high voltages. Furthermore, they should solve the problems related to the continual increase of short circuit power in the network. In this work large area $YBa_2Cu_3O_7$ (YBCO) films are examined with respect to their capability of high-power current limiters. Epitaxially grown, c-axis oriented YBCO thin films on CeO_2 buffered sapphire with critical current densities $J_c > 2MA/cm^2$ are examined with respect to their current limiting properties.

2. Sample preparation

$Au/YBCO/CeO_2$ and $YBCO/CeO_2$ multilayers have been prepared on (1$\underline{1}$02) sapphire substrates with a diameter of 2" by on-axis sputter deposition technique. A detailed report on the preparation technique is given in Ref. 5. Sapphire substrates are chosen due to their surface quality, mechanical stability and heat conductivity (~200W/K at 77K) which is superior to that of other substrates with allow epitaxial growth of YBCO films (e.g. ~1.4W/K at 77K for YSZ). For the deposition of CeO_2 a pressure of 13Pa consisting of Ar and O_2 at a ratio of 4:1 was used. The rf-power on the target was 150W, yielding a sputtering rate of ~0.01nm/s. The sputtering conditions for YBCO were: substrate temperature of ~820°C, 210W dc-power, pressure of 60Pa, and Ar/O_2-ratio 2:1. In this work we have examined 150-175nm thick YBCO films grown on 80 nm thick CeO_2 buffer layer. In some cases a 100nm thick Au shunt has been deposited via dc-sputter technique on top of the YBCO layer. Their transition temperature inductively measured at different parts of the films was $T_{c,off}(ind)$= 88-89K (see Fig.1) with $\Delta T_c < 0.5$ K, the specific resistance above the transition was typically ρ_n=(1.2-1.5)$\mu\Omega$m, and the critical current density J_c(77K,0T)=(2.3-3.3)MA/cm^2. The films are patterned via optical lithography and ion beam etching at 250eV in Ar. In the following chapter we will compare the results of two different thin film SCLD on sapphire, i.e. a 9A YBCO SCLD and a 11A Au/YBCO SCLD with geometrical dimensions: length l=19.7 cm, width w=2mm and thickness of the YBCO layer d_{YBCO}=175nm.

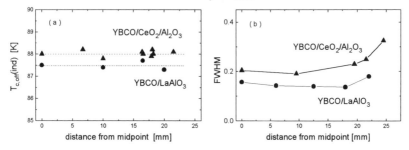

Fig. 1: Spatial dependence of the inductively determined off-set transition temperature and the full width at halve maximum of the YBCO (005) x-ray reflex for a YBCO film on r-cut sapphire and a reference sample on LaAlO₃ revealing the quality and homogeneity of the YBCO-film on sapphire.

3. Experimental results

The YBCO current limiters are characterized via dc and dc/ac puls measurements. Fig. 2 displays the voltage dependence of the current, normalized resistance R/R_n and power for the 9A-SCLD and the Au-shunted 11A-SCLD.

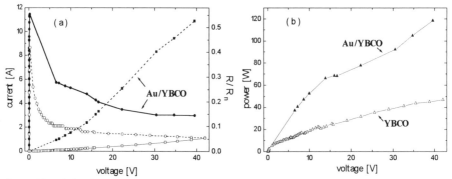

Fig. 2: Plot of the dc current (circles), normalized resistance R/R_n (squares) and power (triangles) as a function of the applied voltage for a bare (open symbols) and a Au-shunted (solid symbols) YBCO strip line on CeO₂/Al₂O₃. The dimensions of the SCLD's are l=19.7cm, w=2mm and d_{YBCO}=175nm.

The voltage dependence of the current is characterized by a steep increase at low voltages. This initial finite slope is an artefact of the 2-point method. Above the voltage at which the critical current is reached, the YBCO 9A-SCLD gradually switches to the normal state finally reaching 5.4% of the normal state resistance R_n at 40V (~2V/cm), whereas the Au-shunted SCLD displays a sudden jump from 11.46A (at 0.18V) to 5.75A (at 6.5V) (values between 0.18V and 6.5V could not be stabilized) and reaches 54% of R_n at 40V. The voltage-current jump can be ascribed to the Au layer, which provides an ideal temperature shunt with a heat conductivity >300W/mK compared to 0.2W/mK for Al₂O₃. The difference in the normalized resistance at high voltages is caused by the lower normal state resistance of the Au/YBCO SCLD, i.e. 25Ω for Au/YBCO and 704Ω for YBCO, respectively. However, an evaluation of the normalized resistance of the YBCO layer $R_{YBCO}/R_{n,YBCO}$ according to a parallel resistance model with $\rho_{n,YBCO}=1.25\mu\Omega m$ yields quite similar values for the Au/YBCO SCLD, i.e. $R_{YBCO}/R_{n,YBCO}=4.2\%$ at 40V.

Puls measurements are executed according to the electrical circuit sketched in Fig. 3. For dc pulses the transformer is exchanged by a dc power supply. R_P indicates the resistances of the connections and contacts. It was typically $R_P \leq 0.5\Omega$ for our measurements.

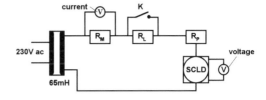

Fig.3: Electric circuit for ac puls measurements.

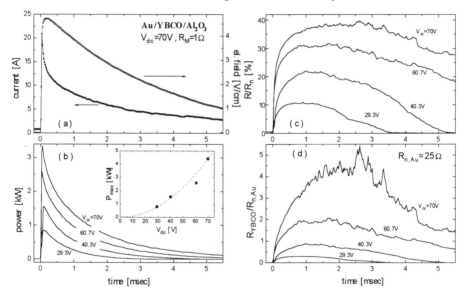

Fig.4: Evolution of (a) current and voltage, (b) power, (c) normalized resistance with $R_n \approx 25\Omega$ and (d) resistance ratio of YBCO and Au layer in time for a dc voltage pulses on the 11A Au/YBCO SCLD. For the evaluation of the resistance ratio in (d) a specific resistance of $\rho_{n,YBCO}=1.25\mu\Omega m$ is assumed. The inset in (b) displays the maximum values of the power which is reached after 40-60μsec together with a quadratic fit.

Fig.4a displays the evolution of voltage and current of the Au-shunted SCLD for dc puls of $V_{dc}=70V$. The resistances are $R_L=50\Omega$, $R_M=1\Omega$, and $R_P<0.5\Omega$. After closing the contact K (at t=0) a sudden voltage increase up to 4.7V/cm is observed, which after 0.2msec decreases gradually within several msec. The resulting current through the SCLD shows a much sharper peak structure. A maximum of 21A is reached after 40μsec and followed by a steep decrease. After 0.4msec the critical current value is surpassed and the current limiting property of the SCLD is established. The power dependence (Fig. 4b) shows a sharp peak with a maximum which is reached after 40-60μsec. The maximum depends quadratically on the applied dc voltage and is of the order of several kW, e.g. 4.4kW for 70V_{dc} (see inset Fig.4b). Fig.4c and d show the development of the normalized resistance for different voltage pulses. The total resistance of the system reaches a broad maximum within the first msec's at 40% ($V_{dc}=70V$), 30% (60.7V), 21% (40.3V) and 11% (29.3V) of the normal state resistance. The steep increase of the resistance is an indication of an extremely high quench velocity and homogeneity of the superconducting layer. The decay of the resistance after a few msec's indicates, that due to the current reduction the temperature in the YBCO layer decreases

622

again. This effect becomes ever more transparent in Fig. 4c, which displays the resistance ratio between the resistance of the YBCO and Au layer. For dc-pulses below 40V the current mainly flows through the YBCO-layer. Above 60V the Au-shunt bears the main current.

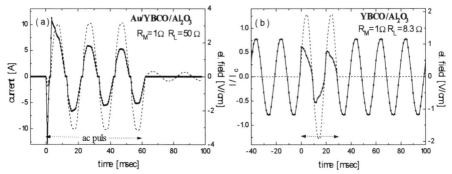

Fig. 5: Demonstration of ac current limitation. In (a) a 60msec puls (3 cycles at 50Hz, $V_{ac,max}$=62V) is applied to the Au/YBCO SCLD, in (b) a 30 msec puls (1.5 cycles, $V_{ac,max}$=38V) is applied to the bare YBCO SCLD.

The ac current limiting properties are demonstrated in Fig.5. The ac measurements are executed with the superconductor in line with a resistive load R_L in the secondary part of a transformer (65mH). The resistive load guaranties a subcritical current during the normal mode. At t=0 the load is short-circuited for a period of 30 or 60msec, respectively. The voltage at the SCLD increases from zero to $V_{ac,max}$=62V in (a) and to 38V in (b). In both cases (Au/YBCO and bare YBCO SCLD) the current is instantaneously limited when the critical current is reached. For instance the Au/YBCO SCLD shows an abrupt decrease of the current at 11A for the first positive half wave. A further decrease of the current is observed for the following half waves. In (b) the immediate recovery of the SCLD after the puls is demonstrated. After 30 msec's the load is added and the original current peaks at $0.76I_c$ are immediately present again. The current limitation is stronger in the case of larger applied ac-voltages due to the increase of power dissipation in the superconducting current limiter. For the case of Au/YBCO SCLD an voltage of V_{ac}=150V leads to a current of ~3.2A.

Our experiments showed, that the thin film SCLD demonstrators with high critical currents are stable even for pulses upto several kW and voltages above 100V. Considering the dimensions of the test devices an upscaling by a factor 20 (i.e. up to 90kW) seems do be feasable without technical modifications of the process. For larger powers extremly thick films might be required and problems with thermal spices or reduced critical properties have to be taken into consideration.

Work in collaboration with Siemens Joint Project 'Development of resistive fault current limiter'. Supported by German BMBF.

References

[1] Tixador P, Brunet Y, Leveque J 1992 *IEEE Trans. on Magnetics*: **28** 446; Tixador P, Leveque J, Brunet Y 1993 *Applied Superconductivity* (ed. H.C. Freyhardt, DGM Verlag) 891

[2] Hara T, Okuma T, Yamamoto T, Ito D, Tsurunga K 1993 *IEEE Trans. on Power Delivery*: **8** (1) 182.

[3] Goren S, Jung G, Meerovich V, Sokolovsky V 1993 *Applied Superconductivity* (ed. H.C. Freyhardt, DGM Verlag) 887

[4] Gaidukov M M, Kozirev A B, Kovalevich L, Samoiliva T B, Soldatenkov O I Superconductivitiy 1994 *Physics, Chemistry, Technology*: **3** (10) 1607-1610

[5] Wang F and Wördenweber R 1993 *Thin Solid Films*: **227** 200

Inst. Phys. Conf. Ser. No 148
Paper presented at Applied Superconductivity, Edinburgh, 3–6 July 1995
© *1995 IOP Publishing Ltd*

Inductive HTS fault current limiter development*

J R Cave, D W A Willén, R Nadi, W Zhu, and Y Brissette

Vice-présidence Technologie et IREQ, Hydro-Québec, 1800 Montée Ste-Julie, Varennes (Québec) J3X-1S1, Canada

Abstract   Superconducting fault current limiters can be designed that utilize the unique property of a superconductor's transition from a superconducting to a resistive state when one or a group of the parameters current density, J, temperature, T, or magnetic field strength, B, exceeds certain threshold values. This transition can be passively triggered by the sudden appearance of the line voltage across the superconducting limiting device in a fault condition. Using this unique superconductor property, in-line "resistive" limiters and also "inductive" or inductively coupled limiters are envisaged. This article describes our ongoing development of inductively coupled fault current limiters, some of the constraints on the superconducting material, and also the operating conditions for the superconductor in this particular application. The synthesis of suitable superconducting materials is an essential consideration for the development of a commercially viable device for utilization in a power utility network.

1.   Introduction

Power utilities are under an increasing demand to provide power that is safe and of high quality. A fault current limiter (FCL) that can act instantaneously and reliably is therefore an attractive apparatus. The unique rapid switching characteristic of a superconductor from the superconducting to the resistive state provides the basis for such a device. However, its selection over other competing technologies will require that the device production cost be as low as possible [1].

Superconducting fault current limiters have been considered for some time and liquid helium based designs have been shown to be feasible [2,3,4]. With the advent of high temperature superconductivity, considerable simplification of the cryogenic system is envisaged (e.g. liquid nitrogen instead of liquid helium). However, the materials properties that are achievable in small samples will have to be realized in larger sizes. Therefore, in order to achieve the goal of a cost-effective high temperature superconducting fault current limiter the superconducting materials properties as well as the limiter design must be developed.

For distribution networks, studies have shown that a clear need for fault current limiters exists [1,2]. There are three suggested locations: on feeders, bus tie positions and secondary windings of power transformers. Some of the most important issues that are left to be resolved are: heat management in a resistive FCL; large iron requirements for inductive and saturated iron core FCLs; and the need for a detailed cost analysis and estimation of what utilities are willing to pay.

2.   Materials requirements and limiter configurations

2.1   Materials parameters

The critical current density, J_c, and the resistivity, ρ_f, after a transition are of importance for this application. Specifically, the product $J_c{}^2\rho_f$ relates to the energy dissipation per unit volume

---

* Joint collaboration project Siemens/Hydro-Québec

624

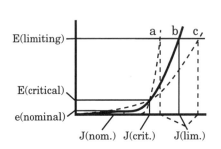
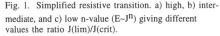

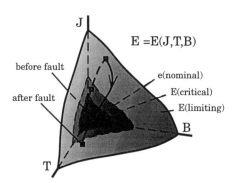

Fig. 1. Simplified resistive transition. a) high, b) intermediate, and c) low n-value (E~J^n) giving different
values the ratio J(lim)/J(crit).

Fig. 2. Schematic trajectory in time of superconductor
during fault, showing complex behaviour of J, T, and B
during the fault event.

and affects the required amount of superconducting material in an SFCL of a given power
rating, while the product $J_c\rho_f$ associates with voltage per unit length which constrains the
superconductor's geometry. For example, bulk Bi-2212 with J_c=1-2 kA/cm^2, ρ_f=3-30 μΩcm,
and normal state resistivity, ρ_n=0.1-1 mΩcm, can be acquired commercially [5].

2.2 Superconducting to resistive transition; E-J-T-B behaviour

The effective values of J_c and ρ_f for a material used in an SFCL depend on loading conditions
such as applied line voltage [6]. It is therefore necessary to understand the dynamic behaviour
of these materials during fault conditions (Fig. 1). A limiter before a fault is current driven, as
the AC line voltage is across the load. A small AC voltage, e, is developed across the limiter
due to the movement of magnetic flux into and out of the superconductor (plus resistance of
the primary winding and iron core losses). During a fault, all of the line voltage appears across
the limiter; this value is in excess of E_c (which defines J_c), and the superconductor enters a
resistive flux-flow state. As the fault is held on, energy is dissipated and the temperature
increases, resulting in increased resistivity. This increases the impedance of the SFCL and the
line current is limited further. The state of the superconductor, therefore, follows a path which
traverses a sequence of equipotential surfaces in the J-T-B space (Fig. 2). These surfaces must
be known in order to fully predict the behaviour of an SFCL. The path that is followed during
a fault depends not only on the superconductor, but also on the specific configuration of the
device.

2.3 Limiter configurations

Figure 3 shows some typical configurations using the superconducting to resistive transition in
an SFCL [3,4,6-16]. The purpose of resistive and inductive shunts in concepts b), e.g. ref.
[3], and c), e.g. [4], are to protect the superconducting element from excessive heating. An

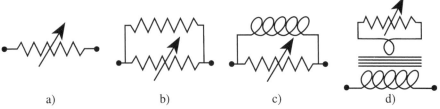

a)   b)   c)   d)

Fig. 3. Configurations of fault current limiters: a) resistive, b) resistive with resistive shunt, c) resistive with
inductive shunt, d) inductive (resistive limitation if $\omega L \gg N^2R$, inductive limitation if $\omega L \ll N^2R$).

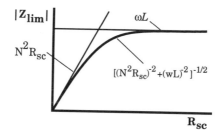

Fig. 4. Schematic of the change in the limiting impedance, $|Z_{lim}|$, of an inductive SFCL with an increase in the resistance R_{sc} of the superconducting switching element.

advantage of the inductive design is the possible elimination of cryogenic current leads.

Hydro-Québec and Siemens are jointly developing FCL technology and evaluating different high-temperature superconducting materials. In this collaboration, Siemens are concentrating on in-line resistive devices (Fig. 3a-c) [7,8], and Hydro-Québec are developing the transformer-coupled inductive concept (Fig. 3d) [6,13].

3. Inductive fault current limiters

Several small-scale prototypes have been built and tested in AC circuits[6, 13]. The impedance of these limiters depends on the effective resistance of the superconductor during fault. As the superconductor's resistance increases, the limiting impedance transits from a resistive regime where $|Z_{lim}| \approx N^2 R_{sc}$ to an inductive regime where $|Z_{lim}| \approx \omega L$. (Fig. 4).

Figure 5 shows the voltage and the current characteristics of a superconducting limiter during a short-circuit fault. This limiter [6] had 273 turns in the primary winding and five melt-cast BSCCO-2212 tubes [5] acting as a screen for the central iron core. The fault is applied after a period of operation at nominal voltage and current of 120Vrms and 32Arms, respectively. It is seen that the limiter acts within a fraction of a cycle, and the fault current is limited to a few times the nominal current. In Fig. 6, the instantaneous power and the cumulative energy dissipated in this limiter is shown. The power is resistive rather than inductive which is due to the relatively low resistivity of the superconducting material[6,13]. This results in considerable energy dissipation in the superconductor which then heats up causing an increased resistance and therefore a decrease in the limited current as seen in Fig. 5.

The number of turns of the primary was then reduced from 273 to 111. Figure 7 shows two faults from a test sequence of this configuration with increased current, I_{nom}=74Arms. In an early fault (a), the current exceeds 500A due to saturation of the iron core. In fault b) the superconducting rings fracture in the first cycle due to the large electromagnetic forces and possibly thermal stresses induced by local heating. This illustrates the important issues of mechanical strength/reinforcement and heat management.

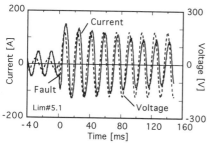

Fig. 5. Current and voltage during a fault for a screened iron-core limiter (N=273) with nominal rating P_{nom}=3.8kVA, V_{nom}= 120Vrms and I_{nom}=32Arms.

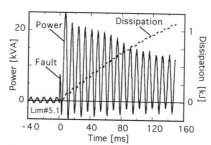

Fig. 6. Instantaneous power and cumulative dissipated energy for the same limiter as in Fig. 5.

626

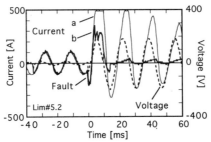

Fig. 7. Current and voltage during a fault for a screened-iron-core limiter with nominal power rating P_{nom}=9 kVA, V_{nom}= 120Vrms and I_{nom}=74Arms (N=111). The current trace "a" describes a fault early in the test sequence. The current trace "b" shows the final test where the five superconducting screening tubes fracture in the first cycle.

4. Conclusions

The initial dynamic behaviour of the superconductor during a fault determines the peak instant current value, which is important for dimensioning of associated equipment such as breakers. The total power dissipation in the superconductor during the fault, the thermal properties (e.g. C_V, κ ...) and the means of heat management will determine the recovery time, which has to be of the order of 1s for most utility requirements. Experiments with small-scale prototypes show that order-of-magnitude improvements in J_C and ρ_f for large size superconductor shapes are necessary for the application of HTS materials to SFCLs. These large geometries must also have reliable mechanical properties that can withstand the large electromagnetic and possibly thermal forces that are associated with fault events in a full-scale power network application.

References

[1] P G Slade, J L Wu, E J Stacey, W F Stubler, R E Voshall, J J Bonk, J Porter, L Hong, 1992 *IEEE Transactions on Power Delivery* vol 7 no 3
[2] R F Giese 1995 *Fault-Current Limiters -- A Second Look* (for the International Energy Agency)
[3] T Verhaege, J P Tavergnier, C Agnoux, C Cottevielle, Y Laumond, M Bekhaled, P Bonnet, M Collet and V D Pham 1993 *IEEE Trans. Appl. Supercond.* vol 3 p 574
[4] D Itoh, K Tsurunaga, T Tada, E S Yoneda, T Hara, K Okinawa, T Ohkuma and T Yamamoto 1992 *Cryogenics* vol 32 p 489
[5] J Bock, S Elschner, and P F Herrmann 1995 *IEEE Trans. Appl. Supercond.* vol 5, no3/4 (ASC'94, Boston)
[6] D W A Willén and J R Cave 1995 *IEEE Trans. Appl. Supercond.* vol 5 no 3/4 (LSC-10, ASC'94, Boston)
[7] W Schmidt 1995 "Preparation of YBCO thick films by pulsed laser deposition for a resistive superconducting fault current limiter" - this conference
[8] G Ries 1995 "Development of resistive HTSC fault current limiters" - this conference
[9] G Bogner 1992 *Concise Encyclopedia of Magnetic & Superconducting Materials,* edited by J. Evetts, p 564
[10] Yu N Vershinin, A R Schnell, R V Schnell, V A Venikov 1979 *Electric Technology U.S.R.R.* no 4 p 91
[11] Yu A Bashkirov, L S Fleishman, T Yu Patsayeva, A N Sobolev and A B Vdovin 1991 *IEEE Trans. Magn.* vol 27 p 1089
[12] L S Fleishman, Yu A Bashkirov, V A Aresteanu, Y Brissette and J R Cave 1993 *IEEE Trans. Appl. Supercond.* vol 3 p 570
[13] J R Cave, D W A Willén, Y Brissette and C Richer 1994 *IEEE Trans. Mag.* vol 30 no 34 p 1895
[14] W Paul, Th Baumann, J Rhyner and F Platter 1994 *IEEE Trans. Appl. Supercond.* vol 5 no 3/4 (LSC-6, ASC'94, Boston)
[15] V Meerovich, V Sokolovsky, G Jung, S Goren 1995 *IEEE Trans. Appl. Supercond.* vol 5 no 3/4 (ASC'94, Boston)
[16] Tixador 1995 *IEEE Trans. Appl. Supercond.* vol 5 no 3/4 (LSC-5, ASC'94, Boston)

Inst. Phys. Conf. Ser. No 148
Paper presented at Applied Superconductivity, Edinburgh, 3–6 July 1995
© *1995 IOP Publishing Ltd*

Design of a superconducting high β linac

H Heinrichs, H Piel, and R W Röth

Fachbereich Physik, Bergische Universität Wuppertal, 42097 Wuppertal,
Germany and Cryoelectra GmbH., 42287 Wuppertal, Germany

Abstract. In this report we introduce the concept of a superconducting H⁻- linac for the European spallation neutron source, ESS. The beam power is 5 MW. We discuss the choice of parameters and compare two versions differing in the final energy: 1.334 GeV and 2.668 GeV. Due to the low operating frequency of 352 MHz the superconducting version offers the possibility to run the low and the high energy part of the linac at the same frequency. Thereby the number of mismatches is reduced. A large ratio of aperture to beam size diminishes the beam spill. The superconducting linac offers advantages with regard to a low radioactivation. The high accelerating gradient reduces the length of the accelerator, the low rf losses reduce the power consumption. It is our aim to show that it may fulfil the main concerns of an accelerator as reliability, availability, low capital and operational costs, as well as hands-on maintenance.

1. Introduction

It is a challenging task to design a high current high β linac for H⁻ ions which is part of the accelerator complex for the European spallation neutron source, ESS.[1] The beam power is 5 MW. Compared to LAMPF the largest proton accelerator up to now the average current is higher by a factor of four and the energy by a factor of 1.7. The avoidance of radioactivation is of decisive importance. We believe that superconducting structures are a good alternative to normal conducting ones.[2] In this report we introduce the concept of a superconducting linac for light ions which would be the first world-wide.

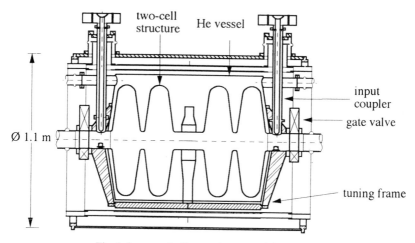

Fig. 1 Superconducting accelerator module

2. General Layout

As the velocity of the particles changes during acceleration, the lengths of the structures have to be adjusted. For the whole energy range about twenty different structures are necessary. We propose a modular design of the accelerator, a cryo module is shown in Fig. 1. It houses two two-cell structures operated at 352 MHz and connected by a tube half a wavelength long, in which the wave cannot propagate because it is in the cut off regime. The frequency of the structures can be tuned by changing their length elastically. A helium vessel which is welded around the structures tolerates this movement. The rf power is fed into the structure via a coaxial input coupler sealed with a ceramic window. At a proposed accelerating field of 10 MV/m 465 kW of peak power have to be coupled to each two-cell structures at the high velocity end of the linac.

Due to the high accelerating gradient of superconducting structures it is still economic to double the energy for the use of H_2^+-ions. Even then the accelerator is shorter than a conventional normal conducting linac with half of this energy. Advantages of this concept are that intense ion sources already exist, that there is no need for funneling, and that the focusing of the beam becomes less critical because the binding energy of H_2^+-ions is higher.[3] The accelerator may easily be designed by replacing the two-cell by four-cell structures. Table 1 summarizes some essential parameters of the accelerators. Version A represent the low, version B the high energy linac.

Table 1 List of accelerator parameters

Version		A	B
Material		Niobium	
Frequency	MHz	352	
Injection energy	MeV	100	
Final energy	MeV	1334	2668
Operating temperature	K	4.3	
Average bunch current	mA	60.8	30.4
Duty cycle	%	6.2	
Effective accelerating gradient	MV/m	8 - 10	
Synchronous phase	deg.	- 15	
Number of cells per structure		2	4
Number of structures		200	
Aperture	cm	10	
Total length of accelerator	m	282	420
Peak rf power for beam acceleration	MW	75	
Dissipated peak rf power in the structures	MW	0.001	0.002
Peak power of klystrons	MW	5 (2.5)	
Number of klystrons		25	25
Peak rf power in one input coupler	KW	465	505

3. Superconducting structures

The structures are proposed to be operated at 352 MHz. There are several reasons for this frequency. It is the frequency of the LEP structures so that a lot of experience regarding to the handling, preparation, and behaviour as well as to an industrial fabrication exists. The superconducting technology is reliable. In the large storage rings like HERA at DESY, LEP at CERN, and TRISTAN at KEK more than 100.000 h of operation are carried out with super-conducting cavities. The operational time of all the superconducting heavy ion accelerators is more than 1 million hours.

Our maximum design value of the accelerating field in the range of 10 MV/m has been demonstrated several times in five-cell 352 MHz structures.[4] The main differences are that the LEP structure has a phase velocity equal to the speed of the light and that it is operated continuously. For pulsed structures the time constant for the field rise time τ is of great importance. If the structure is matched to the rf line in case when the beam is accelerated the klystrons have to be swiched on a time ($\tau*\ln 2$) earlier than the beam arrives at the structure. The average power consumption is increased by about 25 % compared to the power needed if the klystrons would have to be switched on only during the beam on time.

On the other hand the high Q value has an advantage. Within a bunch the beam has to be chopped for a lossless ejection out of the rings. During the notch of the beam which lasts typically 250 ns the fields in the structure increase so that the first buckets of the following beam train are accelerated by a slightly higher field. Due to the high loaded Q-value of a superconducting structure the increase of the fields during the time of the notch is much smaller for a superconducting than for a normal conducting structure. With superconducting structures the required energy resolution better than 10^{-3} can be reached.

According to Slater's theorem deformations of the structure produce a frequency shift. The effect is used for frequency tuning. In addition deformations are produced by the outside pressure of the cooling bath and the inside radiation pressure. The corresponding frequency shift of a five cell LEP structure fixed at both beam tubes but not stiffened on each cell was measured to be 200 Hz when excited to 7 MV/m.[5] A similar frequency shift is expected for the structures described here, it may be calculated that the shift can be reduced by adding stiffening bars on each cell to less than 50 Hz at 10 MV/m, even 10 Hz can be reached. This value is about 1% of the half width. With a suitable low level rf control system the rf klystrons should be locked to the frequency of the structure with beam on.

Table 2 Parameters of superconducting 352 MHz structures with $Q_0=3*10^9$ at E_0T
(low energy version only)

Energy of H$^-$	MeV	100	200	600	1200
Q_L		$3.8*10^5$	$3.1*10^5$	$2.6*10^5$	$2.4*10^5$
$\Delta f_{1/2}$	Hz	458	564	675	720
τ_{field}	ms	0.69	0.56	0.48	0.44
E_0T	MV/m	8	9	10	10

Normal conducting structures usually are optimized with respect to a high shunt impedance. The consequence is a small aperture. Because of it's high Q-value the shunt impedance of a superconducting structure is typically $5*10^4$ times larger than that of a normal conducting one, so there is no strong demand for an increase of the shunt impedance due to a small aperture. We believe that a large aperture affords more safety margin with respect to an avoidance of radioactivation of the whole accelerator. According to the particle dynamics within a bunch, specially if it is space charge dominated, a tenuous halo may evolve.[6] But the beam spill can be reduced by increasing the ratio of structure bore radius to beam radius. In addition it is a duty to avoid all irregularites and mismatches along the beam line. The frequency of 352 MHz opens the possibility to operate the whole linac at the same frequency because the low energy part of the linac, i.e. the RFQ and the drift tube linac, is preferably driven at a low frequency, 352 MHz may be an upper limit.

4. Cost estimation

Based on offers from industrial companies the costs of both linacs are estimated. We find for the capital cost of the low energy version MDM 76 total costs for structures, MDM 90 total costs for rf, MDM 9 for the refrigerator and MDM 41 total costs for the tunnel. It sums up to MDM 216. For the high energy linac we find a sum of MDM 260. These low investment cost are caused by a short length of the linac due to the high accelerating gradient and a small number of klystrons due to the low dissipated rf power. This is also the reason for the low operational costs for ten years which are MDM 191 resp. MDM 204 with 6000 h of operation per year and including klystron renewing.

5. Conclusions

The reliability and availability of the technology of superconducting cavities is proven for high energy applications in the large storage rings like HERA at DESY, LEP at CERN, and TRISTAN at KEK. More than 100.000 h of operation are carried out up to now. Superconducting heavy ion accelerators have collected even much more operational hours. The superconducting linac offers the advantages of a large ratio of aperture to beam radius and a low operating frequency so that there is no frequency jump from the low to the high energy part. This seems to be a key that hands-on maintenance may be possible. In addition the high accelerating gradient reduces the length of the linac, the low rf losses reduce the costs for rf. Therefore capital as well as operational costs are reduced compared with normal conducting linacs.

References

[1]   H.Lengeler, Proposals for Spallation Sources in Europe, *Proc. Fourth Europ. Part. Acc. Conf.*, *EPAC94*, London 1994, p.249-253
[2]   H.Heinrichs, Superconducting Linac High β Section and Cost Comparison, *Proc. Intern. Workshop on High Intensity Proton Linear Accelerator for Future Spallation Neutron Sources*, Witten 1993, p.481-502
[3]   D.Böhne, private communication
[4]   G.Cavallari et al, Status of the RF Superconductivity at CERN, *Proc. 6th Workshop on RF Superconductivity*, CEBAF, 1993, p.49 66
[5]   W.Weingarten, private communication
[6]   G.E.McMichael, B.G.Chidley, and T.Taylor, Considerations for High-Current Linacs, Ed. R.A.Jameson, *Progress Toward Scaling and Optimization Criteria for High-Intensity, Low-Beam-Loss RF Linacs*, Los Alamos Nat. Lab., LA-CP-92-221, p13-35

Inst. Phys. Conf. Ser. No 148
Paper presented at Applied Superconductivity, Edinburgh, 3–6 July 1995

Preparation of YBCO thick films by pulsed laser deposition for a superconducting fault current limiter [1]

W Schmidt[2], P Kummeth, P Schneider[3], B Seebacher,
H-W Neumüller

Siemens AG, Corporate Research and Development, Erlangen and München,
Germany

Abstract. Superconducting thick films of YBCO on polycrystalline zirconia substrates have been prepared by pulsed laser deposition. In short samples (10×20 mm^2) a critical current density j_c of 1.4×10^4 A/cm^2 has been achieved in films with a thickness of $d = 4.5\,\mu$m increasing to 3×10^4 A/cm^2 at $d = 0.9\,\mu$m. The transition to the normal state at 77 K is investigated by applying pulse currents far above j_c.

1. Introduction

The change in electrical resistance during the transition from the superconducting to the normal state is considered to be suitable for the rapid limitation of fault currents in electric power transmission lines [1]. To assess the technical feasibility of fault current limiters (FCL) a project for a 100 kVA model was started recently, within which the preparation of the switching element is one of the major aspects. This project is part of a joint effort to establish high T_c superconductors for power transmission systems [2].

Among the various superconducting materials and the large variety of known preparation techniques, the deposition of YBCO on insulating substrates seems to be the most promising route for producing the required switching elements. The substrates should be insulating to avoid short-circuiting of the high resistance of the superconductor in the normal state. A sufficiently high resistance is essential to effectively limit the fault current. Due to the high power ratings (up to 10 MVA) of FCLs projected for operation in electric power grids, large substrates are required. The choice of substrate material therefore mainly depends on the commercial availability of large sizes. Zirconia ceramic partially stabilized with yttria (PSZ) is available with dimensions up

[1] Work supported by the German Bundesminister für Bildung, Wissenschaft, Forschung und Technologie
[2] Mailing address: Siemens AG, ZFE T EP 4, P.O. box 3220, D-91050 Erlangen
[3] On leave from Forschungszentrum Karlsruhe, Karlsruhe, Germany

to $220 \times 220 \, \text{mm}^2$. It allows the deposition of YBCO without a protecting buffer layer and has a sufficiently matched coefficient of thermal expansion. The $100 \, \text{kVA}$ model e.g. would require a YBCO film with a thickness $d = 5 \ldots 10 \, \mu\text{m}$ covering an area of $200 \times 200 \, \text{mm}^2$ and having an overall critical current density of $j_c = 10^4 \ldots 10^5 \, \text{A/cm}^2$. Physical vapour deposition techniques like pulsed laser deposition (PLD) are appropriate to prepare YBCO films of this quality. PLD has also been proved to operate at very high growth rates of up to $\dot{d} = 870 \, \text{nm/min}$ [3] which is essential to finish deposition within a reasonable time.

As a first step towards a working FCL model, small samples of YBCO on PSZ have been prepared to investigate some of the fundamental aspects, namely polishing of the substrate, optimization of PLD parameters and making of low resistive contacts.

2. Sample preparation

Substrates with $20 \times 10 \, \text{mm}^2$ in size and with thicknesses between 0.1 and 0.25 mm were cut from sintered tapes and plates of fine-grained polycrystalline PSZ. The grain size was $0.2 \ldots 0.3 \, \mu\text{m}$ and the yttria content was 3 mole%. The presence of different phases of zirconia in PSZ had no influence on the superconducting properties. The mechanical stability of PSZ however is much higher than in fully stabilized zirconia. The average roughness of the as-fired surface amounted up to 100 nm measured on a line of $400 \, \mu\text{m}$ length using a stylus with a tip radius of $1.5 \, \mu\text{m}$. Careful grinding and polishing was therefore required to force the growth of highly aligned c-axis orientated grains in the YBCO layer. Afterwards measurements using an atomic force microscope showed a rms-roughness below 5 nm over an area of $4 \times 4 \, \mu\text{m}^2$ but also revealed spikes on the surface. The height of the spikes which are probably due to remains of the polishing media is below 50 nm.

PLD was performed with two different excimer lasers operating at 308 nm (Siemens XP2020) and 248 nm (Lambda Physik LPX300), respectively. While the XP2020 has a practical maximum repetition rate of 10 Hz, the LPX300 can operate at up to 50 Hz, allowing the investigation of high growth rates. The laser beam was focused to a spot of about $3 \, \text{mm}^2$ size by imaging a square shaped aperture with a demagnification factor of about 5. The energy density at the target could be adjusted between 1.5 and 3.5 J/cm^2.

During deposition the oxygen pressure was 0.4 mbar. After deposition the samples were cooled down at an oxygen pressure of 800 mbar. The substrates were loosely positioned on a resistive heater resulting mainly in radiation heat transfer to the sample. The substrate temperature had to be kept in the range of $750 \ldots 820 \, ^\circ\text{C}$ and was controlled by a pyrometer (wavelength 900 nm, emission coefficient 0.75). Compared to a thermocouple the pyrometer read-out was about $5 \ldots 25 \, ^\circ\text{C}$ higher depending on the thickness of the YBCO film. To achieve a good radiation coupling to the heater the rear side of the substrates had to be coated with a black ceramic paint. The heater with the samples is scanned perpendicular to the plasma plume to get a homogeneous thickness distribution.

The samples were patterned into different shapes by wet-etching. Most of the switching experiments [1] were performed using an U-shaped structure, to provide a long strip-line necessary for a sufficiently high limiter resistance. The typical length of this strip-line was 15 mm and the width 4 mm. Current and voltage pads were made by sputtering of Ag followed by an annealing step in flowing oxygen at 600 °C. Current leads were soldered with an alloy melting at 51 °C by gently heating the whole sample and several thin wires were fixed as voltage probes by ultrasonic bonding. Afterwards the overall contact resistance was in the range of $7 \times 10^{-6}\,\Omega\text{cm}^2$ which was sufficient to avoid excessive contact heating with DC currents up to 7 A. A criterion of $1\,\mu\text{V/cm}$ was used for the determination of critical currents.

3. Results

The critical current as a function of film thickness is shown in figure 1. The maximum j_c of about $3 \times 10^4\,\text{A/cm}^2$ was achieved for $d = 0.9\,\mu\text{m}$ with a small decrease to higher thicknesses ($1.4 \times 10^4\,\text{A/cm}^2$ at $4.5\,\mu\text{m}$) and a steep decrease at lower thicknesses. The decrease at low d can be attributed to trenches between groups of closely connected grains in the film surface, which lower the current carrying cross section. The origin of this trenches is not clear at the moment; for higher d they tend to form closed holes in the film as can be observed in SEM micrographs of fractured layers. The decrease of j_c at higher d is probably due to laser droplets accumulating in the YBCO layer and interrupting the c-axis orientated growth. This decrease may also be caused by the variation of the surface temperature of the growing film due to changes in the coefficient of emission [4].

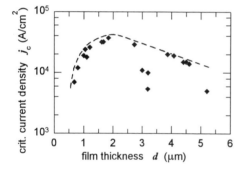

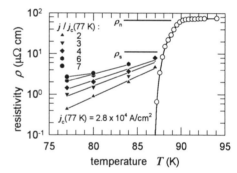

Figure 1. Transport critical current in short samples of YBCO on polycrystalline PSZ at 77 K as a function of film thickness. The scatter in the data is due to various deposition conditions during optimization.

Figure 2. The resistivity ϱ of YBCO as a function of temperature. Below T_c the resistivity was measured using pulsed currents ranging from I_c up to almost $12 \times I_c$.

While j_c determines the cross section of a FCL needed to carry the nominal current, the resistivity ϱ_n above T_c determines the length of the strip line together with the desired limiting resistance. However the time dependent behaviour during the transition from the superconducting to the normal state is determined by the resistivity ϱ_s below T_c at current densities above j_c [5]. We measured ϱ_s by using current pulses of $0.1\ldots1\,\text{ms}$ width at different temperatures below T_c (figure 2). Current densities j of more than $10 \times j_c$ could be applied to our low j_c material before a burn-out of the strip lines occured. The curves $\varrho(j,T)$ have been used as input to simulations of the AC switching behaviour [1].

The samples shown in figure 1 were prepared with a repetition rate of 10 Hz. This resulted in a mean growth rate of about 18 nm/min. Considering the large areas needed for the projected 100 kVA model much higher rates are needed to avoid long deposition times. First experiments with repetition rates up to 50 Hz showed no change in film quality, as far as the transition width of the inductively measured T_c is considered. To confirm this, also a few measurements of the transport critical currents have been performed up to now. In films with a thickness of about 1 μm a j_c of 2.6×10^4 A/cm^2 has been achieved.

4. Conclusions

Deposition of granular low-j_c YBCO thick films by PLD on polycrystalline zirconia seems to be a suitable method for preparing switching elements of high power FCLs. Measurements with pulsed currents show that the samples develop a high resistance at overcurrents below T_c, which leads to a save transition to the normal state. Therefore it seems reasonable that the 100 kVA FCL model can be manufactured using the available preparation techniques.

To achieve this aim a higher current density is desirable either by further optimization of the process or by using a biaxially aligned buffer layer between the YBCO and the substrate. Furthermore the decrease of j_c at higher thicknesses can probably be minimized by stabilizing the temperature of the film surface during deposition.

For FCLs with even higher power ratings however, deposition techniques which can handle areas larger than 200×200 mm^2 should be made available. Therefore magnetron sputtering, thermal evaporation, and plasma flash evaporation are evaluated within the joint project and are considered to replace PLD as soon as the switching elements for high power FCL prototypes have to be manufactured.

Acknowledgements

The authors thank Mrs. W. Ruppert (Siemens AG) for performing the j_c measurements and M. Bauer (Technische Universität München) and J. Markl (Universität Erlangen) for the AFM data.

References

[1] Ries G, Gromoll B, Neumüller H-W, Schmidt W, and Krämer H-P, this conference

[2] Neumüller H-W, this conference

[3] Wu X D, Muenchausen R E, Foltyn S, Estler R C, Dye R C, Garcia A R, Nogar N S, England P, Ramesh R, Hwang D M, Ravi T S, Chang C C, Venkatesan T, Xi X X, Li Q, and Inam A 1990 *Appl. Phys. Lett* **57** 523–525

[4] Westerheim A C, Choi B I, Flik M I, Cima M J, Slattery R L, and Anderson A C 1992 *J. Vac. Sci. Technol. A* **10** 3407–3410

[5] Lindmayer M and Schubert M 1993 *IEEE Transact. on Appl. Superconductivity* **3** 884–888

Inst. Phys. Conf. Ser. No 148
Paper presented at Applied Superconductivity, Edinburgh, 3–6 July 1995
© *1995 IOP Publishing Ltd*

Development of Resistive HTSC Fault Current Limiters

G Ries, B Gromoll., H W Neumüller, W Schmidt, H P Krämer, S Fischer*

Siemens AG, Corporate Research and Development, D-91050 Erlangen, Germany
* Institut für Elektrische Energieanlagen, TU Braunschweig, Germany

Abstract. The aim of this work is to investigate the switching behaviour of a resistive Fault Current Limiter with HTSC elements immersed in liquid nitrogen. The elements consist of YBCO thick films on a YSZ substrate. First results show superior current limiting performance. The steady limiting current is about 3 times the nominal current. The peak fault current is 1/3 of the unlimited peak current and about 10 times the nominal current and is reached within an action time of 0.3 ms. The recovery time is below 0.75 s. Numerical simulation of the switching behaviour shows good agreement with the experiments. These results suggest a resistive FCL-system which is applicable in electric power and distribution systems.

1. Introduction

In an electric power system, the task of Fault Current Limiters (FCLs) is the limitation of mechanical and thermal loads on busbars, insulators and circuit breakers. Superconductive FCLs therefore should work as fast, self restoring fuses. Fields of application are in newly installed networks, in extended networks with increased fault currents, as protector switch for generators and as integrated component of superconductive systems like cables, transformers and SMES. The paper describes the first results of the current limiting performance of a resistive FCL with YBCO switching elements. The work is part of the joint project "HTS for power transmission systems" [1] and part of the joint collaboration between Hydro-Quebec, Canada [2], and Siemens Power Transmission and Distribution/Corporate Research.

2. Preparation of the switching element

Up to now short samples of switching elements have been produced and tested. The YBCO-films were deposited on Yttrium Stabilised Zirconia (YSZ) by pulsed laser deposition. A typical sample is shown in Fig. 1. The thickness and critical current density values cover ranges of 0.5 to 4.5 μm and 5 to 35 kA/cm² respectively [1]. Contacts for current and voltage are made by sputtering silver pads onto the YBCO film and fixing wires by soldering, mechanical pressure or chip-bonding to the pads.

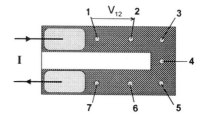

Fig. 1: View of a U-shaped YBCO sample. Substrate YSZ 20*10*0.2mm³

3. Experimental setup

All samples have been mounted on a substrate holder, which is placed in a liquid nitrogen bath at atmospheric pressure. Optical observations of the samples can be made through two windows of 10 cm diameter in the cryostat. The electrical circuit used for the switching experiments is shown schematically in

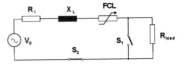

Fig. 2: Limiter equivalent test circuit.

Fig. 2. An AC-voltage of 230V and 50 Hz is transformed to V_0 which can be matched to different FCL-samples. The short circuit reactance X_L and resistance R_i are characteristics of the transformer. By adjusting R_{load} the load current is set to the nominal current $I_0 = I_c/\sqrt{2}$. Switch S_1 makes a short circuit by closing at time $t = 0$ ms whereas switch S_2 opens at $t = 50$ ms thus defining a limiting and holding period. Short circuit make switch S_1 and circuit breaker S_2 are realised by thyristors to allow for exact and phase-locked timing of S_1 and S_2. After 50 ms a small DC-current is driven through the FCL to record the transient resistance $R(t)$ of the superconducting film. By this method the recooling curve and the recovery time of the limiter are determined.

4. Results and discussion

4.1. Switching behaviour

A typical experiment on a sample with a 2.7 μm YBCO-layer and $I_c = 2$ A is shown in Figs. 3a)-d). The limited and unlimited currents are plotted in Figs. 3a),b) on different time scales. It is demonstrated that the limiter reduces the peak fault current I_{pf} to about one third of the unlimited peak current. This is about 10 times the nominal current I_0, whereas the steady state limited current, which has to be interrupted by a conventional breaker, is between $2 \cdot I_0$ and $3 \cdot I_0$. As compared to $23 \cdot I_0$ in the unlimited case this is a drastic reduction by a factor of 10. From Fig. 3b) it can be seen that the action time until the current is effectively limited 0.3 ms.

The evolution of the resistance R of the limiting element is shown in Figs. 3c),d). Here R is normalised to the resistance at 95 K, R_{95K}. The three curves are derived from the voltages recorded between the tabs 1-3, 3-4 and the total length 1-7 respectively, as shown in Fig. 1. We observe a finite resistance $R \approx 0.05 \cdot R_{95K}$ immediately after the short circuit which is typical for the flux flow resistance at $j > j_c$ in our samples. The transition above T_c follows rather abruptly after 0.3 ms. Joule heating causes a further temperature rise up to 225 K after 50 ms. Using the R(T)-curves of the samples R(t) can be correlated to the temperature of the superconductor. Approximate temperature values are given in the figure. We note substantial differences in the resistance between different voltage tabs. A plausible explanation is that in the section 1-3 only part of the superconducting length has gone normal while another part remains superconducting as a result of variations of the local I_c along the sample.

4.2. Recovery time

One of the main advantages of a superconducting FCL compared to a conventional fuse is its self-restoring capability. After the limiting and holding period in case of a short circuit a breaker switches off any further Joule-heating in the FCL element. This allows the FCL to regain its normal state of operation which means a recooling of the superconducting film to temperatures near 77 K. The heat generated in the superconducting film is transferred to the adjacent media which are the substrate on one side and nitrogen on the other side of the film.

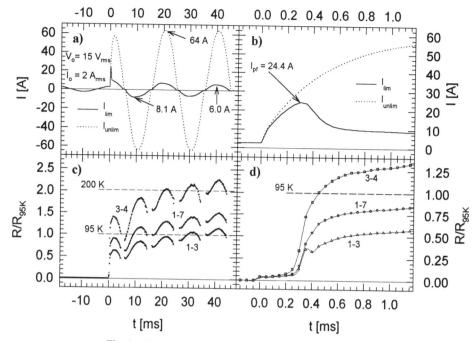

Fig. 4: a),b): Limited and unlimited current as function of time
c),d): Normalised resistance between voltage pads as indicated in Fig. 1 as function of time
b), d): expanded time scale

The recovery time is mainly determined by the heat transfer from the substrate to the surrounding nitrogen by boiling and heat conduction. As explained in the foregoing section the recooling of the superconducting film is measured by a DC-current far below the critical current of the superconductor. Fig. 4 shows typical recooling curves measured at the same sample after limiter experiments with different operating AC-voltages V_0.

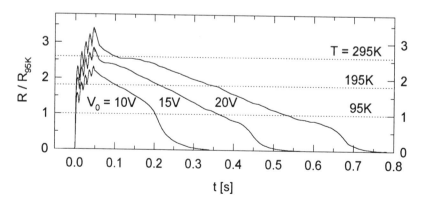

Fig. 3: Heating and recooling of a superconducting FCL during and after a limiting and holding period of 50 ms with transformer voltage V_0 as parameter.

638

As can be seen clearly the recovery time increases with the applied voltage due to the amount of deposited heat. Even in the case of $V_0 = 20$ V with the maximum temperature at the end of the limiting period beyond room temperature the recovery of the FCL takes not more than 0.75 s.

4.3. Simulation

The development of the fault current limiter is supported by parameter analysis, using numerical simulation of the transient heat conduction problem in the switching element, combined with an electric network model. Fig. 5 compares measured and simulated currents I(t) for an experiment similar to that described above using the superconducting resistive characteristics $\rho(j/j_c, T)$ [4] of an investigated sample and the test circuit parameters as model inputs. Calculation and measurement are in good agreement.

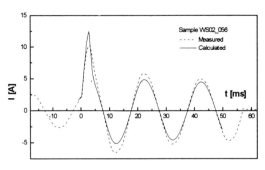

Fig. 5: Comparison of measurement and calculation of limited current as function of time

5. Conclusion and Outlook

The switching behaviour of a resistive FCL consisting of short samples of LN_2-cooled YBCO-films on YSZ substrates has been successfully demonstrated. The peak fault current is 1/3 of the unlimited peak current and about 10 times the nominal current. The action time is 0.3 ms. Investigations of the transient heat transfer between the switching element and liquid nitrogen show that the recovery time is below 0.75 s. The nominal power of the sample is about 40 VA. The next step will be a FCL-demonstrator with a nominal power of 150 VA. A functional model with a nominal power of 100 kVA is planned for 1997.

References

[1] Neumüller H W *Status of the Joint Project HTS for Power Transmission Systems* paper to be presented at EUCAS 1995
[2] Cave J et al *Inductive Fault Current Limiter Development* paper to be presented at EUCAS 1995
[3] Schmidt W et al 1995 *Preparation of YBCO Thick Films by Pulsed Laser Deposition for a Superconducting Fault Current Limiter* paper to be presented at EUCAS 1995
[4] Lindmayer M and Schubert M *IEEE Trans. on Applied Superconductivity* **3 (1993)** 884-888

Transient Analysis Of The HTSC Current Limiter

H. Castro and L. Rinderer

Institut de Physique Expérimentale, Université de Lausanne. CH-1015 Lausanne, Switzerland.

Abstract. We present a theoretical analysis of the transient regime in a HTSC cylinder when a sudden increase in the circulating dc current occurs. First we consider the establishment of the initial operating state with a transport current $I_o < I_c$. Then we study the case of a sudden increase of the transport current to a final value which can be smaller or higher than the critical current. We obtain analytical expressions for the times characterizing this phenomenon and for the power dissipated during the transition.

1. Introduction

A current limiter (CL) consisting on a superconducting cylinder connected in series in a current-carrying circuit is illustrated in figure 1. The operating current $I_o < I_c$ is such that when a large overcurrent occurs the cylinder becomes normal in a very short time, thus forcing the current to a lower value. HTSC CL's are very promising devices that are been commercially explored. However no theory exists, up to now, to predict its transient behavior and to optimize their performance. Our purpose here is to provide such a theory.

We have previously studied the dynamics of flux penetration in HTSC for a magnetic field step[1]. We establish here the equivalence between this problem and a current step in the CL. Then we apply this theory to the CL in the two situations mentioned in the abstract.

2. Theory

The distribution of transport current in a HTSC cylinder is not uniform because of the magnetic flux expulsion. According to Bean's model the current tends to flow through the outer rim of the sample. As I increases the rim shrinks down until the current flows through the whole section of the cylinder for $I \geq I_c$. When a sudden change in current occurs a new configuration must be achieved, as illustrated in figure 2 for the initial (o) and final (f) configurations with $I_o < I_f < I_c$. The transition to the final state requires the penetration of current and magnetic induction deeper into the cylinder. It is the interaction between current and magnetic induction, regardless of their origin, which determines the dynamics of this transition. Therefore this process is equivalent to the one taking place when a magnetic field-step is applied to the cylinder without any transport current.

This phenomenon was previously studied and we obtained the time dependence of the shielding current and magnetic induction inside the cylinder [1] for different field amplitudes. We present here the main results of this theory adapted to the CL, simplified by Bean's approximation, i. e. J_c is assumed constant. More accurate results can be obtained by using the appropriate expression for $J_c(B)$, like an exponential law [2].

640

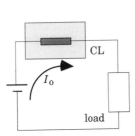

Fig. 1 The Current Limiter (CL) basic configuration in a DC circuit

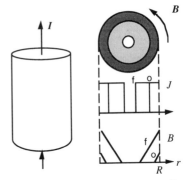

Fig. 2 Current and Induction profiles (i) initial, (f) final configurations

1.1 Establishment of the operating state

We will determine the settling time of the operating state after first application of the operating current I_0. Assuming I_0 to be applied as a step function there will be a corresponding field-step of amplitude H_0 at the surface of the cylinder, given by

$$H_0 = \frac{\mu_{eff} I_0}{2\pi R} \qquad (1)$$

where μ_{eff} is the effective permeability of the granular material and R the cylinder's radius. The magnetic induction will penetrate to a depth prescribed by Bean's model as

$$\delta_0 = \left(\frac{H_0}{H*}\right) \cdot R = \left(\frac{I_0}{I_c}\right) \cdot R \qquad (2)$$

where the second equality arises from a relationship between $H*$ and I_c similar to equation 1.

The flux penetration is described as a flux flow process, obtaining the following implicit relationship for the current density and time [1b]

$$t = \left(\frac{\mu_0 \pi R^3 H_{c2J} I_0}{4\rho_{nJ} I_c^2}\right)\left[-\frac{1}{j} - \ln\left(1 - \frac{1}{j}\right)\right] \qquad (3)$$

where μ_0 is the permeability of vacuum, H_{c2J} the second critical field and ρ_{nJ} the normal state resistivity associated to the Josephson medium and j the normalized current density defined by $j(t) \equiv J(t)/J_c$. Figure 3 is a plot of equation 3 for YBCO at 77 K. The steady state value of the current density is the critical value i. e. $J_c = \lim_{t \to \infty} J(t)$, therefore, we can define the settling time t_{so} by requiring the condition $j \leq 1+\varepsilon$ where ε is a small tolerance number. This leads us to the value

$$t_{so} = \left(\frac{\mu_0 \pi R^3 H_{c2J} I_0}{4\rho_{nJ} I_c^2}\right)\left[\ln\left(\frac{1}{\varepsilon}\right) - 1\right] \qquad (4)$$

Typical values for YBCO-123 at 77K with R = 2 mm are: H_{c2J} = 60 Oe, ρ_{nJ} = 10^{-6} $\Omega \cdot$cm, I_c = 10 A and I_0 = 8A, thus, obtaining $t_{so} \sim 50$ µs for ε = 0.01.

1.2 Current Step

With the operating state completely settled down, let us assume that there is an abrupt increase of current ΔI. As previously stated, this current is accompanied by an increase of magnetic field at the surface, given by equation 1 with ΔI and ΔH instead of I_0 and H_0. We previously observed [1b] that the time for total flux penetration saturates at a minimum value at high fields. Hence, we conclude that when all the weak links are broken the flux vortices travel at a constant (maximum) speed given by

$$v_m = \left(\frac{2\rho_{nJ}I_c}{\mu_o \pi R^2 H_{c2J}} \right) \tag{5}.$$

They will travel at this speed from the cylinder's surface $r = R$ to the initial penetration radius $r_o = R - \delta_o$, in a time $t_o = \delta_o / v_m$. After this point we can analyze the flux penetration as a new problem of an applied field-step to a sample with radius r_o. Now let us apply this result in two possible situations:

Small Perturbation: Let us first consider the case where ΔI is small enough, such that the sample remains in the superconducting state; the required condition is $\Delta I + I_0 < I_c$. Repeating the analysis of the previous case we obtain the same dependence for current vs. time as in equation 3, except for the prefactor which now contains ΔI instead of I_0. Defining the settling time for this part with the same criterion as before and taking into account t_o, as computed after equation 5, we obtain the following approximated expression for the settling time after application of ΔI

$$t_{ss} = \left(\frac{\mu_o \pi R^3 H_{c2J} I_0}{2\rho_{nJ} I_c^2} \right) \left[1 + \frac{\Delta I}{I_0} \right] \tag{6}$$

where ε was taken as 0.01. The final configuration is like the final one (f) in figure 2.

Large Perturbation: Now let us consider an overcurrent that makes the total current overcome the critical value, $\Delta I + I_0 > I_c$. In this case the flux penetration is split in three stages: the first one from the external wall to the initial depth δ_o, the second one from δ_o to the center of the sample and the third one is a further penetration of flux to the center until the equilibrium is reached. The last stage can be interrupted by the transition to the normal state if there is an important generation of heat. This depends upon the thermal conditions of the device, so it will not be discussed here. Instead, we will assume a current excess not so big such that the heat generated is properly removed by the thermal bath. In this case the grains are still superconducting and all the weak links are broken.

The first stage gives the same result as before for t_o. The second stage is treated in the same way as the two previous cases, however, in this case the flux travels from $r = r_o$ to $r = 0$ in a finite time t_1 that is calculated from equation 3. The third stage of flux penetration, solved in ref. 1, gives an exponential decrease of the current density as function of time, with a time constant τ. Therefore the total settling time for the new equilibrium state, which can be either a partial superconducting state with resistive dissipation or the beginning of the transition to the normal state, is estimated as $t_s \approx t_o + t_1 + 3\tau$ obtaining

$$t_s = \left(\frac{\mu_o \pi R^3 H_{c2J} I_0}{2\rho_{nJ} I_c^2} \right) \left\{ 1 - \frac{\Delta I}{2I_0} \ln \left[1 - \frac{6(I_c - I_0)}{\Delta I} \right] \right\} \tag{7}$$

642

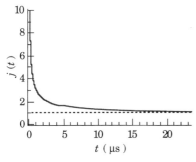

Fig. 3 Current density vs. time (eq. 3)

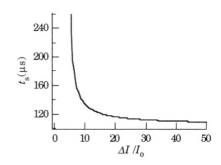

Fig. 4 Settling time vs. ΔI for I_0 = 8A (eq. 7)

where the exact expression was simplified by taking $\varepsilon = 0.01$ and $\mu_{eff} = 1/3$. Figure 4 shows the dependence of t_s on ΔI as given by this equation, for YBCO-123 at 77 K. In the case where $I_0 \approx I_c$, t_s is given by the term in round parenthesis, having a typical value of about 30 μs. In this case t_s is approximated by

$$t_s \approx \left(\frac{\mu_0 R H_{c2J}}{2\rho_{nJ} J_c} \right) \qquad (8).$$

This equation shows that t_s is reduced when the sample's resistivity or the critical current density are increased and is proportional to the sample's radius and to the second critical field of the Josephson medium, which depends on the quality of the weak links.

The heat generated by the CL during vortex penetration can be computed by the volume and time integral of the power dissipated, which according to flux flow [2], is

$$P(t) = \rho_{nJ} \frac{\mu_0 B(r,t)}{H_{c2J}} J(t) \cdot \left[J(t) - J_c \right] \qquad (9)$$

2. Conclusions

A sudden perturbation of the circulating current in the current limiter is accompanied by a perturbation of the magnetic induction. The interaction between current and magnetic induction determines the dynamics of the transition that can be described by flux flow theory and Bean's model. The final state of the system is reached after certain characteristic time which depends on the properties of the Josephson medium: H_{c2J}, J_c, ρ_{nJ}, the size of the device and the amplitude of the perturbation. The present a theory explains the transient phenomena during current perturbations and provide expressions for the characteristic times.

This work was partially supported by the Swiss National Science Foundation.

References

[1] H. Castro PHD Thesis 1995 Experimental Physics Institut of Lausanne University
     H. Castro et al. 1994 Physica C **235-240** 2922
[2] H. Castro et. al 1994 Elvetica Physica Acta **67** 209

Inst. Phys. Conf. Ser. No 148
Paper presented at Applied Superconductivity, Edinburgh, 3–6 July 1995
© *1995 IOP Publishing Ltd*

POSSIBLE APPLICATIONS OF CRYOALTERNATORS IN SHIP PROPULSION SYSTEMS

I.A. Glebov and L.I.Chubraeva

Department of Non-Conventional Electrical Machines,
VNIIElectromash, 196084, St-Petersburg, Russia

Abstract

Analysis of different types of ship propulsion systems, incorporating cryoalternators, and comparison of AC, DC and AC/DC (with frequency converter) types shows advantages of the dual-current ones. The Institute possesses practical experience with the development and testing of a number of superconducting homopolar machines, cryogenerators and a slotless alternator. The homopolar machine designs are based on the excitation system with two superconducting coils and the drum-type rotor. The new carbon-copper brushes for the homopolar motor permit to achieve the current densities above 200 A/cm^2, the developed liquid metal system provides currents up to 10 kA. Development of cryoalternators with slotless armatures (both superconductive and with conventional cooling system) allows to decrease the output electrical voltage without serious demage for the parameters. It helps to develop a propulsion system with a homopolar motor without an intermediate transformer. The results, substantial for future application of these alternators for the ship propulsion systems are being analysed.

1.Introduction

There may be discussed the following versions of ship propulsion systems:
1. DC generator - DC motor - fixed-pitch propeller.
2. AC generator - AC motor - controllable pitch propeller.
3. AC generator - frequency converter - DC motor - fixed-pitch propeller.
The first one is marked by the drawbacks characteristic for the DC commutator machines, the second one is better from the point of view of electrical equipment, but the reliability of controllable-pitch propeller may be inadequate. The last one is the most attractive though relatively expensive.

Problems of application of superconductivity for different types of ship propulsion systems are being investigated alongside with superconductive turbogenerators for the power plants. Main practical achievements were gained in Great Britain, the USA, Japan, Finland and Russia [1-4]. In Russia a number of DC and AC superconductive alternators, including the asynchronous ones, was manufactured and tested [5].

Presented below are the results of development and testing of superconductive homopolar and synchronous alternators by VNIIElectromash. All the investigations were performed at specially equipped test-beds and liquid helium was provided by the cryogenic plants installed in the test-beds.

2.Practical experience of VNIIElectromash

Superconductive homopolar machines. The first homopolar motor manufactured and tested by the Institute in the middle of 70-s had the rating 60 kW. Later on the machine was sufficiently improved. In 1985 there was developed and tested a model homopolar machine with a drum-type rotor with the maximal rating 120 kW, voltage 12 V, rotor current 10 kA, frequency of rotation 2400 rpm. Its general design is presented in Fig.1. The superconducting magnetic system consists of 2 coils manufactured of *Nb-Ti* rectangular wire in copper matrix. The support system of the coils can withstand the action of forces equal to 60 t. The stored energy is 0.56 MJ. The cryostat comprises a helium vessel with a superconducting field winding and an intermediate thermal screen cooled by liquid nitrogen. The helium vessel is suspended with the help of spokes similar to the bicycle wheels ones. The rotor consists of two shells of stainless steel with the channels in between them for the cooling water. The water is being pumped through the shaft. After that a motor with segmented drum multi-turn rotor (4 turns and 2 parallel branches) was developed and tested with the following rated data: U=24 V, n=1200 rpm. The experimental results have shown the motor design is very good for the ship propulsion systems.

The obtained experience was used to develop and start the manufacturing of the homopolar motor of 650 kW, U=220 V, n=420 rpm intended to be a basic model for the ship propulsion systems. The cryostat is positioned inside the rotor, the rotor is also of multi-turn design. The brushes are of copper-carbon type. The rotor and brushes are water cooled. The cryostat comprised a helium vessel, intermediate nitrogen volume. The coolant for the field winding passes through the hollow rotor shaft. Unfortunately at the end of 80-s the financial situation in Russia have changed and the works were stopped because of absence of financial support.

Alongside with these investigations a lot of attention was being paid to the improvement of the sliding contact of the rotor. The works were carried out in two directions: carbon-copper brushed and liquid metal contacts. Among the numerous modifications of brushes with the brush gears the best results were achieved or the brushes with water cooling. The maximum current density obtained under the model rotation was equal to 215 A/cm^2. As for liquid metal, two versions were investigated: based on *Na-K* and on *In-Ga-Pb*. The model with *Na-K*, operating in the inert gas medium provided the current up to 100 kA.

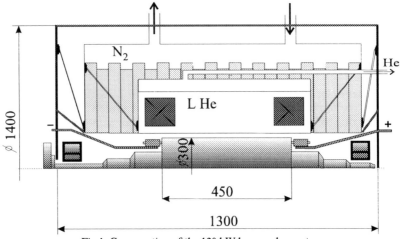

Fig.1. Cross-section of the 120 kW homopolar motor.

The results of investigations reveal that homopolar generator possesses several drawbacks as the alternator applied for a ship propulsion system: it has very low voltage and there may appear very serious problems with the excitation current variation resulting in the increased losses in the shells of the field winding cryostat and support structure, in case they are metallic ones. The cryostat may be non-metallic but our practical experience with non-metallic cryostats shows they need constant pumping out. As for the motors they represent a very simple electrical machine, though their voltage is relatively low (0.5-1.0 kV) and it is necessary to develop a special type of the low-voltage main generator to exclude an intermediate transformer.

Synchronous superconductive machines. Practical experience with the 1.5 and 20 MVA superconductive turbogenerators (Fig.2,a) shows a lot of advantages of such machines when they are installed in the power plants, but reveals as well certain difficulties when discussing the ship propulsion systems. In the accepted for the 20 MVA rotor design an evaporative cooling of the rotor interior by liquid helium. Liquid helium is introduced into the rotor shaft at atmospheric pressure. The flow rate is relatively low (about 4 g/s) and a lot of attention is being paid to the helium transfer coupling with very small gaps between the rotating and stationary elements. Vibrations, which are inevitable on board a ship may cause friction of these parts with subsequent heating and disturb the coolant flow. A much better design for these conditions was proposed by the French specialists , where the transfer coupling is less critical to the vibrations due to larger gaps. But it results in much higher helium flow-rates (up to 100 g/s and even more) [6]. The problem will be simplified substantially in case of HTSC application with liquid nitrogen cooling.

As for the possibility of fast regulation of the excitation current and rating of the generator, demanded for the ship propulsion system, the results of development of high-response turbogenerators shows the problem may be solved successfully by an application of new types of superconductors and by changed approach to the rotor electromagnetic screen system.

Superconductive synchronous motors may be developed as well. Application of fully superconductive alternators will simplify the cooling system of the machine but will arise the problem of a single cryostat design and of rotating shaft seals.

Synchronous slotless machines. The slotless design may be applied to both the main generator and the synchronous motor for the propeller. In the first case the advantages are associated mainly with improved size and weight parameters, increased static stability and low noise of the alternator due to slotless design and helical winding application. Low harmonic contents is important as well. The generator cooling system may be based on the application of insulating liquids and equipped by a heat-pump.

Fig.2. The 20 MVA superconductive and 5 MVA slotless generators.

Conventional generators are very sensitive to the decreased voltage and increased armature winding current value The air-gap winding opens way to the main generator design with decreased armature winding voltage. It turns to be extremely important when discussing the application of superconducting homopolar motors.

In the case of slotless motor design the merits are achieved due to necessity of high overload capability of the drive motor (2-3), especially during storms, and the alternator of slotless design with a relatively big physical air gap is very attractive. In the Institute there was developed and tested a 5 MVA alternator with slotless helical winding (Fig.2,b). The alternator contains a conventional rotor with an indirect air cooling and a slotless stator, meant to be cooled by an insulating liquid at full load. The main parameters and geometrical sizes of the alternator are listed in Table 1.

Table 1

MAIN PARAMETERS OF SUPERCONDUCTIVE AND SLOTLESS . TURBOGENERATORS

Rating, MVA	20	5.0
Voltage (line-to-line), kV	6.3	1.0
Current, kA	1.85	2.9
Frequency of rotation, rpm	3000	3000
Cos φ	0.8	0.8
Rotor OD, m	0.590	0.506
Armature winding cylinder ID, m	0.600	0.525
Ferromagnetic screen ID, m	0.812	0.780
Ferromagnetic screen OD, m	1.600	0.900
Excitation winding active length, m	1.10	1.35
Armature winding active length, m	1.10	1.50
Synchronous reactance, p.u.	0.33	0.81
Flux density in the armature winding, T	0.95	0.35

3. Conclusions

1. It is reasonable to apply superconductivity primarily for the large ships with high ratings of the propulsion system electrical equipment. They are ice-breakers and large transport ships. The rating of the ship propulsion system in this case varies from 20 to 60 MW.

2. Though there exists a possibility of development of a variety of ship propulsion systems and practical experience with the controlled pitch propellers, the main advantages are achieved in case of a dual current one, incorporating an AC turbogenerator, a frequency converter and a DC motor. It operates with a fixed-pitch propeller.

3. Two versions of dual current systems seem the most appropriate: a combination of superconductive turbogenerator and superconductive homopolar motor or of a slotless turbogenerator with conventional cooling and superconductive homopolar motor.

References

1. Appleton A.D. 1975. *IEEE Transactions on Mag.* vol. MAG-11. N 2. p.633.
2. Fox G.R. et al. 1972. *IEEE Publication* 72CH0682-5-TABSC. p.33.
3. Levedahl W.J. 1972. *IEEE Publication* 72CH0682-5-TABSC. p.26.
4. 1989. *Naval engineers journal.* vol.3. p.93-101.
5. Zhemchugov G.A. et al. 1991. *Electrotecnica*, N 9. p.23-29. (In Russian).
6. Glebov I.A. et al. 1982. *CIGRE Session.* Rep.11-14. 8p.

Inst. Phys. Conf. Ser. No 148
Paper presented at Applied Superconductivity, Edinburgh, 3–6 July 1995
© *1995 IOP Publishing Ltd*

Modelling high frequency superconducting converters

O. A. Shevchenko, and H. H. J. ten Kate

Applied Physics Department, University of Twente, P.O. Box 217, 7500 AE Enschede, NL

M. A. Fedorovsky

Institute of Electrodynamics, Nat. Academy of Sciences, Peremoga Pr. 56 Kiev 252680 Ukraine

Abstract.   In the paper a 1 kA thermally switched converter operating at 50 Hz and 4 K is described by means of both analytical and numerical models. The calculated characteristics of the converter and components are supported by measured performance data obtained during full-scale tests. This knowledge is an important step in the development of similar 10 kA devices. New reliable PSPICE circuit simulator models of superconducting components and devices are developed and experimentally verified.

1. Introduction

A superconducting converter is a high current and a low voltage device, which enables both a rectification of an AC current (or voltage) and an inversion of a DC current (or voltage). It consists mainly of a pair of superconducting switches and a superconducting transformer.

During the last years a continuous improvement of such devices is demonstrated but especially for 50 Hz operation a better understanding of the limiting factors is required for further improvement of performance and reliability.

In the paper a thermally switched converter operating at 4 K [1] is characterised both theoretically and experimentally. Both physical and numerical models of the superconducting device and the components are developed. A standard PSPICE circuit simulator is used to build the CAD compatible numerical models.

2. Model of a repetitive superconducting switch

A thermally triggered superconducting gate of a switch [2] is driven by the gate voltage and control heat pulses. The electrical behaviour of the gate is strongly coupled to the thermal and magnetic state.

A new formula to describe the static V-I characteristic of a superconductor in the resistive state is used [3]. Characteristic example for a NbTi gate is given in Figure 1. At lower electrical field the formula is consistent with the commonly used power law.

Normal zones are triggered by the heater during opening of the switch. Their further propagation is determined by the gate current. The velocity of the zones propagation versus the gate current is presented in Figure 2. The measured characteristic agrees with the theory. It is

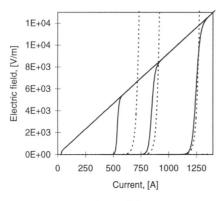 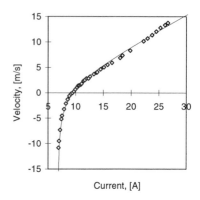

Figure 1 (left). V-I characteristic of the gate conductor at B=0.1 T and temperatures of 9; 7; 6; 4.2 K respectively (from the left to the right). Dashed lines are according to the power law (for T=9 K the line follows the horizontal axis). Solid lines are derived from the new formula.

Figure 2 (right). Velocity of n-s boundaries vs. current for the switch gate in the self-field. Points are measured, while the line is derived analytically. The number of normal zones triggered simultaneously is 24.

used to complete the physical model and to build a 1-D numerical model of a switch.

The diffusion equations are solved by the circuit simulator PSPICE [4] using an equivalent circuit representation. As it was demonstrated before [5], this can be done via an analogy between the electrical, thermal and magnetic processes.

The model requires extended information about the switch. As input data the geometry, the static V-I and normal zone propagation characteristics of the superconductor, the control and cooling conditions, the relevant thermal and magnetic properties are used. Comparing to the known (mainly qualitative) numerical models, the new advanced model describes the activation, propagation, recovery and superconducting modes of the switch quite well. Moreover the model can be easily adjusted to describe new switches such as, for example, HTS switches.

3. Model of a superconducting transformer

The transformer consists of the NbTi primary, Nb$_3$Sn secondary coils, and the iron core [2]. Physical information about the losses versus the applied current and voltage in various operation modes is provided by our experimental data and 2-D simulations.

The standard PSPICE model of an iron core transformer [4] is modified in order to take into account the losses in the SC coils and in the core operating at 4 K. Both current and voltage controlled voltage sources containing the information are added to the model. The self and mutual inductances are voltage dependent [2] and the model allows to calculate their momentary values.

4. Losses in the components

During ramping a magnet, losses in the converter are dependent on both the magnet current and the applied voltage. The main contributions originate from the: repetitive switches S_1 and S_2; SC transformer; persistent mode switch S_3. The loss in the switches S_1, S_2 is dominant and it is calculated assuming that the gate leak- age current is independent of the applied voltage (see Table I); the voltage and the commutation time are both consistent with

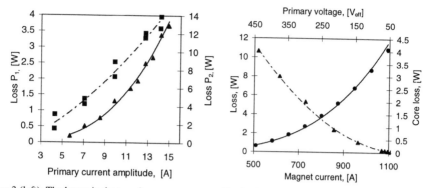

Figure 3 (left): The losses in the transformer components: P_1 -in the separated primary coil, shown by the boxes and dashed line; P_2 -in the coils during the short circuit of the secondary coil, shown by the triangles and solid line; the points are measured; the lines are trends used in the numerical models.

Figure 4 (right): The transformer losses in the converter circuit:-in the coils vs. magnet current (left axis), shown by the circles and the solid line; in the iron core vs. applied voltage (idle mode, 4 K and 300 K [7]), shown by the triangles and the dashed line.

the theory [6]; and that there is no commutation error.

The loss in the transformer consist of the losses in the coils and in the core. Due to the 3-D iron core, a precise calculation of the losses is rather sophisticated. The loss is calculated using scaled experimental data taken from a similar transformer [2]. The losses were measured with a calorimetric method, separately for the iron core (idle mode), for the primary coil, and in the short circuit mode. A summary is given in Figures 3 and 4.

5. Characterisation of a superconducting converter

5.1. Analytical model

The input data of the model [6] are: the operation frequency, the shape and the amplitude of the AC primary voltage; the impedance of the converter during the commutation process; the transformation coefficient and losses of the transformer; the minimum propagation current of the switch gates; the shape of the control pulses, switch recovery times and resistance of the heaters; the impedance of the load. The data for the particular converter considered here, are summarised in Table 1.

The output voltage-current characteristics and efficiency of the converter versus time and versus magnet current, as well as other important values [6] are calculated over the whole range of the parameters covered by the experiment, Figure 5.

5.2. Experiment and comparison

The new converter has a similar design as the one described before [1] with a minor difference in the transformer part. It is investigated within the whole design range of an input voltage up to 240 V, an input current of 7 A, an output voltage of 1.2 V, and an output current of 1 kA. A few characteristics are presented in Figure 5 in solid lines. The characteristics are compared with those calculated analytically on the time scale of the transient process and numerically on the time scale of one operating cycle. Important parameters for the converter operation such as the maximum

650

TABLE 1
INPUT DATA OF THE CONVERTER

Parameter	Value
Operating frequency, [Hz]	50
Transformation coefficient	170.9
Primary voltage amplitude, [V]	55 ⟩ 230
Primary commut. inductance, [mH]	46
Magnet inductance, [mH]	10.7
Supercond. switches $S_{1,2}/(S_3)$:	
Min. propagation current, [A]	9.5/(1)
Recovery time, [ms]	2/(2000)
Heater resistance, [Ohm]	55/(20)
Control voltage amplitude, [V]	16/(6)
Control pulse width, [ms]	0.8/(5)

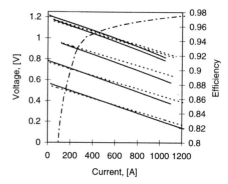

Figure 5: The V-I characteristic of the converter: solid lines are measured, the dotted lines are calculated. The input voltages are 110; 145; 185; 220 and 230 V respectively. The energy efficiency is calculated for the case of 220 V and is shown by the dashed line.

value of the commutation error and the requirements of the control system were determined and verified by simulations during the design stage.

The theory [6] assumes a constant amplitude of the AC voltage applied to the primary coil of the transformer during the ramping procedure. In the experiment a certain voltage drop was present across the room temperature part of the converter. It was changing due to an increase of the primary current. The discrepancy between the theory and the experiment can be eliminated by taking into account this fact.

Conclusions

1. Physical models of a superconducting converter and the components are developed.
2. The equivalent circuit models of several AC superconducting components are built using the standard circuit simulator PSPICE.
3. The experimental model of a SC converter is characterised both analytically and by simulation on the time scale of the transient process and one operating cycle.

Acknowledgement

This work is supported by the Netherlands Organisation for Scientific Research (NWO) and by the State Committee for Science and Technology of the Ukraine.

References

[1] Shevchenko O A, Krooshoop H J G, ten Kate H H J and Fedorovsky M A "Development of 1 kA, 50 Hz Superconducting Converter" 1995 IEEE Trans. on Applied Superconductivity, to be published
[2] Shevchenko O A, *et al.* 1994 Cryogenics **34** ICEC Supplement 745-748
[3] Shevchenko O A, Knoopers H G and ten Kate H H J, to be published in Applied Phys. Lett.
[4] PSPICE 5.2. 1992 The Design Center. MicroSim Corp. 20 Fairbanks Irvine CA
[5] Shevchenko O A, *et al.* 1992 Cryogenics **32** ICEC Supplement 493-496
[6] Shevchenko O A, *et al.* 1994 Adv. in Cryog. Engineering **39** part A 933-940
[7] Strip-wound cut cores. 1987 VAC Publication Vacuumschmelze GMBH Hanau

Inst. Phys. Conf. Ser. No 148
Paper presented at Applied Superconductivity, Edinburgh, 3–6 July 1995

651

On-board linear generator for EDS-MAGLEV vehicles: design optimisation of the SC and induction windings

M Andriollo, G Martinelli, A Morini and A Scuttari

Dept. of Electrical Engineering, University of Padova, Via Gradenigo 6/a, 35131 Padova, Italy

Abstract. The paper describes a method to optimise the winding configuration of the on-board linear generator for EDS-MAGLEV trains. The procedure allows to determine the sizes of the generator coils which maximise the collected e.m.f., with adequate values of the efficiency and vertical stiffness of the magnetic suspension.

1. Introduction

In magnetically levitated trains of electrodynamic type (EDS-MAGLEV) [1], the levitation is produced by the interaction between the magnetic flux of on-board superconducting coils (SCCs) and the currents induced by motion in on-ground levitation coils (LCs); the SCCs also provide the excitation of the linear synchronous motor (LSM) propelling the train. On-board power is required by the cryogenic system and by the auxiliary apparatus. Such power is supplied by on-board linear generators, in which induction coils (ICs) collect e.m.f.s produced by the harmonic flux of the LCs [2]. Auxiliary SCCs increase the collected e.m.f., with lower efficiency of the magnetic suspension due to higher LC currents.

The paper describes a method to optimise the generator configuration with the aim to increase the e.m.f. and to maintain, at the same time, good efficiency and adequate vertical stiffness. Such quantities, related to the geometrical and electrical parameters, are evaluated by means of analytical formulations [3,4] and are taken into account by an objective function to be minimised. The procedure is performed on a PC, more quickly than if numerical methods, such FEM, were used.

2. Coil arrangement

The MAGLEV train has two on-board generators, placed in the front bogie of the head and tail cars. Fig. 1 shows the coil arrangement for one side of the bogie [2]. Coils 1 are the main SCCs of the bogie and produce the flux for the levitation and the LSM excitation; coils 2 are the generator auxiliary SCCs to increase the currents induced in the on-ground LCs; coils 3 are the ICs to collect the e.m.f. due to the field harmonics produced by the LCs. The polarity sequence of the SCCs may be $NnsnsS$ (homopolar) or $NsnsnS$ (heteropolar), the capital letters indicating

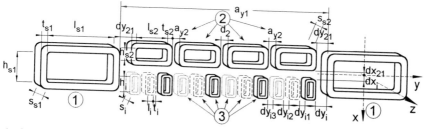

Fig.1 - Coil arrangement and geometrical parameters of the linear generator (one side) [1: main SC coils; 2: auxiliary SC coils; 3: three-phase induction coils, each phase represented by a different line style].

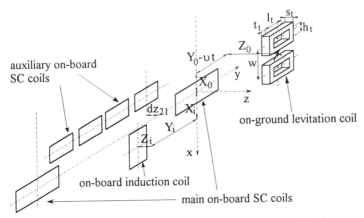

Fig.2 - Geometrical parameters of the 8-shaped levitation coils and reciprocal position between the SC coils, one induction coil and one levitation coil [3,4]. [X_0,Y_0-$\upsilon t,Z_0$: coordinates of the levitation coil; υ: speed; X_i, Y_i,Z_i: coordinates of the induction coil; for sake of clearness, the on-board coils are represented as filiform].

the main coils and the small letters the auxiliary ones. The arrangement corresponds to a synchronous generator with the LCs as field winding and the ICs as armature winding (Fig.2).

3. Calculation of the induced e.m.f.

The calculation of the collected e.m.f. requires a 3D analysis of the time-varying electromagnetic interaction between three coil systems (SCCs, LCs and ICs). With reference to Fig.2 and according to [4], it is possible to evaluate analytically both the current induced in one LC by the SCCs and the mutual inductance between the LC and one IC. By summing up the contributions of all the LCs, the flux linked by the IC can be expressed as a time-varying Fourier series expansion; by derivating the flux, the e.m.f. induced in the IC is then available [4]:

$$e_i(X_i - X_0, Y_i - Y_0, Z_i - Z_0, t) = \sum_{k=1}^{\infty} \omega N^2 \varphi_{it0} L_{iN}(X_i - X_0, Y_i - Y_0, Z_i - Z_0) \sin(kN\omega t - \eta_{kN})$$

with N number of LCs per polar pitch b_y, $\omega=2\pi\upsilon/b_y$ and φ_{it0}, η_{kN}, L_{iN} given in [4]. Finally, by summing up the contributions of the m ICs per phase, it is possible to evaluate the instantaneous phase value of the collected e.m.f. and then the instantaneous and mean values of the rectified e.m.f..

4. Definition of the objective function

In order to optimise the generator performances, it is necessary to increase as much as possible the e.m.f. mean value E_{DC}, maintaining, at the same time, good electromagnetic and mechanical performances. The optimum search is converted into the minimisation of a suitable function of the geometrical parameters of the system. Four items contribute to such function, each expressed in per unit with reference to the starting configuration (indicated with *):

a) the mean value of the rectified e.m.f. $(E_{DC})/(E_{DC})^*=1/f_e$;
b) the efficiency of the magnetic suspension $dr/dr^*=1/f_{dr}$, with dr ratio between the levitation force per bogie F_L and the drag force due to the ohmic losses in the LCs [3]. The car body mass m_C being constant, the bogie mass m_B depends on the coil sizes ($m_B=c_{s1}v_{s1}+c_{s2}v_{s2}+$ $+c_iv_i$, with c suitable coefficients and v the volumes of the SCCs and ICs). Then the required F_L, the corresponding equilibrium vertical displacement $X_0=X_{eq}$ and drag force must be calculated by means of an iterative method for each configuration [3,5];

c) the vertical stiffness of the magnetic suspension $(\partial F_L/\partial X)/(\partial F_L/\partial X)^* = 1/f_{vs}$;

d) the current density in the ICs $(J_i)/(J_i)^*$. The rated power being constant, the generator current is in inverse proportion to the e.m.f.. In order to prevent the current density from exceeding the upper boundary of a given range, a penalty function [5] is defined:

$$g_J = \max\left\{0, k_J(J_i/J_i^* - 1)^{n_J}\right\} \quad J_i/J_i^* = (s_i^* t_i^* E_{DC}^*)/(s_i t_i E_{DC}) \quad k_J > 0, \; n_J = 1,3,5,\ldots$$

The global objective function is:

$$f_{tot} = c_e f_e + c_{dr} f_{dr} + c_{vs} f_{vs} + g_J$$

where the coefficients c_e, c_{dr} and c_{vs} represent the relative weight of each item [5]. Other fixed constraints are: *i)* suitable upper boundaries for the coil sizes; *ii)* constant SCC cross sections; *iii)* constant gap between on-board coils and on-ground LCs. The LC sizes are not changed, as they affect the performances of the whole train and then their definition forms part of the optimisation procedure of the whole system [5]. The adopted multi-variable search technique is based on the discrete step Rosenbrock's method [6].

5. Example of application

A tentative optimisation was performed for both heteropolar and homopolar sequences. The independent variables are 19 geometrical parameters, whose starting and final values are reported in Tab.1 together with other optimised geometrical data; Tab.2 gives other fixed data. The coefficients used in the optimisation are given in Tab.3: they are indicative, since the only aim of the example is to show the method application. The procedure, performed on a 486DX PC, halted after the analysis of ~300 configurations for the heteropolar sequence and ~1500 for the homopolar one. Fig.3 shows the progress of the objective function and its components: its final value is ~0.60. Tab.4 shows the starting and final values of the quantities taken into account by f_{tot}: the rectified e.m.f. (Fig.4), the drag ratio (Fig.5) and the vertical stiffness are increased (the latter 3 times). The optimised performances of both the sequences are similar.

Tab.1 - Optimised geometrical parameters [m] (**Bold**: independent variables)
S: starting values [2,7]; Fhe, Fho: final values for the heteropolar and homopolar sequences

	l_{s1}	h_{s1}	t_{s1}	s_{s1}	l_{s2}	h_{s2}	t_{s2}	s_{s2}	a_{y2}	d_2	dx_{21}	dy_{21}
S	1.020	0.450	0.050	0.080	0.394	0.200	0.050	0.060	0.181	0.181	-0.120	-0.206
Fhe	1.500	0.298	0.150	0.027	0.369	0.508	0.066	0.045	0.148	0.170	-0.118	-0.277
Fho	1.500	0.251	0.150	0.027	0.383	0.588	0.059	0.051	0.167	0.169	-0.112	-0.200

	dz_{21}	l_i	h_i	t_i	s_i	dx_i	dy_i	dy_{i1}	dy_{i2}	dy_{i3}	Z_i	Z_0
S	0.010	0.145	0.300	0.030	0.030	0.020	-0.125	-0.020	-0.020	-0.020	0.055	0.185
Fhe	-0.009	0.150	0.494	0.029	0.020	0.018	-0.189	-0.018	-0.018	-0.018	0.033	0.158
Fho	-0.012	0.145	0.492	0.031	0.020	0.012	-0.120	-0.013	-0.020	-0.019	0.033	0.158

Tab.2 - Fixed system data [2,7]

υ km/h]	b_y [m]	Y_0 [m]	m.m.f.$_{s1}$[kA]	m.m.f.$_{s2}$ [kA]	m	m_C [kg]
500	21.60	0.000	700	500	4	8560

l_t [m]	h_t [m]	t_t [m]	s_t [m]	w [m]	R_t [$\mu\Omega$]	N
0.265	0.255	0.085	0.030	0.435	13.53	48

Tab.3 - Optimisation coefficients

c_{s1} [ton/m^3]	c_{s2} [ton/m^3]	c_i [ton/m^3]	c_e	c_{dr}	c_{vs}	k_J	n_J
200.7	180.6	100.3	0.50	0.25	0.25	5	3

Tab.4 - Design quantities

	X_{eq} [m]	F_L [kN]	E_{DC} [V/turn]	dr	$\partial F_L/\partial X$ [MN/m]	m_B [kg]
Starting heteropolar [2,7]	-0.057	150.0	14.3	48.1	1.9	6720
Final heteropolar	-0.023	173.2	21.3	71.0	5.88	9120
Starting homopolar [2,7]	-0.057	155.0	15.5	49.2	1.9	6720
Final homopolar	-0.022	173.6	21.6	70.1	5.94	9120

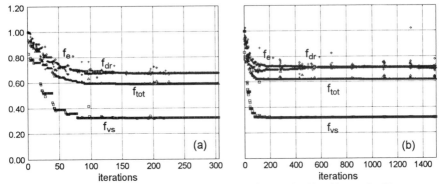

Fig.3 - Progress of the objective function and its components for heteropolar (a) and homopolar (b) sequences.

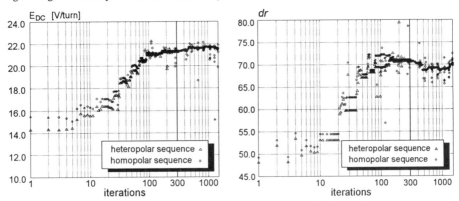

Fig.4 - Progress of the rectified e.m.f..   Fig.5 - Progress of the drag ratio.

6. Conclusions

The paper presents a method to optimise the on-board generator of EDS-MAGLEV vehicles. The procedure is based on the definition of a suitable objective function (depending on the geometrical parameters of the configuration) that can be quickly evaluated by means of a PC, due to the analytical formulation of its components. The function not only takes into account the collected e.m.f., but also allows to optimise magnetic suspension performances, such as the drag ratio and the vertical stiffness, penalised in the starting configuration.

Work supported by the Italian National Research Council (CNR) under the Progetto Finalizzato "Trasporti 2".

References

[1] Tanaka H, "JR Group probes Maglev frontiers", *Railway Gazette International*, July 1990.
[2] Sawano E et al., "Inductive power collection system for MAGLEV vehicles", *Int. Conf. on Speedup Technology for Railway and Maglev Vehicles*, Yokohama (Japan), November 22÷26 1993.
[3] Andriollo A, Martinelli G and Morini A, "General expressions for forces acting on EDS-MAGLEV systems driven by linear synchronous motors", *Int. Aegean Conf. on Electrical Machines and Power Electronics*, Kusadasi, (Turkey), May 27÷29 1992.
[4] Andriollo A, Martinelli G, Morini A and Scuttari A, "Mathematical model of on-board power supply for EDS-MAGLEV systems", *Symposium on Power Electronics, Electrical Drives, Advanced Electrical Motors*, Taormina (Italy), June 8÷10 1994.
[5] Andriollo A, Martinelli G, Morini A and Scuttari A, "Optimisation of the winding configuration in EDS-MAGLEV trains", *14th Int. Conf. on Magnet Technology*, Tampere (Finland), June 11÷16 1995.
[6] Bazaraa M S et al., *Non linear programming - theory and algorithms*, John Wiley & Sons, 1993.
[7] Takao K, "Vehicles for superconducting Maglev system on Yamanashi test line", *Computer in Railways IV (COMPRAIL 94 Proceedings)*, Vol.2, Computational Mechanics Publications, 1994.

Inst. Phys. Conf. Ser. No 148
Paper presented at Applied Superconductivity, Edinburgh, 3–6 July 1995

655

EXPERIMENTAL INVESTIGATION OF CRYOGENIC MOTOR MODEL

L.I. Chubraeva, S.N. Pylinina, V.E. Sigaev and V.A.Tutaev

Department of Non-Conventional Electrical Machines,
VNIIElectromash, 196084, St-Petersburg, Russia

Abstract

Alongside with the development of superconductors for AC applications there have been carried out researches of possible applications of high-purity metals for AC windings. The optimistic results of experimental investigations of new multifilumentary hyperconducting composite wire permit to start the development of an AC armature winding. A simplified disk model of a synchronous alternator was manufactured and tested. It contains a rotor with *Nd-Fe-B* magnets and a stator with the winding of high-purity aluminium. Presented experimental investigations of the model motor were carried out in a temperature range 300-77 K. At 77 K the current density in the aluminium wires is still relatively low. Nevertheless an interesting experimental data was obtained, substantial for the development of high-rated machines. The highest values of armature current density, motor rating and torque may be obtained only at about 20 K. The rating of the model at 20 K may reach 1 kW. The test results of the model machine will be used for the development of cryogenic brushless exciter and different types of cryogenic motors.

1. Introduction

To investigate the principles of operation of cryogenic electrical machines with different winding materials the alternators of simplified disk type seem to be the most appropriate. The one with superconducting rotor and stator windings manufactured of Nb_3Sn was developed and tested in VNIIElectromash in 1990 [1]. It provided initial experimental data for the development of fully superconducting machines, in particular, the peculiarities of operation of synchronous alternator without electromagnetic rotor screen and specific features of AC superconductor transition to normal state. The purpose of development of a new model is the acquiring of further experience for the practical implementation of synchronous alternators incorporating permanent magnets and windings of high-purity aluminium. Synchronous generators with cryogenic cooling and aluminium windings [2] are to have regulated excitation current of relatively high value. In case of application of an independent excitation system there appear the problem of slip-rings. It may be solved with the help of brushless excitation system. A cryogenic motor may find a number of applications as well: it may be a part of the energy storage system or a part of ship electrical equipment [3].

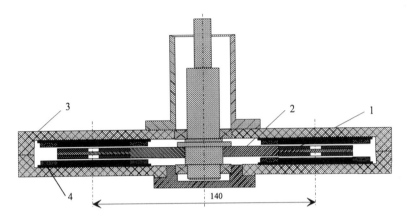

Fig.1. Cross-section of the cryogenic alternator model

2. Motor model design

The scheme of the model machine is presented in Fig.1. It is a 3-phase 8-pole alternator with 16 circular permanent magnets (1) of *Nd-Fe-B* (Russian mark - *He30D6P*) positioned on the rotor (2). The stator winding consists of two groups of 6 circular coils (4) are located on both sides of the rotor. Major sizes of the model and its elements are presented in Table 1.

Table 1

Model outer diameter, mm	250	Magnet thickness, mm	2
Model total height, mm	40	Number of magnet pairs	8
Diameter D_{em} (see Fig.1), mm	140	Stator coil outer diameter, mm	48
Rotor disk diameter, mm	170	Stator coil inner diameter, mm	28
Rotor disk thickness, mm	5	Stator coil height (along the axis),mm	6
Magnet outer diameter, mm	30	Number of stator coils	12

The model is being placed in a helium cryostat. On the cover of the cryostat a 130 W DC drive motor is fixed. The frequency of the rotor rotation is 750 rpm.

3. Investigation of the model elements

Stator winding wire and coils. Alongside with the development of superconductors for AC applications there have been carried out researches of high-purity aluminium and beryllium AC windings. A new step was made when the DC composite aluminium wire technologies were introduced [4]. The one previously developed in Russia and providing manufacturing of multifilumentary strands of hyperconducting aluminium with filaments of 0.3 and 0.1 mm for 50 Hz application was lately improved. The new composite wire of 0.3 mm diameter, containing 49 high-purity aluminium filaments of 0.03 mm in special matrix revealed good current-carrying ability for AC currents up to 800 Hz.

Three modifications of aluminium wires were tested: a twist of 28 insulated filaments 0.1 mm in diameter; a twist of 15 insulated filaments 0.3 mm each; a new composite 0.3 mm wire with the reinforced matrix made of *Al-Mn* alloy with the space factor $K = 0.65$. The samples were tested for DC and AC resistance at varying current density in the range of 77-300 K. The results revealed that the 3-d sample not only had the best strength, but demonstrated the most attractive thermal performance with good current

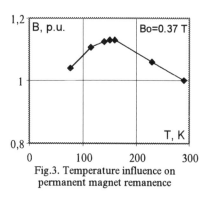

Fig.3. Temperature influence on
permanent magnet remanence

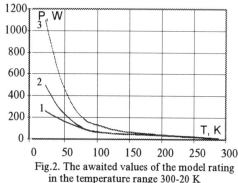

Fig.2. The awaited values of the model rating
in the temperature range 300-20 K

density. However, the 1-st wire has been chosen as the stator coil material and its low mechanical strength has improved by epoxy impregnation. The choice in its favour had only financial grounds.

To study thermal behaviour of the wire in the impregnated winding and to determine the operating current values, the test coils were manufactured and tested.

Permanent magnets for the rotor. Temperature dependence for magnetic flux density was performed close to the magnet surface in the absence of ferromagnetic materials (Fig.2). The investigation shows high stability of the magnets to varying magnetic field (with the frequency up to 400 Hz and the amplitude of the magnetic field up to 25% of the one produced by the magnets) and to thermal cycling 300-77-300 K. The presented curve shows there exists a possibility of regulation of the excitation field be the magnet temperature variation.

4. Model machine test-results

The cryogenic alternator model operating in parallel with a three-phase network was investigated in the generator and motor duties. Phase synchronous and subtransient reactances at 50 Hz are respectively 1.115 Ohm and 0.95 Ohm. In partial units it equals 0.14 p.u. and $x_d" = 0.114$ p.u. (with reference reactance at 300 K - 12.0 Ohm; $x_d = 1.162$ p.u. and $x_d" = 0.95$ p.u. (with reference reactance at 77 K - 1.44 Ohm). The machine reactances will be subjected to further decrease with the temperature and will exceed the possible values for retaining of static stability. Fig.3 presents the curves demonstrating the awaited rating of the model machine at different temperatures (curve 1 - due to decrease of aluminium resistance, curve 2 - due to limits imposed by thermal stability, curve 3 - due to limits imposed by electrical stability. After synchronisation of the model with the network of 12 V, the input power of the driving motor was varied (Fig.4, curves 1 and 2, *P* - rating of the model alternator, *Pa* - rating of the DC drive motor or load). The values of the model rating and current were limited by the DC drive motor feeding. At the line voltage of the network 25 V alternator model investigated as a synchronous motor and generator. The model consumed the reactive power of about 150 W from the network. In generator mode the armature current reached 5A (curve 3) and total rating of the alternator was 230W (curve 4). Upon decreasing of the driving motor input down model changed its duty and started operating as a motor, compensating all the mechanical losses. The total load on the shaft was about 45 W. The drive DC machine changed its mode to the

658

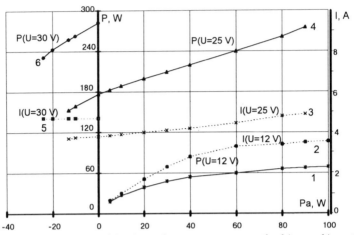

Fig.4. Dependence of the model rating and armature current on the drive machine rating

generator one. When the DC generator output reached 13 W, the model machine fell out of synchronism. The total load of the model shaft was approximately 62 W. If the line voltage network is 30 V (curves 5 and 6) the maximum permissible load on shaft is estimated to be 76 W.

5. Conclusions

1. The optimal temperature for the *Nd-Fe-B* magnets operation is in the range 130-160 K, so they may be cooled by the gaseous cryocoolant and in the thermal scheme of the machine may be positioned behind the armature.

2. Thermal instability of superconducting winding results in its transition to normal state without serious damages if properly protected. It may be vital for the hyperconducting winding because of rapid increase of electrical resistance and attention should be paid to its electrical and thermal design.

3. The alternator with permanent magnets and high-purity metal armature winding is characterised by unstable operation in transient modes because of fast variation of high-purity metal electrical resistance. The rating of the machine may be limited at 20 K by the specific thermal loading and electrical stability.

6. Acknowledgements

The research is supported by the Ministry of Science and Technical Policy of Russia and of the Russian Fund for Fundamental Researches.

References

1. Klimenko E.Yu. et al. 1991. *ICCE-91. Southampton.* 6 p.
2. Oberly C.E. et al. 1993. *Cryogenic Power Syst. Comp. Workshop. USASSDC/Nichols Co.* 7 p.
3. Tixador P. et al. 1994. *ICEM-94. Paris. France.* p.361-366.
4. Oberly C.E. and Ho J. 1991. *IEEE Trans. on Mag.,* Vol. Mag-27. pp.458-463.

Inst. Phys. Conf. Ser. No 148
Paper presented at Applied Superconductivity, Edinburgh, 3–6 July 1995

SUPERCONDUCTING VARIABLE SPEED MOTOR

P. Tixador, H. Daffix*, F. Simon

CNRS - CRTBT/LEG, BP 166, F 38042 Grenoble Cedex 09, France.
* Also GEC ALSTHOM , 3 Av. des Trois Chênes, 90018 Belfort , France

Abstract - The emergence of a.c. NbTi wires makes it possible the design of a three phase superconducting winding for electrical machines. An hybrid structure which associates a superconducting armature and rotating cooled permanent magnets is well adapted to embarked devices due to its lightness and its simple mechanical structure. An experimental 15 kW-750 rpm model has been built and has suffered a large number of tests. The first tests at variable speed are reported. The superconducting armature is fed by a PWM (Pulse Width Modulation) inverter. The quench currents are little reduced compared to 50 Hz operation. The motor may be controlled in speed.

1. Introduction

For a long time it has been impossible to design a three phase superconducting winding since the a.c. losses brought back to room temperature led to a low efficiency. The drastic reduction of the a.c. losses by means of a suitable strand structure [1] enables yet the operation of NbTi superconductors at industrial frequencies. Due to the high current densities allowable in superconductors compared to resistive wires, it is of great interest to utilise superconductors in electrical motors. The superconducting machines outclass the conventional ones in terms of weight, volume and efficiency. Those parameters are crucial for embarked systems and superconducting drives are well adapted. Nevertheless these devices are submitted to high stresses (vibrations, shocks, ...) and the mechanical structure should be suited. This is why a structure combining cooled rotating permanent magnets and a superconducting armature has been proposed [2] and appears as interesting. The cryogenic system does not move. The rotating part consists in cooled permanent magnets which makes the system simple and robust. The cooling of a static winding is not difficult and requires a small cryogenic power. The cryocoolers could be adapted for low ratings. Such a machine happens to be easy to drive.

The fully superconducting solution leads to higher performances in terms of weight and volume for the motor itself but the cryogenic rotating part is mechanically delicate and needs a cryogenic power which is not negligible. So when the motor plus its cooling system are considered the weight and volume balances are more favourable for the hybrid structure at least when the power is not too high. The critical or break-even size is lower for the solution with permanent magnets.

The limited excitation field provided by magnets, even if their cooling at about 150 K enhances the magnetic properties, does not penalise too much the superconductors. The a.c. NbTi critical current densities are very high under low field but decreases rapidly when this latter increases [3]. Moreover a low field is favourable for the a.c. losses. The cooling of the magnets at around 150 K requires practically no power due to the cryogenic environment.

2. Experimental model

To study this hybrid electrical machine a 15 kW, 750 rpm model was built [4] (fig. 1 & table I). This size is of course little representative for a real system but this preliminary stage is indispensable to verify this new conception, to check calculations and designs, to study the electrical, mechanical and thermal behaviours and to point out some possible experimental difficulties. The structure is cylindrical with the magnets rotating inside the armature (fig. 1). They are cantilevered to make their cooling easy. The magnetic circuit is reduced to a rotating FeNi bulk core which support the magnets and to a ring around the helium vessel. It consists in a FeSi sheet stack and confines the flux within the machine. The NbTi three phase windings are immersed in liquid Helium whose feeding is continuous.

Their critical currents amount to about 110 A_{rms}. This figure is lower than the theoretical value deduced from the short sample characteristic. This performance degradation is one problem encountered with the model and may originate from insufficient holding. Nevertheless the critical current is higher than the rated value. The no load voltage is also lower than expected (10 %). The magnet magnetisation was expected to increase more at low temperature and the end effects have been neglected.

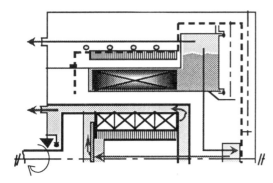

Fig. 1. Schematic cross section of the model.

Table I : Model characteristics.

Power (kW)	15
Speed (rpm)	750
Frequency (Hz)	50
Current (A_{rms})	93
Armature losses (W)	1.3
Dimensions (mm)	
External rotor radius	50
Mean armature radius	72
Magnetic shield ext. radius	88
Active length	200

The machine has successfully suffered a large number of tests as isolated generator and connected to the 50 Hz grid. In addition to the permanent operation transients have been applied in particular three phase sudden short circuits. The rated operation at 15 kW-36 kVA has been reached with a current of 105 A_{rms} and a voltage of 127 V_{rms}. The armature losses have not been measured precisely due to their low value. From calorimetric measurements we have nevertheless estimated the losses of the no load armature to 200 mW. This value is of same order of magnitude than the theoretical calculation (300 mW).

All the tests as motor have been performed at constant speed (750 rpm) fixed by the network frequency (50 Hz). As these machines are designed to be high performance drives it was indispensable to study the operation at variable speed using a variable frequency power supply.

3. Variable speed operation

The motor is then an autopiloted machine : the supply frequency is controlled by the speed itself. For the power supply a PWM (Pulse Width Modulation) voltage inverter is used. It is based on a rectifier followed by a transistor inverter feeding the three phases of the motor. Figure 2 gives the principle schema. It is very classical. The resolver gives the speed and position informations of the rotating part. These quantities with the current values enable to elaborate the orders for the inverter through a current and a speed loop. The inverter applies the three phase armature with voltage steps modulated at high frequency (5 kHz) but due to the armature self inductances the current is close to a sinusoidal wave (fig. 3a).

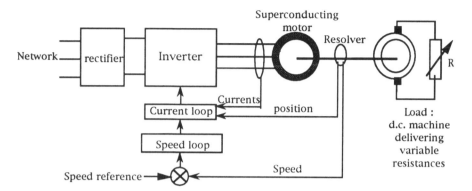

Fig. 2. Principle general schema for the operation at variable speed.

The unknown was the behaviour of the superconducting windings feed through such voltage waves. It is well known that superconducting wires exhibit losses when they are submitted to time varying currents and fields [5]. Even thought a large amount of works has been carried out to understand the loss mechanisms they are not very precisely calculated even under a sinusoidal supply. We feared to get very high losses and in consequence a large quench current degradation. The armature has been supplied with currents up to 85 A_{rms} (fig. 3b). Further quenches occur but the inverter saturates and the current waves become distorted (fig. 3b). The performances are degraded compared to 50 Hz currents (110 A_{rms}) but with a limited effect. If the inverter does not saturate this degradation could be perhaps lower. It has not been possible to measure the a.c. losses. The armature is continuously fed by liquid helium from an external tank. As the flow is relatively high to overcome the cryostat and transfer losses no variations of the flow is recorded when the three phases are supplied. The observations suggest that the a.c. losses are low (< 2 W) even with a inverter.

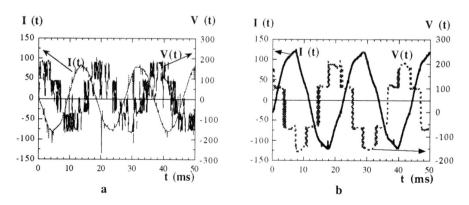

Fig. 3. Instantaneous phase currents and voltages (a : 55.5 A_{rms} b : 85 A_{rms}).

662

4. Variable speed control

The control of a superconducting motor was also experimentally unknown. These machines possess electrical characteristics rather different than conventional ones. In particular their electrical time constants are very high. First studies, based on numerical simulations, have nevertheless shown that the motor could be controlled using suitable adjustments for the current and speed loops. Unfortunately the power converter we used is an industrial one and it is very difficult to modify the loops. So the response to a speed step is bad (fig. 4) with a large excess (50 %) and a low damping of the oscillations. Moreover quenches have been recorded when speed steps were applied (but no explanation has been found for this).

A better adjustment of the correctors should enhance the transient performances. If, instead a step, a ramp is applied the behaviour is more satisfactory. The motor speed follows well the reference (fig. 5 : 50-500 rpm in 2 s) even when the motor is loaded.

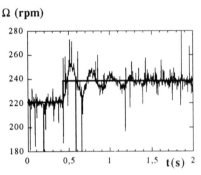

Fig. 4. Speed evolution for a speed step.

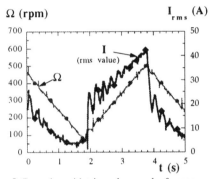

Fig. 5. Operation with triangular speed reference.

Conclusions

A three phase superconducting may be fed through a PWM inverter with a low degradation of the critical currents. The losses do not increase too much but we cannot put forwards figures for instance. The machine may be controlled in speed. This result is very important and promising for the development of superconducting motor as variable speed drives.

Acknowledgements

This study is supported by the DRET (Direction des Recherches et Etudes Techniques). The authors are pleased to thank Messrs Boulbes, Gillia, Leggerri, Pilon and Reynaud for their technical contribution to its work.

References

[1]   P. Dubots et Al., Journal de Physique 45, 467 (1984).
[2]   P. Tixador et Al., Proceedings of MT11, Elsevier, 621 (1990).
[3]   A. Lacaze et Al., IEEE Trans. on Magnetic 27, 2178 (1991).
[4]   P. Tixador, et Al., IEEE Trans. on Applied Superconductivity 3, 381 (1993).
[5]   A. Février, Cryogenics, 185 (1983).

Inst. Phys. Conf. Ser. No 148
Paper presented at Applied Superconductivity, Edinburgh, 3–6 July 1995
© *1995 IOP Publishing Ltd*

STABILITY OF SUPERCONDUCTIVE TURBOGENERATORS WITH SLOW-RESPONSE EXCITATION

I.A.Glebov and L.I.Chubraeva

Department of Non-Conventional Electrical Machines,
VNIIElectromash, 196084, St-Petersburg, Russia

Abstract

Superconductive turbogenerators are characterised by improved static stability as compared to conventional units. Their transient stability may be improved as well. It is mostly evident in case of small reactances of external electrical circuits. The combined-cycle power plants are being equipped with the generators of 200-250 MW unit rating. They are usually situated relatively close to the consumer thus providing a possibility of application of superconductive turbogenerators of slow-response type, which have relatively low manufacturing costs. The alternator, connected with the consumer by a relatively short transmission line will possess necessary level of transient stability. It will need no devices like fast excitation system, AVR or SMES with converter. Evaluations of transient and static stability of non-regulated superconductive units and the ones operating with SMES are being presented. The analysis is being carried out for the 220 MW superconductive machine, which is now under development in VNIIElectromash.

1. Introduction

At present the combined-cycle power plants are being developed, incorporating both gas and steam turbines. They are to be equipped with the generators up to 200-250 MW unit rating. In Russia they are mostly air and water cooled turboalternators.

The superconductive turbogenerators of similar ratings intended for the combined-cycle power plants are to have relatively low manufacturing costs. They are also to possess serious advantages, substantial for the utilities, to be competitive with conventional units. Therefore only superconductive machines with slow-response excitation are worth being discussed, because the high-response ones will be much more expensive. In the Institute there is being under development the 220 MW superconducting turbogenerator of slow-response type [1]. Its design is being based on the experience obtained with the 20 MVA alternator development and investigation [2]. Research of its operation in the network revealed new advantage: possibility of operation with deep under-excitation. Compensation of reactive power represents a serious problem for the utilities of Russia. The other advantages are to follow from the evaluation of the transient and static stability of slow-response type of alternators.

During its operation in the network the superconductive turbogenerator is being subjected to a number of heavy modes, associated with fast changes of the magnetic flux and excitation current density. The most heavy one which the supercon-ducting field winding is to withstand without transition to normal state is the sudden

3-phase short-circuit at the beginning of the transmission line. The mode is a basic one for the theoretical investigations of superconductive alternator transient stability.

2. Transient and static stability of non-regulated superconductive turboalternator

A number of investigations of transient stability problems, carried out in the Institute are based on comparison of theoretical results with those obtained on the electrodynamic model. The parameters of the model are described in [2].

Evaluation of the ultimate value of the turbogenerator rating as concerns transient stability behaviour is carried out in accordance with the area law (Fig.1), enabling to decide whether the alternator retains synchronous operation with the network or is falling out of step [3]. The ultimate value of the transient stability P_t corresponds to the equality of areas A and B.

The duration of oscillations of superconductive alternator rotor with slow-response excitation after the short-circuit depends on damping ability of the rotor circuits. Though protracted oscillations do not decrease the static stability after the first cycle, it is desirable to decrease the continuation of the oscillation process. The latter may be achieved by the increase of the electrical resistance of the rotor electromagnetic screen, thus increasing the penetration of AC magnetic flux in the field winding zone. This may result in a more sophisticated superconducting wire for the turbogenerator field winding.

The equality of areas A and B corresponds to the equation

$$P_t \Delta \delta_k = \int_{\delta_k}^{\pi - \delta_0'} P'_m \sin\delta \, d\delta - P_t(\pi - \delta'_0 - \delta_k), \qquad (1)$$

where P_t - ultimate rating, P'_m - maximal rating after the faulty chain is switched off. It may be further represented as:

$$P_t \Delta \delta_k = P'_m(\cos\delta'_0 + \cos\delta_k) - P_t(\pi - \delta'_0 - \delta_t). \qquad (2)$$

From Fig.1 it follows

$$\sin\delta'_0 = \frac{P_m}{P_{m'}} \sin\delta_0 \qquad (3)$$

with P_m - maximal rating of the initial mode.

Determination of generator current and EMF is carried out in accordance with vector diagrams. It allows to obtain by (2) the limiting transient stability rating, which equals for the 220 MW superconductive turbogenerator P_t=0.59 p.u. while the similar conventional machine is characterised by P_t=0.56 p.u. The higher level of dynamic stability of superconductive unit in spite of its lower mechanical inertia may be explained by fast decay of currents in the quadrature axis of the alternator.

The results of computation of the angle δ variation for the non-regulated alternator are presented in Fig.2 and show that the maximum rotor deviation happens at t=1.2 s. The other curves characterise changes of active P and reactive Q power and generator voltage U. The results refer to relatively long transmission line (540 km).

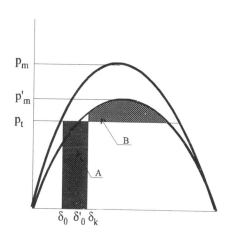
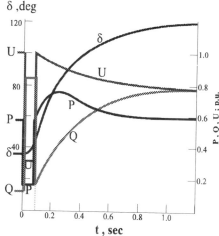

Fig.1. Determination of transient stability limit
by the area law

Fig.2. Transient process of the non-regulated
superconductive turbogenerator

The presented curves show that the increase of the angle δ is followed by substantial increase of the reactive power. The off-loading of the alternator from reactive power short-term changes is possible in case synchronous condensers or static VAR compensators are installed in the beginning of transmission line, thus helping to stabilise the voltage not only at steady-state but during transients as well.

As it was already mentioned, the superconductive alternator has low values of synchronous reactances. Their favourable influence on dynamic stability is revealed when the external circuits also have low reactances. A small rotor angle against the power system and constant transient EMF value are important as well. Analysis of results, obtained on electrodynamic model alongside with subsequent calculations show that the dynamic stability of superconducive unit rated 220-250 MW during the 3-phase sudden short-circuit at the beginning of transmission line is provided in case the length of transmission line is about 250-300 km (the short-circuit is being switched off via 0.14 s, which is a heavy mode for the transient stability). Application of modern protection devices and high-voltage circuit-breakers with SF_6 will decrease this period to 0.1 s and less and improve the dynamic stability, of course.

Therefore if the generator is to provide the necessary stability level and there is no special demand about the equivalency with conventional turboset behaviour in a complicated network, it may be applied at power plants, situated relatively close to the consumer. In this case no expensive additional equipment like fast-response excitation system and AVR will be demanded. Moreover in this case the alternator may have a brushless excitation system, based on a flux pump. The modification of the 220 MW superconductive turboalternator with a flux pump was developed in the Institute.

The static stability of a non-regulated superconductive unit is quite acceptable due to low values of synchronous reactances. The generator has no static stability limit in the under-excited mode, thus providing substantial advantages of superconductive unit as compared to conventional one. Comparison of static stability of superconductive and conventional alternators of 200-300 MW shows that in non-regulated modes the superconductive one possesses 42 % higher static stability limit

666

than a conventional one when operating via a single-chain line and 21 % - via a double-chain line. If the power plant equipped by superconductive units is connected directly to the power grid, the upper limit of static stability is so high (P_m=1.74 p.u.) that it is difficult to use it.

3. Transient stability of superconductive alternator operating with SMES

Preliminary investigations of the problem were presented in [4]. The generator is operating during transients in a non-regulated mode. After the short-circuit is being switched off, the SMES system absorbs the active power and provides additional rotor deceleration during the first deviation. The SMES system an absorb the energy during different time periods: the shorter is the time - the higher is the rating of SMES and converter. The latter is important mainly for the converter. To decrease the size of the stabilising system one has to increase the time of energy absorption. But this time period is not to exceed the time of rotor deviation by the angle π-δ_0' or the generator will fall out of step. The precise determination of the SMES energy must take into account the energy of the rotor deceleration

$$ W = (1 - \frac{B}{A})(P_t - P_E)t_{sc}. \tag{4} $$

In case of 220 MW superconductive unit operating via 540 km transmission line the SMES energy will be about 18 MJ.

One of the main advantages of application of the SMES/converter system as compared to the alternator with high-response excitation is the influence during transients directly upon the generator rating and voltage where as in the second case the influence is provided via an inertia member - superconducting excitation winding.

Conclusion

The presented analysis shows that alongside with the well-known advantages of superconductive turbogenerators some quite new may be discussed:

1. Possibility of stable operation of superconductive turbogenerators intended for the combined-cycle power plants via a relatively short transmission line without application of AVR and fast excitation systems.

2. Practically no limits of static stability and absorbed reactive power (up to the rated value) during under-excited operation.

3. Possibility of development of a built-in brushless exciter, based on a flux pump principle.

Acknowledgements

Research is being supported by the Ministry of Science and Technical Policy of Russia.

References

1. Glebov I.A., Chubraeva L.I. 1994. *RAS News. Power Engineering.* N5. 27--41 (In Russian)
2. Chubraeva L.I., Danilevich Ja.B., Glebov I.A., Mamikonjants L.G. 1994.*CIGRE .* Rep.11-103. SP-6
3. Glebov I.A., Chubraeva L.I. 1993.*EUCAS'93, Gottingen, Germany.* Rep. PDC-02. SP-4
4. KovacsK.P. and Racz I. 1959. *Budapest, Verlag der Ungarishen Academie.* 324--350

Inst. Phys. Conf. Ser. No 148
Paper presented at Applied Superconductivity, Edinburgh, 3–6 July 1995
© *1995 IOP Publishing Ltd*

A versatile and plain thermal equivalent Circuit for the Rotor of a superconducting synchronous Generator

H. Köfler

Institut für Elektromagnetische Energieumwandlung, Abteilung für Elektromaschinenbau, Technische Universität Graz, A-8010 Graz Kopernikusgasse 24, Österreich

Abstract. The paper describes a two mass thermal model of the complex structure of the superconducting rotor of a medium sized synchronous generator. The approach is based on an equivalent network. This circuit contains two masses and is used for monitoring of the generator when in service. Comparison to experimental data and performance data of the rotor calculated with a numerical code are given and show reasonable agreement with respect to the intended purpose.

1. Introduction

Prediction of thermal behaviour of the rotor of synchronous machines with superconducting field windings is of vital interest for all operational states of such machines. There are codes available producing information on Cooldown of the machine that is of primary interest and others for like quenching of the superconducting winding in part or totally with variable depth of accurateness. These calculation codes mainly have been constructed to assure mechanical and electrical integrity at these special operational modes. The elaborated cooldown and quench codes do not fit well monitoring regular operation of the synchronous machines. For this task proven methods in conventional electrical machinery are thermal images like that used to monitor asynchronous motors. The complexity of the images may be reduced to two bodies and two heating sources in the case of the superconducting rotor. Such a two body representation of the rotor gives a versatile and plain description of its thermal behaviour.

2. Theory

Mathematical treatment of electrical machines in general ends with at least three distinct differential equation sets. One set may describe the electrical behaviour of the machine, the other may describe mechanical behaviour and the last may describe the thermal behaviour. The differential equation for a thermal field in a cross section of an infinite arrangement (when length is not considered) is shown below.

$$\rho(\Theta) \cdot c(\Theta) \cdot A(x) \cdot \frac{\partial \Theta}{\partial x} = \frac{\partial}{\partial x}\left[\lambda(\Theta) \cdot A(x) \cdot \frac{\partial \Theta}{\partial x}\right] + S(x,t,\Theta) \cdot A(x) - \alpha(T,x) \cdot U(x) \cdot (\Theta - T) \quad (1)$$

A very common way getting insight into the electrical behaviour of a circuit operates with lumped parameter networks. The lumped parameters are equivalents for the windings and

668

their associated magnetic fields in case of inductances or the associated electric field in case of capacitances. A likely insight into the thermal behaviour of the machine is possible with equivalent networks consisting of lumped parameter elements defined by analogy from thermal constants and thermal field quantities.

Table 1 shows these parameters by conclusion from analogy:

Θ	temperature	°C, K	analogous to	u	electrical potential
$\Delta\Theta$	temperature difference	°C, K	analogous to	Δu	voltage
P	heat $=r\cdot\Theta_\infty$	W	analogous to	i	current
r	therm. resistance $= \tau_{therm}\cdot C$	°/W, K/W	analogous to	R	electrical resistance
Λ	thermal conductance	W/°, W/K	analogous to	R^{-1}	electrical conductance
C	thermal capacity $= m\cdot c$	kJ/°, kJ/K	analogous to	C	electrical capacitance

Table 1

In a direct manner one may reduce the elements of an electrical machine into their thermal equivalents. The build up of the thermal network includes the specific material constants at changing temperature. For this task specific heat of copper is given at three levels of temperature one higher than 60 K, one lower than 60 K and one lower than 20 K (see Fig. 1).

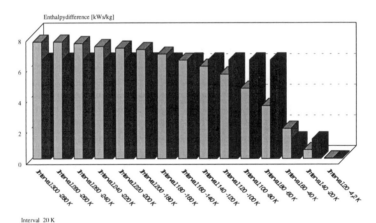

Figure 1: Step representation of enthalpy for use in the thermal network

Modelling of thermal resistances and of heat sources is sensitive too. In the generator heat does not flow as simple as in a uniform body. Looking at the drawing of the rotor of the superconducting generator one sees two main paths that supply most of the heat to the cold central structure of the rotor. One path is the current lead to ambient and the other is the structural link and the hull of the central body. Both paths in normal operation are cooled by gaseous counter flow of the central helium supply. By this heat is intercepted at locations in front of the central cold cavity. The central rotor itself is cooled by continuous supply of liquid helium. The thermal model will model this main part of cooling by an infinite thermal sink the rotor is linked to and the leads and struts as resistances. Heat to be used in the models unfortunately cannot be calculated straight from the design data of the rotor because the main thermal paths are paralleled by numerous side paths. The cross section of the current leads in the machine discussed here is 60 mm^2 over an averaged length of 900 mm. This gives a calculated heat flow to the low temperature region of roughly 10 W. The experimental check resulted in a higher heat value of 82.72W valid at low temperatures in the equivalent circuit. The same statement holds for the combined losses (620.3 W) that apply at the temperatures ranging from 60 to 300 Kelvin.

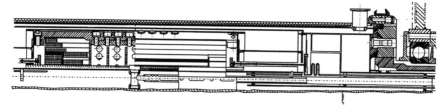

Figure 2 Drawing of the central rotor

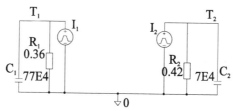

Figure 3 Example of Thermal network for the central rotor

The model shall describe different temperature ranges therefore thermal capacitances particularly and thermal resistances in the model change. Notice in figure 3 the equivalent network that shows two independent branches. One branch contains the very central rotor that is represented from the winding and its thermal capacitance at low temperature (30 to 60 K). The other branch puts together the masses of winding and structural material with the capacitances for the temperature ranging from 60 to 300 K. The very end in the operating temperature of the superconducting generator is dominated by vanishing heat capacities. Heat therefore causes dangerous temperature increase for the superconductors in case of a quench. The equivalent network (right part) can predict these temperatures as well when we use new values for the capacitance and the resistance ($C = 0.008 \cdot 10^4$, R = 0.0148).

3. Results

Experimental readings of various experiments on the thermal behaviour of the superconducting Rotor of SMG are available. An insight into the performance of the simple modelling is given in Figure 4. It shows the temperature of the field winding in the case of heating by self losses to ambient temperature. Specific thermal capacities of the materials involved are far from being constant but the modelling of this temperature course is possible with exponential increase of temperature at two different time constants. Check of the time constants evaluated from the experiment indicated that the proposed averaged heat capacities for the two main rotor components namely the winding and the stainless steel rotor body can be used in these time constants. In the very low temperature region only one time constant is evident. Figure 5 compares the equivalent network results with the calculation of an elaborated numeric code for the thermal behaviour of the rotor [3]. This code has been tested against forecast of quench heating temperatures in coil modules of the exciterwinding of SMG successfully. The current sources can be set to the loss values known from experiments with and without quenches in the commissioning phase of the generator. By this quench detection can be

670

assisted and quenches can be avoided when currents or coolant supply are redesigned according to the predictions of the thermal model. Regular monotoring is facilitated and en-

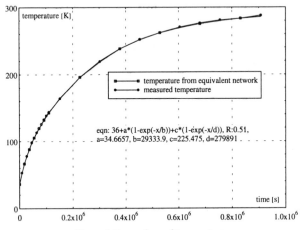

eqn: $36+a*(1-\exp(-x/b))+c*(1-\exp(-x/d))$, R:0.51, $a=34.6657$, $b=29333.9$, $c=225.475$, $d=279891$

Figure 4 Comparison of temperatures

larged from vital quench detection to quench prediction. Normal operation can be reestablished quicker and with less effort than without the prognosis values of the thermal network.

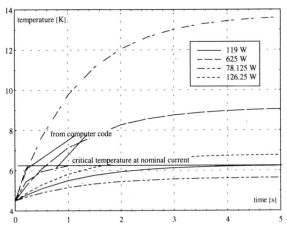

Figure 5 Comparison of quench heating

References

[1] H. Köfler et al, ELIN Zeitschrift **37**, 34, 1/2, 1985. SUSI und SMG ein Entwicklungsprogramm für Synchrongeneratoren mit supraleitender Erregerwicklung.

[2] H. Köfler et al, IEEE Trans. MAG **24**, 800, 1988. Thermal design and operation performance of the rotating cryostat of a superconducting generator.

[3] B. Bodner, Dissertation Technische Universität Graz, Österreich, 1987. Erwärmung und Wärmeabfunr im Rotor eines supraleitenden Synchrongenerators.

Inst. Phys. Conf. Ser. No 148
Paper presented at Applied Superconductivity, Edinburgh, 3–6 July 1995
© 1995 IOP Publishing Ltd

An Active Superconducting Magnetic Bearing

T.A. COOMBS and A.M. CAMPBELL

IRC in Superconductivity, Cambridge University, West Cambridge Site, Madingley Road, Cambridge, CB1 1LN, UK

Abstract. The potential use of $YBa_2Cu_3O_7$ as an active component in a magnetic bearing is being investigated. Measurements are being made of the load bearing capacity and related stiffnesses in comparison to predictions from the critical state model. Although the load bearing capacity is high and increases with the square of the trapped magnetic field the stiffness is low. We report on a novel design concept to overcome this problem.

1. Introduction

Two distinct roles have been identified for superconductors in the construction of magnetic bearings. The first is as a load bearing component the second is as a stabilising component. It has been shown that highly oriented large grain samples of YBCO can trap magnetic fields significantly larger than those which are available from conventional ferromagnetic materials [7]. In addition the ability of superconductors to sustain stable levitation [1] (without active control) has led to bearing designs in which the superconductor is used merely as a stabilising component [2] and permanent magnets provide the levitation force. Tests peformed in vacuo show that bearings constructed using superconductors in either role have an extremely low frictional coefficient[2,5].

There are four problems which must be addressed if superconductors are to be used in bearings. One is flux creep. The second is low stiffness. The third is hysteresis and the final one is the effect of vibrations. Flux creep is a function of the number density and effectiveness of the pinning centres. It is partly a materials problem and as fabrication methods improve then the problem of flux creep should be reduced. The problem of stiffness however is more fundamental. Commonly a bearing will have to bear dynamic loads which are two or even three times the magnitude of the steady state loads and will have to survive crash loads of 5 times the steady state loads. This will have to be achieved in situations (for example turbines) where displacements of 2 or 3 millimetres are unacceptable. There may be some situations in which low or even zero stiffness is desirable [3] where for instance the bearing is acting as a vibration isolator but these are unusual. A bearing is more usually designed to provide one degree of freedom (e.g. rotational) and restraint in one (thrust) or two (journal) axes. One which provided an additional degree of freedom, especially along the loading axis, would generally be undesirable.

Hysteresis will cause the bearing to take up a different position (at the same loading) as the bearing is loaded and unloaded. Moreover the stiffness of the bearing is also dependent on the loading history. Both these problems can be minimised by limiting the dynamic loading so that only minor hysteresis loops are circumscribed but this limits the range of applications to ones in which the loading is constant or nearly constant (such as supporting flywheels).

We have recently identified a fourth possible obstacle to the use of superconductors in magnetic bearings: the effect of vibrations. The effect of vibrations is to induce a cyclic loading on the bearing. Under the influence of cyclic loading a bearing may, under certain circumstances, even collapse as will be shown below. We suggest that the physical

672

mechanism is a random walk mechanism in which repeated pitching of the bearing causes the magnet to advance irreversibly towards the superconductor. This is well described by Brandt[1] who has shown the existence of a range of stable levitation positions at a single given force for the configuration described here. The effect of vibrations is to advance through these positions and alter the bearing gap. Figure 1 shows a typical force displacement curve. the superconductor is initially levitated at a position corresponding to point A. When load is increased the superconductor moves to point B. When the load is again relaxed however the superconductor does not return to point A but instead moves to point C which is at a somewhat smaller gap than represented by point A. If the load were then increased back to that corresponding to point B then the superconductor would simply oscillate between points B and C and the bearing gap would not decrease any further. However the effect of vibrations is to impart

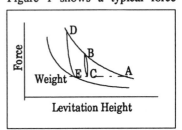

Figure 1 - Typical force displacement curve

energy into the system and the amplitude of the vibrations will build up to a value which is dictated by the dynamic magnifier (which is a function of the stimulating frequency over the resonant frequency and the damping). On the second loop therefore the bearing moves to some point past C and the bearing gap decreases. This continues until the vibrations have built up to a level dictated by the dynamic magnifier.

A superconducting bearing may be represented by a spring damper system in which under small oscillations there is very high Q (low damping). This system has a natural frequency. When the bearing is rotated the rate at which energy will be input into the system will be a function of the ratio of frequency of the rotation relative to natural frequency of the system. The dynamic magnification of the output relative to the input is given by the following equation

$$\Theta_o = \frac{\Theta_i}{[(1 - \omega^2 / \omega_n^2)^2 + 4c^2 \omega^2 / \omega_n^2]^{1/2}}$$

Thus the magnitude of the response depends on c (the damping factor). If c is zero then when $\omega = \omega_n$ the response is infinite. In general when c is small the dynamic magnification increases as ω increases from zero, slowly at first but then very rapidly as ω aprroaches the natural frequency. Beyond the natural frequency the response initially falls off rapidly and then tends to zero as ω tends to infinity. Vibrations will

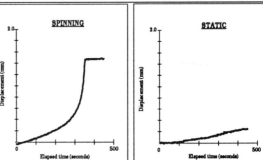

Figure 2 - Decay profiles (without active control) - Spinning case in which vibrations are being induced due to imbalances in the rotor and the decay is largely due to the vibrations and is large enough to cause collapse of the bearing in the time frame of the measurement- Static case in which the decay is due to flux creep.

build up to a steady state value dictated by the dynamic magnifier. During this time the sample will move, since its equilibrium position (the stable levitation height) depends on the maximum travel, and the bearing gap reduces. Once the steady state value has been reached, then the gap will not change any more. As is shown below the gap may actually vanish

If the rate at which energy is put into the system exceeds the damping then the system will tend to advance to some stable point E (figure 1). In the presence of strong enough pinning the magnetisation at E may become negative[1] and the force between magnet and superconductor become attractive. There is no stable levitation position and the bearing will collapse. We have recently observed this effect. Figure 2 shows the decay in the bearing gap which is observed due to vibrations and over the same time scale due to flux creep.

Active control therefore has three functions one is to enable the bearing gap to be set and stabilised at some convenient value. The second is to introduce damping into the system. Finally the third function would be enable the system to work at some point C (Figure 1) while protecting the superconductors from a single large excursion by enabling them to be withdrawn. This enables a catcher bearing to come into play preventing the gap being closed to a level at which the magnetisation becomes negative.

2. Method

In order to provide active control a method must be devised whereby active elements may be included in the system which will respond to changes in load. In an electromagnetic bearing active control is achieved by changing the current in the coils and hence changing the magnetic flux density. In bulk materials where the magnetic field is produced by persistent currents trapped in the superconductor this option is not available. However the available force is a function of the flux density and the area over which it is applied. Thus even if the flux density remains constant then changes in available force may be made by changing the active area of the bearing i.e. increasing the area of overlap between the magnetic components.

Using this method active control may be achieved using bulk materials, low stiffness is no longer a problem and neither are vibrations as the damping in the system may be adjusted in the control loop. In order to determine the characteristics of a system using this method a model has been developed in which the superconductors are replaced by solenoids of equivalent size. Thus as the area of overlap between the magnet and the superconductor increases the radius of the solenoid also increases.

We have developed a prototype actively controlled superconducting bearing which has been reported on in [4] and uses superconductors[6] supplied by Haldor Topsoe . This bearing has enabled the basic principles of an actively controlled superconducting bearing to be investigated. Tests carried out have included both pseudo-static loading (force displacement

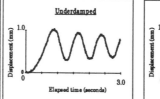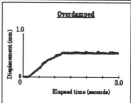

Figure 3 - Response of Actively controlled system to Step input - Underdamped case demonstrates the inherent damping in the system (i.e. that from the superconductor) - Overdamped case shows the response when c has been increased to > 1 using the control loop.

curves)[4] during which the bearing was spun at up to 580 rpm and step tests to examine the dynamic response of the system (figure 3).

In order to further assess the viability of the active controlled method a new test rig is being designed for construction in the IRC. This rig is designed to enable rotation at speeds of greater than 10,000 rpm. A schematic of the planned rig is given in figure 4. The rig is designed to test a superconducting thrust bearing formed by a neodynium iron boride magnet (the thrust plate of the bearing) and four sections of YBCO superconductor. The following are features of this design:

674

1) The shaft itself is supported independently of the superconducting bearing. This simplifies dynamic balancing of the shaft and the design does not need to provide for catcher bearings in the event of the failure of the superconducting bearing (due to loss of liquid N_2 for example).

2) The rotational losses in the bearing may be evaluated by measuring the sideways load on the superconductor rather than having to perform spin down tests which would require an evacuated chamber

3) The bearing gap may be controlled precisely. This will be of significant advantage when measuring the effects of vibration, for example. Vibrations may be simulated by applying a varying current to the electromagnetic loading coils and the variation of load bearing capacity at constant bearing gap investigated.

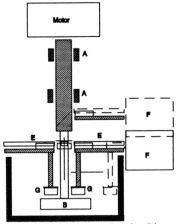

Figure 4 - Advanced Bearing Rig

3. Conclusions

In this paper we have described the factors which influence the design of a superconducting bearing. These are flux creep, low stiffness, hysteresis and vibrations. We have described a method for adding active elements into the system which will help mitigate the deleterious effects of these factors. The method described has three clear potential advantages over a conventional electromagnetic bearing. The first is that the specific loading capacity is higher if superconductors are available which trap fields in excess of the saturation fux of iron. An electromagnetic bearing is limited by the saturation flux of iron (c. 1.5 Tesla). YBCO has already been shown to exceed this figure (e.g. 2.3 Tesla at 77K [1]). The second is that the actuation forces required are lower [4] and the third is that the system is inherently self levelling [4].

A simple demonstrator has been built using the active control method and results obtained showing a PID controller maintaining a constant bearing gap [4] and the response of the system to a step demand. Stable control has been achieved in this rig at speeds of up to 580 rpm (the maximum driving speed). A further rig is under construction which will enable the behaviour of the method at speeds of up to 10,000 rpm and at a range of vibrational frequencies and amplitudes to be examined. Further the rotational losses in the bearing may be measured.

Acknowledgements
Rolls Royce PLC,.J.G. Larsen, Haldor Topsoe Denmark, Sam Brown & Chris Clementsen IRC Cambridge
References
[1] Brandt E. H. 1988 *Applied Physics Letters* 53 No. 16 1554-1556
[2] Chen Q.Y. et al 1994 *Applied Superconductivity* 2 No. 17 457-464
[3] Chu W.K. et al.,"Interaction between high temperature superconductors and permanent magnets and its application",1994 *International Workshop on Superconductivity*, Kyoto, Japan
[4] Coombs T.A et al"An Active Superconducting Magnetic Bearing" 1994 Submitted to the *Proceedings of the Applied Superconductivity Conference*, Boston.
[5] Hull J.R.. et al 1994 *Applied Superconductivity* 2 No. 17 449-455
[6] Larsen J.G. et al 1993 *Physica C* 217 135-145
[7] Weinstein R & Sawh R "An experimental Generator using high Temperature Quasi-Pernament Magnets" 1994 Submitted to the *Proceedings of the Applied Superconductivity Conference*, Boston.

Inst. Phys. Conf. Ser. No 148
Paper presented at Applied Superconductivity, Edinburgh, 3–6 July 1995
© 1995 IOP Publishing Ltd

Magnet Configurations for Superconducting Bearings

P. Tixador, P. Hiebel, E. Hotier
CNRS - CRTBT/LEG, BP 166, F 38042 Grenoble Cedex 09, France.
X. Chaud, R. Tournier, E. Beaugnon
CNRS - EPM/MATFORMAG, BP 166, F 38042 Grenoble Cedex 09, France.

Abstract - Combined with permanent magnets YBaCuO bulk pellets exhibit large electromagnetic forces. These interactions happen to be fully stable due to the specific properties of superconductors. Both materials play an important part. The superconducting pellets should exhibit high critical current densities and its grain structure should be adapted to the magnets. Those are indeed very important for the magnetization of the pellets. The field gradient produced by the magnets is also a term of importance. With the same YBaCuO sample and the same magnet volume the force may vary in a ratio 1:3 and the stiffnesses in a ratio 1:11. This latter is an important parameter when considering stability for example. Different magnet configurations are reported in terms of vertical forces and stiffnesses. A premagnetization has been investigated.

1. Introduction

The superconducting bearings are a promising application of high T_c materials. These are the only materials available to design entire passive stable and efficient magnetic suspensions [1,2]. YBaCuO melt textured materials are the best candidate for an operation at "high" temperature (> 60 K). Whereas YBaCuO faces severe processing difficulties for long length wires, it may be elaborated under the form of bulk pellets of several cm^3 developing high magnetizations [3]. The sample we use are elaborated from $YBa_2Cu_3O_\delta$ (from Rhone Poulenc) with additions of Ag_2O, Y_2O_3 and PtO_2. In order to get a good alignment a magnet field is applied during the melt texturing since the c-axis of the crystal aligns with the magnet field.

2. Magnetic interactions, magnetisation

Combined with permanent magnets superconducting samples exhibit an electromagnetic force which may be expressed in the direction u as :

$$F_u = \iiint_{\vartheta_{sc}} \mu_0\, m(H)\, \frac{dH}{du}\, d\vartheta \approx \mu_0\, M(H)\, \frac{dH}{du}\, V_{sc}^{active} \qquad k_u = -\left(\frac{\partial F_u}{\partial u}\right)_{operating\ point}$$

$$\begin{cases} H : \text{permanent magnet field} \\ V_{sc}^{active} : \text{superconducting magnetised volume} \\ m, M : \text{local and averaged superconductor magnetisation} \end{cases}$$

The force is an important parameter but the stiffness k_u is a term of importance for stability. With the definition above it must be positive in order to get a stable operating point. The stiffness may be different from the slope of the curve force versus distance.

Submitted to external field variations, macroscopic currents are induced inside a superconducting sample to oppose themselves to the field variations in relation to the Lenz law. These currents whose current density is the critical one (Bean model) flow essentially inside in the ab planes of the grains. The critical current along the c axis is much lower than along the ab planes and the grain boundaries act as barriers to the currents. Induced by the magnetization field, the currents tend to reproduce it if they may flow freely in the superconducting sample.

This work is supported by a Brite Euram II Contract N° BR2-CT92-0274.

This "impression" of the magnetisation field is clearly illustrated by figure 1 showing the trapped axial flux mapping of a YBaCuO pellet magnetised by a coil around (fig. 1a) and by an alternating polarity magnet set (fig. 1b). These flux mappings are performed by moving a hall probe half a millimetre above the superconducting sample after its magnetization.

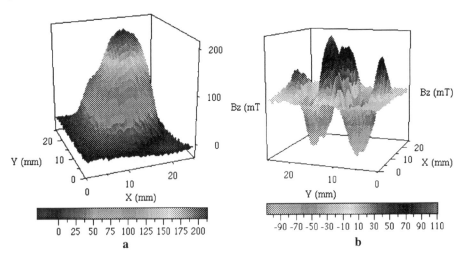

Fig. 1. Remnant field of the same YBaCuO pellet in different magnetization fields.

So in fully penetration conditions the magnetisation in Am^{-1} may be approximately expressed by :

$$M \approx \alpha\, J_c\, \lambda \qquad \begin{cases} \alpha : \text{geometric factor} \\ J_c : \text{critical current density} \\ \lambda : \text{characteristic dimension of the current loops} \end{cases}$$

α depends in particular on the demagnetization field then on the height where the currents flow inside the sample. λ is limited by the grain size otherwise it is fixed by the magnetisation field. For the fig. 1a λ equals the radius of the mono grain sample but λ yields to the pole pitch for fig. 1b. Likewise the penetration field (H_p) is proportional to the product of λ by the critical current density. To develop its maximum magnetization the superconducting sample should be fully penetrated ($H \geq H_p$).

The magnetisation depends little on the grain size if this latter is greater than the dimension of the current loops. Figure 2 shows the remnant field induced by the same alternating polarity magnet set (pole pitch : 5 mm) for two YBaCuO pellets with different grain sizes, 10 mm and 25 mm. The flux mapping shapes are identical whereas the grain structure is different. The largest grain sample exhibits nevertheless a higher remnant field. The critical current densities may be different but the grains are also certainly different inside the sample. If the grains are less deep, the demagnetization field increases and the magnetization is reduced (term α). The flux mapping gives informations above all about the grain structure at the surface.

If the superconductor magnetization is essential for the force, the permanent magnet structure plays a very important part too. It determines not only the field gradient but also in part the superconducting magnetization with competitive influences. To magnetize the whole sample the field must penetrate as much as possible inside the sample. It is contradictory with high field gradients especially for the vertical force.

We have carried out an experimental work to study the magnet configurations. A test bench enables to measure vertical, transverse forces and the trapped flux mapping. It consists essentially in a (2+1) dimension motorised table controlled by a microcomputer which also performs the data acquisition. The superconducting sample is immersed in a liquid nitrogen vessel.

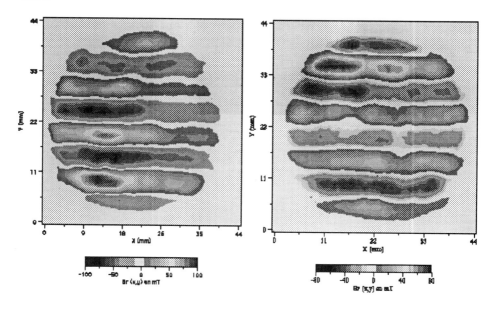

Fig. 2. Remnant field of two YBaCuO pellets 2a : $\emptyset_{grain} \approx 25$ mm, 2b : $\emptyset_{grain} \approx 10$ mm.

Different magnet structures have been investigated. By adding bar magnets of same dimension (5 mm x 15 mm x 50 mm) the different structures are compared in terms of maximum vertical force and stiffness (table 1). The two last structures in table 1 include non magnetic shims between some magnets. The magnet structure shows then a great importance. The suppression of some magnets may even increases the force. Those results show well the competitive effects between a high magnet field gradient and a large magnetized volume. Lower is the pole pitch higher is the field gradient but lower is the magnetization and the magnetized volume since the field decreases rapidly inside the pellet. Similar results have been recorded [4].

For the structure with all the magnets in the same direction, the magnet field may be insufficient to penetrate the sample as the penetration field increases with the size λ of the current loops. This point is certainly a limitation for the levitation force. For large grains with high J_c the magnets become insufficient to magnetize correctly the sample and they limit the force.

Two magnets in repulsion exhibit a high field and gradient. To study and compare such a configuration we have used magnets of square cross section (5 mm x 5 mm) and 25 mm in length. With the same YBaCuO pellet ($\emptyset = 22$ mm) different associations of those bar magnets have been measured (table 2). The last structure is called RMD (Rotating Magnetization Direction [5]). It concentrates the field on only one side and creates so a high field over a larger volume compared to the other structures which are symmetrical. This is the

678

reason why the RMD is the most efficient in terms of maximum force. On the other hand the highest stiffness is achieved by the magnets in opposition.

To increase the active superconducting volume, we have pre-magnetized a YBaCuO pellet with a coil around. The results are gathered in figure 3 and table 3. The pre-magnetization increases by 30 % the maximum force if the effects are adding. In the other case, the performance are on the contrary lowered (factor 2). This enhancement of the performances may be linked to the measured increase (20 %) of the remnant field above a sample magnetized by a coil instead by a permanent magnet. Only the upper part of the superconductor plays a part. Moreover the pellets are not homogeneous in the height. The upper part is the most efficient [6].

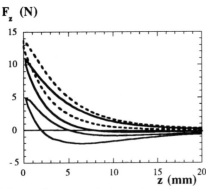

Fig. 3. Influence of a premagnetization on the vertical force cycle.

Table 3
Pre-magnetization influence

Premagnetization	Force (N)	Stiffness (N/mm)
No	10.8	4.4
One direction	13.7	4.5
Opposite direction	4.8	4.7

Table 1 Magnet structure influence
Alternating polarity sets

Magnets 5 mm x 15 mm x 50 mm, $\varnothing_{SC} \approx 40$ mm

Structure	F_z^{max} (N)	k_z^{max} (N/mm)
↑ ↑ ↑ ↑ ↑ ↑	34.3	20
↑ ↑ ↑ ↓ ↓ ↓	53.0	43
↑ ↑ ↓ ↓ ↑ ↑	46.3	34
↑ ↓ ↑ ↓ ↑ ↓	37.1	32
↑ | ↓ | ↑ | ↓ | ↑	43.1	28
↑ | ↑ | ↑ | ↑	25.9	14

Table 2 Magnet structure influence

Magnets 5 mm x 5 mm x 25 mm, $\varnothing_{SC} \approx 22$ mm

Structure	F_z^{max} (N)	k_z^{max} (N/mm)
↑ ↑ ↑ ↑	4.9	1.3
↑ ↑ ↓ ↓	8.7	7.3
↑ ↓ ↑ ↓	9.6	11.3
→ ← → ←	10.4	15
↑ → ↓ ←	12.9	11.2

References

[1]   B.R. Weinberger et Al., Superconductor Science Technology **3**, 381 (1990).

[2]   Bornemann et Al., Applied Superconductivity, **2** (5), 315 (1994).

[3]   P. De Rango et Al., Proceeding of Applied Superconductivity, Göttingen, Oct. 1993, 305.

[4]   M. Komori et Al., J. Appl. Phys. **69** (10), 7306 (1991)

[5]   J.P. Yonnet et Al., Journal of Applied Physics **70**, 6633 (1991).

[6]   P. Hiebel et Al., Cryogenics, vol. 34, ICEC Supplement, 843 (1994).

Inst. Phys. Conf. Ser. No 148
Paper presented at Applied Superconductivity, Edinburgh, 3–6 July 1995
© *1995 IOP Publishing Ltd*

Directional solidification of YBCO cylinders for magnetic levitation applications

J. Mora, X. Granados, V. Gomis, M. Carrera, F. Sandiumenge, S. Piñol, J. Fontcuberta and X. Obradors

Institut de Ciencia de Materials de Barcelona (C.S.I.C.), Campus de la U.A.B., 08193 Bellaterra, Spain

Abstract: $YBa_2Cu_3O_7$ cylinders with 10-20% of Y_2BaCuO_5 and 1% CeO_2 additions have been directionally solidified under a temperature gradient. It is shown that a steady growth regime of domains, typically 1 cm in diameter, is established after polynucleation at the bottom of the cylinders on the substrate interface. The improvement of the crystallinity of these domains is detected by rocking-curve measurements which give $\Delta\omega \approx 0.5$-$0.7°$ far away from the interface. The length of the region where a steady growth proceeds is limited by the liquid loss, which induces an enrichment in unreacted Y_2BaCuO_5 in the upper part of the cylinder, resulting in a polycrystalline structure. The vertical and lateral magnetic levitation forces have been measured and a direct correlation with the size of the domains has been found. We give also evidence for asymmetric lateral force hysteresis loops which reflect the lack of true cylindrical symmetry of the samples.

1. Introduction

Magnetic levitation applications based in YBCO bulk superconducting materials can be nowdays developed because melt processing techniques allow to fabricate big tiles with high critical currents [1-3]. The growth process of these big tiles (2-3 cm in diameter) is, however complex and several techniques, such as top seeding, have been used to control the nucleation processes [4,5]. Directional solidification under a small temperature gradient is the simplest methology to control the growth process and it has been indeed used successfully in the growth of long YBCO bars with small cross sections ($\phi \approx 8mm$) [6,7]. The extension of this methodology to bigger cross sections is not without difficulties and so a full study of the process methodology, in connection with microstructural observations and levitation force measurements is needed. In this work we present preliminary results concerning the growth process of cylinders having 20 mm in diameter, as well as characterize their texture and magnetic levitation force.

2. Directional solidification

Presintered [8] cylindrical pellets with typical height of 30 mm and diameter of 20 mm, and compositions: $A:YBa_2Cu_3O_7$ (123) + 10 wt% Y_2BaCuO_5 (211), B: 123 + 15% 211, and C: 123+20% 211 (powders supplied by SSC [9]), are slowly cooled at 1°C/h from a maximum temperature of 1060°C in a temperature gradient of 10°C/cm. All samples contain 1wt% CeO_2 addition because it allows to optimize the melt viscosity and to refine the resultant 211 precipitate size [6,7]. In this work, only samples nucleated at the bottom are investigated. The alumina substrates are covered by a CeO_2 powder in order to avoid reaction with the ceramic block.

From microstructural observations, using polarized light and scanning electron microscopy, we can conclude that three regions characterize these cylinders (Fig. 1a). In region I above the interface with the substrate, a multinucleaction process occurs. In region II a few domains remain from a growth competition stage (typically 2-4). No major porosity was observed in the samples and the 211 precipitates were distributed homogeneously in this initial region of the crystallization (about 10 mm in height). However, when advancing vertically along the cylinder, a progressive increase of the concentration of 211 phase is observed which perturbs the steady growth of the big domains and, finally, originates new nucleation phenomena (region III). We attribute this perturbation of the growth process to the liquid loss at the interface with the substrate which then impedes the completion of the peritectic reaction at the end of the directional solidification process. It's clear then that only the central 10-20 mm of the cylinder, with a height of 30 mm, can be used to investigate the levitation force of these materials.

It is interesting to note that the inductive critical currents of small pieces (few mm) extracted from single domains have been found to be very similar to those previously published on samples with similar concentration of 211 particles (up to 10^5 A/cm^2 at 77K and zero field) [10].

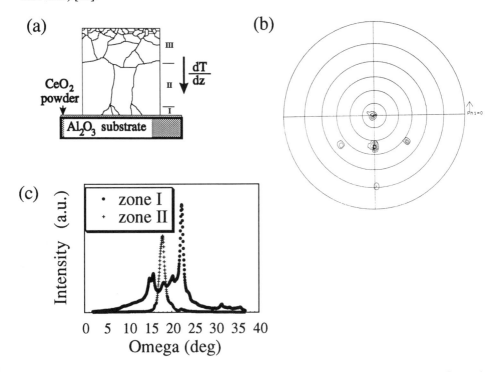

Figure 1: (a) Schematics of the experimental setup and grain growth of YBCO cylinders, (b) Pole-figure of the (005) reflection of the sample containing 15% 211 + 1% CeO$_2$ additions corresponding to regions II, and (c) Rocking-curves of the (005) reflection measured across the central peak in the pole-figure corresponding to zones I and II.

3. Texture

In order to track the evolution of the texture with increasing distance from the base of the cylinders, sample B, giving the higher levitation force, was sliced perpendicular to the axis of the cylinder and examined by x-ray diffraction using the pole-figure and rocking-curve techniques. The (005) reflection was used. In all experiments the whole surface of the sample was illuminated. It was observed that while the positions of the prominent poles are preserved, their widths are narrowed along the height of the samples. Figure 1b shows the pole-figure corresponding to zone II, displaying two prominent peaks, one aligned with the axis of the cylinder and the other 37 degrees apart. Subsidiary peaks appear which tend to align at about 45 degrees from the axis of the cylinder (iso-intensity lines plotted in a square-root scale in order to emphasize the low intensity details). The rocking curves shown in Fig. 1c illustrate the narrowing of the misorientational spread within the grain oriented with c parallel to the axis of the cylinder. The same observation holds for the remaining grains. This results indicate that in the first stage of the nucleation process (zone I), the grains posses a wide orientational spread which reflect the formation of a large number of small angle grain boundaries within each domain. While the number of large grains is essentially maintained, there is a competition between slightly misoriented growth directions within each grain, probably favoured by the axial temperature gradient of the furnace, which finally results in a narrow mosaic distribution of about 0.6 degrees within the independent grains (zone II). Also worth mentioning is the observation of a tendency of the grains to align with their c-axis inclined about 45 degrees to the axis of the cylinder , i.e., with the [103] axis along the temperature gradient. We note that this crystallographic axis corresponds to the diagonal of the pseudocubic perovskite and thus it's very likely that it constitutes a preferred crystallization axis. This crystal orientation has been widely observed by several groups [1,2,6,7] and results from a compromise between an efficient evacuation of the latent heat of crystallization and a fast growth rate. Finally, pole-figures recorded at the top surface of the cylinders indicated no texture. As we mentioned above, inspection by light microscopy indicated an enrichment in 211 phase and nearly polycrystalline material above 2/3 the height of the cylinder, which results from accumulation of unreacted material because of the liquid loss at the interface with the substrate.

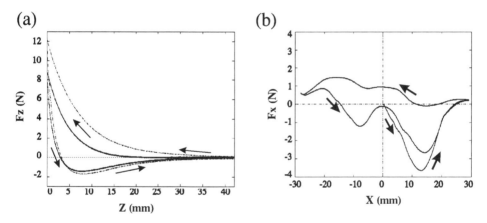

Figure 2: (a) Comparison of the first cicle of the vertical force with the second cicle, and (b) Field cooled (FC) measurement of the transversal force with transversal displacement.

4. Levitation force

The integral vertical levitation force and the transversal force were measured for different samples by using a triaxial force sensor based on strain gauges and having a computer controlled movement along two perpendicular axis. All the materials were tested with the same NdFeB permanent magnet having $\phi=25$ mm, h=20 mm and $B_r=0.3$T. The relative displacement of the superconductor and the magnet was done quasistatically (1.2 mm/s). A typical levitation force loop obtained, after a zero field cooling procedure, is displayed in Fig. 2a. In the figure we also show the hysteresis loop obtained after the first cycle, where it is evidenced that the trapped flux decreases considerably the maximum vertical force. The higher vertical levitation force was observed in the B sample, in agreement with the observation of an smaller number of domains in the x-ray pole-figures. The vertical levitation force of a permanent magnet over a superconductor is the z-component of the vector:

682

$$\vec{F} = \int_V (\vec{M} \cdot \vec{\nabla}) \cdot \vec{B}_{ex}\, dV$$

where B_{ex} is the magnetic induction generated by the magnet and M is the magnetization of the superconductor. If large angle grain boundaries exist among the single-domains, flux penetration occurs and the critical state is established only within the single domains. In this way M_{irr} and F_z are reduced even if Jc remains constant within the domains. This behavior has nicely illustrated by Yamamoto et al [11] by decreasing artificially the effective domain size in bulk superconductors.

Transverse force measurements were also obtained when a lateral displacement was introduced after a field cooling process. A typical hysteresis loop is displayed in Fig. 2b where the magnet stays at 1.5 mm above the superconductor B (15% 211). The stiffness constant corresponding to the linear recuperation force observed near the origin was found to be 300 N/m , which is similar to that observed by other authors in similar geometries [12]. We note, however, that a clear asymmetry is observed in the transversal force hysteresis loop and we found that this asymmetry changes when the direction of the radial displacement within the basal plane of the cylinders is modified. It's clear that this anomalous behavior reflects the absence of cylindrical symmetry associated to the two prominent domains observed in the x-ray texture analysis. We suggest that transversal force hysteresis loops can be used as a sensitive tool to detect the existence of a multidomain structure without uniaxial symmetry.

The absence of uniaxial symmetry in the samples has also been assesed by means of remanence magnetic field profiling with Hall microprobes. Several linear scans were performed along the diameter after field cooling under 0.3T, giving evidence for the existence of large angle grain boundaries.

In conclusion, we have shown that directional solidification long cylinders is a promising area for developing big tiles for magnetic levitation application. However further work must be carried out to avoid the multinucleation process which induces the formation of several domains and reduces the levitation force. Probably a combination of seeding and directional solidification could lead to a strong improvement of the final superconducting characteristics.

Acknowledgements: We are grateful to: CICYT (MAT91-0742), Programa MIDAS (93-2331) and EC-EURAM (BRE2CT94-1011).

References

[1] "Melt processed high temperature superconductors". Ed. M. Murakami. World Scientific Publ. (1992)
[2] K. Salama, V. Selvamanicham and D. F. Lee, Chapter 5 in "Processing and properties of high Tc superconductors". Vol. 1 Bulk Materials. p. 155-211, Ed. Sungho Jin. World Scientific. Singapore (1993)
[3] P. McGuinn, Chapter 8 in 'High temperature superconducting materials science and engineering". p. 345-382. Ed. Donglu Shi. Elsevier Science Ltd. (1995)
[4] L. Gao et al. Appl. Phys. Lett. **64**, 520 (1994)
[5] M. Morita et al in "Advances in Superconductivity III", p.733. Ed. K. Kajimura and H. Hayakawa. Springer (1991)
[6] S. Piñol et al , Appl. Phys. Lett. **65**, 1448 (1994)
[7] S. Piñol et al , IEEE Trans. on Appl. Supercond. (in press); N. Vilalta et al (to be published)
[8] F. Frangi et al, Supercon. Science Technol. **7**, 891 (1994)
[9] Seattle Speciality Ceramics Inc.
[10] V. Gomis et al, in ref. [4] p.373; B. Martinez et al, Phys. Rev. B (in press)
[11] K. Yamamoto et al, Appl. Superconductivity **2**, 487 (1995)
[12] H. J. Bornemann et al, Appl. Superconductivity **2**, 439 (1995)

Inst. Phys. Conf. Ser. No 148
Paper presented at Applied Superconductivity, Edinburgh, 3–6 July 1995
© *1995 IOP Publishing Ltd*

Properties of a Superconducting Magnetic Bearing

P. Stoye[1], G. Fuchs[1], W. Gawalek[2], P. Görnert[2], A. Gladun[3]

[1] Institut für Festkörper- und Werkstofforschung Dresden, D-01171 Dresden, Germany
[2] Institut für Physikalische Hochtechnologie, D-07743 Jena, Germany
[3] Technische Universität Dresden, D-01062 Dresden, Germany

Abstract. Static levitation forces and stiffnesses in a superconducting bearing consisting of concentric ring magnets between two superconducting YBaCuO rings are investigated. In the field cooled state a levitation force of 40 N has been achieved. The axial and radial stiffnesses at the working position have values of 15 N/mm and 10 N/mm, respectively. An arrangement with two bearings supporting a high speed shaft is now under development.

1. Introduction

Passive superconducting magnetic bearings (SMB) using permanent magnets and high temperature superconductors are one of the most promising applications of high temperature superconductors (HTSC). Fields for possible application of SMB are flywheels for energy storage with a large diameter rotor [1], motors of small size [2] for different applications, e.g. liquid gas pumps, and linear transport systems.
A model of a SMB is under construction, where we use melt-textured YBaCuO bulk material produced by a technology reported elsewhere [3]. In order to determine an optimum geometry of the bearing configuration for the motor, levitation forces and stiffnesses in several arrangements of superconducting and permanent magnet rings have been measured.

2. Experimental

The rotor of the SMB (Fig. 1) is built up of two concentric NdFeB ring magnets ($\varnothing$ 60 x $\varnothing$ 43 x 8 mm^3 and $\varnothing$ 39 x $\varnothing$ 26 x 8 mm^3) connected with the shaft. They are placed between two superconducting rings ($\varnothing$ 70 x $\varnothing$ 20 x 10 mm^3), each of them consisting of 6 pellets of YBCO, as shown in Fig. 2. YBCO pellets with nearly the same levitation force values were selected in order to achieve a good rotational symmetry of the magnetization.
Using a strain gauge force sensor and a 3-D translation stage system [5], the levitation force and stiffness between the permanent magnet rings and the superconductor (cooled by liquid nitrogen) as a function of relative position have been measured. The zero field cooled levitation force between one YBCO pellet (30 mm x 10 mm) and a SmCo magnet (25 mm x 18 mm) at a distance of 0.5 mm has a value of 20 N with a maximum deviation of 0.8 N between different pellets [6]. Fig. 3 shows the vertical component of the trapped field in a YBCO ring scanned by a Hall probe at a distance of 1.5 mm from the surface. In the distribution of B_z there are still differences of about 20 % between the segments, mainly because the magnetization peaks within each segment are not located at its centre.

684

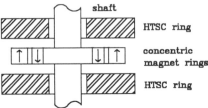

shaft

HTSC ring

concentric
magnet rings

HTSC ring

Fig. 1. Schematic arrangement of the SMB

Fig. 2. Geometry of the YBaCuO ring
($\varnothing$ 70 x $\varnothing$ 20 x 10 mm^3)

3. Results and discussion

The axial levitation forces between superconductor and different magnet configurations have been measured in the zero field cooled state as shown in Fig. 4. During the approach of the magnet to the superconductor the gradient of the magnetic field between HTSC and magnet becomes steeper, thereby inducing an increase of the repulsive force. When the magnet is removed from the superconductor, the force becomes attractive, depending on the amount of the trapped flux inside the superconducting sample. From this plot we derived a maximum repulsive levitation force of 28 N for the small ring, 48 N for the large ring and 72 N for the double ring at zero gap with a maximum axial stiffness of 33 N/mm.

In the field cooled state the double magnet ring is brought close to the superconductor ring before cooling. After cooling the magnet is removed, and then moved again towards the superconductor. In the initial loop a maximum attractive force of 20 N at a vertical distance of 3 mm is achieved caused by the trapped field in the superconductor (Fig. 5).

The radial force has been measured by moving the magnet in a lateral direction to the superconductor and shows a nearly symmetric hysteresis loop (Fig. 6). Near r = 0 the radial magnetic stiffness $\partial F / \partial r$, describing the restoring behaviour of the configuration, has a value of about 10 N/mm.

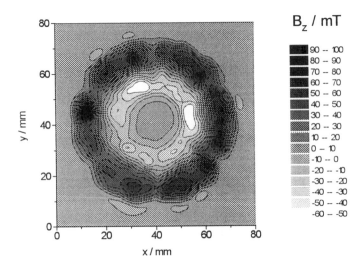

Fig. 3. Contour plot of the vertical component of the trapped field B_z in the superconducting ring.

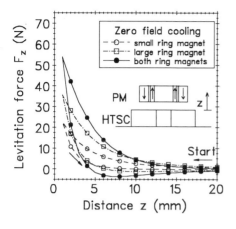

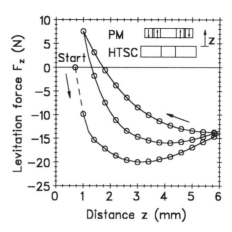

Fig 4. Axial levitation force between magnet ring and YBCO ring versus distance for several magnet configurations (zero field cooling).

Fig. 5. Axial levitation force between magnet ring and YBCO ring versus axial displacement (field cooling).

Measuring the forces in the field cooled state of the SMB configuration according to Fig. 1, the starting gaps between the magnet and the upper and lower HTSC ring were 2 mm and 7 mm, respectively (Fig. 7). When moving down, the magnet is subjected to a force acting upwards, since there is a maximum trapped field in the upper HTSC ring and a small trapped field in the lower HTSC ring. At the maximum upper gap a force of 40 N acts on the magnet. Moving the magnet upwards the force decreases rapidly and finally becomes repulsive.

As a result, this configuration is able to suspend a rotor with a weight of about 25 N in the middle position between the superconductors. Maximum axial stiffnesses of about 15 N/mm can be derived from the minor loops at this position.

At present a high speed motor with two SMB is under construction. As shown in Fig. 8, the shaft carries two double permanent magnet rings and the rotor.

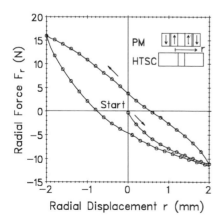

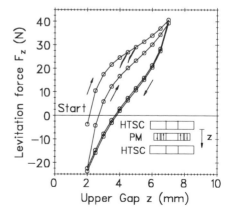

Fig. 6. Radial restoring force between magnet ring and YBCO ring versus lateral displacement (field cooling).

Fig. 7. Loop of axial force in the configuration of double magnet ring between two YBCO rings.

686

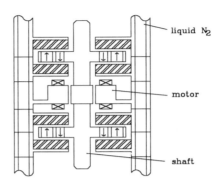

☑ HTSC   [↓] magnet ring

Fig. 8 Scheme of the superconducting magnetic bearing

The stator with the YBCO rings is cooled indirectly by heat conduction. Before cooling, the shaft is lifted up to its highest position and centered by adjustment of mechanical ball bearings at the ends of the shaft. These are also used to protect the motor in case of emergency. As a drive system a three-phase asynchronous motor for high speed rotation has been constructed. It was optimized concerning low losses at liquid nitrogen temperatures .

4. Conclusions

A three-phase asynchronous motor with two SMB is under construction. To optimize the bearing configuration the static forces and stiffnesses in several arrangements of permanent magnet rings and superconducting rings have been investigated. The maximum repulsive force under zero field cooling conditions is 72 N with an axial stiffness of 33 N/mm. In the field cooled state for each bearing a levitation force of 40 N, and maximum axial and radial stiffnesses of 15 N/mm and 10 N/mm, respectively, have been determined. The stiffness values are comparable to other experimental SMB [2,5] of different design. As an advantage over permanent magnet bearings with active control, the SMB provides self stabilization in all degrees of freedom, but its stiffnesses are about one order of magnitude lower than in active permanent magnet bearings.

The forces acting in the bearing configuration with two superconducting rings were found to be able to suspend a rotor with a weight of 25 N in the working position.

This work is supported by Bundesministerium für Bildung und Forschung, contract No. 13 N 6102.

References

[1]   Suzuki T, Suzuki H, Endo N, Yasaka Y, Morimoto M, Takaichi H and Murakami M 1994 Advances in Superconductivity 6 to be published
[2]   Takaichi H, Murakami M, Kondoh A, Koshizuka N, Tanaka S, Fukuyama H, Seki K, Takizawa T and Aihara S 1992 Proc. 3rd Int. Symp. on Magnetic Bearings 1313-1316
[3]   Gawalek W, Habisreuther T, Fischer K, Bruchlos G, and Görnert P 1993 Cryogenics 33 65
[4]   Gladun A, Stoye P, Verges P, Gawalek W, Habisreuther T and Görnert P 1993 Proc. First Europ. Conf. on Appl. Superconductivity DGM Oberursel 973-976
[5]   Bornemann H J, Boegler P, Urban C, Zabka R, Rietschel H, de Rango P, Chaud X, Gautier-Picard P, Tournier R 1993 Proc. First Europ. Conf. on Appl. Superconductivity DGM Oberursel 277-283
[6]   Gawalek W, Habisreuther T, Straßer T, Wu M, Litzkendorf D, Fischer K, Görnert P, Gladun A, Stoye P, Verges P, Ilushin K V and Kovalev L K 1994 Appl. Supercond. 2 456-478

Inst. Phys. Conf. Ser. No 148
Paper presented at Applied Superconductivity, Edinburgh, 3–6 July 1995
© *1995 IOP Publishing Ltd*

Bulk Melt Textured YBCO for Electrical Hysteresis Machines and Levitated Linear Guidance Systems

T Straßer, T Habisreuther, W Gawalek, M Wu, D Litzkendorf and P Görnert

Institut für Physikalische Hochtechnologie e.V., Helmholtzweg 4, D-07743 Jena, Germany

K V Iljushin, L K Kovaljov

Moscow State Aviation Institute, Volokolamskoje shosse, 125871 Moscow, Russia

Abstract. Melt textured YBCO cylindrical blocks are prepared in a batch process.These blocks are employed in a superconducting hysteresis motor and a linear guidance and carrier system. The performance of these systems are presented at liquid nitrogen temperature. Furthermore the quality of the YBCO blocks at low temperatures are compared to at liquid nitrogen temperature and to the quality of the theoretical limit.

1. Introduction

The development of YBCO bulk material has nowadays been progressed in a way that it can be tested and employed to manyfold applications. The batch processed material is suitable for electric motors and generators, HTSC permanent magnets, magnetic field screens as well as for magnetic bearings and linear carrier and guidance systems. For these purposes one uses the typical properties of that material: Screening behaviour to an applied external magnetic field after zero field cooling (zfc) the superconductor. Applying an external magnetic field and field cooling (fc) the superconductor the field is trapped inside due to pinning forces. Furthermore a hysteretic behaviour, that is motion of vortices in the material, can also be used for damping applications. These properties are examined by levitation force measurements.

In the following we present a hysteresis motor with superconducting rotor and a linear guidance and carrier system both developed in the Institut für Physikalische Hochtechnologie, Jena in cooperation with the Moscow State Aviation Institute.

In addition we discuss the levitation properties of our material in the field cooled and zero field cooled state at liquid nitrogen temperature as well as at 17K and compare them to the calculated theoretical limit.

2. Material preparation

The YBCO cylindrical blocks of 30mm or 45mm in diameter and 17mm in thickness are prepared by a precursor powder being composed of $Y_{1.5-1.8}Ba_2Cu_3O_x$ and 1 wt.% Pt addition. This powder is uniaxially pressed to cylindrical shape.

For the modified melt textured growth process the cylinders are placed in a six side heated chamber furnace. After establishing the growth starting temperature the furnace is cooled down with a cooling rate of 0.5K up to 2.0K at a temperature gradient of ∇T=10-20K/cm. Finally the samples are oxygenized by a seperated process.

The multi domain material consists of quasi single crystalline domains with volumes in the range of cm^3, to be observed by polarization photography and trapped magnetic field mapping. The intra domain critical current densities are up to $5*10^4 A/cm^2$ at 77K measured by vibration sample magnetometry.

3. Hysteresis motor

With three blocks of our melt textured material the superconducting rotor (hollow cylinder with 25mm outer diameter, 15mm inner diameter, and 40mm length) of a hysteresis motor had been constructed. The motor was tested in liquid nitrogen for some hours at 3000 rpm and an output power of 80 Watts was reached at 220V/50Hz voltage supply with 1.5A stator current. The mechanical torque was 2250gcm [1]. In comparision with conventional hysteresis motors the specific output power is higher by a factor of 10. The output power is correlated with the material quality i.e. with the zfc levitation force:

Figure 1 shows the output power versus stator current. The bent part up to 1.25A shows the zfc regime. At higher currents the stator field is penetrating the rotor completely and the power gain then

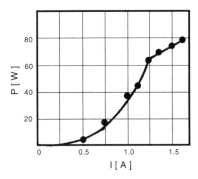

Fig 1: output power versus stator current   Fig 2: output power versus zfc levitation force of a single block

is only linear with increasing stator current. As a consequence of Figure 1 Figure 2 can be understood: It shows the output power versus the zfc levitation force of a permanent magnet on the superconductor (see 5.). The less the penetration of magnetic field into the superconductor the more the zfc levitation force. Thus, by using material of higher levitation force, a higher stator current can be applied with high power gain without reaching the less effective fc-similar region. In Figure 2 we estimate this motor supplied with - already available - samples of 30N levitation force to have more than 100W output power.

4. Linear guidance and carrier system

Magnetic bearings have to be compared to conventional air bearings. With 32 YBCO blocks mounted in a fixed stainless steel dewar a linear guidance and carrier system has been constructed for application relevant investigation (Figure 3). A carrier equipped with four NdFeB permanent magnet rails (length: 40mm, weight: 3.5kg) is placed below the dewar before the system is field cooled to liquid nitrogen. The resulting airgap between superconductors and hanging carrier varies from 4 to 6mm depending on variable cooling distance and carrier load. The carrier can be moved parallel to the rails with very low magnetic friction, for the remaining five degrees of freedom we found high stiffness and damping. At a cooling distance of 1mm between permanent magnets and dewar surface the maximum carrier load is 3kg at a resulting airgap of approximately 6mm. The high guidance precision is being measured by laser interferometry. These measurements are still on the run.

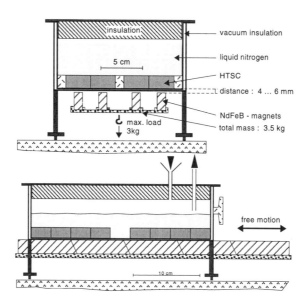

Fig 3: Cross section and side view of the linear guidance and carrier system

5. Levitation properties

The more or less multi domain samples are characterized bei integral levitation force measurements in the fc and zfc state at 77K and recently at 17K. The results are compared to numerical simulations providing the theoretical limits [3]. We obtain the theoretical limits of fc and zfc levitation force as follows:

The theoretical conditions like size of superconductor and permanent magnet, distance, remanence of the magnet are the same as for real measurements. The theoretical superconductor is considered as single domain with infinitely high current density, that is infinitely strong flux pinning.

In the zfc case no magnetic field can penetrate the theoretical superconductor (London penetration depth shall be neglected). This also means that no hysteresis behaviour can be observed due to the missing of flux remaining in the superconductor to provide any attractive force contribution while removing the magnet from the superconductor after approaching it. (In case of a superconductor of infinitely large lateral size the theoretical zfc situation can easily be carried out in experiment by measuring the repelling force of two equal permanent magnets. The results of this experiment show good agreement to the numerical simulation).

In the fc case the frozen magnetic field remains fixed in the superconductor. Again no hysteretic behaviour can be observed because there is no motion of vortices.

The experiments were conducted with samples of 30mm in diameter and 17mm thickness using a SmCo permanent magnet of 25mm in diameter and 15mm thickness with an iron backplate of 3mm thickness. The pole face field of the permanent magnet is 410mT. The field cooling distance between magnet and superconductor and minimum approach distance respectively is 0.5mm.

Figure 4 shows the fc levitation force at 77K (normal line), 17K (shivering line) and the theoretical limit (thick line). As the 77K curve shows a big hysteresis the one in the 17K curve becomes much smaller. The highest levitation pressure (highest levitation force divided by magnet surface) reaches 93% of the theoretical limit (-3.7N/cm²) where there is 65% of the theoretical limit at 77K.In figure 5 we see the zfc levitation forces analogous to figure 4. The 77K curve shows a hysteretic behaviour resulting in a small attractive force whereas almost no hysteretic behaviour can be observed at 17K. The highest levitation pressure of the theoretical limit amounts to 17.7N/cm² which is reached with

690

61% at 17K and 41% at 77K.

The difference in the percentage of the levitation pressure in the fc and zfc case at one temperature is originated in the multi domain properties: In the zfc case the field gradient is higher than in the fc case. The more and smaller the grains in the sample the less the field gradient becomes. Thus the difference of the field gradients of a single domain sample is bigger in the zfc case than in the fc case.

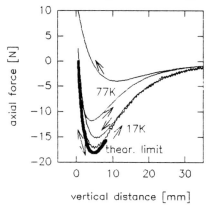

Fig. 4: fc levitation force at 77K (normal line), 17K (shivering line) and theoretical limit (thick line)

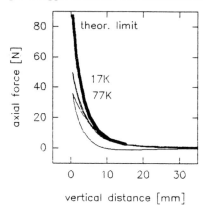

Fig. 5: zfc levitation force at 77K (normal line), 17K (shivering line) and theoretical limit (thick line)

The percentage values are valid for any permanent magnet in a pole field of 0.4 up to 0.5T and not depending on its size and shape. So these values are directive for applications with superconducting bulk material and permanent magnets.

6. Summary

Bulk melt textured YBCO has been used for a superconducting hysteresis motor and a levitated linear guidance and carrier system. In comparision with conventional hysteresis motors the specific output power at 77K is higher by a factor of 10. Very low magnetic friction in on degree of freedom marks out the linear carrier and guidance system whereas we found a high stiffness for the remaining five degrees of freedom. For trapped field and screening applications e.g. permanent magnets, electric motors and generators and magnetic field screens the field cooled levitarion force is a suitable parameter. For selfstabilizing levitation applications e.g. magnetic bearings, linear guidance and carrier systems the field cooled levitation force is a suitable parameter. Temperatures far below 77K will allow a much higher performance of the YBCO bulk material reaching close to the theoretical limit of field cooled levitation forces which is limited by permanent magnets.

This work was suppoted by the German BMFT under contract No.13N6100.

References

[1] L K Kovaljov et al., VII Trilateral German - Russian - Ukrainian Seminar on HTSC, Munich, Germany, Sept. 12-15, 1994

[2] H Weh et al., Symp. on Power Electronics, Electrical Devices, Advanced Electrical Motors, Taomina, Italy, June 8-10, 1994

[3] Simulations performed by H Hupe, A Steingroever, H Pahl, H May, H Weh, Institut für elektrische Maschinen, Antriebe und Bahnen, Hans-Sommer-Str. 66, D-38106 Braunschweig, Germany

Inst. Phys. Conf. Ser. No 148
Paper presented at Applied Superconductivity, Edinburgh, 3–6 July 1995
© *1995 IOP Publishing Ltd*

A comparative study of levitation force and magnetic stiffness of bulk YBCO and YBCO thin films

B. Lehndorff, H.-G. Kürschner and B. Lücke
Institute of Material Science, Dept. Physics, University of Wuppertal, Gauß-Str. 20,
D-42097 Wuppertal, Germany

Abstract. Comparative experiments of levitation force and vertical magnetic stiffness on melt processed, granular and film samples of $YBa_2Cu_3O_{7-\delta}$ are presented. The melt processed and the film samples show a strong hysteretic behavior as expected from their enhanced pinning properties. The vertical stiffness is determined from the slope of minor loops of the force vs. distance curve. When plotted as a function of magnetic field it shows a crossover from $B^{1.5}$ to a B^2-behavior for all samples except for the granular one. This can be explained by flux line lattice effects. In addition a quantitative estimation for the pinning strength of the samples can be obtained from the stiffness measurements. Furthermore a 3-d positioning system is used to map the spatial distribution of levitation force and trapped flux providing a good characterization of sample homogeneity.

1. Introduction

The applicability of melt processed $YBa_2Cu_3O_{7-\delta}$ ceramics in magnetic bearings has been widely demonstrated [1]. Many efforts have been undertaken to optimize the material towards large levitation forces and stiffness [2]. Other possible applications are based on trapped flux magnets. Magnetic flux up to 7 T at 55 K has been trapped in melt grown $YBa_2Cu_3O_{7-\delta}$ samples [3]. Magnetic levitation as well as the effective trapping of high magnetic flux are both strongly correlated to the pinning properties of the melt grown ceramics. Thus the main goal of materials characterization besides the determination of the microstructure is the measurement of levitation force, stiffness and flux distribution, which we present in this paper. In order to obtain a better understanding of the limiting factors for bearing applications we examined the relation between pinning properties and levitation force characteristics. Thus for the first time a quantitative estimate of the pinning properties from magnetic stiffness measurements appears to be possible.

2. Experimental details

The granular sample has been prepared using a conventional sintering process. The melt textured grown (mtg) ceramic has been produced by modified liquid phase processing as described earlier [4]. It has a diameter of 21 mm and exhibits a few 2-5 mm^2 large grains and micron sized Y-211 inclusions as analyzed by electron microscopy (SEM) and energy dispersive x-ray diffraction (EDX). Further details are given in [5]. The $YBa_2Cu_3O_{7-\delta}$ film sample with a diameter of 25.4 mm and a thickness of 500 nm has been prepared by DC-sputtering [6].

A computer controlled setup was used to measure the spatial distribution of levitation force and trapped magnetic flux. Alternatively a strain gauge force sensor or a Hall probe with 200 x 200 μm^2 active area can be attached to the positioning system. Simultaneously the exact position and the force or Hall voltage respectively are read out by the computer. This setup and the measuring procedure is described in detail in ref. [5].

3. Levitation force and vertical magnetic stiffness

The levitation force was measured as a function of the distance between a permanent magnet and the superconductor. Typical force versus distance curves for the three different kinds of samples are shown in fig. 1. The absolute levitation force is highest for the mtg sample compared to the film or the sintered one. The hysteresis is pronounced for the melt grown ceramic and the thin film, whereas the sintered sample not even exhibits an attractive force. This is expected due to the weak pinning of granular samples.

The vertical stiffness η was determined from the slope of minor loops of the main hysteresis curve. The stiffness can be expressed as a function of the external magnetic field, which is applied by the permanent magnet. Interestingly the magnetic field dependence is independent of the history of the sample. In the field cooled (fc) as well as in the zero field cooled (zfc) case the stiffness started with a $B^{1.5}$ dependence and turned over to a B^2 dependence for the mtg and the film sample for fields larger than about 150 mT. The granular sample showed a $B^{1.3}$ behavior over the whole field range. In fig. 2 the stiffness η is plotted as a function of the external field B_{ext} for the sintered pellet, the mtg ceramic and the thin film for the zfc case.

As argued in [7] the pinning parameter α can be calculated from the magnetic stiffness data considering a correlation volume for the flux lines which is of the order of the grain size. The pinning parameter is plotted as a function of the external field in a log-log plot in fig. 3. Also depicted in fig. 3 is the fit to $\alpha = k\, B^n$. The values for α agree well with values obtained from vibrating reed experiments for $YBa_2Cu_3O_{7-\delta}$ samples [8]. This is the first attempt to get quantitative information about the pinning strength from levitation force experiments. The crossover from $B^{1.5}$ to B^2 can be explained by flux line lattice interaction. For small magnetic fields the interaction can be nearly neglected and only single vortex pinning is valid. At fields higher than 150 mT, which corresponds to a vortex spacing smaller than 0.2 µm, the elasticity of the flux line lattice is dominating the forces which leads to the observed square dependence [9]. In the case of the granular sample nearly no flux is trapped within the superconductor and thus the crossover can not take place.

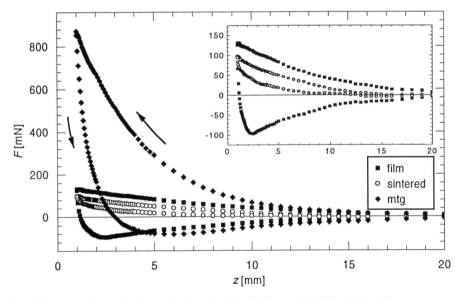

Fig.1.: Force versus distance for the mtg, the sintered and the thin film sample of $YBa_2Cu_3O_{7-\delta}$ (zfc)

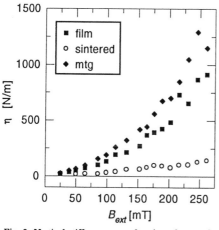

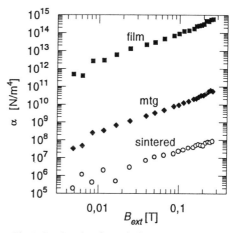

Fig. 2: Vertical stiffness η as a function of external magnetic field B_{ext} for the same samples as in fig. 1.

Fig. 3: Log-log plot of the pinning parameter α as a function of B_{ext}. Lines are a fit to $\alpha = k\, B^n$.

4. Force and trapped flux distribution

Both the force and the trapped flux distribution were measured for the same samples as above. In fig. 4 the spatial distribution of the levitation force is shown for the melt grown ceramic and the thin film. The spatial resolution depends on the radius of the test magnet and the step size which in both cases is 1mm. The scans were made in the zfc state. Both samples exhibit a rather homogeneous force distribution. The thin film has a slightly higher force on one sector. The force distribution of the mtg sample shows a grained structure which corresponds to the physical grain distribution with grain sizes of a few mm.

For the measurement of the remanent field the samples were cooled in the field of a permanent magnet. After careful removal of the magnet and a waiting time of several minutes to reduce relaxation effects during the measurement, the field distribution was measured with the Hall probe and a step size of 1 mm. The results are shown in fig. 5. The trapped flux of the film shows a slightly asymmetric cone. This corresponds to the asymmetric distribution of the levitation force, which leads to the conclusion that the film was not deposited homogeneously. The cone shape is predicted by the Bean critical state model [10]. The mtg sample exhibits several cones, as expected by its grain structure, which overlap mainly due to the final resolution of the Hall scan.

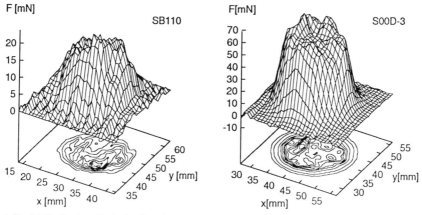

Fig. 4: Spatial distribution of levitation force for the thin film (left) and the mtg sample (right) of $YBa_2Cu_3O_{7-\delta}$.

694

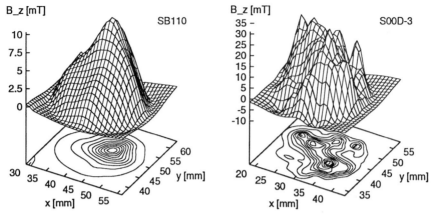

Fig. 5: Trapped flux distribution of the two samples of fig. 4 (left: thin film, right: mtg sample).

5. Conclusions

A comparative study of levitation force and trapped flux distribution as well as vertical stiffness measurements on melt grown, granular and thin film samples of $YBa_2Cu_3O_{7-\delta}$ were presented. The results of the magnetic field dependence of the vertical stiffness reveal the different pinning properties of the samples. While the mtg and film sample show effects of flux line lattice interaction as a crossover from $B^{1.5}$ to B^2 behavior, the sintered one exhibits only very weak pinning and a $B^{1.3}$ dependence. The force and flux distributions give a signature of the grained microstructure of the mtg sample and the homogenety of the thin film. Thus a fairly complete characterization of samples for magnetic bearing applications is possible with our experimental methods. Furthermore the stiffness experiments provide a good means to determine basic physical properties like the pinning parameter α. The comparison of the different kinds of samples show the relatively high levitation force of a thin $YBa_2Cu_3O_{7-\delta}$ film compared to a bulk sample, and also provides a possibility to analyze the homogeneity of film deposition.

Acknowledgment

We grateful acknowledge valuable discussions with A. M. Portis. The thin film sample was kindly provided by H. Schlick.

References
[1]   Moon F C and Chang P Z 1994 *Superconducting Levitation* (Chichester: Wiley)
[2]   Murakami M 1992 *Supercond. Sci. Technol.* **5** 185
[3]   Weinstein R, Ren Y, Liu I, Chen G, Sawh'R, Obot V and Foster C 1993 *6th ISS*, Hiroshima
[4]   Lehndorff B, Kürschner H-G, and Piel H 1995 *IEEE Trans. Appl. Supercond.* **5**
[5]   Lehndorff B, Kürschner H-G, Lücke B and Piel H 1995 *Physica C* to appear
[6]   Müller G et al. 1995 *IEEE Trans. Appl. Supercond.* **5**
[7]   Lehndorff B, Kürschner H-G and Lücke B 1995 *submitted to Appl. Phys. Lett.*
[8]   Ziese M, Esquinazi and Braun H F 1994 *Supercond. Sci. Technol.* **7** 869
[9]   Brandt E H 1993 *Physica C* **195** 1
[10]   Bean C 1964 *Rev. Mod. Phys.* **36** 31

Inst. Phys. Conf. Ser. No 148
Paper presented at Applied Superconductivity, Edinburgh, 3–6 July 1995
© *1995 IOP Publishing Ltd*

Magnetic shielding effects of high-T_c superconducting cylinder by using an amorphous-metal cylinder

M. Itoh, K. Mori*, T. Minemoto*, F. Pavese**, M. Vanolo**, D. Giraudi**, and Y. Hotta***

Dept. of Electronic Engi., Kinki Univ., Higashi-Osaka, Osaka 577, Japan
*Dept. of Computer and Systems Engi., Kobe Univ., Nada, Kobe 657, Japan
**Instituo di Metrologia "G. Colonnetti" (IMGC), 10135 Torino, Italy
***Div. of EMR, Tokin Corp., Kawasaki, Kanagawa 213, Japan

Abstract. The value of maximum shielded magnetic flux density (B_s) was shown to be improved by using two type superimposed cylinders which are the amorphous superimposed and hybrid ferromagnetic superimposed cylinders. For example, the value of B_s for the hybrid ferromagnetic superimposed cylinder by using a Y-Ba-Cu-O (YBCO) thick-film cylinder was found to be about 130 times greater than that of YBCO thick-film cylinder. The values of B_s were analyzed by using several different models, with an important criterion being derived for designing an effective and reliable magnetic shielding for a large-sized vessel.

1. Introduction

The value of the critical current density (J_c) of a high critical temperature (high-T_c) superconducting vessel is an important parameter associated with magnetic shielding. In general, however, it is difficult to fabricate very high-J_c vessels. There is also a practical difficulty with the magnetic shielding when using a high-T_c superconductor. Therefore, it is necessary to improve the magnetic shielding effect of high-T_c superconducting vessel.

In order to simplify the present theoretical analysis, the evaluation of the magnetic shielding effect is limited to a superconducting cylinder in an excitation magnetic flux density (B_{ex}, homogeneous DC magnetic field). The value of maximum shielded magnetic flux density (B_s) is shown to be improved by using two type superimposed cylinders, the amorphous superimposed and hybrid ferromagnetic superimposed cylinders, which are the superposition of an amorphous-metal cylinder over a BPSCCO bulk cylinder and the superposition of hybrid ferromagnetic cylinder over a BPSCCO bulk cylinder, respectively. The hybrid ferromagnetic cylinder is constructed of two cylinders, amorphous-metal and PC (87% Ni) permalloy. The values of B_s for two type superimposed cylinders were theoretically analyzed by use of the several different models as show in Ref. [1]. The experimental values of B_s were found to agree well with theoretical values. The present paper examines also the effects of magnetic shielding within two type superimposed cylinders by use of a YBCO thick-film cylinder.

2. Experimental details

Magnetic shielding effects have been evaluated with all cylinders placed in B_{ex}, applied parallel to axial direction of the cylinders. The effect of B_{ex} on the inner magnetic flux density (B_{in}) within two type superimposed cylinders and the distribution of the maximum shielded magnetic flux density (B_s) within the cylinders were measured with the use of a GaAs Hall device as reported in Ref. [2]. The sensitivity of the measuring system was 100 mV/gauss.

The BPSCCO bulk cylinder (2.9 mm inner radius, 2.6 mm thickness, 30.4 mm length) was manufactured according to the method in Refs. [2] and [3]. Each outside of BPSCCO, amorphous-metal, and permalloy cylinders were wrapped with several turns of fluoroplastic (PTFE) tape, making use of the troidal winding method, in order to avoid sudden temperature changes. After 80 thermal cycles between room temperature (300 K) and the boiling point of liquid nitrogen (77.4 K), the characteristics of BPSCCO, amorphous, and permalloy cylinders, such as shown in Figs. 1 and 2, underwent no significant change in the degree of magnetic shielding. In addition, all characteristics of the BPSCCO cylinder in the present experiment were obtained by zero field coolings. The amorphous-metal cylinder (17.0 mm inner radius, 3.0 mm thickness, 60.0 mm length) and permalloy cylinder (12.5 mm inner radius, 2.0 mm thickness, 100.0 mm length) were degaussed by an AC magnetic field (60 Hz) at room temperature, prior to carrying out the experiments.

3. Results and discussion

Figure 1 shows the characteristics of the magnetic shielding at the center of the innermost BPSCCO bulk cylinder under temperature conditions of 77.4 K. The curves (a), (b), and (c) represent the BPSCCO single-cylinder (open circles), the amorphous superimposed cylinder (solid squares), and the hybrid ferromagnetic superimposed cylinder (solid circles), respectively. In this figure, the points denoted as B_{sa} (=0.98 mT), B_{sb} (=2.90 mT), and B_{sc} (=12.30 mT) express the maximum shielded magnetic flux density (B_s) for the respective curves. It is found that the value of B_{sa} for the BPSCCO single-cylinder is improved by factors of about 3 and 13 for the amorphous and hybrid ferromagnetic superimposed cylinders.

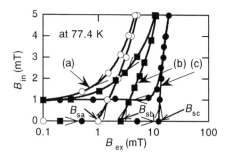

Fig. 1. Typical characteristics of the magnetic shielding at the center of the innermost BPSCCO cylinder at 77.4 K. Curves (a), (b), and (c) represent the BPSCCO single-cylinder, the amorphous superimposed cylinder, and the hybrid ferromagnetic superimposed cylinder, respectively.

Table 1. Experimental and theoretical values (B_{se} and B_{st}) of B_s for the superimposed cylinders by use of a BPSCCO bulk cylinder.

Cylinders	B_{se} (mT)	B_{st} (mT)
A	2.9	3.3
B	10.8	11.0
C	12.3	12.4

Here, A is the amorphous superimposed cylinder, B the permalloy superimposed cylinder, and C the hybrid ferromagnetic superimposed cylinder.

From Refs. [1], [4], and [5], the theoretical value of B_s for the superposition of a ferromagnetic cylinder over a superconducting cylinder, can be written as

$$B_s = B_s{'} \left(|s_1|_{l/r_1 \geq k} + |s_2|_{l/r_1 < k} \right) \tag{1}.$$

In Eq. (1), the shielding factors (s_1 and s_2) are given, following Refs. [1], [4], and [5], for cases in which the end effects of the ferromagnetic magnetic cylinder are neglected and not neglected. Here, l is the length of the magnetic cylinder, k the value of l/r_1 at the crossing point of s_1 and s_2, r_1 and r_2 the inner and outer radii of the ferromagnetic cylinder superimposed over the superconducting cylinder, and $B_s{'}$ the magnetic shielded field modified by r_1.

Table 1 lists the experimental and theoretical values of B_s for the superimposed cylinders by use of a BPSCCO bulk cylinder. The permalloy superimposed cylinder in this table is the superposition of a permalloy cylinder over a BPSCCO cylinder. The theoretical values (B_{st}) were obtained by use of Eq. (1). As shown in Table 1, the experimental values (B_{se}) agree well with the values of B_{st}.

The distribution of B_s along the axial direction of the superconducting BPSCCO bulk cylinder for curves (a), (b), and (c) shown in Fig. 1 are displayed in Fig. 2. Here, the B_{ex} was also applied along the axial direction of the cylinders. The curves (a), (b), and (c) represent the BPSCCO single-cylinder (open circles), the amorphous superimposed cylinder (solid squares), and the hybrid ferromagnetic superimposed cylinder (solid circles), respectively.

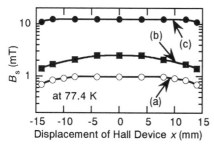

Fig. 2. Distributions of the maximum shielded magnetic flux densities (B_s) along the axial direction of the BPSCCO cylinder at 77.4 K. Curves (a), (b), and (c) represent the BPSCCO single-cylinder, the amorphous superimposed cylinder, and the hybrid ferromagnetic superimposed cylinder, respectively.

Table 2. Experimental and theoretical values (B_{se} and B_{st}) of B_s for the superimposed cylinders by use of a YBCO thick-film cylinder.

Cylinders	B_{se} (mT)	B_{st} (mT)
A	0.12	0.16
B	9.5	9.5
C	10.1	10.4

Here, A is the amorphous superimposed cylinder, B the permalloy superimposed cylinder, and C the hybrid ferromagnetic superimposed cylinder.

The value of the trapped magnetic flux density (B_t) at the center of the superconducting BPSCCO cylinder decreased by approximately 3 % in 1000 seconds under temperature condition of 77.4 K (not shown). The stability of the temporal change for B_t is an important criterion in the process of evaluating superconductor properties for the magnetic shield.

4. Results on YBCO thick-film cylinder

The use of the high-T_c superconducting thick-film is convenient to design large-sized vessels for the magnetic shielding. The silver as a substrate for the high-T_c superconducting thick-film is fine material as shown in Ref. [6], since it does not oxidize during the high-temperature treatment and is fully compatible with YBCO. In general, however, there are some cracks in the YBCO thick-film on re-heating after sintering. The cracks of YBCO thick-film cylinder are eliminated by use of Ni film (100 μm) coated stainless-steel cylinder as a substrate, such as shown in Ref. [7]. Then, a YBCO thick-film cylinder (11.5 mm inner radius, 150 μm thickness, 114.0 mm length) was manufactured according to the method in Ref. [7].

Table 2 lists the experimental and theoretical values (B_{se} and B_{st}) of B_s for the superimposed cylinders by use of a YBCO thick-film cylinder, in much the same manner as in Table 1. The value of B_s for the YBCO single-cylinder was 80 μT. In this results, the amorphous-metal and permalloy cylinders are used the same as in Fig. 1. The experimental values are represented the results after six months, since it is fabricated the YBCO thick-film cylinder. After six months, the value of B_s for the YBCO thick-film cylinder decreased by approximately 50 % under unprotected condition in the atmosphere. The theoretical values (B_{st}) were obtained by use of Eq. (1), and agree well with experimental values (B_{se}) as shown in this table. It is found that the value of B_s for the YBCO single-cylinder is improved by factors of about 1.5 and 130 for the amorphous and hybrid ferromagnetic superimposed cylinders. After 20 thermal cycles between room temperature (300 K) and the boiling point of liquid nitrogen (77.4 K), the experimental value of B_s (=80 μT) for the YBCO thick-film cylinder underwent no significant change in the degree of magnetic shielding.

The distribution characteristics such as shown in Fig. 2 exhibited a similar tendency in the results obtained for the YBCO thick-film, amorphous superimposed, and hybrid ferromagnetic superimposed cylinders (not shown).

5. Conclusion

One of the basic research areas is for the improvement of magnetic shielding effects and the fabrication of large-sized high-T_c superconducting vessels. The present paper has been directed at analysis of two type superimposed cylinders, by using a BPSCCO bulk cylinder and YBCO thick-film cylinder. As shown in Tables 1 and 2, the theoretical values of maximum shielded magnetic flux density (B_s) agreed well with experimental values. It is found that the value of B_s (=B_{sa}) for the BPSCCO single-cylinder is improved by factors of about 3 and 13 for the amorphous and hybrid ferromagnetic superimposed cylinders. Also, the value of B_s for the YBCO thick-film cylinder is improved by about 1.5 and 130 times for the amorphous and hybrid ferromagnetic superimposed cylinders. These results are important criteria fundamental in the design of effective and reliable magnetic shielding for a large-sized vessel. The method used to fabricate of large-sized vessels with greater effective characteristics of magnetic shieldings for wide applications is now investigation by using these results.

Acknowledgements

The authors wish to acknowledge valuable discussion with Dr. Takashi Ohyama. Thanks are due to Dr. Kohji Kajitani for preparing the samples.

References

[1] Itoh M, Ohyama T, Mori K, and Minemoto T 1995 *T. IEE Japan* **115** 696-701

[2] Itoh M, Ohyama T, Minemoto T, Numata K, and Hoshino K 1992 *J. Physics D: Appl. Phys.* **25** 1630-4

[3] Itoh M, Ohyama T, Hoshino K, Ishigaki H, and Minemoto T 1993 *IEEE Trans. Appl. Superconductivity* **3** 181-4

[4] Itoh M, Ohyama T, Mori K, and Minemoto T 1994 *Cryogenics* **34** (ICEC Suppl.) 817-20

[5] Ohyama T, Minemoto T, Itoh M, and Hoshino K 1993 *IEEE Trans. Magnetics* **29** 3583-5

[6] Pavese F, Bianco M, Andreone D, Cresta R, and Rellecati P 1992 *Physica C* **204** 1-7

[7] Vanolo M, Pavese F, Giraudi D, and Bianco M 1995 *Nuovo Cimento* in press

Inst. Phys. Conf. Ser. No 148
Paper presented at Applied Superconductivity, Edinburgh, 3–6 July 1995
© *1995 IOP Publishing Ltd*

Method of vibration damping based on AC losses in bulk high-T_c superconductors

L.S. Fleishman, Yu.A. Bashkirov
Krzhizhanovsky Power Engineering Institute, 19 Leninsky Prospekt, Moscow 117927, Russia

V.A. Malginov
Lebedev Physics Institute, Moscow 117924, Russia

O.L. Poluschenko, N.A. Nizhelskii
Moscow State Technical University, Moscow 107005, Russia

Abstract. Vibration damping due to energy losses in melt-textured YBCO superconductor during permanent magnet cyclic motion has been studied. Both free and forced vibration characteristics have been measured. The magnet-superconductor interaction was found to result in non-linear dynamics and substantial energy dissipation.

1. Introduction

The development of bulk melt processed YBCO superconductors has shown great potential for practical applications such as magnetic bearings and maglev vehicles. The hybrid superconducting magnetic bearing [1] consists of permanent magnets providing high levitation pressure and it is stabilized by bulk superconductors. Back and forth cyclic motion of the magnet relative to a piece of high-T_c superconductor should experience hysteretic forces, with loss of mechanical energy as a result [2]. In this work the vibrating motion of permanent magnets beside a superconducting plate have been studied.

2. Experimental

The measurement system depicted schematically in Fig.1 consists of Nd-Fe-B permanent magnets attached to a steel cantilever beam and a melt-textured YBCO superconducting plate which is mechanically fixed and cooled at 77 K. The magnet motion occurred parallel to the superconducting plate, the separation between them could be as small as 1 mm due to plastic cryostat wall thickness. The vibration characteristics have been measured by means of both free and forced vibration methods. In the latter case the beam was mounted on a vibrating table. The magnet block movement was registered by means of a piezoelectric vibrometer or strain gauge.

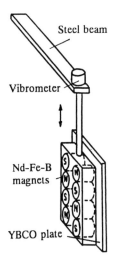

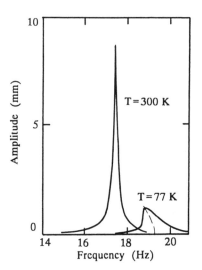

Fig.1. Schematic of the
vibration system

Fig.2. Frequency response before and after cooling
of the YBCO plate

3. Results and discussion

3.1. Forced vibrations

The frequency response of the vibration system in the vicinity of resonance is
shown in Fig.2. When the superconducting plate is at room temperature the
high and narrow resonance peak exists. After cooling the frequency response
changes drastically. The resonance maximum decreases by an order of
magnitude, the bandwidth substantially increases, and the resonance
frequency slightly grows. The response curve becomes asymmetrical having a
steep ascent at pre-resonance frequencies and a flat descent above the
resonance. All these features result from the particular character of
superconductor-magnet interaction. Lateral force and elastic stiffness
between a high-T_c superconductor and levitating permanent magnet were
measured in [3] when the relative positions of the magnet and superconductor
were shifted horizontally by large amounts. The initial displacement
immediately gives rise to a rapidly growing restoring force. It reaches a
maximum, after which it decreases in a nearly linear fashion. The magnetic
stiffness also was proved to decrease at large displacements. This results in the
natural frequency f_0 decrease as the vibration amplitude A grows. The
qualitative $f_0(A)$ dependence is shown by dashed line in Fig.2. The resonance
response is determined by this characteristic line and is a manifestation of the
non-linear dynamics when a spring constant decreases at large displacements
[4].

Damping efficiency can be characterized by the dissipation coefficient Γ
defined as the ratio of the energy dissipation per cycle W to the total elastic
energy E:

$$\Gamma = W/E, \tag{1}$$

If the harmonic driving force is due to vibrating table the dissipation coefficient is given by [4]:

$$\Gamma = 2\pi A_t/A_r, \qquad (2)$$

where A_t and A_r are the table and resonance amplitudes. Using the data from Fig.2: $A_t = 0.075$mm, $A_r(300K) = 8.94$mm and $A_r(77K) = 1.19$mm, we find that $\Gamma = 0.05$ when the superconductor is normal and $\Gamma = 0.40$ when it is in superconducting state.

3.2. Free vibrations

Free vibration measurements enable to observe oscillation waveform and amplitude decay. The strain gauge signals proportional to the beam deflection are shown in Fig.3 at different distances between the YBCO plate and permanent magnet block. Damping decrement increases dramatically when the magnets move directly near the cryostat surface. Since the beam elastic energy is proportional to deflection square, the dissipation coefficient is given by:

$$\Gamma = 1 - (A_{i+1}/A_i)^2, \qquad (3)$$

where A_i and A_{i+1} are any two successive oscillation maxima. The oscillation decay turns out to be exponential for a number of high deflection maxima but the decrement decreases at low deflections. Therefore it would be quite correct to define the dissipation coefficient according to Eq.3 for high amplitude vibrations at different natural frequencies. Fig.4 shows Γ as a function of frequency f for a number of magnet-superconductor separations. The $\Gamma(f)$ functions are found to be fitted well by an inverse proportional law $\Gamma \sim f^{-1}$. Such a frequency dependence is typical for eddy-current loss mechanism. When eddy-current loss takes place the energy dissipated per cycle is proportional to AC field frequency: $W \sim f$ [5]. Using Eq.1 and the beam elastic energy $E = (1/2)M(2\pi f)^2 A^2$, where M being the vibrating mass, we have $\Gamma \sim f^{-1}$.

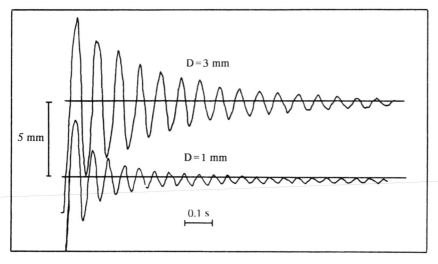

Fig.3. Oscillation waveforms at different distances D between the magnets and YBCO plate

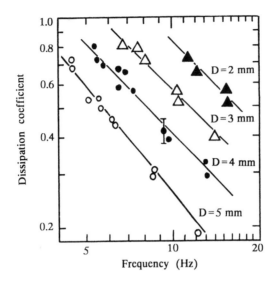

Fig.4. Dissipation coefficient vs frequency at different magnet-superconductor distances D

Eddy-current loss in superconductor is belived to be due to flux-flow accompanied by induced electric field and depends on the magnet velocity. This loss was proved to take place also when a magnetic dipole passed through a hollow YBCO cylinder and was interpreted using the flux-flow resistivity of YBCO superconductors [6].

4. Conclusions

The interaction between permanent magnets and melt-textured YBCO plate has been shown to bring about the non-linear dynamics and effective vibration damping due to flux-flow energy losses.

5. Acknowledgements

This work was supported by the Russian Foundation for Fundamental Research under Project No.93-02-16879 and the HTSC Council under Project No.94014.

References

[1] McMicael C.K. et al. 1992 **Appl. Phys. Lett. 60** 1893-5
[2] Ma K.B. et al. 1993 **IEEE Trans. Appl. Supercond. 3** 388-91
[3] Johansen T.H. et al. 1994 **J. Appl. Phys. 75** 1667-70
[4] Timoshenko S. et al. 1974 **Vibration Problems in Engineering** (New York: Wiley)
[5] Landau L.D. and Lifshitz E.M. 1960 **Electrodynamics of Continuous Media** (Oxford: Pergamon)
[6] Kuze H. et al. 1993 **J. Appl. Phys. 73** 1320-6

Inst. Phys. Conf. Ser. No 148
Paper presented at Applied Superconductivity, Edinburgh, 3–6 July 1995
© *1995 IOP Publishing Ltd*

Dynamic characteristics of iron-cored magnets using high-T_c superconductors

A A El-Abbar, R M Goodall, and C J MacLeod
Department of Electronic and Electrical Engineering, Loughborough University of Technology, Loughborough, LE11 3TU, UK

A M Campbell
IRC in Superconductivity, Cambridge University, Cambridge, CB3 0HE, UK

R G Jenkins and H Jones
Clarendon Laboratory, University of Oxford, Oxford, OX1 3PU, UK

Abstract. This paper describes the dynamic characteristics of high-T_c superconducting magnets with an iron core under different operating conditions. The main objective of this work is to understand the limits of controllability of these magnets for use in controlled applications such as electromagnetic suspensions (EMS) and bearings. Experimental results obtained using laminated and solid iron cores indicated that the ac losses are dominated by eddy currents in the iron cores, and the losses in the superconductor are relatively insignificant.

1. Introduction

As the properties of high-T_c superconducting materials progressively improve, their application to the solution of engineering problems will depend upon a clear understanding of how these properties translate into an engineering framework relevant to the particular application. This paper describes work being undertaken to understand the limits of using iron-cored HTS magnets in control systems. The specific control application is for electro-magnetic "Maglev" systems [1], the requirements for which are directing the work, although the results will be applicable for electro-magnetic actuators in general.

The aim of the current project is to have a controlled HTS magnet supporting a load of 10 kg, but first it is essential to understand the limits of controllability, the magnet design issues, and the dynamic properties of the system with an HTS magnet. The paper describes experimental tests undertaken to achieve this understanding.

2. Magnet design

As reported in ref. [2], dynamic testing was performed on a first experimental superconducting magnet [3] shown in figure1. The results indicated that, while a closed loop control of the excitation of the superconducting magnet was achieved, high flux leakage in the magnetic circuit reduced the effect of changing reluctance on the s/c coils. Figure 2 shows an improved magnetic circuit using the same 4 BSCCO (2212) s/c coils as in the first experimental magnet. The total number of turns is 200, the critical current is 1 A at 77.3 K, and the air-gap of the magnetic circuit is 7 mm. In order to identify the source of ac losses in the magnet, impedance measurements were performed with both laminated and solid iron cores.

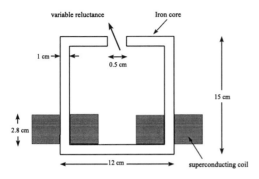

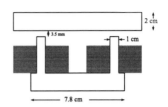

Figure 1 Experimental test magnet

Figure 2 An improved magnetic circuit

Finite element modelling of the two magnetic circuits is shown in figure 3. It is evident from this figure that considerable improvement has been achieved in the modified magnet of figure 2 as compared to the experimental test magnet.

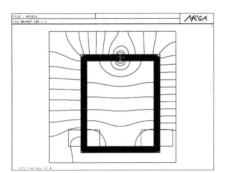

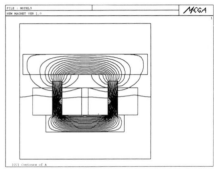

Figure 3 Finite element modelling of the two magnetic circuits

3. Dynamic testing

The ac impedance of the modified superconducting magnet of figure 2 has been measured as a function of frequency (0.1 - 100 Hz) at liquid nitrogen temperature. The test circuit used in these measurements is shown in figure 4. The magnet excitation was provided by a continuous power amplifier, although a switched mode amplifier has also been used. The magnet can either be current or flux controlled in a closed loop manner. The mode of operation of the present testing is to set the current at a nominal value and to inject a sinusoidal signal (of amplitude up to ~ 20% of the nominal value) from a Frequency Response Analyser, which is used to measure the magnitude and phase of the current and voltage variation.

Figure 5 shows typical results for the ac impedance of laminated and solid core at a steady current level of 0.5 A superimposed with an ac signal of amplitude 0.04 A(rms). The ac impedance is dominated by the inductive behaviour in both cases. The resistive component in the laminated core magnet is extremely small for frequencies up to ~ 20 Hz. It is also evident that the lamination of the iron core has resulted in a considerable reduction of the resistive component.

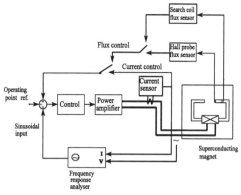

Figure 4   Test circuit for dynamic testing

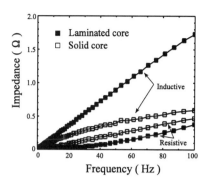

Figure 5 Typical graph of the inductive and resistive components of the ac impedance of the laminated and solid core magnets at I_{dc}=0.5 A and I_{ac}=0.04   A(rms).

4. AC losses

The frequency dependence of the power losses (P= I_{ac}^2 R) in the laminated and solid core magnets is shown in figure 6. Two observations are evident from this figure: firstly, the ac losses are smaller in the laminated core magnet, and secondly, the frequency dependence is different in the two cases. The frequency dependence can be represented by a power law: $P \propto f^n$ with n≈ 1.8 for laminated core and n≈ 1 in the solid core case. These relationships are shown as solid lines in figure 6. It is worth observing that the dc losses in copper coils of the same size would be around 20 mW, and so these losses are tiny by comparison.

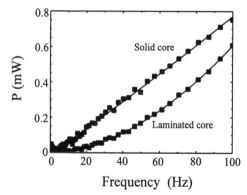

Figure 6 Frequency dependence of the power losses in laminated and solid core magnets at I_{dc} =0.5 A and I_{ac} = 0.04 A(rms).

The ac amplitude dependence of power losses was also studied at different frequencies as shown in figure 7. In both laminated and solid core magnets, the ac amplitude dependence can be described by $P \propto I_{ac}^m$ with m= 2.0 - 2.2. This square-law dependence is an indication that the main contribution to the ac losses is from eddy currents in the iron core. It is anticipated that a power law with different exponents for the frequency dependence of P is due to the fact that skin depth effect in the laminated and solid cores is different. The skin depth (δ) of iron is about 3 -10 mm for frequency range 0.1 - 100 Hz. The fact that δ is much larger than the thickness

706

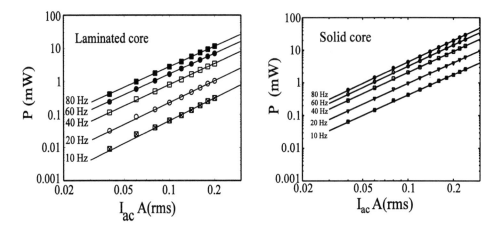

Figure 7 Amplitude dependence of the power losses at I_{dc}=0.5 A.

of the lamination (~ 0.3 mm) and smaller than the radius of the solid core (10 mm) causes a different frequency dependence.

These results indicate that the main source of ac losses is the eddy currents in the iron cores and silver substrate of the tapes. The hysteresis losses in the superconductor are not important factor in the total losses in the system.

5. Conclusions

Tests on iron-cored magnets using HTS coils have been carried out with high performance closed loop current control in order to assess the dynamic effects of variations in amplitude and frequency about different operating conditions. This has demonstrated that the loss in the superconductor itself is not a significant portion of the overall losses, and indicates excellent prospects for the use of such magnets in controlled applications such as Maglev.

Acknowledgement. The authors are grateful to the UK's Engineering and Physical Sciences Research Council for supporting this work (GR/J40089).

References

[1]   Goodall R M and MacLeod C J 1995 Proc. IEEE conf. on control and its applications Albany USA, 28-29 September.

[2]   Goodall R M, MacLeod C J, El-Abbar A A, Campbell A M, Jones H, and Jenkins R G 1994 Proc. Appl. Supercond. Conf. Boston USA 16-21 October.

[3]   Jenkins R G, Jones H, Yang M, Goringe M J, and Grovenor C R M 1994 Proc. Appl. Supercond. Conf. Boston USA 16-21 October.

Inst. Phys. Conf. Ser. No 148
Paper presented at Applied Superconductivity, Edinburgh, 3–6 July 1995
© 1995 IOP Publishing Ltd

4.8T Cryogen Free NbTi magnet, Operated at 5 K with High Tc BiSrCaCuO (2212) Current Leads.

Lisa Cowey, Kevin Timms, Mike Biltcliffe, Peter Cetnik

Oxford Instruments, Eynsham, Oxford, UK

Abstract-A superconducting NbTi magnet has been operated without liquid cryogens, utilising the cooling power of a rare earth enhanced, 2 stage Gifford-McMahon cryocooler. The magnet generated 4.8T in a 9.15cm bore at a current of 100A at a temperature of 5K At this temperature the magnet was calculated to be operating at ~100% short sample; the peak field on the coils was 6.34T and the stored energy 25.3KJ. Base temperature of the system at zero current was 3.7K; base temperature of 5K for the magnet during operation was limited by non optimised thermal anchoring of resistive components in the system. Quench determining critical current at 4.2K was expected to be 145A at peak field of 9.17T, corresponding to a central field of 6.95T. Current was supplied via High Tc BiSrCaCuO (2212) current leads. The system demonstrated long term stability and the magnet, fitted with conventional diode/resistor protection, was successfully quenched without any damage to the system; full recovery from quench was reached within an hour.

1. Introduction

Superconducting magnets, operating in the absence of liquid cryogens, are increasing, both in number and diversity. The use of mechanical coolers makes it possible to dispense with liquid helium as a refrigerant and operate superconductors at temperatures other than the 4.2K helium boiling point. By utilising a refrigeration capacity of a few watts around 10K, Nb3Sn, with Tc of ~18K, has been shown to supply moderate fields at intermediate temperatures. The feasibility of designs for 12K operation of Nb3Sn coils, [1] [2] are now well established. Using Nb3Sn magnets, systems have been run successfully to yield 2T at 14K [3], 4.6T at 11K,[4], 2.0T at 10K[5], 0.8T at 8K [6]. Other example can be found in recent literature. The range of fields is a reflection of the diversity of the individual systems and their envisaged applications, with room temperature bore access ranging from 180mm to 38mm.

The combination of rare earth regeneration material, replacing lead in heat exchangers and ceramic HTc material replacing brass and copper current leads has allowed the development of practical 4K coolers. Using the rare earth regeneration material Erbium-Cobalt-Nickel, 4.2K has been reached using a two-stage Gifford-McMahon (G-M) refrigerator [7] providing 0.95 watts of cooling power at 4.2K, while Erbium-Nickel has been shown to give 1.15W of cooling power at 4.2K [8].

The HTc leads substantially reduce the heat leak to the system both by decreasing the heat conductivity and the ohmic dissipation. Ceramic superconductors have a thermal conductivity some 400 times lower than copper at the same temperatures. A Tc of around 90K reduces the ohmic dissipation to effectively zero below this temperature.

Cooling powers of some 0.5W at 4K are now easily attainable with commercial refrigerators for cooling the magnet and cold stage. The decrease in temperature allows the use of NbTi, Tc ~ 9K. Fields of 6T have now been acheived using NbTi magnets [9]. NbSn/NbTi nested coils were the natural extension and fields up to 10T are now being claimed. A logical progression to HTc magnets operating at elevated temperatures is now being seen with 2.0T in a BiSrCaCuO coil, operating at 30K now demonstated.

708

2. System Design

The system described is shown schematically in Figure 1. It consists of a 2 stage G-M cooler with the magnet supported on the second stage and the HTc current leads anchored between the two stages. A metal shroud encloses the magnet and ensures good thermal contact is maintained between the base of the magnet and the second stage platform. A radiation shield surrounds magnet and leads and a vacuum can encases the entire assembly. Resistor/diode quench protection is mounted onto the second stage. Rhodium-iron temperature sensors are placed on the top and bottom of the magnet casing, on the first stage, the top radiation shield and warmer end of both HTc leads. A 19.6 mm room temperature bore is provided, suitable for VSM or similar lab. scale experiments. The remaining empty bore space is currently being filled with a small HTc magnet, designed to add a further 2T to the central field.

2.1 Cooler

Cooling was supplied by a rare earth enhanced, 2 stage Gifford-McMahon cryocooler. The close cycle cooler provides first stage cooling power of 50W at 50K, and simultaneous second stage cooling power of 0.5W at 4.2K. The ultimate bottom temperature (no load) is below 4K. The first stage acts to cool the magnet radiation shield and to thermally anchor the upper end of the HTc current leads, while the second stage supports and cools the magnet. The first stage approaches a base temperature of 33 K in 3 hours. The second stage reaches base temperature of 3.8 K within 17 hours.

2.2 Magnet

The magnet was constructed using conventional winding methods, proven for NbTi helium cooled magnets; no special arrangements were made during winding to optimise heat conduction from the coil to the cold stage. Several grades of NbTi wire were used and quench determining critical current at 4.2K was expected to be 145A at peak field of 9.17T, corresponding to a central field of 6.95T.

The coil had an inner diameter of 91.5mm, outer diameter of 196mm and a height of 50.8mm. The combined cold mass of magnet and shield was 140Kg and the inductance was 5.06H.

The magnet was thermally anchored to the second stage of the cold head using a clamping arrangement. The temperature of the magnet surround was seen to rise during operation. The temperature increase was shown to be only partly related to the rate at which magnet current was swept. This suggests that it was a product of eddy current heating and resistive heating in the portion of current leads between the HTc ceramic termination and the magnet. The sensors mounted on the bottom and top of the magnet surround read 4.8K and 4.9K respectively just before quench. With the existing arrangement it proved impossible to reduce magnet shield temperature below these values when the current was 100A. At 100A and 5K the magnet was calculated to be operating at ~100% short sample. At this current the peak field on the coils is 6.34T and the stored energy is 25.3KJ.

2.3   Current Leads

The current leads used on this system have been described in some detail in an earlier publication [5]. They are assembled from three discrete sections. From room temperature to the first cooling stage they were composed of brass strands, optimised for currents of 100A in the temperature range 300K to 50K. From the first to second cooling stages they were HTc ceramic rods and from the second cooling stage to the magnet they were a mixture of NbTi and copper strands.

The HTc ceramic section was composed of BiSrCaCuO, (2212) rods supplied by Hoechst. Tc was ~ 90K. Both rods were 12cm long and 8mm diameter. When the central bore field is at 5.0T, the leads experience both self field and magnet stray field. At 100A operating current, the lead temperature is seen to range from 50K at the warmer end, to 5K at the cooler end. The stray field varies from 0.15T to 0.78T in this region. The experimentally determined Jc(B,T) characteristics of the leads shows them to lie safely within the Jc limits of the leads over this temperature and field range. No quench was seen in the leads and no damage to them was detected as a result of a quench in the magnet. The ceramic rods were temperature cycled from room temperature to operating temperature a number of times without any perceived deterioration in their subsequent performance.

3.   System Performance

In early tests of the system, the magnet shroud temperature reached a base of 5K and had risen to 6K when the system quenched at 77A. By improving the thermal contact between shroud and cold stage the measured temperature was seen to fall to a base of <4K, rising with current to 4.9K with an associated increase in quench current to 100A. No further improvement was possible with either further reductions in sweep rate or extended recovery periods between current rises. This suggests that heat input to the second stage was being dominated by the ohmic heating in the final section of the current leads. 100 A is the calculated limiting current for the magnet at 5K. It seems likely that a 4K operating temperature would have seen the central field rise to its expected limit of ~7T.

The modifications made to the system to improve the operating temperature were relatively simple. However, the production of heat during sweep and the requirement for rapid sweep rates in many experimental applications suggests that thought should be given to improving the heat extraction from the inside of cryogen free coils.

The system was deliberatly quenched a number of times. The second stage temperature rose rapidly from ~5K to nearly 40K. The first stage temperature rose towards 50K as the current was increased in the magnet but was not itself adversely effected by the quench. No damage to the current leads or any other part of the system was detected and thermal recovery was complete within one hour. The magnet was subsequently run and quenched several times. No degradation in performance was detected.

710

4. *Conclusions*

The successful operation of cryogen free magnets has now been well proven and documented. Both low Tc Nb3Sn, NbTi and HTc BiSrCaCuO materials have been used to demonstrate fields which are useful for a variety of research and commercial applications.

By developing both the conductors and the coolers together it becomes possible to match superconducting performance and operating temperature in a way which is not fully exploited or optimised when operating temperatures are dictated by the boiling points of available, safe cryogens. Although the engineering designs of such systems require new thought and consideration, there seems to be no intrinsic reason why fields of the magnitude now achieved with conventional wet superconducting magnet should not become available with dry systems.

References

[1] Hoenig, M.O., IEEE Transactions on Magnets 1983 MAG-19 3 880-883
[2] van der Laan, M.T.G. *et al.* Proc. 11th Intl. Conf. on Mag. Tec. (MT11) 1989 1366-1370
[3] Furuyama, M. *et al.* Advances in Cryogenic Engineering Prenum Press, NY 1990 35 625-631
[4] Watanabe K., *et al.* Jpn. J. Applied Physics 1993 32 L488-L490
[5] Cowey *et al.* to appear in Proc. Applied Superconductivity Conference, Boston 1994.
[6] Trifon Laskaris E., *et al.* Cryogenics 1994 24 ICEC Supp. 635
[7] Kuriyama T., *et al.* Advances in Cryogenic Engineering 1994 39 1335-1342
[8] Seshake H. *et al.* Advances in Cryogenic Engineering 1994 37 B 995-1001
[9] Kuriyama T., *et al.* Cryogenics 1994 24 ICEC Supp. 643

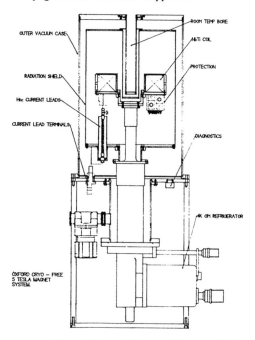

Figure 1. Schematic Diagram of Magnet and Cooler

Inst. Phys. Conf. Ser. No 148
Paper presented at Applied Superconductivity, Edinburgh, 3–6 July 1995
© *1995 IOP Publishing Ltd*

Use of SMES for Compensation of System Perturbations

Prof. Dr.-Ing. E. Handschin, Dipl.-Ing. A. Altmann, Dipl.-Ing. M. Schroeder, Dr.-Ing. Th. Stephanblome, Dipl.-Ing. A. Tromm

Center of Innovative Energy Conversion and Storage (ZEUS), Gelsenkirchen, Germany

Abstract - Due to system disturbances such as short interruptions, voltage dips, flicker and harmonics high sensitive industrial production processes may be affected and subsequently cost intensive production interruptions can occure. In order to cope with such disturbances high power has to be allocated within short times. An active compensator consisting of a SMES and a self commutating power conversion system guarantees high power quality by means of a simultaneous and independent active and reactive power compensation. In this paper specifications for the SMES, the power electronics and the system control are presented. Based on these results such a SMES-system is currently built up at the ZEUS-laboratories.

1. Introduction

An increasing number of industrial production processes and equipments (disturbance sinks e.g. micro processors, power electronics) has become sensitive to deteriorations of voltage quality. These deteriorations of voltage quality can be *voltage fluctuations (flicker)* or *harmonics* caused by disturbing sources (renewable energy applications, customers with special load characteristics) transferred by the electric networks, as well as *outages, short interruptions* or *voltage sags* caused by faults originated in the power systems. Such disturbances bear, in some cases, misoperation and/or damages causing very costly interruptions to industrial processes. Besides it is desirable to achieve a constant active power consumption in order to avoid the effect of a negative sequence system, e.g. the retarding torque of electrical machines [1].

Concerning the problems of system perturbations it is necessary both to reduce or cancel the effect of the disturbance sources on the power systems and to protect sensitive customers (disturbance sinks) from system perturbations in the power systems. The first possibility to reduce such system perturbations on the consumer's side is the enhancement of the short circuit power at the bus where the consumer is connected to the power system. This can be problematic because it has become more and more difficult to get the necessary permission for the reinforcement of existing power systems. The other possibility is the compensation of the system perturbations with passive or active compensators. It is preferable to use an active compensator as it can accomplish multiple tasks due to the customer's requirements mentioned above.

The aim of this paper is to illustrate the engineering of an active compensator consisting of a power conversion system and a system control.

712

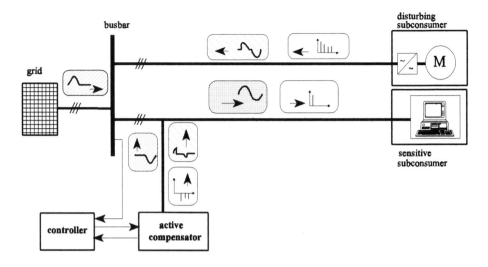

Fig.1: Topology of a system with an active compensator

2. Active Power Line Conditioning (APLC) by Superconducting Magnetic Energy Storage (SMES)

In order to fulfill the requirements mentioned above it is necessary to develop an intelligently controlled power conversion system and to provide a suitable kind of energy storage. Main features of this energy storage system must be:

- *Minimum access time to the stored energy,*
- *A high power rating rather than a high energy rating.*

A SMES with a low energy capacity (micro-SMES) of the High-Critical-Temperature-Type (HT_C-Type) combines the fulfillment of the requirements stated above with a high amount of economical attractiveness. In contrast to low-temperature superconducting materials ($T_C < 4$ K) the HT_C-Type is made of ceramic and operates at T = 20 K or even higher temperatures. That results in a considerable reduction of the rated power of the cryogenic system. Furthermore for the HT_C-SMES the classical quench destroying the coil is impossible and thus excellent performance characteristics and maximum reliability are guaranteed.

To utilize these optimal qualities an appropriate Power Conversion System (PCS) has to be chosen. Though different converter topologies are conceivable, ZEUS pursues a system concept containing three components: a three-phase voltage-source-inverter (VSI), a capacitor as d.c.-link and a voltage chopper, forming the connection to the HT_C-micro-SMES. The used switchable power electronic devices are of the IGBT-Type in order to ensure the possibility of high switching frequencies. The drawback of the VSI-concept, demanding the chopper as additional energy conversion level, is outmatched by a number of advantages, concerning the easy control of the VSI, the additional degree of freedom gained by the voltage chopper and the buffering behaviour of the d.c.-link-capacitor [2].

3. System Control

The controller shown in Fig. 1 must guarantee an event-controlled activation of active and reactive power with respect to the operational requirements defined above. On the other hand the controller has to operate permanently to ensure optimal operating conditions for both, the superconducting coil and the d.c.-link-capacitor. Because of constantly changing operating conditions the performance of the controller has to be optimal within a wide operating area. In order to meet the requirements of such an event driven, robust, multipurpose, multivariable controller, the use of a fuzzy-controller is proposed.

The fuzzy-controller represents a nonlinear mapping from the input to the output variables. The use of the fuzzy-set-theory allows a transparent design of the controller. The fuzzy-controller derives operating conditions from the measured values and selects its appropriate action from heuristic rules which are describing the desired performance of the fuzzy-controller. Because the rulebase can be expanded easily the used fuzzy-control-scheme is of great advantage in case additional control tasks have to be considered.

The fuzzy-controller itself has no dynamic component; i.e. it can immediately perform the correct control action. The fuzzy-set-theory is a generalization of the classical set theory. The degree of membership of a measured value to a fuzzy set is usually described by a weighting factor in the intervall $0 \leq \mu \leq 1$. Fuzzy sets are used to build a model of the linguistic description of the measured quantities (linguistic variables). The numerical values of the measured quantities are mapped according to their degree of membership μ to the linguistic attributes by means of nonlinear membership functions. The performance of the fuzzy-controller is established by the formulation of heuristic rules. The *if-part* of the rules is a logical connection of the attributes describing the linguistic variables.

To illustrate such characteristics of the system controller, the control structure of the busbar voltage $U = U_{busbar}$, the d.c.-link voltage U_d and the superconducting coil current I_L is shown in detail in Fig. 2. The output variables of the fuzzy-controller are the set values for the duty cycle d of the chopper and the control angel α_{VSI} of the VSI. In order to keep up constant availability of the compensator, U_d has to be kept inside the hysteris band between $U_{d,low}$ and $U_{d,high}$ whereas the desired state of the superconducting coil is half charged ($I_L=I_{L,50\%}$). For U_d as well as for I_L certain maximum and minimum values have to be guaranteed. The voltage U is the control value which should be compensated for. Only symmetrical events should be taken into consideration with respect to high transparency of the controller performance. In Fig. 2 the application of one rule to a certain operating point of the entire system is illustrated. The rulebase is completed by a number of rules determinating the controller performance according to the power system's operational requirements, the needs of the superconducting coil and the d.c.-link as well. In order to derive the control action in the inference process the weighting factors of the *if-part* are set to the weighting factors of the *action-part* of each rule. Hence the result is a linguistic description of the output variables from which a numerical value has to be derived for the control action. For this purpose the fuzzy sets of the output variables are weighted with the weighting factors according to the given rules. In Fig. 2 this is shown for the application of one rule. Then the corresponding areas as a result of the application of all other rules are determined. The output variable is determined by calculating the center of gravity of the area that results from the superposition of all areas [3].

714

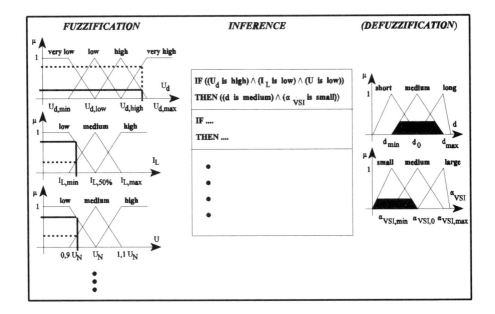

Fig. 2: Fuzzy-controller

4. Conclusions and prospects

For the purpose of compensating system perturbations a SMES-system containing a self commutating converter unit is applied. The use of a HT_c-micro-SMES yields an economic attractive design of the cryogenic system. In order to obtain an event driven, robust and multivariable controller, this paper presents a system control basing on the fuzzy-set-theory.

The work done within the system engineering is basis of a collaboration with American Superconductors Inc. (ASC) concerning a hardware realization of a HT_c-micro-SMES at the ZEUS-laboratories in Gelsenkirchen, Germany at the beginning of 1996. The scope of this project is to simulate the supply situation of certain customers on the flexible network model at ZEUS and to develop concepts for solving their problems by means of a HT_c-micro-SMES. Because of the design of the power conversion system and the HT_c-micro-SMES as well, the results and the concepts obtained from laboratory investigations can be scaled up to customer's real needs.

5. References

[1]   Burke J J et al 1990 *"Power Quality - Two Different Perspectives" IEEE Transactions on Power delivery Vol.5, No. 3, July 1990* 1501-1513

[2]   Kustom R L et al J J 1991 *"Research on Power Conditioning Systems for Superconductive Energy Storage (SMES)" IEEE Trans. on Magnetics* **27** 2320-2323

[3]   Pedrycz W 1989 *"Fuzzy Control and Fuzzy Systems" Research Studies Press Ltd.* Taunton Somerset England

Inst. Phys. Conf. Ser. No 148
Paper presented at Applied Superconductivity, Edinburgh, 3–6 July 1995
© *1995 IOP Publishing Ltd*

Critical current of a Bi-2223/Ag pancake coil

M Lahtinen, J Paasi and Z Han[*]

*Laboratory of Electricity and Magnetism, Tampere University of Technology, P.O. Box 692,
FIN-33101 Tampere, Finland*
[*]*NKT Research Center A/S, DK-2605 Brøndby, Denmark*

Abstract. In this paper we study the current-voltage (I-V) characteristics of a Bi-2223/Ag pancake coil. The measurements were done in self-field in the temperature range 4.2K-108K. The magnetic field dependence of the critical current (I_c) was studied in more detail at 77K in external magnetic field range $B_e=\pm160$mT. At small currents $I<I_c$ the I-V characteristics show temperature dependent resistive tails. In low temperatures and $I>I_c$ the I-V characteristics are excessively slanted, which is caused by current sharing between the Bi-2223 ceramic and the Ag sheath. At 77K the coil self-field is found to have a significant effect on both the $I_c(B_e)$ curves and I-V characteristics.

1. Introduction

The electrical performance of Bi-2223/Ag tapes is usually evaluated using short sample tests. Bi-2223/Ag tapes are now routinely produced in long lengths with reproducible results. Hence, also the evaluation of tape performance in coil samples has become common. The behavior of a coil is, however, different from the behavior of a short tape, mainly due to the larger self-field (B_s) of the coil. In this work we discuss the current-voltage (I-V) characteristics and the critical current (I_c) of a small Bi-2223/Ag pancake coil. Especially, we try to distinguish the behavior of the Bi-2223 ceramic from the effects caused by B_s and sample defects.

2. Experimental

The 13 turn pancake coil was fabricated from a monofilamentary, 107cm long, Bi-2223/Ag tape. The area of Ag in the total tape cross-section 2.8mm×0.1mm was $A_{Ag}=0.22$mm^2. The coil dimensios were: inner diameter 25mm, outer diameter 27.5mm and height 2.8mm. The tape length between the voltage taps was $l\approx100$cm. The calculated self-field of the coil was $B_s=0.63$mT/A in the coil center and the maximum field in the coil winding was 2.6mT/A. As a check of possible short-circuits between the turns, the B_s in the coil center was measured using a Hall sensor. The measured value $B_s=0.60$mT at $I=1$A agreed well with the calculations.

The coil I_c was determined using a voltage criterion $V_c=100\mu V$ over the voltage taps, which corresponds to a uniform electric field $E=1\mu V/cm$ in the coil winding. The temperature dependence of I_c was studied from 4.2K up to the sample transition temperature 108K. In the $I_c(T)$ measurements the coil was cooled by LHe and He vapor in the temperature range 4.2K-75K and by LN_2 and N_2 vapor in the 77K-108K range. In this study, small sample size and slow warming rate were necessary to minimize temperature variations in the sample volume.

2. Results and discussion

2.1 Magnetic field dependence of the critical current

The coil I_c in self-field was 3.1A at 77K. The $I_c(B_e)$ dependence was measured in increasing B_e after zero-field cooling. The measurements were carried out for three B_e and B_s orientations: B_e perpendicular to B_s $I_c(B_e{\perp}B_s)$, B_e parallel to B_s $I_c(B_e{\uparrow\uparrow}B_s)$ and B_e antiparallel to B_s $I_c(B_e{\uparrow\downarrow}B_s)$. In all B_e orientations the $I_c(B_e)$ showed a steep drop in the range $B_e=5mT\text{-}30mT$, Fig. 1. At larger values of the external field, $B_e>80mT$, $I_c(B_e{\uparrow\uparrow}B_s)$ and $I_c(B_e{\uparrow\downarrow}B_s)$ curves overlapped indicating that the effect of B_s on the sample behavior was negligible in this B_e range. However, the $I_c(B_e{\perp}B_s)$ was smaller than $I_c(B_e{\uparrow\uparrow}B_s)$ and $I_c(B_e{\uparrow\downarrow}B_s)$ showing that the Bi-2223/Ag tape was more sensitive to B_e orientation normal to the tape than to B_e parallel to the tape surface.

In low external field, $B_e<30mT$, the effect of B_s was not negligible. In the $I_c(B_e)$ measurement it was most clearly demonstrated as the difference between the curves $I_c(B_e{\uparrow\uparrow}B_s)$ and $I_c(B_e{\uparrow\downarrow}B_s)$ in Fig. 1. The $I_c(B_e{\uparrow\uparrow}B_s)$ curve starts to decrease immediately as B_e is applied whereas $I_c(B_e{\uparrow\downarrow}B_s)$ has a maximum at a nonzero value of B_e at $B_e=4mT$. The maximum is due to the fact that on the inner surface of the coil, where the self-field is the strongest, the total field $B=B_e+B_s$ decreases as B_e increases. Similar behavior in a bulk YBaCuO ring has been reported by Polák et al [1].

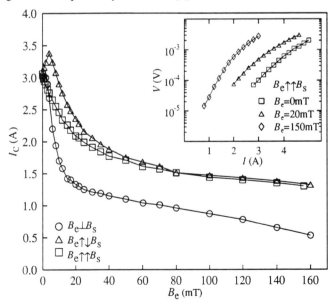

Figure 1. The $I_c(B_e)$ dependence for different B_e orientations, $V(I)$ curves for $B_e{\uparrow\uparrow}B_s$ are shown in the inset.

The self-field has an influence also on the I-V characteristics. At small values of B_e the magnitude and direction of B were nonuniform in the sample volume. Thus the different sections of the Bi-2223/Ag tape operated under different values of local electric field E, because the tape I_c was very sensitive to B. This resulted in slanted $V(I)$ characteristics near I_c, the $B_e=0$mT and $B_e=20$mT cases in Fig. 1 inset. At higher B_e values, B in the coil winding was more uniform leading to steeper $V(I)$ curves, the $B_e=150$mT case in Fig. 1 inset.

2.2 Temperature dependence of the critical current

Fig. 2 shows the measured I-V characteristics (markers) at three temperatures: 77K, 20K and 4.2K. The I-V characteristics have two distinct features. First, at $T=20$K and $T=4.2$K the $V(I)$ curves become more and more slanted as the voltage increases above $V=100\mu$V. Second, all the curves show resistive tails in the voltage range below $V=40\mu$V. The behavior is very similar to that observed by Fukumoto et al. [2] in short Bi-2223/Ag tapes. In the following the relationship of these effects to the temperature dependent sheath resistivity $\rho_{Ag}(T)$ is discussed.

At low temperatures, $T=20$K and $T=4.2$K, the ρ_{Ag} is very small. As I increases, the Bi-2223 core begins to show non-zero resistance and larger and larger fraction of I flows in the Ag sheath (I_{Ag}), in parallel with the current in the Bi-2223 core (I_{SC}). From the tape geometry and published values of $\rho_{Ag}(T)$ for pure Ag [3],[4] we calculate the resistance of the Ag sheath between the voltage taps: $R_{Ag}(77\text{K})= \rho_{Ag}(77\text{K})\times l \,/\, A_{Ag}= 0.25\mu\Omega\text{cm}\times100\text{cm}/2.2\times 10^{-3}\text{cm}^2= 11\text{m}\Omega$. Similarly we obtain $R_{Ag}(20\text{K})=0.19\text{m}\Omega$ and $R_{Ag}(4.2\text{K})=0.091\text{m}\Omega$. In Fig. 2 the solid lines represent $V(I_{SC})$ curves, where the contribution of I_{Ag} has been subtracted from the applied I using the formula $I_{SC}=I-V/R_{Ag}(T)$ [3]. Clearly for $V>100\mu$V the measured $V(I)$ curves at $T=20$K and $T=4.2$K become excessively slanted to the right because I_{Ag} becomes significant when compared to I_{SC}.

At 77K the coil $V(I)$ curve shows a resistive tail below $V=40\mu$V, Fig. 2. The dashed curve next to the I-V characteristic represents the voltage over a $R=12\mu\Omega$ resistor. The resistor

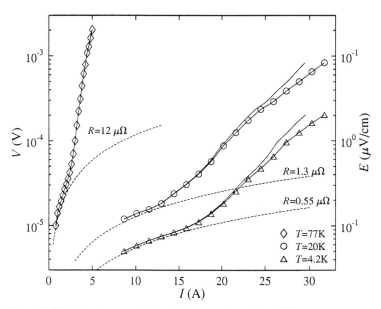

Figure 2. The self-field I-V characteristics of the pancake coil at different temperatures.

value is chosen so that the resistive voltage $V=12\mu\Omega\times I$ matches the resistive tail of the $V(I)$ curve. By a similar matching procedure the resistance values $R=1.3\mu\Omega$ and $R=0.55\mu\Omega$ are obtained corresponding to the resistive tails of the $V(I)$ curves at $T=20K$ and $T=4.2K$. The resistance values R are decreasing with decreasing temperature, which illustrates the strong temperature dependence of the resistive tails.

We believe that the resistive tails are due to transport current flow in the Ag sheath, although the detailed origin of the behavior remains unclear. The current may be forced into the Ag sheath by local defects in the Bi-2223 core or finite current may flow in the sheath due to large surface resistivity at the Ag-ceramic interface.

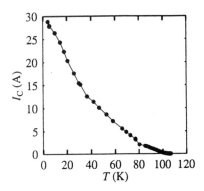

Figure 3. The coil $I_c(T)$ dependence.

Fig. 3 shows the $I_c(T)$ plot determined using the $V_c=100\mu V$ criterion. At this voltage the effects of tape defects and current sharing were small. Hence it is expected that the plot mostly reflects the properties of the superconductor. The $I_c(T)$ plot shows three regions, which are characterized by different slopes of $I_c(T)$ dependence. We believe that the changes in the slope are accompanied by changes in mechanisms limiting the coil current. Starting from $T=108K$, the first change occurs near 80K. Above 80K I_c shows only moderate increase with decreasing T. The change in slope at 80K may be associated with the transition of secondary Bi-2212 phase into the superconducting state, thus resulting in an increase of I_c. Another change of slope occurs near 40K. This change may be related to enhanced flux pinning in low temperatures. We believe that below 40K the role of sample self-field in limiting the coil performance is less important than above 40K.

3. Summary

The experimental results indicate the important roles of sample field B_S and Ag sheath resistivity ρ_{Ag} in the interpretation of pancake coil I-V characteristics. At 77K B_S caused a $I_c(B_e)$ maximum at a non-zero value of B_e, when B_e and B_S were antiparallel. The resistive tails of I-V characteristics at $I<I_c$ were attributed to transport current flow in the Ag sheath. In low temperatures and $I>I_c$ the current sharing between the Ag sheath and the Bi-2223 resulted in slanted I-V curves. As a conclusion, the $I_c(B_e)$ and $I_c(T)$ experiments revealed effects which complicate the optimal design of Bi-2223/Ag magnets.

References

[1] Polák M, Majoroš M, Hanic F, Pitel J, Kedrová M, Kottman P, Talapa J and Vencel L 1989 *J. of Supercond.* **2** 219-233

[2] Fukumoto Y, Li Q, Wang Y L, Sueanaga M and Haldar P 1995 *Appl. Phys. Lett.* **66** 1827-1829

[3] Matthews D N, Müller K-H, Andrikidis C, Liu H K and Dou S X 1994 *Physica C* **229** 403-410

[4] Iwasa Y, McNiff E J, Bellis R H and Sato K 1993 *Cryogenics* **33** 836-837

Inst. Phys. Conf. Ser. No 148
Paper presented at Applied Superconductivity, Edinburgh, 3–6 July 1995
© 1995 IOP Publishing Ltd

Development of Bi-2212/Ag multilayer structures and monolithic pancake coils prepared by tape casting, elastomer processing and composite reaction texturing[†]

M. Chen[*‡], D.R. Watson[*#], A.J. Misson[*], B. Soylu[*], B.A. Glowacki[*#] and J.E. Evetts[*#]

[*]Dept. of Materials Science and Metallurgy and [#]IRC in Superconductivity, University of Cambridge, Pembroke Street, Cambridge, CB2 3QZ, UK
[‡]now at ABB, Corporate Research, CH-5405 Barden-Dättwil, Switzerland

A process has been developed for the synthesis of [Bi-2212+MgO-fibres]/Ag multilayers in bulk form and monolithic pancake coils by co-reacting laminated tapes of Bi-2212, Ag, MgO and Al_2O_3. Green tapes, prepared by tape casting and elastomer processing, were laminated into multilayer structures, subjected to slow burn-off and co-reacted in oxygen to produce densified and well bonded multilayer structures. The high degree of texture observed in the 500μm thick Bi-2212+MgO-fibres layer is induced by specifically aligned MgO-fibres. Multilayers have been processed with various thickness of Bi-2212, Ag and insulating layers with the Ag layer providing current contacts with contact resistivity of $1\mu\Omega cm^2$. The I_c and J_c of the co-reacted Bi-2212 layer are typically 80A and 2200Acm^{-2} respectively at 77K. VSM measurements indicated a J_c in excess of 5.10^4Acm^{-2} at 5K and 12T. Monolithic pancake coils with up to 25 turns have been fabricated with a bore of 1cm and OD of 10 cm. The Bi-2212 occupies over 50% of the total volume of the monolithic coil. The continuous Al_2O_3 layer buffered with Ag insulates the Bi-2212 turns and does not affect the properties of Bi-2212 thus demonstrating an alternative method for fabricating pancake coils.

1. Introduction

Co-reaction of bulk $Bi_2Sr_2CaCu_2O_{8+\delta}$ (Bi-2212) superconductors and Ag is technologically important for processing low resistance high current Ag contacts, for providing an alternate current path should the superconductor quench, for mechanical and environmental protection and for use as a buffer layer between Bi-2212 and various reactive phases. While co-reacting Bi-2212 and Ag is well demonstrated in Ag sheathed wires and thick tapes, bulk processing with Ag is less so because of Ag solution in the Bi-rich liquid and the subsequent creation of a partial melt gradient through the bulk of a sample during reaction. A recent study [1] showed that doping Bi-2212 powder with Ag enables uniform partial melting of Bi-2212/Ag layered composites. This, combined with Composite Reaction Texturing (CRT) [2,3] which uses MgO-fibres to induce texture, allows the fabrication of textured bulk Bi-2212 multilayers.

High T_c pancake coils for high magnetic field applications are being developed primarily with textured Ag clad tapes [4]. The cross sectional area of sheath material and materials for electrical insulation and mechanical integration result in a low engineering critical current density (J_{ce}). An alternative consisting of the synthesis of multilayer composites from green tapes of Ag powder and Bi-2212+MgO-f is presented here. A ceramic insulating layer is introduced and the concept of co-reacting integrated monolithic pancake coils is demonstrated.

2. Experimental procedure

Stoichiometric Bi-2212 (particle size 2-4μm, Hoechst), Ag powder (3.75μm, Degussa), MgO fibres (MgO-f , 5μm dia., 100μm long, made in house), MgO powder (<100nm, Fisons) and Al$_2$O$_3$ (0.5μm, Alcoa) were employed as precursor materials. Two polymer based processing techniques, tape casting (TC) [1] and elastomer processing (EP) [5] were used to produce flexible green tapes with solid loadings of 40-60vol%. [Bi-2212+MgO] tapes (hereafter referred to as BM) were prepared by mixing Bi-2212 with 5wt% Ag followed by 10wt% MgO-f with the relevant binder system. Characteristics of the processing techniques result in MgO-f randomly distributed in the tape plane. Other tapes produced include: Bi-2212 (without MgO-f), Ag, Al$_2$O$_3$ (with 10vol% Bi-2212) and MgO (with 3wt% LiF). Tapes were laminated to produce Bi-2212/Ag, BM/Ag, and BM/Ag/[MgO or Al$_2$O$_3$]/Ag multilayers. Green tapes of BM/Ag/[MgO or Al$_2$O$_3$]/Ag mutilayer was also wound into pancake coils.

Multilayer preforms were subjected to slow organic burn-off and then reacted in pure oxygen on YSZ substrates [1,3,6]. Samples were heated to 890°C with 20min dwell, cooled to 880°C, followed by slow cooling at 4°Chr^{-1} to 840°C with 20hr dwell. Samples were subsequently postannealed at 770°C for 10hr in Ar-2% oxygen followed by furnace cooling.

Microstructures of the co-reacted composites were characterised using scanning electron microscopy. Transport J_c was measured using the standard four probe method at 77K.

3. Results and discussion

3.1. Densification and J_c characterisation of multilayer structures

Co-densification of layered composites requires all layers to exhibit similar densification and shrinkage behaviour to minimise sintering induced stress and distortion, to remain inert to each other and to show small thermal expansion mismatch. Densification studies were carried out on TC tapes and cold pressed pellets, both with green density of 55-60% theoretical. Ag pellets with 5wt% Bi-2212 started to sinter at 600°C, reaching 97% density at 850°C which is below the partial melting temperature (T_m) of Ag doped Bi-2212 in oxygen. BM tapes achieved almost full density when sintered at 10°C above T_m. Sintering aids consisting of 3 wt% LiF in MgO and 10vol% Bi-2212 (containing 5 wt% Ag) in Al$_2$O$_3$ greatly reduced the respective sintering temperatures, achieving around 95% density at 890°C for both ceramics.

Fig.1 contains SEM micrographs of a reacted and thermally etched cross-section of a Bi-2212/Ag multilayer, showing highly densified layers with a sharp interface. More importantly the Bi-2212 layers are highly textured similar to that observed in Ag clad tapes.The BM layers are 200 μm thick with a highly textured microstructure as shown in Fig. 2. The development of such texture is induced by the planar 2D random distribution of MgO-f [2,3].

 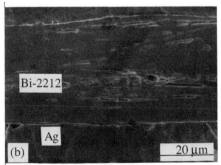

Fig. 1. SEM micrographs of a reacted thermally etched cross section of a Bi-2212/Ag multilayer.

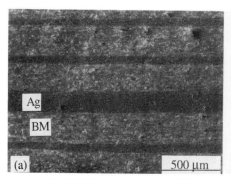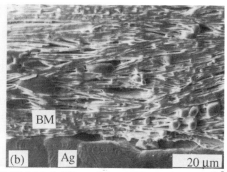

Fig.2. SEM micrographs of a CRT processed [Bi-2212+MgO-f]/Ag multilayered composite.

Fig. 3a shows the microstructure of a BM/Ag/Al$_2$O$_3$/Ag multilayer after CRT processing. The trilayer structure in Fig. 3b indicates that all layers are uniform and highly densified. The BM superconducting layer is buffered by Ag from Al$_2$O$_3$. The layers are bonded and Bi-2212 which occupies over 50% of the whole cross section is highly textured.

All samples showed a T$_c$ onset of 92K. Co-reacted Ag layers have a very low contact resistivity, in the order of 1$\mu\Omega$cm^2 at 77K, thus serving as an effective current contact. The BM/Ag multilayer exhibits a typical transport J$_c$ of 2500 Acm^{-2} in the BM layers, with I$_c$ up to 150A at 1μVcm^{-1} and 77K. In the BM/Ag/MgO(or Al$_2$O$_3$)/Ag multilayer each superconducting layer has a typical I$_c$ of 70-80A at 77K. Both BM/Ag/Al$_2$O$_3$/Ag and BM/Ag/MgO/Ag multilayers show a typical J$_c$ of 2000 Acm^{-2}. The Al$_2$O$_3$ exhibited true insulator properties at 77K while the MgO layers were semiconducting.

3.2. Design and testing of monolithic pancake coils

An ideal pancake coil consists of alternating turns of thick well textured high J$_c$ superconductor and a thin insulator such that the ratio of the cross sectional areas of superconducting to non-superconducting layers is high thereby maximising J$_{ce}$. The difficulty in finding an inert insulator to be co-densified with Bi-2212 led to the development of Bi-2212/Ag/insulator/Ag type coil with Ag as a buffer layer as shown in Fig.4. Two types of pancake coils were developed, coil A and coil B. Coil preforms were made by stacking and joining TC or EP processed green tapes into a multilayer, followed by winding and joining into shape. A co-reacted monolithic coil is shown in Fig.5. The sub-structure of the pancake coil consists of a 500μm thick layer of Bi-2212+MgO-f, two 50-100μm Ag buffer layers and a 100μm Al$_2$O$_3$ layer. The Bi-2212 layer is continuous and electrically insulated between the turns by Al$_2$O$_3$.

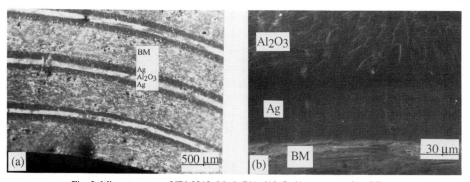

Fig. 3. Microstructure of [Bi-2212+MgO-f]/Ag/Al$_2$O$_3$/Ag co-reacted multilayer.

722

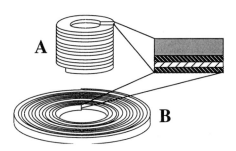

Fig. 4. Schematic of two types of coils showing
Bi-2212/Ag/Al$_2$O$_3$/Ag layers.

Fig. 5. Co-reacted monolithic pancake coil.

These coils are currently undergoing full electromagnetic characterisation however based on a
J$_c$'s of 10^4 and 5x10^4 Acm^{-2} at 20K and 4T and 5K and 12T respectively [3] a 20 turn pancake
coil with a Bi-2212+MgO-f cross section of 5 by 0.5mm can be expected to generate 0.5T and
2.5T respectively under above conditions.

The techniques developed here allow pancake coils of different geometry's to be
processed by co-densification of Bi-2212/Ag/Al$_2$O$_3$/Ag green tapes, followed by CRT
processing to develop bulk texture in Bi-2212. Using these methods, thick cross section coils
can be fabricated with high I$_c$ and J$_{ce}$ and with mechanically integrated turns.

4. Conclusions

A process has been developed for the synthesis of Ag/Bi-2212 multilayer bulk composites and
monolithic pancake coils by co-densification and Composite Reaction Texturing of green tapes
prepared by tape casting and elastomer processing. The multilayer structures were highly
densified with well bonded layers and highly textured Bi-2212 layers. While the texture in the
Bi-2212/Ag multilayer with a 30μm Bi-2212 layer is similar to that of clad Bi-2212 tapes, the
bulk texture in [Bi-2212+MgO]/Ag with superconducting layer around 500μm thick is induced
by the presence of aligned MgO-f. The Bi-2212/Ag interface showed a contact resistivity of
1μΩcm^2, thus providing high current contacts. The co-reacted multilayers containing 500μm
thick Bi-2212+MgO layers show a self-field I$_c$ and J$_c$ of 70-80A and 2000-2500Acm^{-2}
respectively at 77K. The co-reacted pancake coil consists of a 500μm thick Bi-2212+MgO-f
superconducting layer, two Ag buffer layers and an Al$_2$O$_3$ insulating layer. The presence of the
Al$_2$O$_3$ layer does not degrade the Bi-2212 and behaves as a true insulating layer between the
superconducting turns, thus demonstrating a process for fabricating monolithic pancake coils.

References

[1] Chen M and Evetts J E 1995 to be published
[2] Soylu B, Adamopoulos N, Glowacka D M and Evetts J E 1992 *Appl. Phys. Lett.* **25** 3183
[3] Chen M, Glowacka D M, Soylu B, Watson D R, Christiansen J K S, Baranowski R P, Glowacki B A and
   Evetts J E 1995, *IEEE Trans. on Appl. Supercon.*, **5** [in press]
[4] Shimoyama J, Morimoto J, Kitaguchi H, Kumakura H, Togano K, Maeda H, Nomura K and Seido M 1992,
   Jpn. J. Appl. Phys., **31** L163-L165
[5] Watson D R, Chen M, Glowacka D M, Adamopoulos N, Soylu B, Glowacki B A and Evetts J E 1995,
   IEEE Trans. on Appl. Supercon. **5** [in press]
[6] Watson D R, Chen M and Evetts J E 1995, *Supercond. Sci. Technol.*, **8**, 311-316

† Work supported by the European Commission (Brite EuRam II Contract No. BRE2-0208). The authors wish
to acknowledge Dr. W.J. Clegg for providing processing equipment and Dr. A.P. Baker for discussions.

Inst. Phys. Conf. Ser. No 148
Paper presented at Applied Superconductivity, Edinburgh, 3–6 July 1995
© *1995 IOP Publishing Ltd*

Preparation of zone melted YBCO - rods with high current contacts

C. Gross [a,b], **S. Elschner** [a] **and W. Assmus** [b]

a) Central Research, Hoechst AG, 65926 Frankfurt am Main, Germany
b) Phys. Institut, Johann Wolfgang Goethe-Universität, 60054 Frankfurt am Main, Germany

Abstract. YBCO - rods have been manufactured by powder pressing, sintering and subsequent zone melting in a specially designed tube furnace with a steep temperature gradient (100 K/cm in the sample). We discuss different methods of powder pressing (DryBag and CIP), describe the further development of our zone melting process and compare silver contacts which have been produced by various techniques. Rods with sputtered contacts showed lowest contact resistances so far (2.5 $\mu\Omega cm^2$) and carried currents up to 1000 A in selffield.

1. Introduction

YBCO - rods for high current applications should consist of very few domains, at best a single domain. Zone melting of presintered rods has been proved to be a suitable method for producing YBCO - rods with several cm long domains and high current density (j_c = 52.000 A/cm^2 at 77 K, 0 T, 1 $\mu V/cm$) [1-3]. In order to achieve a reasonable growth velocity, a high temperature gradient should be employed during zone melting. This gradient helps to suppress spontaneous seeding in front of the solid - liquid interface and avoids growth of domains with deviating crystallographic orientation. In our furnace, a stable temperature field was established by mounting the heating and cooling elements as close as possible to the sample. The sample (d = 8 mm) is located in a protecting ceramic tube with an inner diameter of 11 mm, which serves as an indirect heater. Because this leaves an average distance of 1 - 2 mm between the sample and the protecting tube, straight YBCO - rods are an essential prerequisite of the zone melting process. Therefore, we investigated the possible advantages of using a DryBag Press instead of a conventional Cold Isostatic Press (CIP) for powder pressing (*see 2.1.*)

For high current applications in electrotechnical devices, e.g. current leads, the High - T_c - rod has to be connected with metallic conductors (usually copper) on both ends. This can be done by putting on silver or gold contacts onto the sample first and soldering it with a low melting solder afterwards. The contacts should not only have a low resistance in order to minimize contact heating under operation, but also have to adhere tightly to the sample, since soldering is inevitably connected with some degree of mechanical stress and might destroy the contacts. Contacts on YBCO - samples with very low resistances (in the order of 10^{-8} Ωcm^2 or even less) have been reported already [4-7]. However, these investigations have been limited to small contact areas (< 0.1 cm^2) and small currents (< 1 A) in most cases. For that reason, the procedures described can not be transferred readily to larger contacts and currents. In particular, measurements of contact resistance should be made with currents comparable with the intended operation current, since a pronounced dependence on the applied current can occur [5]. In this work, contact resistances have been determined with currents up to 450 A (DC) therefore (*see 3.1.*).

2. Sample preparation

2.1. DryBag pressing

In contrast to Cold Isostatic Pressing, the pressing mould is an integral part of the DryBag-Press (see Fig. 1). After filling in the powder mixture, the thick-walled mould is closed by steel stamps on both ends and pressure is applied through an oil chamber. The direction of force is essentially radial. This makes DryBag pressing a convenient tool for elongated samples as rods, but less suited for other shapes (oblate or without rotational symmetry).

DryBag pressed rods came much closer to the desired cylindrical shape than cipped rods. The deviation from the ideal shape could be reduced from more than 1 mm to less than 0.5 mm. This faciliates mechanical treatment (grinding) before sintering, if necessary at all. Although pressure is limited to 170 MPa in our DryBag press (CIP: 300 MPa), a comparable green density (65 % of theoretical) could be achieved. Fig. 2 shows a DryBag pressed and sintered rod before zone melting. Small ceramic pivots have been integrated at the ends of the sample, which simplify fastening in the zone furnace later.

2.2. Zone melting

The sintered rods have been attached to a ceramic support at their lower end and have been mounted vertically in the zone furnace. Afterwards, the furnace has been lowered slowly in order to move a molten zone from the upper to the lower end of the sample (see [2] for details). For shorter rods (< 5 cm) fastening at the lower end is sufficient, since the partially molten zone is mechanically stable enough to carry the recristallized material. For longer rods

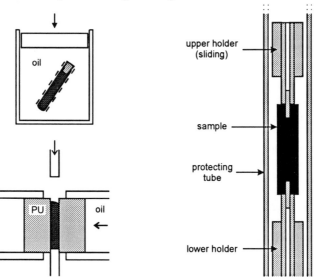

Fig. 1.: left: Principle of Cold Isostatic Press (above) and DryBag Press (below)
right: Sample attachment in the tube furnace (schematic, not to scale)

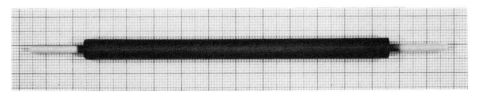

Fig. 2.: DryBag pressed and sintered YBCO - rod before zone melting (length: 11 cm)

an additional fixing of the upper end becomes necessary to avoid contact of the sample with the furnace walls. By use of a sliding holder (*see Fig. 1*), the upper end was stabilized laterally while allowing longitudinal contraction of the sample during zone melting due to remaining pores in the sintered bodies.

2.2. Contact preparation

All contacts have been prepared on zone melted, but not single - domained rods. However, results should be transferable to single - domained material, since the average grain size (1 - 2 cm) was bigger than the typical contact length (1 cm).

Contacts made of *fired silver epoxy* (Demetron E 4037) showed mediocre adhesion on zone melted YBCO, in contrast to reasonable results on polycristalline material [8]. *Thermal evaporation* of silver resulted in poor adhesion, even after an additional annealing step. Contacts produced by *flame coating* stick tightly to the samples and are easy to solder, but reproducibility is bad, since this technique (at least in the used form) involves very rapid silver deposition which is difficult to control. *Sputtering* produced shiny contacts with good adhesion (*see Fig. 3*).

3. Electrical measurements

Samples have been connected with low melting solder (Woods metal with indium addition) to copper leads. A power supply capable of providing 500 A (DC) resp. 1000 A (pulsed) was used. All measurements have been made at $T = 77$ K.

3.1. Contact resistances

In order to differentiate between the resistances of the copper - solder interface on one hand and the solder - superconductor interface on the other hand, two pairs of voltage leads have been employed (*see Fig. 3*). If properly prepared, the first contribution to the over-all resistance is much smaller than the second and the voltages become comparable.

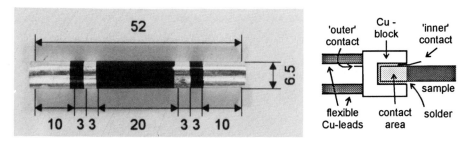

Fig. 3. left: Zone melted YBCO - rod with sputtered silver contacts
right: Connection copper - sample

726

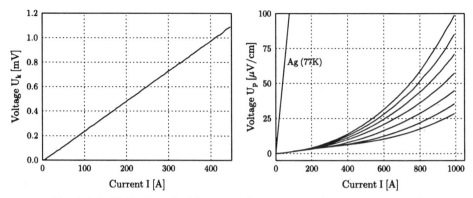

Fig. 4. left: $U_k(I)$ of a rod with sputtered contacts (total contact area: 4 cm²)
right: $U_p(I)$ of a zone melted YBCO rod (cross-section: 0.32 cm²), $B = 0,500,..,3000$ G

Contacts made of fired silver epoxy showed unacceptable high resistances for high current applications. Contacts deposited by thermal evaporation (without additional firing) showed widely varying resistances due to poor adhesion to the samples. Flame coated contacts resulted in a resistance in the order of 5 - 30 µΩcm². Lowest contact resistances could be obtained by sputtering and a subsequent heat treatment (2.5 µΩcm²).

3.2. U(I) - characteristics of the YBCO - rods

The low j_c (100 - 300 A/cm²) reflects the existence of grain boundaries within the sample (*see Fig. 4*). Nevertheless, the rods could carry much higher currents (at least 3000 A/cm²) without any noticeable degradation, since the voltage rises slowly above j_c. This surprising feature is a consequence of the particular microstructure of the material. The sample can be described as a series and parallel connection of large superconducting grains and thin normalconducting barriers, which determine the measured voltage. The weak magnetic field dependence can be explained be assuming that each grain is surrounded by a layer of not exactly stoichiometric material with reduced j_c (compared to the intragrain value) in magnetic fields. The contribution to the over-all resistivity remains low because of the small number of grain boundaries. For $I = 1000$ A and $B = 3000$ G, the volume - specific resistance of the rods was less than 1/10 of silver at this temperature.

References

[1] M. Brand, C. Gross, S. Elschner, S. Gauss and W. Assmus, Proc. EUCAS '93, Göttingen, (DGM, Frankfurt, 1993), edited by H. Freyhardt, p. 369

[2] M. Brand, S. Elschner, S. Gauss, and W. Assmus, Appl. Phys. Lett. **64** (15), 2022 (1994)

[3] C. Gross, M. Brand, S. Elschner, S. Gauss und W. Assmus, *Supraleitung und Tieftemperaturtechnik*, p. 351, publ. by VDI Technologiezentrum, Düsseldorf (1994)

[4] J W Ekin, T M Larson, N F Bergren et al., Appl. Phys. Lett. 52 (1988) 1819

[5] C A Hollin, J S Abell and P W Gilberd, Supercond. Sci. Technol. **8** (1995) 6-14

[6] R Birkhahn, N Browning et al., Appl. Supercond. Vol. 2, No. 1, pp. 67-69, 1994

[7] H Huang and C R M Grovenor, Physica C 217 (1993) 405-417

[8] S Gauss, Hoechst AG, Internal Report (unpublished)

Inst. Phys. Conf. Ser. No 148
Paper presented at Applied Superconductivity, Edinburgh, 3–6 July 1995
© *1995 IOP Publishing Ltd*

An ac superconducting Nb₃Sn coil for use in high magnetic fields

H D Ramsbottom[*], **D P Hampshire**[*], **H Jones**[†] and **D B Smathers**[‡]

[*] Superconductivity Group, Department of Physics, University of Durham, Durham, UK.
[†] The Clarendon Laboratory, University of Oxford, Oxford, UK.
[‡] Teledyne Wah Chang Albany, PO. Box 460, Albany, Oregon, 97321, USA.

Abstract. Many different magnetic measurements require a large ac. magnetic field to be produced in a high background dc. magnetic field. This paper describes the design, construction and performance of an ac. superconducting Nb_3Sn coil . It is wound on a non-magnetic, non-metallic former that is made from silica glass and a machinable glass ceramic. The coil is 216 mm long with an outer diameter of 38 mm and an inner diameter of 18 mm. It is made in two sections, wound coaxially, which are configured in the opposite sense to reduce the coupling between the coil producing the ac. field and the dc. magnet. The coil is able to produce high ac. fields in a 40 mm bore, 17 T magnet. In a dc. field of 12 T the coil can produce 340 mT at 1 Hz and 85 mT at 20 Hz.

1. Introduction

Many magnetic measurements require large ac. fields superimposed onto large dc. fields. For example, in flux penetration measurements on high field superconductors, a high ac. field is required in high dc. fields in order to fully penetrate bulk samples with a high J_C [1].

This paper describes the design, construction and performance of a two component, ac, superconducting Nb_3Sn coil. The coil produces high ac. fields in a 40 mm bore, 17 T superconducting magnet and replaces a NbTi coil which can only operate in fields up to 10 T. [2]. The coil former is non-magnetic and non-metallic which is required for good field homogeneity and reduces eddy current heating and ac. losses.

2. Coil Fabrication

Figure 1 shows the Nb_3Sn superconducting ac. coil. The coil is made from a prototype, multifilamentary, Nb_3Sn wire fabricated at Teledyne Wah Chang Albany. It is wound on a non-magnetic, non-metallic former that consists of a silica tube with 'end-cheeks' that have been made from a machinable glass ceramic. The coil consists of two sections wound coaxially. These are configured in the opposite sense to reduce the coupling with the dc. magnet [3]. It has a maximum diameter of 38 mm so that it will fit in the 40 mm bore of the dc. superconducting magnet. The specifications of the coil are shown in Table 1.

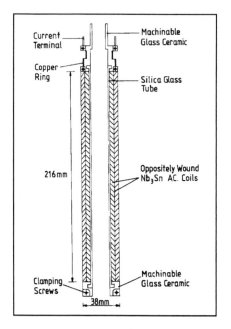

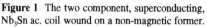

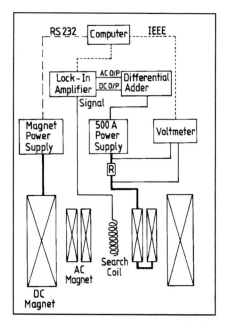

Figure 1 The two component, superconducting, Nb$_3$Sn ac. coil wound on a non-magnetic former.

Figure 2 The external circuit for characterizing the ac. Nb$_3$Sn coil.

After it had been wound, the Nb$_3$Sn ac. coil was inserted into the bore of a copper dc. magnet and the coupling measured between the coil and the magnet. Turns were added or removed from the outer section of the ac. coil to reduce coupling to as close to zero as possible. When the number of turns has been optimized, and the sections are configured in the opposite sense, the coupling was 1000 times less than when they were configured in the same sense. Reducing the coupling is essential to prevent the coil from vibrating in the dc. field and to avoid causing the dc. superconducting magnet to quench.

The coil was then reacted at 650 °C for 200 hours in a Nitrogen atmosphere. The Nb$_3$Sn wires were then soldered to the copper rings, which were connected to the current terminals with 1 mm diameter copper wires. After reaction the coil was vacuum impregnated with resin. During this procedure, the coil was placed in a mould which was sealed in a vacuum chamber. While the chamber was evacuated, heat was applied to drive off moisture and to help the outgassing process. When a good vacuum (< 0.1 torr) has been achieved, resin was admitted to the mould, which was then overpressured to 5 bar of N$_2$ gas to drive the resin into the windings. The resin impregnated coil was then removed from the mould and cured using the 'spit-roast' technique. This involved rotating the coil in a furnace for 50 hours during which time the temperature was increased to 160 °C.

Unfortunately, while removing the coil from the mould, the machinable glass ceramic at the bottom end of the coil was damaged. This was replaced with tufnol. The 'spit-roast' curing ensured that the new 'end-cheek' was properly secured to the coil.

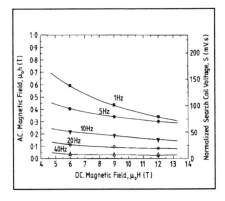

	AC$_1$	AC$_2$	Coil
Inner dia (mm)	21	32	21
Outer dia (mm)	32	38	38
Length of coil (mm)	216	216	216
No. of turns	1250	850	2100
No. of layers	6	4	10
Length of wire (m)	105	95	200
Wire diameter (mm)	0.8	0.8	0.8
Coil const (mT.A^{-1})	16.0	13.5	2.5

Figure 3 The ac. coil performance versus dc. magnetic field as a function of frequency.

Table 1 The specifications of the Nb$_3$Sn coil. The inner section (AC$_1$), the outer section (AC$_2$) and the entire coil.

3. High Field Measurements

An Oxford Instruments 15/17 T (4.2/2 K) superconducting magnet provided the dc. field. The current through the coil was supplies by a 500 A, 5 V power supply. The critical current (I_C) of the wire was measured from dc. up to 40 Hz in dc. fields up to 14 T.

The dc. values of I_C were measured using a standard resistor in series with the coil and determined when the coil quenched. Figure 2 show the experimental arrangement for the ac. measurements. A differential adder was used to offset the programming voltage to the power supply to ensure that the current it supplied was always positive. The power supply is not bipolar. The large smoothing capacitors inside the dc. power supply in parallel with the inductive load of the ac. coil form an LCR circuit. This enabled very large, bipolar currents to flow through the coil. The ac. field in the Nb$_3$Sn coil was measured directly by a calibrated search coil. Both the magnitude and functional form of this voltage is monitored using a lock-in amplifier. The ac. fields quoted in figure 3 were determined when the ac. field was noticeably non-sinusoidal and are typically 10 % below the fields associated with the quench currents.

3. Results and Discussion

At 12 T the dc. critical current is 78 A. This is 10 % lower than the manufacturers specification of 86 A at a criterion of 1 μV.m^{-1}. This reduction in I_C increases to 17 % at 10 T. These differences may be due to several reasons. Local heating in the coil may cause the wire to quench prematurely, the wire may have been damaged during the epoxy resin impregnation or the wire may not have been fully reacted. Experience suggests that supporting the coil on a mandrel during impregnation and ensuring that the epoxy is fully keyed to the former is essential.

The ac. performance of the coil is shown in figure 3. As can be seen, the rms. ac. field falls off with increasing frequency but the coil is still able to produce a field of 85 mT in a dc. field of 12 T. This is at least an order of magnitude better than a dissipative copper coil. NbTi compares more favourably at low fields but is unable to operate in dc. fields above 10 T, which is the upper critical field of the wire. During the ac. measurements no significant Helium boil off or coil vibration were observed, even at high ac. and dc. fields. This is indicative of low coupling and low ac. losses.

4. Conclusions

An ac. superconducting Nb_3Sn coil has been designed and built which can produce large ac fields in large background dc. fields.

It is wound on a non-magnetic, non-metallic former to produce good field homogeneity and reduce eddy currents and ac. losses.

The coil consists of two oppositely wound sections which facilitates use in high field superconducting magnets.

The ac. performance of the coil falls of with increasing frequency but the coil is still able to produce an ac. field of 85 mT at 20 Hz in a dc. field of 12 T.

This coil will enable flux profile measurements to be made on bulk superconducting samples with high J_C in dc. fields up to 17 T.

Acknowledgements

The authors wish to thank R. Hart, G. Haswell and M. Greener for constructing the magnet former, A. Hickman and G. Sheratt for impregnating the magnet, C. Mullaney for building the 500 A power supply, D. Evans for useful discussions and P. Russell for help with the production of the drawings. This work is supported by the EPSRC. GR/J39588 and The Royal Society, UK.

References

[1] Campbell A M 1969 *J. Phys. C* **2** 1492-1501
[2] Ramsbottom H D and Hampshire D P 1995 *Accepted for publication in Meas. Sci. Technol.*
[3] Polák M Pitel J Majorôs M Kokavec J Suchon D Kedrová M Kvitkovic J Fikis H and Kirchmayr H 1994 *ASC '94*